# Valuation: Principles into Practice
## Sixth edition

Edited by R E H Hayward BSc (Est Man) Lond, FRICS

2008

**┌EG┐ Books**

A division of Reed Business Information

Estates Gazette
1 Procter Street, l

First published 1980
Second edition 1984
Third edition 1998
Fourth edition 1992
Fifth edition 2000
Sixth edition 2008
Reprinted in Jan 2009

ISBN 978-07282-0524-6

Cover design by Rebecca Caro
Typeset in Palatino 10/12 by Amy Boyle, Rochester
Printed by Bell & Bain, Glasgow

# Contents and Authors

# Preface to the Sixth Edition

This is the first edition without Bill Rees at the helm. His death in January 2004 left a hole that cannot be filled. The idea for this book was his, he saw to its publication through the first five editions. The Prefaces to the first and fifth editions are reproduced here as they show exactly what his intentions and wishes were, and I hope that this edition is faithful to those.

In 1978 Bill sought out authors to write specialist chapters for *Principles*. He wanted, and achieved, a balance of practitioners and academics — a meeting of town and gown which remains one of the book's strengths. From his invitation to contribute to the book came a friendship that lasted until the day Bill died. His request that I lend a hand with editing the fifth, with a view to taking over thereafter, came out of the blue, and was not a little daunting, coming as it did from the owner of one of the finest intellects in the surveying profession.

Working with Bill was, simply put, sheer delight and enormous fun. We shared the view that we did not have to restrict ourselves to what was necessary to get the book published on time, but it provided an unimpeachable reason to meet for long lunches at 'The Sign of the Pink Moggy' to swap bundles of documents and eat those things which our wives and doctors thought we should not. The Pink Moggy is known to some, I believe, as the Red Lion at Turner's Hill.

So to the sixth edition. It is inevitable that in the seven years since the previous edition was published some of those who contributed will have retired. Of those who have put down their pens several have written for all five of the earlier ones. They deserve long-service awards. That said, fourteen of those who contributed to the fifth have written for the sixth.

Sadly, DC Elgar, the author two chapters in the fifth, and earlier, editions died before that edition was published. My thanks go to his son and daughter, John and Caroline, for agreeing that their father's work could be used in this edition in whatever way the new authors chose. The other contributors who have retired have almost without exception made their work available to their successors, although in the event these chapters have all been completely re-written from scratch. I am grateful to them all for their generosity in making their work available, for their help in the past and, particularly, their active participation in helping me to find their successors.

So, I record a grateful parting Thank You to Richard Baldwin, RE Clark, Charles Goodwyn, John Ratcliffe, Bill Taylor, Ron Westbrook and Brendan Williams. At the same time, a welcome to those who have taken over from them: Charles Cowap, Stephen Crocker, Nigel Dubben, Geoffrey Jackson, John Murdoch, Keith Murray and Chris Thorne; to Simon Galway who has taken responsibility for Chapter 14, with the continuing assistance of Peter Squire, and to Christoper Hart who has added a substantial contribution to the tax chapter. Each is an expert in his field. In expressing my thanks to the authors it is appropriate that I explain that their opinions are confined to their own chapters, and that I should disclaim all responsibility for the views expressed.

Finally, a bit about housekeeping. 'He' in this book refers to the human species and not just to the male variant. Case citations have been confined to the Table of Cases and kept out of the body of the text, where they would either interrupt the flow or clutter up the pages with footnotes.

Richard Hayward
June 2008

# Preface to the Fifth Edition

Since the last edition, three of the contributors have died and nine others have decided not to continue. We are most grateful to those concerned who have allowed successors to use the matter in the last edition and to others who have provided much useful help.

For the first time in the history of this book valuer-contributors are joined by other professionals, chapter 8, Taxation being written by a chartered accountant and a taxation specialist.

The purpose of the book remains as stated in the preface to the first edition, as do the remarks relating to the opinions of the various authors and the disclaimer as to our own views.

As with all such books, the relevant case and statute law may have altered since the chapters were written.

R E H Hayward
W H Rees
November 2000

# Preface to the First Edition

The numbers coming into the surveying profession reading full-time for degrees and diplomas increase year by year. Particularly in the case of such students, it is desirable that the teaching approach should be to guide their reading rather than to dispense facts in lectures. Students following correspondence courses need to supplement these by further reading. In the subject of Valuations, beyond the intermediate state, there is comparatively little reading matter available, other than *Modern Methods of Valuation*. However, in no way does this book attempt to compete with *Modern Methods of Valuation* which covers principles, law and practice; it is intended to complement the latter. There is said by some to be a gap between the principles of valuation as taught for examinations and practice. The purpose of this book is to provide more reading matter at final year level to bridge this gap.

The form of the book, as originally intended, was to have been a series of case studies with notes by way of explanation but during the course of preparation it became apparent that some subjects were better covered by text, hence the chapters vary widely in form. There are also wide differences of approach but in this subject there is frequently room for more than one opinion and this is a not unimportant factor of which a student should be aware. In expressing my thanks to the authors I should explain here that their opinions are confined to the chapter each has written. It is also appropriate that I should disclaim all responsibility for the views expressed. The idea of the book was mine and the choice of authors; my intention in that connection was to try to balance those engaged with teaching with those in practice, a coming together of "town and gown" which is perhaps an all too rare occurrence for the good of the valuer's profession.

I decided to draw the line at the end of November 1979 as far as legislation is concerned; thus no account is taken of the Local Government, Planning and Land Bill.

W H Rees BSc FRICS
March 1980

# A Note of General Application Regarding the RICS Red Book

References are made throughout this publication to the RICS Valuation Standards[1], known colloquially as the "Red Book". Until the mid 1990s the Red Book had limited application outside the field of valuations prepared for financial statements, but now applies to virtually all written valuations. Valuations undertaken in the course of litigation, arbitrations or similar disputes and certain agency or brokerage work are exempt from the mandatory requirements of the Red Book on the grounds that it is often impractical to follow all the requirements completely when undertaking the type of work in question. However, it is generally recognised as best practice to follow the provisions of the Red Book insofar as is practicable.

It should be noted that, unlike *Principles into Practice*, the Red Book does not concern itself with valuation methodology but simply sets rules for the conduct of a valuation instruction. The purpose of these rules is to ensure that clients receive objective advice, in a professional manner that is consistent with internationally recognised standards. It sets a framework for best practice in the execution and delivery of valuations for different purposes, but does not instruct chartered surveyors how to value, nor does it discuss valuation methodology or techniques.

An important point to note about the evolution of the Red Book during recent years has been the increasing significance of the International Valuation Standards. It has long been RICS policy to support the development of a single set of core international valuation standards to provide a common framework for valuers worldwide. Since 2000 there has been rapid development in the scope and quality of the International Standards and the Red Book now adopts the same core standards, applications and definitions that appear in the International Standards.

The distinction between the International Standards and the Red Book should be explained. The International Standards are published by the International Valuation Standards Council[2] which acts purely as a standard setter. It has no individual members and cannot enforce compliance with the Standards it publishes. The RICS is a professional body that has a responsibility under its charter to regulate the conduct of its members. The Red Book incorporates the International Valuation Standards and provides the means of ensuring that RICS members comply with these standards.

As well as being the means of ensuring that chartered surveyors comply with International Valuation Standards, the Red Book also contains national standards, which incorporate additional

---

1     Until September 2007 this was known as the Appraisal & Valuation Standards.
2     Formerly, the International Valuation Standards Committee.

material, or variations form the International Standards, that are required to meet the specific legal or regulatory requirements of particular countries. In the UK there are supplementary rules and guidance relating to valuations for loan security, accounts, share listings, company circulars, taxation, and of pension fund assets and unit trusts.

It is outside the scope of this book to explain the detailed provisions of the Red Book, but readers should be aware of its requirements before putting any valuation principles into practice.

# Biography— William Hurst Rees

Bill Rees was born at Bushey, Hertfordshire in April 1917. Educated at Watford Grammar School, his primary interest was mathematics. He soon showed a mechanical aptitude and throughout his life was interested in clocks and radio receivers.

Bill started his professional life with Salter Rex & Co, studying for his professional examinations by correspondence, while gaining firsthand experience in private practice. He found the correspondence course too much, and after four years he left to become a full-time student at the College of Estate Management in London. His work experience stood him in good stead, enabling him to combine the theory with what he had seen in practice.

The Second World War interrupted his professional career. He joined the Royal Artillery in 1940. He transferred to the Royal Engineers, was promoted to Captain and trained for visual deception (camouflage), the subversive nature of which appealed to his sense of humour. This was his role during the Normandy landings, where he successfully deployed dummy tanks and guns. More deception schemes followed before the Rhine Crossing, after which he was promoted to Major to become liaison officer with the 1st Belgian Engineer Battalion in Germany.

After demobilisation Bill went back to the College of Estate Management as a lecturer. The College inaugurated a programme of full-time study for the University of London BSc (Estate Management) degree, and from 1948–1951, as Head of the Valuation Department, Bill played an important part in its implementation. In 1950 he passed the BSc (Estate Management) degree examination himself.

In 1949 he became joint author, with David Lawrence and Harold May, of the second edition of *Modern Methods of Valuation*, a book that needs no introduction here, and he edited the next four editions.

In 1951, Bill moved to Richard Ellis as head of the valuation department, while continuing to lecture at the College in the evenings. Of interest in the context of *Principles*, he acted for Mrs Harvey in the *Harvey* v *Crawley* case, which gave rise to the concept of Crawley costs. In 1961 he became senior partner of Turner Rudge & Turner of East Grinstead, where he stayed until he became a Member of the Lands Tribunal in 1973, from which post he retired in 1989.

His work as a Member of the Lands Tribunal suited him. His sharp and analytical mind could readily get to the heart of any problem, frequently making the seemingly complex blindingly simple. He was a great believer in the need to apply the correct principles of valuation. He gave a number of leading decisions, notably in connection with the admissibility of ransom value in compulsory purchase cases.

In 1978 he devised and edited the first edition of this book, and edited four further editions, the fifth in conjunction with Richard Hayward. *Principles* was Bill's brainchild and his wish was that it would continue into the future. It will.

Bill was one of the giants of the surveying profession, his innate modesty and kindness providing an effective camouflage. A man of incisive judgement, he cared deeply about professional standards and the quality of education that underpins them and was always happy to visit colleges to meet students. He always had time for people, was infinitely approachable and ever willing to offer comment or advice to those who asked. He was convivial, with a wonderfully mischievous sense of humour. In his spare time he built radios, was enthusiastic about music and opera and was a keen gardener.

In 1941 he married Betty Wight, who worked at the College of Estate Management and who survives him together with their three children, eight grandchildren, and (currently five and rising) great-grandchildren. Their house was, and remains, renowned for its hospitality. Bill died peacefully at home on 6th January 2004, aged 86.

*I am grateful to Martin Hattersley for permission to base this biography on the obituary which he prepared in 2004. REHH.*

This book is dedicated to the memory of its originator and first editor,
Bill, WH Rees BSc (Est Man) Lond, Dtech (Hon), Hon RICS,
Honorary Member of the Rating Surveyors' Association

12 April 1917–6 January 2004
Surveyor, teacher, judge, author and a dear friend and mentor to many

# Table of Cases

# B

# Table of Statutes

# Table of Statutory Instruments

*l*

# Table of Circulars

# Agricultural Property

Once upon a time farms grew crops and reared livestock to feed a grateful nation which knew the dangers of hunger. Farmers were seen as cheery fellows with a deep-rooted attachment to the land, happy with their lot as cultivators and custodians of a strong rural heritage. The job of the agricultural valuer was to advise on the sale and purchase of farms and rural estates, to undertake the annual stocktaking valuation and to see tenants in and out of their tenancies with the myriad calculations which make up a tenant right claim. On one or two days a week the valuer might act as auctioneer in the local mart, and of course would also arrange the special sales of machinery, livestock and implements (deadstock) which would accompany a change in farm occupancy. On hand to provide a broad range of practical advice, the valuer would also deal with the occasional incursion from a new road or pipeline, arrange building work, insurance valuations for the larger estates and claim grants from a generous Ministry of Agriculture.

Now all that has changed. In recent years the farmer began to doubt whether anybody wanted the produce over which so much labour and care was expended and his role as custodian of the countryside has been increasingly questioned. In short, no longer a cheery fellow at all in many cases. Meanwhile, the number of traditional tenancies has declined. Land has been purchased by a new breed of 'lifestyle' buyer. The production of food is no longer the priority and we have forgotten the dangers of hunger. Farms have diversified, with some now resembling a small business development located in a country park. The Ministry of Agriculture is no more, and public administration of the countryside is in the hands of a plethora of public bodies most notably the Department of Environment, Food and Rural Affairs (DEFRA). Harvest festivals cling on, but another annual highlight of the agricultural calendar is the May submission of paperwork to the Rural Payments Agency of DEFRA in order to claim the Single Farm Payment for the maintenance of land on the Rural Land Register in Good Agricultural and Environmental Condition. 'Serious' farming can still be encountered, but it too has changed to an epic industrial scale driven by the quality and environmental imperatives of the major supermarkets.

The work of the agricultural valuer has therefore changed too. Rural surveying firms were among the first to offer brokerage services for milk quotas in the mid 1980s, and they were among the foremost professional groups in helping farmers to cope with the introduction of Arable Area Payments, set-aside, sheep and suckler cow quotas, and IACS — the Integrated Administration and Control System — after the McSharry reforms of the Common Agricultural Policy in 1992. The implementation of the mid-term review of the Common Agricultural Policy in the last two years has been the latest episode this story of increasing complexity in agricultural and rural administration. Professional advice has been needed on the introduction of the Rural Land Register, eligibility for the Single Payment Scheme and historic based entitlements and the annual claims for Single Farm Payment. These changes have also added to the assets that must be appraised by the agricultural valuer: first milk quota, then quotas for sheep and suckler cows before their withdrawal two years ago, and now historic entitlement to Single Farm Payment. Alongside the agricultural revolution of the last decade, there have also been notable changes in the valuation requirements of taxation — Inheritance Tax and its impact on farmhouses in particular. The latest proposals for a Planning Gain Supplement seem likely to be one of the next challenges to agricultural valuers by 2009, in particular the need for 'current use valuations' of redundant agricultural assets which ignore prospects for future development in the form of hope value.

Against this turbulent background, this chapter revisits the fundamental aspects of any agricultural valuation. This covers the qualifications and experience of the valuer, the requirements of a physical inspection of a farm or estate, the desk research which must be undertaken for a complete valuation of a farm, and sources of information with which the valuer should be familiar. The chapter then goes on to look at some more specialised examples of the agricultural valuer's current work.

# The fundamentals

## The valuer

The RICS (Royal Institution of Chartered Surveyors) lays down requirements for the education and training of its members, ensuring that they acquire appropriate practical experience and keep their professional knowledge up to date. The Red Book (*RICS Valuation Standards*) applies to virtually all written valuations and it is considered best practice to follow its requirements. All this ensures that the valuer understands his task, has the confidence borne of experience and is subject to stringent professional regulation. The valuation of agricultural assets is no exception, and most practising valuers will therefore be members of the RICS, and in particular its Rural Faculty.

In addition to RICS membership, many agricultural valuers in England and Wales also belong to the Central Association of Agricultural Valuers and its local branches. This association provides its members with highly focussed technical material, as well as arranging a lively series of local professional meetings, and membership is highly valued by its members. CAAV, like the RICS, also sets a series of rigorous examinations for new entrants. Members can be recognised by the designation FAAV (Fellow of the Association of Agricultural Valuers) after their names. Publications in recent years have covered the emerging changes in the Common Agricultural Policy, changes to agricultural tenancy law and the conduct of rent reviews of farms. Student and probationer membership is available to those wishing to qualify, and anybody with a serious interest in the subject of rural valuation should look into membership.

Like any other valuer, the agricultural valuer needs to be familiar with the five traditional methods of valuation. Although direct comparison is the mainstay of the rural valuer's work, different rural assets may need the investment, residual, profits or depreciated replacement cost methods. Let commercial conversions, barns for redevelopment, specialised leisure enterprises and specialised agricultural buildings would, respectively, all draw on these methods. There has also been a greater acceptance of discounted cash flow approaches in recent years, although more with a view to investment appraisal than the preparation of market valuations. These trends seem likely to continue as the rural economy becomes more diverse. The valuer also needs a sound appreciation of husbandry, forestry and rural construction, as well as a good understanding of the planning system and the increasingly complicated requirements of rural land administration. A good understanding of all forms of tenancy, residential, commercial and agricultural, is also a pre-requisite.

The Red Book includes an Information Paper on 'Rural Property Valuation' (Valuation Information Paper 5), and clearly the valuer should be acquainted with its contents which are reflected later in this chapter.

## The purpose of the valuation

Farms have to be valued for the usual range of reasons: sale and purchase, taxation, compensation on compulsory purchase, loan security are all common examples. Sadly, valuations are also required in connection with divorce settlements, business partnership dissolution, and for probate and taxation purposes following death. Clearly, considerable sensitivity needs to be exhibited by the valuer faced with some of these situations.

## The physical inspection

A comprehensive physical inspection of a farm of any size may take at least several hours, and possibly a lot longer. The valuer will need to be equipped with boots and waterproofs as well as the usual notebooks, recorders, digital camera, tape or laser measure and binoculars. A digital camera is a particularly helpful instrument for a farm inspection. Hygiene precautions — nowadays called biosecurity — have become important again, and the valuer should ensure that clothing, particularly footwear, is scrupulously clean. Clean overalls can be an excellent idea, and on high health status units the valuer should expect to wear boots and overalls provided by the farmer.

It is always helpful to spend some time talking to the current farmer at the start of an inspection, and to note down any information he is willing to share about current stocking and cropping on the farm, houses and cottages, current workforce, eligibility for the Single Payment Scheme, occupation details, participation in environmental or other schemes. This may also be an opportunity to check the existence of any statutory designations which might affect the farm, for example part of the holding may be notified as a Site of Special Scientific Interest, by Natural England, which has recently taken over the functions formerly exercised by its predecessor English Nature. The valuer is likely to be aware of wider designations such as National Park or Area of Outstanding Natural Beauty status from his general knowledge of the area. An important recent development has been the designation of certain areas as 'Open Access' land under the Countryside and Rights of Way Act, over which the so-called right to roam operates. This information can subsequently be verified from the maps published by Natural England, previously by the Countryside Agency. It is important that the valuer should try to form a clear impression of the farm or other business activity taking place, and these data will help to form a professional view on the standard of management of the farm.

In particular, details of all houses and cottages should be noted in the normal way for any residential property. It is helpful to note as much detail as possible about the current occupiers of all residential property, particularly where security of tenure may be an issue. Similar details will also be needed for other buildings, where current use should also be noted and in particular any signs of non-agricultural occupancy or use. The presence of asbestos may be an issue in some farm buildings, and the valuer should be alert to this danger. When appraising the suitability of farm buildings, the valuer should bring a knowledge of the requirements of modern agricultural production to the task. For example, is there adequate storage for waste on a dairy farm (slurry from the cows, washings from the dairy, silage effluent, soiled bedding) and does an arable farm have adequate grain storage against the requirements of modern quality assurance schemes for the marketing of combinable crops (wheat, barley)? Grain stores should be protected against vermin and birds, and lights should be designed so that broken glass cannot fall into the stored grain. Shortfalls should be clearly noted.

A thorough inspection will include an assessment of each field on the holding. Its soil type, current use, access, condition of fences or hedges, availability of water may also need to be noted. Slope and aspect are also important considerations, with gently sloping fields of a south to southwesterly aspect often being the ideal. For some valuations, in particular those needed at the end of a tenancy, it may also be necessary to gather data about recent fertiliser, lime and FYM (farmyard manure) and slurry applications. It is also important to look out for signs of contamination. It is by no means uncommon for farm ponds to have been filled up with a potent mixture of oil drums, chemical containers, rubble and rusty machinery over the years, to be crudely covered with a good layer of topsoil when the opportunity arose. Sites like this are unlikely to be revealed on any contamination registers, but the physical signs may be there in the field (depressed areas, raised areas, vegetation changes, unusual discharges from land drains).

The existence and condition of field drains will also be a matter of note during this part of the inspection. Outfalls from field drains should be clearly visible in the ditches, and the valuer should form a view as to the continuing efficacy of the system (do the drains seem to be running clearly, or are there signs of blockage, ochre contamination and the like). Drainage problems may also be apparent on the surface of the field, in the form of damper areas or patches in the crop which have failed to thrive. Conversely does the farm have access to irrigation, perhaps from its own reservoir and have water pipes been laid to provide an irrigation network? As our climate changes, these considerations may become more important along with the whole question of access to water.

It is useful if the valuer is aware at the outset of any environmental obligations which apply to the farm. For example, there are environmental obligations attached to receipt of the Single Farm Payment and to participation in some of the basic environmental schemes offered by DEFRA — the so-called Entry Level Scheme, for example. Field margins should be left to grass for example, and the valuer should be careful to note any infringements for the office file. This is particularly important when advising a potential purchaser who may be faced with some of the resulting liabilities. There may also be continuing obligations as a result of management agreements under the Wildlife and Countryside Act between the current owner, occupier and Natural England.

More generally the valuer should take in the general layout and amenity of the farm. Is it contained in a ring fence with its own internal tracks and roads, or is it bisected by public roads which provide good field access? Is the electricity supply single phase (240 volts) or three phase (480 volts)? Is the water supply from the mains or a private borehole? Sewerage is nearly always private, but there are exceptions. A mains gas supply is unusual in the countryside, but some farms do use bottled gas. Are there hedges and trees in good condition? Is the overall impression that of a fine country and sporting residence, or a day to day working farm? What use, if any, is made of the sporting interest in the farm (pheasant and partridge in particular on a lowland farm)? Is coarse or game fishing available, and how are these rights used? Much of the interest in the farmland market in recent years has been from the so-called 'lifestyle' buyer, and these considerations may be far more important at the outset than practical matters of agricultural management.

The physical inspection is also an opportunity to note any prospects for development, although these will also need to be the subject of a lot of desk research later. For example, are there potential development plots adjacent to a village or town. Are there older buildings of character that would lend themselves to conversion to residential or commercial accommodation? There has been a very active market in old barns for residential conversion in recent years, and the rent from a conversion for commercial purposes can be a very useful income stream to a farm business. Telecom masts have been described as worth another cottage in rent, so their presence or potential also needs to be noted. Mineral rights can also be very important, and any evidence of the availability of minerals or their extraction should be noted. The presence and condition of woodlands on the property needs to be considered, along with their potential for timber production and their sporting potential. More generally, public access to the property should also be noted, formal and informal, along with any problems arising from fly-tipping, trespass or general anti-social behaviour. For example, farms abutting lay-bys on main roads often suffer badly from litter.

On completion of the inspection the valuer will need to reduce his* field notes to a simple physical description of the farm for the final valuation report, and it should be borne in mind that this may need to be understood by readers unfamiliar with agriculture who have not had the opportunity to see the farm for themselves. It is normal practice to append a plan of the farm, and schedules of fields showing their status and uses, and a similar schedule of the main farm buildings referenced to a simple block plan of the farmyard. The field notes and photographs also need to be stored carefully, along with the valuer's contemporary observations on the office file, for future inquiries arising from the valuation or, in the worst case, against the possibility of claims for professional negligence. They should therefore be clear and unambiguous, and readily understandable by other professional valuers who may need to review them.

## Desk research

A considerable amount of office work may be required both before and after the physical inspection of the farm. Like any valuation, clear instructions must be confirmed fully with the client before proceeding with the work. The requirements of the Red Book apply as much here as to any asset valuation. Particular attention should be paid to Information Paper 5, including the need to identify any special assumptions which might apply to the valuation. Two common points may be an assumption of vacant possession where it is currently not available, and an assumption that planning permission will be obtained for development work.

Careful preparation beforehand will allow the valuer to make the most of the inspection, particularly a less experienced valuer. It is helpful to study as much of the legal documentation pertaining to a farm as possible. This might include deeds or registration documents, in particular with regard to current ownership, the nature of the interest held (freehold etc), the existence of any easements or restrictive covenants, riparian rights, abstraction licences for irrigation water, private water supplies, Rural Land Registration data, history of Single Farm Payment claims, allocated and transferred quota rights, details of environmental or grant schemes and statutory notification concerning Sites of Special Scientific Interest are examples. Documentation will not always be available to the valuer at this stage to cover all these aspects, and Information Paper 5 recognises that a number of matters may need to be established by simple enquiry, with appropriate assumptions reported along with an indication of the need to verify key points.

The existence of sporting, mineral, agricultural, residential and commercial tenancy agreements, or other licence arrangements, should also be reviewed along with grants of easements and wayleaves to statutory undertakers for pipelines or wires running through a farm. In particular the dates on which

---

\*    His/her are used generically throughout the book and apply to both male and female.

tenancies came into effect can be important in considering security of tenure and succession rights. The physical inspection may also prompt the need for further research amongst these sources. For example, properties or land may have been sublet for which no documentation is apparently available. Issues of this kind can become particularly challenging problems in view of the security of tenure enjoyed by some agricultural tenants, and the valuer must be particularly alert to these dangers as they may also have a considerable effect on a valuation. Assumptions may have to be made in order to arrive at a valuation, and these will need to be carefully noted in the report. Again, Valuation Information Paper 5 offers a useful summary of matters that may need to be dealt with by inquiry rather than inspection.

More generally a valuer new to an area can acquaint himself with the types of soil he is likely to encounter by studying the Agricultural Land Classification maps published by the former Ministry of Agriculture. These were published at a scale of one inch to one mile, so their accuracy is limited (the ministry only ever claimed an accuracy to within 80 ha), and land was graded between one and five, with grade one land the best and most versatile. Grade Three land (the predominant grade in the lowlands) was further sub-divided, but these divisions were not shown on the published maps. Land in this category should be capable of growing average yields of cereals, and average to above-average crops of grass. Root crops like potatoes do grow on Grade Three land, but really require land of Grade Two or One to thrive reliably. Of more use may be the local publications available from the Soil Survey of England and Wales, which show at varying scales the soil types to be encountered in different areas. These are well described, allowing the valuer to compare his observations in the field with the range of soil types to be expected in the locality. Close study of this information in the earlier part of an agricultural valuer's career will be worthwhile, in order to develop skill in assessing soil types and appraising their potential.

## Sources of comparable and other market information

As always, comparable transactions of which the valuer has direct knowledge and which relate closely to the subject property in location, character, tenure and valuation date are the best guide. However, the valuer rarely has the luxury of many such comparables, and therefore has to look further afield. Sales particulars will give a good idea of current asking prices, but prices realised may in some cases be significantly different (higher or lower). Some land is still sold by public auction, and so these data are also readily available. The *Farmers Weekly* carries reports of farm sales most weeks in the season, and the *Farmland Market* is published twice a year by *Farmers Weekly* and RICS. This gives a county by county schedule of farm sales, published twice a year. The exact price is not given for every sale, but the summaries indicate an approximate price (for example, 'in the region of' means within 10% of the price quoted). Most of the larger firms of agents publish market commentaries which are readily available from their websites, and the Valuation Office Agency publishes its Property Market Report twice a year which also include a commentary on agricultural land prices based on its own data. The RICS also publishes regular commentaries on the state of the rural property market.

Farmland prices are usually quoted per hectare in official publications. In practice most farmers and agricultural valuers work in acres. One hectare is 10,000 m$^2$ (eg 100 m × 100 m) and one acre is 4,840 sq yds (eg approximately 70 yds × 70 yds). The usual conversion from acres to hectares is 2.471 acres to one hectare. Metric measures can be used alone, but imperial measures should, strictly, always be preceded by the equivalent metric measure. The examples in this chapter are all metric. The analysis of farmland comparables normally involves the allocation of separate values to each house and cottage, different categories and qualities of land, and perhaps to buildings with development potential.

The market in recent years has been characterised by an excess of demand over supply, particularly for quality property with a good house. Lifestyle buyers have tended to dominate the market, their bids fuelled by the general buoyancy in the economy. Farms are usually now sold in lots rather than as complete units, which increases their appeal to neighbouring farmers. Indeed it is often consideration of likely neighbour interest that determines the arrangements of the lots. The neighbouring farmer can expand without increasing overheads in the existing business, and the opportunity to buy conveniently-located land is likely to be seized as a once in a lifetime opportunity. Although less of a feature of the market in the last year or so, farmers who have sold development land have also been important drivers in the market as they have sought new farms. Not only do they wish to maintain their farming interests, but Rollover Relief provides a way to shelter money reinvested in a new farm from Capital Gains Tax.

Farms will often be purchased with a view to direct management of the land by the new owner (in-hand farming). However, for many lifestyle buyers hands-on farming is not a practical proposition. The land may therefore be let, but it is more likely that a contract farming agreement will be arranged with a nearby farmer. These agreements avoid the problems of tenure associated with tenancy agreements, and give the farm owner the status of a working farmer rather than a landlord for a number of purposes (including taxation). The owner of the farm is nominally the farmer and pays for the main farming inputs. A nearby farmer takes on the contracting role, providing labour and machinery and invoicing the owner for these services. A first charge from farm revenues is drawn by the owner — often similar in amount to a rent — and the remainder is divided between the owner and the contractor. This ensures that the owner is exposed to business risk, thus establishing his credentials as a farmer running a business. It also acts as an incentive to the contractor to do a good job. Good advice from an agricultural valuer is essential to ensure that the agreement is workable and the figures add up.

The following simple studies illustrate some of the points discussed above.

## Study 1

Hawthorn Farm, Freehold interest valued with vacant possession.

Hawthorn Farm is a 200 ha holding with one five-bedroomed farmhouse in good condition. It has a modern portal-frame farm building of 500 m². The land is divided between one block of 140 ha of better quality land, Grade Two in the Agricultural Land Classification and the remaining 60 ha is at the lower end of Grade Three. All of the land is registered on the Rural Land Register, and the farmer has established his eligibility for the Single Payment Scheme. The payment entitlement for 2007 is £37,000. This is a combination of historic and regional average entitlement after adjustment for currency conversion, clawbacks for the national reserve, modulation and financial discipline. Local comparables suggest a value for the house of £600,000, for the better land £8,600 per ha and for the poorer land £6,000 per ha. Single Payment entitlement is a separate asset from the land, and recent market activity suggests that it has been trading at 2YP.

The freehold vacant possession valuation of Hawthorn Farm is therefore built up as follows:

|  | £ |
| --- | --- |
| House | 600,000 |
| Better land: 140 ha at £8,600 | 1,204,000 |
| Poorer land: 60 ha at £6,000 | 360,000 |
|  | 2,164,000 |
| But say | 2,200,000 |
| *Add* SPS Entitlement: £37,000 × 2 YP | 74,000 |

Note: The Single Payment Entitlement is kept as a separate asset because it is a separate and identifiable asset.

The availability of a range of traditional farm buildings with planning permission for residential conversion could easily add another £100,000 to the value of Hawthorn Farm in many areas. This value might be more accurately tested by a residual valuation although some market evidence would suggest that this would not act as a constraint on the price of a suitable property. An increase of value of this magnitude may, however, be offset by some loss of value from the house in the anticipation of nearby development.

## Study 2

Hawthorn Farm: freehold interest subject to a 1986 Act Tenancy

The physical facts are the same as study 1. Here, however, Hawthorn Farm is subject to a tenancy under the terms of the Agricultural Holdings Act 1986. This Act governs most tenancies of agricultural land granted before 1 September 1995, including statutory successions to such tenancies. Tenants have considerable security of tenure as there is a very limited range of reasons for which the landlord can serve a valid notice to quit. Rent reviews are covered by the statutory provisions of Schedule 2 to the Act and tenancies granted before August 1984 carry statutory succession rights for up to two generations. The rent passing is £31,000, which can be analysed as £125 per ha for the land on average (reflecting the availability of the building) and £6,000 pa for the house. Although it has been four years since a rent review, and therefore either landlord or tenant could demand another rent review, it is considered unlikely that a review would lead to much change in the rent.

Using the investment method the farm would be valued as follows:

|  | £ |
|---|---|
| Net Rent | 31,000 |
| YP in perpetuity at 4% | 25 |
| Value of freehold interest | 775,000 |

Note that the gross rent has been capitalised in this study, which is a common approach in the valuation of these interests although in practice landlord's outgoings (building repairs, insurance, management fees) could easily be absorbing 40% of rent. It is usual in these cases also to compare the valuation with the equivalent freehold interest with vacant possession (VP). In this case, the investment value is only 36% of the VP value, just within the range normally encountered of 35 to 50%. Given the lack of direct evidence for an appropriate discount rate in some parts of the country, valuers will sometimes value freehold interests subject to these tenancies directly by a discount from VP value. However, even where this is the primary approach it is still worth checking the yield which such a valuation would imply.

In checking the adequacy of the rent passing for a valuation like this, regard should be had to the provisions of the 1986 Act. Rent reviews can take place every three years, and are initiated by the service of a notice of arbitration in accordance with section 12 of the Act. An arbitrator can be appointed by agreement (before the term day from which the new rent will run), or by the RICS President (on application before the term day, although the date of the appointment itself may be later). In practice most rent reviews are settled by agreement, but having regard to the directions to arbitrators as to rent properly payable which are set out in Schedule 2 of the 1986 Act. These direct that the rent should be that payable by a prudent and willing tenant to a prudent and willing landlord taking into account 'all relevant factors'. Extensive guidance on these rent reviews has been published by RICS and CAAV, and an update is expected later in the year to take account of Single Farm Payment issues and other recent developments. It is necessary to consider the rent payable for comparable holdings, suitably adjusted for scarcity and marriage value (although there is no direction to disregard these factors in setting the rent for the subject holding). Regard must be had to all the terms of the tenancy and there are a number of statutory disregards, for example tenant's neglect, government contributions to the cost of landlord's improvements and high farming, ie the adoption of special systems of husbandry. The review has to consider an ordinarily competent tenant farmer rather than the super-farmer or laggard who may be in actual occupation. Schedule 2 also directs the arbitrator to have regard to the 'productive capacity' and 'related earning capacity' of the holding, meaning the cropping and stocking plans which a typical farmer might follow and the margins he might expect to make from them, but no indication is given as to the weight which these factors should have. In practice, little weight may be put on these factors where a farm has a number of other attractive features, eg houses and cottages. Latent value, for example the suitability of part of the farm for a caravan park, may also be reflected if such a use would be permitted under the tenancy agreement.

Study 3 shows how to value the freehold when it is subject to a farm business tenancy. The same approach, of valuing a reversion to vacant possession value, will be relevant to a farm let under the 1986 Act when there are good reasons to expect a 1986 tenancy to terminate at a determinable time in the near future. This might be because of an elderly tenant who has no eligible successors, or because the landlord can serve an incontestable notice to quit for one of the limited range of reasons set out in the Act. The most common of these is a Case B Notice to Quit, where the landlord has obtained planning permission for an alternative use.

## Study 3

Hawthorn Farm: freehold interest subject to a Farm Business Tenancy

The physical facts are again the same as study 1, but this time the farm is let under the terms of the Agricultural Tenancies Act 1995, ie as a farm business tenancy. This legislation is much less restrictive on landlords than the Agricultural Holdings Act 1986, and the new law was introduced with effect from 1 September 1995 in the hope of encouraging a revival of farm lettings and to accommodate the growing need for all farmers to consider diversification of their businesses. There is no statutory security of tenure beyond the agreed term, other than a requirement for a minimum notice to quit of one year for tenancies of more than two years duration. Rent reviews still take place on a three year cycle, but the parties can agree their own basis of rent. In the absence of specific terms, rent reviews are based on market values. Rents under Farm Business Tenancies therefore tend to be higher than traditional tenancies because of the absence of restrictions like those imposed by Schedule 2 of the Agricultural Holdings Act.

The rent in this study is, therefore, £54,000 pa, which can be analysed as £12,000 pa for the house, and £210 per ha on average for the land. There is little evidence if any for the choice of an appropriate discount rate, but professional opinion seems to be settled at about 6 or 7%. This valuation will assume that Hawthorn Farm has been let on a five year term, which will be terminable by the service of notice sometime during year four. Failing such notice, the tenancy will continue from year to year until a year's notice to quit is served by either party (the notice expiring on the term date of the tenancy).

|  | £ pa | £ |
|---|---|---|
| Net Rent | 54,000 | |
| YP 5 years at 6% | 4.2124 | |
| | | 227,470 |
| Reversion to | | |
| Vacant possession value | 2,164,000 | |
| PV £1 in 5 years at 6% | 0.7473 | |
| | | 1,617,157 |
| | | £1,844,627 |

Value of freehold interest, say £1,850,000

Note that the reversion here is to VP value, because of the ability of the owner to crystallise this value at the end of the term if required. In practice, it might be questioned as to why an owner who has let now would wish to sell with vacant possession in five years time. However, the fact remains that if the objective was to obtain the best price for the property a sale with vacant possession would be the way to achieve this. There may also be some scope for discussion over the choice of discount rate on the reversion, as this has been applied to a future VP value rather than a reversionary rent. A higher risk might be attached to the prospects for VP values than continuing rents, thus depressing the YP for this part of the calculation. This valuation approach is relatively untried in practice, as the first farm business tenancies were let on relatively short terms while practitioners developed confidence with the new concept. Therefore any owners contemplating the need for a sale would not find themselves encumbered with a farm business tenancy. However, terms have been getting longer and it is only a question of time before such interests have to be valued for compulsory purchase, probate, Inheritance Tax or on an unexpected sale. The value itself is much closer to the VP value than study 2, because of the effect of the reversion and the higher rent during the term. This has more than offset the higher discount rate.

# The impact of diversification

The valuation studies above have all assumed that the farm is a straightforward holding, its land and buildings devoted to agricultural production. There has been an increasing trend for farmers to diversify into other enterprises due to the economic pressures on agriculture. Grants have been available to encourage this although the Rural Enterprise Scheme closed to new applicants in June 2006 and has not been replaced with a new scheme yet.

Diversification in the farm business will not always add to the value of the property, as the new enterprise may be very personal to the proprietor in view of special skills or knowledge of the market. The effect of this would be that a purchaser would not be able easily to replicate the same business. It can be useful to think of a new enterprise in terms of the management time it requires, the expertise needed to run it and its potential profit. An enterprise which requires little management time, limited expertise and the potential of a reasonable profit for example, is likely to add significantly to the value of the property. Some new enterprises on farms, however, are very demanding in expertise and management time, and may offer only modest profits. In this case, the enterprise is very personal to the owner and is unlikely to add significantly to the value of the property.

A good example of an enterprise that would add to the value of property would be conversion of buildings to commercial uses for letting. Once the conversion is completed, day to day management requirements are likely to be modest. Significant expertise is required to plan and oversee the conversion work, and to find new tenants but once a tenant has taken up occupancy the expertise to manage the site from day to day becomes modest. Reasonable returns can be expected from a fully let conversion. Therefore the value of the property would normally be expected to increase.

The valuation approach may be based on comparables, but is just as likely to draw on one of the other traditional methods of valuation. An office conversion, for example, would be valued using the investment method of valuation. Other types of enterprise might have to be valued using a profits approach, or a modification of it.

Before finalising a valuation of a diversified farm, it is important to consider a range of additional factors. The existence of planning consent is important as many developments have arisen in an *ad hoc* fashion, sometimes without the formality of a proper planning consent. Rural planning consents may also have been made subject to special conditions. For example, new dwellings on farms will almost certainly be subject to a special occupancy condition. In the case of agricultural dwellings, the typical condition restricts occupation to somebody engaged in agriculture in the vicinity, or retired from agriculture, or their dependants. Although there is relatively little market evidence for the impact of such conditions on value, a widely held view among valuers is that a clause like this can depreciate values by 25 to 40%.

It is also appropriate to 'stand back' from the property and to consider whether the diversified business has changed the fundamental nature of the property. Is it still a farm with another enterprise? Or has it become a different business, with the remnant of a farm? If the latter, a different overall valuation approach may be indicated.

A number of other regulations may also be important to the value of a diversified farm. For example compliance with Building Regulations, specific Health and Safety requirements including Fire Regulations and some sector-specific rules. An example of the latter would be the local authority licensing of riding establishments. It may not be necessary for the valuer to check all these aspects in detail, but clear assumptions may have to be made and reported as to the existence of the necessary consents.

The degree of specialisation required, and management time, have already been mentioned. Local competition and viability should also be considered. How easy would it be, for example, for another farmer to set up the same enterprise nearby?

Tenanted farms raise further questions over diversification. For example, Schedule 2 of the Agricultural Holdings Act 1986 requires that tenant's improvements are excluded from consideration of the rent properly payable for the holding. In the unreported case of *Tummon* v *Barclays Bank* the enterprise in issue was a caravan park. The decision made clear that:

> You have to leave out of account the fact that the improvement has been done by the tenant ... but take into account the possibility that the new tenant might ... carry out the improvement and so get much better value out of the farm.

# The valuation of agricultural tenancies

One of the more challenging valuations to arise from time to time concerns the valuation of a tenant's interest in an agricultural tenancy, in particular 1986 Act tenancies. It is standard practice for these tenancies to be subject to an absolute bar on assignment or subletting. Where the tenant is free to assign, the landlord may seek arbitration under the Act as to the terms of the tenancy and the arbitrator can impose a suitable clause. It is not unknown for a tenant who finds himself in the happy position of occupying under an assignable tenancy to transfer his interest to a limited company, so as to secure its perpetual continuation.

Therefore there is no real market in agricultural tenants' interests. However, these interests do have to be valued from time to time. In particular, the need can arise in connection with Inheritance Tax and Capital Gains Tax, on compulsory purchase and on the dissolution of a business partnership or divorce. In the absence of market evidence, the Lands Tribunal has had to consider a number of approaches.

The legal basis for a valuation on compulsory purchase will be section 20 of the Compulsory Purchase Act 1965 in most cases, which entitles short term tenants to the value of their unexpired term or interest, or the Inheritance Tax Act 1984 (as amended) for an Inheritance Tax valuation. Chapter 8 discusses tax valuations in more detail, but suffice to note here that the valuation is also a market valuation for which it may be necessary to consider a hypothetical sale by a hypothetical vendor. One effect of this contortion is that the actual vendor may be considered as a potential purchaser, and it is this argument which has been used successfully to establish that it is appropriate to establish a value on an un-assignable tenancy for tax purposes.

*Wakerley* v *St Edmundsbury Borough Council* [1977] is widely mentioned in textbooks concerning the valuation of tenancies for compulsory purchase. However, the case report itself has little to offer. The claim in contention was formulated on the basis of the farm profits capitalised over the period for which the tenant was expected to remain in occupation had the compulsory acquisition not occurred. A nominal award of £100 was made for the value of the tenant's interest because both valuers had omitted it from their claims as a separate item. As the rest of the award consisted of a lifetime's expectation of profits for the tenant it may be wondered whether the tenant had not already been fully compensated for the loss of his interest in the tenancy, but this does not seem to have occurred to the Tribunal. Wakerley and other cases do make it clear, however, that regard should be had to the likely duration of the tenancy in the absence of compulsory purchase in assessing the value of the tenant's interest.

A common approach to the assessment of compensation under section 20 of the Compulsory Purchase Act is to consider the vacant possession premium which arises in respect of a tenanted farm. Effectively this premium is split 50:50 between landlord and tenant, as study 4 shows.

## Study 4

Hawthorn Farm: acquisition of four hectares under compulsory purchase.

Four hectares of the farm already described are to be acquired under compulsory purchase powers. The holding is tenanted under the Agricultural Holdings Act 1986, with strong prospects of succession to the tenancy by the eldest of the farmer's three children. There is therefore little prospect of an early reversion to the landlord.

|  | £ |
|---|---|
| Value of 4 ha with vacant possession | 34,400 |
| Value of 4 ha subject to tenancy | 15,000 |
| Vacant possession premium (VPP) | 19,400 |
| Value of tenant's interest at 50% × VPP | 9,700 |

*Note*

In addition to the compensation for the value of the land taken, those affected are entitled to claim for Rule 6 items (Disturbance and other matters not related to the value of the land), including the cost of professional fees. Please refer to Chapter 8 for further details.

Section 20 of the Compulsory Purchase Act 1965 also allows a claim for 'any just allowance' which would be made to an outgoing tenant, covering for example traditional tenant right items.

The Agriculture (Miscellaneous Provisions) Act 1968 entitles the tenant to payment of four years' rent in respect of the land lost. As this involves an element of double-counting with the value of the tenant's interest, the amount of the rent payment is deducted from the value of the interest and claimed separately. This preserves the tax-free status of the rent payment, whereas the value of the tenant's interest may be subject to Capital Gains Tax.

Another approach was established in *Walton v Inland Revenue Commissioners* [1994], where the Lands Tribunal accepted that the prospect of early reversion to vacant possession was speculative and remote. A sophisticated profit rental approach was adopted, demonstrated by study 5.

## Study 5

Hawthorn Farm: valuation of tenancy for partnership dissolution.

|  | £ pa | £ |
|---|---|---|
| Estimated arbitration rent award | 31,000 | |
| Rent payable | 25,000 | |
| Profit rent, bottom slice | 6,000 | |
| YP 1 year 9 months at 5% | 1.60 | |
|  |  | 9,600 |

| | | |
|---|---:|---:|
| Tender Rent | 35,650 | |
| Estimated arbitration rent award | 31,000 | |
| Profit rent, top slice | 4,650 | |
| YP 3 years at 12% | 2.40 | |
| | | 11,160 |
| Value of tenant's interest | | £20,760 |

The profit rent has been divided into two slices. The 'bottom slice' is the difference between the rent currently payable (reduced from the previous study for the purpose of this one) and the rent which would be set under Schedule 2 of the 1986 Act on arbitration. Tender rents for 1986 Act lettings were always considerably higher than arbitrated rent levels because of the scarcity of available farms to let, so a second slice of profit rent is found from the difference between tender rents and arbitrated rents. Both slices of profit rent are then capitalised. The bottom slice is capitalised until the next rent review, and the tender rent is capitalised until the first review that could occur with a new tenant. The two slices are not entirely compatible here in terms of their timing, but nevertheless this was the preferred approach in the *Walton* case. *Walton* concerned a tax valuation, but the same approach has since been endorsed by the Court of Appeal in *Greenbank* v *Pickles* [2001] for the valuation of agricultural assets on a partnership dissolution.

# Agricultural value

One of the latest challenges in professional practice concerns the availability of Agricultural Property Relief from Inheritance Tax. Inheritance Tax is payable on death and on certain lifetime transfers of property, on the market value of assets. The rate is currently 40% of all value above a nil-rate threshold (currently £300,000) which is reviewed every year. The definition of market value is set out in section 160 of the Inheritance Tax Act 1984. Agricultural Property Relief is available in respect of in-hand and let farms under section 115 of the Act, but it is important to note that the relief is restricted to 'agricultural value' under the Act. The concept of agricultural value assumes a perpetual covenant that restricts the use of the land to agriculture. Therefore value for other uses has to be excluded. A series of recent cases before the Commissioners of Taxation and the Lands Tribunal has explored these issues in some depth, as a result of which a greater onus is emerging on valuers to provide a very full appraisal in connection with potential claims for Agricultural Property Relief.

The first step is to identify 'agricultural property' for the purposes of this relief. As far as land is concerned this is straightforward. Dwellings, woodland and buildings can also be recognised as agricultural property if they are 'ancillary' to the occupation of land, and if they are of 'character appropriate' to it. Five tests have been postulated to judge 'character appropriate'. The factors to consider are the size of the farm in relation to buildings and dwellings, the type of farming undertaken, and the so-called 'elephant test' ('you know it when you see it'). The fourth test is to ask whether the property is a farm with a house, or a house with land (a 'gentleman's residence'), and the final test is to look at the history of agricultural production and the length of association between the farmhouse and the farming of the land. There must have been continuous occupation for the purposes of agriculture in the two years before death for in-hand land, and for seven years for let land.

The rate of relief is 100% in the case of in-hand land and lend let on tenancies since 1 September 1995, and 50% for land let before 1 September 1995. This date was set as the introduction of the new Farm Business Tenancies, although the detailed wording of the legislation means that succession tenancies under the Agricultural Holdings Act 1986 which have been awarded since 1 September 1995 also benefit from the higher rate of relief.

In the recent *Antrobus* and *McKenna* cases a number of issues have been tested, as a result of which it is advisable for valuers to ensure they receive very clear instructions from executors and their solicitors which should cover the need to identify agricultural property clearly, to set out both market valuations and agricultural valuations, and to identify those dwellings and buildings which are 'ancillary' and of 'character appropriate' to agricultural land. If the valuer has access to any information about continuous occupation during the periods of two or seven years, this may also be offered or if any evidence is apparent from the valuer's inspection this may also be relevant. Such valuation reports were often provided quite informally, as probate valuations. It is now clear that they are full Red Book valuations, and they should be described as Inheritance Tax Valuations as the term probate valuation is now taken to imply a deliberate under-valuation.

## Study 6

Hawthorn Farm: Valuation of the freehold interest with vacant possession for Inheritance Tax purposes

The entire property appears upon inspection to be agricultural in nature. There is no evidence of non-agricultural use in the modern portal-framed building and the house and buildings are of reasonable size for the holding having regard to its size, current cropping and stocking and in comparison with other farms in the area. The farm has been occupied as one unit for over fifty years by the same family, and although pleasant in appearance the house is visually dominated by the portal-framed building.

The entire holding is therefore considered to consist of agricultural property, with the buildings and house ancillary to the occupation of agricultural land. The house and buildings are therefore considered to be of 'character appropriate' to the property.

The market value for Inheritance Tax purposes is considered to be £2.2 m (from study 1). The agricultural value of the property has been assessed as follows:

|  | £ |
|---|---|
| House at 70% OMV (as *Antrobus*) | 420,000 |
| Better land (from study 1) | 1,204,000 |
| Poorer land (from study 1) | 360,000 |
| Agricultural value | £1,984,000 |

The valuation of the buildings has been reflected in the land value, and it will be seen that the market value of the house has been reduced by 30%. This reflects the effect of the perpetual covenant assumed by the Inheritance Tax Act, restricting occupation to the purposes of agriculture. The discount applied here follows the outcome of the second of the two *Antrobus* cases, where the Lands Tribunal took the view that this was the appropriate discount for a substantial farmhouse in Warwickshire within reasonable commuting distance of Birmingham and Worcester. The evidence for the amount of the discount was however, fairly thin so considerable caution should be exercised in applying it elsewhere. For example, a farm in the commuter belt of any major conurbation may be subject to a significant difference between market and agricultural value, whereas less attractive property in more remote or less popular areas may not justify a discount at all. Due consideration should also be given to any agricultural occupancy planning conditions when establishing market value.

In this study, Agricultural Property Relief will be given as 100% of the agricultural valuation, ie £1.984 m, leaving a difference of £216,000 liable to Inheritance Tax. Business property on the farm, ie crops in store, tractors and machinery, and livestock will also need to be valued, and will be relieved by Business Property Relief which is also available at 100% in most cases.

# Concluding comments

Agricultural valuation has always been a challenging area of practice, and these challenges continue to change and grow. This chapter has offered an introduction to some of the many factors which must be considered when valuing agricultural property, taking the reader from consideration of some of the fundamentals which apply to any farm valuation to some of the more specialised agricultural valuations which are not currently documented very thoroughly elsewhere. If it ignites an interest in this area of practice in a young valuer or surveyor, or serves as a warning to the unwary it will have served its purpose. A reading list is offered for those wishing to pursue the topic further, which will also provide well-documented coverage of a number of valuation examples which this chapter has sought to complement.

© Charles Cowap 2008

# Further Reading

*A Schedule of Time Limits for Action by Valuers*, CAAV (2006)
> Helpful summary of time-limits which apply to notices under the Agricultural Holdings Act 1986, Agriculture Act 1986 (milk quota compensation), Agricultural Tenancies Act 1995, Arbitration Act 1996 and the Dairy Produce Quota Regulations 2005.

*Agricultural Tenancies: the 2006 Reforms and Update*, CAAV (2006)
> Deals with the impact of CAP reforms on agricultural tenancy matters.

*Agricultural Valuations: a Practical Guide* Williams, RG. 3rd ed, Estates Gazette (1998)
> An invaluable 'how to do it' book although this edition is inevitably starting now to show its age. Not all practitioners agree with all the approaches advocated, but full of examples of different types of valuation which will at least get you started on the task in hand. An essential companion for students and probationers in mainstream agricultural practice, especially if you plan to take the CAAV examinations.

*Appraisal and Valuation Standards* (The Red Book) Royal Institution of Chartered Surveyors
> Essential reference for practising valuers as infringement is a disciplinary offence for RICS members. New version launched in 2003, which is fundamentally different in layout and approach from previous editions. In particular see Information Paper 5: Valuation of Rural Property which was prepared by the Rural Valuation Group of RICS.

*Dilapidations on the end of a tenancy*, CAAV (2007)
> Companion paper to the 2006 publication on end of tenancy compensation.

*End of Tenancy Compensation under the Agricultural Holdings Act 1986*, CAAV (2006).
> Up to date guidance on valuation matters which must be addressed at the end of a traditional agricultural tenancy.

"Entitlements — a valuer's perspective" O'Brien, J. *Bulletin of the Agricultural Law Association*, Issue 42, Autumn 2005, pp7–9 (Agricultural Law Association)
> An early perspective on the valuation of single payment entitlements, considering a DCF approach.

*Excel for Surveyors*, Bowcock, P & Bayfield, N. Estates Gazette (2000)
> Very useful for its relevance to valuation theory, and covers some basic aspects of spreadsheet use as well.

*Just what is a farmhouse? Recent developments in agricultural property relief from Inheritance Tax and their implications for professional practice*, Cowap, C. RICS (2007)

RICS Roots Rural Research Conference, London, 17 April 2007.

Examines the recent *Antrobus* and *McKenna* cases and assesses in detail the implications for professional practice and instructions, setting out a series of steps which should satisfy the requirements of both estate executors, the Capital Taxes of HM Revenue and Customs and District Valuers.

*Modern Methods of Valuation of Land, Houses and Buildings*, Johnson, T; Davies, K; Shapiro, E. 9th ed Estates Gazette (2000)

Good all-round introductory text for basic theory as well as a range of applications.

*Rent Reviews under the Agricultural Holdings Act 1986*, CAAV and RICS (2000) RICS Information Paper; CAAV Numbered Publication 178

Comprehensive guidance on rent reviews under the 1986 Act. A new edition is expected in September 2007 which will incorporate guidance on the treatment of single Farm Payments.

*Single Payment Entitlement trading, transfer and valuation issues (England region)* RICS Rural Faculty Briefing Note, RICS (2006)

RICS Guidance prepared by the Rural Valuation Group on various aspects of the new Single Payment Scheme.

'Tenant farmers out in the cold' Cowap, C. *RICS Land Journal*, January 2007, pp14–16, RICS

Reviews, with worked examples, the new Owner and Occupier Loss payments introduced under the Planning and Compulsory Purchase Act 2004. There are particular complications for annual agricultural tenancies under the 1986 Act, and all farming claims need to consider the alternative of a tariff-based payment for land and buildings as an alternative to an occupier's claim based on 2.5% of compensation claimed.

*The valuation of agricultural tenancies: art or artifice?* Cowap, C. RICS (1997)

Paper presented at RICS Roots Rural Research Conference, Reading University, January 1997 London: RICS

Detailed consideration of the valuation of agricultural tenancies for compulsory purchase, taxation and other purposes. Includes details of Lands Tribunal valuation decisions.

*The Valuation of Rural Property* Prag, PAB. Packard, Chichester (1998)

All round general introduction to the valuation of rural property.

*Valuation, Special Properties and Purposes*, Askham, P. (ed) Estates Gazette (2003)

Two chapters on milk quotas and forestry respectively may be particularly useful for anybody faced with one of these highly specialised valuations.

# Useful case references

## *Valuation of agricultural tenants' interests*

*Agricultural Mortgage Corporation plc v Woodward* [1994] EGCS 98
*Anderson v Moray District Council* [1978] Scots Law Times 37
*Baird's Exors v Commissioners of Inland Revenue* [1991] 1 EGLR 201; [1991] 09 EG 129 & 10 EG 153
*Dawson v Norwich City Council* [1979] 1 EGLR 204
*Gooderam v Dept of Transport* [1995] RVR 12
*Greenbank v Pickles* [2001] 1 EGLR 1; [2001] 09 EG 230

*Layzell* v *Smith Morton and Long* [1992] 1 EGLR 169; [1992] 13 EG 118

*Runcorn Association Football Club Ltd* v *Warrington & Runcorn Development Corporation* [1982] 2 EGLR 216

*Wakerley* v *St Edmundsbury Borough Council* [1977] 1 EGLR 158

*Walton (Ex'cr of JH Walton dec'd)* v *Inland Revenue Commissioners* [1994] RVR 217, LT

*Walton (Ex'cr of JH Walton dec'd)* v *Inland Revenue Commissioners* [1996] RVR 55, CA

## Agricultural Property Relief from Inheritance Tax

*CJF Arnander, DTM Lloyd and MM Villiers, Ex'crs of David McKenna (dec'd) and Lady Ceciliar McKenna (dec'd)* v *HMRC* SPC 00565 23.10.2005

*Dixon* v *IRC* [2002] STC (SCD) 53

*Higginson, John Sidney (Ex'crs of)* v *CIR* [2002] STC (SCD) 483

*Korner* v *CIR* (1969) 45 TC 287

*Lloyds TSB (personal representative of Rosemary Antrobus dec'd)* v *Commissioners of Inland Revenue (Antrobus No 1)* [2002] STC (SCD) 468; SC/3062/2002, 9.10.02

*Lloyds TSB Private Banking plc (personal representative of Rosemary Antrobus dec'd)* v *Peter Twiddy (IR Capital Taxes) (Antrobus No 2)* [2005] DET/47/2004, 10.10.2005; [2006] RVR 138

*Pissaridou (HM Revenue & Customs)* v *Enid Rosser (personal rep of Phillips dec'd)* [2005] TMA/40/2005

*Rosser, Enid Meriod Amelia* v *CIR* SpC 368 [2003] SSCD 311

*Starke & Another (exec'rs Brown dec'd)* v *IRC* [1994] STC 295; [1995] STC 689, CA

*Williams, R* v *HMRC* [2005] SWTI 1682

# Residential Properties

Historically residential property, as opposed to commercial property, has been subject to legislative restrictions. The normal economic rules of demand and supply, and freedom of contract, have been curtailed and there is therefore an entirely different valuation approach to that which applies to other types of property. Unlike the market for commercial property, there are two separate and distinct markets, purchase and rental, and there is no fixed correlation between them. All valuations must be carried out with an eye on the vacant possession value of the property and, except in the very rarest cases, this is the maximum value which can be put on a property, no matter how high or how low a yield it would seem to give.

The vacant possession value for the purposes of this chapter may be defined as the price at which a house or flat could be sold to an owner–occupier in the open market, based on the definition of market value as laid down in the *RICS Valuation Standards* (the Red Book), the underlying assumption being that this is the maximum value. It is therefore made on the assumption that the property is in such a state of repair and of such a nature as to be capable of being mortgaged to a building society or similar lending institution. This definition is applied to a single unit of accommodation and not to an estate of houses or to a block or blocks of flats taken as a whole. The reason for this restricted definition is that there should be a distinction both from properties which require emendation (in terms of conversion, repair or change in design) and from properties which require a wholesale as opposed to a retail consideration.

The Rent Act 1977 (as amended by the Housing Act 1980), the Housing Act 1988 and the Housing Acts 1996 and 2004 are the principal Acts of Parliament giving security of tenure to tenants and limiting the rent payable in respect of residential property, whether furnished or unfurnished (provided that the letting comes within the ambit of the Acts). There are several types of tenancy under the Acts, viz:

1. tenancies created prior to January 1989:
   (i) regulated
   (ii) secure
2. tenancies created after January 1989:
   (i) assured shorthold
   (ii) assured.

Irrespective of whether a tenancy was created prior to January 1986 a regulated tenancy can only exist if the flat or house had a rateable value below certain limits, ie below £1,500 in Metropolitan London and £750 elsewhere, as at March 31 1990, whether furnished or unfurnished. As rateable values are no longer the basis for local taxation, these RV limits will now remain constant with no future changes. There will be no new RV limits or any substituted system, because a tenancy which is entered into after the commencement of the Housing Act 1988 cannot be a regulated tenancy (except in the special cases set out by section 34 of that Act) and no regulated tenancy can exist where the rental value is in excess of £25,000 pa. The Act came into force in January 1989.

The four types of tenancies referred to above are briefly described as follows.

## 1. Regulated

This is a tenancy to which the Rent Act 1977 applies. It provides security of tenure for the tenant and up to two successors. It also restricts the rent which can be charged by a landlord to a 'fair rent', which is fixed by a rent officer or, on appeal, by a rent assessment committee; a successor spouse to the tenant has the same protection as the tenant, but a non-spouse successor becomes an assured tenant (see later). Within the definition of fair rent there is the assumption that the supply of residential accommodation within the area is in approximate balance with the demand for such accommodation. This assumption removes the 'scarcity factor' from the normal working of the market and thus ensures a rent which is lower than the open market rent, except in locations where no scarcity of rented accommodation exists.

This type of tenancy is becoming scarce as they must have been created more than 18 years ago and so the tenant(s) are usually elderly. Thus with every passing year the prospect of vacant possession or a transfer to an assured tenancy becomes greater.

## 2. Secure

A secure tenancy is one which exists in the public sector and is governed by sections 28–61 of the Housing Act 1980 and the Housing Act 1985. Tenants in the public sector now enjoy security of tenure, subject to certain exceptions. Because this type of property has no effect on the valuation of residential property in the open market no more will be said of it.

## 3. Assured shorthold (AST)

These were the creation of section 20 of the Housing Act 1988 and were up to 1997 tenancies:

(i)   granted for a term certain of not less than six months
(ii)  in respect of which there was no power for the landlord to determine the tenancy at any time earlier than six months from its beginning and
(iii) in respect of which a notice was served by the person who was to be the landlord on the person who was to be the tenant stating that the tenancy was to be an assured shorthold tenancy.

All residential tenancies granted after February 28 1997 are now assured shorthold tenancies (ASTs) unless there is a written agreement which unambiguously creates an Assured Tenancy (Housing Act 1996); thus there is now no need for a pre-tenancy notice. Because of this change, it is necessary to check the date when the original tenancy came into existence. If the tenancy was created before February 28 1997 then, in the absence of a written agreement and pre-tenancy notice, such a tenancy would either be a regulated tenancy or an assured tenancy. The fact that a tenancy was renewed after 1997 does not make it an AST if the first tenancy did not comply with the above specified three rules. In effect, once an assured or regulated tenancy then always one.

In summary therefore all tenancies entered into before January 1989 would be assured tenancies unless specifically created as an Assured Shorthold Tenancy and there was a pre-tenancy notice and unless the other conditions given at the beginning of this section were complied with. If the same tenancy was created after 1997, then such a tenancy would automatically be an assured shorthold Tenancy unless there was a written agreement specifically making it an assured tenancy.

In this type of tenancy the tenant will not have security of tenure and only a limited degree of control of the rent; the latter would not help the tenant very much because he could be evicted within a short term. However, the rules relating to the stamping of documents must be followed if an order for possession is to be obtained. Possession can only be obtained through the courts, if not given voluntarily, and the Prevention from Eviction Act 1964 still applies with criminal sanctions if not followed; illegal eviction carry dire consequences both in terms of fines, custodial sentences and damages.

## 4. *Assured*

These are the creation of section 1 of the Housing Act 1988 and, briefly stated, are tenancies where the tenant has full security of tenure but there is a limited control on the rent. They resemble commercial tenancies, which are within the control of Part II of the Landlord and Tenant Act 1954, and as stated above must now be specifically created.

The various types of tenancies will be considered in greater detail later in this chapter. However, in view of the statement given above that all valuations must be carried out with an eye on the vacant possession value of the property, it is clear that it is the security of tenure provisions which have the greatest impact on the valuation of a property if it is subject to a tenancy. It should also be borne in mind that there are a number of other tenancies which are outside the control of either the Rent Acts or the Housing Act 1988 and these include company lettings, holiday lettings, service tenancies, licences and those where there is a large measure of service.

A further factor which distinguishes residential property from commercial property is the different taxation treatment which is given to it. An owner-occupier is specially favoured from the point of view of capital gains tax, as any increase in the value of his principal home is free from taxation. It is worth noting the expectation of capital growth on the sentiment of persons in the market for residential accommodation. It is usually this which determines whether they will become tenants or owner-occupiers and it also drives the buy-to-rent market.

The cost of interest of a 100% mortgage on a £100,000 house (interest only) at say 6.0% is £6,000 pa. Assuming that the house increases in value over 10 years by an average of 5% pa compound, the value of the house after 10 years will be £162,890. Therefore the total net cost over 10 years ignoring tax will be as follows:

|  | £ |
|---|---|
| Annual cost, £100,000 at 6% | 6,000 |
| Amount of £1 pa for 10 years at 6% | 13.18 |
| Total: interest | 79,085 |
| *Less*: Capital Gain | 62,890 |
| True Net Cost | 16,195 |
| Annual equivalent of true net cost (£16,195 ÷ 13.18) = | £1,229 pa |

A tenant of the same property will not make any gain over the period of the tenancy. Such a tenant should therefore only be prepared to pay a rent for the property which equates to an average of £1,229 pa over the 10-year period if he is to be in the same position as his neighbour who purchased the house with the aid of a mortgage. Such a rent would be totally unacceptable to an investor, since over the 10-year period he would see an average yield of only 1.23%. This would obviously be unacceptable at times of interest rates at say 5–6%, unless the investor could obtain vacant possession at the end of the 10-year period. If the latter were possible then the investor's average yield over the 10-year period would be 6.23% before tax. The investor's after tax position is complicated by the fact that part of the capital gain will be tax free being that part above the rate of inflation or will be reduced by tapering relief. The suggested rate is also patently too low as it suggests too much of a bargain for the short term tenant and thus the demand/supply equation would push this rent up.

Before considering this matter further, some comment must be made on the state of the housing and mortgage market at the time of writing this chapter, which is the winter of 2006/07, and the historical context in which it is set.

A recession overtook the housing market throughout the United Kingdom in about the Spring/Summer of 1989. Capital values fell and, with the exception of one or two false starts, they continued to fall until about the middle of 1996. The fall in values from the peak to the troughs was an average of approximately one-third although in some areas, the fall was much greater; London and the south east were the hardest hit.

The housing market began a recovery from about the beginning of 1996 and significant price rises were recorded in the period 1997–2000 (1999 being a boom year). This recovery was due to a combination of very low interest rates and the general recovery in the economy as a whole. Government also became very aware of the importance of a strong housing market for the stability of the general economy, and hence there is now a general political philosophy to the effect that the housing market must be maintained with both main parties wishing to see a gentle but sustainable increase, year on year, of the market as a whole. The government changed in the spring of 1997 and the then new Labour government passed control of interest rates to the Bank of England with the result that interest rates have ceased to be a political weapon but have become the basis of control of inflation. Over the last many years these have been running at historically low levels resulting in mortgages becoming cheaper and the lending institutions increasing their multiples of earnings. The general result has been that house prices have continuously increased (with only minor set-backs). The fall of interest rates combined with the strength of the housing market has also led to a considerable increase in investor interest, now called the buy-to-let market, which has further led to a strengthening of the market.

The underlying hypothesis of this chapter is that mortgage interest rates will remain within the range of 4–6% over the course of the foreseeable future and that house prices will rise at a moderate and sustainable rate. This has been the trend since the second world war during which time the market has seen many peaks and troughs and gyrating interest rates. The rate of increase in value has varied according to location within the country, but the average increase between the end of 1973 and the end of 2006 for all properties in the United Kingdom has been approximately 8.5% pa compound. A more detailed picture, based in the Nationwide Building Society UK House Prices Index is as follows:

| Years | House Prices | (Cost of Living) |
|---|---|---|
| 1983–1988 | 16.28% pa compound | (4.65%) |
| 1988–1993 | 2.53% pa compound | (5.65%) |
| 1993–1997 | 5.40% pa compound | (2.86%) |
| 1997–1999 | 11.09% pa compound | (2.48%) |

| Years | House Prices | (Cost of Living) |
|---|---|---|
| 1999–2006 | 14.51% pa compound | (3.67%) |
| 1983–2006 | 8.50% pa compound | (3.75%) |

It would appear from the above that an assumption that house prices will continue to rise over the foreseeable future, with perhaps short-term periods of stagnation or even falls, is a reasonable hypothesis, but over the long term there is no guarantee that the rate of increase will outpace the cost of living.

The percentage of households which owned their own houses, as opposed to renting them, rose continuously from the turn of the century until about 1989 when it got very close to 70%. The reasons for this increase were a combination of the prospects for large capital gains, the reduction in both the public and private sector supply of housing to rent, and the easy availability of mortgage finance. The second of these was caused by the right to buy policy of the Thatcher Conservative government (not rescinded by the subsequent Labour government) which reduced the number of local authority units (the proceeds could not by statute be re-invested in the building of new council housing) and the private sector's concern over security of tenure and rent control legislation. Tenants and potential tenants were encouraged to buy by freely available mortgages of up to 100% of the purchase price at low interest rates. The percentage has fallen slightly over recent years due to the scars left by the early 1990's recession when many new owner occupiers lost all their capital as a result of foreclosures and the growth in the availability of housing to rent following the growth in the buy-to-rent sector.

As stated above, the bubble burst during 1989 and in the subsequent years this led to the creation of a new term in the housing vocabulary: negative equity. The high percentage of loans taken by many and rising interest rates, combined with falling prices led to many owner-occupiers having mortgages which were greater than the capital value of their property. Many homes were repossessed by the banks and building societies and many families became disenchanted with the concept of owner occupation. For an unprecedented period of years, many owner-occupiers found themselves unable to move because they could not sell their properties and this led to a restriction on the mobility of labour at a time when unemployment was increasing rapidly.

For the first time in at least two generations (and possibly three), many people, both sophisticated and not sophisticated, see the rental market as providing a more satisfactory home than the owner-occupier market. Some families will never recover from the trauma of losing their home and their capital, and will hereafter wish to rent rather than buy; others believe that the advantages of mobility and the opportunity of capital growth in other investment markets overrides the disadvantages of renting.

The changes in the law concerning security of tenure and rent control has also led to the expansion of the private sector market for letting, so that the demand for good quality housing can now be met by the supply of such housing, albeit as a short term lettings only. The economics of renting, as opposed to buying, as set out above still apply, so that a significant part of the return to the investor will remain as capital gain as opposed to just rent. The likely net result is that the headline rent will be lower than the opportunity cost of capital but the cost of renting over any significant period of time will still be greater than the annualised cost of purchase.

There will also be the social factor, which is that in all but a few cases lettings will be on a short term basis, and thus the tenants will not have security of tenure. This in turn will prevent them from investing in their own homes, so that for the average property there will be the social disadvantage of living with somebody else's fixtures and fittings in so far as this applies to kitchens, bathrooms etc. The properly advised investor will not allow tenants to have security of tenure because of their need to retain the ability to make capital gains by disposals, and these will only be fully available by the sale of vacant properties. This situation will only change if residential tenants are prepared to take on long term rental commitments with frequent rent reviews, and where such tenants are of good covenant. In the view of the writer, this is unlikely to happen because the principal advantage of renting, as opposed to buying, is the benefit of mobility which would be lost if a long term commitment had to be taken.

It will therefore be clearly apparent that the rent which can practically and economically be paid by a tenant will be significantly below the amount which an owner–occupier can afford to pay by way of interest on a mortgage. At this level of rent no landlord can afford to let unless he is guaranteed that he can obtain vacant possession after the letting term (in order to secure a capital gain).

The situation is, however, entirely different if vacant possession is not obtainable at the end of the term. We will assume that a higher rent than £1,229 pa could be extracted from a tenant, notwithstanding that this would make renting more expensive than purchasing if vacant possession values rose by 5% pa compound. For the purposes of this example, we will assume a rent of £5,000 pa on a £100,000 home and we will also assume that this rent would rise annually in accordance with increases in the cost of living, which we will take at 3% pa. On this basis, the rental value after 10 years would have increased to £6,720 pa.

The true performance of this investment, provided that vacant possession was available at the end of the 10-year period, would have been an average running yield at a purchase cost of £100,000 at 5.86% plus capital growth of 5% giving a total average yield of 10.86%. However, if vacant possession was not available at the end of the term, all that would be receivable at that time would be the increased rent of £6,720 pa, which would really be an unacceptably low yield having regard to the commercial risks involved on non full repairing lettings, but also, in all probability, a capital loss both in real and in nominal terms. (No allowance has been made so far for landlord's outgoings and the rents utilised have been assumed to be net rents.)

Without the ability to sell the property with vacant possession at the end of the first lease, the yield on the investment is probably unacceptable, as the capital growth which was taken into account in estimating the yield when comparing it with a gilt-edged stock (which is risk free) would no longer be present. It is true that the rent would increase over the term but this would only mitigate the loss, not reduce it.

There are, therefore, two conclusions from the above analysis:

(a) security of tenure if given to a tenant is disruptive of capital value, as an acceptable return on the investment can only be obtained if there is a reversion to vacant possession value
(b) that owing to the existence of capital growth the average cost of occupation over any period is significantly lower than the interest which is actually being paid. The result of this is that tenants should only pay a rent which is considerably lower than what at first view would be a fair return on the value of the property they occupy.

It will be appreciated from the above that for all property which is saleable with vacant possession to an owner-occupier the valuation approach must be geared to the differential between the vacant and investment values; it must thus be concerned with marriage values. If the £100,000 house, which was referred to earlier, was let at, say, £2,500 pa, it would be illogical to consider that the investment value is only £25,000 (applying, say, a 10% return), because the size of the gap between that figure and the vacant possession figure of £100,000 is an incentive for the parties to merge their interest and share the profit. This will be considered in detail later in this chapter.

# Valuation on the basis of vacant possession

Owner–occupation in the UK rose from about 30% of dwellings in 1945 to a little under 70% in 2006. The reason for this growth was a combination of the effects of:

(i)   restrictions on private renting (which led landlords to sell off their properties rather than offer them to let)
(ii)  greater affluence and
(iii) the appeal of owner-occupation to the young, the unmarried and those with higher incomes who could not obtain rented accommodation from local authorities.

The prices realised in the market place are as a result of:

(i)   the demand and supply for housing
(ii)  levels of affluence and
(iii) government policy to encourage the sale of council and housing association property.

There are no exact limits to the category of property which is capable of sale with vacant possession to owner–occupiers because it is a matter which has very much to do with personal taste, and at its extremes includes any property capable of being lived in — from a sprawling country estate to a dilapidated shack. In practice, the lending policies of the financial institutions tend to limit the compass of the category although it does vary for different parts of the country.

Whether a property is indeed saleable to an ordinary purchaser, and the price it would obtain are matters for local knowledge. A valuer from outside the area must take extreme care when valuing a property in an area of which he has no intimate knowledge; while there are general trends of value, the specific will override the general.

The principal factors for determining the value of a property are the following:

(i)   location
(ii)  accommodation
(iii) the nature of the housing unit
(iii) the state of repair, appearance and quality of finish
(v)   the quality and quantum of the fixtures and fittings
(vi)  the potential of the neighbourhood in terms of general improvement and the potential of the property to be improved and modernised and
(vii) plot size.

All the above points are really self-explanatory and are well known to those who practice in the house agency sphere. Some locations are more fashionable than others owing frequently to historical or current associations with the rich and famous, also owing to matters such as accessibility to transport facilities, shopping and recreational facilities and other community advantages.

There is a great deal of snobbishness in the housing market which is reflected in the fact that some postal districts are more fashionable than others, and likewise with telephone exchanges, especially in the past, when the exchanges had names rather than numbers. Other locational factors which influence value are travelling distance from the centre of the nearest major city and the attractiveness of the local townscape and of the local surrounding area. In most towns there is a right side and wrong side of the tracks, and similar properties will have different values according to which side they are on. The political complexion of the local authority coupled with the level of council tax is also relevant.

The quantum of accommodation is clearly important and usually the larger the property the higher the value. There is, however, no exact correlation between value and size; therefore, if a house has 20% more accommodation it does not mean that it is 20% more valuable. In many areas there is an optimum size of a house or flat which commands the highest price per m². For example, a three-room flat with kitchen and bathroom/WC having a gross internal floor area of 60 m² might be worth, say, £120,000 but a four-room flat in the same block having a gross internal area of, say, 70 m² may be worth only £134,000; thus the value has fallen from £2,000 per m² to just under £1,915 per m² as a result of the increase in size. The reason for this is that as three-bedroom flats (ie four rooms) are more suitable for family occupation (as opposed to occupation by young married or elderly occupants) they are not as suitable as, say, a three-bedroom semi-detached house, and hence demand shifts from the flats to the houses, with a resultant loss in value. Similar results may well be found when comparing three-bedroom houses with four-bedroom houses, but in reverse, so that the former is less valuable per m² than the latter. It is not possible to set out any rules, since the effect of accommodation will vary according to area and housing type.

The nature of the housing unit also causes variation in value per m², so that a three-bedroom semi-detached house is less valuable than a three-bedroom detached house of exactly the same size and in exactly the same location. Thus detached houses have a higher value per unit of area than semi-detached houses, which in turn have a higher value than terraced housing. Other similar influences on value are historical associations, prettiness (in the case of cottages) and appearance of grandeur (in the case of large houses) coupled with the size of the garden which accompanies the house. With flats, the influencing factors are whether they are purpose built or conversions and whether they have lifts to the upper floors. The size and grandeur of the entrance hall, the extent of services offered and provisions for security also have a large part to play.

As a general rule, it is not possible to value residential property by applying to it a local price per m². Unlike offices, which may have a local value of £2,500 per m² being the product of £200 per m² pa capitalised at 8%, it is not really possible to say that flats in the given area have a capital value of say £2,000 per m². Houses and flats in the United Kingdom are not usually sold on this basis and are consequently not devalued on this basis. This is because there is no uniform unit of comparison. Thus offices are valued on a net usable floor space basis and this basis is defined in the RICS Code of Measuring Practice with modern offices being in clear space. Residential units are not in clear space but are divided by non-demountable walls. The value of the unit will be determined by the scale of the rooms, the quantum of bathrooms, the outlook from the rooms and the general feel of the unit on first approach. Small variations in size will not affect the value except within the immediate development (being the block of flats or the housing estate). The general unit of comparison is a similar property taken overall and then adjusted to reflect the factors already discussed. However in some markets, particularly in the Far East, housing units are sold on a price per m² and this applies when United Kingdom property is sold to that market. This only applies in very particular locations within the United Kingdom market, which in this case means within very particular locations within the Central London market. Properties should only be valued in the way that the market approaches valuation, so that the unit of comparison for shops is so much per m² Zone A; for offices it is so much per m² of usable space, and for factories and warehouses, it is so much per m² of gross internal space. Thus, for houses and flats, the 'currency' is so much per unit.

Having said that, residential property should not be valued at a rate per m² basis. This is a convenient approach in some cases such as for valuations under the Leasehold Reform Legislation where tenants' improvements have to be disregarded and it is not always possible to inspect all of the flats in a collective enfranchisement situation. Floor plans may be available to show the different sizes of the flats in the building and applying rates per m$^2$ may be the only logical approach to the valuation calculation, the appropriate rate being obtained from comparables analysed on this basis.

This valuation method is also frequently used by developers to arrive at a Gross Development Value before detailed plans of the proposed development have been drawn up, instead of a valuation by comparison with existing properties of similar size and configuration located in comparable locations, but clearly derived from the same.

Before most house purchasers actually buy a house they have the property surveyed in order to ascertain its structural condition and its state of repair. Surveyors are asked to quantify the actual costs of necessary repairs and purchasers then seek a deduction from the purchase price to reflect this cost. Where the defect is major, such as the need for a new roof, there is clearly some betterment to the property and the deduction allowed will frequently be less than the actual cost of the works. However, in many cases the reverse will be true because of the degree of upheaval and mess necessitated by the repairs. The major repairing items which affect value are settlement and heave, the extent of any dampness (both rising dampness and penetrating dampness), the extent of any rot (both dry rot and wet rot, although the former is the more relevant) and the condition of the roof and services.

Coupled with the actual condition of the property is the appearance which the property gives on being shown. The lay purchaser is influenced by a house which looks as if it has been well maintained, with fresh decorations and the appearance of having been modernised. If he feels that he can virtually move in without carrying out a great deal of work he will pay a higher price than if the reverse is true. The importance of this aspect is shown by the American market where the agent arranges viewing days (with the owners absent) and where they 'improve' the property by putting in flower displays around the house and generally giving it the best beauty treatment that they can. The structural survey may then prove that the appearance of being in good repair is superficial only and that there are major defects which require remedying, owing possibly to an excess of DIY zeal.

The first price agreed between vendor and purchaser is of the property as first shown and the final price will depend on the result of the survey. The writer has known many cases where a dream house has been turned into a nightmare because rising damp and dry rot have been concealed behind panelling, walls have been removed without supporting RSJs (rolled steel joists) being provided, and extensions constructed without building regulation consent and without adequate foundations. In the latter case the deduction in value following the survey will almost certainly exceed the actual cost of the works to allow for the fact that the house will be uninhabitable while the builders work within the property.

Where works of modernisation and decorations have been correctly carried out the actual standard of these works will influence the final value of the property, although taste has a great deal to do with the final answer; taste and expenditure do not always go hand in hand. Stated at its simplest, a house with central heating will have a higher value than a house without it unless the heating system itself is obsolete and requires replacing. A house with double glazing will have a higher value than a house with single glazing although not necessarily by the amount of the extra cost of this provision. A house or flat with attractive cornices, good quality doors, and high-quality kitchen and bathroom fittings will have a higher value than a house or flat with lower quality and this really speaks for itself. However, if the owner's taste is unusual in choice of colours or materials then this could well nullify the extra costs which have been expended in providing that particular standard of finish.

The aspect of a property is frequently seen as an important factor by purchasers. Thus south facing gardens are often seen as important selling points even though the converse of this is that the rooms on the north side of the house get no sunlight whatsoever. The same considerations with regard to sunlight apply to flats as well as houses. Balconies and roof terraces are also of great importance: a roof terrace can add 10% to the value of a flat and a balcony between 2.5% and 5% although these figures are not hard and fast. Likewise, a flat with its own garden will be 10–15% more valuable than a standard flat with shared gardens. In assessing the value of a garden, terrace or balcony orientation will be very important.

Finally, the potential of the neighbourhood and also of the house itself will be a determining factor in the valuation of the property. Areas change from being unfashionable to fashionable and the chances of such an event happening should be taken into account when valuing the property. The reverse is also true and, consequently, civic proposals often have a great deal to do with value. If there are local authority proposals to build a large comprehensive school on a green field site of high natural beauty this will reduce the value of the houses where views will be altered. Likewise if there are major road proposals or similar future schemes which the buying public may apprehend as being adverse.

The reverse, of course, is also true and proposed improvements in communications will increase the value of property in an area which might come within a more reasonable commuter distance of a major city, whereas at the present time it is beyond such a reasonable distance. Similarly, the potential of the property itself to be increased in size or improved will also play a part in determining its value.

The one further consideration which is vital in the valuation of a residential property is its mortgageability. Many of the lending societies limit their lending to post-first world war purpose-built houses or flats of standard-built construction in good structural condition with pitched tiled or slated roofs. Some will accept flat roofs or flats provided by conversion or above shops, but the more the property varies from the standard the more difficult it will be to mortgage and therefore to sell, and therefore the lower its relative value. It is absolutely vital that the nature of the construction of the property is such that it will conform with the general policies of the lending institutions. The more unusual or experimental the nature of the construction, the more difficult it will be to mortgage the property and consequently the more difficult it will be to sell it, thus resulting in a lower value. A simple example is provided by houses with thatched roofs, which are more difficult to mortgage than houses with pitched tiled roofs and consequently more difficult to sell; the result is that these have a lower value than adjacent houses of similar size but of more conventional construction unless their 'prettiness' moves values in the opposite direction. With flats a further factor is the length of lease notwithstanding the leasehold reform legislation. The lending institutions still require minimum lease terms and a lease may still need to be extended in order to secure a satisfactory mortgage.

The difference between residential property and commercial property is that no profit in economic terms is derived from the use of residential property; it has a functional use in providing shelter but it also has an esoteric function in providing enjoyment of ownership and occupation to its possessor. In view of the latter, it is like any other possession in that it has a subjective value to its owner. Consequently, the more the property can be enjoyed by its possessor the higher its value. The exercise of the function of the valuer is not purely mathematical any more than the valuation of a picture or sculpture is a mathematical exercise.

In so far as houses are concerned, the usual rule is that the larger the plot (ie the bigger the garden), the more valuable the house. Where the house is set in 'grounds' extra acreage will not have as great an effect on the value as the initial acreage. Thus there will be little difference in value between a house set in 10 acres and a house set in 11 acres, but there will be a considerable difference in value between a house set in 1 acre and a house set in 2 acres. This proposition is put on the basis that there is no development value. Where there is development value, then clearly every additional square metre will have value because of the potential to redevelop in due course.

No consideration has been given above to whether the property is freehold or leasehold or the effects of the latter tenure on the value. This will be dealt with later in this chapter as also will the effect of the level of service charges, particularly in relation to flats.

# Freehold properties let on regulated tenancies

It should be noted that the law concerning succession to a statutory tenancy has been amended by section 39 of and Schedule 4 to the Housing Act 1988. A tenancy which passes to a spouse will continue as a regulated tenancy, but otherwise the tenancy which passes is an assured tenancy; the occupancy period in order to qualify to succession with the predecessor in title is increased to two years. The questions which must be asked by the valuer when valuing properties let on regulated tenancies are as follows:

1.  Is there a single letting of the property or is there more than one tenancy?
2.  Is there a single occupier-tenant for the whole of the property so that on the death or departure of that tenant vacant possession will be obtained?
3.  What is the age of the tenant(s) and what is his/her (or their) state of health?
4.  Can the property be sold in its existing state with vacant possession or will it need modification or improvement? If the later, what is the likely cost?
5.  Are there successors to the statutory tenant? If so, will there be a transmission to a spouse (thus continuing the regulated tenancy) or to a member of the family (which would therefore be a succession to an assured tenancy)?

## Study 1

A semi-detached house in a good suburban location with a vacant possession value of £160,000 is let to a regulated tenant at £3,500 pa exclusive of rates (just registered).

*1(a)  Value of landlord's interest as a pure investment*

|  |  | £ pa | £ pa |
|---|---|---:|---:|
| Rent reserved |  |  | 3,500 |
| *Less:* Outgoings: repairs | | 500 | |
| insurance | | 340 | |
| management | | 175 | 1,015 |
| Net Income |  |  | 2,485 |
| YP in perpetuity at, say, 8% |  |  | 12.5 |
|  |  |  | 31,062 |

Say, £31,000

This is the value to an investor on first view, but it is, of course, much too low, for reasons that will follow. The value of £31,000 is the apparent value to an investor and not to a sitting tenant. Between the investor and the sitting tenant there is a considerable marriage value notwithstanding the non-open market saleability of the tenant's interest. The tenant possesses a limited state of irremovability which has the effect of preventing the real value of the property being realised.

*1(b)  Consideration of marriage value*

|                                             | £       | £      |
|---------------------------------------------|---------|--------|
| Vacant possession value                     | 160,000 |        |
| Investor's value (on first view)            | 31,000  |        |
| Marriage value (MV)                         | 129,000 |        |
| Say 50% of MV                               | 0.5     | 64,500 |
| Investor's value                            |         | 31,000 |
| Potential value (59.69% of VP value)        |         | 95,500 |

Net yield (Net Income as return on the investor's value plus share of MV)
= 2.6% (£2,485 pa ÷ 100  £95,500)

The potential value must now be looked at in detail, as its yield is considerably below that possibly expected for what appears to be a fairly poor investment, bearing in mind the commercial banks' minimum lending rates. The disadvantages of the investment are, first, that there are landlord's outgoings which might in some years take a large part (or even all) of the income and, second, that the annual rate of growth of the income is limited, as it is not truly related to economic factors (the new rent is determined by a rent officer or rent assessment committee, who can be influenced by non-economic factors). The advantages are, first, that the income is secure, since its non-payment will give early vacant possession (which would be a windfall), and, second, that there is a long-term rate of capital growth which far outstrips most other normal investment.

Assuming that the underlying value of the property increases over 10 years from £160,000 to £260,000 (an increase of 5% pa compound) and that vacant possession is obtained after 10 years, the capital growth would equate to 10.36% pa compound before tax if the property was purchased for £95,500. If inflation is also running at 3% pa average over the 10-year period, the net return to a 40% tax-paying investor is as follows:

(i)   initial income yield 2.6% — this will be subject to increase every two years by re-registration, subject to the rent cap imposed to keep fair rents in line with inflation

(ii)  capital gain before tax = £164,500 (ie sale price of £260,000 less original cost (£95,500). The annual equivalent of the gain is 10.53%.

Marriage value usually envisages a deal between the landlord and the tenant. In this case it is taken to mean the effect of a potential deal done immediately. The value achieved by this calculation is used to calculate the return to the investor if the deal is not done and he waits 10 years for vacant possession.

The above analysis is based on assuming the price at which a sale to the tenant could be made to arrive at a value equal to just under 60% of the VP value, and using this price to consider the long term return to the investor if a sale is not made to the tenant. As a result of this analysis the valuation approach for investments of this kind is therefore to take a percentage of the vacant possession value of the property. The percentage taken will depend on the answers to the questions raised at the beginning of this section. It is rare for the percentage to exceed 70% and the usual percentage taken in average circumstances is 40%–60%, but the valuer must use his own judgment to decide on the appropriate percentage. The general rule is that the longer the anticipated delay until vacant possession can be obtained, the lower the percentage of vacant possession value. If the rental yield after taking say 50% of VP is unusually high, then the value would increase until a more usual yield is achieved. It is the prospect to early vacant possession that is the usual determination of the ultimate percentage and while this is usually related to the age and health of the tenant(s) it may also relate to the prospect of possession through rehousing.

A sale to a sitting tenant would justify a higher price because the true marriage value can be calculated using an investment value of say 60% of VP value thus giving a value of 80% of the VP value. The discount to the sitting tenant will be greater the lower the prospect for obtaining vacant possession by natural means.

# Freehold properties let on unfurnished tenancies with fair rent and with multiple tenancies

During the inter-war period, and for a decade or so after the second world war, it was quite common for houses to be let in floors with the bathroom and WC shared by the tenants. There was no self-containment between the floors and cooking facilities were provided in the smallest room on each floor. Thus a standard three-bedroom semi-detached house would often be occupied as two two-room flats each with a kitchen and sharing bathroom, WC and garden. The statistical chance of obtaining full vacant possession is less than for single tenancy houses.

There is a market for part-possession houses, as some owner–occupiers see the let portion as a way of helping to pay for the mortgage and there is, of course, a considerable discount on the full vacant possession value. The extent of the discount is difficult to estimate; it will vary according to location and type of property and the age of the remaining tenant(s).

Such property is of considerable appeal to young, first-time buyers who are prepared to put up with the disadvantages of sharing but can buy at a discount to the full VP value and also benefit from the additional income; the actual discount will depend on the age and type of the tenants and whether they have any children living with them. However, these properties may be difficult to mortgage.

The property is also of interest to an investor who might use the vacant portion to rehouse a tenant from another part-possession property which he owns elsewhere.

## Study 2

A three-storey house with vacant possession of the ground and first floors and the second floor let to a regulated tenant at a rent of £2,000 pa (recently registered). The accommodation comprises two rooms and a back-addition room on the ground floor, three rooms and a back addition (containing bathroom and WC) on the first floor, and three rooms on the second floor. The second-floor tenant uses the bathroom and WC on the first floor but the back addition could be extended to provide a bathroom and WC for the second floor and thus also provide a self-contained unit — see Note (1).

|  | £ | £ |
|---|---|---|
| Vacant possession value of ground and first floors if made into self-contained maisonette, Say |  | 160,000 |
| Value of second-floor flat (say 50% of assumed VP value of, say, £100,000), Note (2) |  | 50,000 |
|  |  | 210,000 |

| *Less:* | £ | £ | £ |
|---|---|---|---|
| (i) cost of improvement work, (3) say | 40,000 | | |
| (ii) fees on improvement work | 4,000 | | |
| (iii) legal and agent's fees | 4,200 | | |
| (iv) interest on (i) and (ii) (4) | 1,320 | | |
| (v) developer's profit, (5) say | 31,500 | | |
| | | 81,020 | |
| Residue, being value of the house, acquisition costs and interest until sale of improved building | | 128,980 | |
| PV £1 in 6 months at 6% | | 0.971 | |
| | | 125,240 | |
| *Less* 2% allowance for acquisition costs | | 2,456 | |
| | | | 122,784 |

Value of house, say £122,750

*Notes*

(1) The tenant's consent to improvements is required, but the courts have power to authorise modernisation works if the tenant does not lose accommodation which he reasonably requires. The improvements will increase the fair rent to, say, £2,500 pa.

(2) Some rent rebate may be necessary during the course of the works because of disturbance to the second-floor tenant.

(3) No allowance has been given in this example for improvement grants and the costs are assumed to include VAT.

(4) Short-term finance cost has been taken at 6% averaged over six months.

(5) The developer's profit has been taken as 15% of the final estimated entirety value to reflect his risks and to justify the exercise fully.

(6) The residual amount of £128,980 is calculated before providing for the costs of acquisition, such as stamp duty, legal costs etc., and before interest. These latter costs are deducted as end allowances and are in addition to the costs allowed for in (i)–(v).

# Freehold tenancies let on assured shorthold tenancies

The basis of these tenancies is that the tenant will not have security of tenure. They have to be let for a fixed term of not less than six months, there must be no power for the landlord to determine the tenancy at any time earlier than six months from the beginning of the tenancy and if the original tenancy was created before February 28 1997 there must be a notice served in the prescribed form before the tenancy is entered into stating that the tenancy is to be a shorthold tenancy; if these formalities are not followed the tenancy becomes an assured tenancy. The situation is reversed for all tenancies created after February 28 1997 and these are therefore assured shorthold tenancies unless specifically agreed otherwise in writing. In the case of a grant of an assured shorthold tenancy, section 21 of the Housing Act 1988 provides that a court shall make an order for possession of the dwelling-house subject to the landlord's giving not less than two months' notice that he requires possession of the dwelling, given before or at any time after the date on which the tenancy was due to come to an end. The tenancy may be allowed to continue without affecting the right of the landlord to obtain possession provided that a minimum of two months' notice is given that possession is required.

The only restriction on rent is provided for by section 22 of the Act, under which a tenant may make an application to a rent assessment committee for a determination of the rent which, in the committee's opinion, the landlord might reasonably be expected to obtain under the assured shorthold tenancy. The rights of the rent assessment committee are limited to amending the rent being paid, but they can only reduce the rent being paid under the assured shorthold tenancy in question if it is significantly higher than the rent which the landlord might reasonably be expected to be able to obtain had it been an assured tenancy.

This rent assessment committee function is somewhat academic, since if the rent determined by the committee is unacceptably low the landlord will determine the tenancy at the first available opportunity. The existence of an assured shorthold tenancy is a mandatory ground for possession by the court.

As stated at the very beginning of this chapter, the maximum value of any property in usual circumstances is the vacant possession value. The question therefore to be asked is: 'What discount from the vacant possession value is appropriate because of the existence of the assured shorthold tenancy?'. The answer lies in the length of the tenancy and the quality of the tenant.

Notwithstanding the nature of the tenancy it is illegal for a landlord to evict a tenant from the premises without a court order. Consequently delays could be experienced in obtaining possession since no action for possession through the courts can be commenced until after the landlord's notice has expired. It could well take two to three months (longer in some areas) to obtain a hearing date from the court and at least another 28 days for actual possession to be obtained. Usually the court will specify a possession date 28 days after the initial court hearing and a further delay can then be experienced, as a court officer will be required in order actually to obtain possession if a tenant does not go voluntarily.

Because of the above, it is reasonable to allow a minimum period of six to nine months after the expiration date of the tenancy in order to obtain actual possession. During this period rent should be paid by the tenant, but in many cases it might in fact be difficult to obtain the rent.

In view of the above, the maximum value which can be placed on a property let on an assured shorthold tenancy should be 90%–95% of its vacant possession value. However, if it is the buyer's intention to continue letting the property 100% of the VP value will be achievable. Therefore, if the property is more of interest to owner occupiers there would be a discount, but not otherwise.

Many of these properties may be classified as Houses in Multiple Occupation (HMO's). These must be registered and planning permission will be required to convert these into self contained flats for sale; in inner city locations planning consent will not usually be granted.

Purchasers of properties for letting on assured shorthold tenancies should also be aware of the outgoings which landlords must meet in relation to repairs, insurance and management, including annual inspections to ensure the safety of gas and electrical installations and the need to comply with the HHSRS (Housing, Health and Safety Rating System), improvement notices and prohibition orders. Landlord's net income is therefore frequently less than 70% of the gross rental income.

# Freehold properties let on assured tenancies

The only difference between an assured tenancy and a regulated tenancy is that the rent payable under the assured tenancy is an open market rent and not a fair rent. There is no rent officer or committee to interfere with the freely negotiated rent between the landlord and the tenant and the tenancy can provide for a review of the rent, with any mechanism that the parties desire in order to fix the rent on review.

However, for the purposes of securing an increase in the rent under an assured tenancy, the landlord must serve on the tenant a notice in the prescribed form proposing a new rent to take effect at the beginning of the new period of the tenancy specified in the notice, being a period beginning not earlier than:

(a)  in the case of a yearly tenancy — six months; in the case of a tenancy where the period is less than one month — one month; and in any other case a period equal to the period of the tenancy
(b)  except in the case of a statutory periodic tenancy, the first anniversary of the date on which the first period of the tenancy began
(c)  if the rent under the tenancy had previously been increased by virtue of a notice, the first anniversary of the date on which the increased rent took effect.

Where the parties cannot agree the rent payable, the matter is referred to a rent assessment committee which, after taking due notice of any improvements or dis-improvements effected to the premises by the tenant, shall determine the rent at which they consider that the dwelling-house concerned might reasonably be expected to be let in the open market by a willing landlord under an assured tenancy, ie an open market rent.

It is therefore clear that the rent payable under an assured tenancy should be higher than that payable under a fair rent, unless there was little or no scarcity of accommodation to let with security of tenure, but the prospects of obtaining possession are no different than with a regulated tenancy, except that the higher rent and more readily available alternative accommodation could cause an increase in the lessee's mobility.

The majority of these tenancies have come into existence by the failure of landlords to create assured shorthold tenancies correctly, usually the pre-notice of lack of security was not served; this can no longer happen. These tenancies now only come into existence because of successions to regulated tenancies to non-spouses and by tenants holding over after the expiration of long tenancies under Part I of the Landlord and Tenant Act 1954, or where tenants are rehoused under the Rent Act. There is no logical reason to create an assured tenancy in the normal market as the creation of a situation where a tenant obtains security of tenure is usually destructive of capital value.

In exactly the same way as with regulated tenancies, the greater the prospect of vacant possession the higher the percentage of vacant possession value which will be applicable to the property. There is little rent incentive for a tenant to remain in occupation and therefore this will usually allow a value of between 75% and 85% of the vacant possession value to be taken.

In all cases where tenants have security of tenure there will always be landlord's outgoings in respect of repairs, insurance and management. More and more legislation is being passed to protect tenants and, therefore, where investors are borrowing, regard must be had to complying with costs arising out of the Gas Safety (Installation and Use) Regulations, the Electrical Equipment (Safety) Regulations and improvement notices or prohibition orders issued under the Housing, Health and Safety Rating System (HHSRS) created by the Housing Act 2004. Failure to comply with orders and notices can lead to fines of up to £20,000.

## Freehold 'break up' valuations

Up to now we have considered single properties with one or more tenants on short term tenancies. Many residential investments comprise blocks of purpose-built flats. The valuation of these is a combination of investment analysis as in study 1 and development appraisal. In a block of any size, a number of flats may become vacant over the course of a year. It is usual either to sell the block unbroken with possession of a number of flats or to 'break the block up' (meaning selling on long leases). Partially broken blocks are less attractive to investor/dealers as the statistical probabilities have been altered by the part disposals and consequently the yield requirement on the balance must be higher. The more vacant flats there are, the greater the potential for general block improvement, which will increase the average sales price of the individual flats.

The general considerations already discussed apply to flat break-up as they do to individual units. The flats must be mortgageable and capable of smooth management. There must be an acceptable management structure involving a single company which is responsible for repairs, insurance, services, etc, and with which each lessee covenants in respect of payment for a share of the costs; the landlord must similarly covenant in respect of unsold flats. It should be noted that freehold flats are difficult to mortgage because of the problems of enforcing positive covenants and therefore a break-up is best effected by the grant of long leases (over 60 years). This has been changed by the introduction of a commonhold system which will allow the sale of freehold flats, but to date few, if any, such schemes have been created.

The block should be in good repair before the break-up is commenced because the buying public are (and should be) wary of involving themselves in a large service charge commitment for works which should have been carried out years before. Similarly, sitting tenants would not be too happy for their landlords to escape existing repairing obligations by passing these on to them as buyers. It must be borne in mind that if a sitting tenant buys his flat on a mortgage he will, from an income point of view, be financially worse off for many years than if he had continued renting his flat, ie his interest and service charge expenditure will be considerably greater than his rent at the beginning of his mortgage term. The viability and profitability of a break-up scheme will depend on the number of tenant sales.

The sale of the block as a whole will be affected by the Landlord and Tenant Act 1987, which requires an option to be given to the tenants and any sale can be held up for up to seven months. The Act provides that a sale cannot be made to an outside purchaser for less than the block is offered to the tenants. This Act was honoured more in the breach than otherwise and therefore the Housing Act 1996 has brought in penalties for non-observance. There is also a special scheme for sales by auction where the tenants need not bid but may then pre-empt a purchase at the sales price.

## Study 3

A tenant rents a flat at £5,000 pa including £800 pa for services. The vacant possession value is £160,000 for a 999-year lease at £125 pa. The sitting-tenant price is £128,000.

|  |  | £ pa |
|---|---|---|
| *Present expenditure* | rent | 4,200 |
|  | services | 800 |
| Rent fixed for 2 years |  | 5,000 |
|  |  |  |
| *Proposed expenditure* |  |  |
| Assuming 100% (interest only) mortgage at 5.5% mortgage interest |  | 7,040 |
| Services (Note 1) |  | 1,200 |
| Ground rent |  | 125 |
| Annual cost |  | £8,365 |

Note

(1)  The service charge will be higher than the service element included in the rent because this will include repairs and insurance.

Assuming that rent and all costs rise by 5%p.a. compound, but that interest charges remain constant, it will be about 20 years before expenditure is equated. The compensation for this is the immediate capital profit of £32,000 and the added psychological advantages of ownership, although this is not without worry in view of rising service charges.

## Study 4

The value of the freehold interest in a block of 12 self-contained flats on three floors is required. Three flats are vacant and nine are let to regulated tenants at £4,450 pa each, including £750 for services (just registered). The block is in good repair and the vacant flats are worth £150,000 each if sold on 99-year leases with £50 pa ground rents.

| | £ pa | £ pa | £ |
|---|---|---|---|
| *Let flats*: | | | |
| 9 at £4,450 pa | | 40,050 | |
| | | | |
| *Less*:   services: 9 at £750 | 6,750 | | |
|      repairs at, say, £500 per flat | 4,500 | | |
|      management at, say, 5%+ VAT | 2,353 | | |
|      insurance, say | 1,450 | 15,053 | |
| Net Income | | 24,997 | |
| YP in perpetuity at 3% | | 33.33 | 833,150 |
| (approx 60% of vacant possession value) | | | |
| | | | |
| *Vacant flats*: | | | |
| 3 at £150,000 | | 450,000 | |
| 3 ground rents at £50 pa × 12 YP | | 600 | 450,600 |
| | | | 1,283,750 |
| | | | |
| *Less* | | | |
| (i)  legal and agent's fees at 3% on sales + VAT (4) | | 15,862 | |
| (ii)  profit at 15% on sales (1) | | 67,500 | 83,362 |
| | | | 1,200,388 |
| Less interest for, say, 6 months at 6%, but say | | | 35,000 |
| | | | 1,165,388 |
| | | | |
| Value of freehold interest, say (2) | | | £1,165,000 |

Notes

(1)  The profit is required in the same way as any wholesaler requires a profit because of the risk of obtaining the estimated prices, delays, etc.
(2)  The average price per flat is just over £97,000.
(3)  The yield is really immaterial and results from the let flats being valued at just over 60% of VP value.
(4)  Fees and profit are only calculated on the immediate sales of the three flats with VP. The real profit from the investment will come from future sales of let flats when they become vacant.

The above example may be made more complex by having a variety of tenants at different rents and the need to estimate the relevant fair rents. Furnished and assured shorthold lettings should be considered separately from unfurnished regulated lettings and also from assured lettings as the appropriate yield (ie percentage of VP value) will be different. Potential decreases of rent must be considered because of possible registration, together with the cost of putting the block and the individual flats into a saleable state.

Some blocks, because of their location or design, are not capable of being broken up and must be considered as pure investments. If this is the case, then the yield must be much higher because of the absence of the potential to the vacant possession profit. If the state of repair is poor, then the cost of refurbishment and modernisation can be taken as an end-allowance together with the financing costs thereon. Many schemes have failed because insufficient allowance has been made for financing the refurbishment expenditure which might be required, either to the exterior and common parts or to the interior of flats which are going to be sold off immediately. In order to ensure the viability of a break-up, this expenditure must be recovered from the initial sales or the financing cost must be deducted from the general income, or a combination of both.

It is very unusual to consider a modernisation/break-up without a reasonable percentage of vacancies at the beginning, as in study 4, since it is very risky to assume that large numbers of sitting tenants will buy; they very often do not have the resources and are too old to obtain a mortgage. It is therefore better to wait until there is a good vacant possession base.

In many developments there are several blocks. A break-up might best be effected by modernising and selling one block at a time. Sitting tenants can be moved by court order to suitable alternative accommodation to assist in making individual blocks totally vacant. If a sales campaign to sitting tenants is mounted then the purchasers should be made aware of the developer's future intentions as to expenditure, so that they do not buy cheaply in a run-down break-up and then find themselves called upon for further large capital sums to meet their share of the improvement expenditure, which they might not be able to afford. It would be better to effect the improvements first and then to sell to the sitting tenants, who will no doubt need larger mortgages.

In considering the value of any block regard must also be had to its development potential. Can, for example, the roof space be developed or can further units be built in the grounds?

In some break up situations some of the flats have already been sold off and those leases may now be sufficiently short to allow for some assumption as to further income being derived from surrenders and re-grants. Under the provisions of the Leasehold Reform, Housing and Urban Development Act 1993 all lessees have the right at any time once they pass the ownership test to require a lease extension of 90 years and the reduction of their ground rent to a peppercorn. The price at which the lease extension will be sold will include half the marriage value (where leases exceed 80 years unexpired). There is also a right for collective enfranchisement by long lessees although there are qualifying tests in so far as the relevant percentages are concerned. This will be considered in the next chapter under leasehold reform and will not be considered further here.

## Leasehold properties

The entire analysis so far has been concerned with freehold properties, but many residential properties are leasehold. Consequently, the above analysis must be amended to take account of the unexpired term and the fact that purchasers of the leasehold interest in individual flats can exercise rights under the leasehold reform legislation. If the property is a house, then the leasehold value is in fact the freehold value less the cost of enfranchisement with allowance for a margin to reflect the costs and delays involved.

Chapter 3 deals in detail with the Leasehold Reform Act 1967, its subsequent amendments and the Leasehold Reform, Housing and Urban Development Act 1993, but it is worth mentioning at this time that a leaseholder of an entire block of flats is thought to be unable to obtain lease extensions for all of the individual units where not sub-let to a qualifying lessee, he also cannot collectively enfranchise. However, such a leaseholder can grant to himself individual leases of all non-qualifying flats and obtain lease extensions for these after two years; there will however be tax consequences in so doing. Therefore in study 4 the value of the leasehold interest would be the value of the freehold less the cost of extending all of the leases.

Referring back to the analysis, so far it will be seen that the rate of interest effectively adopted in capitalising the income to arrive at a freehold value must depend on the possibility of early possession. In a leasehold situation it used to be that it must depend on the probability of obtaining vacant possession before the end of the lease. This is no longer the case, because of the landlord's ability to obtain a lease extension at any time. When valuing freehold reversions to ground-rented long leases, regard must be had to the potential for enfranchisement by the lessee or his successor in title. A view must be taken as to future enfranchisement possibilities.

It will be necessary to study Chapter 3 in order to calculate the enfranchisement price applicable under the different circumstances that apply. With the exception of houses having a rateable value in London of less than £1,000 and elsewhere of less than £500, the enfranchisement price where leases are less than 80 years will be the value of the landlord's interest plus half the marriage value and this could amount to a very significant sum. Only in respect of the houses below the rateable value limits will the enfranchisement price be confiscatory so far as the landlord is concerned.

## Study 5

This study is in respect of a head lease with 60 years unexpired at a ground rent of £600 pa, using the same information as in study 4 with all the flats in hand. It assumes that the value of the head lease is effectively the same as the freehold, less the cost of acquiring the 12 lease extensions, each of 60 years, under the Leasehold Reform, Housing and Urban Development Act 1993.

|  |  | £ | £ | £ | £ |
|---|---|---|---|---|---|
| 1. | *Value of freehold as before* leases not extended (see study 4) |  |  |  | 1,170,000 |
|  | less say 1% for flats not being freehold following sale (Note 1) |  |  |  | 11,700 |
|  |  |  |  |  | 1,158,300 |
| 2. | *Cost of obtaining 12 lease extensions* of 90 years at a peppercorn rent: |  |  |  |  |
|  | (i) *Value of interest after leases extended* (Note 2) |  |  |  |  |
|  | (a) Let flats: 9 × £150,000 × 60% |  |  |  | 810,000 |
|  | (b) Vacant flats: 3 at £150,000 |  |  |  | 450,000 |
|  |  |  |  |  | 1,260,000 |
|  | (ii) *Value of existing leases* say 80% of extended lease values (Note 3) |  |  | 1,008,000 |  |

|  | | £ | £ | £ | £ |
|---|---|---|---|---|---|
| (iii) | *Diminution in value of freehold interest* | | | | |
| (a) | Before leases extended (Note 4) | | | | |
| | *Term* | | | | |
| | 12 Ground Rents at £50 | 600 | | | |
| | YP 60 years at 7% | 14.04 | 8,424 | | |
| | *Reversion to* (Note 5) | | | | |
| | 12 flats at £151,500 | 1,818,000 | | | |
| | PV of £1 in 60 years | | | | |
| | at 5% (Note 6) | 0.0535 | 97,263 | | |
| | | | 105,687 | | |
| (b) | After leases extended | | | | |
| | *Reversion* | | | | |
| | 12 flats at £151,000 | 1,818,000 | | | |
| | PV £1 in 150 years at 5% | 0.00066 | 1,200 | | |
| | *Diminution in value* | | | 104,487 | |
| | | | | | 1,112,487 |
| (iv) | *Marriage value* | | | | 147,513 |
| | Share at 50% | | | | 50% |
| | | | | | 73,756 |
| 3. | *Cost of obtaining extension* | | | | 104,487 |
| | *Cost of lease extensions* | | | | 178,244 |
| 4. | Value of leases after extensions | | | 1,583,000 | |
| | *Less* Cost of obtaining lease extensions | | | 178,244 | |
| | Net value before costs (Note 7) | | | 980,056 | |

*Notes*

(1)   Each new lease will be for 150 years at a peppercorn rent, but it is considered that these will have a combined value of 1% less than had the final interest been freehold.

(2)   The extended lease value of nine of the flats is reduced by the fact that they are subject to under-lettings at regulated rents with the tenants having security of tenure. If the lettings were to Assured Tenants then the ratio would be higher, and if to assured shorthold tenants then no discount would be applied.

(3)   See Chapter 3 to explain the ratio of 80% for the 60 year lease. This percentage is not guaranteed or fixed and will depend on the relevant evidence.

(4)   The 1993 Act requires a calculation of the diminution of the landlord's interest. This being the case, there will be a reversion to a capital value after the expiration of the current term plus 90 years to be the diminution in value as there would be a small review after the lease is extended.

(5)   The reversion is often taken at a higher value than the extended lease because the freeholder is not required to sell a lease of the current term plus 90 years, he is free to sell any length of term or even a share of the freehold.

(6)   This yield has been determined by the Lands Tribunal in *Cadogan* v *Sportelli* (see Chapter 3) and it excludes hope value. The exclusion has been taken to the Court of Appeal and this valuation may need to be amended after that judgement has been given.

(7)   This value of £980,056 does not take into account the costs to the head lessee of obtaining the lease extensions or the risks that the actual costs will prove to be higher than anticipated, or there might be a change in the law. The costs could amount to £15,000 plus VAT and an end allowance for risk could be taken at anything between 10 and 25%.

# American style 'lofts'

This is a comparatively new style of property which has been recently been coming in to the market following the obsolescence of some types of commercial property. The basic concept is that a large clear space is sold with planning permission for use as residential accommodation, but that space is not fitted out as a flat as such. The accommodation is serviced by the provision of mains water, sewerage, electricity and gas, but the purchaser fits out the accommodation to suit his own needs. The usual feature of accommodation of this kind is that the space is large and frequently of above average height, which allows for the construction of mezzanine floors or even duplex apartments within the shell.

This type of accommodation is generally seen most in the disused dock areas of the major cities where old multi-storey warehouses have been converted to residential use. This use is now spreading to city centres where office blocks, telephone exchanges etc, are being converted into residential use.

Because of the popularity of this kind of unit, consideration should be given to the conversion of obsolescent commercial buildings to residential use. The valuation method will of course be the residual method. This is discussed in Chapter 13.

# Non-owner–occupier-type properties

In general, there are three types of properties in this category: (i) properties in very poor repair or in areas scheduled for redevelopment; (ii) houses in multi-occupation; (iii) artisan dwellings.

## (i) Freehold properties in very poor repair or scheduled for redevelopment

There comes a point in the life of a property when the cost of repair may approach the value of the property in good repair. In such a case, the maximum vacant possession value is the site value less the cost of demolition. In addition, a landlord can be forced to do repairs to the property under the Public Health Acts and the Housing Acts and, in most cases, by a civil action for breach of covenant to repair. Consequently, there will be a capital loss situation in any investment. Local authorities are now much more active than they have been and are serving notices under Schedule 9 to the Housing Act 1957 requiring major works of repair to be carried out. If the works are not done then they may enter and do the works themselves, recovering the costs from the rents. Grants are often available for the work.

### Study 6

A terrace house is let to a single tenant at £1,560 pa. It has been badly neglected and will now cost £17,000 plus VAT to put it into an acceptable and habitable state of repair, in which condition it will have a fair rental value of £3,500 pa. The house is let to a couple who are 45 years old and is worth £45,000 with vacant possession.

|  |  | £ pa | £ pa | £ |
|---|---|---|---|---|
| *Fair rent in repair* |  |  | 3,500 |  |
| *Less:* | annual repairs | 400 |  |  |
|  | insurance | 225 |  |  |
|  | management | 205 | 830 |  |

|  |  | £ pa | £ pa | £ |
|---|---|---|---|---|
|  | Net Income |  | 2,670 |  |
|  | YP in perpetuity at 10% |  | 10 |  |
|  |  |  |  | 26,700 |
| *Less:* | necessary repairs |  | 17,000 |  |
|  | supervision fees at 10% (1) |  | 1,700 | 18,700 |
|  | Value (2) |  |  | £8,000 |
| or: |  |  |  |  |
|  | *Present fair rent in disrepair* |  | 1,560 |  |
| *Less* outgoings: |  |  |  |  |
|  | repairs | 500 |  |  |
|  | insurance | 225 |  |  |
|  | management | 92 | 817 |  |
|  |  |  | 943 |  |
|  | YP in perpetuity at 20% (3) |  | 5 |  |
|  |  |  |  | 4,715 |
| *Less:* | essential repairs |  |  | 4,000 |
| Value |  |  |  | £715 |

*Notes*
(1)  The 10% yield reflects the non-probability of early possession.
(2)  The value after the costs is say 40% × VP = £18,000 so that the net yield is 14.83% but there is a capital loss as the cost is £26,700.
(3)  The 20% yield reflects possibilities of notices under the Public Health Acts, etc.

The house in its present state is virtually worthless, but if it is put into a full state of repair has a cost which is greater than its market value. This is compensated for by a high yield and the prospect of significant gain, albeit a long time in the future if the tenants remain in occupation for their lifetime.

This valuation could be affected by the availability of local authority grants and up to 90% of the cost of works could be obtained. If this is the case then the works are worth doing and the value of the property will be considerably increased.

If, because of the disrepair, the tenant does not pay his rent, then possession may be obtained and a profit made, but this is a doubtful occurrence as the tenant would be advised to press for the repairs via the local authority or to sue on the landlord's deemed repairing covenants.

For property of this type there is also always the threat of compulsory purchase as a property which is unfit for human habitation (see Chapter 9).

## (ii) Houses in multiple occupation

This is a similar category to studies 2 and 3, but the properties were not usually of a sufficient quality or of satisfactory design to allow part owner-occupation either by conversion (into all or part self-contained units) or by sharing or because the local planning authority will not give consent for change of use to a single dwelling or for self-containment. Consequently, there may be little logic in leaving part vacant in the hope of full vacant possession since there is a security risk from squatters and there will be substantial outgoings in the form of repairs, insurance and rates. (NB conversion of two dwellings into a single dwelling does not require planning permission.)

Such property is the poorest possible type for investment. There is very little prospect of capital profit because there is little or no vacant possession market and consequently there is almost no option but to re-let when a part becomes vacant. This type of property is also politically sensitive, since the tenants are usually the poorer members of society who need statutory protection. Rent levels are often geared to Social Security payments and therefore suspect to political interference.

These have to be registered as a matter of law under the Housing Act 2004. The general assumption is that a reversion to a single dwelling, or conversion into self-contained flats for sale, will not be permitted as planning permission for change of use will be required and will be refused. Therefore these properties must now be valued as investments with no assumption of a profit to be generated by obtaining vacant possession. This will of course depend on location at the demand and supply of such properties.

They can either be valued by reference to gross or net income. Current sales analysis indicates a YP on gross income of up to 15 which would indicate a net yield of as little as 6%. Such yields can only be justified by the low rates of interest currently being charged in the UK and by the hope of obtaining planning consent to convert into self contained flats for sale. These low yields also only apply in high quality locations. The usual yields will be much higher with gross YPs of 11 or more. The gross to net income should take into account landlord's outgoings as well as the need for more repairs than in single lettings and to allow for voids and bad debts. If interest rates rise then the value of these properties will fall unless there is a compensating rise in rental income. However, the rental income will be geared to wages and inflation generally and not to underlying rises in the capital values of VP houses and flats.

## (iii) Artisan dwellings

This is the name given to purpose-built blocks of houses and flats built around the turn of the century by the predecessors of today's housing associations, usually charitable trusts; good examples are the blocks built by the Guinness and Peabody Trusts. At the time they were built, these properties were a major advance on the 'working class' housing then existing.

By today's standards, however, these artisan blocks are old fashioned and lack modern amenities. They were built to a very high density with blocks being close together and often five storeys high, without lifts. They tend to have open landings and staircases. Some of the better blocks can be, or have been modernised, to acceptable standards so that 'break-ups' can be effected. Others, by reason of their design or location, can never be made acceptable owner-occupier-type properties although they can be provided with standard amenities. Consequently, they provide a purely investment type of property and the yield must reflect a pure income situation. Regard must be given to the state of repair and the level of outgoings on insurance, services and management. Finally, regard must be had to potential obsolescence, since it is hoped that, in due course, this type of housing will be demolished as not being of a sufficiently high standard in our enlightened age.

These blocks will be valued in the same way as houses in multiple occupation but will usually require a higher yield.

If a freehold interest in this category of property is one of the poorest investments available, then a leasehold interest must be the worst of all as there is also the problem of dilapidations. The covenants of the lease must be considered in valuing the lease and if there are the usual forms of repairing covenant then the property may be a liability rather than an asset. An appropriate sinking fund must be used. The leasehold in the block can be effectively converted to a near freehold in the same way as shown in study 5.

# Furnished property

This type of property does not really warrant separate treatment as there is no substantial difference today between a furnished and an unfurnished letting, since both types are subject to the same rent and security of tenure controls. The maximum value is the vacant possession value.

© EF Shapiro 2008

# Leasehold Enfranchisement

The Leasehold Reform Act 1967 was a measure to protect certain occupiers of houses holding under long leases at low rents. It is now extended to cover flats held on long leases and non occupying leaseholders, while the low rent test has been removed in the majority of cases. It was the culmination of an argument since the latter part of the last century that although the land in such circumstances may belong in equity to the freeholder, the buildings should belong to the leaseholder. Leasehold enfranchisement, or the principle that certain residential leaseholders should be able to expand their interest in premises, was the policy of both major political parties in 1966, but the basis for enfranchisement was the subject of controversy and still can be.

The original principle of enfranchisement adopted in the 1967 Act was that the tenant might either take a further 50-year term of ground lease at a so-called modern ground rent subject to review after 25 years, or purchase the landlord's interest. In the latter case the enfranchisement price payable was the value of the landlord's interest subject to the tenant's unexpired term of lease, the right to a further 50-year ground lease at a modern ground rent and the present value of the premises at the end of the 50-year extension. This principle only applies to lower value houses in the now widened setting of enfranchisement.

The 1967 Act was amended by section 82 of the Housing Act 1969, which requires that any additional bid that a tenant might make be excluded from the enfranchisement price. Section 118 of the Housing Act 1974 extended the right to enfranchisement to a further range of interests subject to different rules for assessing prices in the higher rateable value bands. It also made the amendment that for enfranchisement qualification purposes rateable values may be adjusted to discount the value of any tenant's improvements. The Leasehold Reform Act 1979 closed the loophole of creating intermediate interests to increase the cost in the enfranchisement price provisions of the 1967 Act and, this was confirmed by the House of Lords in *Jones* v *Wrotham Park Settled Estates* [1979].

The Housing Act 1980 made further amendments to the 1967 Act, reducing the original five years' residential occupation requirement to three years and correcting and improving the 1974 Act provisions regarding rateable value adjustments. It also included within the protection of the 1967 Act tenancies terminable on death or marriage to close an avoidance loophole. It also extended transitional relief to tenants paying modern ground rents on lease extensions and introduced a formula approach for those enfranchising against minor superior tenants. Section 142 provided for the referral of valuation disputes in the first instance to leasehold valuation tribunals drawn from rent assessment panels, with an appeal right to the Lands Tribunal.

The Leasehold Reform, Housing and Urban Development Act 1993 introduced provisions for collective leasehold enfranchisement by tenants of certain flat blocks, amended the enfranchisement rules for houses under the 1967 Act, provided for Estate Management Schemes for newly enfranchisable estates and the right to a management audit of service charge accounts. The 1993 Act enables certain occupying leaseholders to acquire collectively the freehold, and any other superior interests, in a block of flats where they hold long leasehold interests. The individual flat leaseholders also each have the right to be granted a new lease of their flat with an additional 90 years added to the existing lease term free of ground rent, whether or not collective enfranchisements rights have been exercised.

The 1993 Act provisions are intended to help flat leaseholders where the lease terms are getting too short to be saleable or to be suitable security for lending. They are also aimed at helping tenants to improve the management of their blocks, through the rights to service audits and management codes. The audit right enables detailed investigation as to the appropriateness and efficiency of management functions and service charge expenditure. The Secretary of State is given power to approve codes of management practice designed to promote improved standards and while there are no direct sanctions relating to the operation of a code such may be used as evidence in proceedings.

Tenants of houses in higher rateable value bands were excluded from the right to enfranchisement under the Act of 1967 but now the value limits are removed by the 1993 Act, although the right to the alternative of a 50-year lease extension at a section 15 modern ground rent is excluded for these newly qualified houses. An additional low rent test qualification was added by the 1993 Act whereby at the commencement of the tenancy, or before 1 April 1990, the tenancy should be at either no rent or a rent of no more than two thirds of the letting value (if a pre 1963 tenancy) or two thirds of the rateable value (for post 1963 tenancies). For such newly qualified houses the valuation assumptions in section 9(1A) of the 1967 Act are to be made requiring the parties to share any marriage value and it is not to be assumed that the tenant has any right to carry on in possession at the end of the tenancy. The 1993 Act also excludes certain houses let by charitable housing trusts from the right to enfranchisement.

The Housing Act 1996 made a number of changes to the 1967 Act, including: (i) removing the requirement for a low rent test where the tenancy is for an original term of over 35 years and the tenancy is not excluded; (ii) certain costs under section 9(4) or 14(2) were put into the jurisdiction of a leasehold valuation tribunal; (iii) providing compensation for the postponement of the termination of a tenancy as a result of ineffective claims to have the freehold under the 1967 Act; (iv) extended time-limits for applications under the 1993 Act, Chapter IV of Part I.

The Commonhold and Leasehold Reform Act 2002 introduced further amendments to the 1967 Act, which came into force in England on 26 July 2002. These included: (i) the general abolition of the residence test; (ii) a two year residence test in some circumstances; (iii) exclusion of certain business tenancies; (iv) for claims to have the freehold abolition of the low rent test where the tenancy is for an original term of 35 years or less; (v) the right of personal representatives of a tenant to make a claim; (vi) right to the freehold where a tenant has already extended under the 1967 Act; (vii) excluding any marriage value where a tenancy exceeds 80 years and limits landlord's marriage value, if payable, to one half of that; (viii) absent landlord cases to be dealt with by the County Court and leasehold valuation tribunals; (ix) new leasehold valuation tribunal procedures and restricted rights of appeal from them.

It should be noted that residential tenants, whether or not qualifying for enfranchisement, generally still have the alternative, but probably less beneficial, protection of Part I of the Landlord and Tenant Act 1954. This effectively entitles them to a protected tenancy at the expiry of the long lease.

Valuations of landlords' and tenants' interests in residential property must take account of the effect of enfranchisement rights and may be specifically required for the assessment of enfranchisement price or lease extension terms. They must take account of the statutes and the precedents which interact to influence the rights to enfranchisement and the basis and terms for it. There have been numerous references to the Lands Tribunal since 1967 on issues under the Act and, since 1980, many before leasehold valuation tribunals as the initial reference point for disputes. Many cases have been heard by the courts too under the provisions but the basis for the valuations required is still not always clear. The importance of surveyors understanding the consequences of the tests for qualification under the Act, for example, was stressed in a breach of duty of care case. In this the valuer did not remind the client of the consequences under the Act of agreeing to a rateable value in excess of £1,500 and was held to be negligent: see *McIntyre* v *Herring Son & Daw* [1988].

The principal precedents on enfranchisement terms include: *Kemp* v *Josephine Trust Ltd* (1970); *Farr* v *Millersons Investments Ltd* (1971), in which the Lands Tribunal described the generally recognised approaches including three alternative means of calculating the modern ground rent; *Official Custodian for Charities* v *Goldridge* (1973), in which the Court of Appeal disapproved of the adverse differential approach; *Norfolk* v *Trinity College, Cambridge* [1976] and *Lloyd-Jones* v *Church Commissioners for England* [1982], the first two cases under the additional valuation rules of section 9 (1A) introduced by the 1974 Act.

The decision in *Pearlman* v *Harrow Keepers and Governors* [1978] dealt with tenant's improvements and qualification for enfranchisement under the 1974 Act amendments. On the issue of what is a house for enfranchisement purposes, a majority decision of the House of Lords in *Tandon* v *Spurgeons Homes Trustees* [1982] held that the tenant of a shop in a parade occupied together with living accommodation above qualified even though the shop part was also a protected business tenancy. The Duke of Westminster has taken a number of cases to the Court of Appeal and one to the House of Lords over the issue of what is a low rent in the setting of the qualification provisions of the Act: see studies 1 and 2 below.

*Arbid* v *Earl Cadogan* (2005) concerned six appeals to the Lands Tribunal regarding the deferment discount rate to apply to valuation reversions in determinations of price under section 9(1C) of the 1967 Act. The Lands Tribunal, in four of the cases, determined the discount rate at 4.75%/4.5% rather than the 6% rate applied by the LVT, resorting to the finance markets to assess deferment rates. The cases concerned valuable houses in Chelsea and the Tribunal sets out an objective approach to the discount rate. Tribunal decisions do not, however, establish coventions or precedents and the cases do not establish a hard framework for a reduced discount rate in all cases. The Lands Tribunal, in the case of *Earl Cadogan and Cadogan Estates Ltd* v *Sportelli* [2006] specified rates which it believed should be applied in all but exceptional cases, again referring to the finance markets to assess deferment rates. The Court of Appeal in *Earl Cadogan* v *Sportelli* [2007] dismissed appeals against the Lands Tribunal's decisions. It was confirmed that hope value was not a permissible element in valuations under Schedules 6 or 13 of the 1993 Act for collective enfranchisement or lease extensions. Marriage value was specifically provided for in the 1993 Act with a 50% allocation where the lease has less than 80 years to run. The Court of Appeal also confirmed that it was appropriate for the Lands Tribunal to lay down guidelines as to the deferment rate to apply in future cases to promote consistent practice in land valuation. In future cases the deferment rate adopted by the Lands Tribunal would be the starting point, although evidence might be called to show that a different rate should be applied. This confirmed the deferment rates for houses at 4.75% and for flats at 5% in those Central London cases.

Other helpful cases include *Maryland Estates Ltd* v *63 Perham Road Ltd* (1996), which dealt with a nominee purchaser's acquisition of the freehold reversionary interest in a house converted into four flats. *Becker Properties Ltd* v *Garden Court NW8 Property Company Ltd* (1997) dealt with similar issues. *Trustees of the Eyre Estate* v *Saphir* (1998) dealt with the enfranchisement price under 9(1C), yield evidence and the duty of the Lands Tribunal, and Marriage Value. In *Cadogan Estates Ltd* v *26 Cadogan Square Ltd* [2007] the Court of Appeal held that the 1993 Act did not confer a head lessee of mixed use properties the right to lease extensions for the residential elements.

The provisions of the Housing Acts, which granted public sector tenants the right to buy long leases of flats, granted wider rights than those which applied to long-leased flats in the private sector originally excluded from the leasehold reform provisions. The Landlord and Tenant Act 1987 conferred on groups of tenants of flats certain rights of first refusal to acquire their landlord's reversion and the Leasehold Reform, Housing and Urban Development Act 1993 further extended these provisions with the right to collective acquisition of a landlord's freehold reversion of blocks of flats or for individual occupying tenants to have an additional 90 years added to their lease terms.

Sections 89–93 of the Housing Act 1996 introduced amendments to the Landlord and Tenant Act 1987 to deal with contractual arrangements structured to avoid the rights of first refusal, granted to certain flat tenants in the 1987 Act and made it an offence for a landlord not to comply with the provisions of Part I of the 1987 Act. Schedule 6 to the 1996 Act introduced revised procedures for a landlord to serve offer notices on tenants with rights of enforcement by certain flat tenants against a purchaser.

A proposition was made at (1979) 249 EG 31 — "Even freeholders may have human rights" by Harry Kidd — that the Leasehold Reform Act 1967 was contrary to the European Convention on Human Rights as it can allow the expropriation of property unjustly and without fair compensation, other than in the public interest. Following this line of reasoning the principle of the enfranchisement price provisions was questioned by the European Commission on Human Rights and before the European Court of Human Rights at the instigation of the Trustees of the Duke of Westminster. The application was rejected and no breach of the convention was held to have occurred. It is now thought unlikely that any challenge could be brought to the Leasehold Reform legislation under the Human Rights Act 1998.

Following the abolition of the domestic rating system by the Local Government Finance Act of 1988, the Local Government and Housing Act of 1989 conferred powers on the Secretary of State to make consequential amendments to the 1967 Act. By SI 1990 No 434 and SI 1990 No 701, a new section 1 (1)(a) of the 1967 Act was inserted to deal with houses qualifying for enfranchisement after 1 April 1990, where the house had no rateable value on 31 March 1990. In such cases, and where the lease commences after 31 March 1990, a formula is to be applied to compute an R value, which was not to exceed £25,000 for the house to qualify. These provisions are, to a degree, superseded by the effect of the 1993 Act's amendments to the qualification rules for houses.

Leasehold valuation tribunals have jurisdiction to determine the terms of acquisition of interests under the 1993 Act, to approve estate management schemes and, under the Act of 2002, to deal with absent landlord cases. This jurisdiction is to be exercised by rent assessment committees acting as leasehold valuation tribunals. Various regulations made under these provisions deal with the detailed practices and procedures under the 1967 Act and its amending provisions. Following the conferment of new jurisdiction on leasehold valuation tribunals by the Acts of 1985, 1987, 1996 and 2002, with a right of appeal to the Lands Tribunal, the Lands Tribunal (Amendment) Rules 1997 (SI 1997 No 1965) and 1998 (SI 1998 No 22) now require leave to appeal to be obtained from the leasehold valuation tribunal.

Reference to the Act and sections of statute in this chapter are to the Leasehold Reform Act 1967 unless otherwise stated. The *Handbook of Leasehold Reform*, Sweet and Maxwell (1988) but currently out of print, is a loose-leaf manual on the subject. It includes digests of many earlier cases of note, including court decisions, Lands Tribunal determinations and leasehold valuation tribunal determinations, many of the latter of which are unreported. For unreported cases referred to in this chapter which are dealt with in the *Handbook of Leasehold Reform*, references to the digests contained therein are given. HBLR LVT 108, for example, refers to the leasehold valuation tribunal case of *Booer* v *Clinton Devon Estates* (1989) which is found digested as LVT 108 in the handbook. Unreported Lands Tribunal determinations are referred to with the Tribunal reference such as LRA/39/1997. A useful website *www.lvtbulletin.com* provides copies of LVT decisions, while Lands Tribunal decisions and guidance is on *www.landstribunal.gov.uk.*

## Study 1 Enfranchisement of a long term of lease

This illustrates enfranchisement qualification and price assessment under the Act for an interest in a house held with a long term of lease unexpired.

### Study Facts

A terraced house in the Midlands, built 50 years ago, is held on lease with 75 years' unexpired term at a rent of £15 pa. The rateable value of the house was £140 at 23 March 1965. The present tenant has occupied the house for the last seven years as assignee of an earlier tenant. An estate of freehold ground rents in the same locality totalling £100 pa with reversions in 80 years was recently sold as an investment at auction for £950. An enfranchising tenant on the same estate, with a 78 years' unexpired term, recently paid 10.5 YP for his freehold on your advice.

### Qualification

Prior to 1 November 1993, the Act applied only to houses with a certain rateable value and within other financial limits. There were six rateable value tests to apply in assessing whether interests qualify to enfranchise. Compliance with any one would permit enfranchisement. The tests were:

(a)    was the rateable value not more than £200 (£400 in Greater London) on 23 March 1965 or at the commencement of the tenancy or
(b)    was the rateable value not more than £500 (£1,000) in the valuation list on 1 April 1973 or
(c)    was the rateable value not more than £750 (£1,500) on 1 April 1973 and was the tenancy created before 18 February 1966 or
(d)    if the rateable value in the above cases was greater than the limits given, would the figure be within the limits were any tenant's improvements to be discounted from the rateable value or
(e)    where the provisions of the Act would apply to a tenant of a house, but for the financial limit being exceeded, the same right to acquire the freehold applies if the limits of rateable value of £750 (£1,500 in Greater London) were not exceeded or
(f)    if a tenancy was entered into on or after 1 April 1990 and the house and premises had no rateable value at 31 March 1990 the R value does not exceed £25,000 under the following formula:

$$R = P \times \frac{1}{1 - (1 + I)^{-T}}$$

where P is any premium paid for the lease or 0 if no premium was paid.
I is 0. 06
T is the term of years of the lease.

Since 1 November 1993 the rateable value limits have effectively been abolished for those acquiring a freehold under the Act. They remain relevant for qualifying to have an extended lease or to determine the correct approach to the enfranchisement price.

The rateable value at the appropriate day (see section 1 (4) and (5) of the Act), in this case 23 March 1965, was less than £200. The house is held on a long lease of over 21 years at a low rent (1) which was less than two-thirds of the rateable value at the appropriate day. The premises appear to be within the definition of a house (2) and have been occupied (3) by the enfranchising tenant as his only or main residence for the last two years(4) The tenant qualifies to serve notice on the freeholder desiring to have the freehold or an extended lease (5) (6).

If the rateable value was more than £500 (£1,000 in Greater London) and the tenancy created after 18 February 1966 or the R value for a post 31 March 1990 lease is greater than £16,333 the question of different assumptions under section 9 (1A) and a payment of marriage value could apply in this case: see *Lowther* v *Strandberg* (1985). If the tenant had qualified with a house with a rateable value of over £750 (£1,500 in Greater London) or under the 1993 Act's additional qualification test he would only qualify to have the freehold under the assumptions in section 9(1C) and not an extended lease at a modern ground rent.

## Analysis of comparables

The auction result indicates a market investment yield on these ground rents of 10.5%. The enfranchisement settlement shows a discount rate of 9.5%. From this information a reasonable enfranchisement capitalisation rate is assumed as 10%. Note that the yields applied are for illustration purposes only and changing markets can render alternative approaches applicable.

*Valuation for enfranchisement price* (7)

| Term | £ pa | |
|---|---|---|
| Ground rent | 15 | |
| YP in perp (8) at 10% (9) | | 10 |
| Enfranchisement price (10) (11) | | £150 |

*Reversion*
Of nominal value only (8)
*Plus landlords fees & costs* (12)

Notes

(1)  The full rules on qualification are in sections 1 to 4 of the 1967 Act as amended by section 118 and Schedule 8 to the Housing Act 1974, section 141 and Schedule 21 to the Housing Act 1980 and sections 63 to 68 of the Leasehold Reform, Housing and Urban Development Act 1993. Section 4 defines a low rent in certain cases as being less than two-thirds of the letting value and in *Manson* v *Duke of Westminster* [1981] the Court of Appeal held that the letting value for this purpose should include the annual equivalent of the premium paid: see also *Collin* v *Duke of Westminster* [1985] and *Johnston* v *Duke of Westminster* [1986], which both involved the issue of low rent. Tenancies not at a low rent under the section 4(1) test of the Act may now qualify under the test in section 4(A) introduced by the 1993 Act. This test covers tenancies at zero rent or those at rents satisfying the new tests in relation to rateable values or letting values set out in section 4 (A).

(2)  A house is defined in section 2 and while it does not include horizontally divided flats themselves it can include a whole house which is converted into flats or part of a building divided vertically as long as there is no part of other premises above or below: see article at (1974) 229 EG 1165 "Leasehold Reform Act 1967: Meaning of 'House'" by W A Leach also *Peck* v *Anicar Properties Ltd* (1970), *Wolf* v *Crutchley* (1970), *Baron* v *Phillips* [1978], and *Tandon's* case (*supra*). In *Cresswell* v *Duke of Westminster* [1985] a terraced house with accommodation over a side access passage was held to be within the definition of a house for the purposes of section 2 (2). See also *Malpas* v *St Ermin's Property Ltd* [1992] where the Court of Appeal confirmed that a house converted horizontally into two flats could be enfranchised where the respondent held the long lease of the whole premises and occupied part.

(3)  Section 1 allows enfranchisement rights to a tenant "occupying ... as his residence". In *Duke of Westminster v Oddy* [1984], a tenant who held the leasehold interest in a house as a bare trustee for a company was held not to be a tenant with rights under the Act even though he occupied the house with his family.

(4)  For enfranchisement qualification section 1(1)(b) originally required occupation for the last five years or for periods amounting to five years in the last 10. Schedule 21, 1(1) to the 1980 Act amended each period of five to three years, ie three out of the last 10, and the 2002 Act generally abolished the residence test after 26 July 2002. An investor leaseholder could therefore now also enfranchise.

(5)  The form of notices to be used by a tenant under the Act is set out in regulations updated from time to time.

(6)  In *Oliver v Central Estates (Belgravia) Ltd* [1985] a house in London had a rateable value of £347 on 23 March 1965 but £1,347 on 1 April 1973. The LVT held that the rateable value at the relevant time was the 1973 figure and the enfranchisement price fell to be assessed under section 9 (1A). In *MacFarquhar v Phillimore* [1986] the Court of Appeal considered cases where the rateable values of two London houses exceeded £1,500 on 1 April 1973 but subsequent proposals were agreed to reduce these retrospectively. It held that the altered rateable values were the relevant values for the purposes of the 1967 Act and the tenants qualified to enfranchise. The distinction between the relevance of rateable values for qualification purposes and for the purpose of determining the basis for enfranchisement price assessment should be noted: see *Oliver*'s case (*supra*). This is set out in section 9 of the Act.

(7)  The valuation date is in this case the date of the tenant's notice: sections 9 (1) and 37 (1).

(8)  The reversion to the modern ground rent in 75 years can be ignored although ones as far distant as 56 years have been valued: see *Gordon v Lady Londesborough's Marriage Settlement Trustees* (1974) and *Uziell Hamilton v Hibbert Foy* (1974). In *Collins v Howell-Jones* [1981] a leasehold valuation tribunal capitalised a 50 year term of a £5 pa ground rent and then the reversion to the modern ground rent deferred 50 years both at 7%. leasehold valuation tribunals have valued reversions after long terms of lease, including one after 72 years in *Wright v Hawkins* (1989) and 79 years in *Jong v Peachey Property Management Co Ltd* (1981) and *Pemberton v Alders Ltd* (1982). If the rateable value was over £500 and the approach adopted in the *Lowther* case (*supra*) were to be applied, then the reversion could be sufficiently significant to be included here. In *Cummings v Severn Trent Water Authority* (1985) a ground lease with reviews to current market value, the first in 14 and a half years and then after a further 30 and 60 years, was dealt with by reverting in perpetuity to a section 15 rent in 14 and a half years.

(9)  The 10% rate in this case is supported by the all-important comparables, but for an up-to-date review of the issues and considerations. *Earl Cadogan and Cadogan Estates Ltd v Arbid* (2005) and related cases and also *Cadogan v Sportelli* (2006) where deferment rates of 4.75% and 5% have been determined at the Lands Tribunal for high value London houses and flats and confirmed in the Court of Appeal. See *Earl Cadogan v Sportelli* (*supra*). The Tribunal preferred financial market evidence and made explicit adjustments for growth and other factors.

(10)  See similar decisions in *Jenkins v Bevan-Thomas* (1972) (10% basis), *Barber v Eltham United Charities Trustees* (1972) (10% basis), *Janering v English Property Corporation Ltd* (1977) (11% basis), and *Ugrinic v Shipway (Estates) Ltd* (1977) (9% basis). An exception was *Cohen v Metropolitan Property Realisations* [1976] where 7% was adopted for a 59 year term. In *Yates v Bridgwater Estates Ltd* (1982) (LVT) a leasehold valuation tribunal valued a £3.62 pa ground rent receivable for 971 years at £10, a capitalisation rate of interest of 36.2%. Leasehold valuation tribunal determinations for long lease terms in the last few years have typically been at between 5% and 10% discount rate with a number of determinations as high as 15%, 17.5%, 19% and even 50% where the rent was very low and the term very long. The recent *Cadogan* decisions previously referred to may alter rates generally to be adopted.

(11)  In its decision in *Re Castlebeg Investments (Jersey) Ltd*'s Appeal [1985], the Lands Tribunal accepted the landlord's unchallenged evidence of long-term ground rents selling at 16 years' purchase, 6.5% and more, when linked with an obligation on tenants to insure the premises through the landlord's agency: see *Lynch v Castlebeg Investments (Jersey) Ltd* [1988] in which the LVT capitalised the potential insurance commission at 6.5 years' purchase as against 14.28 years' purchase for the ground rent. In *Wells v Hillman* (1987) insurance commission rights were reflected by an extra two years' purchase, while in *Pabari v Calthorpe Estate Trustees* (1990) the insurance commission was capitalised at just over three years' purchase. See also *Calthorpe Estate Trustees v*

*Pabari* [1991] at the Lands Tribunal and *Blackstone Investments Ltd* v *Middleton-Dell Management Co Ltd* (1997). In *35 Dennington Park Road Management Ltd* v *Maryland Estates Ltd* (1997) unreported, the LVT considered insurance commission was earned by virtue of an insurance agency not from ownership. It may be better reflected in the yield adopted. In *Divis* v *Middleton* (1983), a ground rent with fixed rent increases in 23 and 56 years and to full value in 89 years was capitalised at 7%.

(12)   Section 9(4) of the Act requires, *inter alia*, an enfranchising tenant to bear the landlord's costs of any valuation of the house and premises. In *Naiva* v *Covent Garden Group Ltd* [1994] the Court of Appeal confirmed that the landlord's costs of an LVT reference were not recoverable.

## Study 2  A lease extension of 50 years

This further considers enfranchisement qualification together with the assessment of a modern ground rent on a 50-year lease extension to a house held with a short term unexpired. Extensions on houses are rare in practice, as they may not give the tenant such a good deal, but they have to be assumed in enfranchisement cases where a tenant wishes to buy the freehold of a house with a rateable value of less than £500 (£1,000 in London) under the section 9(1) original provisions of the Act. A tenant wanting a 50-year extension and falling in the £500 to £750 rateable value band (£1,000 to £1,500 in London) would also qualify to take such an extension at a modern ground rent under section 15 and not on the less favourable assumptions to be adopted for assessing enfranchisement price under section 9 (1A) which would apply if the tenant wished to take the freehold. For such lessees extension may be more beneficial: see article at (1983) 268 EG 876 at p978 and further in Studies 7, 8 and 9.

### Study Facts

A 98-year-old house in London is held with one-year unexpired term of lease at a rent of £25 pa. The rateable value of the house at 23 March 1965 was £380 and at 1 April 1973 was £980. The lessee has held the interest for the last 10 years during which time he has used the ground floor as his main residence. A small area of the ground floor has been sublet as a betting office and the upper floors have been sublet as unfurnished flats. Both sublettings contravene the headlease covenants. The house is a listed building with an indeterminate but reasonable future life. The adjacent houses do not qualify to enfranchise but the plot might have some redevelopment potential. The plot area is 800 m² and evidence shows that the house if improved might sell for £750,000 freehold with vacant possession. Sites in this area are worth about 40% of the freehold vacant possession value of such houses.

### Qualification

The house is within the £400 London rateable value limit at 23 March 1965, the appropriate day, in this case under section 1 of the Act and the low rent level. Although the house has been used for other purposes it may still qualify for enfranchisement. (1)

*Assessment of modern ground rent for extended lease* (2)

| | £ | £ pa |
|---|---|---|
| Standing house approach: (3) | | |
| Entirety value (4) (5) | 750,000 | |
| Site value at, say, 40% (6) | 300,000 | |
| Section 15 rent at 7% (7) | 0.07 | |
| | | 21,000 |
| *Less:* factor to reflect possible repossession rights, say 10% (8) | | 210 |
| Section 15 modern ground rent (9) (10), reviewable after 25 years | | 20,790 |

Notes

(1) See the Court of Appeal cases *Harris* v *Swick Securities Ltd* (1969) and *Lake* v *Bennett* (1969). In *Baron* v *Phillips* (*supra*) the subletting of the ground-floor shop part led to the loss of enfranchisement rights. In *Tandon* v *Spurgeons Homes Trustees* [1982] a shop in a parade with a flat above occupied by the retailer was held to be a house. In *Methuen-Campbell* v *Walters* [1978] the Court of Appeal held that the enfranchisement right did not extend to 1.6 acres of paddock demised with the premises: see also *Gaidowski* v *Gonville & Caius College, Cambridge* [1976].

(2) Where a tenant requires, and is within the value groups qualifying for, an extended lease the new rent is fixed not earlier than 12 months before the original lease termination date and the tenant bears all costs: section 15 (2)(b). In *Burford Estate & Property Co Ltd* v *Creasey* [1986] a leasehold valuation tribunal rejected a tenant's claim that acceptance by the landlord of the old ground rent for four years of the extended lease implied agreement of the old rent figure as the revised section 15 rent for the first 25 years of the new lease. Section 15 rents have also been determined on lease extensions reported in *Eckert* v *Burnett* (1987) and *Duke of Norfolk* v *Brandon* (1988).

(3) The standing house approach is one of the three generally accepted ways of arriving at the section 15 modern ground rent and stems from *Kemp* v *Josephine Trust Ltd* (1970) but the Lands Tribunal heavily criticised it in *Miller* v *St John Baptist's College, Oxford* (1977) and in *Embling* v *Wells & Campden Charity's Trustees* [1978]. It has stated that the approach should be used only where there is no relevant evidence of a market in residential development land. Many cases referred to the Lands Tribunal and leasehold valuation tribunals have led to determinations using the standing house approach where there was no relevant evidence of land values on the assumption otherwise required.

(4) Entirety value has become a way of expressing the full freehold vacant possession value in this context. There may not be evidence on the basis required by the Act but there should normally be evidence to support the entirety value: see *Carthew* v *Estates Governors of Alleyn's College of God's Gift* (1974).

(5) As the modern ground rent is to be the iletting value of the site for the uses ... of the existing tenancy, other than uses which by the terms of the new tenancy are not permitted, (section 15 (2) (a)), there may be some argument here as to the basis for the entirety value: see *Lake* v *Bennett* (1971), where the best entirety value for the house was the value as two maisonettes, the most profitable use permitted by the lease. In *Kingdon* v *Bartholomew Estates Ltd* (1971) the basis for entirety value was taken to be mixed commercial and residential use. In *Barrable* v *Westminster City Council* (1984), the entirety value was taken at the higher value as if the house was converted into three flats, even though it was used as a single dwelling at the date of enfranchisement.

(6) The approach may be acceptable where the house is likely to remain standing for the foreseeable future. This figure was adopted in *Carthew*'s case (*supra*) but the percentage accepted will depend on locality, building costs, site attributes, evidence etc. In decided cases the figure has ranged between 11.5% and 50% of entirety value and the percentage to be adopted depends very much on location. In the Kensington and Hampstead areas of London, for example, 40% has often been applied and 50% was determined by the Lands Tribunal in *Cadogan Estates Ltd* v *Hows* [1989]. In North and East London figures of 27.5% to 30% have been used while in South Wales, the Midlands and the North figures of between 20% and 30% are more usual. In Sheffield site value proportions as low as 11.5 % and 13 % have been determined by leasehold valuation tribunals in the cases *Duke of Norfolk* v *Bell* (1982) and *Duke of Norfolk* v *Brandon* (1988). In the *Embling* case (*supra*) the Lands Tribunal was critical of the valuers' arbitrary approach to the percentage for site value.

(7) This percentage will depend on the evidence available and decisions by the Lands Tribunal have ranged between 6% and 8%, but 7% seems to have become the generally accepted figure. In *Windsor Life Assurance Company Ltd* v *Buckley* (1995) the Lands Tribunal applied 7% to a term ground rent of £45 for 70.5 years and 6.5% to the section 15 reversion. The Lands Tribunal in an appeal by *Mrs JB Taylor* (1998) applied a 6.5% return to compute the section 15 rent for a reversion in 65.5 years and capitalised that at 6.5%. The term ground rent was discounted at 7%. Yields of 6% and 6.5% have more recently been applied by the tribunals on the basis of evidence particularly in prime central London areas. However, see also the *Sportelli* and *Arbid* cases (supra) where lower yields at 4.5% and 5% have been applied by the Lands Tribunal, and now confirmed by the Court of Appeal.

(8) Section 17 reserves to the landlord the future right to possession for redevelopment where a tenant takes an extended lease. In *Carthew*'s case (*supra*) it was accepted that the possibility of repossession might reduce any potential rental bid under section 15.

(9) Section 15 modern ground rents calculated as percentages of fair rents were held to be inappropriate in *Carthew*'s case (*supra*).

(10) An alternative method might be the new for old approach see Study 5.


## Study 3  Enfranchisement price with a short lease unexpired — Section 9(1)
## Assumptions

This study examines the possible enfranchisement price on the facts assumed in Study 2, as the tenant in that case must look at the alternative merits of buying the freehold rather than taking the 50-year lease extension. The approach adopted takes account of the reversion to the full value of the house and premises after the assumed 50-year extension has expired. This follows the decision of the Lands Tribunal in *Haresign* v *St John the Baptist's College, Oxford* (1980), but whether such a reversion is built into the valuation will depend on the significance of it and the strength of evidence to justify such an approach. As the rateable value at the relevant time, the date of the tenant's notice to have the freehold, was not more than £1,000 the price will be under the assumptions required by section 9(1), which are the original ones excluding any marriage value.

*Valuation for enfranchisement price assessment*

| Term | | £ pa | £ |
|---|---|---|---|
| Rent receivable | | 25 | |
| YP 1 year at 7% (1) | | 0.935 | |
| | | | 23 |
| Reversion to section 15 rent figure | | | |
| adopted as assessed in Study 2 | | 20,790 | |
| YP 50 years at 7% | 13.8 | | |
| PV of £1 in 1 year at 7% (3) | 0.935 | 12.90 | |
| | | | 268,191 |
| Reversion to standing house | | | |
| value (2) in 51 years | | £750,000 | |
| PV £1 in 51 years at 7% | | 0.032 | |
| | | 240,000 | 240,000 |
| Enfranchisement price of the freehold interest | | | £508,191 |
| But say | | | £508,200 |

Notes

(1) Although 7% has been adopted in numerous LVT determinations since 1980, in *Haresign*'s case (*supra*) 6% was adopted in discounting a term rent for three years and the reversion to the section 15 rent. 7% was adopted only for the reversion to the standing house value at the end of the notional lease extension. However, in the case of *Lowther* v *Strandberg* [1985], the Lands Tribunal approved the adoption of 9% throughout the valuation. The figure finally adopted will depend on the relative strength of any supporting evidence. In *Speedwell Estates Ltd* (1998) the Lands Tribunal applied 7% to a term rent fixed for 38 years and 6.5% to the reversionary section 15 rent. Recent Lands Tribunal decisions previously referred to may alter rates generally to be applied. See *Sportelli* and *Arbid* (*supra*) where lower deferment rates of 4.75% and 5% have been determined by the Lands Tribunal and now confirmed by the Court of Appeal.

(2) In *Haresign*'s case (supra) it was argued successfully that, as the residue of the contractual term was so short, the three-stage basis of taking into account the landlord's reversion to freehold possession at the end of the

lease extension was sufficiently material to be included: see also, for example, *Lowther*'s case (*supra*). In *Ball* v *Johnson* (1973) a reversion to a development site after the 50-year lease extension was allowed for where the house had additional land with it not likely to remain as garden indefinitely. In *Griffith* v *Allen* (1989) a leasehold valuation tribunal determined a price reflecting a reversion to standing house value after a combined term and 50-year lease extension giving a total period of 84 years. A discount rate of 7% was used for the term and 50 years' extension, while 8% was applied to the *Haresign* reversion.

(3) Dependable evidence is often difficult to obtain. In *Letorbury Properties Ltd* v *Ivens* (1982) the sale of a freeholder's interest at auction a few days after the enfranchisement notice was held to be the most dependable market evidence.

## Study 4  Enfranchisement against two superior interests and alternative approaches to section 15 rent

This illustrates an approach to assessing enfranchisement price for a house held with a short term of lease but with two superior interests against which to enfranchise. The house has a limited life but the freeholder wishes to impose covenants on the freehold which will restrict the future use and thereby the value of the premises.

The house outside London is on a 0.2 ha site and is held with a five-year unexpired term of sublease at £10 pa from the head leaseholder who holds an eight-year unexpired term from the freeholder at a rent of £5 pa. The occupying subleaseholder qualifies to enfranchise and has recently served valid notices to acquire both superior interests. The rateable value was less than £500 (£1000 in Greater London) on 31 March 1990 and the enfranchisement price falls to be assessed under section 9(1) of the Act. It has been agreed that the freeholder will impose covenants in the conveyance to the effect that not more than one house be erected on the site, that any new design and layout be within certain constraints, and that there be various domestic limitations. The existing house, after much-needed improvements, would have an entirety value of £350,000. Comparable sites have sold for £300 per m² for flat development, planning permission for which is readily forthcoming in the neighbourhood. Large single house plots with planning permission but subject to restrictive covenants such as those to be imposed are worth about £750,000 per ha.

*Valuations for enfranchisement price assessment* (1)

*Section 15 modern ground rent* (2)

| | £ | £pa | |
|---|---|---|---|
| Cleared site approach: | | | |
| 0.2 ha restricted use value (3) | | | |
| 0.2 ha £750,000 site value per ha | | 150,000 | |
| Section 15 rent at 6% | | 0.06 | |
| | | | 9,000 |
| Or | | | |
| Standing house approach: | | | |
| Entirety value | | 350,000 | |
| Site value at 30% (4) | | 0.3 | |
| | | 105,000 | |
| Section 15 rent at 6% | | 0.06 | |
| | | | 6,300 |

Assume that the cleared site approach is the
most appropriate in this case (5) and the
section 15 modern ground rent is £9,000 pa

*Enfranchisement price of the freehold interest*

| Term | £ pa | £ |
|---|---|---|
| Rent receivable | 5 | |
| YP 8 years at 6% | 6.2 | 31 |
| | | |
| Reversion to section 15 rent | 9,000 | |
| YP in perp (6) at 7% deferred 8 years | 8.3 | |
| | | 74,700 |
| Enfranchisement price of freeholder's interest | | £74,731 |
| But say | | £74,750 |

*Enfranchisement price of the headlessee's interest*

| Term | £ pa | £ |
|---|---|---|
| Rent receivable | 10 | |
| Rent payable | 5 | |
| | | |
| Net income | 5 | |
| YP for 5 years at 7% and 4% (Tax at 40%) (7) | 2.6 | |
| | | 13 |

| *Reversion* to section 15 rent | | 9,000 | |
|---|---|---|---|
| Rent payable | | 5 | |
| Net income | | 8,995 | |
| YP 3 years at 8% and 4% | | | |
| (Tax at 40%) | 1.6 | | |
| PV £1 in 5 years at 8% | 0.68 | 1.1 | |
| | | | 9,895.00 |
| | | | £9,907.50 |
| Enfranchisement price of the headlessee's interest (8) (9) (10), Say | | | £9,910.00 |

Notes

(1) Provisions on enfranchisement by subtenants are in section 5(4) and Schedule 1. The reversioner, in this case the freeholder, acts for all the superior interests in dividing up the total enfranchisement price: see *Goldsmiths' Company* v *Guardian Assurance Co Ltd* (1970); *Hameed* v *Hussain* (1977); *Nash* v *Central Estates (Belgravia) Ltd* (1978); *Burton* v *Kolup Investments Ltd* (1978); *Mortiboys* v *Dennis Fell Co* (1984) and *Pilgrim* v *Central Estates (Belgravia) Ltd* [1986]. The leasehold valuation tribunal determination in *Booth* v *Bullivant* (1987) dealt with enfranchisement against the freeholder and head leaseholder by a tenant with a 970-year term. It awarded a nominal £10 to each of the freeholder and head leaseholder for incomes of £1.15 and £2.35 pa receivable for 970 years.

(2) In *Farr* v *Millersons Investments Ltd* (1971) it was suggested that the valuer should use one main approach to the section 15 site value and one of the other accepted approaches as a check. As the life of the property is unsure, the so-called new for old approach may have been appropriate here: see Study 5.

(3) Section 10 deals with the restrictions which may be imposed on the freehold title and in the event of dispute these matters may be referred to a leasehold valuation tribunal with an appeal to the Lands Tribunal. Restrictions on the freehold title to be conveyed may reduce the enfranchisement price but the rights to be assumed in assessing the section 15 rent must be borne in mind: see *Buckley* v *SRL Investments Ltd* (1970); *Peck* v *Hornsey Parochial Charities Trustees* (1970); *Hulton* v *Girdlers Company* (1971); *Grime* v *Robinson* (1972) and *Barrable* v *Westminster City Council* (1984) for examples of covenants being imposed on the freehold titles conveyed: see note (5) to Study 2 (*supra*).

(4)  This percentage is assumed but would have to be justified by reference to supporting evidence: see note (6) to Study 2 (*supra*).

(5)  See the Lands Tribunal's decision in *Farr*'s case (*supra*).

(6)  The effect of the notional 25-year rent review and the reversion after 50 years may be ignored in the calculation: see *Farr*'s case (*supra*) but see also *Haresign*'s case (*supra*).

(7)  This interest is valued on a conventional dual rate tax adjusted basis as it is a short leasehold interest. Although differential risk rates are applied with 6% to the term and 7% to the reversion in this study, in many tribunal decisions the same rate has been applied to both the term the reversion and in recent cases a lower discount rate to the reversions where these will have review provisions. Wholly net of tax valuations were rejected in *Perrin v Ashdale Land & Property Co Ltd* (1971) The tax rate is a negotiable issue depending on timing, length of lease unexpired and current rates.

(8)  Marriage of the head leasehold and freehold interests prior to enfranchisement might increase the total enfranchisement price in this sort of case.

(9)  The House of Lords in *Jones* v *Wrotham Park Settled Estates* [1979] confirmed the validity of a freeholder creating an intermediary headlease which increased the total costs of enfranchisement from £300 to £4,000. This loophole was closed by the Leasehold Reform Act 1979, which requires that the enfranchisement price cannot be artificially increased by transactions involving the creation of new intermediary or similar interests or the alteration in the terms of the lease after 15 February 1979.

(10) In this case had the head leaseholder a profit rent of not more than £5 pa and a reversion of not more than one month the enfranchisement price would be on a formula basis as set out in Schedule 21 (6) to the Housing Act 1980. The formula is:

$$\text{Price (P)} = \frac{\text{Profit Rent (R)}}{2.5\% \text{ Consols Yield (Y)}} - \frac{R}{Y(\ -\ Y)^n}$$

and it capitalises the term profit rent at the 2.5% Consols yield at the valuation date. The formula counts any part of a year as a whole year. There is no appeal to a leasehold valuation tribunal on the enfranchisement price in such cases: see *Afzal* v *Cressingham Properties Ltd* (1981).

## Study 5  Enfranchisement price, comparables and the adverse differential issue — section 9(1) Assumptions

This considers further the methods of arriving at site value in the original extended lease at a modern ground rent enfranchisement hypothesis of section 9 (1) of the Act and examines the now generally discredited concept of the adverse differential. It also looks at the role of enfranchisement settlements as evidence and the use of other risk rate approaches.

Assume a large residential estate, the subject of much enfranchisement with most of the houses held on ground leases expiring shortly at rents of £15 pa each. The landlord's valuers have recently settled several enfranchisement claims at figures of about £115,000 for five year unexpired terms. The frontage, depth and amenity of the plots vary although the plot areas are all very similar. The landlord's valuers have established a comprehensive approach to site values on the estate and have developed techniques to adjust prices to allow for minor differences between the plots.

The study considers enfranchisement price from both the landlord's and the tenant's point of view in order to draw out various matters which might be the subject of negotiation. The interest concerned qualifies for enfranchisement within the less than £500 (£1,000 in Greater London) rateable value bands and the price falls to be assessed under the assumptions of section 9(1). The same assumptions apply following the abolition of domestic rating to new houses which had no rateable value at 31 March 1990 and the R value calculated under section 1(1) (a) (ii) is no more than £16,333. The lease has an unexpired term of five years at £15 pa. The house is agreed to have a limited life and the plot has a frontage of 10 m, a depth of 40 m and an area of 400 m².

*Enfranchisement price using arguments to the landlord's advantage*

| Term | £ | £ pa | £ |
|---|---|---|---|
| Rent receivable | | 15 | |
| YP 5 years at 6% (1) | | 4.2 | |
| | | | 63 |
| Reversion to section 15 rent | | | |
| Cleared site approach: (2) | | | |
| 10 m frontage at landlord's adjusted | | | |
| rate per m frontage for plots of | | | |
| 400 m², 10 m at £12,000 (3) | 120.000 | | |
| Adjustments to site value: (4) | | | |
| (i)   for better location/amenity + 15% | 18,000 | | |
| (ii)  10 m deeper than average plot + 10% | 12,000 | | |
| | | | |
| Site value Say | 150,000 | | |
| Section 15 rent at 6% | 0.06 | | |
| | | | |
| Section 15 rent | | 9,000 | |
| YP in perp at 6% deferred 5 years (5) (6) | | 12. 5 | |
| | | | £112,500 |
| | | | £112,563 |
| Landlord's view of enfranchisement price (7) | | | |
| Say | | | £112,500 |

Notes

(1)    While rates of 6% or 7% were generally adopted for the term following *Farr*'s case (*supra*) the lower rate favours the landlord. More recent decisions previously referred to in *Sportelli* (*supra*) and *Arbid* (*supra*) may alter rates generally applied.

(2)    The cleared site approach is used as the house has a limited life.

(3)    A similarly scheduled approach to site values on a large estate was used by the landlord's valuers in *Siggs* v *Royal Arsenal Co-operative Society Ltd* (1971). The Lands Tribunal was noncommittal, but accepted the approach in that case.

(4)    These adjustments are given solely by way of illustration and could only be used in practice with clear and careful justification from market driven evidence.

(5)    The section 15 rent has been recapitalised at the same 6%. This follows *Official Custodian for Charities* v *Goldridge* (1973) in which the Court of Appeal disapproved of the adverse differential. This differential was earlier adopted by the Lands Tribunal to take account of section 82 of the Housing Act 1969 and its exclusion of the tenant's bid for the freehold in assessing enfranchisement price: see *Farr*'s case (supra) and many subsequent cases where site values were decapitalised at about 6% and then recapitalised at 8% giving an adverse differential of 2%. In *Grainger* v *Gunter Estate Trustees* (1977) an attempt to use a larger adverse differential was rejected by the Lands Tribunal. In *Wilkes* v *Larkcroft Properties Ltd* [1983], the Court of Appeal held that evidence of en bloc sales of ground rents did not justify a claim that the adverse differential should be applied as a matter of law. The court's decision also provides a useful examination of the nature of evidence required to support a decision by the Tribunal.

(6)    In *Siggs* case (*supra*) it was held that the large estate landlord might bid for leaseholds as they came on to the market in order to reap marriage value. This enabled the larger estate to argue against the adverse differential and apply one common rate in arriving at the section 15 rent reflecting the marriage by sale incentive. The Court

of Appeal disapproved of this argument in the *Goldridge* case (*supra*). The degree to which the *Sportelli* (*supra*) and *Arbid* (*supra*) cases will influence the discount rates and the deferment rates to be adopted remains to be seen. The lower deferment rate cases did not relate to section 9(1) assumptions, but to a section 9(1A) case.

Enfranchisement settlements should be adjusted before use as evidence in order to reflect the *Delaforce* effect. This is the extra amount a tenant might be willing to pay in a negotiated settlement to avoid the worry, risk and costs of litigation: see *Delaforce* v *Evans* (1970) and *Ugrinic*'s case (*supra*). In this study settlements are given at about ££115,000 and it therefore might be argued that there has been a £2,500 *Delaforce* effect allowance in this price. In *Wilkes*'s case (*supra*) the Court of Appeal held that the Lands Tribunal did not err in declining to make a deduction for the *Delaforce* effect.

*Enfranchisement price using arguments to the tenant's advantage*

| Term | £ | £ pa | £ |
|---|---|---|---|
| Rent payable | | 15 | |
| YP 5 years at 8% (1) | | 4 | |
| | | | 60 |
| Reversion to section 15 rent | | | |
| New-for-old approach: (2) | | | |
| Sale price of new house on site | 480,000 | | |
| Less: development costs | 380,000 | | |
| Section 15 site value | 100,000 | | |
| Or | | | |
| Sale price of new house on site | 480,000 | | |
| Site value at 20% | 0.2 | | |
| Section 15 site value | 96,000 | | |
| Adopt lower | 96,000 | | |
| Section 15 rent at 8% | 0.08 | 7,680 | |
| YP in perp at 10% (3) deferred 5 years | | 6.2 | 47,616 |
| Tenant's view of enfranchisement price | | | |
| Say | | | £47,676 |

Notes

(1) Early cases involved higher rates when interest rates were high. See *Patten* v *Wenrose Investments Ltd* (1976) 6.8% was accepted at all stages of the valuation and in *Lowther*'s case (*supra*) 9% was adopted throughout for houses in Holland Park, London. Recent decisions such as *Sportelli* (*supra*) referred to may alter rates generally applied.

(2) The new-for-old approach is an alternative where the house has a limited but indeterminate future life: see *Gajewski* v *Anderton* (1971) and *Farr*'s case (*supra*). The figures here are assumed and would need to be supported with evidence.

(3) Although the Court of Appeal disapproved of the adverse differential in the *Goldridge* case (*supra*), there have since been cases before the Lands Tribunal when it has been accepted. The Court of Appeal left the matter open if the concept could be justified with evidence and reason: see *Lead* v *J & L Estates Ltd* [1975] and *Perry* v *Barry Marson Ltd* (1976).

*Possible form of valuation for a negotiated settlement in this study*

| Term | £ | £ pa | £ |
|---|---|---|---|
| Rent reserved | | 15 | |
| YP 5 years at 7% (1)(2) | | 4.1 | |
| | | | 62 |
| Reversion to section 15 rent | | | |
| Site value Say | £105,000 | | |
| Section 15 rent at 6% (1) | 0.06 | | |
| | | 6,300 | |
| YP in perp at 7% deferred | | | |
| 5 years | | 12.45 | 78,435 |
| Enfranchisement price (3) | | | 78,497 |
| | | | 78,500 |

Notes

(1)  For the settlement figure calculation a rate of 6% has been adopted but this would of course need to be supported by dependable market evidence derived from the financial and money markets and consistent analysis of market transactions and settlements: Recent decisions previously referred to may alter rates generally applicable.

(2)  See also the Lands Tribunal and Court of Appeal decisions in *Arbid* v *Earl Cadogan* [2005] (*supra*) and *Earl Cadogan* v *Sportelli* [2007] (*supra*) where deferment rates of 4.75% and 5% applied by the Lands Tribunal have been confirmed as guidelines for future cases.

(3)  No reversion to the standing house after 55 years has been included as was in *Haresign*'s case (*supra*), as the house has been stated to have a very limited future life and such an approach would be inconsistent with the cleared site and new-for-old approaches adopted.

## Study 6

### *Rateable value adjustment for qualification to discount tenant's improvements*

This examines further qualification for enfranchisement or extension under section 1(4A) of the Act introduced by section 118 and Schedule 8 to the Housing Act 1974 as amended by section 141 and Schedule 21 to the Housing Act 1980.

A bungalow outside London is held on the residue of a long lease with a 10-year unexpired term at a ground rent of £12 pa. The premises had a rateable value of £300 at March 23 1965, one of £805 at 1 April 1973 and has a gross external area (GEA) of 225 m². The present occupier purchased the premises in 1970 and at once built an extension with a GEA of 25 m². The previous owner had installed central heating and erected a double garage in 1968. The value of the bungalow now, freehold and with vacant possession, would be £525,000 but without the tenant's improvements would be only £325,000. Apart from the rateable values the tenant is assumed to be otherwise qualified under the provisions of section 1(1) and 1(5) of the Act, ignoring the 1993 Act extensions. The rateable values after adjustment to discount the tenant's improvements may not exclude qualification.

### *Qualification for enfranchisement*

The tenant did not qualify under the original 1967 rateable value limit and at first sight does not qualify under the amendments of the Housing Act 1974 as the rateable value was over £750. The tenant may, however, use section

1(4A) of the Act, contained in Schedule 21 to the Housing Act 1980, which enables a tenant who is otherwise qualified to claim a reduction (1) in the notional rateable value for enfranchisement purposes to exclude the annual value of any tenant's (2) improvements (3) (4). The tenant might also qualify under the 1993 Act amendment removing rateable value limits but on a less favourable enfranchisement price basis because the 1993 Act basis omits the right to the 50-year extension of the lease.

*Adjustment to the rateable value under section 1 (4A) of the Act and Schedule 8 to the Housing Act 1974*

|  |  | £ | £ |
|---|---|---:|---:|
| Rateable value at 1.4.1973 (5) |  | 805 |  |
| Gross value at 1.4.1973 |  | 1,000 |  |
| Analysis of gross value | 225 m² at £3.67 |  | 825 |
|  | Central heating |  | 80 |
|  | Double garage |  | 95 |
|  | Gross value |  | 1,000 |
|  |  |  |  |
| Assessment without tenant's improvements |  |  |  |
| (exclude 25 m² extension) | 200 m² at £3.67 |  | 734 |
|  | No central heating |  | — |
|  | No garage |  | — |
|  | Gross value |  | 734 |
| Less: statutory deductions |  |  | 151 |
| Adjusted rateable value without tenant's improvements (5) (6) (7) |  |  | £583 |

Therefore, the lessee will qualify to enfranchise but the enfranchisement price will be under the additional section 9(1A) introduced in 1974 as the rateable value is over £500 (7) (8). For houses on leases granted after March 1990 which had no rateable value at March 31 1990, the R value under section 1(1)(a)(ii) of the Act should be less than £16,333 for qualification under section 9(1). To qualify under section 9(1A) the R value should be between £16,333 and £25,000.

Notwithstanding the above process the tenant might alternatively claim qualification under section 1(A) of the Act, introduced by the 1993 Act, which confers in certain circumstances the same rights to have the freehold as apply to those within the earlier financial limits. In such circumstances no right to an extended lease can be claimed and the enfranchisement price is on the additional basis set out in section 9(1C) introduced by the 1993 Act.

Notes
(1)  The tenant first serves a notice on the landlord requiring him to agree to the nature of the improvements to be discounted and proposing a figure for the reduced rateable value. The form of notice is set out in Schedule 8 to the Housing Act 1974 and the tenant must specify the improvements and works concerned. The tenant must bear the reasonable costs of the landlord's investigations of the works of improvement claimed on notices, see Schedule 21 (8) to the Housing Act 1980. Failing agreement between the parties the county court may determine the extent and nature of the improvements to be taken into account and the valuation officer may be required to determine the reduced rateable value excluding the value of the improvements.
(2)  The provisions regarding the adjustment of rateable value apply to the improvements of both the current and any previous tenants: see Schedule 8 (1) to the Housing Act 1974.
(3)  The time scale for the procedures under Schedule 8 is generally mandatory and a tenant may only make a single application under them for that reason: see *Pollock* v *Brook-Shepherd* [1983]. In *Arieli* v *Duke of Westminster* [1984] the Court of Appeal overruled a county court decision to refuse an extension of time to originate an application to the court under Schedule 8. In *Johnson* v *Duke of Devonshire* (1984) a 12-day extension of the time to refer the matter to the court was held to be within the judge's discretion.

(4) Improvements may include an addition, for example a garage. Whether the construction of a replacement house is an improvement is not clear but the Court of Appeal in *Pearlman* v *Harrow School Keepers and Governors* [1978] decided that the installation of central heating in this context was an improvement.

(5) There has been some uncertainty as to the date by reference to which rateable values are to be considered and adjusted: see article at (1983) 266 EG 187 — "Leasehold Reform Act — Notional reductions in rateable value" by Nigel T Hague QC — *Pollock's* case (*supra*), and *Mayhew* v *Free Grammar School of John Lyon* [1991]. It has been held to be the date of agreement or the date of the certificate issued by the District Valuer *John Lyon's Charity* v *Vignaud* [1992].

(6) The rating valuation approach illustrated is likely to be adopted by the valuation officer where the landlord and tenant cannot agree to the appropriate reduction.

(7) See *Woodruff* v *Hambro* [1991] where a tenant who had surrendered a long lease and taken a new one was precluded from applying for a reduction in rateable value in order to qualify for enfranchisement.

(8) There is no appeal against a valuation officer's certificate of adjusted rateable value. In *R* v *Valuation Officer for Westminster and District, ex parte Rendall* [1986] a tenant argued that the valuation officer's certified adjusted rateable value was incorrect, it being more than £1,500, and sought judicial review. This was dismissed by the High Court and the Court of Appeal.

(9) As the notional rateable value after adjustment would be over £500 and under £750 the enfranchisement price should be assessed under the assumptions in section 9(1A) of the Act, introduced by section 118(4) of the Housing Act 1974.

## Study 7 Enfranchisement price where the rateable value is over £500 (£1,000 in London) — the "Norfolk" approach — section 9(1A) assumptions

This illustrates the assessment of enfranchisement price under the provisions of section 9(1A), which do not assume a modern ground rent for the notional 50-year lease extension and do not preclude any marriage value from a tenant's bid. Using the facts adopted in Study 6 and with the rateable value reduced to £583 the tenant now qualifies to enfranchise but at a price to be assessed under the less favourable assumptions introduced by section 118 of the Housing Act 1974 and first interpreted by the Lands Tribunal in *Norfolk* and developed further in *Lloyd-Jones* (*supra*) and many cases subsequently: see also notes (4) and (5).

*Enfranchisement price assessment*

*Valuation of lessor's interest excluding marriage value (1)*

| | £ pa | £ |
|---|---|---|
| Term | | |
| Rent receivable | 12 | |
| YP 10 years at 7% | 7 | |
| | | 84 |
| Reversion to a protected tenancy under the Landlord and Tenant Act 1954 (2) | | |
| Assume rent Say 7% of the unimproved (3) standing house value (4) of ££325,000 | 22,750 | |
| *Less*: outgoings, say 15% | 3,412 | |
| | | 19,338 |
| YP in perp at 7% (4) deferred 10 years | | 7.26 |
| | | 140,394 |
| Value of lessor's interest excluding marriage value | | 140,478 |
| (see note (5) for an alternative approach) | | |
| Say | | 140,500 |

*Valuation of the lessee's interest excluding marriage value*

| Term | £ pa | £ |
|---|---|---|
| Annual value of house, say 10% on the full freehold vacant possession value (4) of £525,000 | 52,500 | |
| Less: rent payable | 12 | |
| Profit rent | 52,488 | |
| YP 10 years at 8% (4) and 2.5% (tax 35%) (6) | | 4.6 |
| | | 241,445 |
| Value of lessee's interest excluding marriage value, Say | | 241,450 |

(No value has been put on the reversionary right as a reversion in 60 years may be too distant to be significant)

*Apportionment of marriage value to arrive at enfranchisement price (7)*

| | | £ | £ | £ |
|---|---|---|---|---|
| (i) | Value of lessor's interest exclusive of marriage value | | | 140,500 |
| (ii) | Assessment of lessor's share of marriage value | | | |
| (a) | Freehold vacant possession value | £525,000 | | |
| | Less: lessee's improvements | £200,000 | | |
| | | | 325,000 | |
| (b) | Value of lessor's interest exclusive of marriage value | | 140,500 | |
| (c) | Value of lessee's interest exclusive of marriage value | 241,450 | | |
| | *Less*: value to the lessee of his improvements (8) | 73,600 | | |
| | | | 167,850 | |
| (d) | Total value of freehold and leasehold interests unmarried (b + c) | | 308,350 | |
| (e) | Gain on the marriage of the interests ignoring the lessee's improvements (a–d) | | 16,650 | |
| | Lessor's share of the gain Say 50% (9) | | 0.05 | |
| | | | | 8,325 |
| (iii) | Enfranchisement price (i) + (ii) | | | £148,825 |

Notes

(1)	The study broadly follows the apportionment of marriage value basis used by the Lands Tribunal in *Norfolk* v *Trinity College, Cambridge* [1976]. The principle confirmed in that case was that section 9(1A) of the 1967 Act does not exclude the extra value of the tenant's bid where the enfranchisement price is assessed under the section 118 provisions of the Housing Act 1974. The price therefore includes any extra amount that a tenant might pay for the value increase arising on the marriage of the freehold and leasehold interests.

(2)	The principle in section 9(1A) (b) is that, instead of reverting to a modern ground rent, enfranchising tenants revert in this enfranchisement hypothesis to a protected tenancy under the Landlord and Tenant Act 1954.

(3)	The valuation must ignore the value of any tenant's improvements, section 9(1A)(d). It seems that the word improvement here has a wider meaning than for the rateable value reduction provisions in Schedule 8 to the 1974 Act but would exclude maintenance and renewals. They must, however, have added to the value.

(4)	The assumptions here are illustrative only and clearly will be open to negotiation. In *Lloyd-Jones* v *Church Commissioners for England* [1982], the second reported case under the different valuation assumptions introduced by the 1974 Act, it was held that such an approach was inappropriate in London as it was almost unknown for tenants of particular types of house to continue in occupation on a protected tenancy. Tenants, it was stated, normally surrender and renew leases or purchase the freehold. In *Lowther* v *Strandberg* [1985] the effect of the tenant's bid on a Holland Park, London W14, estate was considered and the *Lloyd-Jones* approach was adopted even though the reversion was not for another 81 years. A discount rate of 9% was adopted throughout the valuation, but lower rates currently prevail. See also *Vignaud* v *Keepers and Governors of the Free Grammar School of John Lyon* [1996], and *Earl Cadogan* v *Sportelli* (*supra*) where much lower deferment rates were determined and Lands Tribunal determinations confirmed as a starting point for consistent valuations.

(5)	In the *Lloyd-Jones* case (supra) the Lands Tribunal accepted the strong settlement evidence of the landlord's valuers and dismissed, as being out of touch with reality, the tenant's view of the reversion to a fair rent. The approach adopted for the landlord's reversion was simply to deduct 10% from the freehold vacant possession value to reflect the risk of a tenant claiming a tenancy under Part I of the Landlord and Tenant Act 1954.

*The* Lloyd-Jones *case approach to the value of the landlord's interest excluding marriage value would be*:

| Term | £ | £ |
|---|---|---|
| As before | | 84 |
| Reversion to unimproved vacant possession value (6) (7) (8) | 325,000 | |
| Less: for risk of tenant claiming a tenancy under Part I of the Landlord and Tenant Act 1954 | | |
| 10% (9) | 32,500 | |
| | 292,500 | |
| PV of £1 in 10 years 7% | 0.51 | |
| | | 149,175 |
| | | £149,259 |
| Say | | £149,250 |

In this study the enfranchisement price might be little different but in the *Lloyd-Jones* case it increased the enfranchisement price by £68,000.

(6)	The lessee's term has been valued on a conventional dual rate tax-adjusted basis. No value has been placed here on the lessee's right to a further protected tenancy although as there may be a loss in value to the lessor there is an argument for such a further element in the valuation.

(7)	This is an apportionment approach similar to that used in *Norfolk*'s case (*supra*).

(8) This has been found by taking an annual value of £16,000 for the improvements and capitalising it on a dual rate tax-adjusted basis for the lessee's term unexpired.

(9) The percentage applied in the *Norfolk* case (*supra*) but valuers must be cautioned against simply following the approaches adopted in other cases as leasehold valuation tribunal and Lands Tribunal decisions are not necessarily good evidence of value for other cases. Market evidence is likely to be of equal or even greater relevance.

## Study 8 Further rateable value adjustments as a means of reducing enfranchisement price

This study considers the effect of obtaining a further reduction in the rateable value to below the £500 (£1,000) figure in order to enable the enfranchisement price to be assessed under section 9(1) which may be more favourable to the tenant.

The facts are as in Studies 6 and 7 except that on detailed survey of the house and premises it becomes apparent that part of the bungalow dates from 1900 and part from 1922. The gross external area of the premises as originally constructed in 1900 can be shown to be 160 m². The value of the freehold interest without these further improvements is assumed to be £275,000.

### *Qualification for enfranchisement*

*Adjustment to the rateable value under sections 1(4A) and 9(1B) of the Act*

|  | £ |
|---|---|
| Gross value at 1.4.1973 (as before) 1,000 | 1,000 |
| Gross value without tenant's improvements | |
| 160 m² at £3.67 | 587 |
| No central heating | |
| No double garage | |
| Gross value | 587 |
| Less: statutory deductions | 126 |
| Adjusted rateable value without tenant's improvements (1) | £461 |

The tenant qualifies to enfranchise but as the rateable value was less than £500 (outside Greater London) the enfranchisement price will be assessed under the original section 9(1) assumptions of the 1967 Act.

*Enfranchisement price assessment* (2)

| Term | £ | £ pa | £ |
|---|---|---|---|
| Rent reserved | | 12 | |
| YP 10 years at 7% | | 7 | |
| | | | 84 |
| Reversion to section 15 rent | | | |
| Standing house approach (3) | | | |
| Entirety value | 525,000 | | |
| Site value at 25% | 131,250 | | |
| Section 15 rent at 7% | 0.07 | | |
| | | 9,188 | |
| YP in perp at 7% deferred 10 years | | 7.3 | |
| Enfranchisement price (4) | | | |
| This compares with £149,250 in Study 7 (5) say | | | £67,072 |

Notes

(1)  In the leasehold valuation tribunal case of *Effra Investments Ltd* v *Stergios* (1982) a house in London had been converted into flats taking the rateable value over £1,000. Although the tribunal gave the tenant time to take action to reduce the rateable value under Schedule 8 to the Housing Act 1974 he did not do so and the tribunal fixed a high enfranchisement price under the provisions of section 9(1A), closely following the principles of the *Norfolk* decision.

(2)  In this case the value of any tenant's bid, which might include a marriage value component, is excluded by section 82 of the Housing Act 1969.

(3)  This approach was considered more fully in Studies 3 and 4.

(4)  It can be seen from this enfranchisement price that it may well be in the lessee's interest to try to ensure that the rateable value is reduced to a level which brings the enfranchisement price within the original 1967 Act valuation rules and excludes the marriage value component and the resulting higher enfranchisement prices adopted under section 9(1A) in the *Lloyd-Jones* and *Norfolk* cases (*supra*).

(5)  The enfranchisement price assumptions for premises in the £500 to £750 (£1,000 to £1,500 in London) rateable value bands or with an R value of between £16,333 and £25,000 are those of section 9(1A). If a tenant in that band of value served a notice to take an extended lease instead, he would be entitled to such at a modern ground rent for 50 years notwithstanding the rateable value level. Subject to the repossession rights of section 17 (see Study 12) he might avoid having to pay the marriage value element yet would still obtain a very valuable interest: see article at (1983) 268 EG 876, 978 and Study 9 (*infra*).

## Study 9 Houses in the higher rateable value bands: extending a lease as an alternative to, or prior to, buying a freehold

The Lands Tribunal in *Hickman* v *Phillimore Kensington Estate Trustees* [1985], confirmed by the Court of Appeal in *Mosley* v *Hickman* [1986], determined that a tenant of a house in the higher rateable value band who had first extended the lease for 50 years at a section 15 rent could enfranchise at a price reflecting that extended lease. Section 23 of the Housing and Planning Act 1986 subsequently amended section 9(1A) of the Act to require that no rights to such an extension be assumed and that if a lease has been extended it is to be assumed to terminate at the original term date. The alternative of lease extension was also examined at (1983) 268 EG 876, 978, where it was shown that in some cases it might be potentially better value for a tenant than enfranchising under the assumptions of section 9(1A). This alternative is still available to otherwise qualified tenants in the higher rateable value band who wish to expand their interest under the Act by a lease extension but without any capital outlay.

This study assumes a house in London with an 1 April 1973 rateable value of £1,300 held on a ground lease with 10 years to run. The ground rent is £50 pa and the tenant is assumed to qualify under the provisions of the Act, as amended, the house is agreed as having an unencumbered freehold value of £750,000 at the relevant date.

### Lease extension

The tenant of the house can claim under section 14 to have an extended lease of a further 50 years in addition to the present term and such a lease would be in substitution for the existing one. This new lease would be for 60 years, the first 10-year term at the existing rent and the further 50 years at the section 15 rent. The section 15 rent would not be fixed until the last year of the original lease, in this case in nine years time. The new lease would normally provide for a rent review, in this case in 35 years time: section 15(2). The tenant will still have the right to serve a notice to have the freehold up to the time when the original lease would have expired, in this case during the next 10 years: section 16(1)(a). Such an arrangement would ensure that the tenant retains a quite valuable interest at quite a modest cost, in present value terms, without having to borrow any capital. However, the nature of the section 15 rent could mean that the interest is less marketable than it would be if freehold.

*Valuation of the tenant's interest after extension*

| Term of existing lease | £ pa | £ |
|---|---|---|
| Full annual value, say 6% on £750,000 | 45,000 | |
| Less: ground rent | 50 | |
| Profit rent | 44,950 | |
| YP 10 years at 6% and 2.5% (tax 35%) (1) | 4.99 | |
| | | £224,301 |
| Reversion to section 15 rent and extended term (2) | | |
| Full annual value as before | 45,000 | |
| Section 15 rent based on site value as 30% of the entirety of £750,000 = £250,000 | | |
| Section 15 rent at 6% of £250,000 (3) | 15,000 | |
| Profit rent | 30,000 | |
| YP 50 (4) years at 6% and 2.5% (tax 35%) deferred 10 years at 10% | 7.137 | |
| | | 214,110 |
| Total value of tenant's extended interest | | 438,411 |
| Say | | £438,500 |

Notes
(1) All the component figures would need justification by reference to market evidence.
(2) While the extended lease will in due course become a wasting asset it might appeal to a tenant who simply wishes to remain in occupation and has no intention, or wish, to realise the best financial return from a sale of the house.
(3) Extension of the lease would limit the cost of remaining in occupation to the present value of the liability to pay the section 15 rent. The discounted net present cost today should be compared with the present value of the interest obtained and the cost of an outright purchase of the freehold (*supra*).
(4) It must be remembered that a tenant who has taken an extended lease can be the subject of proceedings for possession for redevelopment under section 17, subject to the compensation right for the value of the extended lease: see Study 12.

## Study 10 Higher Rateable Value Band Enfranchisement price if notice to have the freehold is served after the lease has been extended

If the leaseholder had served notice to have the freehold after the lease was extended, but before 7 November 1986, the enfranchisement price was assessed "on the assumption that the vendor was selling ... subject to the tenancy". The tenancy existing would be the extended one at a section 15 modern ground rent, not the type of tenancy normally to be assumed under section 9(1A) for houses in the higher rateable value band. The tenant was not, however, excluded from the notional market in which the freehold is being sold and a marriage value bid could have been included in an enfranchisement price calculation.

Following the amendment of section 9(1A)(a) by the Housing Act of 1986, introduced to overturn part of the effect of the *Hickman* decisions, the assumption to be applied in computing the enfranchisement price is that even if the lease has been extended at section 15 rent it is to be assumed that it will terminate on the original term date. In such circumstances the enfranchisement price might be computed as in Study 7.

## Study 11 Enfranchisement price for a house qualifying beyond the applicable financial limits in section 1–section 9 (1C) approach

Under amendments to the Act introduced by sections 63 to 66 of the Leasehold Reform, Housing and Urban Development Act 1993, a new subsection 9(1C) was added which provides that for houses now qualifying under the provisions of the new sections 1A, 1AA and 1B the enfranchisement price is computed in accordance with section 9(1A) with the additional assumptions of section 9(1C). In any such case where there is marriage value to be taken into account, the share of marriage value to which the tenant is entitled shall not exceed 50%. The new section 9(A) provides additional elements of compensation to the landlord where a tenant qualifies under sections 1A or 1B to include:

(i)    any reduction in value of any interest of the landlord in other property resulting from the acquisition;
(ii)   any loss or damage which results, to the extent that it relates to the landlord's ownership of any interest in other property;
(iii)  any loss of development value, which means any increase in value attributable to the possibility of demolishing, re-constructing or carrying out substantial works of construction on the whole or a substantial part of the house or premises.

## *Facts*

A detached Victorian six bedroomed house in a north London suburb with a large garden with various tenant's improvements. The house had rateable values of £530 at 1 April 1963 and £2,200 on 1 April 1973. It is held with 53 years unexpired of a 85-year lease at a ground rent of £200 pa with a fixed uplift of the ground rent every 15 years to the sum by which 0.25% of the capital value of the unimproved house exceeds £200 pa. The tenant qualifies under section 1A(1) of the Act by the extension of the qualification rules beyond the previous rateable value limits. The assumptions to apply to assessment of the price are those in section 9(1A) as amended by 9(1C). The ground rent will be reviewed to £2,500 in three years time. The freehold vacant possession value is £1,250,000, but the tenant's improvements contribute £200,000 to that value. The value of the leaseholder's present lease is agreed to be £725,000 excluding the value of improvements.

| | | £ |
|---|---|---|
| *Term* | | |
| Current ground rent | £200 pa | |
| YP 3 years @ 6% (1) | 2.67 | £534 |
| | | |
| *Revised Ground Rent* | | |
| Reviewed Rent | £2,500 pa | |
| YP 50 years deferred 3 years @ 6% | 13.23 | £33,084 |
| | | |
| *Reversion* | | |
| Vacant possession value | £1,250,000 | |
| Less tenants improvements | £200,000 | |
| | £1,050,000 | |
| | | |
| PV of £1 in 53 years @ 6% | 0.046 | £48,300 |
| Current value of freehold | | £81,918 |
| | | |
| *Marriage Value* (1) | | |
| Freehold VP value excluding improvements | | £1,050,000 |

| *Less* | | |
|---|---|---|
| Current value of freehold | £81,918 | |
| Plus current value of lease | | |
| excluding improvements | £725,000 | |
| | | £806,918 |
| | | |
| Marriage Value | | £243,082 |
| Freeholders share 50% | | £121,541 |

| *Enfranchisement Price under section 9(1C) of the 1967 Act* | |
|---|---|
| Value of freehold | £81,918 |
| + Freeholders share of Marriage Value | £121,541 |
| | £203,459 |
| | |
| *Enfranchisement Price* (2) Say | £203,500 |

*+ Any loss or reduction of value of other
interests or loss of development value (3)*

Notes
(1)    This approach follows that applied by the Lands Tribunal in *Re John Lyons Charity* (1996). See also the Lands Tribunal in *Trustees of the Eyre Estate* v *Saphir* (1999). The Court of Appeal's decision in *Cadogan* v *Sportelli* [2007] (*supra*) may have a significant effect on the deferment yields to be used as the yields adopted by the Lands Tribunal are now to be adopted as a starting point.
(2)    See also the Lands Tribunal determination in *John Lyons Charity* v *Brett* (1998) and *Sharp* v *Cadogan Estates* (1998).
(3)    These are the additional areas of compensation rights specifically added to the landlord's entitlement in such cases under section 9(A) where the right arises under section 1A or 1B.


## Study 12 Premises with redevelopment potential and the repossession rights under section 17

This examines the assessment of enfranchisement price where the house and the premises have development potential. The problem is how to take account of the landlord's rights to repossession for redevelopment under section 17 of the Act which are available where a tenant has taken advantage of the rights for a lower valued house to have a 50-year lease extension at a section 15 rent. The principles have not been widely tested before the Lands Tribunal or the courts, probably because of the fact that the gains available through marriage via enfranchisement are sufficient to ease successful negotiations. In the leasehold valuation tribunal determination in *Booer* v *Clinton Devon Estates* (1989) the tribunal reflected the redevelopment right for six plots, but included in the enfranchisement price the full value of a reversion to a section 15 rent. It should be noted that for houses qualifying under section 1A and paying a price assessed under section 9(1C) the landlord specifically has to be compensated for any foregone development value under section 9A.


### Facts

An obsolete house on a 0.7 ha plot in a provincial city is held with a six-year unexpired term at a ground rent of £25 pa. The house had a rateable value of £195 at 23 March 1965 and the tenant qualifies to enfranchise under the enfranchisement price assumptions in section 9(1) of the Act requiring the assumed 50 years lease extension at a section 15 rent. The value of the plot with planning permission for the best feasible redevelopment would be £250,000 freehold with vacant possession. The value of the site restricted to use for one dwelling would be £55,000 and the entirety value of the house improved would be £120,000, both figures being freehold with vacant possession. The lessors are willing to sell the freehold interest without restrictions on redevelopment on the title and have accepted the lessee's notice of enfranchisement. The lessee requires a valuation of his current interest as he is now proposing to sell to a developer.

## Valuation of the lessee's interest for sale (1) taking account of the landlord's redevelopment rights under section 17(2)

| | £ pa | £ | £ |
|---|---|---|---|
| *Cost of enfranchisement* | | | |
| Term | | | |
| Rent payable | 25 | | |
| YP 6 years at 7% | 4.8 | | |
| | | | 120 |
| | | | |
| *Reversion* to full redevelopment value | | 250,000 | |
| *Less*: section 17 and Schedule 2 compensation. The value to the lessee of a 50-year lease at a section 15 modern ground rent. (2) | | | |
| Annual value of house, Say (3) | 12,000 | | |
| | | | |
| *Less*: section 15 rent based on 7% of £55,000 | 3,850 | | |
| | | | |
| Notional annual profit rent | 8,150 | | |
| YP for 50 years at 7% and 2.5% (tax 35%) | 12.39 | | |
| | | | |
| Schedule 2 compensation in 6 years' time (3) | | 100,979 | |
| Value of reversion less compensation | | 149,021 | |
| PV £1 in 6 years at 7% (4) | | 0.666 | |
| Present value of lessor's reversionary rights under the enfranchisement hypothesis | | | 99,248 |
| | | | |
| Total enfranchisement price (5) | | | 99,368 |
| | | | |
| Add landlord's and tenant's enfranchisement fees and costs (6) Say | | | 2,000 |
| | | | |
| Total enfranchisement costs (7) | | | 101,368 |
| Say | | | 101,500 |
| | | | |
| Full development value of the site (as given) | | | 250,000 |
| Less: costs of enfranchisement | | | 101,500 |
| | | | |
| Value of lessee's interest for sale (8) (9) | | | £148,500 |

Notes

(1)  A lessee may assign his lease with the enfranchisement rights as soon as a valid notice of enfranchisement has been served: see section 5.

(2)  Section 17 reserves to the landlord of a dwelling, where the lessee has taken an extended lease under the Act, the right to repossession for redevelopment. Here it is assumed in assessing enfranchisement price that the tenant has an extended lease; it would appear logical to incorporate in the valuation the notional right of the landlord to repossession. This right is available not earlier than one year prior to the original term date and the landlord must compensate the tenant under Schedule 2 for the value of a 50-year lease at a section 15 modern ground rent.

(3)  The basis for and approach to these figures is open to argument and negotiation.

(4)  The deferment allows for the assumption that the lessor cannot obtain possession until the end of the term, when he will then have to pay the lessee compensation under Schedule 2.

(5)  If the rateable value were over £500 (£1,000 in London), so that the enfranchisement price fell to be assessed under the *Norfolk* v *Cambridge* and *Lloyd-Jones'* approaches (supra), the marriage value component might add further to the enfranchisement price and reduce the value of the lessee's interest, even though the assumptions under section 9(1A) preclude the notional right to repossession under section 17 and assume a reversion to a registered rent rather than a modern ground rent.

(6)  The tenant is responsible for landlord's reasonable legal and valuation costs incurred in connection with the enfranchisement: see section 9(4).

(7)  No cases from the Lands Tribunal or the courts on the points covered in this study have been reported. The Lands Tribunal briefly considered a related section 17 matter in *Cottingham-Mundy* v *Dover Borough Council* (1971), while a leasehold valuation tribunal did consider the matter in *Booer* v *Clinton Devon Estate* (1989).

(8)  If the lessee assigned his interest before agreement as to the enfranchisement price, negotiations might subsequently delay development. Therefore the full redevelopment value of the site might be deferred for a suitable period to allow for agreement.

(9)  A valuation in this case of the lessee's interest but ignoring the section 17 redevelopment rights gives a very different answer. The total enfranchisement costs on a cleared site approach under section 9 (1) would be lower. It can be seen that this section 17 approach is very much a landlord's argument.

## Study 13 Flats: Individual right to acquire a new lease under the Leasehold Reform, Housing and Urban Development Act 1993 — 90 years lease extension

Section 39 of the Leasehold Reform Housing and Urban Development Act 1993 grants, to qualifying tenants on long leases who have been the registered leaseholder of their flats at the Land Registry for two years, the right to surrender and take a new lease for 90 additional years at a peppercorn rent. The immediate landlord may not always be the competent landlord: see section 40 of the 1993 Act. A premium is payable to the landlord, or to each landlord where there is more than one, assessed under Schedule 13 to the 1993 Act as amended by the 2002 Act, and the new lease is on the terms of the existing lease with modifications as appropriate under the provision of section 57 of the 1993 Act. While a further right to a new lease may be exercised, none of the statutory security of tenure provisions will apply (section 59). The tenant has to pay the various and reasonable costs of the landlord under section 50 and the landlord retains the right to apply to the court for possession for redevelopment either during the final 12 months of the original term or during the last five years of the new lease. In such circumstances the landlord would have to pay compensation for the value of the tenant's interest under Schedule 14 of the 1993 Act.

The premium payable in these circumstances is the total of the reduction in the value of the landlord's interest and the landlord's share of marriage value as set out in Schedule 13 to the 1993 Act. Key valuation assumptions include an open market value without the tenant seeking to buy, excluding the right to the new lease and any tenant's improvements.

## Facts

A third-floor flat in a well located London mansion block with four bedrooms, three living rooms and a net internal area of 250 m$^2$ (gross internal area 272 m$^2$). The flat was held on a lease from the freeholder for 27 years with seven years expired at a fixed ground rent of £150 pa. The tenant has been accepted to qualify to have the new lease following notice under section 42 of the 1993 Act and it is agreed that the landlord should receive a premium for the value of a lease for a further 90 years at a peppercorn rent. The vacant possession value of the new lease on the flat is agreed at £700,000, or about £2,574 per m$^2$ of the gross internal area, with an uplift of 5% if it was freehold. There is a well run service charge arrangement for the block.

*Valuation of the Premium Payable by the tenant in accordance with Schedule 13 to the 1993 Act*

| | | | |
|---|---|---|---|
| (a) | *Diminution of the landlord's interest* | | |
| | Current Interest (1) | | |
| | Existing lease 7 years unexpired | | |
| | Ground rent | £150 pa | |
| | YP 7 years @ 6.5% | 5.49 | £824 |
| | | | |
| | Reversion to capital value | | |
| | Lease for 90 years agreed value | | £700,000 (2) |
| | Add for freehold interest say 5% | | £35,000 |
| | Say | | £735,000 |
| | | | |
| | But less for Landlord and Tenant | | |
| | Act 1954 rights/risk of not | | |
| | obtaining possession | | |
| | Say 40% leaving | | £441,000 |
| | PV of £1 in 7 years @ 6,5% | | 0.6435 | £283,786 |
| | | | £284,610 |
| | Deduct | | |
| | Value of freeholder's future interest after | | |
| | lease extension — Capital value Say | £441,000 | |
| | in 90 years (as above) | | |
| | PV of £1 in 97 years @ 6.5% | 0.0022 | £981 |
| | Diminution in landlord's value | | £283,805 |
| | | | |
| (b) | *Marriage Value* | | |
| | Tenant's interest under the new lease | £700,000 | |
| | Landlord's reversionary interest | £981 | £700,981 |
| | | | |
| | *Less* Tenant's existing leasehold | | |
| | interest agreed at say | £200,000 | |
| | Landlord's existing interest | £284,610 | £484,610 |
| | Gain on marriage | | £216,371 |
| | Landlord's share (4) (5) Say 50% | | 0.50 | £108,186 |
| | Total premium payable for extended | | |
| | lease of flat plus fees & costs of landlord | | £391,991 |
| | Say | | £392,000 |

Notes

(1) See *Cadogan Estates Ltd* v *McGirk* (1998)) where a similar approach was determined by the Lands Tribunal and a range of comparables were considered with the issue of whether the lease value should be discounted at the end of the existing term to reflect the loss of statutory protection.

(2) The appropriate capitalisation and discount rates may be rather lower depending on wider financial market considerations: see *Arbid* and *Sportelli's* cases (*supra*).

(3) In *Goldstein* v *Conley* (1998) the Lands Tribunal made a further deduction of 40% from the freehold unencumbered value to reflect the potential impact of the Landlord and Tenant Act 1954 rights on this element of the computation. A 72.5% marriage value share as applied in *Shahgholi* (below) was also applied in this case. Following revisions under the 2002 Act, Marriage Value is now to be split 50%/50% in all such cases.

(4) In *Cadogan Estates Ltd* v *Shahgholi* (1998) the Lands Tribunal accepted compelling evidence that the landlord's share of marriage value would have been struck at 72.5% not at the convention of 50% which is now generally applicable, in a reference concerning a long leasehold flat in a six-storey period house.

(5) See also *Re Grosvenor Estate Belgravia* (1999) which concerned, inter alia, the quantification of the loss in value of the freeholder's interest in the whole building caused by a tenant's new lease right. A 6% discount rate was applied and a 30% adjustment for the statutory tenancy right.

## Study 14 Flats: Collective Enfranchisement under the Leasehold Reform, Housing and Urban Development Act 1993

### Legal background

These provisions, sections 1 to 38, enable qualifying tenants to buy the freehold and intermediate leasehold interests in a block of flats through a nominee purchaser. The right extends to appurtenant property, including garages, outhouses and gardens and applies to self contained buildings or parts of buildings where there are two or more flats owned by qualifying tenants and at least two thirds of the total number of flats in the block are owned by qualifying tenants. Self-contained buildings are those structurally detached or vertically divided and capable of separate development. Excluded are buildings where more than 10% of the gross internal floor areas, excluding common parts, are non residential or the building has no more than four units and has a resident landlord.

Qualifying tenants are those of flats on long leases, with those on business leases and tenants of charitable housing trusts specifically excluded. The leaseholder of the flat must have also been registered at the Land Registry for a period of 2 years prior to serving notice. Long leases include ones over 21 years in length, perpetually renewable ones and continuation tenancies under Part I of the 1954 Act.

The processes for a qualifying group of tenants to combine and create the nominee purchaser and progress collective enfranchisement are complex and the reader is referred to sections 11 to 30 of the 1993 Act. The provisions relating to determination of price and costs of enfranchisement are in sections 32 and 33 and Schedule 6 to the Act and the owner of any interest will have a lien for any other outstanding debts following collective enfranchisement. The costs of the process are in general born by the nominee purchaser, who is effectively all the enfranchising tenants.

The price for the freehold is the open market value of the freeholder's interest, the freeholder's share of marriage value and compensation for other losses resulting from enfranchisement. The valuation assumptions are that the title is subject to the existing leases and that the leaseholders have no right to buy collectively or to acquire a new lease. Tenant's improvements are excluded and there are anti avoidance provisions. Marriage value is 50%. In addition compensation for other loss or damage to the freeholder arising from the acquisition is to be paid to include:

(i) reductions in the value of other interests of the freeholder

(ii) loss of development value, even if that could have been reduced, for example by the freeholder taking a leaseback.

## Facts

The nominee purchaser, established by qualifying tenants, wishes to acquire the freehold of a block of 20 identical self-contained flats outside London all held with 70 years unexpired of 99-year leases all at equal £100 pa ground rents, totalling £2,000 pa fixed for the balance of the leases. The lessees reimburse the lessor for the cost of insurances while services are effectively operated by the lessees through the nominee purchaser company and the freeholder has no ongoing service or management responsibilities. 16 of the 20 leaseholders are participating tenants. At the date of the nominees' notice the open market values of each existing 70-year leasehold interest in a flat is agreed to be £50,000. The freehold reversionary interest in the whole block is agreed to be currently worth £26,000.

*Collective Enfranchisement Price for the Nominee Company* (1)

| | | |
|---|---:|---:|
| Value of existing leases of participating tenants | | |
| 16 × £50,000 | £800,000 | |
| | | |
| *Value of Virtual Freehold* | | |
| Say total value of leases plus 5% (2) | £840,000 | |
| Add for elimination of ground rent | | |
| £1,600 pa × 7 YP (3) | £11,200 | £851,200 |
| | | |
| *Less* | | |
| (i)   Value of existing leases | £800,000 | |
| (ii)  Value of existing freehold | £26,000 | £826,000 |
| | | |
| Marriage Value | | £25,200 |
| Freeholder's share 50% (4) | | £12,600 |
| | | |
| Collective Enfranchisement | | |
| Price | | |
| (i)   Value of existing freehold | | £26,000 |
| (ii)  Share of Marriage Value | | £12,600 |
| | | £38,600 |

*Plus* Any reduction in the value of the interests in property held by the freeholder arising from the acquisition and any loss in the development value.

*Plus* The freeholder's fees and costs.

Notes
(1)   The nominee purchaser can subsequently grant new or extended leases to the flat occupiers on terms to be agreed between themselves. The funding and organisation of the nominee purchaser would have to be established, together with the various occupying tenants' joint agreement to underwrite the costs, whether abortive or otherwise prior to commencing.
(2)   See *Maryland Estates Ltd* v *Abbature Flat Management Company Ltd* (1998) where an adjustment of 5% was determined by the Lands Tribunal; also *Maryland Estates Ltd* v *63 Perham Road Ltd* [1997].
(3)   See *Maryland Estates* (*supra*).
(4)   The marriage value amount under Schedule 6 to the 1993 Act was originally a minimum of 50% or such greater sums as can be proved with relevant evidence: see *Cadogan Estates Ltd* v *Shahgholi* (1998) (*supra*) where 72.5% was determined. Following revisions by the 2002 Act, Marriage Value is now to be split 50%/50% in all cases. It should be noted that provisions of the Commonhold and Leasehold Act 2002, yet to take effect, require the

creation of a 'right to enfranchise company' prior to collective enfranchisement to replace the nominee purchase. There is, at present, no indication as to when these provisions are to come into effect. See also the *Arbid* and *Sportelli* cases (*supra*) where the marriage value rules and deferment rates to be applied were reconsidered.

## Study 15 Flats: Collective Enfranchisement under the Leasehold Reform, Housing and Urban Development Act 1993: Shorter Lease Terms Unexpired

A freehold purpose built seven-storey block of 13 flats in London is occupied by tenants all on 99-year leases with 20 years unexpired at the agreed valuation date. Nine of the leases are held by qualifying tenants who wish to collectively acquire the freehold through a nominee company. The ground leases are at fixed rents of £50 pa each and the vacant possession values of each current leasehold flat are agreed at £250,000 on a new 175-year lease but subject to the current unexpired lease only £165,000. There will be an agreed loss in value to the freeholder of about £100,000 to an adjacent development site, the access to which will be lost by the sale of the block.

*Value of Freeholder's Interest in accordance with Schedule 6 to the 1993 Act* (1)

| | | | |
|---|---|---|---|
| *Participating tenants' leases* | | | |
| Ground rents × 9 | | £450 | |
| YP for 20 years @ 8% (2) | | 9.818 | 4,418 |
| | | | |
| *Reversion in 20 years* | | | |
| 9 flats worth £250,000 each | | £2,250,000 | |
| *Less* risk of LTA 1954 rights | | | |
| Say 40% (3) reduction | | £1,350,000 | |
| PV of £1 in 20 years @ 8% | | 0.214 | £289,640 |
| | | | £294,058 |
| | | | |
| *Non Participating Tenants' Leases* | | | |
| Ground Rents × 4 | | £200 | |
| YP for 20 years @ 8% | | 9.818 | £1,964 |
| | | | |
| Reversion in 20 years | | | |
| 4 flats worth £250,000 | | £1,000,000 | |
| *Less* risk of LTA 1954 rights | | | |
| Say 40% (3) | | £600,000 | |
| | | | |
| PV of £1 in 20 years @ 8% | | 0.214 | £128,728 |
| | | | £130,692 |
| | | | |
| Total Freehold Interest (£294,058 + £130,692) | | | £424,751 |

*Marriage Value*
*New Nominee Freehold Value*

| | | | |
|---|---|---|---|
| (i) | Nine new long 125 year leases at nominal rents in participating flats. | | |
| | 9 @ £250,000 | £2,250,000 | |
| (ii) | Four existing non-participating flats. | | |
| | Existing freehold values | £130,692 | £2,380,692 |
| | of ground rents and reversions | | |

*Current Position*

| | | | |
|---|---|---|---|
| (a) | Existing freehold interest in participating flats | £294,058 | |
| (b) | Existing leasehold interests of participating flats 9 @ £165,000 | £1,485,000 | £1,779,058 |
| | Gain on Marriage | | £601,634 |
| | Freeholder's share of marriage value 50% | | £300,817 |

*Premium Payable*

| | | |
|---|---|---|
| (i) | Value of freehold interest pre-enfranchisement | £424,751 |
| (ii) | Freeholder's share of marriage value | £300,817 |
| (iii) | Loss in value of freeholder's adjacent development plot agreed at say reflecting cost of alternative access | £100,000 |
| | | £825,568 |
| | Total premium payable (i) + (ii) + (iii) | |
| | Say | £825,600 |

*Plus* the freeholder's related fees and costs (4)
(Cost of process £98,415 per flat for the 9 participants (5)).

Notes
(1)   See *Becker Properties Ltd* v *Garden Court NW8 Property Co Ltd* (1997) in which the Lands Tribunal explores hope value of premiums from non participating tenants.
(2)   The discount rates will need to be suitable to reflect current precedents and arguments. In *Becker Properties Ltd* (*supra*) discount rates of 9% were accepted but for longer lease terms albeit with rising ground rents. Again, care must be taken in the approach to capitalisation rates to reflect recent LVT and Lands Tribunal determinations following the *Arbid* and *Sportelli* cases (*supra*).
(3)   This allowance is for the risk of the tenant remaining in occupation as a protected tenant under the 1954 Act: see *Goldstein* v *Conley* (1998) (*supra*).
(4)   See also *Kemp and Siegfried* v *Myers* (1998).
(5)   The participants as freeholders would have the potential to extract premiums from the non participants for lease extensions as the term dates for their leases draw closer.

## Study 16 Rights of occupiers of buildings in two or more flats to acquire their landlord's interest under the provisions of Part 1 of the Landlord and Tenant Act 1987 as amended by the Housing Act 1996

The Landlord and Tenant Act 1987, as amended, provides, *inter alia*, the right of tenants to have first refusal on a relevant disposal of any premises containing two or more flats, where those flats are held by qualifying tenants, and the number of flats held by such tenants exceeds 50% of the total number of flats in the premises. The rights of such tenants are confined to privately owned blocks and to exercise these rights the following conditions must be satisfied:

(i)   That the tenant occupies a flat under a tenancy which is neither a protected shorthold, a business tenancy under the Landlord and Tenant Act 1954 Part II, a tenancy terminable on cessation of employment or an assured tenancy.

(ii)   The landlord must be proposing a relevant disposal of an estate or interest.
(iii)  The premises must contain two or more flats held by qualifying tenants.
(iv)   The number of such flats must exceed 50% of the total number of flats in the premises.
(v)    The premises may consist of the whole or part of a building.

There are a number of exemptions from the provisions of the 1987 Act which would not, for example, apply if any part of the premises were intended to be occupied other than for residential purposes and those parts exceeded 50% of the internal floor area. Resident landlords are excluded from the operation of the 1987 Act as are a number of exempt landlords including local authorities, housing associations and urban development corporations.

## Facts: Shop with three flats above built 80 years ago

The freehold of the whole premises is held by a private property company. The shop is let separately from the flats on a five-year lease at a recently fixed rack-rent of £5,500 pa net. Two of the flats are occupied by tenants on long leases each with 27 years to run at £10 pa net with a separate reviewable service charge. One flat is let on a tenancy at a registered rent of £1,350 pa including a reviewable service charge.
   The internal floor areas are as follows:

| | |
|---|---|
| Shop | 153 m² |
| Common parts | 32.5 m² |
| Each flat | 57.6 m² |

The landlord is proposing to offer the freehold interest for sale.

## Possible effects of the 1987 Act

The three flats, each apparently occupied by a qualified tenant, exceed the two required and 50% of the flats in the premises appear to be within the Act. The non-residential part appears to be insufficient to disqualify the premises as the internal floor area of the shop is only 47% of the whole after having disregarded the common parts.

$$\frac{153m^2}{(172\ m^2 + 153\ m^2)} = 0.47$$

If one of the flats was let as offices then the premises would not be subject to the 1987 Act as more than 50% of the internal floor area would be occupied for other than residential purposes. The landlord is not exempt or resident and therefore before selling on the open market should serve offer notices under section 5 of the Act, as amended, on the three qualifying tenants of the flats. This notice will need to define the estate to be sold, the premises, the consideration required and give the tenants the statutory time to respond. Two out of the three would need to vote to accept the offer for them to be able to proceed and they would then need to arrange to organise and fund the purchase. If the tenants were not to proceed with an acceptance notice in the prescribed period the landlord may, during the following 12 months dispose of the protected interests for a price not less than and on other terms corresponding to those offered to the tenants. The landlord might offer the shop, as well as the flats, to the qualifying tenants of the flats, who might have to fund a pure investment property purchase of the shop. Alternatively, the landlord could choose to offer the tenants the reversionary interest in only the flats on long leasehold, although this might conflict with the good management of the shop and its future investment performance. A landlord's proposed sale might be delayed by the revised 1987 Act procedures introduced by Schedule 6 of the 1996 Act even if the tenants did not in the event proceed with the purchase. If the freehold of the premises were owned by a local authority or new town development corporation, however, it could be disposed of immediately.

## Procedure and tactics

The cost of purchasing the shop as well as the reversionary interest in the flats could deter the flat tenants from accepting the offer or from negotiating an agreement. If the landlord failed to follow the offer notice procedure and sold the interest to a new landlord, the qualifying tenants could, if the majority agree, serve a purchase notice on the new landlord. If, following the purchase notice, the parties fail to agree terms a leasehold valuation tribunal could be required to determine the estate or interest to be disposed of and other terms under section 12. A tenant may argue before a tribunal that the shop part should not be part of the interest to be disposed of.

### Possible approach to the valuation of the landlord's interest under the 1987 Act

This is on the basis that the landlord is disposing of its interests in the whole premises.

|  | £ pa | £ | £ |
|---|---|---|---|
| Shop part |  |  |  |
| Rent income net | 5,500 |  |  |
| Less: management | 300 |  |  |
| Net income | 5,200 |  |  |
| YP in perp at 8% | 12.5 | 65,000 | 65,000 |
| *Two flats occupied on long leases* |  |  |  |
| Term |  |  |  |
| Rents received 2 @ £10 | 20 |  |  |
| YP 27 years at 10% (1) | 9.24 | 185 |  |
| *Reversion* to two flats to be sold on |  |  |  |
| new 99-year leases 2 @ £45,000 | 90,000 |  |  |
| *Less*: factor to reflect the chance of |  |  |  |
| tenants opting to continue under Part I |  |  |  |
| of the Landlord and Tenant Act 1954 |  |  |  |
| Say 20% (2) | 18,000 |  |  |
|  | 72,000 |  |  |
| PV of £1 in 27 years at 12% (1) | 0.047 | 3,384 | 3,569 |
| *One flat let on a registered rent* |  |  |  |
| Rent received |  | 1,350 |  |
| *Less*: outgoings |  |  |  |
| services | 200 |  |  |
| repairs | 150 |  |  |
| insurance | 100 |  |  |
| management | 105 | 555 |  |
| Net income |  | 795 |  |
| YP in perp at 14% |  | 7 | 5,565 |
| *Plus* an amount to reflect the possible sale |  |  |  |
| to the sitting tenant of a long lease or the sale |  |  |  |
| of the flat with vacant possession (2) |  |  |  |
| Say |  | 6,000 | 11,565 |
| Total value of the landlord's interest (3) |  |  | 80,134 |
| Say (4) (5) |  |  | 80,000 |

Notes
(1) The discount rates to be adopted would need to be drawn from justifiable current comparable market evidence and no strong reliance should be placed on those used in this example.
(2) The factors and amounts to be allowed for to reflect the possible sale of flats with vacant possession will depend on the age profile, the financial and family circumstances and the intentions of the sitting tenants as these factors will influence the market's view of such opportunities.
(3) The consideration figure to be inserted on any offer notices may need to be higher than this in order to provide a reasonable margin for balanced negotiation. The key elements are the shop value and the less-clear marriage value element. In *Twinsection Ltd* v *Jones* (1997) the Lands Tribunal in dealing with a purchase by qualifying tenants considers the rights of tenants over gardens and amenity land and changes since the tenant's purchase notice.
(4) Given the high proportion of value attributable to the shop element it is unlikely that the leasehold occupiers of the flats would wish to, or be able to, fund the purchase of the landlord's interest and they probably would not serve an acceptance notice allowing the landlord to sell his interest the market.
(5) See *Davis* v *Stone* [1992] a determination of the purchase price under these provisions by a Leasehold Valuation Tribunal which also concerned the terms for the conveyance.

## Study 17 Collective enfranchisement of a Victorian house in five flats under section 24 of the 1993 Act as amended by the Commonhold and Leasehold Reform Act 2002.

### *The Facts*

A large house in a leafy conservation area outside London was converted into five self-contained flats 21 years ago. Four of the flats are held by leaseholders on the balance of 99 year leases at £50 pa ground rents, two of whom are in occupation and two are investors. The freeholder of the house and premises resides in flat 5. All leaseholders have rights over the gardens and common parts, but the freeholder has also retained the basement as storage. Three of the leaseholders are participating in the purchase and have appointed a nominee purchaser and valid notices and counter notices have been served. The freeholder has agreed and is entitled to take a leaseback of his occupied flat for 99 years at a peppercorn rent. All the occupiers contribute one fifth of common parts maintenance. The notice to have the freehold was served two years ago (1) when the leases all had about 78 years to run. The value with very long leases at peppercorn rents of flats 1, 2, 3 and 5 would be £170,000, while flat 4 would be £210,000. The tenants served notice when there was less than 80 years to run on each lease (2) and will have to pay some marriage value.

### *Calculation of Freehold Acquisition Price to the Nominee Purchaser*

*Value of Freeholder's Present Interest*
*Flats 1–4*
*Term*

|  |  | £ | £ | £ |
|---|---|---|---|---|
| Flat 1 |  |  |  |  |
| Ground Rent | £50 pa |  |  |  |
| YP 78.03 years @ 7% | 14.2129 |  |  |  |
|  |  | 711 |  |  |
| Flat 2 |  |  |  |  |
| Ground Rent | £50 pa |  |  |  |
| YP 78.08 years @ 7% | 14.2132 |  |  |  |
|  |  | 711 |  |  |

|  | | £ | £ | £ |
|---|---|---|---|---|
| **Flat 3** | | | | |
| Ground Rent | £50 pa | | | |
| YP 78.11 years @ 7% | 14.2133 | | | |
| | | 711 | | |
| | | | | |
| **Flat 4** | | | | |
| Ground Rent | £50 pa | | | |
| YP 78.00 years @ 7% | 14.2128 | | | |
| | | 711 | 2,844 | |

*Reversion*:

|  | | £ | £ | £ |
|---|---|---|---|---|
| Flat 1 | | | | |
| Flat value | 170,000 (4) | | | |
| PV £1 78.03 years @ 7% | 0.0051 | | | |
| | | 867 | | |
| | | | | |
| Flat 2 | | | | |
| Flat value | 170,000 | | | |
| PV £1 78.08 years @ 7% | 0.0051 | | | |
| | | 867 | | |
| | | | | |
| Flat 3 | | | | |
| Flat value | 170,000 | | | |
| PV £1 78.11 years @ 7% | 0.0051 | | | |
| | | 867 | | |
| | | | | |
| Flat 4 | | | | |
| Flat value | 210,000 (4) | | | |
| PV £1 78.03 years @ 7% | 0.0051 | 1,071 | 3,672 | |

| | | | | |
|---|---|---|---|---|
| Value of Freeholder's present interest in flat 1–4 | | | | 6,516 |

|  | £ |
|---|---|
| Flat 5 Reduction in Freeholder's Value | |
| Value of Current Freeholder's Value Say | 170,000 |
| Value of Interest following leaseback* | |
| Say 98% of freehold | 166,600 |

Lease granted to current freeholder on valuation date assumed to be a 99 year lease at a peppercorn rent for the whole term and no restriction on underletting or assignment

| | | | | |
|---|---|---|---|---|
| Reduction in value of Freeholder's present Interest in Flat 1 (Difference between value of current interest and value of interest following leaseback (5) | | | 3,400 | 3,400 |
| | | | | 9,916 |

|  | £ | £ | £ |
|---|---|---|---|
| *Marriage Value flats 1–4* (6) | | | |
| Value of freehold interests  1 @ £210,000 | | 210,000 | |
| 3 @ £170,000 | | 510,000 | |
| | | 720,000 | |
| Value of leasehold interests | | | |
| Say Freehold less 2% £70,000 – £14,400 | | 705,600 | |
| Marriage value | | 14,400 | |
| 50% to freeholder | | | 7,200 |
| *Basement Price* (7) | | | |
| Income (for use as storage) | | | |
| 641 sq ft @ £2 per sq ft | | 1,282 | |
| YP perp deferred 1 year at 11% (8) | | | |
| Value of Freeholders present interest in Basement | | 8.19 | 10,500 |
| Total Value of Freeholder's present interest | | | 27,616 |
| Total freehold acquisition price | | | |
| Say | | | £27,600 |

*Plus* landlord's reasonable legal and surveyors fees and costs

Notes

(1) Originally under these provisions the valuation date was the date of agreement or the date of the leasehold valuation tribunal's decision. The Act of 2002 has amended this and the valuation date would be the date of the valid initial notices, thereby fixing values and the term unexpired for valuation purposes.

(2) If the tenants had more than 80 years to run on their leases at the date of the notice they would not have had to pay marriage value. Here they were two years late and will have to pay 50% of the marriage value.

(3) Capitalising the term at 7% reflects LVT approaches prior to the *Arbid* and *Sportelli* cases (*supra*). Lower yields may well become generally applicable following the Court of Appeal decisions in the *Sportelli* cases (*supra*), given the revised status of Lands Tribunal determinations. The different length of lease terms reflects the actual position given different lease commencement dates.

(4) The value of freehold or very long leasehold flats at nominal ground rents would need to be determined on the basis of relevant market evidence in the locality at the valuation date.

(5) This reflects the loss to the freeholder who is agreeing to give up the freehold for a long leasehold. Some LVT's have simply applied a simple percentage reduction to reflect the lesser attractiveness of a 99 years lease. Here 2% has been applied, but market evidence would be important.

(6) Loss of marriage value at 50% has to be paid to the freeholder as all the leaseholders failed to serve their notices before their leases went below 80 years unexpired term. The position would be further complication where some leaseholders had served their notices with 80 years and more to run.

(7) The basement has some value to the freeholder and here an approach to valuing the loss of the freeholder in transferring it to the nominee purchaser is demonstrated.

(8) This high discount rate reflects the risky and speculative nature of the basement element, but would depend on market evidence.

# Further reading

Aldridge, T M *Leasehold Law*, Oyez Longman.

Aldridge, T M *Rent Control and Leasehold Enfranchisement*, Oyez.

Barnes, D M W *The Leasehold Reform Act 1967*, Butterworths.

Hague, N T *Leasehold Enfranchisement*, Sweet & Maxwell (4th ed, 2003)

Hubbard, C C and D W Williams, *Handbook of Leasehold Reform*, Sweet & Maxwell (1988)

Wellings, V G Woodfall *Landlord and Tenant*, Sweet & Maxwell.

MacGilp J and G Fox, RICS/Institute of Housing, *Leasehold Reform, Housing and Urban Development Act 1993 — A guide to Part I and Part II*, 1994.

The Leasehold Advisory Service *www.lease-advice.org*

*www.landstribunal.gov.uk*

*www.lvtbulletin.com*

# The Landlord and Tenant Act 1954

The Landlord and Tenant Act 1954 Part II together with the Landlord and Tenant Act 1927 gives security of tenure to tenants of business premises and a right to possible compensation at the end of their tenancies. The 1954 Act was amended by the Regulatory Reform (Business Tenancies) (England and Wales) Order 2003 (SI 2003 No 3096) as from 1 June 2004 and this chapter reflects these changes. The security under the 1954 Act is afforded by:

(a) automatic continuance of the tenancy notwithstanding expiry of the term at common law (section 24)
(b) compelling a landlord who desires possession to establish one or more of the grounds listed in section 30
(c) giving the tenant a right to apply for a new tenancy.

## Conditions for security of tenure to apply

Before these rights accrue, the tenancy must be one to which the 1954 Act applies, such conditions being outlined in section 23(1), which provides:

> ... applies to any tenancy where the property comprised in the tenancy is or includes premises which are occupied by the tenant and are so occupied for the purposes of a business carried on by him or for those and other purposes.

In practice, therefore, first there must be a tenancy, so that licences are excluded. In *Street v Mountford* [1985] the House of Lords held that where residential accommodation is granted for a term at a rent with exclusive possession the grant is a tenancy. While this decision relates to occupation of residential accommodation, it is submitted that it is just as applicable to business premises. In *Onyx (UK) Ltd v Beard* [1996] the court held that the absence of a provision for the payment of rent in the agreement pointed to the fact that the arrangement was not of a commercial character and raised the question of whether it might be a licence.

Second, the premises must be occupied by the tenant for the purposes of a business. Usually, this will not give rise to any problems, as occupation by an agent, for instance a manager, will suffice, as in *Cafeteria (Keighley) Ltd v Harrison* (1956). It has been suggested, in *Bagettes v GP Estates Ltd* (1956), that a tenant who occupies premises for the sole purpose of subletting parts of the building is outside the scope of the 1954 Act, so that the right to renew does not enure in favour of a tenant who has totally sublet: see *Narcissi* v *Wolfe* [1960]. In *Graysim Holdings Ltd* v *P&O Property Holdings Ltd* [1996] the House of Lords held that as a matter of principle a tenant under a business tenancy cannot sublet part of the property to a business subtenant and at the same time continue to occupy a holding comprising all the property contained in the demise to him. The definition of holding in section 23(3) excludes property not occupied by the tenant indicating that two persons (other than when acting jointly) cannot be in occupation for the purposes of the 1954 Act at the same time. In *Linden* v *Department of Health and Social Security* [1986] flats which were occupied by persons employed by a district health authority which, in turn, was exercising the functions on behalf of the Secretary of State were held to be occupied for the purposes of a government department. As a result of section 23(1A), (1B) occupation or the carrying on of a business (a) by a company in which the tenant has a controlling interest; or (b) where the tenant is a company, by a person with a controlling interest in the company shall be treated for the purposes of section 23 as equivalent to occupation or, as the case may be, the carrying on of a business by the tenant. It is to be noted that there is no equivalent provision for limited liability partnerships.

Finally, business is defined in section 23(2) as including

a trade, profession or employment and includes any activity carried on by a body of persons, whether corporate or unincorporate

This has been held to include the activities of a tennis club and the storage of goods in a lock-up garage: see *Bell* v *Alfred Franks & Bartlett Co Ltd* [1980] (see also *Groveside Properties Ltd* v *Westminster Medical School* [1983]). The term will obviously include such premises as shops and offices. However, it should be noted that if a tenant carries on a business in breach of a prohibition in the lease, the protection of the 1954 Act does not apply to the premises unless the landlord has consented to, or acquiesced in, the breach: section 23(4).

## Excluded tenancies

If the tenancy comes within section 43 of the 1954 Act it will not be subject to the security of tenure provisions. In general, the Act does not apply to:

(i)   agricultural holdings but may apply to a field used for horse-riding lessons — see *Wetherall* v *Smith* [1980]
(ii)  mining leases
(iii) service tenancies
(iv)  short leases which, with certain exceptions, are for terms not exceeding six months
(v)   extended leases as there is no right to renew an extended tenancy granted under section 16(1) of the Leasehold Reform Act 1967
(vi)  tenancies at will — see *Manfield & Sons* v *Botchin* [1970] 2 QB 612
(vii) tenancies granted by exempt bodies: sections 57–60 of the 1954 Act.

Following the introduction of the Landlord and Tenant (Licensed Premises) Act 1990 the position of licensed premises with regard to 1954 Act protection is as follows: the effect of section 1(1) of that Act is that section 43(1)(d) of the 1954 Act ceases to have effect in relation to any tenancy entered into on or after July 11 1989 otherwise than in pursuance of a contract made before that date.

If the tenancy in question was entered into before July 11 1989 and continues in existence until July 11 1992 it will be protected by the 1954 Act thereafter. In such a case a notice served under section 24(3)(b) does not have effect: section 1(2) of the 1990 Act.

Section 1(3) of the 1990 Act provides that in the circumstances envisaged by section 1(2) the landlord or the tenant can serve certain notices as if section 43(1)(d) had already ceased to have effect, namely:

(i)   a section 25 notice specifying as the date of termination July 11 1992 or any later date
(ii)  a section 26 notice requesting a new tenancy beginning not earlier than July 11 1992
(iii) a section 27 notice stating that the tenant does not desire his tenancy to be continued.

# Automatic continuance under section 24

Under section 24 a tenancy to which the Act applies cannot come to an end save in the manner provided by the Act, so that if the procedures laid down in sections 25 and 26 are not strictly adhered to the tenancy continues in force indefinitely. Section 24 provides that a tenancy to which the Act applies shall not come to an end unless terminated in accordance with the provisions of the Act and either the landlord or the tenant under such a tenancy may apply to the court for a new tenancy. Neither the landlord nor the tenant may make an application under section 24(1) if the other has made such an application and the application has been served. Neither the tenant not the landlord can make an application if the landlord has made an application for a new tenancy or termination of the current tenancy under section 29.

Section 24 also expressly reserves several common law methods of terminating a business tenancy. For instance, the right to forfeit the tenancy or any superior tenancy is also provided for as well as the ability of the tenant to serve a notice to quit.

# Termination of business tenancy

The renewal or termination procedure may be commenced either by the landlord serving a section 25 notice to terminate or by the tenant serving a section 26 request for a new tenancy. There are alternative forms for the landlord's section 25 notice, namely

(i)  where the landlord is willing to grant a new tenancy and
(ii) where the landlord is unwilling to grant a new tenancy.

## (a) Section 25 notice

If the landlord is willing to grant a new tenancy he should serve a notice in the prescribed form which is specified as Form 1 in Schedule 2 to the Landlord and Tenant Act 1954 Pt2 (Notices) Regulations 2004 (SI 2004 No 1005). This form of notice advises the tenant of the termination of the tenancy, contains the landlord's proposals for the grant of a new tenancy, advises that either party can make an application to the court and specifies the date for the termination of the tenancy. If the landlord is unwilling to grant a new tenancy he should serve a notice in the prescribed form which as specified as Form 2 in the 2004 Regulations. This states that an application to the court will be opposed by the landlord, states what grounds in section 30 will be used by the landlord, advises that the tenant can apply to the court for a new tenancy and specifies the date for termination of the current tenancy. The form is given by the competent landlord who is the first superior landlord having an interest that will not come to an end within 14 months by effluxion of time (section 44). Such a notice must state the date on which the current tenancy will come to an end and such a date must not be earlier than the date on which the tenancy would expire either by notice to quit (periodic tenancy) or by effluxion of time (fixed-term tenancy).

The section 25 notice must be given not more than 12 months and not less than six months before the termination date specified in the notice (section 25(2)). In computing the date of service of the section 25 notice or a tenant's counternotice the general rule is that only one day of the notice period is excluded and not both the date of service and the date of expiry: *Hogg Bullimore & Co* v *Co-operative Insurance Society Ltd* (1984). Since the changes brought into effect On June 1 2004, a tenant no longer has to serve a counter-notice to a landlord's section 25 notice.

If a 21-year lease contained a break clause giving the landlord an option to terminate the lease every seven years, what would be the effect of section 25 on the provisions in the lease? In *Weinbergs Weatherproofs Ltd* v *Radcliffe Paper Mill Co Ltd* [1957], it was stated that the service of an ordinary break clause notice would not satisfy section 25, but may enable a section 25 notice to be served thereafter, while in *Scholl Manufacturing Co Ltd* v *Clifton (Slim-line) Ltd* [1966] it was held that a single notice may, in certain circumstances, operate both the break clause and satisfy section 25.

It should be noted that the provisions of section 25(3) and (4) (the 'not earlier than' provisions) do not apply where the tenancy has already expired or been terminated at common law and is continuing under section 24: *Lewis* v *MTC (Cars) Ltd* (1975). In such an example, the notice under section 25 can be given at any time to expire not more than 12 nor less than six months later than the date on which it is given.

## (b) Section 26 request

For a tenant's section 26 request to be valid, the tenant's current tenancy must have been granted for either a term of years certain exceeding one year or a term of years certain and thereafter from year to year. However, a request cannot be made after the landlord has served a section 25 notice. Should the landlord fail to reply, he will lose his right to oppose a new tenancy.

# Application to the court

Either party may apply to the court. This can be made as follows:

(i)   by a tenant who has made a section 26 request or who has received a section 25 notice
(ii)  by a landlord who is willing to grant a new tenancy or
(iii) by a landlord who is opposed to the grant of a new tenancy and has either served a negative section 25 notice or served a counter-notice to the tenant's section 26 and is making the application for termination.

As both the landlord and tenant are empowered to make an application to the court, dual applications are avoided by section 24(2A). A similar provision is contained in section 24(2B) which prevents the making of an application by either party where the landlord has already made an application under section 29(2) to terminate. Once a section 25 notice has been given to a tenant or a section 26 request has been given to the landlord, the relevant time-limits become linked to the statutory period. The significance of these time-limits is that a court cannot entertain an application made under section 24(1) or section 29(2) if it is made after the end of the statutory period. The statutory period is defined in section 29(2) by reference to the section 25 notice or the section 26 request. Where the landlord has served a section 25 notice the statutory period is the period ending on the date specified in his notice for termination of the current tenancy. Where the tenant has served a section 26 request, the statutory period is the period ending 'immediately before the date specified in his request'. In the case of a section 25 notice the beginning of the statutory period is the date on which the notice is given to the tenant. In the case of a section 26 request, the beginning is the date two months after the giving of the notice to the landlord, or if the landlord serves a counternotice under section 26(6), immediately after the counternotice has been served. An application made too early cannot be entertained by the court; section 29(3). After the landlord's section 25 notice or tenant's section 26 request, but before the end of the statutory period, the landlord and tenant may agree that an application to the court may be made before the end of the period specified in the agreement which will expire after the end of the statutory period: section 29B(1).

It is of the utmost importance to note that recourse to the court to determine the terms of the new tenancy is to be made only where there is disagreement. Section 28 provides that where the landlord and the tenant agree to the grant of a further tenancy of the holding, the current tenancy will continue until that date but will not be a tenancy to which the 1954 Act applies. To satisfy section 28 the agreement must be in writing and there must be agreement on all the material terms: see *Derby & Co Ltd v ITC Pension Trust Ltd* [1978]. In *RJ Stratton Ltd v Wallis Tomlin & Co Ltd* [1986] an agreement for the purposes of section 28 was held to be a binding contractual arrangement enforceable by the parties at law. Application to the court within the strict time-limits is essential to protect the tenant's interest.

# Landlord's grounds of opposition

Whether the landlord states his opposition in his own section 25 notice or in his counternotice to the tenant's section 26 request he will have to confine himself to the seven grounds outlined in section 30(1), paras (a) to (g), which are as follows:

(a) *Failure to repair.* In order to be able to rely on this ground the landlord must prove that the state of repair of the holding is such that the tenant should not be granted a new tenancy. There is a discretion in the court as to the degree of disrepair: see *Lyons v Central Commercial Properties Ltd* [1958]. In *Eichner v Midland Bank Executor & Trustee Co Ltd* (1970) the court was of the opinion that it was entitled to consider the whole of the tenant's conduct in relation to his obligations and was not limited to the landlord's grounds.

(b) *Persistent delay in paying rent.* It is clear from *Horowitz v Ferrand* [1956] that the arrears of rent do not have to be sustained, but if the reason for the delay no longer applies, the landlord may not succeed under this ground: see also *Hopcutt v Carver* (1969). In *Hurstfell Ltd v Leicester Square Property Co Ltd* [1988] the Court of Appeal was of the opinion that the crucial question of whether the arrears were likely to recur was one for the judge at first instance.

(c) *Breaches of other obligations.* This ground includes substantial breaches of the tenant's obligations under the current tenancy or any other reason connected with the tenant's use or management of the holding. In grounds (a), (b) and (c) the whole of the tenant's conduct may be considered: see *Eichner* (*supra*).

(d) *Alternative accommodation offered.* Where the landlord offers alternative accommodation to the tenant such accommodation must be provided or secured by the landlord. The terms of the alternative accommodation must be reasonable having regard to the current tenancy and be suitable for the tenant's requirements, enabling him to preserve goodwill.

(e) *Better return if let or sold as larger unit.* Where the competent landlord is a superior landlord (and not the immediate one) and the tenancy in question is a subtenancy the landlord may oppose a new tenancy on the ground that by letting all the premises contained in the head lease together more rent could be obtained.

(f) *Intention to demolish or reconstruct.* This ground will apply where the landlord establishes that on the termination of the current tenancy he intends to demolish or reconstruct the premises comprised in the holding (or a substantial part of it) or to carry out substantial work of construction that he could not reasonably so do without obtaining possession. In *Betty's Cafés Ltd v Phillips Furnishing Stores Ltd* [1959], it was held that the landlord must have this intention at the date of the proceedings and he will fail if major obstacles still lie ahead of him, eg no planning permission obtained. Further, in *Fisher v Taylors Furnishing Stores* [1956], it was stated that the motive for effecting the work was immaterial.

   The type or amount of work needed to satisfy ground (f) is a question of fact and degree. In *Housleys Ltd v Bloomer-Holt Ltd* [1966], the court held that demolishing a garage and wall was sufficient. Demolition without reconstruction is also sufficient. In *Botterill v Bedfordshire County Council* [1985] the Court of Appeal held that infilling by removal of topsoil, depositing waste and replacing the topsoil was not reconstruction within section 30(1)(f). Under section 31A the tenant may still be able to obtain a new tenancy if he agrees to the inclusion in the new tenancy of terms giving the landlord sufficient access to undertake the work or that he is willing to accept a tenancy of a smaller part of the holding. In *Mularczyk v Azralnove Investments Ltd* [1985] the Court of Appeal rejected the contention that the works proposed could be undertaken under the provisions of section 31A(1)(a) because the landlord could not reasonably carry out the works without obtaining possession of the holding and without interfering for a substantial time or to a substantial extent with the business user of the land. Further, the tenant may be able to show that the current tenancy contains an access clause and the work falls within the terms of that clause: see *Heath v Drown* [1973].

(g) *Intention to occupy himself.* The landlord can establish this ground where he proves that on the termination of the current tenancy he intends to occupy the holding for the purposes, or partly for the purposes, of a business to be carried on by him there, or as his residence. In *Re Crowhurst Park, Sims-Hilditch v Simmons* (1973), it was held that the requirement of paragraph (g) was satisfied where the landlord intended to use the premises for partnership purposes. In *Westminster City Council v British Waterways Board* [1984], it was held that a tenant council could not protect its occupation by intimating its refusal of planning permission to the landlord water board.

   Some problems occur in practice regarding the intention to occupy where the landlord concerned is a company forming a group of companies. In such a case, section 42(3) provides that the intention is to be construed as including intended occupation by any member of the group for the purposes of a business to be carried on by that member. Finally, it should be noted that the court must be satisfied that the landlord has a sufficient intention and it will enquire into the matter as under para (f): see *Gregson v Cyril Lord Ltd* (1962). In *Europark (Midlands) Ltd v Town Centre Securities plc* [1985] the Court of Appeal accepted the following as showing a firm and settled intention on the part of the landlords:

(i)   minutes of board meetings

(ii)  evidence of quotations received from the supplier of equipment and

(iii) an affidavit from the landlords' property director.

# Terms of a new tenancy granted by the court

Sections 32 to 35 provide guidance to the court when it is called to determine any of the terms of a tenancy. Given a successful application by the tenants, the court is bound to order the grant of a new tenancy (section 29). As a general rule, the following limits will apply in so far as the parties have failed to agree.

## *Property to be comprised in new tenancy (section 32)*

As a general rule, the new tenancy must comprise the premises as they stood at the date of the order, except that where the landlord has sought to oppose a new tenancy on ground (f) the court may grant a tenancy of part of the holding under section 31A.

## *Duration of a new tenancy (section 33)*

The length of the new tenancy shall be as agreed between the landlord and tenant or, if decided by the court, it shall not exceed 15 years. It is clear that the duration of the old lease is a relevant factor as in *Betty's Cafés* (*supra*) where a 14-year term was ordered but was reduced to a five-year term by the Court of Appeal, so that the court will rarely grant a term longer than the original lease. Where the landlord fails to establish any of the grounds (d) to (g) but persuades the court that he is likely to be able to satisfy the grounds in the near future, the court may order a short-term lease: see *Upsons Ltd v E Robins Ltd* [1956].

In view of the comments concerning the interrelationship of section 25 and a break clause it is significant to note that the court can insert a break clause into a new tenancy: see *McCombie v Grand Junction Co Ltd* (1962). An interesting example of the exercise of this power took place in *Adams v Green* [1978], where the landlord owned a block of 12 shops in one of which the tenant asked for a new lease. The landlord wished to have a break clause inserted which would enable him to determine the lease after giving notice. The Court of Appeal granted a 14-year lease with a break clause included. In *CBS (United Kingdom) Ltd v London Scottish Properties Ltd* [1985] the court stated that while it was perfectly fair and proper for the landlord to seek to maximise the value of his investment, the court has to decide what is reasonable in the circumstances and the matter ought to be decided with fairness and justice. As the tenants were in the process of moving to a new location the court granted a 12-month term, while the landlord had argued for a 14-year term. (Contrast with the decision in *Charles Follett Ltd v Cabtell Investments Ltd* [1986].)

## *Rent under new tenancy (section 34)*

Such rent is to be determined by the court having regard to the terms of the tenancy (other than those relating to rent) at which the holding might reasonably be expected to be let in the open market by a willing lessor, there being disregarded:

(a) any effect on rent of the fact that the tenant has or his predecessors in title have been in occupation of the holding

(b) any goodwill attached to the holding by reason of the carrying on thereat of the business of the tenant (whether by him or by a predecessor of his in that business)

(c) any effect on rent of an improvement to which this paragraph applies

(d) in the case of a holding comprising licensed premises, any addition to its value attributable to the licence if it appears to the court that having regard to the terms of the current tenancy and any other relevant circumstances the benefit of the licence belongs to the tenant.

The working of section 34 needs some explanation as it touches on other areas of landlord and tenant law concerning, for example, rent review. It is clear that the terms of the lease may have a direct bearing on the rent. The point was emphasised in *Charles Clements (London) Ltd* v *Rank City Wall Ltd* [1978], where the court rejected an attempt by the landlord, as a means of raising the rent, to force on the tenant a relaxation of a covenant limiting user which would have been of no benefit to the tenant. But what tenant must the court have in mind? Section 34 directs the court to the willing lessor. The answer to the meaning of this phrase was given in *FR Evans (Leeds) Ltd* v *English Electric Co Ltd* [1978] where Donaldson J was of the opinion that 'the willing lessee is an abstraction — a hypothetical person ... He will take account of similar factors but he too will be unaffected by liquidity problems, governmental or other pressures'. In *Dennis & Robinson Ltd* v *Kiossos Establishment* [1987] the Court of Appeal was of the opinion that under such an assumption the strength of the market is a matter for the valuer. On the other hand, it is suggested that the open market rent may be affected by the profitability of the tenant's particular business. Whatever the position on profitability, the amount of the previous rent is not a factor for determination of the new rent. Further, section 34(3) allows the insertion of a rent review clause and in *WH Smith & Son Ltd* v *Bath City Council* (1984) (unreported) local factors affecting the pattern of the rent review were held to prevail over other factors. In recent years the question of the inclusion of a rent review clause in the new lease at renewal has been considered by the courts on numerous occasions.

## Other terms of the new tenancy (section 35)

It is in determining the new rent and other terms of the new tenancy that the courts have faced problems in recent years. This problem was highlighted in *O'May* v *City of London Real Property Co Ltd* [1982], where the lease was of office premises for a term of five years. Under the lease the landlord was responsible for maintenance and repairs. When the lease expired, the landlord agreed to offer the tenants a new lease but wished to change it into a clear lease by which responsibility for maintenance and repairs was to be on the tenants. An additional service charge was to be levied in return for which the tenants would be compensated by a reduction in the proposed new rent. The tenants applied to the court for a new tenancy. At first instance, Goulding J enunciated four tests in deciding whether a particular landlord's proposals were justified:

1. Has the party demanding a variation of the terms of the current tenancy shown a reason for doing so?

2. If the party demanding the change is successful, will the party resisting it in principle, be adequately compensated by the consequential adjustment of open market rent under section 34?

3. Will the proposed change materially impair the tenant's security in carrying on his business or profession?
4. Taking all relevant matters into account, is the proposal, in the court's opinion, fair and reasonable as between the parties?

The House of Lords agreed with the first three tests but stated that the fourth test of discretion should be applied in each of the preceding stages. It thus seems that, to justify a change in any of the other terms in a lease, the landlord will have to satisfy the tests laid down in the *O'May* case: see also *Gold v Brighton Corporation* [1956].

## Contracting out of the 1954 Act

A landlord and tenant may agree that the provisions of sections 24 to 28 of the 1954 Act shall be excluded from the tenancy: section 38A(1). The procedure to be adopted does not require the parties to make an application to the court. The agreement to exclude sections 24 to 28 is void unless the landlord serves on the tenant a notice in the form (or substantially in the form) set out in Schedule 1 of the Regulatory Reform (Business Tenancies) (England and Wales) Order 2003. The following conditions must be met:

(i) the notice must be served on the tenant not less than 14 days before the tenant enters into the tenancy to which it applies or (if earlier) becomes contractually bound to do so
(ii) if (i) is met, the tenant must, before he enters into the tenancy, make a declaration in the form set out in Schedule 2 to the 2003 Order. This confirms that the tenant has received the notice and accepts the consequences
(iii) if (i) is not met, the notice must be served on the tenant before the tenant enters into the tenancy to which it applies, or, if earlier, becomes contractually bound to do so, and the tenant must before that time make a statutory declaration in the form set out in Schedule 2 of the 2003 Order
(iv) the agreement under section 38A(1), or a reference to the agreement, must be contained in, or endorsed upon, the instrument creating the tenancy. There are similar provisions for agreements to surrender.

# Rent reviews and business tenancies

The problem of the implementation of a rent review clause may occur in a business tenancy at two stages, namely, at any time during the currency of the lease and/or on renewal. In most cases, the rent review or reviews will take place during the currency of the lease and will require careful consideration. Draftsmen now recognise that there are several essential matters that must be stipulated in such a clause, namely:

• stipulations as to time in the service of notices
• interval of rent review
• formula and machinery for determining the new rent
• provisions in event of disagreement.

## Service of notices

It is frequently the case that the rent review procedure is activated by the service of a trigger notice which commences the rent review process. In most cases, the notice will be served by the landlord or the tenant and the main problem that has arisen in this area is whether a failure to keep strictly to the timetable set in the rent review clause will result in the landlord losing its rights to a reviewed rent. In *United Scientific Holdings Ltd* v *Burnley Borough Council* [1977], the House of Lords laid down the general rule that time was not of the essence in the service of notices in a rent review process but stated that there could be exceptions to this general rule. It is possible to make time of the essence by stating expressly in the lease that it should be so, but time can also be made of the essence where there is an interrelationship between the rent review clause and some other clause in the lease. Such was the case in *Al Saloom* v *Shirley James Travel Service Ltd* [1981], where an underlease contained both a break clause and a rent review clause. The last date on which the landlord could serve the rent review notice was the same as that on which the tenant could give notice exercising his option to determine the underlease. In these circumstances, it was held that the presence of the break clause has the effect of making time of the essence. The question of what constitutes unreasonable delay has arisen on several occasions. In *H West & Son Ltd* v *Brech* [1982] a delay of 18 months was not sufficient to affect the landlord's rights. In Oliver LJ commented:

> ... But I know of no ground for saying that mere delay, however lengthy, destroys the contractual right. It may put the other party in a position, where, by taking the proper steps, he may become entitled to treat himself as discharged from his obligation: but that does not occur automatically and from the mere passage of time ...

In *United Scientific* (*supra*) the House of Lords recognised that a contra-indication may make time of the essence for the service of a rent review trigger-notice. What amounts to a contra-indication is the subject of dispute. In *Henry Smith's Charity Trustees* v *AWADA Trading & Promotion Services Ltd* [1984] the clause contained an elaborate time schedule and provided that the rent stated by the landlord be deemed to be the rent if the tenant's counternotice was not served in time. The Court of Appeal held that where the parties had not only set out a timetable but had provided what was to happen in the absence of strict compliance with that timetable, the general rule was rebutted. The Court of Appeal reached a different conclusion in the interpretation of a deeming provision in *Mecca Leisure Ltd* v *Renown Investments (Holdings) Ltd* [1984]. In *Greenhaven Securities Ltd* v *Compton* [1985], the rent review clause provided that if the parties had not, within a 15-month time-limit, agreed on a arbitrator or made an application for the appointment of one, the new rent should be a sum equal to the old rent. Goulding J distinguished the decision in *Mecca* and held that the default provision constituted a contra-indication. The opposite view was taken in *Taylor Woodrow Property Co Ltd* v *Lonhro Textiles Ltd* [1985], where the court noted that in *Henry Smith's* the deeming provisions were two-way while in *Mecca* they were one-way, so that time was not of the essence in the service of the counter-notice.

## Interval of rent review

It is clearly important for the interval of rent review to be expressly stated so that, for example, on a 21-year lease the rent review clause may become operative in the seventh and 14th years. Where it is not so, the rent review clause runs the risk of being inoperable. In *Brown* v *Gould* [1972] the option for a new lease was for a term of 21 years 'at a rent to be fixed, having regard to the market value of the premises at the time of exercising the option'. The court held that if no machinery was stated for working out the formula, the court will determine the matter itself. A liberal approach to the construction of an option to purchase was adopted in *Sudbrook Trading Estate Ltd* v *Eggleton* [1983].

## Formula and machinery

The formula and machinery has, of necessity, a direct relationship with the interval of rent review. Valuers are faced with enough problems on rent review without the clause adding to those problems by failing to define the rent on review. Such was the case in *Beer* v *Bowden* [1981] where the clause provided only that the rent should be the fair market rent for the premises. In the particular circumstances of the case the Court of Appeal stated that the rent should be the fair market rent for the premises. See also *Thomas Bates & Son Ltd* v *Wyndham's (Lingerie) Ltd* [1981], where a term was implied that the rent was to be that which is 'reasonable as between the parties'. By way of contrast, in *King* v *King* [1980] the court refused to look at the defective rent review clause from a reasonableness viewpoint.

Some novel provisions on rent are being considered by draftsmen seeking to increase the landlord's benefit as in *Bovis Group Pension Fund Ltd* v *GC Flooring & Furnishing Ltd* (1982), where the clause provided for the new rent to be assessed by reference to the rent that could be obtained if the premises were let for office purposes and the court stated that it was to be assumed that the building had planning permission for office use notwithstanding that no such permission had, in fact, been granted. Similarly, in *Pugh* v *Smiths Industries Ltd* [1982] it was held that the rent review should be on a literal construction of the lease where the formula provided that the presence of the review clause should be disregarded in calculating the new rent: see also *Lister Locks Ltd* v *TEI Pension Trust Ltd* [1982]. The converse situation applied in *GREA Real Property Investments Ltd* v *Williams* (1979), where it was decided that the effect of improvements on a rent review of premises in shell-form only, had to be disregarded. If no such improvements disregard clause is present, the tenant will have to pay increased rent on his own improvements (contrast with the situation on a lease renewed under section 34 of the 1954 Act). If a strict user clause is present in the lease, this may also have a dampening effect on the new rental level as in *Plinth Property Investments Ltd* v *Mott Hay & Anderson* [1979]: see also *Law Land Co Ltd* v *Consumers' Association Ltd* [1980]. In *British Gas Corporation* v *Universities Superannuation Scheme Ltd* [1986] Browne-Wilkinson V-C said that the correct approach in these circumstances was as follows:

(a) words in a rent exclusion provision which require all provisions as to rent to be disregarded produce a result so manifestly contrary to commercial common sense that they cannot be given literal effect

(b) other clear words which require the rent review provision (as opposed to all provisions as to rent) to be disregarded, must be given effect to, however wayward the result and

(c) subject to (b), in the absence of special circumstances it is proper to give effect to the underlying commercial purpose of a rent review clause and to construe the words so as to give effect to that purpose by requiring future rent reviews to be taken into account in fixing the open market rent under the hypothetical letting.

The interpretation of the *British Gas* guidelines was considered in *Co-operative Wholesale Society* v *National Westminster Bank plc* [1995] in relation to the question of whether deductions should be made to headline rents to reflect concessionary or rent-free periods other than for fitting out when determining the open market rent after the expiry of the rent-free period. The Court of Appeal held that a clause which excludes the assumption that an hypothetical tenant would have the expense of moving in and fitting out is more in accordance with the presumption in favour of reality than one which does not, under the guidelines enunciated in *British Gas Corporation v Universities Superannuation Scheme Ltd* [1986]. On the other hand, a clause which deems the market rent to be the headline rent obtainable after a rent-free period granted simply to disguise the fall in the rental value of the property is not in accordance with the basic purpose of a rent review clause. Such a clause enables the landlord to obtain an increase in rent without any rise in property values or fall in the value of the money by reason of changes in the way the market is choosing to structure the financial packaging of the deal. Therefore, in the absence of unambiguous language, a court should not be ready to construe a rent review clause as having this effect.

## Disagreement

A well-drafted rent review clause should always provide for a procedure in the event of disagreement between landlord and tenant on the new rental level. It should be made clear whether reference to an arbitrator or independent expert is desired. In most cases, if the parties cannot agree on whom should be appointed, the President of the Royal Institution of Chartered Surveyors is the person most frequently requested to appoint someone.

# Compensation

Under the amendments to the 1954 Act brought into force in June 2004, section 37 has been replaced by sections 37(1), 37(1A), 37(1B) and 37(1C) by providing for three different sets of circumstances in which a tenant is eligible for compensation on the termination of a tenancy. The following are the key issues to consider.

(a) First compensation case is where the tenant applies for a lease renewal but the court is precluded from granting a new lease only by one or more of certain of the section 30(1) no fault grounds.

(b) Second compensation case is where the landlord opposes renewal only on one or more of certain of the section 30(1) no fault grounds.

(c) Third compensation case is where the landlord opposes renewal only on one or more of certain of the section 30(1) no fault grounds and the tenant does not make an application under section 24(1) (or the landlord does not make an application under section 29(2)) or such application is made and is subsequently withdrawn.

(d) The amount of compensation is dependent upon the inter-relationship of section 37 of the 1954 Act and the Landlord and Tenant Act 1954 (Appropriate Multiplier) Order 1990 (SI 1990, No 363). Where the date of service of the landlord's section 25 notice or section 26 counternotice is on or before 31 March 1990, the appropriate multiplier is 3. Where the date of service of the landlord's section 25 notice or section 26 counternotice is on or after 1 April 1990, the appropriate multiplier is 1. Compensation amounting to twice the appropriate multiplier is payable where during the whole of the 14 years immediately preceding the termination of the current tenancy, the premises have been occupied for the purposes of a business carried on by the occupier or for those and other purposes or, during those 14 years there was a change in the occupier of the premises and the person who was the occupier after the change was the successor to the business carried on by the person who was the occupier immediately before the change. It should also be noted that there are transitional provisions contained in Schedule 7 to the Local Government and Housing Act 1989. Where the transitional provisions apply, 31 March 1990 is deemed to be the date for the determination of the rateable value and not the date (after 1 April 1990) of the landlord's section 25 notice or section 26(6) counternotice. The transitional provisions apply in the following circumstances:

(i)   the tenancy concerned was entered into before April 1 1990 or was entered into on or after that date in pursuance of a contract made before that date and
(ii)  the landlord's notice under section 25 or, as the case may be, section 26(6) is given before April 1 2000 and
(iii) within the period referred to in section 29(3) for the making of an application under section 24(1), the tenant gives notice to the landlord that he wants the special basis of compensation provided for by this paragraph and
(iv)  there must be a rateable value shown on the valuation list as at 31 March 1990.

Where the conditions are met, the appropriate multiplier is 8 times the rateable value on 31 March 1990.
   Where the transitional provisions apply, compensation amounting to 16 times the rateable value on 31 March 1990 is payable (section 37(3) of the 1954 Act) if:

(i)   during the whole of the 14 years immediately preceding the termination of the current tenancy, the premises being or comprised in the holding have been occupied for the purposes of a business carried on by the occupier or for those and other purposes or
(ii)  during those 14 years there was a change in the occupier of the premises and the person who was the occupier immediately after the change was the successor to the business carried on by the person who was the occupier immediately before the change.

## Compensation for improvements

The Landlord and Tenant Act 1927 as amended by the Landlord and Tenant Act 1954 Part III provides that on quitting a holding at the termination of the lease a business tenant (section 17) may in certain cases claim compensation from the landlord for improvements carried out by the tenant or his predecessor in title (section 1).
   In order to qualify for compensation, the tenant must follow the statutory procedure strictly. Prior notice of the proposed work must be served on the landlord giving an opportunity for him to object, or to elect to undertake the work in consideration of a reasonable increase of rent, or such rent as the court determines (section 3(1)). If the landlord objects, and no agreement is reached, the tenant may apply to the court for a certificate that the work is a proper improvement. This will be granted if the work adds to the letting value of the holding, is reasonable to its character, and does not reduce the value of any other nearby property belonging to the landlord. Where no notice of objection is served, or a court certificate is obtained, the tenant may proceed with the work. A further certificate of completion must be obtained from the landlord or the court (section 3(6)). Any increased rent under a renewed lease should not include the value of improvements carried out by the tenant or his predecessors in the business within the currency of the lease now ending or within the previous 21 years whichever is the longer. The 21 years is measured back from the date of the notice for a new tenancy. (section 34, Landlord and Tenant Act 1954, as amended).
   The tenant's claim for compensation must be made within the appropriate time-limits (section 47). The amount is the lesser of the net addition to the value of the holding resulting directly from the work

or the reasonable cost of carrying out the work at the termination date less an allowance for obsolescence (section 1(1)). As the basis of the compensation is the value of the improvements to the landlord, the amount may be reduced if the landlord intends to alter or change the use of the premises, and no compensation is payable if the premises are to be demolished (section 1(2)). If disputed, the amount may be determined by the court (section 1(3)).

© Del W Williams 2008

# Industrial and Distribution Properties

## Introduction

As a valuation field, industry contains not only the greatest range of sizes compared with shops, office and other commercial premises, but also by far the greatest number of different types of use. At one extreme the valuer will find a small workshop, possibly on an upper floor of an old building where only one or two people work, perhaps in the rag trade with a sewing machine. At the other extreme are the car assembly plants sometimes occupying many square kilometres. The valuer may have to consider properties that range from the fully automated plants incorporating a great deal of sophisticated computerised equipment to ancient workshops where the processes have not changed for centuries.

Before embarking on the valuation of any property the valuer should first familiarise himself with the current edition of the RICS's *Valuation Standards* (The Red Book). In *A Valuer's Guide to the Red Book* (Andrew Cherry, published by the RICS) readers are advised 'The best starting off point is to assume that the Red Book applies to all valuations and then look at the limited number of exceptions'. Valuers should heed that advice. Since the first edition of the Red Book in the mid 1970s, the bases of valuation have evolved being expanded in the 1990s to include such bases as Estimated Realisation Price which were dropped in the 5th ed. The section entitled the bases of valuation gives the valuer a choice of just Market Value and Existing Use Value for non-specialised properties, or Depreciated Replacement Cost for specialised properties. Although no longer treated as a universal valuation basis, existing use value is retained as a basis for valuations for financial statements in the UK where UK generally accepted accounting principles apply. Valuers should be conscious that the bases of valuation are kept under constant review by the profession and will continue to evolve as accounting and lending standards are harmonised across international jurisdictions.

The Red Book requires the valuer to stipulate in the terms of engagement agreed with the client the basis or bases of value that will be reported. Sometimes the basis will be specified by law or prescribed in a legal agreement. In practice, in most instances, the valuer will need to guide the client and the client's professional advisers.

The valuer of industrial property in the UK should be familiar with the Value of Plant and Machinery to the Business (as per the Red Book) and ensure that the basis of valuation adopted for the land and buildings is compatible with that for any plant and machinery, particularly where the two elements are to be valued by different valuers.

Valuations of property for company accounts or other financial statements which have regard to the land and buildings as investments must be on the market value basis. Where the property is owner occupied in the UK as part of a going concern it will usually be valued on either a market value basis or an existing use value basis. For owner-occupied specialist buildings, however, a depreciated replacement cost approach may be more appropriate. The depreciated replacement cost basis will certainly be appropriate where plant and machinery forms a major integral part of a building, such as a cement works.

# Types of industries

While not necessarily directly germane to the mechanics of valuation, an understanding of the economic categorisation of industry will help in the valuer's appreciation of the context in which an industry has evolved and sometimes point to the determinants of that industry's location. The four basic categories, which should not be confused with Planning Use Classes, are:

(i)  *Primary industries*
     These are concerned with extracting material direct from the land or the sea and do not involve any processing or fabrication of a finished product. Mining and quarrying come into this category.

(ii)  *Secondary industries*
      This includes the bulk of manufacturing industries from food and drink to chemical and mechanical engineering, printing and publishing.

(iii)  *Tertiary industries*
       These provide services which are basically oriented towards the retail market. Such industries as construction, transport and communications are included in this group.

(iv)  *Quaternary industries*
      These include research establishments and those concerned with the provision of information or expertise.

# Influences on location

A wealth of economic and demographic information is now available from both local authorities and private research organisations. Examination of such information will help the valuer understand what has influenced the location of industry historically and the factors influencing current location decisions.

Clearly, primary industries can only be located at the source of raw materials. On the other hand, the locational requirements of secondary industries are much more complex. Some are mainly located with reference to the market which they serve, such as bakers, while others are strongly tied to raw materials, such as brickworks. The majority are, of course, intermediate between these extremes. Traditionally, service industries were almost always located where those services were required — at the market. However, the revolution in communications technology in the late 1990s and early 2000s has reduced the necessity for this. Warehouses serving the traditional distribution industry were located to serve local or sub-regional markets. The 1990s saw the advent of dedicated distribution parks strategically located close to the motorway network designed to complement on a national scale the highly sophisticated stock control systems of both the retail and manufacturing sectors.

Quaternary industries are very often more market oriented and businesses may be located to serve the needs of just one large customer or to be near the source of a specialist skills base such as on one of the science parks associated with a university.

Given an understanding of the general locational needs of different types of industry, what factors influence the choice of any particular location? For established businesses, the location of the factory will often be due to historic reasons which bear little relation to the current market circumstances. However, in deciding to relocate a factory or warehouse, a business is likely to consider at least some of the following.

- The availability of labour, be it skilled or unskilled and, increasingly, its cost. The cost and availability of housing will be part of the equation.
- The availability of land and buildings. For smaller businesses the prior availability of suitable advance factories and warehouses or of second-hand stock will be highly influential in helping the relocation to be achieved more quickly than would be possible if land had to be acquired and a bespoke building constructed. For larger operations, for which a purpose built plant is the only likely option, the availability of land will be a major determinant. Given the competition from alternative uses, particularly housing, the availability of large sites is relatively limited.
- The quality of communications in terms of effective access for the workforce, access to raw materials, access to markets for the finished product, and access to ports and airports. Proximity to a motorway junction remains a major influence, but other factors such as drive times from the docks, for those who are import/export oriented, or proximity to airports for those with overseas business and interests will all play their part.
- For some industries the provision of services will be fundamental. Typically this requirement will be for water, but may include sewage disposal facilities, gas or electricity.
- Environmental and legislative considerations will also have an influence. A business where there is a risk, for instance, of chemical seepage may seek sites with advantageous geological conditions and avoid sites close to watercourses, while businesses depending heavily on road transport will consider the current law prescribing the hours that a truck driver may drive between rests.

From a valuation perspective, recognition of these influences is important, particularly when considering generally comparable properties in, say, neighbouring towns.

# Government grants

Encouraging industry and influencing its location has long been an instrument of economic policy and at any one time government (in various guises) may offer both grants and tax concessions. Grants can take the form of direct cash incentives to assist with the construction of buildings or cash incentives to train staff. These incentives could be obtained from the European Union, central government or local government. While the availability of grants varies from time to time, they have proved to be an effective means by which government has influenced the establishment and location of industry. Enterprise Zones, with attractive planning and rating regimes along with tax benefits for investors, have been designated in the past to promote the case for specific, usually run-down, areas.

The valuer should seek information as to the availability of these grants and designations and also be conscious of the distortions that they may have on the market.

# Regulatory regime

A steady stream of both domestic and European legislation dealing with environmental and health and safety issues imposes on the valuer an obligation to keep abreast of rules that apply both to sites and to new and older buildings. Attention to the issues surrounding the Environmental Protection Act 1990 and the Environment Act 1995, and familiarity with the Factory Acts, are important as is knowledge of planning legislation. It would be unrealistic, however, to expect any but the most experienced of valuers to be familiar with the regulations specific to each industry and a valuer may need to agree limitations with the client and qualify the valuation accordingly.

In preparing a valuation, the valuer should be cognisant of the effect that general health and safety rules will have on the maintenance of the building and of the impact of Construction, Design and Management Regulations (CDM) on costs, particularly when compared to other buildings being relied upon as comparables. The valuer should also be familiar with the requirements of the Disability Discrimination Acts of 1995 and 2005 and check that the property complies.

# Building specification

The industrial property market is dominated by the general purpose industrial building which has evolved as a consensus building suitable as an envelope for a wide range of manufacturing and distribution purposes. The characteristics of buildings for industry and distribution converged in the 1970s and early 1980s so that the buildings were identical regardless of the use to which they might be put under planning controls. This has resulted in the ubiquitous institutional 'shed'. As industry had to react to markets that changed ever more rapidly, owner/occupied buildings became less common and companies started to rely on the provision of standard buildings built as 'advance' units by development corporations, institutions and property companies. As a low interest rate environment became established in the early 2000s, it became more attractive to (particularly smaller) companies to own their own buildings but this has not impacted on the design of the majority of sheds except for the very large distribution units which are dealt with separately below.

When considering comparables, the valuer should be alert to the differences between properties, the most important of which are set out below.

(i)   *The site*
How accessible is the site? Are there obstacles en route from the main arterial roads? Low bridges, for instance? Is the access to the site suitable for large vehicles? Is the road adopted? Is it properly lit? To what use is neighbouring land put? If neighbouring land is residential, are there any restrictions on working hours or noise emissions?

(ii)  *Site cover*
The footprint of the building (being the area of the ground floor of the building measured on a gross external basis in accordance with the RICS's Code of Measuring Practice) divided by the area of the site gives the site cover. (Plot ratio is the ratio of the gross external area of all floors to the site area and is used in town planning.) While the utility of the land not built upon is of more interest to the occupier than the site cover itself, the valuer should be aware of current institutional thinking as to an optimum site cover ratio. At present a site cover of between 35 and 40% would be deemed optimum. Care must be taken where there is a high office content at first-floor level as the site cover ratio may mask an over-development. A site cover of over 50% is likely to result in congestion.

(iii)  *Eaves height*

In recent years the institutional 'standard' for eaves height has risen along with increased floor loading requirements. While 5 m was an acceptable standard in the mid-1970s, a height of 6 to 7 m is now required for even smaller buildings, rising to 12 m or more for large distribution facilities. The Code of Measuring Practice has a diagram distinguishing between external eaves height, internal eaves height and clear internal height. The difference between the latter two could be half a metre so valuers should take care to compare like with like.

(iv)  *Floor loading*

Many occupiers could manage with much less than the institutional requirement for a floor loading of between 30 and 40 kN/m$^2$. The design 'rule of thumb' is 2.5 kN/m$^2$ per metre of eaves height. The institutional rationale for their apparent over specification is the need for their buildings to be able to accommodate the possible requirements of as wide a range of potential occupiers as possible without the need for those occupiers to strengthen the floors for specific uses.

(v)  *Office content*

Offices with a separate entrance and their own toilets are usually provided with the standard institutional shed. While there is a debate as to the optimum office content, 10% of the total area of the factory is the accepted norm, considerably less would not be unusual for very large properties.

(vi)  *Car and lorry parking*

Providing the site cover ratio is low enough, adequate car and truck parking facilities should be easily accommodated on the site. Care should be taken where the site cover is high to ensure that trucks can manoeuvre safely up to the loading doors. Manoeuvring and parking areas for trucks should be in concrete.

(vii)  *Fit-out*

Advance units are usually fitted out to the lowest standard that the developer believes he can get away with. This will usually include fitting out the offices with carpets, central heating, lighting and suspended ceilings, a roller shutter or sliding loading door to the factory premises and some landscaping to the perimeter of the site. A higher specification such as electrically operated loading doors, heating and lighting to the factory and perhaps dock levellers may also be included.

(viii) *Shape*

Modern general purpose buildings tend to be rectangular with the length rarely more than two and a half times the width. A very long and narrow building may, however, have more attractions to certain occupiers. The optimum office arrangement is usually considered to be a pod to the front or side of the building. The clearer the interior is of internal columns, the better.

(ix)  *Size*

Where comparing evidence of rents per m$^2$ (or still, usually, rents per sq ft!), the valuer should consider carefully the size of the building from which the evidence comes.

(x)  *Energy consumption*

With increasing concern as to the efficient use of energy, buildings that are energy efficient are likely to attract a premium. For instance, the ratio of translucent roof panels to solid ones was traditionally 10%. A higher ratio will allow in more light and thus save on the cost of artificial lighting. Modern buildings may have a BREEAM (Building Research Establishment Environmental Assessment Method) rating and where possible the ratings of buildings used as comparables should be compared.

(xi) *Insurance risk rating: contamination and deleterious materials*

Certain buildings may be more expensive to insure than others that seem to be very similar. The valuer should be aware of not only those materials that were historically used in buildings and are now known to be dangerous, asbestos being the most obvious example, but also of the possible dangers from more recent materials and components. For instance, in certain circumstances composite panels, commonly used in cladding, have been found by insurers to represent a greater fire risk than other forms of cladding.

By their nature many industrial buildings will be sited on industrial sites established long before contamination became an issue. The valuer should consider the historic usage of the site and advise on the need for an environmental search if appropriate. Problems may arise from past processes on the site, for instance from spilt chemicals, and some research into previous uses may be desirable. Although not strictly contamination, Japanese Knot Weed can cause considerable damage to buildings and valuers should be aware of where it is most likely to appear. The risk of flooding should also be addressed.

# Large sheds

While clearly large distribution sheds have evolved from the 1970s institutional shed, they represent something of a market of their own and valuers should be cautious about using per square foot comparables taken from smaller units and applying them to major distribution facilities.

Site cover can be rather higher, although there will be regional variables, up to about 45% in South East England. All units will require extensive concrete hard standing for trucks — and perhaps a 50 m deep yard. Height to underside of eaves will be circa 14 metres and dock levellers at one per say 1,000 m² will be required plus two level access loading doors. Floor loadings of about 50kN/m² are the norm.

# Bespoke buildings

Whether a bespoke building can be valued by reference to evidence from transactions involving institutional sheds will depend upon the extent that the specification is at variance with the institutional norm. Some buildings will be so specialist that they do not bear comparison. These buildings may, however, have a value to occupiers in the same industry somewhat in excess of the value that a depreciated replacement cost valuation would suggest. This is particularly true for the new technology industries for which the immediate availability of a modern fully equipped plant with, for instance, fully tested clean rooms would often outweigh the advantages of a purpose built facility for which there could be several months' or years' delay. Specialist buildings will often contain a much greater proportion of plant than would be found in an advance factory and the valuer should bring to the attention of the client the potential value of any capital allowances this plant may attract.

# Plant and machinery

As indicated above, not only are there an enormous number of different industrial processes, but these are often tied up with detailed locational requirements, such as the need for pure water or a particular quality in the labour force. The valuation of the range of plant required for specific operations requires highly specialised skills and a knowledge that goes beyond the expertise needed to value land and buildings. A plant and machinery valuer may not be familiar with every industrial process, especially if it is of an unusual or specialised type but experience will give the valuer an understanding of how to assimilate the nature of a specific plant. By enquiry and research, the valuer will be able to establish inherent obsolescence, which is an important factor in such valuations. Almost always, much of the associated plant and equipment in a factory is common to many others. This applies, for example, very much to the chemical industry, where a complex looking process may appear to represent an insoluble conundrum. On detailed examination, however, this will break down into a series of pipes, pumps, condenser tanks, filtration plants and so on, leaving only a few unfamiliar items which need to be researched.

Clearly, a plant valuer must have the ability to recognise what he sees in all the major branches of industry. He must have at his fingertips records giving the present day cost of all the plant he is likely to come across, and know where to obtain such information about unusual items.

The valuation of plant and machinery is an increasingly specialised field (and one where the valuer usually practising in the land and buildings sector should be wary). None the less, the property valuer should be familiar with plant which normally forms a component of industrial buildings. Guidance Note 2 of the Red Book cautions that the boundary between the category of plant and machinery that usually forms part of the building services installations, and which would thus usually be included in the land and buildings valuation, can be difficult to distinguish from the process plant. Plant such as the heating installations, air conditioning and ventilation are clearly part of the building. Items such as crane gantries, which would usually pass with the building on sale or letting but which are really fitted in order to help with a specific process, would however need special consideration. The valuer should ensure that the appropriate adjustments have been made for the plant when considering the comparables. Some plant will add value while some will reduce it. Particular care should be taken when pipes and ducts are to be left *in situ*. The cost of cleaning chemical and gaseous contaminants can be very high.

# Valuations for financial statements

(See also Chapter 16.) The valuer is now regularly required to update the asset side of the balance sheet. This enables the correct present day values to be substituted for the existing values, which may well have been calculated on the basis of historic cost.

The Red Book is now prescriptive as to the bases on which valuations for financial statements must be made and valuers should refer to UKPS 1: Valuations For Financial Statements. Market value and existing use value are the appropriate bases for non-specialised property in the UK while for specialised property, the depreciated replacement cost is appropriate. Such properties that meet, or approximate to, standard institutional buildings can usually be valued to market value or existing use value by the conventional investment method or by direct comparison with prices paid per m$^2$, where such information can be obtained. When utilising the investment method of capitalising passing and estimated future income streams be aware that investment yields are usually quoted as net of a notional purchaser's acquisition costs (Stamp Duty Land Tax, legal and surveyors' fees and VAT).

Valuers will be familiar with the definition of market value:

The estimated amount for which a property should exchange on the date of valuation between a willing buyer and a willing seller in an arm's length transaction after proper marketing wherein both parties had acted knowledgeably, prudently and without compulsion.

Existing use value is the appropriate basis for properties '... owner occupied by a business or other entity.' Its definition includes that of market value, but with additional assumptions that:

The estimated amount for which a property should exchange on the date of valuation between a willing buyer and a willing seller in an arm's length transaction after proper marketing wherein both parties had acted knowledgeably, prudently and without compulsion, assuming that the buyer is granted vacant possession of all parts of the property required by the business and disregarding potential alternative uses and any other characteristics of the property that would cause its Market Value to differ from that needed to replace the remaining service potential at least cost.

The Red Book also comments that, 'Where the property is of a type that is commonly traded in the market, with no higher value for an alternative use and no unusual features that could restrict marketability there would normally be little material difference between market value and existing use value.' Indeed where there is a difference, the Red Book requires valuers to provide a valuation on both bases and explain the difference. In other words, the result of an existing use valuation should be to establish, by reference to market transactions and/or analysis, the price a business would have to pay to replace a property if it were deprived of it.

Valuers should be aware that the investment and owner occupied markets for sheds can produce very different results. Consider two 10,000 sq ft modern fully specified sheds adjacent to each other on an estate and with no characteristics to distinguish one from the other except that one, unit A, is occupied by a tenant with 10 years to run on a full repairing and insuring lease at market rent while the other, unit B, is being offered to the market with vacant possession. An investment method valuation might produce a value as follows for unit A:

|  | £ pa | £ |
|---|---|---|
| Rent | 60,000 | |
| YP perp at 6.5% | 15.385 | |
| Gross value | | 923,077 |
| Allow for purchaser's costs of 5.725% (÷ by 1.05725) | | |
| Net value: | | 873,092 |

but say £875,000
This equals £87.50 per sq ft in capital value.

A valuer would be justified in adopting the same approach to arrive at the value of unit B. However, because developers are usually shy of fragmenting their developments by selling some units to owner occupiers and letting others for sale to investors, vacant freehold buildings may rarely become available in the local market. As a result, prospective owner occupiers may well compete for the rare opportunity and be prepared to pay a premium for the opportunity:

The owner occupier's offer might be: 10,000 sq ft at £95 per sq ft          £950,000

One of the reasons for the acquisition may well be that the business owner wants to put the property into a Self Invested Pension Plan (SIPP), holding the freehold in the plan and letting it to the business. The valuer for the SIPP then carries out an investment method value and the purchaser finds that the property is worth less than was paid for it. Valuers should not be afraid of explaining this conundrum to their clients.

The Red Book definition of depreciated replacement cost, appearing in the Glossary is 'the current cost of reproduction or replacement of an asset less deductions for physical deterioration and all relevant forms of obsolescence and optimization' which readers may find a little obscure. Elsewhere the Red Book refers to the FRS (Financial Reporting Standard) 15 definition which is:

> The measure of the cost, or revalued amount, of the economic benefits of a tangible fixed asset that have been consumed during the period. Consumption includes the wearing out, using up or other reduction in the useful economic life of the tangible fixed asset whether from use, effluxion of time or obsolescence through either changes in technology or demand for goods and services produced by the asset.

Depreciation is thus a measure of the extent to which the asset is worn out where the definition of wearing out is much wider than we would normally understand from the term. It follows that the valuer must, for this basis, separate the land from the buildings in his calculations. Very often the valuer will find that it is just due to the buildings being so special that the existing use value basis is precluded and that the land can be valued by reference to comparables. Problems may, however, occur where there is a *sui generis* planning use. In such circumstances earlier editions of the Red Book directed that

> the approach should be to value the land with the benefit of an assumed planning permission for a use, or a range of uses, prevailing in the vicinity of the actual site.

The approach clearly required values to make a broad range of assumptions that could not be easily tested. The 6th edition of the Red Book is thus more circumspect and states that where potential for an alternative use can be identified but "the value for that use cannot be reliably determined without significant research" then "a simple statement that the value for the site for a potential alternative use may be significantly higher than the value derived from using the DRC method will be sufficient."

To establish the future economic life of buildings, depreciation should be considered under two heads:

(i) *economic obsolescence*: this is the valuer's measurement of the physical obsolescence of the premises, taking into account age, present condition and future costs in use, particularly maintenance and repair.

(ii) *functional obsolescence*: this involves a consideration of the suitability of the premises for their present use and their likely efficiency if continued in that use by the undertaking, or in some other use if such were reasonably conceivable.

Valuers might also consider the possibility of what might be called *statutory obsolescence*.

The basic methodology requires the valuer to establish the cost of replacing the existing buildings, including fees and fit out, (the gross current replacement cost) and to then depreciate that figure to reflect the value attributable to the remaining economic working life of the building. The valuer should also take into account environmental and planning policies when considering the appropriate depreciation factor. Where there have been technological changes, it may be appropriate to consider a modern substitute building of a smaller size but equivalent capacity

The efficacy of the depreciated replacement cost basis is dependent upon the business having adequate potential profitability with regard to 'the value of the total assets employed and the nature of the operation' (The Red Book).

# Industrial building allowances and capital allowances

It is because the tax status of each of the buyers and sellers that comprise 'the market' varies so widely that tax, except in respect of sinking fund calculations, is not taken into account in valuations. Indeed most valuations will include a caveat such as:

> No account has been taken in our valuation of any liability for tax (including Value Added Tax) on either the rental income from the property (if any) the notional sale prices or any gains which may be realised on disposal

It is not the role of the valuer to calculate the quantum of any allowances that may be available and it will almost always be desirable for a specialist to advise. However the valuer should bear in mind that these allowances might have an impact on a sufficiently significant part of the market to affect value. Equally it is not the task of this chapter to detail tax calculations but simply to demonstrate the principles.

## Capital allowances

These are available on plant and machinery (P&M). Regulations do not distinguish between the P&M that is normally part of the building (in the case of sheds this would include roller-shutter doors, dock-levelers, space heating, lighting and the like) and plant and machinery that is associated with the process carried on in the building — conveyors, presses etc. However Guidance Note 2 in the Red Book — Plant and Machinery — gives a clear explanation of this distinction as it affects the valuation of property.

Plant and machinery attracts a writing down allowance of 25% pa on a reducing balance basis and valuers should be aware that this is different from the straight line depreciation adopted by many businesses for accounting purposes. An example of how this works is given at the end of this section. However, if a business buys a piece of plant for £100, it may claim an allowance of £25 in the first year (25% of £100), £18.75 in the second year (25% of £75) and so on.

## Industrial building allowances

Industrial buildings allowances may be claimed for buildings used for qualifying purposes which include manufacturing; subjecting goods to a process; or storing goods or materials used in manufacturing or processing. Storage will meet the qualifying purpose criterion if it is for goods or materials which are to be used in the manufacture of other goods or materials; goods which are to be subjected to a process in the course of a trade; goods which have been manufactured or processed or subjected to a process but have not yet been delivered to any purchaser; goods on their arrival in the UK from a place outside the UK.

A building used for a qualifying purpose has a 25 year tax life and attracts a writing down allowance of 4% pa on a straight line basis on the qualifying expenditure which is based on the construction cost of the building. Construction costs will include preparing and levelling land but not the cost of the land itself. Architect's fees are included, legal fees are not.

## Study 1

Take for example a new industrial building of 5000 $m^2$ with a 12% plant and machinery content and just occupied for a qualifying use. It will attract industrial building allowances (IBAs) and, from the date of first occupation, will have a tax life of 25 years. It will also attract capital allowances on the plant and machinery. While unlikely to be as simple as this in practice, the total allowances that could be claimed would be calculated as follows:

|  | £ |
|---|---:|
| Cost of construction including plant and machinery and fees | |
| Say 5,000 $m^2$ constructed at £500 per $m^2$ | 2,500,000 |
| *Less* Cost of plant and machinery say | 300,000 |
| Qualifying cost for IBA | 2,200,000 |

In the year of first claim the allowances claimed would be:

|  | |
|---|---:|
| 25% of £300,000 | 75,000 |
| 4% of £2,200,000 | 88,000 |
| Total allowance claimed | 163,000 |

For a company paying 30% corporation tax, this will represent a saving of £48,900 in tax payable.

|  | £ |
|---|---:|
| In the second year the calculation would be: | |
| 25% of £225,000 (ie £300,000 – £75,000) | 56,250 |
| 4% of £2,200,000 | 88,000 |
| Total Allowance claimed | 144,250 |

giving a tax saving of £43,275 in that year.

In a market dominated by gross funds (in other words those investors that are not required to pay tax), the availability of these allowances is of no immediate relevance. However, in a market where investors typically borrow a substantial proportion of the purchase monies, the relevance of the allowances becomes apparent.

## Study 2

The freehold building in example 1 is let at £75 per $m^2$

|  | £ pa | £ |
|---|---:|---:|
| Annual rent | 375,000 | |
| YP in perp at 6.5% | 15.3846 | |
| Gross value | | 5,769,225 |

| To allow for notional purchaser's costs: | |
|---|---:|
| Divide by 1.05725% to give net value of | 5,456,822 |
| Assume that the investor has borrowed 55% of this sum | 3,001,252 |
| At a fixed rate of 6.25% the interest cost would be £187,578 pa | |

Assuming the investor has no other income, the tax calculation in the second year will be:

| | £ pa |
|---|---|
| Gross income | 375,000 |
| Less interest | 187,578 |
| Taxable income before allowances | 187,422 |
| Less capital and industrial building allowances | 144,250 |
| Taxable Income | 43,172 |

This is a necessarily simplistic example to illustrate the point, but shows how a building used for a qualifying purpose might be worth considerably more to one class of investor than to another.

Where an industrial building allowance is not claimed in any one year, it remains in the allowance pool until the 25 years has expired and can be utilised as an allowance in equal parts over the remaining years of the original 25 years.

## Study 3

Thus, if the above building were owned from the start by a pension fund that did not pay tax and the above sale was 21 years after it was first used for a qualifying use, then the calculations would be:

| | £ |
|---|---|
| Capital allowance | 75,000 |
| IBA £2,300,000 / (25 – 21) | 575,000 |
| Total allowances | 650,000 |

An investor could thus shelter £650,000 of taxable income.

Editor's Note: These examples were prepared prior to the 2007 Budget which proposed substantial changes to both industrial building allowance and capital allowances.

# Offices

Office buildings are invariably valued by the investment method, which is the determination of the present value of the right to receive, and liability (if any) to pay, a future sum or a series of future sums by discounting them at a compound rate determined by direct comparison with yields obtained on other investments, ie it is the determination of the value of an interest by the capitalisation of rental income.

The attention of all surveyors carrying out any valuation is drawn to the *Valuation Standards* (the Red Book) published by the Royal Institution of Chartered Surveyors.

The methodical assembly of all the information required is of paramount importance to ensure that no relevant factors are overlooked. How this is done is illustrated in this chapter by considering the valuation of various interests in an office building. The following are introductory matters.

The freeholder (A) built on the site an office building which was completed and let in 1959 to a tenant whose lease expired in 1989. He then re-let the building to tenant (B) for a term of 75 years. B has subsequently underleased the building to various occupying tenants. A valuation is required as at 2007 of the following interests in the building:

- the freehold
- the head leasehold
- the freehold in possession, ie the value to the freeholder if he was entitled to receive the rack-rents receivable by the head lessee.

The building is assumed to be in a first-class location in the West End of London. The accommodation comprises a basement used as storage with offices on the ground and six upper floors. The floors were leased in open plan form and these areas formed the basis for the apportionment of service charges (see later). Tenants have installed their own internal partitioning. The building is centrally heated by two gas-fired boilers which were renewed two years ago, and this system is in good working order. The upper floors are served by three lifts, all of which have been renewed within the last two years. The sixth floor of the building was constructed three years ago by J Ltd, the occupational tenants, at their own expense.

The first essential in valuing any building is to read the various leases and to prepare a synopsis of the relevant details. The importance of this cannot be expressed strongly enough.

# The leases

## The headlease

The freehold is owned by A, who has leased to B on the following terms:

| | |
|---|---|
| Term: | 75 years from 1989. |
| Initial rent: | £40,000 exclusive of all outgoings. |
| Rent reviews: | At the end of the 25th (2014) and 50th (2039) years. |
| Basis of rent review: | £40,000 pa plus 5/8th of the annual rack-rental value (Note 1) of the building at the date of review in excess of £64,000 pa subject to such figure being not less than the rent being paid at the date of review. The rent on review is to be assessed having regard to the terms of the head lease and on the assumption that the building is vacant and available for letting in accordance with market practice at the date of review (Note 2). |
| Insurance: | The tenant is required to insure against all risks at full replacement cost to include fees and for three years' loss of rent. |
| Repairs: | The tenant is required to carry out all repairs and decorations internally and externally and to maintain all plant and equipment (eg lifts and central-heating installations) in sound working order. |

Notes

(1) It is customary to estimate the rental value at the date of review on the basis of rental values current at the date of valuation and this will need to be undertaken in order to arrive at the amount of the head rent on review.

(2) On this basis it must be assumed that the building is let to its best advantage. It could be assumed to be let as a whole or by floors or by parts of floors. Some head leases put restrictions on how a building may be let, ie no more than one tenant per floor or not more than a given number of tenants in the building as a whole. It is also to be assumed that if current market practice is to let on the basis of five-year rent reviews the tenant will let on that basis.

## *The underleases*

The synopsis of the underleases is best prepared in schedule form as shown in Schedule I. Although it may be purported that all sublettings are in accordance with a standard form of underlease, all leases should be perused, as variations from the standard are frequently to be found. As the building has been let to a number of tenants, their superior landlord B will endeavour to recover from them a proportion of the cost of the outgoings he has to bear. These will be recovered by service charges but the liability for payment by individual occupiers may not be identical in all cases. The method of apportioning service charges should also be noted. It is usual in the case of office buildings to base the apportionment on the ratio of the floor area occupied by a particular tenant to the total area of the building and therefore it is important to note the definition of floor area (if any) in the underleases. Many leases which intend to recover 100% of the cost of services provided when a building is fully let, do not in reality do so. Where in respect of any tenancy the landlord does not recover the whole of the apportioned service charge a deduction from rent will have to be made in respect of the amount of service charge which is irrecoverable. It is advisable to record on this schedule the area of each subletting, the floor on which it is situated, the specific use and the estimated rental value on reversion.

## Schedule I
## Schedule of underleases

| Tenant (1) | Floor | Use | Floor area | Rent paid £ pa | Service charge | Rent review/ end of lease | Estimated Reversionary Value, £ pa |
|---|---|---|---|---|---|---|---|
| C Ltd | Basement | Storage | 371 | 1,000 (4) | No provision for recovery of the cost of heating or lifts | End of lease 2039 No rent reviews | 26,000 |
| D Ltd | Ground | Offices | 492 | 2,500 (4) | No provision for recovery of the cost of heating or lifts | End of lease 2039 No rent reviews | 220,000 |
| E Ltd | First | Offices | 492 | 93,000 | Full recovery | Rent reviews 2008 and 2013 End of lease 2018 | 220,000 |
| F Ltd | Second | Offices | 492 | 106,000 | Full recovery | Rent reviews 2009, 2014 and 2019 End of lease 2024 | 220,000 |
| G Ltd | Third | Offices | 492 | 240,000 (2) | Full recovery | Rent reviews 2010 and 2015 End of lease 2021 | 220,000 |
| H & Co | Fourth | Offices | 492 | 93,000 | Full recovery | Rent review 2008 End of lease 2013 | 220,000 |
| J Ltd | Fifth | Offices | 492 | 93,000 | Full recovery | Rent reviews 2008 and 2013 End of lease 2018 | 220,000 |
| J Ltd | Sixth (3) | Offices | 232 | Nil (Tenant's improvements) | Full recovery | End of lease 2018 | 103,000 |

Notes on Schedule I

(1) Companies C, D, E and F are all substantial public companies. G Ltd is a public property development company which has lately been in financial difficulty and its covenant strength is doubtful. H is a firm of apparently successful solicitors and the two senior partners are guarantors under the terms of the lease. J Ltd is a foreign-based oil exploration company and for the last two years it has disclosed substantial losses in its annual accounts.

(2) The lease to G Ltd is subject to upward-only rent reviews and the rent from this floor, which was negotiated six years ago, is now considered to be in excess of the current market rent by about 25%. There may therefore be no increase in rent on review in 2010.

(3) This floor was constructed by J Ltd at its own expense in 2003. There is no provision in the lease requiring them to construct this floor and the lease provides that on review improvements carried out at the expense of the tenant are to be ignored in assessing the rent.

(4) The rents payable under these leases are fixed until the expiration of the terms in 2038 (31 years) and the tenants do not contribute towards the cost of lifts or heating. Owing to likely further inflation, the head lessee will receive a steadily declining income on this account.

# Cost of services

The headlessee B is responsible for all outgoings and these need to be carefully analysed in order to calculate the landlord's residual liability in those cases where the apportioned cost is not fully recoverable from the underlessees. A detailed analysis of the service charge is shown in Schedule II. It will be noted that there is no provision for the recovery of any sum in respect of the renewal of plant or equipment. Where this is the case a deduction will need to be made at the end of the valuation to allow for obsolescence where renewal is imminent. The landlord has, however, renewed the boilers and lifts recently, so that no deduction is required in this case. The cost of services varies from building to building and where actual figures are not available estimates must be built up by pricing the individual services supplied.

For more information on service charges valuers should refer to the RICS *Code of Practice for Service Charges in Commercial Property* published on 1 April 2007.

## Schedule II

Schedule of items included in the service charge and analysis of the actual cost per m² of providing these services

| Item | Cost per m² £ |
|---|---|
| Staff wages | 8.60 |
| Rent for caretaker's flat | 1.55 |
| Cleaning of common parts | 2.80 |
| Electricity | 2.85 |
| Central heating of building (1) | 10.50 |
| Heating maintenance contract (1) | 2.95 |
| Building and third party insurance | 5.50 |
| General maintenance | 16.50 |
| Lift maintenance (1) | 2.25 |
| Gardening | 0.20 |
| Managing agent's fee (2) | 6.76 |
| Auditor's fee | 0.40 |
| Total cost of the provision of services per m² | £60.86 |

Notes on Schedule II
(1)   Cost of lifts and central heating: it is necessary to calculate these separately in respect of the basement and ground floor as they are not recoverable from the underlessees.

| | |
|---|---|
| Central heating | £ 10.50 |
| Heating maintenance contract | £ 2.95 |
| Lift maintenance | £ 2.25 |
| | £15.70 |

| | Area in m² |
|---|---|
| Basement | 371 |
| Ground floor | 492 |
| | 863 at £15.70 per m² = £13,549 |

(2)   Managing agent's fees based on 12.5% of the total cost of services and not on a percentage of rent.

# Reversionary rental values

Where parts of the building are let at rents which do not reflect current rental values these will have to be estimated. Such rents on reversion should be calculated on open market rental values as at date of valuation and not on the estimated rental value at the date of review, on the assumption that there will be an increase owing to inflation. The best comparables for this purpose will be recent lettings or sublettings which have taken place in the building itself. As a check on their validity, recent lettings in comparable buildings in the vicinity should be examined. Schedule III illustrates the methods used in the devaluation of comparables. More comparables would be required in practice. From a consideration of the devalued comparables the valuer will determine his basic current market valuation rates per m². This determination will be based not entirely on mathematical calculations but also on the valuer's judgement based on experience. In the example, he has decided on £70 per m² for the basement storage and £445 per m² for the office floors.

On any rental and service charge analysis the valuer should be careful not to mix metric and imperial measurements. For example, rents are most likely to be given in imperial terms.

## Schedule III

### Schedule of underleases

| Address | Tenant | Floor | Area m² | Service charge | Date of letting | Rent paid | Adjustment £ pa | Adjusted rent m² /£pa |
|---|---|---|---|---|---|---|---|---|
| The subject property | K Ltd (1) | Part 2nd | 186 (1) | Full | Letting just agreed Lease to 2019. Rent reviews 2009 and 2014 | 70,000 = 376.34 m² | See Note (1) | 309.01 |
| The subject | L Ltd (2) | Basement | 371 | No provisions for the recovery of the cost of heating or lifts. | Lease just agreed for 5 years. | 32,000 = 86.25 m² | For service charge see Note (2) | 70.56 |
| A similar property | M Ltd (3) | 4th Floor | 550 | Full | Agreed 1 year ago Lease for 20 years to 2026. Rent reviews 5 yearly. | 270,000 = 490.90 m² | See Note (3) | 445.41 |

Notes on Schedule III

(1)  F Ltd (see Schedule I) have retained 218 m² of office accommodation for their own use. In order to provide access to the office suite let to K Ltd and to its own accommodation, a common access passage has been partitioned off from the area originally leased by F Ltd. This has an area of 88 m².

As the basis for valuation is the total internal floor area for each floor exclusive of lavatories, lift lobbies and staircases, the net rent per m² of the sublet portion needs adjustment to allow for a proportionate share of the internal corridor as follows:

Rent of subletting          £70,000 pa

Net area of subletting =     $(492 - 218) - 88 = 186$ m²
                             £70, 000 ÷ 186 m² = £376.34 per m²

Adjustment:             Total net area of occupied parts

                        Total area of subletting

                        218 m² + 186 m²

                        492 m²

                        = 0.8211 × £376.34 = £309.01 per m²

(2)   C Ltd (see Schedule I) have let the entire basement area at a rack-rent, but otherwise on the same terms on which they themselves hold it. This rent therefore reflects the fact that the tenant has no liability to reimburse the landlord with a proportion of the cost of heating, maintenance of heating or of the cost of providing lifts. An adjustment to the rent will have to be made to take this into account. Rent £32,000 pa *less* £5,825 (irrecoverable cost of heating and lifts) = £26,175 pa (£70.56 per m²).

(3)   This floor has been partitioned by the landlord to provide individual offices. It is estimated that the partitioning adds approximately £45.50 per m² to the value of this accommodation. In the case of the subject property, no partitioning has been provided by the landlord and therefore this must be deducted for the purpose of comparison.

## Inspection of the building

The collection of data may take some time but having put the work in hand an inspection of the building should now be made. It is not proposed to go into detail on matters to be noted on inspection, but the following are the principal ones:

- location — general and particular
- general description, construction and condition of the building
- services: eg lifts, central heating and public services available
- degree of obsolescence of building and services with particular reference to renewals for which the head lessee may be liable
- floor areas should be checked against plans if available, otherwise a survey will be necessary. The measurement of any building should be in accordance with the Code of Measuring Practice issued by the Royal Institution of Chartered Surveyors. For offices it is the net internal area that is required
- floor loading capacity. Ideally should be able to support a uniformly distributed super imposed load of 4 kN/m² plus 1 kN/m² for partition loading.

### Enquiries of public authorities

Information should be obtained on town planning consents, any established use rights and building regulation consents. Local authority development plans should be inspected to see whether any developments are projected in the vicinity and the highway authority should be consulted on road proposals as either of these matters might affect the potential value of the property.

## *Assessment of the rental data (Schedule IV)*

Having collected all the relevant data it is necessary to analyse and co-ordinate them. Since the valuation will be arrived at by capitalisation of net income, the reversionary rental values of all underleases will need to be calculated and thereby the increased income at each rent review is ascertained. This has been summarised in Schedule IV to give the total rent receivable by the head lessee at each rent review year based on current rental values. From these totals will be deducted the head rent and any irrecoverable service charges. The head rent falls to be reviewed in 2007 and is calculated as follows:

## Calculation to determine rent to be paid on review under headlease rack-rental value of building

|  | Area m² | Rent per m² | Rent £ pa |
|---|---|---|---|
| Basement storage (2) | 371 | 70 | 26,000 |
| Ground to 5th floor offices | 2,952 | 445 | 1,313,640 |
| Totals | 3,323 | | £1,339,640 |
| | | | |
| Rack-rental value | | | 1,339,640 |
| Less: Head Rent | | | 64,000 |
| | | | £1,275,640 |

Revised head lease rent = 5/8 (as per head lease) × £1,275,640 + £40,000 = £837,275 pa

## Schedule IV

Notes

(1)  The sixth floor is excluded from the calculation as this is a tenant's improvement which the lease states is not to be taken into account for rent review purposes.

(2)  The rent review is based on the rack-rental value and not on the rent received.

A further deduction to arrive at net income has been made in respect of that part of the income arising from the rent of the third floor which is in excess of current rental value. This unsecured income will be valued separately.

   For more information on landlord and tenant matters and lease renewals and rent reviews rents see Chapters 4 and 15 respectively.

## Schedule IV

Schedule showing the present and future net income at current rental values receivable by the head lessee

|  | 2007<br>£ pa | 2008<br>£ pa | 2009<br>£ pa | 2014<br>£ pa |
|---|---|---|---|---|
| Basement (1) | 1,000 | 1,000 | 1,000 | 1,000 |
| Ground floor (1) | 2,500 | 2,500 | 2,500 | 2,500 |
| First floor | 93,000 | 220,000 | 220,000 | 220,000 |
| Second floor | 106,000 | 106,000 | 220,000 | 220,000 |
| Third floor | 240,000 | 240,000 | 240,000 | 240,000 |
| Fourth floor | 93,000 | 220,000 | 220,000 | 220,000 |
| Fifth floor | 93,000 | 220,000 | 220,000 | 220,000 |
| Sixth floor | – | – | – | – |
| Total rents receivable | 628,500 | 1,009,500 | 1,123,500 | 1,123,500 |
| *Less:* irrecoverable service charge | 13,549 | 13,549 | 13,549 | 13,549 |
| Total net rents receivable | 614,951 | 995,951 | 1,109,951 | 1,109,951 |
| *Less:* rent received for 3rd floor<br>in excess of rental value | 20,000 | 20,000 | 20,000 | 20,000 |
|  | 594,951 | 975,951 | 1,089,951 | 1,089,951 |
| *Less:* head lease rent | 40,000 | 40,000 | 40,000 | 837275 |
| Net rents receivable | 554,951 | 935,951 | 1,049,951 | 252,676 |

### Note on Schedule IV

(1)  The reversionary incomes in respect of the basement, ground and sixth floors have been ignored because they are so long deferred. In the case of the basement and ground floor the underlease still has 32 years unexpired while in the case of the sixth floor, despite the fact that no notice has been served under the Landlord and Tenant Act 1927, the tenant will still have the benefit of these improvements if a new lease is granted by virtue of the Law of Property Act 1969.

Two factors affecting the capitalisation rate at which income is valued are the frequency of rent review and the quality of the covenant of the tenant. It is therefore advisable to analyse the rental data to illustrate the position. This has been done in Schedule V.

## Schedule V

Analysis of existing and future rental income from the property and analysis of rent received by frequency of review

|  | £ pa | % |
|---|---|---|
| Rent received at present subject to five-year reviews | 625,000 | 99.5 |
| Rent fixed for 32 years | 3,500 | 0.5 |
|  | 628,500 | 100 |

Analysis of rental value of property by frequency of rent reviews

|  | £ pa | % |
| --- | --- | --- |
| Open market rental value of parts subject to five-year reviews | 1,100,000 | 75.9 |
| Open market rental value of part where rent fixed for 32 years | 246,000 | 16.9 |
| Open market rental value of tenants' improvements | 103,000 | 7.2 |
|  | 1,449,000 | 100 |

Analysis of rent received from underlessees according to the covenant value of tenants

|  | £ pa | % |
| --- | --- | --- |
| Major public companies | 202,500 | 32.4 |
| Public company in financial difficulties | 240,000 | 38.0 |
| Private partnership | 93,000 | 14.8 |
| Foreign-based company | 93,000 | 14.8 |
|  | 628,500 | 100 |

The capitalisation rates to be applied will be ascertained by analysis of market transactions. It is important that transactions are analysed on the same methods as are adopted for the valuations. Office property is but one form of investment which is competing for investment funds, and its performance must be measured against other investments in terms of current yield, rental and capital growth. Other sources of investment whose current performance can readily be ascertained from the *Financial Times* are fixed interest investments such as Government stock, debentures and also equities.

Yields are constantly varying, thus affecting the popularity of particular types of investment, so that valuers must keep themselves up to date with the investment market in general.

Valuation is not, however, an exact science: it is the valuer's personal interpretation of the market in relation to the particular property being valued which is important. It has been assumed in the valuations which follow that the current yield obtainable on prime freehold office investments is of the order of 5%. Prime is not easy to define, but in simple terms it means a well-built modern building, fully let at current market rents, subject to five-year upward-only rent reviews to tenants whose covenant is blue chip. Yields adopted for the valuation take account of the variations in the property being valued compared to the prime freehold investment. Adjustments to the yield will need to be made in respect of age, location, quality of building and the frequency of rent reviews. Although it is customary to adopt values current at the date of valuation in estimating reversionary values, the extent to which a property is a hedge against inflation will depend on the frequency of the rent reviews. Corresponding yields for different frequencies of rent review at varying growth rates are to be found in *Donaldson's Investment Tables*[1].

---

1    Available from Donaldsons, London.

## Valuation of the freehold interest subject to the headlease

| | £ pa | £ |
|---|---|---|
| Present Net Income | 40,000 | |
| YP in perp at 6% (1) | 16.67 | 666,800 |
| | | |
| Increase in rental income received in 2014 | | |
| (revised headlease rent £837,275 *less* £40,000 | 797,275 | |
| YP in perp deferred 7 years at 7% (2) | 8.9 | 7,095,748 |
| Value of freehold subject to the headlease | | 7,762,548 |
| *Less*: purchaser's costs at 5.7625% (3) | | 422,945 |
| | | £7,339,603 |

Value of freehold interest subject to the headlease, say £7,340,000

Notes

(1)  This part of the income is fixed for the full term of the head lease, so it has been valued in perpetuity. Assuming that long-dated gilts are showing a return of 6%, this investment, though long dated, has the advantage that (a) there is an ultimate reversion to a central London freehold, (b) there is a substantial marriage value (see below), so justifying a yield in line with long-dated gilts.

(2)  The estimated additional income receivable in seven years is fixed for 25 years. The reversion to full market value is so long deferred and so uncertain, with the building 100 years old at the end of the lease, that it has been ignored and the reversionary income receivable in 2014 valued in perpetuity. The increased income is also less secure than the basic ground rent because it is dependent on rental values in six years' time, hence the yield has been adjusted to 7%.

(3)  Purchaser's costs in acquiring the interest being valued are deducted so as to give the actual rate of return to the investor, based on his total capital outlay, which will include costs of acquisition. The costs comprise 4% stamp duty on purchase, 1% agent's fee and 0.5% legal fees plus VAT on agent's and legal fees. The percentage deducted needs to be altered as changes occur, particularly on stamp duty which at the time of writing (2007) is at 4% for values above £500,000. Also note the costs of purchase relate to the end value. To arrive at the end value divide the gross value by 1.057625 to give value net of costs. Do not take 5.7625% of the gross figure as this will inflate costs.

An examination of Schedule IV will show that the income receivable by the head lessee falls into three parts. Owing to the substantial increase in head rent in 2014 only £178,825 (referred to below as Minimum income) is secured for the full term of the head lease after deducting the non-recoverable service charge and the excess of rent payable over rental value for the third floor (Schedule 1 Note (2)). Of this income £3,500 pa cannot be reviewed until 2039 (32 years hence), but the balance is subject to five-year rent reviews. Nevertheless, the head rent payable on review in 2014 represents about 77% of estimated net rent receivable at that date after deducting the excess rent of the third floor and the non-recoverable service charge expenditure.

The second part of the income is that received only until the head lease rent review in 2014 (referred to in the valuation as 'Excess income received before review in 2014'). This income is receivable for only seven years and its capital value must therefore be amortised over the seven-year term. This type of income pattern is unattractive to an institutional investor who favours a steady income growth over a period, whereas in this case the income received from the property falls substantially in seven years' time. Consequently, while this income is reasonably well secured a relatively high rate of interest, namely 10%, is used.

Finally, there is the excess income over rental value in respect of the third floor, payable by a tenant whose covenant is doubtful. This income has therefore been valued at a high rate (12%) and amortised over the remainder of the term using the taxed sinking fund rate.

## Valuation of head lessee's interest

|     |       |                                                  |       | £ pa    | £          |
| --- | ----- | ------------------------------------------------ | ----- | ------- | ---------- |
| (a) |       | *Minimum income* (see Schedule IV )              |       |         |            |
|     | (i)   | Years 2007/2014                                  |       |         |            |
|     |       | Net rent                                         |       | 252,676 |            |
|     |       | YP 7 years at 7% and 4% (tax at 40%)(1)          |       | 3.56    | 899,527    |
|     |       |                                                  |       |         |            |
|     | (ii)  | Years 2014/2064                                  |       |         |            |
|     |       | Net rent (2)                                     |       | 252,676 |            |
|     |       | YP 50 years at 8% and 4% (tax at 40%)            | 11.00 |         |            |
|     |       | PV £1 in 7 years at 8%                           | 0.58  | 6.38    | 1,612,073  |
|     |       | Sub-total 1                                      |       |         | £2,511,600 |

Notes
(1)  For sinking funds tax at 40p in the £ is the usual rate adopted by valuers but some valuers use 30p in the £ to reflect rate of corporation tax. The rates of interest and tax are interdependent.
(2)  Although the net income is the same as in the first seven years it is now top slice income subject to the payment of a head rent of about 77% of the total income.

|     |       |                                                  |       | £ pa    | £          |
| --- | ----- | ------------------------------------------------ | ----- | ------- | ---------- |
| (b) |       | *Excess income received before review in 2007*   |       |         |            |
|     | (i)   | Years 2007/2008                                  |       |         |            |
|     |       | Net rent (£544,951 – £252,676)                   |       | 302,275 |            |
|     |       | YP 1 year at 8%                                  |       | 0.93    | 281,116    |
|     |       |                                                  |       |         |            |
|     | (ii)  | Year 2008                                        |       |         |            |
|     |       | Net rent (£935,951 – £252,676) (1)               |       | 683,275 |            |
|     |       | YP I year at 8%                                  | 0.93  |         |            |
|     |       | PV £1 in 1 year at 8%                            | 0.93  | 0.86    | 587,617    |
|     |       |                                                  |       |         |            |
|     | (iii) | Years 2009/2014                                  |       |         |            |
|     |       | Net rent (£1,049,950 – £252,676) (1)             |       | 797,274 |            |
|     |       | YP 6 years at 8%                                 | 4.62  |         |            |
|     |       | PV £1 in 2 years at 8%                           | 0.86  | 3.97    | 3,165,177  |
|     |       |                                                  |       |         | £4,033,911 |

Pannell fraction — which is the adjustment to allow for amortisation over seven years.

$$\frac{\text{YP 7 years at 8\% and 4\% (tax at 40\%)}}{\text{YP 7 years at 8\%}} \quad \frac{3.436}{5.206} = 0.66$$

0.66 × 4,033,911 = £2,662,381

An alternative approach would be by the use of a double sinking fund.

(c)  *Rent received in excess of rental value for third floor (for alternative methods of valuing excess rents see end of chapter)*

|  | £ pa | £ |
|---|---|---|
| Rent | 20,000 | |
| YP 14 years at 8% and 4% (tax at 40%) | 5.84 | 116,800 |

| *Total value of head lessee's interest* | £ |
|---|---|
| (a) | 2,511,600 |
| (b) | 2,662,381 |
| (c) | 116,800 |
| Total value | 5,290,781 |
| *Less*: purchaser's costs at 5.7625% | 288,270 |
| | £5,002,511 |

Value of Head lessee's interest, say £5,000,000

Note: All rents have been reviewed to current rental values by 2010 except for the basement and ground floor. As these are not subject to a rent review for 32 years the value on reversion is negligible.

## Valuation of freehold interest in possession

The figures of net income have been extracted from Schedule IV.

|  | £ pa | £ |
|---|---|---|
| Current rental income 2007 | 594,951 | |
| YP in perp at 6.25% | 16.0 | 9,519,216 |
| | | |
| *Reversion* in 2008 | | |
| Net rent | 975,951 | |
| *Less*: current rent | 594,951 | |
| Increase in rent | 381,000 | |
| YP in perp deferred 1 year at 6.25% | 15.06 | 5,737,860 |
| | | |
| *Reversion* in 2009 to (1) | 1,089,950 | |
| *Less*: current rent 2008 | 975,951 | |
| Increase in rent | 113,999 | |
| YP in perp deferred 2 years at 6.25% | 14.17 | 1,615,366 |
| | | 16,872,442 |
| | | |
| *Excess rental income received for third floor* | | |
| Rent | 20,000 | |
| YP 14 years at 7% and 4% (2) (tax at 40%) | 6.21 | 124,200 |
| | | 16,996,642 |
| *Less*: purchaser's costs 5.7625% | | 926,067 |
| | | 16,070,575 |

Value of Freehold interest in possession, say £16,070,000

Notes
(1)  The rental value of fifth-floor tenant's improvements has been ignored, since the tenant has the benefit of these improvements for another 12 years under his existing lease and for the term of any subsequent lease or 21 years whichever is longer (Law of Property Act 1969). Similarly the rental value of the basement and ground floor has been ignored as these areas are subject to leases without review for an unexpired term of 32 years.
(2)  Although this is in the nature of short leasehold interest it is unlikely to be of interest to a gross fund, who above all want a secure income, and consequently it has been valued using a tax adjusted table.

The freehold subject to the headlease is not an attractive purchase for an institution because of the infrequency of the rent reviews. Equally, the head leasehold interest is unattractive because of the sharp fall in income in seven years' time. The valuation of the freehold subject to the occupation leases illustrates the considerable increase in value by amalgamating the freehold and head leasehold interests. The difference between the value of the freehold in possession and the value of combined freehold and leasehold interests valued separately is known as the marriage value.

|  | £ | £ |
|---|---|---|
| Value of freehold in possession |  | 16,070,000 |
| Value of freehold subject to head lease | 7,340,000 |  |
| Value of head lessee's interest | 5,000,000 |  |
|  | 12,340,000 | 12,340,000 |
| Marriage value |  | £3,730,000 |

It would be usual that, at some stage, either the freeholder or the head lessee would endeavour to effect an amalgamation by buying out the other at a price equivalent to the value of his own interest together with somewhere around 50% of the marriage value, the final outcome depending on the negotiating skills of the parties.

# Valuation of rents in excess of open market rents

During the 1990s it was not unusual for valuers to come across situations where a property had a rental stream in excess of the open market rent, usually due to rental values falling. While in 2007 we are in a period of rental growth, market conditions can change and valuers need to know how to deal with such changes.

To the landlord this means that the rental investment is not adequately secured by the property, and the additional income above the open market rent is at risk, and the continuation of this excess rent is dependent on the financial strength of the tenant. Should the tenant become bankrupt then the additional rent will no longer flow from the property, especially given the recent changes in the law on the assignment of leases.

For many years, the simple way of dealing with this was for valuers to apply a higher rate and capitalise over the remainder of the term using a sinking fund rate. This is the method adopted earlier in the chapter.

However, the method currently being adopted by the market is what is known as the short cut Discounted Cash Flow (DCF) approach.

In this method, the valuer assesses the expected growth in rent from the property, and applies this to the open market rent (as opposed to the passing rent) to ascertain initially at which review or lease termination (whichever is the earlier) the current open market rent has grown to be equal to, or in excess of, the passing rent. Having ascertained this review date, the passing rent is capitalised until the appropriate rent review date or lease renewal date at the top-sliced discounted rate. From the review date onward, the anticipated rental income is capitalised at the appropriate yield for the property, discounted at the equated yield.

An example is as follows:

| | |
|---|---|
| Office area | 500 m² |
| Passing rent £400/m², ie | £250,000 pa |
| Assessment open market rent £300/m | £200,000 pa |
| Over-rented element | £50,000 pa |

| | % |
|---|---|
| Review pattern 5-yearly — lease 3 years in existence | |
| Assumed all risk yield | 6 |
| Equated yield | 10 |
| Expected growth is rent | 5 |

At the first review, in two years' time, assuming an expected growth of 5% pa, the open market rent will be £165,375 pa. This is still below passing rent.

At the second review, in seven years' time, again assuming rental growth of 5% pa, the open market rent will be £211,064, which is above the passing rent. This means the equated yield is adopted until the review in seven years' time.

## Valuation

| *Next 7 years* | | £ pa | £ |
|---|---|---|---|
| Passing rent | | 250,000 | |
| YP 7 years at 10% | | 5.39 | 1,347,500 |
| | | | |
| *Estimated rent at review 7 years* | | | |
| Net rent, say | | 281,420 | |
| YP in perp at 6% | 16.67 | | |
| PV £1 in 7 years @ 10% | 0.513 | 8.55 | 2,406,141 |
| | | | 3,753,641 |
| *Less*: costs of purchase at 5.7625% | | | 204,518 |
| Value | | | £3,549,123 |
| But say £3,550,000 | | | |

It should be pointed out that in the DCF approach, different growth rates could be used for each year to reflect the valuer's judgement on anticipated growth in rents for the property. Deciding on growth rates is not always easy. If in doubt, it is recommended that the valuer uses the growth rates for the economy that are forecast by the Treasury.

# Retail Properties

The valuation of retail properties for non-statutory purposes is affected to only a limited extent by legislation, the principal enactments being Part II of the Landlord and Tenant Act 1954 (as amended by the Law of Property Act 1969 and revised in June 2004), Part I of the Landlord and Tenant Act 1927 and the Offices, Shops and Railway Premises Act 1963. Reference may also need to be made to the Use Classes Order 1987 and General Permitted Development Order 1995.

The wide variation in retail types, including the isolated 'corner' shop, the departmental store of 30,000 m$^2$ occupying an island site in a premier shopping position in a major city, and the out of town superstore, probably provides the valuer with a greater test of his expertise than is generally encountered in the valuation of other types of commercial property.

With few exceptions, retail outlets, whether freehold or leasehold, owner-occupied or held for investment, are normally valued on an 'investment' basis, ie by the capitalisation of rents and rental values, and it is the assessment of rental value to which the valuer has initially to apply his mind.

## Rental value

### Situation

In arriving at retail rental values the location is of paramount importance. In an established town or city centre the pattern of trading will usually be readily assessed from inspection and from commercially produced street plans which are available for most centres giving the names of all the traders. Retail valuers commonly identify the section of a street which commands the maximum rental value as the '100% position' and from analysis of rentals elsewhere in the street are able to describe the less valuable sections in percentage terms, ie the '90% position', '75% position' etc. The more general expressions of 'prime', 'secondary,' or 'tertiary' are also often used.

Where there are long, continuous parades of shops the points at which rental values change may be difficult to identify and may in fact be constantly moving. Where parades are interrupted by side roads or by non-retail frontages such 'breaks' often become the divisions between sections of different rental values.

The level of rents achieved in the 100% position will largely depend on demand and the importance of the centre concerned and the rental value in the prime position in one town may be several times greater than the rental value for a similar-sized unit in the peak position in a smaller town even though the occupier of both units is the same multiple retail company.

In practice the identification of the rental value pattern has become increasingly difficult in recent years. There has been a great deal of development activity in the shopping field and in some cases existing shopping centres have been entirely redeveloped. In other cases new centres have been built adjacent to existing centres and have caused a radical distortion of the previous trading pattern. In the latter cases inspection alone may not be sufficient to identify those positions which have increased or decreased in value and, where evidence is lacking or is inconclusive, discussions with traders and owners of new schemes may be necessary. Established regional centres such as the Metro Centre at Gateshead, Meadowhall at Sheffield or Lakeside at Thurrock have undoubtedly affected levels of rents in competing towns. Increasingly sophisticated research techniques are also being employed to estimate the catchment area likely to be served by a centre and the income groups and spending habits of the potential shoppers.

Known future developments may have to be taken into account. For instance a large proposed food store with a surface car park may have the effect of increasing the value of shops close to the new property, possibly at the expense of shops previously of higher value.

Identification of trends can be important. New Bond Street, W1, and Marylebone High Street, W1, are examples of London streets which have changed in value dramatically in recent years because they have become centres for the sale of jewellery and designer goods respectively. Similarly the change to 'speciality' shopping in South Molton Street, W1, produced a rapid escalation in rental value in the late 1980s. Even established shopping streets such as Oxford Street, W1, can undergo changes which produce a greater-than-average variation in rental values over a short period. It should be borne in mind that changes in rental value brought about by special demand might be short term only, which could quickly be reversed. Local, regional and national trends can vary considerably both in rate and in direction, and it is this irregular movement of rental values which makes it difficult to construct indices which are anything more than a general guide.

Tourism can materially affect the turnover of shops in central London and in historic towns in the regions. Changes in the economic climate in the country of origin of visitors, or in currency exchange rates, can have a substantial and sometimes very rapid effect on turnover, and consequently on rental value. In some cases the effect may be too short-lived to influence rental values; if prolonged, however, rents agreed in times of peak tourist booms may not be maintained subsequently. The '9/11' terrorist bombings in September 2001 had an immediate impact on London retail rents as air travel from the US became less popular and London was perceived as a possible future target for terrorists. Oxford Street in Central London is a particularly good example of such rental movement and it is one of the most volatile locations in the UK, relying to a significant extent on tourist trade from both home and abroad.

All potential tenants will first make an estimate of the turnover likely to be achieved in the retail unit in question, and from the 'mark up' (ie the amount by which the sale price exceeds the cost) appropriate to their particular merchandise they will then calculate the rent they can afford as a residual of all other operating costs. The process is not wholly objective, and a wish to be represented in a particular location, or the prospect of above-average turnover growth, can be among factors which override purely actuarial considerations. Rents related directly to turnover are common in the USA, but there are still relatively few cases in the United Kingdom — the shopping centres at Eldon Square, Newcastle Upon Tyne, the Metro Centre, Gateshead and the Harlequin, Watford, being some of the major examples. Even these rents are usually linked to a minimum base figure which may be 80% of the open market rent. Elsewhere, owners and occupiers in the UK are concerned at the uncertainty of rent/income which a turnover rent introduces.

# Size and type

The traditional shopping street in the smaller town often started life as a series of terraces of houses which were progressively converted into shops, and this origin accounts for the limited size of many of the shops in the older centres. The tendency in recent years has been for tenants to demand larger units, and new developments will now contain a significant number of larger shops. It will be necessary for the valuer to know the level of demand for the various sizes of shops, otherwise the interpretation of rental evidence can be misleading. For instance, demand for small shops suitable for 'niche' operators in a particular location may produce rental values which cannot be safely applied to larger shops, even by zoning or other discounting methods. Similarly, there may be a special demand for large units, particularly in major locations. The special demand might exist, however, for only a limited time. The kind of problem which often faces a valuer is the valuation of, say, an old established store which has a very large ground floor in a good position in a multiple shopping street. If it is known that one or more of the major space users at present occupying badly planned premises in an inferior position are in the market for better premises, the rent which the valuer estimates might be received may be in excess of rents previously achieved. Conversely, if all the major space users are already locally represented in adequate premises, the valuer might decide that there would be limited demand for the unit as a whole and arrive at his rental value by dividing the property into viable units, making allowance for the cost of conversion in the capital value.

The ability to display goods in an efficient and attractive manner is an important factor in retailing, and a major retail company can spend as much on finishing and fitting a unit as on the cost of the structure. Consequently, the internal layout has to be taken into account in assessing rental value. Irregular shapes, low ceiling heights, badly sited staircases, differences in floor levels, intrusive stanchions and beams and such like can adversely affect value.

Parking restrictions and pedestrianisation make off-street loading facilities increasingly important. Lack of rear loading facilities can make shops in heavily restricted areas unlettable for certain trades, eg supermarkets, where large and frequent deliveries are necessary.

Although it has been customary for corner shops with display windows on the return frontage to be regarded as having a greater value than non-corner units, the difference in value is largely a matter of judgment in the particular circumstances. In some cases any additional value could be completely offset by the cost to the tenant of maintaining display space, and subsequent interference to the internal layout. A corner unit at the junction between a primary and a good secondary shopping street would, however, normally be expected to command a higher rent than a similar unit without a return frontage, although the difference has reduced in recent years.

The majority of retailers require staff or stockroom facilities within their units either at the rear, in the basement or on an upper floor. The analysis of rents of lock-up shops, shops with basements and shops with basements and/or upper floors in a shopping centre will usually provide the valuer with a guide to the value of the ancillary accommodation, but experience is necessary when building up a composite value, eg of a shop with a basement and first floor, when rental evidence of directly comparable properties is not available. Attempts are sometimes made to value basements and upper floors on the basis of a proportion of the value of the ground floor. In practice, however, the market is usually too erratic for any reliable analysis of this nature to be made and it cannot be over-emphasised that this type of mathematical analysis plays virtually no part in the process by which a tenant arrives at his estimate of the rent he thinks he can afford to pay. The retail valuer should guard against the tendency to substitute theory for experience in cases where direct evidence is not available.

It is customary in the case of existing retail units to calculate floor areas on a net basis, ie excluding lavatories, staircases, lifts, landings, escalators, plant rooms etc. In new developments, however, where units are being let in 'shell' form (ie with no wall or ceiling finishings and where the tenant is to install lavatories etc) it is not uncommon for letting areas to be quoted on a 'gross internal' basis. The definitions of gross internal area and net internal area given in the Code of Measuring Practice issued by the Royal Institution of Chartered Surveyors should be followed in all cases.

Although zoning of ground floors plays an essential part in rating valuations and in the preparation of valuations for Lands Tribunal or court cases it should be borne in mind that in the open market the zoning system has to be adapted to suit the facts and should not be used on a pre-conceived basis. Ideally, the depth and number of zones should be arrived at by analysis of known rents, but the frequent absence of open market lettings has resulted in a tendency for the adoption of 'standard' zone A depths. Over-reliance on standard depths, particularly when used in conjunction with the analysis of secondary evidence, such as rent review settlements, can produce misleading results. Mathematical precision is not a feature of the valuation of retail properties for open market rental purposes and even between skilled valuers the margin of opinion may be surprisingly wide because of the imperfections in the market.

The conventional zoning basis is to divide the ground floor into sections of equal depth and to assume that each section or zone is worth 50% of the preceding zone, the zone on the street frontage (zone A) having the 100% value. When dealing with deeper shops, however, the 'halving back' process normally stops after four or five zones. To continue halving back can produce what may appear to be unrealistically low values for the rear zones, and ground floor sales rates, even at the rear, should generally always exceed ancillary and storage rates elsewhere in the shop. With very deep shops, however, the reality might be that a zone A value derived from the analysis of rents of smaller shops may be of little or no relevance and an alternative approach, such as a valuation on an 'overall' basis, may be more appropriate.

Irregular shaped shops where, for instance, the unit widens at the rear or extends sideways behind the front zone or zones present difficulties for the zoning process and to avoid arriving at an excessive rental value it may be necessary to regard areas at the side of the rear zones (often referred to as 'masked areas') as if they were part of one of the deeper zones (eg a masked area in zone B being regarded as if it were in zone C).

Irregularities and disabilities are also frequently dealt with by application of 'end' allowances, usually in the form of percentage discounts applied to the aggregate rental value or the rental value of the part suffering that disability. Where arbitrary allowances have to be made to arrive at realistic rental values, however, it is frequently an indication that over reliance has been placed on the zoning system and that the search for comparable properties needs to be widened or other approaches adopted.

One adjustment which is frequently made to the ground floor rental value of a retail shop is for its size, often referred to as a 'quantum' discount. This is where the shop unit being valued is much larger than the properties from which the comparable evidence is being drawn. The discount is a reflection of the reduced demand for the larger property, and therefore the valuer must be aware of the demand profile for the location in question before making the adjustment. Quantum discounts are usually in the range of 0% to 10%, and have tended to reduce in recent years as the demand for larger units from the major retail multiples has increased.

A frequently adopted variation to applying a progressively halved zone A value is to calculate the area of each zone and then to apply the 'halving back' process to the area of each zone, so that the area of zone B is reduced by 50%, that of zone C by a further 50%, *et seq*. The chosen zone A value is then applied to the aggregate of the adjusted areas. The resulting total area is referred to as the area 'in terms of zone A', usually abbreviated to ITZA. The ITZA approach is useful for comparison purposes but in some cases can be misleading as it is possible for two shops to have identical ITZA areas despite having widely differing area distributions between the various zones. In practice it is usually advisable to state both the ITZA area and the actual zone areas.

It cannot be over emphasised that zoning is only a tool and that the acid test is whether or not the rental value which has been arrived at is one which is achievable in the open market.

## Study 1

A rental valuation is required for lease renewal purposes. The shop has a frontage of 6.5 m, a depth of 22.5 m and a net internal area of 133.1 m$^2$.

Evidence is available of the following current transactions:

Shop 1: Lease recently renewed at £30,000 pa. Similar position. Frontage 6.5 m, depth 15 m. Net internal area 90.5 m$^2$.

Shop 2: New letting at £44,000 pa. Better position. Frontage 7.2 m, depth 25.1 m. Net internal area 173.6 m$^2$.

Shop 3: New letting at £37,200 pa. Similar size and position but better property. Net internal area 129.2 m$^2$.

Shop 4: Lease renewed six months ago at £34,000 pa Identical to the subject unit.

Analysis of the rents of the four comparables shows that by adopting two 7.5 m zones and a remainder and 'halving back', a reasonably consistent zone A figure can be deduced:

| | | Area in m$^2$ | £ | £ pa |
|---|---|---|---|---|
| Shop 1: | Ground floor: | | | |
| | Zone A | 46.3 | 439 | 20,326 |
| | Zone B | 44.2 | 439/2 | 9,702 |
| | No remainder | – | – | – |
| | | 90.5 | | 30,028 |
| Say £30,000 pa | | | | |
| | | | | |
| Shop 2: | Ground floor: | | | |
| | Zone A | 52.5 | 459 | 24,097 |
| | Zone B | 52.5 | 459/2 | 12,049 |
| | Remainder | 68.6 | 459/4 | 7,872 |
| | | 173.6 | | 44,018 |
| Say £44,000 pa | | | | |
| | | | | |
| Shop 3: | Ground floor: | | | |
| | Zone A | 50.1 | 466 | 23,347 |
| | Zone B | 39.8 | 466/2 | 9,273 |
| | Remainder | 39.3 | 466/4 | 4,578 |
| | | 129.2 | | 37,198 |
| Say £37,200 pa | | | | |

|  | | Area in m² | £ | £ pa |
|---|---|---|---|---|
| *Shop 1:* | *Ground floor:* | | | |
| *Shop 4:* | *Ground floor:* | | | |
| | Zone A | 48.2 | 421 | 20,292 |
| | Zone B | 45.1 | 421/2 | 9,494 |
| | Remainder | 40.2 | 421/4 | 4,231 |
| | | 133.5 | | 34,017 |

Say £34,000 pa

Shop 2 is in a better position than the subject property and should have a higher zone A value. Shop 3 is a better shop and will also have a higher zone A value. Shop 4 gives a good indication of the rental value six months ago but is now low. The analysis suggests a zone A value of £445 per m2 for the subject property and this is supported by shop 1.

*Valuation*

| | Area in m² | £ | £ pa |
|---|---|---|---|
| Ground floor: | | | |
| Zone A | 47.2 | 445 | 21,004 |
| Zone B | 43.6 | 445/2 | 9,701 |
| Remainder | 42.3 | 445/4 | 4,706 |
| | 133.1 | | 35,411 |

Say £35,400 pa

*Notes*

(1)   It has been assumed that the lease terms are similar in each case. In practice, adjustments may have to be made, eg for differing repairing liabilities to produce comparable rents. The analysis will also include any basements or other floors included in the leases.

(2)   While this example is based on 7.5 m zones, the great majority of valuers devalue rents based on 20 ft (6.1 m) zones. Despite the metrication of many of our measurement bases, the zoning of retail units remains stubbornly on an imperial basis.

# Restaurants

In recent years there has been a significant increase in the number of restaurant outlets, although these are mainly located in secondary non-retail locations due to problems in obtaining planning consents in the main retail pitches, and the higher rental costs involved in these locations. Many traditional restaurants which were a feature of most high streets have disappeared owing to their inability to produce profit margins comparable with those of other multiple retail traders. Where conventional restaurants do exist in retail locations they do not normally present valuation problems as their rental values will be dictated by retail rents in that location, assuming there is no restriction on general retail use.

The valuation of restaurants, particularly those in secondary locations, normally requires a considerable degree of specialised knowledge on the part of the valuer. In some markets, central London for instance, the ground floor seating area is valued at a single overall rate, and various percentages of that rate are applied to other parts of the property, depending on whether they are used for seating or kitchen/ancillary purposes.

# Food stores

Food stores are valued differently from normal shops, and open market rental transactions are rare. The tendency has been for size requirements to increase, with great importance being placed on the availability of car parking facilities. In out of town locations, food stores with a total area of over 10,000 m² are not uncommon. The relationship between gross area and sales area varies between retailers but ancillary space (storage, preparation, etc) equivalent to about 25%–35% of the ground floor sales area will usually be required. Early supermarkets were predominantly food based, but the operators now provide an increased range of merchandise such as hardware, clothing, and furniture. Changes in operational requirements tend to be relatively rapid and it is important that the valuer should be aware of current developments.

Normal rental valuation practice for purpose-built food stores is to apply an overall rate to the total gross internal area. It is obviously essential that the same approach should be used for analysis.

Purpose-built supermarkets, particularly the larger ones, have become a specialised category of retail property as their size and location have made comparison with normal retail properties increasingly difficult. It is now generally accepted that, in the absence of evidence of rental values of similar sized units in the same location, comparison will have to be made with the rents in other locations, perhaps even on a national basis. Very large food stores, ie of the type operated by Tesco Extra, which are often over 10,000 m² in size, will sometimes attract a premium rent above the tone of evidence for smaller stores. This is because such properties are more readily able to dominate the local catchment population, and offer a wide range of profitable non-food lines.

# Departmental and variety stores

Departmental and variety stores present particular problems, as their number is limited and as with food stores there is rarely any direct evidence of comparable open market transactions available. The majority of the very large older stores are owner-occupied, the tenure being either freehold or long leasehold, and where stores do change hands it is more often than not as part of a transaction involving the sale of the owning company. Some rental evidence is found from sale and leaseback transactions and rack-rented stores are occasionally found in new schemes. There is, however, a relatively large body of rent review evidence for variety stores in the range of 3,000 to 10,000 m² total area. Many of these reviews are based on previous rent review evidence in the region, a direct result of the scarcity of open market transactions. As a result it is a commonly held view that the better stores are under-rented, ie reviewed at less than their true open market value, while the reverse is true for poorer stores where, in reality, little demand may exist.

Size is important and, as a reflection of poorer demand, it is usually the case that larger units attract lower rates per m² than small units.

It should be appreciated that whereas the proprietor of a normal sized retail shop will usually stock merchandise which is both limited in range and chosen as far as possible to produce the maximum profit margin, a departmental or variety store will usually offer a very wide range of merchandise with varying profit margins and may even operate some sections at a loss in order to be able to offer a full service. As a general rule the smaller and faster moving lines are sold on the ground floor and the larger and slower moving lines, such as furniture, on the upper floors or in the basement. In practice, an analysis of turnover by departments can show that, in terms of profitability, the basement and the first floors of a multi-floor store are worth more in relation to the value of the ground floor than would be the case if a similar analysis were made of the turnover of a conventional shop. The analysis will also probably show, however, that the value of the floors above first-floor level, despite service by lifts and escalators, falls much more rapidly than would be indicated by the analysis of normal sized retail properties, and that fourth and fifth floors, if used for retailing, may have a minimal value in profitability terms.

The extent to which profitability should be taken into account is a matter of experience but it is common for store operators to think of rent in terms either as a percentage of overall turnover or as an overall rate per m². In arriving at his final answer the valuer might have to use a combination of a conventional overall rate valuation, an estimation of rent as a proportion of turnover and an estimate of the alternative use value of the premises. The alternative use value may be of little consequence in the case of a modern, purpose-designed store but may produce the only viable valuation in the case of an old, 'off pitch' store. The alternative use value is sometimes referred to as the 'break up' value where the only practical alternative is the division of the ground and other floors into smaller units with appropriate provision being made for the costs of conversion.

Department and variety stores are valued on the basis of an overall rate applied to the total gross internal area, although sometimes smaller stores are valued on a floor by floor basis. The valuer uses his skill to adjust this rate for size, location, number of floors and layout. A common topic of discussion is the size threshold over which a retail unit should be valued on an overall, rather than zoning basis. Much depends on the comparable evidence and size of ground floor, but generally stores with over 1,000 m² at ground floor level are valued on an overall basis.

Care should be taken as to whether a store is to be valued on a fitted or shell basis. Valuing fittings (eg escalators, lifts, and plant) can produce a significant increase in the rental value, often in the order of 10% of the total rent. Existing use valuations are generally carried out on a fitted basis, as are rent reviews following a sale and leaseback transaction. Many rent review clauses do, however, direct a shell assumption, or provide a list of fittings to be valued.

## Study 2

A valuation is required for rent review purposes of a department store on the fringe of a good multiple shopping position. The store was purpose-built in 1925 but has been modernised and is adequately served by lifts and escalators. The lease provides that for the purpose of the rent reviews the rental value shall be that of a department store and that the store should be valued as fitted.

The store comprises ground floor, basement, and four upper floors; the ground floor has a frontage of 30 m and a depth of 100 m. The gross internal area of the entire store is 12,000 m².

|  | m² | £ | £ pa |
|---|---|---|---|
| Total GIA | 12,000 | 60.00 | 720,000 |
| Plus 10% for fittings |  |  | 72,000 |
| Total |  |  | 792,000 |

Say, £790,000 pa

*Notes*

(1) Evidence of similar sized stores in similar locations suggests a range of £55 to £70 per m².
(2) Evidence of rent reviews of fitted stores suggests that 10% is the appropriate adjustment.

## *Retail warehouses*

Retail warehouses are generally found in out of town locations, either as solus units (usually DIY), as clusters of two or more units, or on fully fledged purpose built retail parks containing as many as 20 units varying in size from 500 to 10,000 m². They are valued on an overall rate basis applied to the total gross internal area. Such matters as demand, parking, access, prominence, size, competition, loading facilities and design all play their part in determining the rental value. Size is a big issue, as the demand for smaller units is often substantially higher than for larger ones. Open market evidence is often quite plentiful and this is very much a sector for the specialist valuer. Many retail warehouses only have a 'bulky goods' planning consent which restricts demand. Some have open retail consents which can substantially increase their value where demand exists from high street retailers. Fosse Park in Leicester and the Fort at Birmingham are good examples of the latter category.

# Lease terms

Where a retail property is valued subject to a lease, the terms of the lease will have to be taken into account in the valuation. Landlords will almost always seek to pass the responsibility for all outgoings to the tenant, and first lettings of new properties will provide for the tenant to be responsible for rates, insurance and all repairs. In the case of entire properties, the tenant will usually be required to undertake all repairs; lettings of parts of properties can require the tenant to pay a proportion of the expenditure incurred by the landlord in repairing the property. Many leases of parts of older properties still provide for the tenant to be responsible for internal repairs only, and where such leases are renewed the provisions of the Landlord and Tenant Act 1954 make it difficult for the landlord to succeed in widening the repairing covenant (*O'May* v *City of London Real Property Co Ltd* [1982]).

Leases of shopping centres frequently provide for the payment of an annually reviewable service charge by the tenant to cover all outgoings, including normal services, special services such as security staff and the maintenance of large areas of common parts such as service roads. The charge is usually related to floor areas but is sometimes linked to rateable values. In covered centres with air conditioning and heating, the service charges can be high and in some centres it is possible that the level of these charges may adversely affect rental value when the rents are due for review.

The frequency of rent reviews is a factor which has a major effect on the value of properties subject to existing leases. The norm for the past 15 to 20 years has been for rents to be reviewed at five-year intervals, previously review periods of seven, 14, 21 or even 35 years were common.

Rental valuations may now still have to be undertaken on properties which are the subject of leases which were granted in the early days of rent reviews with these longer intervals. The assessment of the rent to be paid at the review dates can cause difficulties as there is likely to be no evidence of new lettings with a review frequency longer than, say, five years and very little information as to settlements in similar cases. It is generally assumed that a tenant would pay a higher rent for a lease with a term of, say, 14 years without review than for a lease with reviews at five or seven yearly intervals. A rule of thumb developed in the 1980s that 1% pa should be added to the rent for each year over a normal five yearly cycle. Therefore a 14 yearly cycle would attract a 9% increase. This percentage reduced as prospects for rental growth diminished in the 1990s, but the level of increase is back to 0.5% to 1% pa, depending on the prospects for rental growth for the type and location of the property concerned.

Leases which provide for the payment of a rent based on a percentage of turnover are sometimes encountered, mainly in fairly modern shopping centres. The most common form of percentage rent clause provides for the payment of a base rent and/or the payment of an additional rent based on turnover. Where there is a base rent there is usually also a base turnover, eg the lease might stipulate a rent of £50,000 pa plus an additional rent of 7% of the turnover in excess of £1,000,000 pa. The base rent is often related to a percentage (often 80%) of the open market value of the unit, and may be subject to review at intervals. The valuer may be called upon to assess both the initial base rent and the turnover percentage at the review dates and this calls for a knowledge of the relative profitability of different trades as the percentage of turnover will vary from trade to trade.

The alienation provisions in a lease should always be carefully considered. Usually a tenant is allowed to assign his lease, or to underlet the whole property. For larger properties, a prohibition on underletting part may be seen by the tenant as onerous. For example, where a variety or department store has a broad frontage to the high street, a restriction on underletting individual shop units may result in an adjustment to the rental value, if the comparable evidence is drawn from properties where no such restrictions are in place. Underletting is often prohibited in shopping centres where the landlord wishes to maintain control of any new lettings

Another important issue is the length of the unexpired term which is being adopted or assumed in the rental valuation. Lease lengths have reduced significantly in recent years, and most high street shop unit retailers are now content to take five or 10 year terms, whereas in the 1980s 20 or 25 year terms were the norm. For larger properties such as department stores, food stores, retail warehouses etc, the occupier will generally require a longer lease. Valuers should be keenly aware of the length of term which is to be assumed in the valuation, and should make a rental adjustment if it is outside the normal range.

User clauses in retail leases vary between fairly open users and those limited to a particular trade, and can have a considerable bearing on rental value. It can be argued that a limitation to a specific trade will always restrict the potential market and will, therefore, adversely affect rental value. In practice the effect will be less where there is likely to be sufficient demand from other operators in the specified trade to generate the full open market rental value. Where the specified trade is one for which there is likely to be little demand if the shop was offered in the open market with the same restriction, the adverse effect can be considerable. In the case of shopping centres, however, it is sometimes found that a tenant mix policy, which restricts certain types of trade to specified units, provides an advantage as occupiers will be assured that direct competitors will not be allowed to trade elsewhere in the centre.

## Study 3

The leases of three similar shops in the same parade are due for renewal. Rental valuations are required by the landlord. The position has improved considerably since the leases were granted 14 years ago and Marks & Spencer and Waitrose have opened new stores in the adjoining block. Originally there was a mixture of private traders and small multiples, but recent lettings and assignments have been to major multiple companies. Evidence suggests that the open market rental value of each of the three shops is £100,000 pa.

Shop 1:  let to a coffee shop, with no user restriction in the lease.

Shop 2:  let to a CD store and the lease contains an absolute prohibition against change of use. Although the tenant operates three other shops and trades efficiently, the maximum rent which he could afford to pay for this branch is £75,000 pa and this is supported by audited accounts.

Shop 3:  let to a mobile phone operator with a restriction against assigning or subletting to any trade already represented in the parade.

*Valuations*

Shop 1 is straightforward and a valuation of £100,000 pa can be reported.

Shop 2 presents the valuer with a problem. The tenant's trade is one which in recent years has been in decline due to internet sales and 'downloading'. If the CD operator decides to seek a renewal of his lease he will argue that the restriction adversely affects rental value, and in the event of court proceedings will probably succeed in having the restriction continued in the new lease (*Charles Clements (London) Ltd* v *Rank City Wall Ltd* [1978]). Evidence of the effect of such a restriction is unlikely to be available, in which case the valuer will probably advise that the landlord may have to be prepared to accept a rent lower than £100,000 pa (tenants' accounts are not, however, generally admissible as evidence of rental value — (*QH Barton Ltd* v *Long Acre Securities Ltd* (1981)).

Shop 3 is also not straightforward. If there are a number of trades not represented in the parade, the valuer may take the view that the restriction does not affect the rental value, and report a figure of £75,000 pa. If, however, most of the potential uses are already present, the valuer may believe some reduction is necessary, but not of the same magnitude as that of shop 2. The valuer may even consider, on the other hand, that the use has a particular premium value if there is good demand for it.

In the case of new shopping centres in good locations where all the units are marketed simultaneously, there is generally little or no variation between the rents achieved for units of similar size and in similar positions, regardless of whether or not there are user restrictions in the leases. In such cases there are grounds for arguing that a similar result should be achieved when reviewing rents or renewing leases. Provided there is not an excess of the less profitable use, the semi-monopoly status mitigates the effect of the user restriction. In small local parades restrictions can enhance the value of individual units.

By their nature, retail properties are more likely to have alterations and improvements made to them than other types of commercial property. The retail valuer must always, therefore, take steps when dealing with properties subject to leases to identify alterations which have been made by the tenant or his predecessor which could rank as improvements for the purposes of section 34 of the Landlord and Tenant Act 1954. Where qualifying improvements have been made and the purpose of the valuation is to advise the landlord or the tenant as to the rental value for renewal of the lease, the property has in effect to be valued as if the improvement had not been carried out. Where an improvement was made during the currency of the expiring lease, and regardless of the length of that lease and of the date of the improvement, the effect on rent of the improvement is to be disregarded for the duration of the subsequent lease (the duration being at the discretion of the court with a maximum of 15 years). Where an improvement was made during a previous lease, and more than 21 years prior to the date of service of notice for a new tenancy, the tenant loses the right to have the improvement disregarded.

## Study 4

A 35-year lease is about to end and the tenant requires a rental valuation prior to negotiating a new tenancy. The building comprises a shop with two upper floors. Present floor areas are: ground floor 97 m$^2$; first floor 92 m$^2$; second floor 78 m$^2$. Twenty-five years ago the tenant, at his own expense, extended the ground and first floors at the rear, the net additional area on each floor being 11 m$^2$. He also removed the original staircase between the ground and first floors and replaced it with a new staircase in the extension. The original staircase occupied an area of 5 m$^2$ on the ground floor and 3 m$^2$ on the first floor. Current rental values are: ground floor (valued overall) £250 per m$^2$: first floor £45 per m$^2$; second floor £30 per m$^2$.

*Valuation*

| | Area in m² | £ | £ pa |
|---|---:|---:|---:|
| Second floor | 78 | 30 | 2,340 |
| First floor | 78 (92 – (11 + 3)) | 45 | 3,510 |
| Ground floor | 81 (97 – (11 + 5)) | 250 | 20,250 |
| | | | 26,100 |

Say, £26,000 pa

*Notes*

(1)   For the purpose of the study it has been assumed that the same rate per m² can be applied to the area of the original and of the extended ground floor; in practice different rates may be applicable.
(2)   The tenant is entitled to the benefit of his improvement during the currency of the new tenancy up to a maximum term of 15 years.

It is sometimes suggested that the value in the unimproved state should be increased to reflect an element of potential, possibly by ascribing a notional ground rental value to the area occupied by the improvement, if it extends the built area. It is also sometimes permissible to value a property in its improved state, if it can be shown that any reasonable tenant would have carried out those improvements. For instance a tenant may, at his own cost, construct a new staircase to improve access to an ancillary floor, where previously access may have been difficult, or possibly non-existent. The valuer may consider valuing the property in its improved state, although the cost of the improvement must, of course, be taken into account. It is worth bearing in mind that this approach was generally rejected in the case of *Iceland Frozen Foods Plc* v *Starlight Investments Ltd* [1992], although this judgement has been much criticised (see *Lewisham Investment Partnership Ltd* v *Morgan* [1997]).

It should be appreciated that the provisions of section 34 of the Landlord and Tenant Act 1954 do not apply to rent reviews. It is common practice for wording either identical or similar to that of section 34 to be incorporated in rent review clauses, but in the absence of such wording improvements carried out by the tenant need not be disregarded when assessing rental value for review purposes (*Ponsford* v *HMS Aerosols Ltd* [1978]).

Where a rent review clause is silent on the subject of improvements an anomaly can arise in that the tenant may have to pay the full rental value for the property in its improved state at the rent review date(s), but on termination of the lease may be entitled to a new lease at a rent disregarding the improvements.

Where a retail unit is let in shell form and the lease is not granted in consideration of the tenant completing the fitting out, the works, if extensive, may qualify as improvements. This can cause problems for the valuer at the end of the lease (or at the review date(s)); there could well be no evidence of the rental value of units let in shell form (*GREA Real Property Investments Ltd* v *Williams* [1979]).

In recent years the interpretation of leases has been a major source of litigation. It is essential that the valuer should be aware of current court decisions, particularly in respect of repair, user, alienation, improvements, length of assumed term, and rent reviews.

# Capital value

The rate of interest at which the rents and rental values of retail units are to be capitalised can vary considerably over a relatively short period and will depend on the demand for retail investments at the time of the valuation. Where a property is let, the demand will be primarily from the investment market and the interest rate will be derived from experience and from the analysis of transactions involving similar properties. In the case of vacant units there may be competition from potential owner occupiers and the valuer will need to identify situations where the investment or the occupational market is in the ascendancy in any particular case.

The primary factors which decide the yields likely to be acceptable to investors are the location, age, and condition of the property, the calibre of the tenant, the tenure, and the terms of the lease, and the likely future rate of growth of rental values. The lowest yields occur when there is an optimum combination of these factors.

Location is important. The majority of investors favour retail properties situated in centres where there is an established and stable pattern of trading, and where a majority of the traders are well known retail companies. As location deteriorates, yields rise and different classes of investors may be involved. Traditionally, the highest rents and lowest yields, and consequently the highest capital values, are to be found in the cities and the larger towns. Retail warehouse and food store yields move on a more national basis, and do not necessarily reflect the strength or otherwise of the nearest town centre.

Maximum security of income is usually associated with the financial standing of the tenant. Units let to major retailers with large numbers of branches provide the highest level of demand and the lowest yields. Since the larger multiple retailers have branches in centres of widely differing size and importance, the particular combination has to be taken into account in assessing value.

A feature of the retail property market is the considerable variation in the type and age of the properties concerned. They can range from units in newly erected shopping centres to converted period buildings, and yields may have to be adjusted accordingly. Although the repairing liability imposed on the tenant by the lease may make a lack of routine repairs of little concern to the landlord, the presence of more serious defects can adversely affect the likely yield.

Investors pay the highest prices for properties where the outgoings and liabilities of the landlord are at a minimum and where the rental income is at a maximum. The lowest yields are obtained from properties let for optimum terms (ideally 10 to 25 years for most retail properties if they can be found with leases of that length) at full open market rents and on modern leases which provide for rent reviews at frequent intervals and for the tenant to be responsible for all repairs and outgoings.

© CJJ Osmond 2008

# Taxation

Note: The following studies have been prepared in accordance with legislation in force at the date of the assumed transaction. Taxation law is subject to annual change. Readers should therefore check for any amendments in relevant legislation before using these studies for current transactions.

The following abbreviations are used in this chapter:

| | |
|---|---|
| CGT | capital gains tax |
| FA | Finance Act, and unless stated otherwise the Finance Act 1998 |
| HMRC | Her Majesty's Revenue and Customs |
| ICT Act | Income and Corporation Taxes Act 1988 |
| IHT Act | Inheritance Tax Act 1984 |
| ITTOI Act | Income Tax (Trading and Other Income) Act 2005 |
| SDLT Act | Stamp Duty Land Tax Finance Act 2003 |
| TCG Act | Taxation of Chargeable Gains Act 1992 |
| VAT Act | Value Added Tax Act 1994 |

This chapter deals with the computation of tax on a sum or gain of a capital nature arising out of land. Such an amount will be treated for tax purposes as either capital or income and taxed in one or more of the following ways:

1. as a chargeable gain under CGT or, in the case of a company, corporation tax on the chargeable gain
2. as the profit or gain from a trade or adventure in the nature of trade and taxed as trading income
3. as a gain of a capital nature arising out of the disposal or development of land and taxed as trading income
4. as a premium on a short lease and taxed as property income
5. as the value transferred under a chargeable capital transfer made on or within seven years of death and subject to inheritance tax
6. on the acquisition of property within the UK and subject to stamp duty land tax.

The computation of the amount subject to tax involves both calculation and valuation. In some cases the taxable amount is found by calculation only, eg the profit from a trade in land; in others, the taxable amount can be found only by a combined process of calculation and valuation, for example the computation of the chargeable gain under CGT. In other cases the discounted value of a rent, or the amount of chargeable consideration in a land transaction. This chapter is concerned with the group of computations which require valuation and deals with CGT, stamp duty land tax and inheritance tax. The object is to explain the valuations and computations necessary to produce the amount of gain or value assessable to tax. Calculations showing the amount of tax payable have been omitted.

## Open market value

The statutory provisions governing valuations for taxation are summarised under each of the taxes, but they have a common base, market value. This is defined in similar terms for each of the taxes and has received judicial interpretation which may be generally applied. Following agreement between HMRC and the RICS (and others) in 2000 it, the basis of valuation, is now included in the RICS *Valuation Standards* [the Red Book] as Guidance Note 3 [UK GN 3]. The basis of valuation includes certain assumptions, these assumptions seek to prevent the minimisation of value, therefore, can lead to significant variation in the levels of value achieved as opposed to using a conventional open market value approach. The transaction, or sale, which is deemed to have occurred at the statutory point in time, which varies from tax to tax, is a hypothetical one, between a hypothetical vendor and a hypothetical purchaser. In this respect reality is not allowed to defeat the statutory purpose — *Lynall* v *IRC* (1972). It is summarised below.

A sale in the open market assumes a hypothetical seller who is a free agent, not anxious to sell without reserve, but one who would sell only at the best possible price that is obtainable. It assumes a hypothetical purchaser with knowledge of the condition and situation of the property and all surrounding circumstances. The market is open and includes every possible purchaser, including a special purchaser who would bid more because the property has a special value to him. A special purchaser, who must be real, solvent and known to the market, *Walton's Executors* v *IRC* (1996) CA, would not necessarily need to make only one bid more than the general price to secure the property but may be forced to bid higher by speculators hoping to buy the property and sell it on to him at a higher price. It is the existence of the special purchaser and his general impact on the market's perception of value that is the key; it is not assumed that the special purchaser himself secures the property, but another wishing to sell on to him. The price which a property is expected to fetch in the open market is found by reference to the expectations of properly qualified persons who have fully informed themselves of all relevant information about the property, its capabilities, the demand for it and the likely purchasers *IRC* v *Clay* [1914]; *Glass* v *IRC* [1915]; *Lynall* v *IRC* [1972].

The open market is not a purely hypothetical market, exempt from restrictions imposed by law, but the actual market and any restrictions on either seller or purchaser may be taken into account, *Priestman Collieries* v *Northern District Valuation Board* [1950]. However, as in the case of an agricultural tenancy, reality, contract or statutory bar cannot defeat the intentions of the taxing statute, *Lynall* v *IRC* (1972). The property must be assumed to be capable of sale, even if it is subject to restrictions on sale or actually un-assignable. However, the purchaser must take the property subject to any such restrictions and adjusts his price accordingly, *IRC* v *Crossman* [1937]; *Baird dec'd* v *IRC* [1991]. The sale assumes a hypothetical seller and buyer of the property, but the valuation must reflect any actual third parties with interests in the property, for example on the valuation of a share in a joint tenancy the intentions of the actual landlord and the other actual joint tenant are relevant, *Walton dec'd* v *IRC* [1996].

Market value must be assessed on the assumption that the property is sold in its actual state of repair at the date of valuation, but that the vendor would arrange for the property to be sold in the most advantageous manner without incurring unreasonable expense of time and effort in making the arrangements. In the case of a landed estate this may involve dividing the property into easily established lots for sale if this would produce a higher price than selling the estate as a single unit, *Duke of Buccleuch* v *IRC* [1967]; *Earl of Ellesmere* v *IRC* [1918]. If two or more property interests are being valued for the same taxable event, it may be necessary to assume that separate interests are sold together if that would produce a higher price, for example where the deceased owned freehold farmland and a share in the partnership owning a farming tenancy over that land, *IRC* v *Gray (Exor of Lady Fox dec'd)* [1994]. Alternatively, where a single asset only is the subject of valuation, it cannot be assumed to be aggregated with other assets owned by the taxpayer, for instance if freehold property is let to a company controlled by the freeholder, the interest is to be valued as being subject to that tenancy and not as if it were an unencumbered freehold, *Henderson* v *Karmel's Executors* [1984].

The above illustrates a significant difference between valuation for Inheritance Tax purposes and valuation for CGT purposes. Although in *Gray* and *Karmel* the facts are similar, for the purposes of IHT on death it may be assumed that all the vendor's assets are on the market at one and the same time, and therefore may be aggregated. For CGT purposes it is assumed that only the asset subject to the valuation requirement, which in *Karmel* was a rebasing to 31 March 1982, is on the market. So aggregation with other assets, which may be owned by the real vendor, but cannot be presumed to be owned by the hypothetical vendor, may be included for IHT but not CGT. No allowance is given for hypothetical costs of sale which might be incurred by the hypothetical vendor: *Duke of Buccleuch* v *IRC*, above.

A valuation should take into account any inherent possibilities or prospects attaching to the property at the date of valuation, for example, hope value for future development or the possibility that a lease might be surrendered, *Raja Vyricherla Narayana Gajapatiraju* v *Revenue Divisional Officer, Vizagapatam* [1939]. An apparent method for ascertaining this type of value has been adopted by the Lands Tribunal in *Kirkwood* v *Boland* (2002) LT, where planning consent had not been achieved, but a probability or possibility of it being achieved existed. The approach is effectively to value the property with an assumed consent, then adjust for the costs of achieving the assumed consent, then discount as appropriate, for time and risk, to reach market value. Where the valuation is actually made after the valuation date, however, one may not take account of events which occurred after that date and which showed how those possibilities or prospects in fact turned out; for example, the subsequent grant of planning permission or the surrender of the lease. The valuation must reflect those events as possibilities and not as the certainties they have become by the time the valuation is made, *IRC* v *Marr's Trustees* [1906]; *Gaze* v *Holden* [1983]. However, where there are properties on offer in the market, which are similar to the subject property but have not yet been sold, a valuer may have regard to this secondary evidence, and then cite the subsequently achieved value as supporting his view reached at the appropriate time.

For property held jointly, either as tenants in common or as beneficial joint tenants, where the object of the trust for sale is to enable the parties to have the proceeds of sale, the normal rule for valuing one of the joint interests has been to allow a reduction in value of 10%, *Cust* v *CIR* [1917]. Where the trust, under which shares in property are held, was created for the purpose of continuing occupation by the joint owners, the court would be unlikely to order a sale to benefit one of them. In that case the discount should be 15%, *Wight* v *IRC* [1982]. In *Charkham* v *IRC* [1997] (unreported), the Lands Tribunal held that for undivided minority shares in investment property the hypothetical purchaser would discount the open market vacant possession value for the uncertainty of whether the court would exercise its discretion to order a sale under a section 30 application under the Law of Property Act

1925) and for the uncertainty regarding the liability for the costs of the action, rather than valuing or a capitalised yield basis. The same discount should apply to different sized minority holdings in the same property, and that the discount to be applied should take into account the particular factors affecting each property at the valuation date, with the result that for one property the discount was determined at 15%, and for others 20% and 22.5%.

Where the Trusts of Land and Appointment of Trustees Act 1996 applies, this may affect the discount applicable. A beneficiary with an interest in possession who is also in occupation may have a statutory right to continued occupation, *Arkwright* v *IRC* [2004]; and a minority holder may not be able to enforce sale.

## *Evidence of value — valuation methodology*

In taxation valuation cases evidence of value is drawn from transactions that occur in the actual market. As a matter of fact the best primary evidence is actual transactions or market evidence although in the case of historic valuations this poses it own problems. Evidence of value can be obtained from many sources; the Valuation Office Agency (VOA), which is an agency of HMRC, has a database of evidence collected from a range of confidential sources, and sometimes uses evidence of valuations agreed or determined in other cases. In *Newman* v *Hatt* [2001] the Lands Tribunal Member reviewed the usability of evidence obtained from confidential sources. From the private sector perspective from 1 April 2000, any person is entitled, subject to certain conditions and on payment of a fee, to inspect and make copies of Land Registry entries, including the price paid on the last change of proprietorship, Land Registration Act 1998, section 1(1) and the Land Registration (NO3) Rules 1999 The VOA guidance manuals state that while settlements in comparable cases constitute secondary evidence, they can be used where there is insufficient primary evidence for open market transactions.

Settlements in tax cases are usually made in the context of a particular taxpayer's affairs, and are often not a true reflection of market value. There are sometimes disputes as to the use or applicability of hindsight evidence in making historic valuations. In a large number of taxation cases a valuer is required to prepare a valuation after the statutory date. In an IHT case this may be relatively shortly after the date of death, but in a CGT case this may be an ascertainment of value as at 31 March 1982. Generally speaking the valuers make their valuation based on that information which would have been available to them as at the date in question. However, where there is evidence, albeit of a secondary nature, of properties being offered for sale in the market as at the statutory date and where the subsequent transaction, say within one or two months after the statutory date bears out the validity of the asking price, then a valuer may be entitled to take the view that the subsequent values obtained bore out or supported the opinion formed on or at the statutory date. Scaling back subsequently achieved prices or attempting to inflate prices achieved forward in time, based on the use of indices, is to be avoided as it is for valuers to use their skill and judgment based on the available evidence to arrive at an opinion of value at the statutory point in time.

Methods of valuation are not pre-determined by statute or by case law, while transactional evidence may reflect the application of a variety of valuation methods, in practice this does not affect the comparability of the price realised in every case. In practical terms, property assets are valued or appraised by whichever method is most appropriate for that type of property.

Special classes and categories of assets will be appraised in different ways, using different techniques, because of the way in which the market values. A presumption of *Lynall* v *IRC* [1972] was that the vendor — when advertising the property — makes such information available to the purchasers of that asset as they would expect to receive, or be made available, in normal market transactions. It should be borne in mind therefore, in this area of valuation there is no presumption of a restricted or closed sale between connected persons. There is also the presumption that information would be made available to the potential purchaser that might not be made available to the public at large. For example, where trading information which involved throughput or turnover figures, which might be considered commercially sensitive, were germane to the quantification of the valuation, while these might normally only be made available at a later stage of the sales process, they are assumed for the purposes of valuation for taxation purposes to have been generally made available to the market as a whole.

## Apportionments

In certain circumstances apportionments are required for taxation purposes. This is particularly common for Stamp Duty Land Tax (SDLT) purposes and for the ascertainment of taxable value in cases involving principal private residence relief for CGT purposes. However in many areas of property valuation for taxation purposes it is necessary to apportion value. Most taxing statutes refer to a just and reasonable apportionment, this is necessarily subjective and each case will be considered on its own merits, as stated in HMRC statement of practice, SP 1/04. In relation to SDLT, it is suggested therein that an apportionment might be on the basis of the percentage areas quoted in a planning application, where appropriate, or alternatively floor space relating to respective uses.

There is a general requirement that apportionment for taxation purposes should be on a just and reasonable basis. It should be remembered that this has equal application both to the taxpayer and HMRC. There is no particular method of apportionment laid down in statute; however, the objective is to arrive at the contribution which each part makes to the sum to be apportioned, whether that sum is a reflection of open market value or actual sale or consideration in an arm's length transaction. Apportionment by area is considered to be only appropriate where a value is spread evenly throughout the land as was stated *obiter* in *Salts* v *Battersby* [1910]. This case dealt with a method of apportioned rent, and, in that circumstance, Darling J stated that the correct approach was to be value, and not area based.

> It seems to be clear from the authorities that what you have to regard is not the bare acreage of the severed portions of the land in demise, but their relative value. You find the same principal running through them all. The county court judge was of the opinion that the proper way to apportion rent was to have regarded the yardage and the yardage alone. On that count I think he was wrong. If it can be shown that land is of equal value throughout, no doubt the apportionment must be on the basis of yardage. But yardage cannot be a sufficient test of the relative value by itself; and here, so far from there being evidence that there was an equal value throughout, evidence was the other way.

Apportionment where one values the constituent parts is generally known as rateable apportionment and generally this is used in some form of formula approach which can ascertain the individual value of any of the component parts of the transaction. In the case of *Bostock* v *Topham* [1997], in upholding the special commissioners decision the court suggested that the approach adopted did no more than apply the principal of the value based rateable apportionment where there is a part disposal, and was appropriate to the circumstances of the case. In all apportionments it is necessary to consider what it is that the apportionment seeks to deliver. Generally speaking for taxation purposes the reason for the apportionment will be to identify and in some cases isolate, the amount of taxable consideration or value. Therefore it is necessary to ensure that the value being ascertained is the value required for the statutory purpose.

## *UK Guidance Note 3*

Currently Guidance Note 3 (UKGN3) of RICS Red Book contains the basis of valuation to be adopted for taxation purposes as set out above. In general, HMRC expect to receive properly worked through valuations, carried out on the correct basis, by competent valuers, in much the same way as other valuations are carried out in accordance with the directions and recommendations of the manual.

# Capital Gains Tax

The Finance Act 1965 introduced capital gains CGT in respect of chargeable gains accruing to a person on the disposal or assumed disposal of assets after 6 April 1965. The statute law on CGT is now contained in the Taxation of Chargeable Gains Act 1992 (TCG Act 1992) as amended by subsequent Finance Acts. CGT is payable by individuals and by trustees. Chargeable gains generated by a partnership, less allowable losses, are apportioned to the partners and assessed on them as individuals. For an individual, chargeable gains less allowable losses in excess of the annual exemption limit are subject to the lower rate of income tax and the higher rate where applicable. Thus, chargeable gains are treated as the top slice of the taxpayer's income. Trusts are taxable at the special rate applicable to trusts. Companies do not pay CGT, but a company's chargeable gains (calculated as for CGT) are included in the profits and charged to corporation tax (TCG Act 1992 section 1(2) and the Income and Corporation Taxes Act 1988 section 6(3) (ICT Act 1988)).

A gain of a capital nature arising out of the disposal of land, which is not specifically exempt, will not automatically suffer CGT. Two hurdles must first be overcome before it does so. First, the disposal must not be a trading disposal or an adventure in the nature of trade. If it is, then it will not be subject to CGT, but the profit or gain will be taxed as trading income. Second, the disposal must not fall within the anti-avoidance provisions of section 776 of the ICT Act 1988. This is a widely drawn section designed to tax as trading income a gain of a capital nature arising out of the disposal of land which was held as stock, or was acquired or developed with the object of realising a gain.

CGT applies to the disposal or assumed disposal of land. The term disposal is not defined, but includes a sale, assignment, gift, and grant of a lease at a premium, grant of an option, receipt of a capital sum, settlement of land on trustees, and the demolition or destruction of a building. The conveyance or transfer of an asset by way of security, eg a mortgage, is not a disposal for the purposes of CGT.

In general terms up to 5 April 1998, a CGT computation comprised the deduction from the consideration for the disposal of certain expenditure, such as the cost of acquisition, improvements and costs of disposal, and then the further deduction of an allowance for inflation by reference to the Retail Prices Index (RPI) (indexation) since 31 March 1982 or the date of acquisition if later, subject to variations proposed in 2008 which are dealt with below. Where a property was owned on 6 April 1965, only the gain after that date is taxable. Where a property was owned on 31 March 1982, the market value at that date can be substituted for the earlier acquisition cost to reduce the gain. In practice, therefore, the gain on the disposal of a property acquired before or after March 1982 will usually be restricted to the real gain since that date or the later date of acquisition. Study 2 which follows, however, shows the alternative methods of calculating the chargeable gain, although in practice it will usually be obvious that re-basing the gain to the March 1982 value will produce the lower gain without the need for further calculations. Therefore the subsequent studies use re-basing of the gain to the March 1982 value.

The Finance Act 1998 changed the CGT regime for disposals by individuals and trustees. The CGT regime for disposals realised by companies is unaffected by the 1998 Act and so the remarks in the rest of this paragraph do not apply to them. For the disposal after 5 April 1998 of an asset acquired before the 1 April 1998, the allowance for inflation only goes up to April 1998, although for individuals and trustees this relief is expected to cease after 5 April 2008. For assets acquired from 1 April 1998 onwards no allowance is given for inflation. Instead, for disposals after 5 April 1998 and up to 5 April 2008, the chargeable gain may be reduced by taper relief. In appropriate circumstances this reduces the amount on which tax is charged in proportion to the number of whole years of ownership. Qualifying business assets benefit from more favourable taper relief than other assets up to 5 April 2008. The notes to Study 1 give an example of the effect of taper relief on a chargeable gain and to the impact of the 2008 changes.

The rules in relation to CGT are expected to be subject to significant change for disposals from 6 April 2008 following the announcement in the pre-budget report of October 2007. As a generality for individuals and trustees the rate of CGT is proposed to reduce to 18%, which is a major change to the regime up to 5 April 2008 where gains were taxed at the individual's marginal rate of income tax. Other major changes announced affecting individuals and trustees are the withdrawal of taper relief, the withdrawal of the indexation allowance, and the compulsory use of the valuation at 31 March 1982 for assets held at that date. The measures first introduced in the pre-budget report of October 2007 have already been subject to significant variation with the announcement in January 2008 of a relief on disposal of a business commonly referred to as entrepreneur's relief. This relief will be available in respect of gains made on disposal of all or parts of a business, or gains made on disposal of assets following the cessation of a business, by certain individuals who are involved in running the business. There are two basic rules, first, the qualifying period for holding the assets will generally be only one year, and, second, the relief applies to the first £1,000,000 gains that qualify for relief. This relief will be allowed on a cumulative lifetime basis for the taxpayer. Entrepreneur's relief reduces the gains liable to CGT by four ninths, resulting in an effective 10% tax rate. The legislation is intended to achieve the effective 10% rate which was available on the disposal of certain business assets after a two-year qualification period under the taper relief regime applying up to 5 April 2008.

Companies are liable to corporation tax in respect of their chargeable gains, and it is expected that they will be unaffected by any of the CGT changes announced in the pre-budget report of October 2007 above. Accordingly companies will continue to benefit from indexation of the base cost of their assets, and their chargeable gains will be assessed at their marginal rate of corporation tax.

A CGT computation often requires the incorporation of values, depending on the circumstances of ownership and disposal. Various valuations may be required. Those most usually needed are market value at the date of disposal, eg where there is a gift or disposal otherwise than by way of a bargain at arm's length, Studies 2 and 6. Examples of market value at 31 March 1982 for the purposes of re-basing the gain to that date or calculating the indexation allowance may be found in Studies 2 and 4. Other valuations that may be required include: the value of the interest retained following a part disposal (for incorporation in the part disposal fraction), Study 3; the capital value of a rent-charge or income (where the consideration for the disposal comprised or included such an annual sum), Study 5; and the notional premium for the gratuitous variation or waiver of lease terms.

The statutory definition of market value for CGT purposes is contained in section 272 TCG Act 1992. It is the price which the property might reasonably be expected to fetch on a sale in the open market, no reduction being made for the assumption that the whole of the property holding has been placed on the market at one and the same time. Where a valuation is made on a disposal, the date of that disposal is the date of the contract and not the conveyance or, in the case of a conditional contract, the date when the condition is satisfied, section 28 TCG Act 1992.

Where a property is subject to any interest or right by way of security, it shall be assumed, both on acquisition and disposal, that the property is not subject to such security, section 26(3) TCG Act 1992. In certain circumstances, where a person making a disposal retains a contingent liability, no allowance for this is made in the valuation. If the liability subsequently becomes enforceable then a retrospective adjustment is made, section 49 TCG Act 1992.

Any appeal concerning a dispute as to the value of any land in the United Kingdom for the purposes of CGT or corporation tax on a chargeable gain is to be heard by the Lands Tribunal under section 46D Taxes Management Act 1970, although appeals regarding apportionments may be heard before the Special or General Commissioners.

There is a procedure for a taxpayer to agree a valuation for CGT purposes with HMRC after a transaction has occurred and before the time-limit for submission of the relative tax return, IR Capital Gains Manual paragraphs 16600 *et seq*. This is so that the taxpayer can file the self-assessment tax return incorporating agreed figures where necessary. The procedure is to complete and submit form CG34 to HMRC with details of the disposal and the context in which the valuation is required. Details to be submitted include the CGT computation showing the valuation used and a copy of any professional valuer's report supporting the valuation. The HMRC office will refer any valuation of land, unquoted shares, or goodwill to the relevant specialist valuation office, normally via the Land Portfolio Valuation Unit based in Coventry. HMRC consider the valuation and either agree it or enter into negotiations with a view to agreeing it. They require a minimum 56 days prior to the filing deadline to review a valuation, and they do not guarantee to do this by the filing deadline. Accordingly, it is better to submit the information as early as practicable following the transaction. The procedure is not available to consider pre-transaction valuations.

## Study 1

This study sets out the basic framework of a CGT computation where the property disposed of was acquired after 31 March 1982. The notes also cover taper relief, and rollover relief on sale and on compulsory purchase.

In March 1984 the freehold interest in a shop was bought for £200,000 by Andrews Ltd. The incidental costs of purchase were £7,500. In June 1985 £10,000 was spent on improvements and the company sold the property in December 2007 for £800,000. The incidental costs of sale were £12,500.

All references are to the TCG Act 1992, the approach set out applies to corporate bodies, both post the FA 1998 amendments and post the 5 April 2008 amendments.

| *CGT computation for Andrews Ltd* | £ | £ |
|---|---:|---:|
| Consideration for disposal | | 800,000 |
| *Less*: Allowable expenditure: | | |
| Consideration for acquisition plus incidental costs (section 38(1)(a)) | 207,500 | |
| Enhancement expenditure (section 38(1)(b)) | 10,000 | |
| Relevant allowable expenditure | 217,500 | |
| Incidental costs of disposal (section 38(1)(c)) | 12,500 | 230,000 |
| Unindexed gain (section 53(2)(a)) | | 570,000 |

*Less*: Indexation allowance (sections 53, 54, and 55(2)) by reference to relevant allowable expenditure only:

Consideration for acquisition plus incidental costs £207,500 multiplied by

$$\frac{RD - RI}{RI}$$

where

RD = RPI for month in which disposal occurs
December 2007 = 210.9

RI = RPI for month of acquisition
March 1984 = 87.48

$$\frac{RD - RI}{RI} = \frac{210.90 - 87.48}{87.48} = 1.411$$

£207,500 × 1.411         292,782

Enhancement expenditure £10,000 multiplied by

$$\frac{RD - RI}{RI}$$

where

RD = RPI for month in which
disposal occurs = 210.9

RI = RPI for month in which
expenditure incurred (June 1985) = 95.41

$$\frac{RD - RI}{RI} = \frac{210.9 - 95.41}{95.41} = 1.210$$

£10,000 × 1.210       12,100     304,883
Chargeable gain       £265,117

*FA 1998 changes applicable to individuals and trustees*
If the property had been owned by an individual, Mr Bolland, and the above disposal had occurred in December 2007, and the property qualified as a business asset because it was used by Mr Bolland for the purposes of his trade throughout his period of ownership, the CGT computation would be modified as follows:

Unindexed gain (as above) for Mr Bolland       570,000

*Less*: Indexation allowance, restricted to the period to April 1998
(section 122 FA 1998):

Consideration for acquisition plus incidental costs £207,500 multiplied by

$$\frac{RD - RI}{RI}$$

where

| | | | |
|---|---|---|---|
| RD | = | RPI for April 1998 = | 162.60 |
| RI | = | RPI for March 1984 = | 87.48 |

$$\frac{RD - RI}{RI} = \frac{162.60 - 87.48}{87.48} = \quad 0.859$$

£207,500 × 0.859        178,243

Enhancement expenditure £10,000 multiplied by

$$\frac{RD - RI}{RI}$$

where

| | | | |
|---|---|---|---|
| RD | = | RPI for April 1998 = | 162.60 |
| RI | = | RPI for month in which expenditure occurred = | 95.41 |

$$\frac{RD - RI}{RI} = \frac{162.60 - 95.41}{95.41} = \quad 0.704$$

| | | |
|---|---|---|
| £10,000 × 0.704 | 7,040 | 185,283 |
| Chargeable gain | | 384,717 |

*Tapered at*: (9 qualifying holding years), 75% relief    .
See below regarding taper relief                              0.25
Chargeable gain                                          £96,179

If Mr Bolland had his annual CGT exemption for 2007/08 of £9,200 otherwise unused and so available to off set against the above chargeable gain, and he was a higher rate income tax payer so that the rate of tax chargeable on the gain was 40%, the CGT payable would be £34,792.

*FA 2008 Changes*
If the above disposals occurred in May 2008, apart from any change in the available indexation allowance due to further expiry of time, the calculation for Andrews Limited, above would remain the same.

However for Mr Bolland, as an individual, the tax calculation is expected to change significantly. The calculation to the point of the unindexed gain remains the same.

| | | |
|---|---|---|
| Unindexed gain | 570,000 | |
| *Less* Entrepreneur's relief (4/9th) | 253,333 | |
| *Less* Annual Exemption (2008/2009 assumed) | 9,200 | |
| | | 307,467 |
| CGT @ 18% | | 0.18 |
| Tax payable | | £55,344 |

This compares to the liability of £34,792 under the CGT regime applicable in 2007/08, see above.

# Taper relief

To calculate the amount of taper relief available on the disposal of a business asset, the period from 5 April 1998, or the date of acquisition if later, is measured to establish the number of whole years of qualifying holding period. For the disposal of business assets after 5 April 2002 taper is applied at the rate of 50% after one whole qualifying holding year. After two years the maximum taper of 75% is applied so that only 25% of the indexed gain is chargeable. Section 2A TCG Act 1992 subsections (5) and (8), as modified by section 66 of the Finance Act 2000 and section 46 of the Finance Act 2002.

In Study 1 the business asset was owned by Mr Bolland on 5 April 1998 and so qualifies for at least two years of qualifying holding period for taper relief purposes.

Had the asset not been used for business purposes by Mr Bolland, but instead had been an investment, for example being let as a dwelling to a third party, non-business assets taper relief would have applied. As Mr Bolland had owned the property on 16 March 1998, he would have a qualifying holding period of 10 whole years by the date of disposal in December 2007, giving him taper relief of 40% and resulting in 60% of the indexed gain being chargeable (see below).

For non-business assets, taper is applied at the rate of 5% from the third whole qualifying holding year after 5 April 1998, increasing at 5% for each additional whole year of the holding period up to a maximum of 10 years. Non-business assets held on 16 March 1998 are deemed to have a minimum of one qualifying holding year for taper relief, section 2A TCG Act 1992 subsections (5) and (8) as amended. Accordingly, for non-business assets, the earliest date a disposal may benefit from taper relief is 6 April 2000. The maximum taper relief for non-business assets is achieved after 10 whole years of qualifying holding period at which point the taper will be 40%, so that only 60% of the indexed gain will then be chargeable.

For individual or trustee investors in properties not used for business purposes the impact of the new rules are beneficial for those who have acquired assets since 1998. Prior to 6 April 2008 a taxpayer needed to own the asset for 10 whole years to reduce the effective maximum rate of tax to 24%, now the effective rate of tax is 18%. However the new regime may be detrimental to those who have owned assets since before April 1998 as the loss of indexation relief may more than outweigh the lower tax rate applicable in the new regime, depending on the relative value of the base cost to the disposal value and the length of time the asset has been owned before April 1998.

For individual or trustee business asset owners the proposed new entrepreneur's relief applicable from 6 April 2008 is allowed on a cumulative lifetime basis. Therefore the relief is a personal one rather than a statutory deduction for each transaction.

# Roll-over relief for assets used in a trade

If the property disposed of has been used exclusively for the purposes of a trade or profession, a gain on disposal may be rolled-over if the taxpayer reinvests the proceeds of sale in other qualifying capital assets for the purposes of a trade, provided these are acquired within the period running from one year before to three years after the disposal. The effect of the roll-over is that the gain on the disposal is not charged to tax then, but instead the acquisition cost of the replacement asset is reduced for CGT purposes by the gain rolled-over. The result is that the taxation of the gain is deferred until the replacement asset is itself disposed of, although it may be possible to roll-over again if the qualifying conditions are then met, sections 152 to 159. Qualifying furnished holiday letting properties can also benefit from this roll-over regime, section 241. Note that the gain rolled-over is not reduced by taper relief.

If the asset disposed of has not been used exclusively for the purposes of the trade throughout its period of ownership, either physically or in terms of time, apportionment will be required between the part so used and the remainder, with the roll-over provisions only applying to the part so used, section 152(7) and (8). If the proceeds are only partly reinvested within the period allowed, the part not reinvested in excess of the indexed base cost will be chargeable to CGT, subject to any taper relief available.

Properties used for qualifying furnished holiday letting qualify for roll-over relief: see TCG Act 1992, section 241 (30) and (3A), and ITTOI Act 2005 section 323 to 325 for the definition of qualifying furnished holiday letting.

## Roll-over relief and compulsory purchase

Roll-over relief is available when a gain arises on land being acquired by an authority exercising, or having, compulsory powers, provided the landowner took no steps to market the property, and he reinvests the consideration in other land, section 247. Offering to sell property to an acquiring authority is regarded as taking steps to sell — those affected have to be careful with this: the property must not be offered to the authority, but one may indicate a willingness to sell if asked to do so. The reinvestment time-limits and the provisions for taxing part of the gain if part only of the consideration is reinvested apply as explained above. This form of roll-over relief applies to investors as well as those carrying on a trade or profession, except that it excludes dwellings qualifying for principle private residence relief, section 248.

### Study 2

This study sets out the basic framework of a CGT computation where a property was acquired before 31 March 1982 and disposed of before 6 April 2008.

The freehold interest in a commercial building was bought in June 1967 for £8,000. The incidental costs of acquisition were £300. In June 1974 the purchaser spent £1,000 on improvements and in December 1979 legal costs of £750 were incurred in successfully defending the title to the property. The building was let on full repairing and insuring terms, and on 31 March 1982 the lease had three years unexpired at a rent of £1,500 pa exclusive. The rack-rental value was £2,500 pa.

In February 2007 the purchaser transferred the freehold to his son for a consideration of £120,000. The incidental costs of disposal were £600. At the date of disposal the property remained let on a full repairing and insuring lease which had twelve years unexpired, at a rent of £15,000 pa exclusive with five-yearly rent reviews: the next review was in four year's time. The rack-rental value was £16,000 pa.

All references are to the TCG Act 1992.

This transaction is between connected persons, section 286(2): it is, therefore, treated as otherwise than by way of a bargain at arm's length, and market value must be substituted for the actual consideration, sections 17 and 18.

The following example reflects the situation prior to the implementation of the new CGT regime from 6 April 2008 announced in the pre-budget report of October 2007.

| (i)   Market value at disposal (section 272) | £ pa | £ |
|---|---|---|
| Rent on lease | 15,000 | |
| YP 4 years at 7% | 3.4 | 51,000 |
| Rack-rental value | 16,000 | |
| YP perp deferred 4 years at 7% | 10.9 | 174,400 |
| | | 225,400 |
| Market Value at date of disposal: February 2007, say | | £225,000 |

| (ii)   Market Value at 31 March 1982 (sections 35, 55 and 272) | £ pa | £ |
|---|---|---|
| Rent on lease | 1,500 | |
| YP 3 years at 7.5% | 2.6 | 3,900 |
| Rack-rental value | 2,500 | |
| YP perp deferred 3 years at 7.5% | 10.7 | 26,750 |
| | | £30,650 |
| Market Value at 31 March 1982, say | | £31,000 |

| *CGT calculation* | Cost | OMV March 1982 |
|---|---|---|
| | £ | £ |
| Deemed consideration for disposal (market value) | 225,000 | 225,000 |
| Less Incidental costs of disposal (section 38(1)(c)) (note 1) | 600 | 600 |
| | 224,400 | 224,400 |
| *Less*: cost: | | |
| Consideration for acquisition plus incidental costs (section 38(1)(a)) | 8,300 | |
| Enhancement expenditure (section 38(1)(b)) | 1,000 | |
| Legal costs defending title (section 38(1)(b)) | 750 | |
| Relevant allowable expenditure | 10,050 | |
| *Less*: OMV March 1982 | | 31,000 |
| Unindexed gain (section 53(2)(a)) | 214,350 | 193,400 |

*Less*: Indexation allowance (sections 53, 54, and 55(2)):
by reference to higher of relevant allowable expenditure
and OMV at March 1982 = OMV at March 1982 £31,000
multiplied by

$$\frac{RD - RI}{RI}$$

where

RD = retail price index (RPI) for month in which
disposal occurs or April 1998 if earlier = 162.60

RI = RPI for March 1982 = 79.44

| $\dfrac{RD - RI}{RI} = \dfrac{162.60 - 79.44}{79.44} =$ | 1.047 | 32,457 | 32,457 |
|---|---|---|---|
| Indexed gain | | 181,893 | 160,943 |
| Lower gain is used: | | | £160,943 |

If the taxpayer has already made an election for all his assets to be re-based to 31 March 1982 values, only the market value at March 1982 would be used (section 35(5) TCG Act 1992).

Following the proposed changes to CGT announced in the pre-budget report of October 2007, the option of choosing original base cost rather than the 31 March 1982 valuation, as shown above, is expected to be abolished, and using the valuation at 31 March 1982 will be compulsory. Also indexation allowance will no longer available to individuals or trustees. However, on the assumption that the premises qualify for the proposed entrepreneur's relief, the following calculation applies, where the disposal is post 5 April 2008 by an individual or by trustees:

|  | £ |
|---|---:|
| Market Value at date of disposal say August 2008 | 225,000 |
| *Less*: Incidental costs of disposal (section 38 (1)(c) | 600 |
|  | 224,400 |
| *Less*: valuation at 31 March 1982 (see above) | 31,000 |
|  | 193,400 |
| *Less*: Entrepreneur's relief at 4/9ths | 85,955 |
|  | 107,445 |
| Less annual exemption (2008/2009 assumed) | 9,200 |
| Taxable gain | 98,245 |
| CGT payable at 18% | £17,684 |

If the asset had not qualified as a business asset the entrepreneur's relief would not have been available, in which case the tax payable would be £33,156. The same result would arise if the asset would qualify for entrepreneur's relief but the taxpayer had already used his lifetime £1 million allowance of entrepreneur's relief on other disposals.

If the disposal had occurred before 6 April 2008 and it had qualified for business assets taper relief, the maximum tax payable would be £12,414:

|  | £ |
|---|---:|
| Indexed gain as above | 160,943 |
| *Less*: Taper relief at 75% | 120,707 |
|  | 40,236 |
| Less: annual exemption (2008 / 2009 assumed) | 9,200 |
| Taxable gain | 31,036 |
| CGT payable at maximum rate of 40% | £12,414 |

Note
(1)   The actual costs of disposal must be used (section 38(1)(c)) even though market value is substituted for the disposal proceeds as the vendor and the purchaser were connected parties.


## Study 3

This study illustrates the CGT computation where there is a part disposal of a freehold interest by the grant of a lease not exceeding 50 years at a premium. It also illustrates the relationship between CGT and income taxed as property income: Schedule 8, paragraph 5(1) TCG Act 1992 and section 34(1) ICT Act 1988 for corporation tax, and section 277 ITTOI Act 2005 for income tax.

George purchased the freehold interest in a commercial building in July 1986 for £65,000. His incidental costs of acquisition, including stamp duty, were £3,300. He subsequently let the whole property on full repairing and insuring terms at a rack rent until the tenant vacated in December 1996.

In January 2007 George granted a new full repairing and insuring lease for 15 years at a fixed rent of £13,250 pa exclusive and a premium of £80,000. The rack-rental value was £22,500 pa. His incidental costs of this part disposal were £2,500.

All references are to the TCG Act 1992 except where indicated otherwise.

*Market value of the freehold reversion after the part disposal (section 42(2)(b) and Schedule 8 paragraph 2(2))*

|  | £ pa | £ |
|---|---|---|
| Rent on lease | 13,250 | |
| YP 15 years at 15% See note 1 | 5.8 | 76,850 |
| Rack-rental value | 22,500 | |
| YP perp deferred 15 years at 7.5% | 4.5 | |
| | | 101,250 |
| | | £178,100 |
| Say | | £178,000 |

This amount is B in the part disposal fraction $\dfrac{A}{A + B}$ (see below).

*CGT computation*

*Apportionment of premium (Schedule 8 paragraph 5(1), and section 34 ICT Act 1988 or section 277 ITTOI Act 2005)*
The premium on a lease granted for not more than 50 years is subject to both CGT and property income tax. The part of the premium subject to CGT is calculated as shown below. The premium is apportioned, and from this are deducted the apportioned costs of acquisition adjusted for indexation:

|  | £ | £ |
|---|---|---|
| Total Premium | | 80,000 |
| *Less*: Part of premium taxed as property income: | | |
| Total premium | 80,000 | |
| *Deduct*: 1/50th for each complete year of the term other than the first: 14/50 × £80,000 | 22,400 | 57,600 |
| Portion of premium treated as consideration for CGT part disposal | | £22,400 |

*Chargeable gain calculation*
The grant of a lease at a premium is a part disposal of the landlord's interest (Schedule 8 paragraph 2(1)).

|  | £ | £ |
|---|---|---|
| Consideration for part disposal (premium, as above) | | 22,400 |
| *Less*: incidental costs of disposal (section 38(1)(b)) | | 2,500 |
| | | 19,900 |
| *Less*: allowable cost: | | |
| Cost of purchase | 65,000 | |
| *Plus*: Incidental costs of purchase | 3,300 | |
| | 68,300 | |

Apportioned to the part disposal by the fraction (section 42)

$$\frac{A}{A + B}$$

where

A = consideration for the part disposal (see note 2)
= £22,400 in numerator but
£80,000 in denominator (Schedule 8 paragraph 5(1))

B = market value of property
Not disposed of = £178,000

| | | | | |
|---|---|---|---|---|
| $\dfrac{A}{A + B} = \dfrac{22{,}400}{80{,}000 + 178{,}000} =$ | | 0.087 | $\underline{0.087}$ | |
| Apportioned allowable cost | | | 5,942 | $\underline{5{,}942}$ |
| | | | | 13,958 |

*Less*: Indexation allowance, (sections 53, 54 and 55(2))
Calculated by reference to allowable cost

$$\frac{RD - RI}{RI}$$

where

RD = retail price index (RPI) for month in which
disposal occurs or April 1998 if earlier = 162.60
RI = RPI for July 1986 = 97.52

| | | | |
|---|---|---|---|
| $\dfrac{RD - RI}{RI} = \dfrac{162.6 - 97.52}{97.52} =$ | | $\underline{0.667}$ | $\underline{3{,}963}$ |
| Chargeable gain | | | £9,995 |

The portion of the premium which is not taxed under CGT (£57,600) is taxed as property income. From 6 April 2008 the above calculation would apply to a corporate body; an individual or trustee would not receive indexation allowance and would be subject to the proposed new rules so that the chargeable gain would be £13,958.

Notes
(1) High capitalisation rate used to reflect the fact that a fixed rent for 15 years would be unattractive to a prospective purchaser.
(2) The part disposal fraction A ÷ (A + B) in section 42 is modified in the case of a part disposal of property by the grant of a lease not exceeding 50 years for a premium. In such a case, part of the premium is taxed as property income (section 34 of the ICT Act 1988 or section 277 of the ITTOI Act 2005). The numerator A is the premium less the amount taxable as income, whereas in the denominator "A" is the full premium received (Schedule 8 paragraph 5(1) TCG Act 1992).

## Study 4

This study illustrates the CGT liability arising on the disposal by assignment of a short lease for a premium.

In December 1977 Jack took an assignment of the head lease of a shop which had 45 years unexpired at a fixed rent of £100 pa exclusive. He paid the assignor £20,000 and incurred incidental costs of £1,000. Jack spent £15,000 on improvements to the property in March 1980. He sublet it on a full repairing and insuring lease for 25 years from 25 March 1982 at a rent of £9,000 pa exclusive, with rent reviews every five years.

On 26 March 2007 Jack disposed of his head lease for £250,000. His incidental costs were £7,500.

All references are to TCG Act 1992.

| (i)   Market Value at 31 March 1982 (sections 35, 55 and 272) | £ pa | £ |
|---|---|---|
| Rent receivable | 9,000 | |
| *Less*: rent payable | 100 | |
| Profit rent | 8,900 | |
| YP 40.75 years at 9% and 4% (taxed at 40%) | 9.4 | 83,660 |
| Market value, say | | £83,500 |

| *CGT computation* | | £ | £ |
|---|---|---|---|
| Consideration for disposal | | | 250,000 |
| *Less*: Allowable expenditure: (note 1) | | | |
| Calculated by reference to market value at 31 March 1982 (note 2) | | 83,500 | |

*Reduced by*: Amount written off
(Schedule 8 paragraph 1(3) and (4) and section 53(2)(b) and (3)):

$$£83,500 \times \frac{P(1) - P(3)}{P(1)}$$

where

P(1) = % derived from table for duration of lease at
deemed acquisition at 31 March 1982
40 years 9 months = 95.895

P(3) = % derived from table for duration of lease
at disposal 15 years 9 months = 63.491

Amount written off:

$$\frac{95.895 - 63.491}{95.895} = 0.338$$

£83,500 × 0.338 = £28,223

| Reduced amount = | | | |
|---|---|---|---|
| Market value on 31 March 1982 | | 83,500 | |
| *Less*: Adjusted allowable expenditure | | 28,223 | |
| | | 55,277 | |
| *Plus*: Incidental costs of disposal (section 38(1)(c)) | | 7,500 | 62,777 |
| Unindexed gain (section 53(2)(a)) = | | | 187,223 |

*Less*: Indexation allowance (sections 53, 54 and 55(2)):

Calculated by reference to relevant allowable expenditure (as above), £55,277    55,277

Multiplied by

$$\frac{RD - RI}{RI}$$

where

RD = RPI for month in which disposal occurs
or April 1998 if earlier =      162.6

RI = RPI for March 1982 =      79.44

$$\frac{RD - RI}{RI} = \frac{162.6 - 79.44}{79.44} =$$      1.047      $\frac{1,047}{57,875}$      57,875

Chargeable gain      £129,348

In the case of a post 5 April 2008 disposal the above example remains the same for corporate bodies, but for individual and trustees there would be no indexation relief. Therefore the chargeable gain would be £187,223, subject to a possible further deduction of 4/9ths for entrepreneur's relief if the asset so qualified.

Notes
(1)  A lease is treated as a wasting asset when the unexpired term does not exceed 50 years at the relevant time, ie at acquisition or disposal (Schedule 8 paragraph 1(1)). The allowable expenditure in respect of such a lease is written off over the duration on a reducing basis in accordance with the table in Schedule 8 paragraph 1.
(2)  Where the acquisition value (whether actual or deemed), or any enhancement expenditure, used in computing the unindexed gain is amended by any enactment (for example, the short lease curved line restriction of allowable expenditure in Schedule 8 paragraph 1), indexation is calculated on the amended amount (section 53(3)).
(3)  Again we have the variation of approach between a corporate entity and an individual, in the above example an individual would not receive indexation allowance.

## Study 5

This study illustrates the CGT computation where there is a part disposal of a leasehold interest by the grant of a sublease at a premium out of a head lease with less than 50 years unexpired.

In May 1988 Fred granted Henry Ltd a full repairing and insuring lease of a shop for 35 years at a fixed rent of £100 pa exclusive plus a premium of £40,000. The company's incidental costs were £1,250.

In January 2007 the company sublet the shop to Thomas on a full repairing and insuring lease for 10 years at a fixed rent of £250 pa exclusive plus a premium of £200,000. The company's incidental costs were £2,500.

The CGT liability for Henry Ltd is to be calculated.

All references are to the TCG Act 1992 except where indicated otherwise.

*Notional premium on sublease to Thomas if the rent under sublease (£250 pa) had been the same as the rent under the head lease (£100 pa) (Schedule 8 paragraph 4(2)(b)).*

|  | £ pa | £ |
|---|---|---|
| Premium under sublease |  | 200,000 |
|  |  |  |
| *Add*: additional premium if the rent under the sublease had been the same |  |  |
| as the rent under the head lease (ie the capital value of the profit rent): |  |  |
| Rent received from sublease | 250 |  |
| *Less*: Rent paid under head lease | 100 |  |
| Net Profit Rent | 150 |  |
| YP 10 years at 12% and 4% (taxed at 30%) | 4.2 | 630 |
|  |  | 200,630 |
| But say |  | £200,650 |

| *CGT computation* | £ | £ |
|---|---|---|
| Consideration for part disposal (premium) |  | 200,000 |

Less: part of premium taxed as rental income (Schedule 8 paragraph 5 (1) and section 34 ICT Act 1988):

| Premium | 200,000 |
|---|---|
| *Deduct*: 1/50th for each year of the term other than the first: 9/50 × £200,000 | 36,000 |
|  | 164,000 |

*Less*: allowance for premium paid on grant of head lease, being the amount chargeable to income tax on Fred for granting the head lease (section 37 ICT Act 1988): note 1

£40,000 − (34/50ths × £40,000) = £12,800

Multiplied by

$$\frac{\text{duration of sublease to Thomas at grant}}{\text{duration of head lease to H Ltd at grant}} = \frac{10}{35}$$

| Allowance 10/35 × £12,800 = | 3,657 |  |
|---|---|---|
| Part of premium chargeable as income | 160,343 | 160,343 |
| Part of premium treated as consideration for CGT part disposal |  | 39,657 |

*Less*: Allowable expenditure:

| Consideration for acquisition plus incidental costs (section 38(1)(a)) | 41,250 |
|---|---|

Apportioned to part disposal by fraction (Schedule 8 paragraph 4(1) and (2)(b)): note 2

$$\frac{P(1) - P(3)}{P(2)}$$

where

P(1) = % derived from table for duration of
head lease at grant of sublease
(16 years 4 months) =                                          64.901

P(2) = % derived from table for term of
head lease at grant of that lease
(35 years) =                                                      91.981

P(3) = % derived from table for duration of
head lease at termination of sublease
(6 years 4 months) =                                          32.601

$$\frac{P(1) - P(3)}{P(2)} = \frac{64.901 - 32.601}{91.981}$$                                    0.351

                                                                                                                14,479

*Less*: for reduction in premium on sublease due
to increased rent over headrest (Schedule 3
paragraph 4(2)(b)):

$$\frac{\text{Premium paid on sublease}}{\text{Notional Premium on sublease}} = \frac{200,000}{200,650} =$$                    0.997

Relevant allowable expenditure                                                        14,436

*Plus*: Incidental costs of part disposal (section 38(1)(c))                  2,500                16,936

Unindexed gain (section 53(2)(a))                                                                              22,721

*Less*: Indexation allowance (sections 53, 54, and 55(2)):

Calculated by reference to relevant allowable expenditure only            14,436
multiplied by

$$\frac{RD - RI}{RI}$$

see note 3 where

RD = RPI for month in which disposal occurs
— January 2007 =                                                    201.60

RI  = RPI for May 1988 =                                          106.20

$$\frac{RD - RI}{RI} = \frac{201.6 - 106.20}{106.20} =$$                          0.898                12,964

Chargeable gain                                                                                                  £9,757

As Henry Ltd is a company, the chargeable gain is subject to corporation tax, section 1(2) TCG Act 1992. Part of the premium (£160,343) is taxed as income in the company's corporation tax computation. Indexation Allowance also remains available post 5 April 2008.

Notes
(1)    Part of the premium paid by Henry Ltd to Fred (ie £3,657 out of £40,000) on the grant of the head lease is set-off against the portion of the premium received on the grant of the sublease by Henry Ltd to Thomas which is taxable as income in the hands of Henry Ltd (ie £164,000 out of £200,000), to leave the net amount chargeable to income on Henry Ltd at £160,343, section 37 ICT Act 1988.

For the period while Henry Ltd remained in occupation and used the shop for the purposes of its trade, it will have been able to claim an annual deduction when computing its corporation tax liability on its trading profits. As the headlease was granted for a 35 year term, the allowance is 1/35th of the amount chargeable to income tax on the landlord Fred (section 87 ICT Act 1988), which in this case is £12,800 – 35 = £366 pa. (The premium received by Fred on the grant of the 35 year headlease was £40,000. 34 – 50 was taxable on him as capital, ie £27,200, and the rest was taxable on him as income £12,800).

For a company carrying on a property rental business, rather than a trading business, there is a similar relief for a deduction in the company's rental income tax computation, section 37(4) ICT Act 1988.
(2)    The normal part disposal fraction (see Study 3) does not apply where the part disposal is the grant of a sublease out of a lease with less than 50 years unexpired, Schedule 8 paragraph 4(1).
(3)    Because Henry Ltd is a company it continues to benefit from indexation after April 1998, but it does not qualify for taper relief, sections 1(2) and 53(1A) TCG Act 1992.

# CGT and dwellings

Relief from CGT on the disposal of a taxpayer's private residence is provided under sections 222 to 226 TCG Act 1992, as amended by FA 1996, Schedule 20 paragraph 59 (2). Section 222(1) provides relief for a gain realised by an individual in so far as it is attributable to disposal of an interest in a dwelling house or part of a dwelling house which is or has been, or has at any time in the taxpayer's previous ownership been, his only or main residence, or land which a tax payer has for his own occupation and enjoyment with that residence as its gardens and grounds up to the permitted area.

Section 222 (3) and (4) of the TCG 1992 provides that the permitted area shall be 0.5 of a hectare (1.236 acres) or such larger area as is required for the reasonable enjoyment of the dwelling house as a residence, having regard to its size and character. This is an area, which is of increasing interest to HMRC, particularly following certain remarks made in the case of *Lloyds TSB Private Banking plc (personal representative of Antrobus, deceased)* v *Peter Twiddy* [2004]. This case, which is commonly known as *Antrobus (2)*, is dealt with in more detail in the IHT section of this chapter. However, during the case, evidence was given which identified a class of purchaser of large houses with additional land surrounding them under the term lifestyle purchasers. It was stated that whereas this class of purchaser might pay say £1,000,000 for a house with 1 acre of land, for the same house with two or three acres of land he might be prepared to pay a significantly higher sum, possibly as much as 50% more. This led HMRC to the view that, potentially, a significant amount of tax might be at stake in this type of case, and vendors who have not considered their affairs prior to disposing of their principal private residence are often surprised by the attitude of HMRC which comes to their notice one or two years later.

There has been significant litigation in this area over the years and it is now clear that a residence may include a flat or a house and in certain circumstances may also include a caravan in the grounds of a property under renovation or construction: see *Makins* v *Elson HMIT* (1997) and *Moore* v *Thompson* (1986). It does not however include a partly built building which is, as at date of disposal, uninhabitable. The issue of what is or what is not garden and grounds is a question of fact. The situation is judged as at the date of disposal of the residence itself, *Varty HMIT* v *Lynes* (1976). This can, for example, have the effect that a garden sold at the same time as, or before, the sale of the qualifying dwelling to which it is attached will qualify relief. But the same garden, if sold after the qualifying dwelling is sold, will not qualify for relief because it is not, at that time, occupied with the residence.

Generally land surrounding a qualifying dwelling and in the same ownership is considered the grounds of the residence unless it is used for another purpose. This will include land which has traditionally been part of the grounds of the residence, but where it has been used for another purpose or separated away from the main residence, perhaps to facilitate a sale, it will no longer be considered to form part of the garden and grounds. The land itself does not have to be contiguous to the subject property, as was held in the Special Commissioner's case of *Wakeling* v *Pearce* (1995). The question of whether or not a larger area is required for the reasonable enjoyment of the dwelling house is much more difficult, and it has been held that this is a question of what is objectively required. This is not the same as that which the individual might require, this concept was recently considered in the case of *Longson* v *Baker* (2001). This case approved the HMRC position, which is essentially that required is to be equated with necessary. The concept originally dates from the case *Re Newhill Compulsory Purchase Order 1937, Payne's Application* (1938). Some commentators believe the Longson case to be wrongly decided upon on the facts, as the *Newhill* case related to land in the same ownership as the garden and grounds of the main residence but not actually forming part of them. There are two concepts concerning the residence and its garden and grounds which it is necessary to understand. First, there is the concept of curtilage which is only used to determine the extent of the dwelling house. Curtilage, generally speaking, includes the residence itself and any buildings attached to it which are used in connection with it, and are not separately used for a non-qualifying purpose such as a business or trade. Therefore, buildings which are let will not be considered to form part of the dwelling itself. From an HMRC perspective, garden is taken to mean an enclosed piece of ground devoted to the cultivation of flowers, fruit or vegetables. This definition is obviously subject to modification where no defining enclosure exists. The concept of grounds is also taken to mean enclosed land surrounding or attached to a dwelling or other building serving chiefly for ornament or recreation. Yet again the concept of enclosure appears in the HMRC Manuals but may not be appropriate in all cases.

Adjustments fall to be made where the property was not occupied by the taxpayer during the entirety of his ownership and may have been used by, or let to, other persons. However, the qualifying dwelling will always be treated as if it were the relevant individual's only or main residence in the last 36 months of its ownership. Therefore a property which is vacated, perhaps during a protracted period of sale, by the taxpayer will still be considered his principal private residence if sold during this extended period. There are also concessions in ESC D49 concerning absence from the property during which building work or renovation takes place.

Generally speaking on the disposal of an interest in a property to a taxpayer's spouse or civil partner then the latter's period of ownership is taken to include the period of ownership of the taxpayer who made the disposal. Where an individual lives with a spouse or civil partner the relief is restricted to one residence. Where part of a qualifying property does not qualify for the exemption because it is used exclusively for the purposes of a trade, business, profession or vocation, then an apportionment will be required to identify the taxable part of the property.

Relief is not available if the acquisition of an interest in a qualifying dwelling was made wholly or partly for the purposes of realising a gain from the disposal of it. Similarly, relief is not available in relation to any gain which is attributable to any expenditure which was incurred wholly or partly for the purpose of realising a gain from a disposal, section 224(3) TCG Act 1992.

Where a qualifying dwelling is acquired to provide a home, it is unlikely that this will have been done in order to realise a gain from it. Just because the purchaser also hoped profit from subsequent sale is not enough in itself to remove the relief, *Jones* v *Wilcock* (1996). From a practitioner's point of view, it is useful to follow the steps set out in the VOA manuals, first to determine the permitted area and if this less than the whole area, the amount of relief it is potentially able to receive. The VOA's five basic steps are as follows:

- step 1, determine the entity of the dwelling house, ie, which building qualifies for relief under section 222(a)
- step 2, determine the extent of garden or grounds, ie, which land is occupied with the dwelling house and can be described as garden or grounds
- step 3, determine the size of the permitted area, ie, if the garden or grounds are in excess of 0.5 ha how much of the land is required for the reasonable enjoyment of the dwelling house as a residence
- step 4, determine the location of the permitted area ie which part of the garden or grounds would be the most suitable for occupation and enjoyment with the residence
- step 5, apportion the proceeds of the disposal and the acquisition cost between the part of the property qualifying for relief and the remainder.

Although these steps are set out from the perspective of HMRC, they form a useful guide in ascertaining what is and what is not required. Generally speaking, land which is put to a particular purpose, such as a swimming pool or tennis court, will be included within the permitted area. Valuations in connection with apportionments for this purpose can be complex. However, when making an apportionment, one must consider that the entirety of the land and buildings are in the ownership of the taxpayer, and that land outside the permitted area is not to be considered as being severed from, or a separate asset to, the main property. Therefore, it cannot be assumed that an area of land in excess of the permitted area will be isolated or land-locked, it must be assumed that access to it can be gained via the permitted area if required. This to a large extent defeats the concept of part of the land being able to exert a ransom price from the potential development of other parts, but this could only occur if the various portions of land were in separate ownerships.

It is common practice that when ascertaining the area required for the reasonable enjoyment of the dwelling, regard will be had to the size of the plots of land of comparable properties. At one point the VOA sought to argue that no considerations of value could entertain this concept; however, this no longer forms part of their instructions. In order to decide on the nature of comparable property, it is proper to have regard not only to its size and type but also to the value category into which such a property might reasonably fit. A useful measure can also be the relative size of the dwelling in relation to the overall size of the plot; and generally it is accepted that larger properties require larger areas of land for their reasonable enjoyment. Therefore it is common to see larger areas of land being accepted as reasonably required with properties in a non-urban environment rather than those in heavily developed areas.

# Inheritance Tax

The FA 1975 abolished estate duty and introduced a new tax, now known as inheritance tax (IHT). The legislation for IHT was consolidated in the IHT 1984, which has been amended by subsequent Finance Acts. References to legislation in this part of this chapter are to the IHT Act 1984 unless indicated otherwise.

Subject to certain exceptions, the tax applies to gifts and other gratuitous transfers of property, whether actual or deemed, made during a person's lifetime if they were made after 26 March 1974, and to the deemed transfer on death after 12 March 1975. IHT is based on the value transferred by a transfer of value. This is, broadly, a disposition whereby the transferor's estate is reduced in value, unless it is excluded because it is not intended to confer a gratuitous benefit and is an arm's length transaction between unconnected persons; or is a disposition such as might be expected to be so made. Thus a gift, a transfer at undervalue, a deliberate omission to exercise a right, or the creation of a settlement may be transfer of value. Permitting someone else the use of assets for less than full consideration may result in a transfer of value.

On death, a person is deemed to have made a transfer of value of all his remaining estate immediately before his death. This is chargeable to IHT, but not to CGT.

In the pre-budget report of October 2007, the Chancellor announced a transferable IHT nil rate band (NRB). Although this concession is welcome, it requires the keeping of accurate historic records. The measure will allow any unused NRB on a deceased spouse's or civil partner's death, whenever they died, to be available to increase the NRB of the surviving spouse or civil partner, providing the survivor's death occurs on or after 9 October 2007. This applies when the NRB of the first spouse or civil partner to die was not fully used in calculating the IHT liability of their estate. When the surviving spouse or civil partner dies, the unused proportion of the NRB of the first to die is added to the survivor's NRB. The unused portion of the NRB is computed in percentage terms rather than an actual monetary amount left unused. The relief only applies in computing IHT on death; chargeable lifetime transfers will be taxed solely with reference to the transferor's personal NRB.

For example, assume that a husband died in 2002/03, when the NRB was £250,000, and that none of his NRB was used (because his will provided that his entire estate was left to his surviving spouse), and therefore his estate was exempt from IHT. Accordingly, 100% of his NRB is available to carry forward for use by his surviving wife on her death, provided this is on or after 9 October 2007. Assume that when the wife dies, in 2010/11, her NRB on her death is £350,000. Her NRB would be increased by 100% to £700,000.

If in the above case the husband had left a legacy to his children of £150,000 on his death, with the remainder to his surviving spouse, he will have used up 60% of his NRB (£150,000 ÷ £250,000). Therefore, if the NRB on his surviving spouse's death is — as above — £350,000, then that will be increased by 60% to £560,000.

The Finance Act 1986 introduced the potentially exempt transfer (PET). This is a transfer of property (eg a gift) made between individuals, or into a settlement in which the beneficiary has an interest in possession; or into an accumulation and maintenance trust; or into a trust for the disabled; or the disposal or termination of a beneficial interest in possession in settled property under certain circumstances. If the transferor lives for at least seven years after the date of the PET it will be exempt from IHT, section 3A(4).

The scope of a taxpayer's ability to make a potentially exempt transfer into a settlement has been restricted with effect from 22 March 2006 by FA 2006. Since that date, IHT has been chargeable on most lifetime transfers of value into settlements, and to all transfers of value on or within seven years of death, subject to any exemptions applicable. If a PET becomes a chargeable transfer because the transferor does not survive for seven years, it is necessary to take into account any earlier chargeable transfers made in the seven years before that chargeable transfer in order to establish the IHT rate applying to it. This is explained below. Bringing such earlier chargeable transfers into account for this purpose does not of itself alter their taxability.

Certain lifetime gifts do not qualify as being potentially exempt from IHT, but are chargeable transfers subject to IHT at half the rate applicable on death. Up until 22 March 2006 the main category of transfer, which did not qualify for PET status, was a transfer into a discretionary settlement, where no beneficiary has a right to the income as it arises, unless it was a transfer into an accumulation and maintenance settlement as defined in section 71. The Finance Act 2006 changed the rules so that most new lifetime gifts of capital into settlements made from 22 March 2006 will be chargeable transfers, apart from certain specific types of settlement for a disabled person, or for a bereaved minor, section 3A.

IHT is chargeable on the loss to the transferor by his transfers of value, as they occur throughout each rolling seven-year period on a cumulative basis, except to the extent that the transfer is an exempt transfer. On death, the deceased is deemed to have made a transfer of value of all his remaining property. His death may also trigger a tax liability on PETs made within the previous seven years, which thereby become chargeable transfers, and this may increase the tax rate applicable to subsequent chargeable lifetime transfers.

For example, taxpayer Roland gave away £3,000 each year in order to use up his annual IHT exemption but otherwise his only transfers were, first, a gift of £290,000 to his son on 1 February 2003, and second on 1 March 2004 when he settled £200,000 on discretionary trust for his grandchildren. He died on 1 April 2008.

The gift on 1 February 2003 was initially a PET and so was ignored when the IHT position was calculated for the chargeable lifetime transfer of 1 March 2004, with the result that the £200,000 settled was within Roland's IHT Nil Rate Band for 2003/04 of £255,000.

However, on Roland's death (on 1 April 2008) the £290,000 gift made on 1 February 2003 becomes chargeable because it was made within seven years of his death. The gift still does not attract IHT as it is covered by the Nil Rate Band applicable on Roland's death in 2007/08, ie £300,000. However, as it is now chargeable it is taken into account in recalculating the IHT due on the property settled on 1 March 2004. The result is that tax becomes chargeable on the trustees:

|  | £ | £ |
|---|---|---|
| Chargeable lifetime transfer |  | 200,000 |
| Nil rate band applicable at date of death | 300,000 |  |
| *Less* Chargeable transfers in the seven years before death | 290,000 |  |
| Nil rate band remaining |  | 10,000 |
| Taxable on death |  | 190,000 |
| IHT at 40% |  | 76,000 |
| Less taper relief — transfer made between four and five years before the death so tax is tapered to 60% of the untapered tax (see note 8 to Study 6 below) — IHT payable |  | £45,600 |

There is now only one table of tax rates for IHT applicable to the deemed transfer on death. This table also applies to any PETs, which become taxable by falling within seven years of death, and to other chargeable lifetime transfers made within seven years of death. The rate of tax payable on such lifetime transfers is subject to taper relief, which mitigates the liability if the transfer was made between three and seven years before death (see above and note 8 to Study 6 below). This taper relief for IHT should not be confused with taper relief for CGT (see Study 1).

Where a transferor does not survive seven years after the transfer and the market value of property included in the lifetime transfer has fallen between the date of transfer and date of a sale at arm's length to a third party, or, if later, the transferor's death, relief may be claimed, section 131(2). The relief is given by deducting the reduction in market value from the value of the lifetime transfer, and charging that reduced amount to IHT at the date of death. The relief does not alter the deceased's total of cumulative transfers for the purposes of computing IHT on other transfers. In the case of an interest in land, sections 137 and 138 adjust the value to be used for this purpose on the sale or later death to adjust for changes in tenure or other factors, for the reduction in the term of a lease which had less than 50 years to run at the date at the date of the lifetime transfer, or for compensation received for any reduction in value.

Questions of valuation arise whenever a transfer of value is made or is deemed to have been made. The statutory definition of value is set out in section 160 as being the price which the property might reasonably be expected to fetch if sold in the open market at the time the transfer occurred, no reduction being made for the assumption that the whole of the property is to be placed on the market at one and the same time. This definition is very much the same as the statutory definition for CGT; the case law illustrating the appropriate valuation methodology is applicable to both taxes.

Note, however, that for actual transfers of value, as opposed to deemed transfers such as on death, the valuation required is the loss to the transferor's estate resulting from the transfer, section 3(1). In some cases the loss to the transferor will exceed the value of the property transferred owing to the consequent depreciation in value of the transferor's retained estate. For example, if the transferor owned a property outright and then gave half his interest to his son, the value of the half retained may well be less than 50% of the value of the whole. Here valuations will be required of the whole of the transferor's estate, both before and after the disposal. While the general rule requires that a transfer of value is calculated by valuing the whole of the transferor's estate before and after the transfer, in practice, where the property transferred forms a clearly separate unit with no implications for the rest of the transferor's estate, it seems to be readily agreed that by valuing the property transferred it will also be possible to ascertain the loss to the transferor's estate for IHT purposes.

The valuation of a transferor's estate can be affected by related property. This is defined in section 161 as property comprised in the estate of the transferor's spouse or civil partner, or property which is, or has been in the preceding five years, held by a charitable trust, political party, housing association, or qualifying public body as a result of an exempt transfer by the transferor or his/her spouse or civil partner after 15 April 1976. If the value of any property in the transferor's estate is less than it would be if it were the appropriate portion of the aggregate value of that and the related property, then its value must be taken to be the appropriate portion of that aggregate. Related property can take various forms, such as jointly held property, successive interests in property, parts of a set, and property with close physical proximity. For example, if a husband owned a block of land with poor access worth say £20,000, and his wife owned an adjoining block which on its own was worth £10,000, but which gave good access to both blocks so that together they would be worth £70,000, then on a transfer by the husband, the related property provisions would apply. The value transferred would be the appropriate portion of £70,000. This is calculated on the proportionate values of the separate assets valued as such, ie £20,000 ÷ (£20,000 + £10,000). The loss to his estate from the transfer would be £46,667, ie two thirds of £70,000. Thus, in related property situations, it is necessary to value the separate assets as such, and the aggregate value as well.

A person with a beneficial interest in possession in settled property, which arose before 22 March 2006 or in other limited circumstances following the FA 2006 (ie a settlement for a disabled person or for a bereaved minor child), is treated as owning the property in which the beneficial interest subsists, section 49. This may have an effect on the value of both the settled property and of other property forming part of the person's estate for IHT purposes. However, if that interest in possession comes to an end during the beneficiary's lifetime, he is deemed to make a transfer of value equal to the value of the trust's assets in which his interest subsisted, section 52. A beneficial interest in settled property is one where the beneficiary has the right to present enjoyment of income as it arises, without the need for any decision by the trustees, *Pearson v IRC* [1980].

Most new interests in possession arising on or after 22 March 2006 will not result in the trust property being treated as forming part of the beneficiary's estate for valuation purposes, except in the special cases of a settlement for a disabled person, or for a bereaved minor.

In determining the value of a person's estate, his liabilities are to be taken into account, provided they were imposed by law — such as rates — or to the extent that they were incurred for consideration, such as mortgage, section 5(3) and (5); but a liability in respect of which there is a right of reimbursement is taken into account only to the extent that reimbursement cannot reasonably be expected to be obtained, section 162(1). A liability is valued at the time of the chargeable event and, if it is a future liability, it is discounted, section 162(2). A liability which is an encumbrance on a particular property is taken to reduce the value of that property for IHT purposes, section 162(4). The discounting of future liabilities or receipts is common to both IHT and CGT but is not applicable to SDLT.

Where a residential property has to be valued which was acquired under the right to buy provisions of the Housing Act 1980, and is therefore subject to a liability to repay the discount if it is sold on within five years of purchase, its market value will be the amount which would be received on a hypothetical sale, subject to the obligation which would fall on the hypothetical purchaser to repay the discount in the event of a disposal by him within the specified period; but on the assumption that the hypothetical sale did not itself give rise to the obligation to repay the discount — see *Alexander v Inland Revenue Commissioners* [1991].

Most importantly, in the case of an actual transfer of value, the loss to the transferor's estate due to the transfer of property takes into account the IHT payable on the transfer when valuing his estate immediately after the transfer, section 5(4). Thus the net value transferred must be grossed up by the amount of IHT in order to arrive at the value transferred, see Study 6. This amount then takes its place in the aggregate of chargeable transfers during the rolling seven-year period. Other taxes borne by the transferor as a result of the transfer, for example income tax or CGT, are ignored in computing the loss to the transferor's estate.

If, however, the transferee bears the transferor's liability for IHT, then the loss to the transferor's estate is diminished, reducing the liability to IHT, section 5(4). If the transferee is an individual and bears the transferor's liability to income tax or CGT on the transfer, the chargeable transfer is reduced by the amount of tax so borne, section 165.

Where, under a contract, the right to dispose of any property has been excluded or restricted — for example where an option has been granted over the property — then, on the next transfer, that exclusion or restriction has to be ignored when valuing the property except to the extent that consideration was given for it. However, allowance will be made for that part of the value transferred which is attributable to the exclusion or restriction, if the contract imposing the exclusion or restriction was itself a chargeable transfer, section 163.

Special rules apply to the deemed transfer on death. For example, any change in the value of all or part of the deceased's estate due to his death shall be taken into account in the valuation as if it had occurred before death, section 171. Reasonable funeral expenses may be deducted in computing the value of the deceased's estate, section 172.

On an appeal to the Special Commissioners or the High Court against a determination by the Board of Inland Revenue concerning an IHT matter, any question as to the valuation of land in the United Kingdom is to be referred to the Lands Tribunal, section 222(4).

Where the value of an asset has been ascertained for the purposes of charging IHT on death, that value is taken as being the market value at that date for the purposes of CGT, section 274.

Special reliefs apply in respect of agricultural property, business property and woodlands necessitating special valuation rules. These are explained below.

## Agricultural Property Relief for IHT

The reader is also referred to Chapter 1 regarding this subject.

100% relief is given for transfers of qualifying agricultural property occurring after 9 March 1992 where, immediately before the transfer, the transferor had the right to obtain vacant possession within the next 12 months, section 116(1) and (2), or where he had held the property since before 10 March 1981 (subject to the conditions specified in section 116(3)). 100% relief is also due if the land is let on an agricultural tenancy that commenced on or after 1 September 1995, which is when the Agricultural Tenancies Act 1995 came into force.

Inland Revenue Concession F17 permits the transfer of tenanted farmland to be regarded as qualifying for 100% agricultural relief on the vacant possession basis if it carries the right to vacant possession within 24 months of the transfer, or if it is valued at an amount broadly equivalent to vacant possession notwithstanding the terms of the tenancy, as may happen if there is related property affecting the valuation.

If qualifying agricultural property does not meet the requirements above for 100% relief, it will obtain relief at 50%.

The relief is given by reducing the value transferred attributable to agricultural property, before the application of any available exemptions.

To qualify for agricultural relief, the property must have been occupied by the transferor for agricultural purposes for a minimum of two years up to the date of transfer, or owned by the transferor for a minimum of seven years up to the date of transfer and occupied (by anyone) for agricultural purposes during that period, section 117. For the relief to apply to a PET which becomes chargeable because the transferor has died within seven years of the transfer, the transferee must have owned the property continuously from the date of transfer to the date of the transferor's death and it must have remained qualifying agricultural property, section 124A. There are provisions for allowing the disposal and replacement of agricultural property without loss of relief, sections 118 and 124B. Relief can also be claimed on a transfer of shares in a company where the transferor held more than 50% of the voting rights in the company immediately before the transfer, and the company's property included qualifying agricultural property. The relief is given against the agricultural property element of the value of the shares, section 122.

Agricultural property is defined in section 115(2), (4) and (5) as agricultural land in the United Kingdom, the Channel Islands or Isle of Man, together with cottages, farm buildings and farm houses that are of a character appropriate to the property; and includes woodlands, and buildings used for the intensive rearing of livestock or fish, provided their occupation is ancillary to the occupation of agricultural property. The breeding and rearing of horses on a stud farm qualifies as agriculture. From 6 April 1995 land and buildings used for short rotation coppice will qualify as agricultural property, FA 1995, section 154; and from 26 November 1996 land in habitat schemes also qualifies, section 124C.

The relief is given on the agricultural value of the property. This is taken to be the value of the property as if it were subject to a perpetual covenant prohibiting its use otherwise than as agricultural property, section 115(3). In the case of a farmhouse this was held to exclude the additional amount a lifestyle purchaser would pay to outbid a working farmer bidding to purchase the agricultural property and buildings as a commercial farm — *Lloyds TSB Private Banking plc (personal representative of Antrobus deceased)* v *Peter Twiddy* (2004), see below.

There has been a series of cases examining whether a house qualifies as a farmhouse for the purposes of agricultural property relief, and if so whether it is of a character appropriate to the property. The necessary characteristics were reviewed in *Lloyds TSB (personal representative of Antrobus, deceased)* v *IRC* [2002], commonly called *Antrobus 1*. This looked at whether the house was appropriate in size, content and layout with the farm buildings and the farmland, drawing on the definition of a farmhouse laid down in the income tax case *IRC* v *Korner* [1969], and how long the house had been associated with the agricultural land. In *Higginson's Executors* v *IRC* [2002] a hunting lodge of considerable size on a 63 acre agricultural estate was held not to qualify as a farmhouse. A former farmhouse occupied by a retired farmer where the farming activities were carried elsewhere did not qualify — see *Rosser* v *IRC* [2003]. In *Arnander* v *Revenue & Customs Commissioners* [2006] a grand house on an estate was held not to be a farmhouse as the farming was all carried out by contract farming arrangements so that the farming was not managed from the house.

While the above cases have further defined the limitations on what type of property qualified as a farmhouse for the purposes of agricultural relief, a subsequent case — *Lloyds TSB Private Banking plc (personal representative of Antrobus)* v *Peter Twiddy* [2004] apparently increased the limitation further. This case is commonly called *Antrobus 2*. This case followed *Antrobus 1*, above, at which the Special Commissioner decided that the property, Cookhill Priory, qualified as a farmhouse for the purposes of agricultural relief. The matter was referred to the Lands Tribunal for a decision as to the appropriate level of agricultural value. The Tribunal's decision introduced further restrictions regarding the definition what was or was not a farmhouse. As this was not properly a matter for the Tribunal to rule on, the remarks may possibly be *obiter* rather than *ratio decedendi*. These remarks placed a narrow construction on the meaning of a farmhouse. In reaching their decision the LT found three main pointers.

1. The farmhouse must be of a character appropriate to the property. If the property was indeed such that it could be expected to be occupied by a lifestyle farmer, what character of house is appropriate and how would this be determined? In the Tribunal's opinion the correct approach is that the land as agricultural land (and in particular the nature and profitability of the agricultural operations on it) should be commensurate with the house. If this is so, it can only be because the concept of a farmhouse is that of a dwelling for a working farmer who requires a suitable house to support his working life.

2. Development of the legislative position since the introduction of estate duty had seen a progression from relief for country estates to one for working farms. Particular significance was attached to the exclusion, on the introduction of CTT, of a mansion house from the definition of agricultural property, so emphasising, in the Tribunal's view, the operational components of agricultural units. This apparently ignores the legislative changes made in 1976 and 1981 and, subsequently, the abolition of the concept the test for working farm relief in 1984. The exclusion of mansion houses was to be seen as excluding those occupiers who might perhaps be in overall control of the farm but were not undertaking it on a day to day basis.

3. The fact that cottages and farm buildings also attract relief under Part 3 of the definition of agricultural property was thought to support the Tribunal's view of the concept of the farmhouse as being the dwelling of the person who manages the farm on a day to day basis.

Much of the Tribunal's reasoning was apparently aimed at excluding the potential bid of a person considered to be a lifestyle farmer. This has also been considered under the concept of principal private residence relief for CGT purposes. The lifestyle purchaser is one who wishes to occupy a substantial dwelling with a large area of land and who might conduct farming operations, as one, but not perhaps the main, financial activity. Evidence was given in *Antrobus 2* to the effect that this category of purchaser would bid for the property up to its full market value, and not be concerned by the restriction placed upon it by the existence of section 115(3), occupancy in connection with agricultural purposes, requirement.

While much of the Tribunal's reasoning in *Antrobus 2* was aimed at the exclusion of the bid of the so-called lifestyle purchaser, this concept, which has also been mentioned in connection with principle private residence relief, is that there is a category of purchaser who wishes to secure an enhanced area of land with a substantial residence, and if the land is large enough may consider pursuing farming. However, farming will not be the main activity of such a lifestyle purchaser, whose main source of income will derive from other pursuits. While the Tribunal had seen no evidence of farmhouses sold in the open market subject to a perpetual agricultural covenant, it found on the evidence before it that a discount of 30% from the property's full market value was appropriate. It also found that if the bid of the lifestyle purchaser had been accepted as admissible, then the discount might have been reduced to 15% or perhaps even lower.

This has created considerable difficulty when valuing farmhouses for the purposes of agricultural relief, particularly where the property is being passed down through the generations for agricultural purposes without actually being sold, which was the case in *Antrobus 2*. The narrow construction of the definition of farmhouse adopted is difficult to reconcile with what was said by Lord Upjohn in *IRC v Korner* (1969) — an income tax case, "that a farmhouse should be judged in accordance with ordinary ideas of what is appropriate in size, content and layout, taken in conjunction with the farm buildings and the particular area of land being farmed, and not part of a rich man's residence."

The Tribunal's decision in *Antrobus 2*, with its reference to the concept of "a farmhouse being the chief dwelling house attached to a farm, the house in which the farmer of the land lives"; and answering its own question "who is the farmer of the land for the purposes of the definition of section 115(2)?" with "in our view it is the person who lives in the farmhouse in order to farm the land comprised in the farm and farms the land on a day to day basis" imports another test: that of day to day operation of the farming activity. This test, which is apparently to be ascertained in relation to the day to day operations of the farm, will have significant impact on those farmers who, while they control the operation of farming, employ farm managers to run their own farms; or those who are elderly farmers who have retired or partly retired; or those who are farmers whose main source of income comes from outside agriculture; or those farmers who, by operating through contractors, retain little involvement in day-to-day farming operations. This may be of of great detriment to those seeking to retain agricultural operation of the land from generation to generation without seeking to dispose of core property after a death.

In valuing agricultural property for IHT, farm cottages occupied by farm workers are to be valued ignoring any value attributable to their suitability for occupation by anyone else, section 169. A cottage occupied by a retired farm worker may benefit from this provision subject to certain conditions (Inland Revenue Concession F16).

A liability secured on any property reduces the value of that property for IHT purposes, section 162(4). Accordingly, a liability secured on property qualifying for agricultural relief will reduce the value of the property before the application of the relief. If the value of the property exceeds the agricultural value, for example if it is enhanced by development potential, the secured liability will be apportioned rateably between the agricultural value and the non-agricultural value (IR Inheritance Tax Manual paragraph 24073).

In the case of a working farmer occupying agricultural property for which there is some additional non-agricultural value, for example hope value for development, the excess value will not be covered by agricultural relief but may instead be covered by business property relief, see below.

## Woodlands regime for IHT

A special regime for woodlands is provided for by section 125, but it is only applicable on the death of the taxpayer and cannot relieve IHT on lifetime transfers. If woodlands relief is to be claimed, the land must be in the UK and the deceased must have owned it for the five years prior to his death, or he must have become beneficially entitled to it otherwise than for valuable consideration — for example by gift or on inheritance — in which case the five-year rule does not apply. An election can be made within two years of the death for the value of the trees and underwood to be left out of account when computing the IHT on the deceased's estate; the value of the underlying land remains chargeable in the deceased's estate.

Woodlands which qualify for agricultural relief as being occupied with and ancillary to an agricultural holding do not qualify for woodlands relief, but automatically receive agricultural property relief. Commercially managed woodlands which are not part of an agricultural holding can qualify for business property relief if the qualifying conditions are met (see below). Since both these reliefs are more favourable than woodlands relief, a claim under section 125 is rare.

If an election has been made, but the timber is disposed of before a subsequent death, then on a sale for full consideration the net sale proceeds, or in any other case the net value of the timber at that time, is charged to IHT as the top slice of the deceased's estate at the rates of IHT in force at the time of the disposal. If the deceased could have claimed business property relief instead of woodlands relief on death, the value charged to IHT is reduced by 50%. Net proceeds or net value are defined as the proceeds or value after deducting such expenses of disposing of the timber, and the costs of replanting within three years of the disposal, as are not allowable as deductions for income tax, sections 127(1) and 130.

## Business property relief for IHT

Business property relief applies where relevant business property is transferred, subject to the various qualifications outlined below. The relief is given by reducing the value transferred that is attributable to qualifying business property, before the application of any available exemptions.

100% relief is given on the transfer of a business by a sole trader, or of an interest in a business (for example, a partnership share), or of shares in a company which is not listed on a recognised stock exchange, or of securities in such a company which by themselves — or with other shares owned by the transferor — gave him control of the company immediately before the transfer, section 105(1) (a) to (bb). For a transfer of unquoted shares prior to 6 April 1996, the transferor needed to control 25% of the voting power to benefit from 100% relief, otherwise the rate of business relief was 50%.

Relief of 50% is given on the transfer of land, buildings, machinery or plant owned by the transferor which immediately before the transfer was used for the purposes of a business run by a partnership of which he was a member, or by a company which he controlled; and also to such items owned by a settlement in which the transferor was beneficially entitled to an interest in possession, and which immediately before the transfer were used by a business carried on by the transferor. It also applies to shares or securities listed on a recognised stock exchange which gave the transferor control of the company immediately before the transfer, section 105 (1)(cc) to (e).

Prior to 10 March 1992 the rates of relief were respectively 50% and 30%.

A partner on retirement from a partnership ceases to qualify for business property relief in respect to his capital account (*Beckman* v *IRC* [2000]) as he no longer owns business property but merely a money debt due to him from the partnership.

Where agricultural property relief, and/or woodlands relief (see above) apply to a transfer of value which also qualifies for business property relief, they are applied in priority and business property relief only applies to any residual value, section 114.

The business property must be owned by the transferor for a minimum of two years immediately preceding to the transfer, section 106. There are provisions allowing for the disposal and replacement of business property without loss of relief, section 107. Where control of a company is required to qualify for 100% business property relief, see above, this is defined as having more than 50% of the voting rights, section 269. Where shares are related property in relation to the transferor (see above), their voting rights are aggregated with his for the purpose of determining whether he had voting control immediately before the transfer, section 269(2).

Business is defined as including a business carried on in the exercise of a profession or vocation, but excludes a business not carried on for gain, section 103(3). A group holding company can qualify provided its subsidiaries themselves carry on qualifying businesses, section 105(4). A business wholly or mainly involved in holding investments, or dealing in land, buildings, or shares or securities (other than a market maker or discount house carrying on trade in the UK) is disqualified, section 105(3) and (4). Thus claims for business property relief were denied for letting industrial property units (*Martin* v *CIR* [1995]), and for letting furnished flats (*Burkinyoung* v *CIR* [1995]). There have been five cases on caravan park businesses; the earlier cases were reviewed by the Court of Appeal in *IRC* v *George* (*executors of Stedman, deceased*) [2004]. The court held that it is necessary to look at the activities of the business in the round to see if it is wholly or mainly an investment holding business or not.

The business of property development for resale (as opposed to property investment) does qualify for the relief.

For a lifetime transfer of business property made within seven years of the death to benefit from business property relief, it is necessary for the transferee to continue to own the property throughout the period from the transfer to the death of the transferor. Also, the property must be used for a qualifying business property relief purpose by the transferee, section 113A. Disposal and replacement with other qualifying business property is permitted, section 113 B.

Section 110 defines the value of a business for the purposes of business property relief as being the value of the assets used in the business (including goodwill), reduced by the aggregate of any liabilities incurred for the purposes of the business. The Special Commissioners reviewed whether assets and liabilities arising from trading contracts were used in the business or not, and following *Van den Berghs Ltd* v *Clark* [1935] concluded they were not (*Hardcastle (the executors of Vernede, deceased)* v *IRC* [2000]). Assets with no nexus to the business are excluded, even if a guarantee for business purposes is secured thereon: *IRC* v *Mallender (executors of Drury-Lowe, deceased)* [2001]. HMRC consider that income tax is a personal liability of the proprietor or partner and so is not deductible from the value of the business (Valuation Office Manual paragraph 25153).

In the case of a sole trader, a transfer of an asset used in the business, but not the business itself nor a part thereof, will not qualify for the relief as it will not fall within any of the headings qualifying for relief listed in section 105.

In the case of an interest in a partnership or company which otherwise qualifies for business property relief with respect to the transferor, section 112 denies relief for excepted assets. These are defined as either not having been used wholly or mainly for the purposes of the business throughout the two years preceding the transfer, or throughout its period of ownership if shorter, section 112(2) and (5), or not being required at the time of the transfer for future use for those purposes. While section 112 applies to a sole trader, assets not used in his business will have been disallowed by section 110. In the case of any land or building where part only has been exclusively used for the purposes of the business, the value of the part so used qualifies for relief and the remainder is disqualified as an excepted asset. The value of the whole asset must be apportioned between the two parts as may be just, section 112(4).

## Heritage property exemption, and maintenance funds

There is a special regime for heritage property under sections 30 to 35A. This provides that a claim for conditional exemption from IHT may be made in respect of the deemed transfer on death; or on a lifetime transfer provided the transferor has been beneficially entitled to the property throughout the six years ending with the transfer, either personally or together with his spouse or civil partner. Where the claim is accepted, the value attributable to the heritage property is treated as conditionally exempt and is left out of account in computing the IHT chargeable on that transfer, section 30.

To qualify as heritage property, the Treasury, on a claim being made to them, must decide whether to designate the property under section 31 as being an object or collection, land, or building that are of outstanding national, scientific, historic or artistic interest. They will require an undertaking to preserve the property and secure reasonable public access.

Following conditional exemption being obtained in respect of a transfer, when the transferee himself subsequently transfers the property either on his death or by a chargeable lifetime transfer (for example a PET which he does not survive by seven years) the conditional exemption is reviewed. Provided the Treasury obtain a satisfactory undertaking in the new circumstances, the designation remains in force and the transfer is itself conditionally exempt. If no satisfactory undertaking is obtained on the transfer or disposal, or if the Treasury determine that the undertaking has not been observed in some material respect, a chargeable event occurs, section 32. IHT is then charged on the market value of the property, or proceeds of sale on a disposal at arm's length, by reference to the circumstances of the relevant person. The relevant person is the last person who made a conditionally exempt transfer, unless there were two or more such conditionally exempt transfers in the last 30 years, in which case HMRC may select whichever they choose. If the relevant person is alive at the date of the chargeable event, the tax is assessed at lifetime rates. Otherwise the rates on death apply. The liability falls on the person who would or did benefit from the disposal proceeds, section 33.

There is a code under section 27 and Schedule 4 for providing exemption for a transfer of value to a settlement in order to fund the maintenance, repair, preservation, or provision of public access to land or to a building designated by the Treasury under section 31. Provided the settlement requires that for at least six years the fund must only be applied for the maintenance etc of the designated property, or, to the extent it is not so used, that it is retained in the settlement, subsequent distributions of the fund to another such settlement, to a charity, to a heritage body, or to the settlor, his spouse or widow are exempt unless the entitlement was purchased. There are provisions for charging IHT where the conditions for an exempt transfer out of the settlement are not met.

## Study 6

This study illustrates the IHT grossing up computation for a chargeable lifetime transfer where the transferor bears the tax; the computation where a PET becomes subject to tax because the donor dies within seven years of making the transfer; and it also deals with business property relief, agricultural relief, CGT hold-over relief for a gift of business assets and the relationship between IHT and CGT on gifts. The facts are as follows.

On 1 January 2000 Tim made his first lifetime transfer, being a cash gift of £152,000 to his daughter.

On 2 February 2001 Tim settled on discretionary trusts the freehold factory he owned, which was occupied by his 100% owned non-listed manufacturing company for the purposes of its trade. The discretionary trust did not qualify as an accumulation and maintenance settlement under section 71 and Tim and his wife were permanently excluded from being beneficiaries. As at the date of the transfer the freehold factory was valued at £500,000, and it was still held by the trust on Tim's death, at which point Tim still owned the manufacturing company. Tim paid the IHT chargeable on the lifetime transfer of the freehold factory, but he made a claim under section 165 TCGA 1992 to hold over the capital gain arising on the transfer.

On 3 March 2004 Tim made a gift to his son Simon of his freehold interest in a farm. Tim's late father had died in September 1984 leaving him the freehold reversion in the farm at which time it was let on an Agricultural Holdings Act tenancy to a tenant in his mid 30s. The probate value on his father's death in 1984 was agreed at £190,000. The farm remained let continuously. Tim's legal costs for the gift were £12,500. The gift was valued at £480,000 (vacant possession value would have been £1,400,000).

Tim died on 30 April 2007. At that date Simon still owned the farm and it had increased in value since Tim had given it to him.

All references are to IHT Act 1984 except where indicated otherwise.

*IHT computation*

The gift on 1 January 2000 was a PET with no liability to IHT initially. The transferor, Tim, survived seven years so the gift became permanently exempt from IHT for all purposes.

The transfer of the factory on 2 February 2001 was a chargeable lifetime transfer. Tim paid the IHT so the chargeable transfer had to be grossed up (see note 4 below). On Tim's death within seven years of the transfer the IHT is recomputed at the rate applicable on death on the table of IHT rates in force at the date of death, Schedule 2 paragraph 1A.

The gift on 3 March 2004 was a PET and accordingly was initially assumed to be exempt from IHT, section 3A (5). On the death of Tim within seven years of the gift, it is established as being a chargeable transfer, section 3A (4). IHT is charged at the rate applicable on death on the table of IHT rates in force at the date of death.

IHT Computations on the lifetime transfers

|  | £ | £ |
|---|---:|---:|
| IHT on chargeable lifetime transfer on 2 February 2001: |  |  |
|  |  |  |
| Value of freehold factory | 500,000 |  |
| *Less*: Business property relief — 50% (note 1) | 250,000 |  |
|  | 250,000 |  |
| *Less*: Annual exemption — 2000/01 (note 2) | 3,000 |  |
|  | 247,000 | 247,000 |
| *Add*: Chargeable transfers in previous 7 years (note 3) | Nil | Nil |
|  |  | 247,000 |
| *Less*: Zero rate band for 2000/01 |  | 234,000 |
|  |  | 13,000 |

| | | |
|---|---:|---:|
| *Gross up* — transferor paying the IHT — lifetime rate 20% | | |
| — multiply £13,000 by $\dfrac{100\%}{100\% - 20\%}$ (note 4) | 3,250 | £16,250 |
| | | |
| IHT on £16,250 at the lifetime rate — 20% | | £3,250 |
| Cumulative total of chargeable transfers carried forward | 250,250 | |

On the death of the transferor within 7 years the IHT must be
recomputed (section 7(4))

| | | |
|---|---:|---:|
| Additional Tax due on death (note 5) | | £Nil |

IHT due on PET 25 March 2004:
Initially exempt, but becomes a chargeable transfer as Tim
does not survive for 7 years. Therefore the IHT position
is recomputed on death:

| | | |
|---|---:|---:|
| Transfer of value (note 6) | | 480,000 |
| *Less*: Agricultural relief — 50% (note 7) | | 240,000 |
| | | 240,000 |
| Annual exemptions   2003/04 | 3,000 | |
|                               2002/03 (note 2) | 3,000 | 6,000 |
| Chargeable transfer | | £234,000 |

IHT payable
Cumulative chargeable transfers in previous 7 years:

| | | |
|---|---:|---:|
| 1 January 2000 Gift of cash — ignore as exempt, the transferor having survived 7 years (note 3) | Nil | |
| 2 February 2001 — as above | 250,250 | 250,250 |
| 25 March 2004 Chargeable transfer as above | | 234,000 |
| | | 484,250 |
| *Deduct*: Zero rate band (note 5) | | 300,000 |
| | | £184,250 |
| IHT at 40% on £184,250 | 73,700 | |
| *Less*: Taper relief — reduction to 80% of full rate (note 8) | 14,740 | |
| IHT payable (note 9) | | £58,960 |

Notes

(1)   The factory was used immediately before the transfer by a trading company controlled by the transferor and thus qualifies for 50% business property relief, section 105(1)(d).

(2)   Each individual has a £3,000 annual exemption for IHT per fiscal year, section 19. To the extent the annual exemption is not used it can be carried forward for one fiscal year only. As Tim made a transfer in the preceding fiscal year, 1999/00, which exceeded the annual exemption for that year and thus used it up, there is no annual exemption to bring forward to use against the transfer in 2000/01. HMRC practice is that if the transfer exceeds the exemption, only the balance is a PET or chargeable transfer (IR Inheritance Tax manual paragraph 14141).
      The value transferred is reduced by business property relief before the annual exemption is utilised.

(3)   The gift on 1 January 2000 is more than seven years before the transferor's death and thus converts from being a PET to being an unconditionally exempt transfer and is not taken into account in reassessing the cumulative total of chargeable transfers in respect of any later transfers, section 3A(4).

(4) The value transferred by a lifetime transfer is the reduction in the transferor's estate resulting from the disposition which includes the IHT payable by him; thus the value of the chargeable transfer must be grossed up, sections 3(1) and 5(4).

(5) The table of tax rates used for the re-computation is the lower of that applicable at the date of the transfer and at death, Schedule 2 paragraph 1A. The IHT rate table at the date of death in 2007/08 is lower; its zero rate band is £300,000. This exceeds the cumulative total of chargeable lifetime gifts at 2 February 2001, £250,250, so the recomputed charge on death is £Nil. However there is no refund for the tax paid originally, section 7(5).

(6) The incidental costs of the transfer are left out of account if borne by the transferor; if borne by the transferee they reduce the value transferred, section 164.

(7) Agricultural property relief applies in this case as the transferor had owned the freehold reversion for more than 7 years. Agricultural property relief is at 50% as the transferor's interest in the property did not carry the right to vacant possession within twelve months, section 116(2), or 24 months (IR concession F17). The transferee owned the freehold reversion continuously from the date of gift to the date of the transferor's death and it continued to be occupied by the tenant for agricultural purposes.

(8) As the gift of the farm occurred more than three but less than four years before the death, the rate of tax is subject to taper relief which reduces it to 80% of the full rate. Taper increases as the period from the transfer to death lengthens: three to four years — 80% of the full rate, four to five years — 60%, five to six years — 40%, six to sven years — 20%, section 7(4).

(9) The transferee is liable to pay the IHT on death, section 204(8), and consequently there is no grossing up when a PET becomes a chargeable transfer due to the death of the transferor within seven years thereafter. The IHT paid by the transferee may be deducted in computing any future chargeable gain on disposal of the property, provided a hold-over claim was made in respect of the gift by the donor and donee under TCG Act 1992 section 165(1) (TCG Act 1992 section 165(10) and (11)). See note 12 below.

As the IHT is attributable to land which qualified for agricultural relief, the tax may be paid by annual interest free instalments over 10 years, sections 227 and 234.

*CGT computation on the gift in 2004*

| | £ | £ |
|---|---:|---:|
| Deemed consideration for disposal, market value (note 10) | | 480,000 |
| *Less*: Allowable expenditure: | | |
| Base Cost, ie probate value (note 11) | 190,000 | |
| Incidental costs of disposal (section 38 (1)(c)) | 12,500 | 202,500 |
| Unindexed gain (section 53(2)(a)) | | 277,500 |

*Less*: Indexation allowance (sections 53, 54, and 55(2)) by reference to relevant allowable expenditure only:

Relevant allowable expenditure £190,000 multiplied by

$$\frac{RD - RI}{RI}$$

where

RD = RPI for month of April 1998 = 162.60
RI = RPI for month of acquisition September 1984 = 90.11

$$\frac{RD - RI}{RI} = \frac{162.60 - 90.11}{90.11} = 0.804$$

| | | |
|---|---|---:|
| £190,000 × 0.804 | | 152,760 |

| | |
|---|---:|
| Indexed gain | 124,740 |
| *Less*: Taper Relief — 20% (note 12) | 24,948 |
| Chargeable Gain (notes 13 and 14) | £124,740 |

Notes

All references are to TCG Act 1992 except where indicated.

(10) This transaction is between connected persons, section 286(2), it is therefore treated as otherwise than by way of a bargain at arm's length, and market value, section 272, must be substituted for the actual consideration, sections 17 and 18.

(11) Where the value of an asset has been ascertained for the purposes of charging IHT on death, that value is taken as being the market value at that date for the purposes of CGT, section 274.

(12) The property has been let throughout Tim's ownership from prior to 17 March 1998 to its disposal on 3 March 2004, a holding period of 6 whole years including the bonus year for the period 17 March 1998 to 5 April 1998. Its use in that holding period did not qualify it as a business asset. Accordingly, taper relief for CGT is calculated using the non-business asset rate of taper relief. A six year holding period gives 20% taper relief.

Had the asset been held until after 6 April 2004, its letting to an individual for use for the purposes of his trade would qualify the asset as a business asset for taper relief, and Tim would gradually have acquired an increasing entitlement to some business asset taper relief.

(13) The chargeable gain on the gift could be held-over by a claim being made by the donor and donee since the property qualified for agricultural property relief for the purposes of IHT. In that case the donor would be treated as making a no-gain no-loss disposal for CGT, and the donee's acquisition would be treated as being at market value less the held-over chargeable gain ignoring taper relief, section 165(1) and (4) as extended by Schedule 7 paragraph 1. If the claim is made, the donee's base cost would be £355,260 (ie market value £480,000 less the indexed gain £124,740 = £355,260).

(14) If no hold-over claim is made, the CGT liability would be the transferor's liability. Any CGT paid by the transferor is ignored for IHT purposes in computing the value transferred by the transfer of value, section 5(4) IHTA 1984.

# Value Added Tax and property

Note: Unless otherwise stated, all references in this part of the chapter are to the VAT Act 1994.

## *An overview of VAT liability*

For VAT purposes property is divided into three broad categories and the VAT treatment varies for each of them:

1. qualifying buildings: dwellings, relevant residential buildings and relevant charitable buildings
2. commercial buildings and civil engineering works and
3. bare land.

## *Qualifying buildings: dwellings, relevant residential buildings and charitable buildings*

Qualifying residential buildings are dwellings (Schedule 8, Gp 5, Nt (2)) and relevant residential buildings which are communal residential buildings such as homes for the elderly, hospices, children's homes (Schedule 8, Gp 5, Nt (4)).

Qualifying charitable buildings are those used for a relevant charitable purpose, ie charitable non-business purposes or as a village hall or similar buildings providing social or recreational facilities for a local community (Schedule 8, Gp 5, Nt (6)).

## Construction (goods and services)

The construction of qualifying buildings is zero-rated. However, the zero-rating for relevant residential and relevant charitable buildings is restricted to supplies to the person who intends to use the building for the relevant purpose and subject to that person issuing a certificate to the supplier confirming the use (Schedule 8, Gp 5, Items 2 and 4, and Nt(12)). HMRC operate an Extra Statutory Concession for relevant charitable buildings allowing zero-rating where any non-qualifying use is less than 10%.

Zero-rating also covers the construction of a self contained annex to an existing relevant charitable building provided that the annex is capable of functioning independently, and the main access to the annex is not through the existing building or vice versa (Schedule 8, Gp 5, Nt (17)).

Where a building is enlarged or extended and that enlargement or extension creates an additional new dwelling, the separate disposal of which is not prevented by any planning or similar consent, the construction goods and services are zero-rated.

Sub-contractor's services are zero-rated only for the construction of new dwellings.

Special rules apply for protected buildings (see below).

The reduced rate of VAT applies to supplies of goods and services in the course of certain works to dwellings and relevant residential buildings (Schedule 7A, groups 6 and 7). The works which qualify for relief at the reduced rate are as follows:

1. converting premises into:
   * a single household dwelling
   * a different number of single household dwellings
   * a multiple occupancy dwelling, such as bed-sits or
   * a relevant residential building.
2. renovating or altering the following buildings that have not been lived in for two years:
   * a dwelling or
   * a relevant residential building.

Supplies of goods and services in the course of the conversion of a non-residential building into a new dwelling or a relevant residential building are zero-rated when supplied to a relevant housing association, subject to the housing association issuing a certificate to confirm the qualifying use in the case of a relevant residential building.

Apart from the exceptions mentioned above, standard-rating applies to any work done to an existing building, including repair and maintenance, and any alteration, extension, reconstruction, enlargement or annexation of an existing building.

## Sale

The first grant of a major interest (freehold or lease exceeding 21 years) by the person constructing a qualifying building is zero-rated (Schedule 8, Gp.5, Item 1). For relevant residential and relevant charitable buildings, zero-rating applies only if the recipient provides a certificate to the grantor confirming the use of the building solely for the qualifying purpose (Schedule 8, Gp 5, (Nt 12)). HMRC operate an extra statutory concession for relevant charitable buildings allowing zero-rating if any non-qualifying use is less than 10%.

A person converting a non-residential building or part of a building to create a new dwelling(s) or a building for relevant residential purposes can zero-rate the first grant of a major interest in the building (Schedule 8, Gp 5, Item 1). Zero-rating applies when the building (or part) being converted (Schedule 8, Gp 5, Nt (7)):

- has never been used as a dwelling or number of dwellings or as a relevant residential building or
- in the 10 years immediately before the sale, the building (or part) has not been used as a dwelling or number of dwellings or as a relevant residential building.

Zero-rating for conversion to a relevant residential building applies only if the recipient provides a certificate to the grantor confirming the use of the building (Schedule 8, Gp 5, Nt (12)).

## Study 7

The conversion of a non-residential building to residential

|  | £ | VAT | |
|---|---|---|---|
| Purchase of old redundant hotel | 170,000 | nil | |
|  |  |  |  |
| Conversion: |  |  |  |
| 2 retail shops | 40,000 | 7,000 | (17.5%) |
| 4 flats | 80,000 | 4,000 | (5%) |
|  |  |  |  |
| Professional fees | 30,000 | 5,250 | |

Shops and flats to be let

Notes
(1) The intended short letting will be exempt, with no VAT recovery on expenses
(2) Election to waive exemption on property will make the letting of the shops standard-rated to facilitate VAT recovery on expenses relating to the shops. Election will not be effective for flats
(3) Consider the zero-rated grant of a major interest in the flats to a separate entity (which will then short let them) to enable VAT on the expenses relating to the flats to be reclaimed.

The sale of a dwelling, relevant residential or relevant charitable building which does not fall within the scope of zero-rating is exempt from VAT.

## Letting

All lettings (other than a lease exceeding 21 years by the person constructing) of dwellings, relevant residential or relevant charitable buildings are exempt from VAT.

## Protected buildings

Protected buildings are, broadly, listed buildings and scheduled monuments which are also qualifying buildings (Schedule 8, Gp 6, Nt (1)).

Approved alterations to protected buildings are zero-rated, but repairs and maintenance are standard rated. In most cases an approved alteration means an alteration to a protected building for which listed building consent is both needed and has been obtained from the appropriate local planning authority prior to commencement of the work. Special rules apply to churches, buildings on Crown or Duchy land and scheduled monuments (Schedule 8, Gp 6, It 2 and 3 and Nt (6)).

The grant of a major interest by a person substantially reconstructing a protected building is zero-rated (Schedule 8, Gp 6, It 1).

## Change of use

VAT must be accounted for if, within 10 years of the completion of a relevant residential building which has been zero-rated by certificate, the whole, or part of the building is put to a non-qualifying use (Schedule 10, paragraph 1). Where there is continued occupation but a change of use, a deemed standard-rated self supply arises. Any sale or letting for a non-qualifying use is standard-rated.

Similar rules apply for a relevant charitable building, however, from 21 March 2007, where zero-rating was obtained under the extra statutory concession for less than 10% non-qualifying use, no self-supply charge applies if there is a change of use that was not anticipated when zero-rating was obtained under the concession.

### Study 8

The change of use of a new building from residential use

| 1 April 01 | Zero-rated construction (by certificate) | |
|---|---|---|
| 1 April 07 | Sale of freehold OR<br>Letting (all or part of building<br>for use as hotel | standard-rated |

Note: The sale or letting for a non-qualifying use is within 10 years of obtaining zero-rating for the construction of the building, therefore the supplies are standard-rated.

## Holiday accommodation

Special rules apply to holiday accommodation (Schedule 9, Gp 1, Nts (11) and (13)) and (Schedule 8, Gp 5, Nt (13)).

## Commercial buildings and civil engineering works

Commercial buildings and civil engineering works do not qualify for zero-rating or the reduced rate of VAT. These include offices, retail premises, industrial buildings, such as factories and warehouses, roads, drains and airfields.

## Construction

The construction of commercial buildings and civil engineering works is standard-rated.

## Freehold sale

The VAT treatment of the freehold sale of a commercial building or civil engineering work depends on whether the building/work is new or old. For VAT purposes new means less than three years from the date on which the building is completed. The date of completion is the date the certificate of practical completion is issued, or the date the building is fully occupied — whichever happens first. All freehold sales which take place within the three year period are standard-rated. Freehold sales of buildings/ works over three years old are exempt from VAT (Schedule 9, Gp 1 It 1). However, an option to tax can be made (see 2.).

## Letting

All letting (including the sale of long leasehold) of commercial buildings and civil engineering works is exempt from VAT (Schedule 9, Gp 1 It 1). However, an option to tax can be made — see below regarding option to tax).

## Bare land

The freehold sale and all letting of bare land is exempt from VAT (Schedule 9, Gp 1 It 1), subject to certain specific exceptions. However, an election to waive exemption can be made — see below. The exceptions to the general exemption for land which are taxed at the standard-rate are specified in Schedule 9, Gp 1 (a) to (n) and include:

- grant of any interest, right or licence to take game or fish, other than the freehold sale of land which includes the sporting or fishing rights
- provision of pitches for tents or of camping facilities, including caravan pitches on seasonal sites
- granting of facilities for parking a vehicle
- grant of a separate right to fell and remove standing timber.

Mixed supplies can arise when freehold land is sold including new civil engineering work which does not cover the whole of the land. The consideration should be apportioned on a fair and reasonable basis, eg cost.

# Option to tax

The election to waive exemption (option to tax) was introduced in 1989 for land, civil engineering works and commercial buildings (Schedule 10, paragraph 2). To be effective, all options to tax must be notified to HMRC in writing within 30 days of the option to tax being made. If previous exempt supplies of the property have been made, it is necessary to obtain HMRC's prior written permission to opt to tax, unless one of the automatic permissions detailed in notice 742A applies. As a general rule, once an option to tax is made and notified to HMRC, supplies of the land or property become taxable, thus allowing the VAT incurred on expenses relating to the supplies to be reclaimed as input tax.

## Study 9

Option to tax — Freehold sale

| | With Option | | Without Option |
|---|---|---|---|
| | Cost | VAT | Cost (VAT inclusive) |
| | £m | £ | £m |
| Building purchase | 2.0 | 350,000 | 2.350 |
| Refurbishment costs | 1.0 | 175,000 | 1.175 |
| | 3.0 | 525,000 | 3.525 |
| Sale of Freehold | (3.5) | 612,500 | (3.5) |
| Profit/(Loss) | 0.5 | | (0.025) |

Special rules apply for options to tax made by members of a VAT group. Broadly an option to tax by one group member binds the other members of the group.

The gift of a property on which an option to tax has been exercised is a standard-rated supply.

An option to tax can only be revoked, subject to certain conditions (including obtaining HMRC's permission in some instances), within six months of being made or after 20 years. All revocations must be notified to HMRC.

An option to tax cannot apply for certain supplies, for example land sold to a housing association or an individual for construction of a dwelling, and property intended for use as a dwelling, or solely for relevant residential or relevant charitable purposes. If the vendor/landlord of a mixed use building has opted to tax, the option applies only to the non-qualifying element. For example, in the case of a shop with a flat above, only the shop element will be standard-rated, the option will not apply to the selling price or rent attributable to the flat.

An anti-avoidance measure was introduced with effect from 19 March 1997 to restrict the option to tax in certain circumstances where the property will be used other than for taxable purposes, Schedule 10, paragraph 2(12) to (17). The dis-application of the option affects freehold sales, leases, assignments, surrenders and any other commercial property supplies to which the option to tax could apply.

Under the provisions, an option for taxation is dis-applied in respect of a grant if, *at the time of the grant*:

1. the development is, or is expected to become, a capital item for the purposes of the capital goods scheme, either for the grantor, a person to whom the development is transferred or a person treated as the grantor and
2. it is the intention or expectation of the grantor or the person treated as the grantor or the person responsible for financing the grantor's development, that the building will be occupied by them or a person connected to them and
3. the person occupying the development will be doing so for other than eligible purposes.

For someone to be in occupation of the development for eligible purposes they must be occupying it for the purpose of making mainly taxable supplies, or for other supplies which entitle them to credit for their input tax. Mainly means substantially more than half.

## Study 10

Option to tax — dis-application

1. Developer constructs a doctors' surgery costing £500,000 plus VAT.
2. Doctors pay £100,000 premium. This is used by the developer as part funding for the development.
3. Developer lets the building to the doctors.

Notes

(1) Building is a capital goods scheme asset of the developer (because its value is £250,000 or more).
(2) The doctors are providing finance and occupying for an exempt use.
(3) Option to tax would be dis-applied.
(4) Take account of the non-recovery of VAT on the developer's costs when setting the rental value.

## Study 11

Option to tax — supplies *not* affected by dis-application

1. *Pension fund*
   - purchase/construction
   - let to a fully taxable connected party for own occupation (ie the connected tenant is able to reclaim all VAT on expenses).

2. *Property investment company*
   - purchase/construction
   - let to an unconnected third party for exempt use (provided the third party is not involved with the financing of the development).

# Whether or not to opt to tax

The main factors which affect the decision on whether to opt to tax are:

- the amount of VAT on expenses which will be an irrecoverable cost if an option to tax is not made
- the VAT status of the recipient of the supply and
- the wording of an existing lease.

If the tenant of a building is to be a business within the VAT exempt sector, such as a bank, building society, insurance company, or educational institution, any VAT charged on the rent will be an additional cost to the tenant, who will have a partial or total restriction on the recovery of the VAT. In some situations it may be possible to reach agreement for the landlord not to opt to tax in return for an increase in the rent payable to compensate him for the VAT on expenses which he is unable to recover. The same principle applies to freehold sales. The landlord needs to be certain that the short term advantage of opting to tax does not create long term problems, for example when he wishes to sell the freehold of the property. It should also be remembered that opting to tax will involve compliance costs associated with VAT registration.

The wording of an existing lease may affect the landlord's decision on whether to opt to tax. If an existing lease is silent on VAT, the landlord can add VAT to the rent or service charges payable, unless the lease specifically prevents it, section 89. If the lease contains a provision preventing the landlord charging VAT in addition to the rent and service charges, he may still opt to tax, but will have to account for VAT out of the actual rent and service charges payable.

The following are some examples of tenants that will have a restriction on the right to reclaim VAT on rent and service charges:

- financial institutions eg banks, building societies, finance houses
- financial agents eg stock brokers, mortgage brokers
- insurance companies, agents and brokers
- universities and private schools
- pension funds
- betting shops
- charities
- private hospitals
- non-profit making sports clubs
- undertakers.

# Stamp Duty Land Tax

Note: Unless otherwise stated, all references in this part of the chapter are to the FA 2003.

SDLT was introduced on the 1 December 2003, its predecessor stamp duty was partly repealed and now only impacts on the transfer of shares, securities and partnership interests. The legislation is complex and the paragraphs below only deal with those practical aspects of the tax likely to be of interest to valuers. Practitioners seeking more detailed information are referred to *Hutton and Anstey* on SDLT, *The Stamp Duty Land Tax Handbook* by Hart and Johnson (on which the following is based) or any of the informative publications by Patrick Cannon, barrister. The tax is convoluted, and to quote Patrick Cannon "In addition to complex legislation, practitioners have also had to contend with the compliance system of enormous complexity and detail, coupled with shortcomings in its administration".

SDLT is charged on land transactions. A land transaction is any acquisition of a chargeable interest, section 43. This is its major difference from its predecessor stamp duty, which was a tax on documents. A chargeable interest means "an estate, interest, right or power in or over land in the UK", section 48(1)(a). It therefore includes freehold and leasehold interests and their equivalent in Scotland. Interests which are wholly exempt are:

- a security interest, for example a mortgage
- a licence to use or occupy land
- tenancies at will
- advowson, the right of presentation to a beneficiary
- franchisees which are grants by the crown, for example the right to hold a market fair or take tolls
- manors.

Most land transactions involve the parties entering into a contract which is completed by the actual conveyance of the interest of the purchaser or lessee. A payment made at the contract stage which is not the substantial amount does not trigger the payment of SDLT. The tax becomes due on completion or when the balance of the purchase price is paid. In SDLT there is the concept of substantial performance, where a contract is effectively actioned prior to completion. There are two circumstances when this can occur. First, where the purchaser or a person connected with a purchaser takes possession of the whole or most of the subject-matter of the contract and receives, or becomes entitled to receive the rents or profits. Second, if the purchaser is allowed to take possession after the contract but before completion, possibly to carry out work to the property. If a contract is substantially formed then this is the date at which the payment of SDLT becomes due. There is a significant number of reliefs from SDLT detailed in statute and in the regulations. The basic premise is that unless a specific relief is granted by some statutory or regulatory provision, then the transaction will be subject to SDLT.

Chargeable consideration for SDLT purposes is any consideration in money or moneys worth given directly or indirectly for the subject-matter of the transaction by the purchaser or by a person connected with him. This connection is specified in ICT Act 1988 section 839 per FA 2003 Schedule 4, paragraph 1. In simple form this connection is either by blood or by business.

Where transactions can be considered to be linked for the purposes of SDLT, the consideration is then aggregated and tax is due on the higher amount. SDLT rates are stepped, with the rates of tax applying to the entirety of a transaction once the threshold is crossed. The concept of a linked transaction is to prevent the fragmentation of acquisitions of property interests to take advantage of a lower rate band. Consideration is detailed in Schedule 4, and includes the assumption of a debt, or in certain circumstances works done as a consequence of the contract, either on the land which is the subject of the contract or other land.

The primary legislation for SDLT is the FA 2003 as but various regulations and statutory instruments also apply, making this a difficult and fragmented piece of legislation to follow. Where a land transaction involves an exchange, each transaction is taxed as if it were distinct or separate from the other, section 47.

A major interest in land for SDLT purposes means a freehold or a lease for a term of years absolute, whether subsisting at law or in equity, section 117(2). While the effective date of a transaction for SDLT is the date of completion, as explained above, there are several exceptions to the concept of substantial completion contained within the complex rules that apply to options, rights of pre-emption, agreements for leases and the like. It is beyond the scope of this chapter to deal with these in any detail.

There are two major areas which are met with in practice: the granting of leases and sales of property. This chapter concentrates on these. With regard to Value Added Tax, any VAT chargeable in respect of a lease or chargeable on the consideration as part of a sale is taken to be part of the chargeable consideration. This differs from position under the old stamp duty provisions. Where VAT becomes chargeable by virtue of an election to tax made after the effective date of the transaction, it will not be considered that VAT forms part of the transaction. When calculating the charge for consideration to be taken as rent on the grant of a lease, the following do not count as chargeable consideration:

- an undertaking by the tenant to repair, maintain, insure or let the premises
- a tenant's undertaking to pay service charges, maintenance, repairs ensuring for management costs
- rent or performance guarantees
- penalty rents
- costs incurred by the tenant exercising a statutory right to be granted a new lease
- obligations to bare the landlord's costs incidental to the grant of a lease
- any other obligations that the tenant takes on that does not effect open market rent.

# Leases

When taking a new lease the lessee will be liable to pay SDLT on the consideration payable. In England, Wales and Northern Ireland, a lease means "an interest or right in or over land for a term of years (whether fixed or periodic)", section 120 and Schedule 17A paragraph 1. It also means "a tenancy at will or other interest or right in or over land terminable by notice at any time," which is curious given that a tenancy at will and a licence to use or occupy land are not chargeable interests. A lease in Scotland has its usual meaning, Schedule 17A, paragraph 19.

In most cases the consideration will include the rent payable, although in some instances leases are taken where the rent is nil or a peppercorn. In addition to the rent, the lessee may pay a premium, which is a capital sum, or there may be two or more such sums payable at specified times in the future.

The calculation of SDLT payable on premiums is the same as that for other capital payments. Reverse premiums, capital payments by lessors to lessees for taking a lease, are not subject to SDLT, Schedule 17A paragraph 18. As regards rent, SDLT is payable on the capital equivalent of the rent, calculated as the discounted amount, or net present value (NPV). The calculation is made in accordance with section 56 and Schedule 5.

The present value of the right to receive a sum in the future is determined by discounting it for the waiting period in the usual way at the appropriate discount rate (the temporal discount rate), Schedule 5 paragraph 8.

In the case of SDLT, the discount rate is determined by the Treasury. At the time of writing the rate is 3.5%. The rent payable, which is to be discounted, is the rent reserved under the lease. Where the rent is inclusive of other matters, such as insurance, or rates, or repairs, without apportionment, the whole sum is treated as rent, Schedule 17A paragraph 6. However, this is without prejudice to the provision for a just and reasonable apportionment under Schedule 4, paragraph 4. In any event, the tenant's obligations for liabilities such as repairs, insurance, service charges and other non-rent matters, are not taken to be chargeable consideration, Schedule 17A paragraph 10. It is in the tenant's interest for the rent to be stated separately from any other payment.

The calculation of the NPV (V) is made by applying the formula found in paragraph 3 of Schedule 5:

$$V = \sum_{i=1}^{n} \frac{r_i}{(1 + T)^i}$$

where

$r_i$ is the rent payable in year i
i is the first, second, third etc year of the term
n is the term of the lease and
T is the temporal discount rate

The NPV is the relevant rental value for the calculation of the tax payable.

The calculation is the same as conventional valuations using the YP single rate, at 3.5%. Apart from applying the figures from the YP Single Rate Table, as is done in the examples below, the figure can be obtained from the Revenue website *www.inlandrevenue.gov.uk/so/bull_.news_flyers_sdlt.htm*. In this part of the chapter deferred YPs have been calculated using the subtraction of YP method rather than the more usual approach of using the PV £1 as a multiplier. The subtraction method only works for single rate YPs. It should be noted that one starts with the YP for the entire period (not the period of actual receipt), from which is deducted the YP for the period of deferral.

Where the rent is subject to variation dependent on a future contingent event, the possibility of deferring payment until the outcome of that event is known, under section 90, is not available — see section 90(7).

For example, if a lease of a warehouse is granted at a rent reflecting its use as a warehouse, but with a provision for a higher rent to be paid should planning permission be granted for retail use, SDLT will be assessed on the higher rent as stated, or a reasonable estimate of the rent payable if it is not stated. A provision for rent to be varied in line with changes in the retail price index is not treated as a rent subject to variation dependent on a future contingent event, Schedule 17A, paragraph 7.

Where a rent is to be reviewed to reflect the prevailing market value at the time of the review, any review which comes into effect up to the end of the first five years of the term of the lease is incorporated in the calculation of the net present value, Schedule 17A, paragraph 7: see the example below. This requires the lessee to predict the changes in rental value over the period up to the review date, presumably on the advice of a valuer. The forecast rent is then incorporated in the net present value calculation. If, at review, the rent is higher than was forecast, increased SDLT becomes payable on the increase in the net present value, calculated by re-computing the original net present value but adopting the actual rent as reviewed. If lower, the same process is followed, with a refund of SDLT becoming due on the reduction in the net present value.

Circumstances may arise where a lessee takes a lease where a rent review is will fall due five years from a date which is itself earlier than the commencement date of the lease ("a specified date"), perhaps the preceding quarter day. If this specified date is three months or less before the commencement date, and if there is a rent review five years after the specified date, it is assumed that the review date is five years after of the beginning of the term, Schedule 17A, paragraph 7A. This removes the need to forecast rent.

Rent reviews which are effective after five years or more are not taken into account in determining net present value.

## Study 12

Lease, only rent payable, rent review within the first five years.
A residential property is let for 12 years at a rent of £12,500 pa, subject to upward only rent reviews after four and eight years.

|  | £ pa | £ |
|---|---|---|
| Initial rent | 12,500 | |
| *Calculate predicted rent* after 4 years — assume rental value increases of 3% pa | | |
| Initial rent | 12,500 | |
| Amount of £1 for 4 years at 3% | 1.1255 | |
| Predicted rent | 14,610 | |
| | | |
| *Calculate NPV* | | |
| Initial rent | 12,500 | |
| YP 4 years at 3.5% | 3.6731 | 45,914 |
| | | |
| Rent after 4 years | 14,610 | |

| | | | |
|---|---|---|---|
| YP 12 years at 3.5% | 9.6633 | | |
| *Less* YP 4 years at 3.5% | 3.6731 | | |
| YP 8 years deferred 4 years at 3.5% | 5.9902 | 5.9902 | 87,515 |
| NPV | | | 133,429 |
| *Less*: Exempt amount | | | 125,000 |
| SDLT at 1% on | | | 8,429 |
| SDLT payable | | £84 | |

Note: The rent review after five years is ignored. The calculation will be repeated when the rent is fixed at the first review. It is proper to round off rental estimates if required.

The length of the term of a lease is a key factor in calculating the net present value. Where the lease is for a fixed term, the term is taken as the contractual term specified in the lease or, if shorter, the period from the date of the grant of the lease until the end of the contractual term. The right of any party to determine or renew a lease, such as a break clause or an option to extend, is ignored in determining the term, Schedule 17A, paragraph 2.

## Study 13

Lease, only rent payable
A commercial property is let for 10 years at a rent of £100,000 pa

| | £ pa | £ |
|---|---|---|
| Rent | 100,000 | |
| YP 10 years at 3.5% | 8.3166 | |
| NPV | | 831,660 |
| *Less*: exempt amount | | 150,000 |
| SDLT at 1% on | | 681,660 |
| SDLT payable | £6,816 | |

## Study 14

Lease, only rent payable after a rent free period
A commercial property is let for 10 years at a rent of £100,000 pa, with an initial rent free period of six months.

| | | £ pa | £ |
|---|---|---|---|
| Rent    Year 1 | | 50,000 | |
| YP 1 yr at 3.5% | | 0.9662 | 48,310 |
| Rent    Years 2–10 | | 100,000 | |
| YP 10 years at 3.5% | 8.3166 | | |
| Less YP 1 yr at 3.5% | 0.9662 | | |
| YP 9 years deferred 1 year | 7.3504 | 7.3504 | 735,040 |
| NPV | | | 783,350 |
| Less: exempt amount | | | 150,000 |
| SDLT at 1% on | | | 633,350 |
| SDLT payable | | £6,333 | |

## Study 15

Lease, rent payable plus an initial premium
A flat is let for 99 years subject to a premium payable of £180,000 and ground rent of £100 pa which doubles after 33 years and 66 years.

|  | £ pa | £ |
|---|---|---|
| Rent | 100 | |
| YP 99 years at 3.5% | 27.6234 | |
| NPV | | 2,762 |

Note: The increase in rent, being payable after five years, is ignored because one only has to have regard to foreseeable increases during the first five years. The net present value will be taxed separately from the premium to determine the SDLT payable, in this case nil as below the threshold of £125,000.

An option to extend a lease is ignored in determining the length of the term. This allows for some tax mitigation. If a lessee wishes to take a lease for, say, 20 years, the initial payment of SDLT will be reduced if the lease is for a term of, say, 12 years, with an option to extend the lease, (or to take a new lease), for eight years. The landlord would probably require cross options. It might be possible for the initial lease to be for a term where the rent has a net present value below the exempt limit. The second payment may be higher if rents increase, but it is unlikely that the lessee would be worse off in discounted cash flow terms.

The renewal of a fixed term lease is treated as a new lease commencing on the expiry of the previous lease. This applies in particular to business leases granted under Part 2 of the Landlord and Tenant Act 1954 or leases under the Business Tenancies (Northern Ireland) Order 1996 (SI 1996/725 (NIS)). However, if a fixed term lease is not brought to an end, so that the lessee remains in possession, the lease is treated as being extended (tacit relocation in Scotland).

For SDLT purposes, the lessee is treated as taking a new lease from the start of the fixed term lease for a period 12 months longer than the original lease. If the situation has not changed after 12 months, it is assumed that a further lease for 12 months has been taken, and so on until a new fixed term lease is taken, Schedule 17A, paragraph 3. These 12 month extensions will steadily increase the old net present value so that although, initially, it was within an exempt level band of net present value, the cumulative adjustments could, over time, render the net present value taxable.

If, at the end of the fixed term, the lessee is liable to pay an interim rent, but this has not been agreed or may not be known until determined by the court, the provision relating to contingent events will apply.

## Study 16

Lease expiry, lessee remaining in possession after the expiry of a fixed term.
A lessee occupies shop under a lease for a term of five years which has expired, but continues to pay rent of £30,000 pa which the landlord accepts. To ascertain if additional tax is payable, we now assume the lease was originally granted for six years.

|  | £ pa | £ |
|---|---|---|
| Rent | 30,000 | |
| YP 6 years at 3.5% | 5.3286 | |
| NPV | | 159,858 |

*Compare original position*

| | | |
|---|---|---|
| Rent | 30,000 | |
| YP 5 years at 3.5% | 4.5151 | |
| NPV | | 135,453 |
| Increase in NPV | | 12,203 |

SDLT liability, £Nil, tax payable on excess above £150,000

## Study 17

Lease expiry, lessee remaining in possession after the expiry of a fixed term.
As above, the tenant paying £30,000 pa, but the landlord has applied to the courts for an interim rent under the lessee occupies shop under Landlord and Tenant Act 1954.

| *Predicted interim rent* | | £ pa | £ |
|---|---|---|---|
| Estimated current FRV | | 40,000 | |
| Rent payable at end of lease | | 30,000 | |
| Interim rent, predicted | | 35,000 | |
| YP 6 years at 3.5% | | 5.3286 | |
| NPV | | | 159,858 |

| *Calculate NPV* | | | |
|---|---|---|---|
| Rent | | 30,000 | |
| YP 5 years at 3.5% | | 4.5151 | 135,453 |
| Plus Reversion to rent of | | 35,000 | |
| YP 6 years at 3.5% | 5.3286 | | |
| Less YP 5 years at 3.5% | 4.5151 | | |
| YP 1 year deferred 6 years at 3.5% | 0.8135 | 0.8135 | 28,473 |
| NPV | | | 163,926 |

Note: If the predicted interim rent is not actually determined at £35,000, then the NPV will be recalculated, with consequential adjustments to the tax liability of the lessee.

# Agreements for leases

An agreement for a lease is commonly entered into where the lessee is to carry out development. The agreement provides that the lessee will carry out specified work, which can range from alterations to an existing building to the construction of a major development. On satisfactory completion of the work, the lease attached to the agreement comes into effect.

For the purposes of SDLT, the term of the lease commences from the date of substantial performance of the agreement and ends on the contractual termination date of the lease. A contract will be substantially performed when the lessee takes possession of the property under the terms of the agreement for a lease. When the lease is subsequently granted, it is treated as a surrender and re-grant for the variation of the term of a lease, Schedule 17A, paragraph 12A.

# Indefinite term

Where a lease is granted for an indefinite term, it is treated as a lease for 12 months, in the same way as a lessee who remains in occupation after the expiry of a fixed term. Examples given in Schedule 17A paragraph 4 include a periodic tenancy or other interest terminable by a period of notice, a tenancy at will, or any other interest terminable by notice at any time.

# Variations

Where a lease is varied so as to increase the amount of rent payable, the variation is treated as the grant of a new lease, the rent payable being the increase in the rent. This does not apply to an increase in rent already provided for in the lease, Schedule 17A, paragraph 13, nor to a lease where stamp duty was paid — a lease commencing before 1 December 2003, Schedule 17A, paragraph 9(4).

Where a lease is varied so as to reduce the amount of rent payable, the variation is treated as an acquisition of a chargeable interest by the lessee, Schedule 17A, paragraph 15A(1). However, if no consideration is paid by the lessee for the variation of the lease then there will be no tax payable.

## Study 18

Variation of rent payable under a lease

A lease of a warehouse was granted for 15 years at a rent of £120,000 pa without review. The lease has eight years to run. The lease is to be varied by allowing non-food retail use and the lessee has agreed to pay a revised rent of £200,000 pa.

|  | £ pa | £ |
|---|---|---|
| Rent payable | 200,000 | |
| Less: Old rent | 120,000 | |
| Additional rent | 80,000 | |
| YP 8 years at 3.5% | 6.8740 | |
| NPV | | 549,920 |
| Less: Exempt amount | | 150,000 |
| SDLT at 1% on | | 399,920 |
| SDLT payable | £3,999 | |

Where the period of a lease is extended, this is treated as the grant of a new lease for the extended term. This new lease is treated as a transaction which is linked to the original grant.

Where the period of a lease is reduced, this is treated as the acquisition of a chargeable interest by the lessor, Schedule 17A, paragraph 15A(2). As in the case of the reduction in the amount of rent payable, if the lessor receives no consideration for the reduction in the term, then there is no consideration on which tax will be payable.

## Study 19

Variation of both the rent payable under a lease and of its term
As in Study 18, but the lease is to be extended for a further seven years.

| | | £ pa | £ |
|---|---|---|---|
| Next 8 years | | | |
| Rent payable | | 200,000 | |
| Less: Old rent | | 120,000 | |
| Additional rent | | 80,000 | |
| YP 8 years at 3.5% | | 6.874 | 549,920 |
| *Following 7 years* | | | |
| Rent payable | | 200,000 | |
| YP 15 years at 3.5% | 11.5174 | | |
| Less YP 8 years at 3.5% | 6.8740 | | |
| YP 7 years deferred 8 years at 3.5% | 4.6434 | 4.6434 | 928,680 |
| NPV | | | 1,478,600 |
| *Less*: exempt amount | | | 150,000 |
| SDLT at 1% on | | | 1,328,600 |
| SDLT payable | | | 13,286 |

Where a lease is surrendered in return for a new lease between the same parties, SDLT is payable on the rent reserved in the new lease. Any value in the lease surrendered is ignored (Schedule 17A, paragraph 16). However, any rent that was payable under the surrendered lease is deducted from the rent reserved under the new lease in determining the net present value, similar to a variation of a lease in respect of rent, Schedule 17A, paragraph 9.

## Study 20

Surrender and re-grant of a lease
As in Study 19, but the lessee surrenders the existing lease and takes a new one for 15 years at a rent of £200,000 pa. The NPV will be calculated as has been done in Study 18.

The sale or an assignment of an interest, whereby the vendor or assignor takes a lease of the premises, the subject of the deal (commonly called sale and leaseback), is not uncommon, particularly where the owner wishes to raise capital to use in the business. For example, a multiple retailer will sell the freehold interest in a store, where the yield is, say, 5%, to invest the proceeds in the business, where the return might be 15%.

In these cases, the purchaser or assignee will pay SDLT. However, no SDLT will be payable in respect of the new lease, subject to certain conditions. These are that the premises which are sold are the same as the demised premises in the lease, that the consideration is in the form of cash or the release or assumption of a liability, and that the parties to the sale and the lease are the same. This does not apply if the parties are companies within a group, Schedule 17A.

If the consideration is partly rent and partly some other consideration, such as a premium, if the net present value of the rent does not exceed £150,000, no tax is payable on the rent consideration. Similarly, if the rent does not exceed £600 pa, and the other consideration does not exceed £150,000, no tax is payable on the other consideration. Where the rent does exceed £600 pa, there is no 0% band, so that other consideration of up to £150,000 attracts tax at 1%, Schedule 6, paragraph 5.

## Resolution of rent review

As explained above, a rent review within the first five years of the term is treated as a contingent event. This requires an estimate of the likely level of rent which will be agreed on review to be adopted when calculating the NPV. When the rent review is settled, the original net present value needs to be re-calculated if this rent is different from the predicted rent. This will lead to a consequential further payment, or a repayment, of SDLT. In practice, when acting for a lessee in respect of a rent review, the valuer should be made aware of the predicted rent adopted at the start of the lease so as to appreciate the SDLT consequences of agreeing a rent at a different level.

## Resolution of contingent event

Where the net present value has been calculated in relation to a future contingent event within the first five years of the lease, the net present value will be recalculated once the actual outcome is known, in the same manner as for the resolution of a rent review.

## Turnover rents

If the rent payable under a lease is a turnover rent (the rent being a percentage of the lessee's turnover), or a rent which is part fixed and part turnover, clearly the amount of rent that will be payable under the lease will be unknown at the commencement of the lease, which is when the NPV is required.

In order to calculate the NPV, an estimate of the likely level of rent that will be achieved is made, and these predicted rents are included in the calculation.

If the lease is for a term exceeding five years then, at the end of the fifth year, the highest rent paid in any continuous period of 12 months in those first five years is identified, and this is adopted as the rent payable for the balance of the term, Schedule 17A, paragraph 7. A revised calculation is then made of the NPV with consequential further payments or repayments being made. The revised calculation adopts the rents actually paid in the first five years, and the highest 12 month figure for the remaining period.

### Study 21

Lease with turnover rents.

1. *Start of lease*

   A shop is let in September 2004 for 15 years at a fixed rent of £10,000 pa and an additional payment of 5% of the turnover of business conducted from the shop.

   The lessee's business plan is based on an anticipated annual turnover of £1,000,000 after the first start up year, when the turnover will be less. If successful, turnover should increase at around 10% each year once the business is established.

   It would be unrealistic to project these assumptions into annual payments for each of the 15 years of the term. The rent, based on the target turnover of £1,000,000 will be £50,000 (5%) plus £10,000 = £60,000. It would seem to be reasonable to adopt this rent, or perhaps something higher, to reflect optimistic outcomes

| *Hence*: | £ pa | £ |
|---|---|---|
| Predicted rent, say, | 60,000 | |
| YP 15 years at 3.5% | 11.52 | |
| NPV | | 691,044 |
| *Less*: Exempt NPV Band | | 150,000 |
| Taxable NPV | | 541,044 |
| SDLT at 1% | 5,410 | |

2.  *End of the fifth year*

    At the end of the fifth year, the business records show that, in the 12 month period starting in the June of the fourth year up to the end of May in the fifth year, turnover was £1,400,000. This produces a rent of £70,000 + £10,000 = £80,000 pa, the highest rent in any 12 month period. In practice, identifying the highest rent over any 12 month period would depend on how frequently the turnover figures are produced. It should be noted that the 12 month period means any period of 12 consecutive months — it is not referenced to the years of the lease or the years of the calendar.

    The rent actually paid in the first five years was:

| Year 1 | £25,000 |
|---|---|
| Year 2 | £53,000 |
| Year 3 | £60,000 |
| Year 4 | £60,000 |
| Year 5 | £74,000 |

*Hence, recalculate NPV as follows*:

| | £ pa | £ NPV | |
|---|---|---|---|
| *Year 1* | | | |
| Rent paid | | 25,000 | |
| YP 1 year at 3.5% | | 0.9662 | 24,155 |
| | | | |
| *Year 2* | | | |
| Rent paid | 53,000 | | |
| YP 2 years at 3.5% | 1.8997 | | |
| *Less* YP 1 year at 3.5% | 0.9662 | | |
| YP 1 year deferred 1 year | 0.9335 | 0.9335 | 49,475 |
| | | | |
| *Years 3 and 4* | | | |
| Rent paid | | 60,000 | |
| YP 4 years at 3.5% | 3.6731 | | |
| *Less* YP 2 years at 3.5% | 1.8997 | | |
| YP 2 years deferred 2 years | 1.7734 | 1.7734 | 106,404 |
| | | | |
| *Year 5* | | | |
| Rent paid | | 74,000 | |
| YP 5 years at 3.5% | 4.5151 | | |
| *Less* YP 4 years at 3.5% | 3.6731 | | |
| YP 1 year deferred 4 years | 0.842 | 0.842 | 62,308 |

*Years 6 to 15*

| | | | |
|---|---|---|---|
| Rent paid | | 80,000 | |
| YP 15 years at 3.5% | 11.5174 | | |
| Less YP 5 years at 3.5% | 4.5151 | | |
| YP 10 years deferred 5 years | 7.0023 | 7.0023 | 560,184 |
| NPV | | 802,526 | |
| *Less*: Original NPV | | 691,044 | |
| Additional NPV | | | 111,482 |
| Additional SDLT at 1% | | £ 1,114 | |

## Variable rents

It is rare, nowadays, for a rent under a lease to be a fixed rent unless it is a short lease, say up to a term of five years. Rents are varied in several ways. One is by rent review, commonly upward only to the prevailing market rental value. A second is by pre-determined increases set out in the lease. A third is variation based on changes in the retail price index.

In the case of rent review and fixed increases, any changes that will take place after the fifth year are ignored. Instead, the highest rent paid in any continuous 12 month period is adopted as the rent payable for the balance of the term after five years. As shown above where there was a rent review after four years, this produced a new rent which was the highest rent in the first five years, and was therefore adopted as the rent for the remainder of the term.

Where a lease is granted with five yearly reviews, the initial rent, or the first annual rent paid after any rent free period, is taken as the rent payable throughout the term. An extension to this is where there is provision in the lease for a rent review, expressed as following five years after a date which falls within the three months before the lease terms commences, usually the preceding quarter day, so producing a rent review in the final quarter of the fifth year. In such circumstances, the first five years of the term are taken to end on the rent review date, Schedule 17A to FA 2003, paragraph 7A. Where the variation is through changes in the retail price index, these changes are ignored, Schedule 17A, paragraph 7(5).

As has been stated, changes to rents after the fifth year through rent review or stepped rent provisions or increased turnover are left out of account in calculating NPV. However, there is provision for rent increases to be brought into the SDLT net if the increase in rent is abnormal. The HMRC introduced this provision as an anti-avoidance measure to prevent lessees exploiting the ignoring of rent changes after the fifth year. As this requirement may give rise to the need for a further return it is dealt with later in this chapter.

## Acquisition of lease from SDLT exempt assignor

Various bodies are exempt from the liability to pay SDLT. Hence, if such an exempt body takes a lease, no SDLT is payable by that lessee. If the exempt body later assigns the lease, the assignee is acquiring a lease where no SDLT has been paid on the rents, including the future rents that the assignee will pay. If the assignee is not itself an exempt body, then SDLT will be payable on the net present value of the remaining rents, in addition to any SDLT payable on any capital payment for the lease. The assignment is treated as if it were the grant of a new lease for the unexpired term of the actual lease and on the same terms, Schedule 17A, paragraph 11.

## Study 22

Assignment by an exempt assignor.

A charity takes a lease of a shop to use as a reception for its charitable purposes. The lease is for 15 years at an initial rent of £60,000 pa with a rent review in year five. The rent is reviewed to £90,000 pa. The charity assigns the lease after eight years for £200,000 to a multiple retailer. It is assumed that the assignee has taken a new lease for seven years at an initial rent of £90,000 pa.

| *SDLT payable* | £ | Tax £ |
|---|---|---|
| Consideration | 200,000 | |
| SDLT at 1% | 0.01 | 2,000 |
| Rents | 90,000 | |
| YP 7 years at 3.5% | 6.1145 | |
| NPV | 550,305 | |
| *Less*: Exempt amount | 150,000 | |
| Taxable NPV | 400,305 | |
| SDLT at 1% | 0.01 | 4,003 |
| SDLT payable | | £6,003 |

Note: Strictly, the rent payable at the next review should be predicted as it is within the first five years of the notional new lease. As the intention is to illustrate the effects of assignments by exempt assignors, it is assumed that no change in rent is predicted.

# Land transaction returns

There are circumstances where a Land Transaction Return (LTR), a further LTR, or a supplementary LTR is required. An LTR is made on one of a variety of statutory forms. The primary form (SDLT 1) indicates where positive information or returns are required. Also dealt with below are the different computations potentially needed in these circumstances, linked transactions and transactions between spouses and civil partners.

## *Basic provisions*

Section 76(1) of FA 2003 states: "In the case of every notifiable transaction the purchaser must deliver a return (a land transaction return) to the Inland Revenue before the end of the period of 30 days after the effective date of the transaction."

The return must include a self-assessment of the tax that, on the basis of the information contained in the return, is chargeable in respect of the transaction, since the FA 2007 the tax payable does not have to accompany the LTR but payment of the amount chargeable must be made within the 30 day period — section 76(3). The Inland Revenue also has power, by statutory instrument, to vary or shorten the 30-day period, section 76(2). It is now a regulatory requirement that where an SDLT 1 is completed and the codes 3 or 4 are entered at question 3, signifying that the transaction comprises of mixed (commercial and residential) or commercial property, then a supplementary return on form SDLT 4 is automatically required.

While it might be thought a simple matter to decide the effective date, different situations where an LTR is required, each have their own specific requirements.

## Notifiable transactions

Notifiable transactions are defined in section 77 as set out below.

a.  The grant of a lease for a term of seven years or more where it is granted for chargeable consideration.
b.  The grant of a lease for less than seven years if either the chargeable consideration consists of, or includes, a premium in respect of which tax is charged at a rate of 1% or higher, or where the chargeable consideration consists of or includes rent in respect of which tax is chargeable at a rate of 1% or higher (or where the tax would be chargeable in the absence of a relief). (In this respect one takes into account consideration not covered by the £125,000 threshold for sales, or the £150,000 threshold for commercial transactions. In the case of residential property situated in a Designated Disadvantaged Area, the applicable threshold is £150,000.)
c.  Any other acquisition of a major interest in land is notifiable unless it is exempt from charge under the provisions of Schedule 3. However, if the transaction is residential and the consideration is less than £1,000, it is considered *de minimis*, section 77(3)(b).
d.  Any other acquisition of a chargeable interest where the consideration is taxable, or would be taxable but for a relief.

## Private finance initiative lease and leasebacks

A private finance initiative (PFI) project involves the sale or lease of land by a public body to a private sector body which then leases back the land, or underlets it, to the public body. The treatment of PFI projects is set out in Schedule 4 paragraph 17, which was inserted by the SDLT (Amendment of Schedule 4 to the Finance Act 2003) Regulations (SI 2003 No 3293). These transactions also include, where appropriate, public private partnerships (PPP). These regulations provide that where the transaction is a qualifying transaction: "neither the lease-back to the public body nor carrying out of works nor provision of services is chargeable consideration for the transfer or grant of the lease by the qualifying body, or for the transfer of the Surplus Land" (in other words tax will generally be charged only on any cash premium or rent paid by the private sector supplier).

To qualify as a PFI or a PPP transaction, the transaction must be between a private sector body which is the supplier, and a public sector body, to which the supplies are made. Section 66 contains an extensive list of those public bodies, including central and local government, health, and planning authorities, which qualify for this relief, together with certain bodies concerned with higher or further education contained in paragraph 17. Hence, a primary care trust would be capable of entering into a PFI/PPP transaction, but a group of doctors or dentists, acting together in a partnership, would not constitute a public sector body, and therefore would be unable to qualify for the relief. It is possible for a person to be prescribed to enjoy the relief by Treasury Order. All PFI/PPP transactions are notifiable, therefore the private sector supplier would make an LTR at appropriate points, and the public sector body would self-certify on the relevant return.

## Assignments

The assignment of a lease, if deemed to be the grant of the lease at the time of the assignment, would be a notifiable transaction (for example, an assignment of the lease held by a charity to an SDLT taxpayer). Also, an assignment where there is consideration chargeable at a rate of 1% or higher, or would be but for any relief, is a notifiable transaction. An agreement for lease is not a notifiable transaction. However, if substantial performance occurs, then it becomes a notifiable transaction and the agreement for lease is effectively treated as if it is the commencement of the lease. Similarly, where an agreement for lease is assigned, if substantial performance occurs, either by the assignor or assignee, then the point of the substantial performance is treated as the commencement of the lease for SDLT purposes.

Where substantial performance occurs, an LTR is required within 30 days. If any consideration is paid by the lessee to the lessor (or assignee to assignor) following occupation, this sum is potentially taxable. Therefore, if the consideration payable is taxable then the procedures described above must be followed. If no taxable consideration is payable then the self certification procedures must be followed. When the actual lease commences a further return is required on form SDLT 1.

For the purposes of SDLT, a lease cannot start before the date it is granted — following the decision in *Bradshaw* v *Pawley* [1980], where it was said "no actual term of years can be created until the lease has been executed and so the grant has been made". This is of course at variance with the doctrine of substantial performance which is effectively an anti-avoidance measure. In practice, an agreement for lease where occupation of the property is granted to the future leaseholder is usually used where works have to be carried out to the satisfaction of the landlord before the granting of the occupational lease. There may be a licence agreement during this period and the payment of a licence fee. Paragraph 9 and 9A of Schedule 17A work to give relief in relation to the consideration already taxed under the initial LTR. The mechanism is the same as that adopted in the situation where a tenant holds over, which is dealt with below.

Where the assignment is that of a lease subject to a relief under section 57A (sale and leaseback), Parts 1 or 2 of Schedule 7 (group, reconstruction and acquisition relief), section 66 (transfers involving public bodies), Schedule 8 (charities relief), or regulations mentioned in section 123(3) of the Act (interaction with stamp duty relief); then the assignment is treated as if it were a lease granted by the assignor. The lease, which in the hands of the assignor was subject to relief, is treated in the hands of the assignee as if it is the grant of a new lease for the remainder of the original term. An LTR is therefore required within 30 days of the assignment.

## Leases for indefinite terms

Schedule 17A, paragraph 4, deals with the treatment of leases for an indefinite term. Such leases are treated, in the first instance, as if they were for a fixed term of one year. If the lease continues after the end of the initial deemed term of one year, it is then treated as if it were a lease for a fixed term of two years. Thereafter, if the lease then continues beyond the end of the second deemed term it is then treated as if it were for a lease for a fixed term of three years, and so on. This means that, while at the initial grant of the lease, the transaction may not be notifiable, or require the completion of an LTR, over time the lease and total rents may accumulate to a point where the total consideration becomes taxable. The requirement in Schedule 17A, paragraph 4, is that when this situation occurs the purchaser (or tenant) must complete an LTR in the normal way, and return it to the Inland Revenue together with the appropriate amount of tax calculated on a self-assessment basis.

## Study 23

A residential property, not in a Designated Disadvantaged Area, is let for one year at £28,250 pa.

In year one, the NPV of the rent passing is below the threshold of £125, 000 .

In years two, three and four the rent has accumulated to a total of £113,000 and tax is not payable.

However, by the end of the fifth year of the term, the total rent payable will have accumulated, by the end of that year, to some £141,250 which, at a discount rate of 3.5% (YP 5 years at 3.5% = 4.5151) has a NPV of £27,000 × 4.5151 = £127,552.

This therefore brings the letting above the threshold, an LTR must be returned and SDLT is payable.

The time at which the leasehold threshold is crossed is the point at which the tax is calculated, and, therefore, any increase in the rate of tax since the initial commencement of the occupation would impact on the amount of SDLT payable, as would any reduction. Therefore, in the above, the LTR would fall to be filed within 30 days of the commencement of the fifth year.

## *Holding and holding over prior to the grant of new lease*

There are specific rules in relation to the situation where a tenant holds over while negotiating the grant of a new lease. These are set out in paragraph 9 of Schedule 17A. Paragraph 9A was introduced by the FA 2006, section 164, and paragraph 3(1) of Schedule 25. First, the situation where a tenant holds over on a year to year basis is dealt with, then where a tenant holds over while in negotiation leading to the grant of a new lease. As shown below, Schedule 17A paragraph 3(2) also provides for the payment of additional tax and the making of an additional LTR where a fixed term lease comes to an end, with the tenant remaining in occupation and paying rent. Where SDLT was payable on the initial rent during the term then, as the deemed term extends, so further SDLT becomes payable. A tenant in this situation should consider whether it is in his interests, from an SDLT perspective, to seek a further lease rather than pay SDLT on an accumulating basis. This may not be the case. In Study 25, where a new five year lease would have a net present value below £150,000, but as a six, seven or eight year lease it might have a net present value above the threshold on a cumulative basis, it is clearly beneficial to seek a new lease, particularly where the rent payable remains at a similar level.

## Study 24

A lease is granted subject to the Landlord and Tenant Act 1954. It commences on 1 January 2004 at £150,000 pa for five years.

As the lease is for five years, the NPV is £677,265 and SDLT of £5,272 is payable.

An LTR is required within 30 days of 1 January 2004.

The tenant remains in occupation at the end of the term paying the same rent. The lease is re-notified and recharged as a lease for six years.

The NPV is now £799,290 (YP 6 years at 3.5% = 5.3286) and SDLT of £1,220 is due (£6,492 less £5,272 paid in 2004).

An LTR is due within 30 days of 1 January 2009 as the effective date of the deemed six-year lease.

If the tenant continues to hold on after 1 January 2010, the process repeats, and so on.

Where a lease comes to an end by effluxion of time it is common for the tenant to hold over, usually at the existing rent, while the terms and conditions of a new lease are agreed. This is an area of some complexity from the perspective of SDLT and the following paragraphs are based on a technical business brief issued by HMRC following the FA 2006. Where a period of holding-over is followed by the granting of a new lease, it is common for the new lease to have as its start date the express end date of the original lease, thus covering the period of holding over of the original lease. It may also be that the new, increased, rent is payable for this period under the new lease. For SDLT return of such a renewable lease cannot start before the date it is granted (per *Bradshaw* v *Pawley*, see above). To avoid the double taxation of a person who has held over, and made a return and paid SDLT for this period, but who then takes a new lease backdated to start from the end of the old one — ie overlapping the period of holding-over — the SDLT for the new lease is calculated from the end, not the beginning, of the holding-over period that has already gone by. (FA 2003, Schedule 9, paragraphs 9 and 9A.) One is, of course, immediately in breach of the rules because the liability ought to have been declared within 30 days of the start of the new lease, except that the start date may have been backdated a couple of years!

This is despite the fact that the original lease may legally have finished the day before the renewal lease is granted. Where there is an increased rent for the holding-over period, if the increased rent during the hold-over period is paid for the grant of the new lease, HMRC will treat it as a premium for the new lease and tax it as such. If it is paid for occupation under the original lease, it is taxable as rent under the original lease. The following are examples as set out by HMRC.

## HMRC Example 1

The express term of a business lease came to an end on 13 November 2010 but the tenant continued in occupation while negotiating terms for a renewal lease. The original lease was subject to an annual rent of £150,000. The renewal lease is granted on 1 March 2012, having terms from 1 December 2010 at an increased annual rent of £350,000.

The original lease was granted or treated as granted prior to 1 December 2003, the period of holding over, as an extension of the original lease, does not need to be notified at all.

If the increase in rent during the holding over period (£249,315 in total) is paid for the grant of the renewal lease, it will be taxed as a premium for the grant of renewal lease together with any tax due on the NPV of the renewal lease. Only one LTR has to be made notifying the grant of the renewal lease, if tax is payable or the term of the renewal lease is seven years or more.

If, however, the increased rent during the holding over period is paid for occupation under the original lease, it is not taxed as a premium. Instead the new lease will need to be notified and tax paid if tax is payable or the term of the renewal lease is seven years or more. Notification would have to be given of the increased rent under the original lease (by virtue of Paragraph 13 to Schedule 17A of FA 2003) treating the rent increase as the grant of the lease. As the duration of this deemed lease is known (15 months) the NPV can be calculated to be some £239,272. (This is the total of £200,000 discounted by 3.5% and £49,315 discounted by 3.5% twice). This deemed grant would not be linked with the original grant but it will be notifiable with tax due of £892, within 30 days of 1 March 2012, this being the effective date of the grant of the renewal lease. As a term end date is one day prior to the effective date, this transaction is to be sent to the Birmingham Stamp Office. Had there been no tax to pay it would not have be a notifiable transaction.

If the original lease was granted on or after 1 December 2003, the period of holding over, as an extension of the original lease, will need to be notified if the NPV of the rent for each extra year means that more tax is payable. Therefore if the lease started on 1 December 2003, the NPV of the original lease would have been £917,181. On 1 December 2010, the lease is treated as extended by one year to 30 November 2011 and the NPV now becomes £1,031,093. Extra tax is due and notification would have to be made by 31 December 2010. On 1 December 2011, the lease is again treated as extended by a further year to November 2012. This results in an NPV of £1,141,152 and a LTR has to be made, as more tax is due.

If the increased rent during the holding over period of some £249,315 is paid for the grant of your lease, it would be taxed as a premium to be granted for the renewal lease together with any other premium (if any) and on the NPV of the renewal lease. The rent for the first year of the renewal lease (some £350,000) will be reduced by the period of "double counting" of rent of 275 days from 1 March 2012 to 13 November 2012, known as the overlap period. The reduction will therefore be £150,000 divided by 365 and multiplied by 275, being some £113,013. The amount to be entered on the NPV calculator for year 1 of the renewal lease will therefore be £236,987 (£350,000 – £113,013). Only one LTR has to be made notifying the grant of the renewal lease if tax is payable or if the term of the renewal lease is seven years or more.

If a lease is specifically extended during its term this extension is treated as surrender or re-grants. SDLT is therefore payable on the NPV on the re-granted lease, with the rent element reduced during the overlap period by the rent already taken into account for SDLT purposes.

## HMRC Example 2

The express term of a business lease ends on 30 November 2010. In 2008, the tenant negotiates with the landlord to extend the term of the lease by 5 years to 30 November 2015. This new lease is granted on 1 December 2008. The rent remains the same at an annual amount of £200,000.

- If the original lease was granted or treated as granted before 1 December 2003, there is no overlap relief for the rent as it has not been subject to SDLT. The new lease for seven years has an NPV of £1,222,908 and should be notified on an LTR.
- If the original lease was granted on or after 1 December 2003, this is subject to SDLT and the rent used to calculate the NPV of the original lease in the overlap period, 1 December 2008 to 30 November 2010, is available for overlap relief. Therefore, the rent to take into account of the new lease in the first and second years is zero, as overlap relief is the same as the rent due. The NPV of the rent for this lease is therefore £842,969.

It is legally possible for a tenant to enter into a reversionary lease. This is a lease, which takes effect sometime in the future when the current lease expires. The SDLT treatment of these leases depends on whether they are actually granted or whether an agreement of such a lease is entered into. Where a reversionary lease is granted, notification by way of an LTR and any tax are due within the 30 days after the grant. Therefore, where a reversionary lease is granted on 1 December 2005, which has a term from 1 December 2010 to 30 November 2020 at an annual rent of £100,000 the NPV, is £831,660. The first years calculation for NPV is £100,000 discounted by 3.5% not a £100,000 discounted by 3.5% six times. This is because the calculation of the NPV only starts the first year of the term of the lease (paragraph 3 of Schedule 5), which in the case of a reversionary lease is the date the lease commences, 1 December 2010 in this example, and not the date when it is granted.

This compares to the case where an agreement for a reversionary lease is entered into. The agreement is a contract subject to section 44 of the FA 2003. Although the tenant may be in occupation of the property, this will be by virtue of the current lease and does not count as substantial performance of the agreement for a lease, because the occupation is not by virtue of the agreement. When the current original lease determines, the agreement for a lease would be substantially performed, if the reversionary lease has not already been granted as part of the determination of the original lease. Therefore SDLT will only be payable at that time, and the lease, or substantial performance of the agreement, would become notifiable. Therefore, where an agreement for a lease is entered into on 1 December 2005 for a lease to be granted from 1 December 2010 until 30 November 2020 at an annual rent of £100,000, the NPV is again £831,660 but the tax is only payable by 31 December 2010.

It should be noted that the HMRC automated process for the calculation of NPV is unable to deal with reversionary leases, since the correct start date of the lease is in the future. In these cases the LTR should be completed in the appropriate way and then sent to the Birmingham Stamp Office with a covering letter explaining the exact circumstances.

## Variable or uncertain leases

Schedule 17A, paragraph 8, stipulates that, where the rent payable under a lease is variable or uncertain, for example a turnover rent, then, at the end of the first five years of the term of the lease, or the point at which the rent becomes ascertainable, a further return must be made to the Inland Revenue. Where a lease is assigned, any responsibility for the making of any supplementary LTR passes to the assignee (Schedule 17A, paragraph 12(1)). This includes the need for a further LTR in cases where a contingency is fulfilled, or in consequence of a later linked transaction, or where rent becomes ascertainable. An assignee, therefore, needs to ensure that all relevant information is made available by the assignor so that responsibilities are both identified and capable of being fulfilled. Where a lease is varied so that either the term or the rent increases, then a supplementary LTR will be required if more SDLT is payable. Schedule 17A paragraph 13 states that any increase in rent due to a variation of the lease is deemed to be the grant of a new lease. Rent increases due to existing contractual arrangements, such as rent reviews, do not come within this provision. An increase in the term of a lease is considered a linked transaction, section 108, FA 2003 and an additional LTR is required, section 81A and Schedule 17A, paragraph 3.

## Surrenders and reduction of rent or lease term

Where a lease is granted in consideration of the surrender of an existing lease (between the same parties), the surrender is not treated as chargeable consideration for the grant of the new lease. Also, the grant of the new lease does not count as chargeable consideration for the surrender (Schedule 17A, paragraph 16). Neither is the release of obligations under the lease in relation to such a surrender treated as chargeable consideration. These are tenant's obligations, as set out in Schedule 17A, paragraph 10, such as repairing obligations, rental guarantees, penal rents for breach of obligations etc.

Schedule 17A, paragraph 15A, states that where a lease is so varied as to reduce the amount of rent payable, the variation is treated as an acquisition of a chargeable interest by the lessee. Similarly it states that, where a lease is varied so as to reduce the term, the variation is treated as the acquisition of a chargeable interest by the lessor. In each case the effective date will be the date of the deed of variation and an LTR, together with any SDLT due, must be filed within 30 days.

While it is apparent what the consideration for the transaction will be where the lessee makes a payment to the landlord for a reduction in rent, or a landlord makes a payment to a tenant to reduce the outstanding term, and to gain an early surrender, if no payment is made by the lessor to the tenant then there will be no consideration to be brought into account. What is apparent, however, is that the existence of a condition of the lease to allow for upward and downward rent reviews will not be affected by this provision. Nor will an existing condition of the lease allowing for a break clause to be triggered by landlord or tenant. While these are ignored for the purposes of ascertaining the initial length of the lease, they still form part of its contractual terms.

# *Uncertain or unascertained rents*

Where a lease contains provisions for the adjustment of the rent by a rent review, from a specified date or dates that fall within the first five years of the term, the consideration is calculated at the original effective date on the basis that the contingent sum will be payable. Therefore, where the amount is not known, a reasonable estimate is used. For increased rents that become payable on review after the fifth year of the lease, they are (generally) ignored for the purposes of the original LTR. For turnover rents and the like, the highest rent payable in any 12 month period of the previous five year term is used in the recalculation of SDLT.

In this situation there are requirements for either the making of a further LTR together with the payment of any additional SDLT due, or the claiming of a refund within the appropriate time-limits. In the case of a rent review within the first five years there are three points at which an LTR potentially becomes due (Schedule 17A, paragraphs. 7, 7A and 8). First, there is the date at which the rent is ascertained; second, failing that, the end of the fifth year of the lease. Third, where the rent is ascertained after the end of the fifth year, a further LTR is required.

So, in the case of a review in the third year of a lease, it is the point at which the rent is agreed that becomes the effective date, unless still not ascertained at the end of the fifth year of the lease, when an LTR is required, with a further LTR on the later ascertainment of the rent.

## Study 25

A lease is granted for five years, with rent increases to be based on turnover.
  An initial reasonable estimate (Schedule 17A, paragraph 7) is included in an LTR.
  After five years a revised LTR is submitted based on highest rent paid in any 12-month period.
  Additional SDLT is payable (or refundable) depending on estimated and actual rent ascertained.

## Study 26

1.  A lease is granted as above, but for five years.
    An initial reasonable estimate for first five years is made, from which is determined the highest rent likely to be payable in first five years.
    The LTR includes NPV calculated for seven-year term.
    After five years, an amended LTR is required with revised NPV calculation submitted.
    Rent is determined from highest amount actually paid in any 12-month period in first five years.
    SDLT is payable on seven year lease calculation, with credit for original SDLT paid, or a refund is made.

2.  A seven year lease with rent review is granted on 1 January 2005.
    A seven-year lease with rent review at year four is granted on 1 January 2005.
    Initial rent is £100,000 pa.
    Estimated rent at year four is £125,000.
    Rent agreed on review £150,000.
    (a)  Assume that the review was concluded on 31 December 2008. The following LTRs would be required:
         By 31 January 2004, with NPV based on £100,000 (years 1–4) and estimated rent of £125,000 (years 5–7).
         By 30 January 2009, an LTR with NPV based on £100,000 (years 1–4) and £150.000 (years 5–7).

    (b)  Assume that the review was concluded on 1 July 2008. The following LTRs would be required:
         By 31 January 2004, with NPV of £100,000 (years 1–4) and estimated rent of £125,000 (years 5–7).
         By 31 July 2008, an LTR with NPV based on £100,000 (years 1–4) and £150,000 (years 5–7).

(c)   Assume the review was concluded on 1 July 2009 after dispute.
An estimate on 30 January 2009 would be required (say £140,000) as the five year point is reached at 31 December 2008. The following LTRs would be required:

- By 31 January 2004, with NPV based on £100,000 (years 1–4) and estimated rent of £125,000 (years 5–7).
- By 30 January 2009, with NPV based on £100,000 (years 1–4) and re-estimated rent of £140,000 (years 5–7).
- By 31 July 2009, with NPV based on £100,000 (years 1–4) and known rent of £150,000 (years 5–7).

## Abnormal increases in rent

Schedule 17A, paragraphs 14 and 15, set out the conditions under which a rental increase can be considered abnormal. Basically, the date of review in relation to these provisions falls at or after the end of the fifth year of the lease. These have been subject to significant change introduced by the FA 2006, section 164 and Schedule 25 paragraph 8(1), which reduced the six steps envisaged in the original legislation to three. The question the taxpayer must ask at the point of review is whether or not the abnormal increase provisions are in point. These provisions will not affect rent reviews prior to December 2008. The legislation originally provided that, where the annual rent payable after year five increased by more than 5% plus RPI, over and above the rent used in the original LTR, then such increases were regarded as abnormal and would be treated (deemed) as the grant of a new lease. It can be seen that if the RPI increased at say 3% pa then an increase of 50% over say five years would trigger the provisions.

The provisions of paragraph. 15 to Schedule 17A of FA 2003 have been modified by the introduction of a formula which requires the application of three steps.

- Find the start date.
- Find the number of whole years between the start date and the date on which the new rent first becomes payable.
- Test the new rent against the formula R × Y /5 where R is the rent previously taxed and Y is the number of whole years.

The excess rent must be greater than the product of the formula, this is a slightly complex way of stating if the rent increases by more than 100% then it will be treated as abnormal.

Assume a lease for 21 years is entered into on 1 January 2004 (the start date) subject to reviews at three yearly intervals. At the first review in January 2007 no additional (abnormal increase) LTR is required. However, in January 2009 the rent passing, as agreed under the review as at 2007, must be tested to see if it can be considered abnormal. Then, following each review in 2010, 2013, 2016, 2019 and 2022, the rent must be retested to see if, with reference to the original LTR submitted in January 2004, or to the last occasion when there was an abnormal increase, the rent increases can be considered and a fresh LTR may be required with additional SDLT.

The LTRs generally required, however, in the context of this lease, would follow those in example above, the abnormal rent increase calculations providing an additional requirement. While these provisions seem to be more logical from a valuation standpoint than the initial ones because they no longer assume that rents should move within an RPI plus a set percentage driven framework. In practical terms they require the re-evaluation of rental payments at regular stages throughout the term of the lease. Therefore, beneficial variations to a locality, which change value patterns in terms of rents are not, apparently, to be taken into account, only increases in rental value. Effectively, this allows for the additional payment of SDLT if market forces drive rents above a ceiling.

The deemed grant of the new lease is taken as being made on the date the first increased rent becomes payable, and the new lease is taken as being for the, then, unexpired term of the lease. Therefore, if the review of 2010 in the situation envisaged above brought the abnormal rent increase provisions into play, then the deemed lease would be for a term of 15 years. It should also be appreciated that, if the rent increases again during the course of the deemed lease, and it can be judged abnormal by reference to the rent used in the calculation of the net present value in the most recently returned LTR, then a further lease can be deemed to be granted. There is no additional five year rule for the deemed leases as they are past the point of the initial five years of the lease, the transactions being treated as linked transactions to the originating lease.

A further consideration is the meaning of Schedule 17A, paragraph 14(5). It refers to the date on which the increase in rent first becomes payable. This will, unless the lease otherwise provides, possibly be the date of the review or the date on which the rent increase is agreed. Normally, a reviewed rent becomes payable from the quarter day following the date of the new rent being determined, but calculated from the review date In this situation the specific provisions of the lease define the effective date. This leads to the interesting situation that a taxpayer may need to make an LTR within 30 days following the review date after the fifth year of the term, where he believes that the review could lead to the creation of a deemed lease under the abnormal increase provisions after estimating the amount of rent payable on review.

As a rent review, seeking a significant uplift in rent, may lead to extended negotiations and even litigation, the level to which the rent may rise must be estimated. It is also unlikely that deferment of tax (under section 90 of the FA 2003) may be sought as the rent, while uncertain, falls to be paid prior to the date of the LTR (from the first quarter day) rather than at least six months hence. In addition to the abnormal rent LTR, any other LTR requirements, as set out above, also apply.

## Linked transactions

Linked transactions are defined in section 108 of the FA 2003, which states transactions are linked for the purposes of this part of the Act if they form part of a single scheme, arrangement or series of transactions between the same vendor and purchaser, or in either case, persons connected with them. Where two or more transactions have the same effective date they may be notified using a single LTR, the rate of tax being defined by the sum of the chargeable considerations paid. There is a certain amount of confusion surrounding the area of linked transactions, mainly stemming from the SDLT manual at SDLTM 30100, where it states: "It is a question of fact whether or not transactions are linked. A purchaser will need to make a full examination of all the circumstances leading to the transactions before completing their LTR. Just because two transactions are between the same purchaser and seller does not necessarily mean they are linked. The transactions will be linked, however, if they are part of the same deal."

First let us look at what comprises a linked transaction. If a purchaser acquires an entire property by way of two or more transactions, at the same time, from the same vendor, say a flat over a shop, this would clearly be a linked transaction. Where a purchaser acquires, say, three flats from a developer and receives a discount, perhaps by buying off plan, this would clearly form part of the same deal and be a linked transaction. What though is the situation where a purchaser acquires two flats, individually, in the same block from the same vendor and receives no discount? The HMRC view is that if the transactions have the same effective date they are linked, or at least this is what their initial view was, although the wording of section 108, together with the wording of the SDLT Manual (which does not have the force of law) would indicate the possibility of a different view.

Similarly, if a husband and wife purchased two individual flats from the same developer, independently, but at the same date, by virtue of being connected persons these transactions would be linked. The key here is probably the effective date, together with the parties' intentions at the point of the first transaction. Where individual properties are purchased over time without the intention of the transactions being linked as part of a series, then factually they would be separate transactions. For example, if one spouse purchased a flat and the other spouse subsequently purchased another, in the same block, then it would be a matter of fact that the transactions were not linked Similarly if an investor purchased a shop property and subsequently purchased the living accommodation above, perhaps on it falling vacant, it would be hard to see this as a linked transaction unless there was some form of agreement in place between vendor and purchaser at the point of the initial transaction.

The taxable amount, and the applicable rate, are judged from the sum of the chargeable consideration. HMRC's penalty notice forms use the word total consideration, in connection with linked transactions, this has led some HMRC officers, mistakenly, to argue that non-chargeable consideration can be aggregated together with chargeable consideration in setting the applicable rate. This view is wholly wrong as only chargeable consideration is taxable to SDLT, and therefore capable of aggregation.

## Transactions between spouses and civil partners

There is no automatic relief for transactions between spouses or civil partners, even when associated with marriage, divorce and civil partnership. Various different scenarios may arise, as set out below, however as SDLT is charged where an interest in land is transferred for consideration, it should be borne in mind that consideration will include:

- any cash payment
- any assumption of liability to pay a mortgage
- the liability assumed is taken to be a proportion of the outstanding mortgage corresponding to the proportion of the share of the property acquired.

For example, say a house is valued at £180,000. The transferring partner has equity of £90,000 and there is an outstanding mortgage of £90,000. The transferee partner pays a cash sum equivalent to half the equity and acquires a 50% share in the property. The consideration is therefore the cash payment of £45,000 plus 50% of the outstanding mortgage, totalling together some £90,000. As this is below the SDLT threshold of £125,000 there will be no tax to pay. However, details of the transaction must be returned using an LTR. For this purpose, joint tenants are treated as if they each owned 50% of the property.

The situation is different where a couple are divorcing, dissolving a civil partnership or splitting up and wish to transfer the property from their joint names into the name of one partner. Where such a transaction is effected in pursuance of a court order or an agreement between the parties in connection with divorce, nullity of marriage, judicial separation, or the dissolution of a civil partnership it is exempt (Schedule 3). In this case the transaction can be self-certified to the Land Registry using form SDLT 60 and no LTR is required. Otherwise SDLT will be charged as set out above.

For example, a house is valued at £350,000. The partners have an equity of £250,000 and there is an outstanding mortgage of £100,000. The transferee partner pays a cash sum equivalent to 50% of the equity and acquires sole ownership of the property. The consideration is therefore the cash payment of £125,000 plus 50% of the outstanding mortgage, totalling some £175.000. Therefore tax is payable at 1%, a sum of £1,750. An LTR must be made in the usual way.

# Chargeable consideration

"The amount of tax chargeable in respect of a chargeable transaction is a percentage of the chargeable consideration for the transaction", section 55(1). The determination of the chargeable consideration in respect of a purchase is explained below. Chargeable consideration is defined in the FA 2003, section 50 and Schedule 4.

## Single payment

In the majority of cases the purchase of an interest in a property is the payment of one agreed amount, which is the chargeable consideration. It may be that a part payment is made on exchange of contracts, but that is part of the agreed amount. SDLT is payable on the agreed amount. As stated, the agreed amount paid determines the tax payable, so that, for example, where the transaction is a gift with no consideration, there is no SDLT payable, Schedule 3, paragraph 1, although market value may be substituted for the operation of other taxes. The one exception is where the purchaser is a company purchasing from a connected vendor, when market value will be adopted if it is different from the actual consideration, section 53 of FA 2003.

The position is the same where payment is made by someone connected with the company purchaser. If the price paid includes value added tax (VAT) because, for example, it is a new non-residential building, or the vendor has opted to tax, SDLT is payable on the VAT inclusive price, but not otherwise. Unlike the old stamp duty, if no VAT is included but the price could be subject to VAT after the effective date of the transaction, by virtue of a later election to opt to tax, the potential VAT payment is left out of account, Schedule 4, paragraph 2. On the other hand, if there is a single bargain, but separate consideration is given, or purports to be given, for various elements (which would consequently reduce the tax payable), tax is payable on the aggregate consideration, Schedule 4, paragraph 4.

## Payment other than cash

In those cases where the consideration is part in money and part in some other form, or indeed wholly in another form (money's worth rather than money), the market value of the non-monetary consideration at the effective date of the transaction is the value of the consideration, Schedule 4, paragraph 7. Where the non-monetary consideration is the provision of services, the value is the amount that would have to be paid in the open market to obtain those services, Schedule 4, paragraph 11.

For example, if a residential developer purchases land for development and agrees, in addition to paying a sum of money for the land, to sell back one of the houses to be built to the vendor at cost, the market value of buying at cost is added to the price of the land for SDLT purposes. Say the market value of the house is £300,000 and the vendor will have to pay only £160,000 to buy it, a figure around £140,000 would be added to the money paid by the developer for the land. The vendor, when buying back the new house, will pay SDLT on £160,000.

Where a transaction includes a requirement for the vendor or someone connected to the vendor to carry out work, the value of such work will be added to any other money payment by the purchaser to determine the chargeable consideration. The value of the work is the sum that would have to be paid in the open market to have the work carried out, Schedule 4, paragraph 10. This is principally an anti-avoidance measure to stop the loss of tax by a scheme whereby a developer sells some land, typically a plot of land for the building of a house, on the basis that the developer will then build the house for the purchaser. In the absence of this anti-avoidance measure, the tax payable would be calculated only on the price of the land (which may, perhaps, have been sold at less than its market value).

The measure does not apply where the works are carried out after the effective date of the transaction and are to be carried out on the land the subject of the transaction, or on other land owned by the purchaser or someone connected to the purchaser, and it is not a condition of the transaction that the works are to be carried out by the vendor or someone connected to the vendor.

In a press release in April 2004 (SDLT — sale of land with associated construction, & c, contract), the Inland Revenue stated that

> we have been asked how to determine the chargeable consideration for SDLT purposes where V agrees to sell land to P and V also agrees to carry out works (commonly works of construction, improvement or repair) on the land sold. Our view is that the decision in *Prudential Assurance Co Ltd* v *IRC* [1992] applies for the purposes of SDLT as it does for stamp duty. This is because the basis of the decision was the identification of the subject-matter of the transaction and this is as relevant for SDLT as it is for stamp duty.

Where the sale of land and construction contract are, in substance, one bargain then, following *Prudential*, there must be a just and reasonable apportionment.

## Phased payments

Where all or part of the consideration is payable at an agreed date after the transaction, no discount is given for such postponed payments, Schedule 4, paragraph 3. For example, if some development land is purchased under a single contract as a single transaction and payment is agreed to be spread over three years, by four equal payments of £500,000, one at the time of the transaction and three further payments at the end of years one, two and three, SDLT will be payable on 4 x £500,000, £2 m, even though the value to the vendor, and the cost to the purchaser, is less than £2 m.

## Contingent payments

Some transactions are structured to provide for further payments to the vendor contingent on a future event taking place. The treatment of the consideration in such cases is set out in section 51 of the FA 2003. Typically a payment is made at the time of the transaction, with provision for a further payment on, say, receipt of planning permission. The further payment may be fixed in the agreement leading to the transaction, or may be a figure which is to be determined at the time of the contingent event. For example, a farmer sells land to a developer for £100,000 with a further payment to be made of £900,000 on the grant of planning permission for residential development.

However, under section 90 of the FA 2003, if the further payment will be made, or may be made, more than six months after the effective date of the transaction, the purchaser may apply to the HMRC to defer payment of tax on the contingent sum.

Alternatively, the farmer sells the land to a developer for £100,000 with a further payment to be made of 90% of the market value of the land if planning permission is granted, at the time it is granted. In this case a reasonable estimate of the further payment must be made and added to the payment of £100,000, the whole being treated as chargeable. It is likely that a purchaser, in such circumstances, might seek advice from a valuer, who would need to check whether offering estimates of future values is covered by his professional insurance. This type of case is one where consideration deferment under section 90 is likely to be applied, as it depends on a future contingent event to trigger the additional payment.

## Debt as consideration

In some transactions, the purchaser may include in the terms an allowance involving debt, which is reflected in the overall consideration. For example, the purchaser may release the vendor from a debt owed to the purchaser, or may take over responsibility for a debt owed by the vendor, such as taking over the vendor's outstanding mortgage.

In such circumstances, the amount of debt released or assumed is part of the chargeable consideration, Schedule 4, paragraph 8. However, if the addition of this outstanding debt takes the chargeable consideration above market value, the chargeable consideration is limited to the market value. Where the debt is secured on the property which is the subject-matter of the transaction, and two or more persons hold an undivided share in the property before or after the transaction, the debt is apportioned in the proportion of the undivided share to the whole property. Joint tenants are treated as holding equal undivided shares of the property.

## Exchanges

When a transaction involves the sale of a property where the vendor also purchases a property from the purchaser, the chargeable consideration in respect of each purchase is the market value of the property being purchased, Schedule 4, paragraph 5. So, if A sells Blackacre to B in return for the purchase of Greenacre from B, the chargeable consideration on which B pays SDLT for the purchase of Blackacre is its market value, while A pays SDLT on the market value of Greenacre.

This applies where the subject-matter of the sale or purchase is a major interest in land (basically a freehold or leasehold interest or its equivalent in Scotland). If none of the transactions is in respect of a major interest, the chargeable consideration is the amount or value of any consideration given for the acquisition other than the disposal. If the acquisition is the grant of a lease, at a rent, the chargeable consideration is that rent. In certain instances, where a house building company acquires from the purchaser of one of its new houses the interest in the house owned by the purchaser, the company is not liable to pay SDLT on its purchase.

## Payments for chattels

It is not uncommon for items other than the property itself to be included in a transaction, for example furniture, carpets and curtains with a house, or plant and machinery, and fixtures and fittings with a commercial property. Since SDLT is payable in respect of a chargeable land transaction, it follows that items other than the property itself should be left out of account. Since the payment structure of SDLT is illogical, throwing up high marginal rates of tax as the consideration moves from one band to another, there is a temptation to inflate the price paid for chattels so as to keep the consideration for the property itself in a lower band. It is impossible to say how widespread this practice is, but the HMRC have made it clear that they intend to attempt to stop it.

# Compulsory Purchase

The following abbreviations are used in this chapter:

| | |
|---|---|
| LCA 1961 | Land Compensation Act 1961 |
| CPA 1965 | Compulsory Purchase Act 1965 |
| LCA 1973 | Land Compensation Act 1973 |
| ALA 1981 | Acquisition of Land Act 1981 |
| T&CPA 1990 | Town and Country Planning Act 1990 |
| P&CA 1991 | Planning and Compensation Act 1991 |
| TCGA 1992 | Taxation of Chargeable Gains Act 1992 |
| P&CPA 2004 | Planning and Compulsory Purchase Act 2004 |

## PART I: PROCEDURAL MATTERS
## Historical summary

The consolidation of compulsory purchase and compensation procedures began with the Lands Clauses (Consolidation) Act 1845 which set out a code designed to regulate the procedure, compensation and any other disputed matters likely to arise between owners and acquiring authorities. Under this legislation, all compensation was assessed on the basis of value to owner.

A major review of the principles relating to the assessment of compensation resulted in the Acquisition of Land (Assessment of Compensation) Act 1919. This introduced several modifications; principal of which were the introduction of the six rule governing the assessment of compensation and the alteration of the general basis of valuation from value to owner to market value. The six rules of that Act were re-enacted in section 5 of the LCA 1961 and this Act was made applicable to all cases of compulsory acquisition.

The CPA 1965 consolidated the earlier 1845 Acts in so far as procedural and other matters relating to compulsory acquisition were concerned while the ALA 1981 lays down a detailed procedure for the making and confirmation of a compulsory purchase order applicable to acquisitions by local authorities and in respect of transportation matters.

The LCA 1973 addressed a number of perceived deficiencies apparent in the assessment of compensation as at the time of its enactment while the T&CPA 1990, the P&CA 1991 and the P&CPA 2004, although directed mainly at planning matters, included various compensation provisions.

Thus most acquisitions will now involve the rules of the 1965 Act, the procedure of the 1981 Act and the compensation provisions of the 1961, 1973, 1990, 1991 and 2004 Acts. No problems then!

Finally it should be noted that the TCGA 1992 makes provision for rollover relief for taxation purposes where a chargeable gain arises as a consequence of a compulsory acquisition. For further details regarding this, please see Chapter 8.

# Commencing the process

There are two alternative ways that the acquiring authority can implement the compulsory purchase process; by service of notice to treat or by general vesting declaration (GVD). Each has advantages and disadvantages for both the authority and the claimant. The notice to treat procedure generally gives greater flexibility to all, however title to the land to be acquired cannot be secured by the authority until the compensation is either agreed or determined by the Lands Tribunal. On the other hand, under the GVD procedure, title to the land passes to the authority on the date of vesting, but this procedure tends to offer less flexibility.

## Notice to treat

Until 1991, there was no statutory time-limit governing the life of a compulsory purchase order where the notice to treat procedure was to be used, although there were a number of case law decisions where it was held that due to the passage of time the compulsory purchase powers had been abandoned and so could not be implemented. This uncertainty changed in 1991. Section 67 P&CA 1991 inserted a new section 2(A) in the CPA 1965. This provides that a notice to treat shall cease to have effect after three years unless possession has been taken, compensation has been agreed, awarded, paid, or paid into court or the question of compensation referred to the Lands Tribunal. The parties may agree, however, to extend the three year period. Compensation is payable for any loss or expenses occasioned by the serving of the notice to treat and its ceasing to have effect, section 2(C) including interest 2(E).

The P&CA 1991 further provides that possession of the land to be acquired must be taken within three years of the date of service of the notice to treat. The combined effect of this legislation is therefore that a compulsory purchase order has a life expectancy of a maximum of six years.

The date of service of notice to treat does not fix the date for valuation. The owner of a compensatable interest must therefore continue to ensure that its interest remains fully protected. For example, it should continue to insure, maintain and protect the property from vandalism: see *Lewars v Greater London Council* (1981) and *Blackadder v Grampian Regional Council* [1992].

However, compensation can be claimed only in respect of such interests in the land as exist at the date of the notice to treat and the acquiring authority's total burden cannot be increased by the creation of fresh interests after that date if done with the object of increasing the compensation: see *Mercer v Liverpool, St Helens & South Lancashire Railway* [1904]. This early case law has now been overtaken by statute. Section 4, ALA 1981 provides that in assessing compensation, the Lands Tribunal may ignore any interest created or work done if it is satisfied that either was done only with the object of increasing the compensation payable, even where either was done prior to the date of service of notice to treat.

For many years it was the case that no claim for compensation, particularly in respect of disturbance, was sustainable prior to the date of notice to treat. However, following the decisions in

*Prasad* v *Wolverhampton Borough Council* (1983) CA and *Director of Buildings and Land (Hong Kong)* v *Shung Fung Ironworks* [1995] PC, claims for losses incurred prior to the date of Notice to Treat may be claimed in certain circumstances; see Disturbance later in this chapter.

Following the service of a notice to treat the owner may be forced to sell and the acquiring authority obliged to purchase the land in question, but the authority has a six-week statutory period in which it may withdraw the notice to treat. This period runs from the receipt of the claim (section 31(1) LCA 1961). In the absence of a valid claim, a notice to treat may be withdrawn within six weeks after the decision of the Lands Tribunal (section 31(2)).

Compensation for any loss or expenses occasioned by the giving and withdrawal of the notice is payable under section 31(3) but this will be limited if it is withdrawn under section 31(2) and the claim was made late. It was held in *R* v *Northumbrian Water Ltd, ex parte Able UK Ltd* [1995] that the water company had an unqualified right to withdraw a notice to treat under section 31(1) within six weeks of a claim even though it had entered into possession of the land, but it should be noted that if the notice to treat had been withdrawn under section 31(2) the end result might have been different. In *Williams* v *Blaenau Gwent Borough Council* (1994) the Lands Tribunal decided that they had jurisdiction to award compensation where the authority had decided not to proceed with the acquisition but the owners had agreed to the withdrawal of the notice to treat upon payment of the proper compensation.

## General vesting declaration

This is effectively an expedited procedure whereby the acquiring authority can secure clean title to an interest in land faster than would be the case if the Notice to Treat procedure was used. The governing legislation is the Compulsory Purchase (Vesting Declarations) Act 1981.

The acquiring authority must first serve a notice of intention to use such powers. The notice must be published in a local newspaper. Additionally, within a reasonable period of time following publication of the notice in the press, notice must be served on those affected by it, except those with minor tenancies.

Not earlier than two months after the notice of intention is published, unless every party affected agrees in writing to a shorter time scale, the acquiring authority can then make the GVD in the form prescribed by the Compulsory Purchase of Land Regulations 1994 (SI 1994 No 2145). Notice of the making of the GVD must be served on every person affected, except minor interest holders, and upon any other person that responded to the notice of intention.

The GVD will take effect on the date specified in the notice advising of the making of the GVD provided that date is not less than 28 days after the making of the GVD.

Under section 4 CPA 1965, a GVD must be exercised within three years of confirmation of a CPO. In *Co-operative Insurance Society Ltd* v *Hastings Borough Council* [1993] it was decided that the Notice of Intention was merely a warning to ensure that the powers were subsequently exercised properly. The GVD should come into effect within the three year time-limit measured from the date of confirmation.

# Date of valuation on compulsory acquisition

With effect from the enactment of the P&CPA 2004, the date of valuation has been codified by statute. Section 103 P&CPA 2004 inserted a new section 5A into the LCA 1961 which states that in all cases except those to which rule 5, section 6 LCA 1961 (see later) applies, the valuation date is the earlier of:

- the date when possession is taken
- the date that compensation is agreed between the parties
- the date when compensation is determined by the Lands Tribunal.

This legislation effectively enacted what had been understood to be the case since 1968 following the decision in *West Midland Baptist (Trust) Association (Inc) v Birmingham Corporation* (1968) CA.

When possession is taken will be a question of fact in every case. In *Courage Ltd v Kingswood District Council* [1978] it was decided that the council had taken possession where it had entered the land and carried out certain site works which increased the value although a key to the entrance gate was not handed over by the claimants until seven months later.

The date when compensation is determined by the Lands Tribunal follows the decision in *C&J Seymour (Investments) Ltd v Lewes District Council* [1992]. The date of valuation was held to be the last day of the tribunal hearing. This followed the Court of Appeal decision in *W&S (Long Eaton) Ltd v Derbyshire County Council* [1975].

For rule 5 cases the *West Midland Baptist Trust* position is maintained in that the valuation date is the earliest date on which the work of reinstatement could reasonably have been expected to start.

Section 103 P&CPA 2004 also provides that in assessing market value under rule 2, the valuation must be made as at the valuation date. This clarifies the application of the decision in *Bwyllfa & Merthyr Dare Steam Collieries (1891) Ltd v Pontypridd Waterworks Co* [1903] where it was held that hindsight evidence could be used to determine the compensation payable in connection with a claim for disturbance, Lord Macnaghten saying in that case, "Why should the arbitrator listen to conjecture on a matter which has become an accomplished fact? With the light before him, why should he shut his eyes and grope in the dark?" It is important to note that the *Bwyllfa* decision only applies to the assessment of compensation for disturbance.

# Time barring of claims

Where the notice to treat procedure had been adopted it was not appreciated, until 1998, that any time-limit applied to the making of a reference to the Lands Tribunal. However, in *Hillingdon London Borough Council v ARC Ltd* [1998] CA, it was held that in accordance with section 9 of the Limitation Act 1980, a claim for compensation must be referred to the Lands Tribunal within six years of date of entry on the land unless by reason of its conduct the authority was not entitled to rely on that section. Subsequent cases in respect of the application of the Limitation Act have examined the effects of such as estoppel with mixed results for the various parties, each case depending upon its individual circumstances.

The position is reasonably clear in a case where the notice to treat procedure has been used. Unless the acquiring authority has agreed in writing to waive the Limitation Act point, the claimant should ensure that a reference has been made to the Lands Tribunal to determine the compensation within the period of six years of the date of possession.

The position is even clearer where the GVD procedure has been used. Section 4 CPA 1965 provides that compulsory purchase powers must be used within three years of those powers becoming operative; that is within three years of the date that notice of the confirmation of the order is given. It was held in *Co-operative Insurance Society v Hastings Borough Council* [1993] that a notice of intention under section 3 of the Compulsory Purchase (Vesting Declarations) Act 1981 did not amount to the exercise of the powers. However even this relatively clear legislation may be open to challenge. In *Co-operative Wholesale Society v Chester le Street District Council* [1996] it was held, based on the conduct of he parties, that a reference to the Tribunal made after the six-year period of limitation referred to in section 10(3) of the Compulsory Purchase (Vesting Declarations) Act 1981 was not time barred. The President found that two issues fell to be decided. First, whether the time-limit could be waived either by agreement or by estoppel (and held that it could be) and, second, whether the conduct of the authority in continuing negotiations well after the cut-off date amounted to waiver. The appropriate test will be found in Lord Denning's judgment in *Amalgamated Investment & Property Co Ltd v Texas Commerce International Bank Ltd* [1982] CA.

# PART II: THE VALUATION RULES
# Compensation for land taken — the statutory valuation rules

Section 5 LCA 1961 contains six sub-sections each containing one of the "six rules" which relate to the assessment of compensation, as follows.

**Rule 1** "No allowance shall be made on account of the acquisition being compulsory"

The rule was introduced in 1919 to halt the practice of adding 10% to the assessed compensation to acknowledge the compulsion of the acquisition. However the LCA 1973 saw the introduction of home loss and farm loss payments while the P&CPA 2004 saw the introduction of basic and occupier loss payments; all effectively bonus payments to reflect the compulsory nature of the acquisition! The introduction of these payments must beg the question as to whether this rule now has any real meaning!

**Rule 2** "The value of land shall, subject as hereinafter provided, be taken to be the amount which the land if sold in the open market by a willing seller might be expected to realise"

The classic statement of this rule is perhaps to be found in the decision in the case of *Raja Vryricherla Narayana Gajapatiraju v Revenue Divisional Officer Vizagapatam* (1939) otherwise and more helpfully known as the *Indian* judgment.

> The compensation must be determined, therefore, by reference to the price which a willing vendor might reasonably obtain from a willing purchaser. The disinclination of the vendor to part with his land and the urgent necessity of the purchase to buy must alike be disregarded.

The rule allows for potentialities and hope value in so far as they would be reflected in market value: In *Ali v Southwark London Borough Council* (1977) the fact that the tenants were close friends of the claimant landlord and there was an understanding that they would vacate if he wished to reoccupy was decided by the tribunal to be of no relevance and compensation was awarded on an investment value approach, presumably as the market would have considered there to be a high degree of risk of the understanding not coming good. However, in the case of *Sullivan v Broxtowe Borough Council* (1986) LT, it was held that the claimant was entitled to take into account the potentiality of obtaining a home improvement grant when assessing his claim for compensation. Similarly in the case of *John David v Lewisham London Borough Council* (1977) it was held that compensation could be based on vacant possession value if the occupier was a licensee, presumably to reflect the possibility of recovering possession.

**Rule 3** "The special suitability or adaptability of the land for any purpose shall not be taken into account if that purpose is a purpose to which it could be applied only in pursuance of statutory powers or for which there is no market apart from the requirements of any authority possessing compulsory purchase powers"

This wording has applied since 1991 as a result of amendments enacted by Schedule 15, para 1 P&CA 1991.

This rule was originally enacted to remove the prospect of additional value being paid due to some inherent physical quality which the land possessed and was worded ostensibly to effect this result. For example, if land was to be acquired for the purpose of creating a reservoir, the purchase of farmland comprising a valley would be ideal. Considerable expenditure would be saved through not having to construct flank retaining embankments and in the absence of this rule, that saving in cost could have been translated into land value.

Over the years however it was successfully argued that the rule also applied to ransom strips, ie areas of land that possess a value due to their providing access and so releasing latent value in other land, with the result that it became commonplace for acquiring authorities to acquire access to backland suitable for development at nominal cost for the necessary access. Perhaps the worst example of this application of rule 3 was a case where a claimant was only awarded £1 as compensation for a strip of land that held the key to releasing 2 acres of residential building land for development.

An important point, which seems to have been overlooked in the earlier cases, is that the exclusion of the special suitability over-bid only applies where its existence is wholly dependent upon statutory powers (ie the land could not be put to the use without them) or is wholly due to demand from bodies possessing compulsory purchase powers. In the ransom strip cases it was eventually pointed out in evidence that commercial developers commonly buy access rights at high prices in the open market.

Thus it was that from around 1984 onwards, the Lands Tribunal began to give more pragmatic decisions where ransom strips were in issue, so reflecting the commercial reality of their value. In various decisions, the tribunal took into account the cost that might be incurred in purchasing an alternative parcel of land to provide the necessary access, the merits of competing accesses and even imported hypothetical speculators to justify the value of an access until the point was reached where it could no longer be said that the rule applied to such cases and the 1991 amendment to rule 3 completed the process.

However, even the revised wording of rule 3 might still have created difficulties for ransom strip owners had it not been for decisions in two lengthy litigations.

*Hertfordshire County Council* v *Ozanne* [1989] concerned the compulsory acquisition of land for a highway improvement. The improvement would facilitate the development for residential purposes of a large area of land in a different ownership once a statutory road closure order had been made in respect of the existing public highway which was on land that was not part of the reference land. On the eventual appeal to the House of Lords [1991] it was held that statutory powers conferred on the Secretary of State to order the stopping up of a highway on land that was not part of the land being acquired could not justify the application of rule 3 to the land which was being acquired from the claimant. Rule 3 applied only if the statutory powers relating to the purpose for which powers were sought related to the use of the land being acquired.

In *Batchelor* v *Kent County Council* [1990] the Court of Appeal ruled that rule 3 could apply only if the land had a special suitability or adaptability. The case involved a potential ransom strip. The Lands Tribunal had found that although the most suitable access was on the order land, this was not the only possible access as there were other alternatives available at the time. While the reference land might therefore have been the most suitable, "most suitable" did not correspond with "specially suitable" and consequently the prefatory words of rule 3 were not satisfied so as to exclude ransom value: See also *Wards Construction (Medway) Ltd* v *Barclays Bank plc* (1994) a further reference of the same case where Barclays Bank plc was substituted in place of Mr Batchelor.

The meaning of special suitability was examined by the House of Lords in *Waters* v *Welsh Development Agency* [2004], and the interpretation given it by the courts in the previous cases was much criticised. Over the years the phrase had come to mean almost uniquely suitable and there was a general view that rule 3 had become of little application. The House ruled that this very narrow interpretation was far too restrictive and in the light of that ruling the judgment in *Waters* from now should be regarded as the primary source of judicial guidance. Further cases on rule 3 must now be expected, but however suitable the land may be, the exclusion of the extra value only applies if its existence entirely depends on statutory power or the sole interest of authorities having compulsory powers for its release.

**Rule 4** "Where the value of land is increased by reason of the use thereof or of any premises thereon in a manner which could be restrained by any court, or is contrary to law, or is detrimental to the health of the occupants of the premises or to the public health, the amount of that increase shall not be taken into account"

Section 4 P&CA 1991 addresses time-limits on enforcement action and introduced a new section 171B to the T&CPA 1990. Where there has been a breach of planning control consisting of the carrying out without planning permission of building, engineering or mining operations or similarly where there has been a change of use of any building to use as a single dwelling-house, no enforcement action may be taken after four years from completion of the operations or, in the latter case, of the change of use. In the case of any other breach of planning control, the period is 10 years from the date of the breach. Section 10 P&CA 1991 also substituted new sections 191 to 194 of the T&CPA 1990 replacing established use certificates with certificates of lawful use or development. Section 10 (2) of the P&CA 1991 now provides a new section 191(2) T&CPA 1990 stating that uses and operations are lawful if no enforcement action may be taken. Furthermore, the possibility of planning consent being assumed under sections 14 to 17 of the LCA 1961 must also be borne in mind when considering unlawful uses.

In the light of the above provisions it is clearly impossible to treat an established use under the T&CPA 1990 as being contrary to law within the meaning of rule 4.

However, *Hughes* v *Doncaster Metropolitan Borough Council* [1991] HL, concerned land used without planning permission but immune from enforcement action. The point at issue was whether the disturbance claim should be reduced on account of the use of the land not being lawful. It was held that the use could not be taken to be contrary to law within the meaning of rule 4 and was therefore outwith the scope of it. Compensation for disturbance was therefore payable in full. The House of Lords stated that an Act could not be used to take away private property rights without compensation unless it specifically authorised this, and held that although in practice the value of the land and disturbance were separately assessed, the courts had consistently maintained the principle that the value of the land taken and compensation for disturbance were inseparable parts of a single whole in that together they made up the value of the land to the owner. The P&CA 1991 has inserted a new section 191(2) into the T&CPA 1990 which renders lawful any use or other breach of planning control that is beyond the reach of enforcement action.

**Rule 5** "Where land is, and but for the compulsory acquisition would continue to be, devoted to a purpose of such a nature that there is no general demand or market for the land for that purpose, the compensation may, if the Lands Tribunal is satisfied that reinstatement in some other case is *bona fide* intended, be assessed on the basis of equivalent reinstatement"

Although there are earlier cases concerning this rule, the case of *Sparks* v *Leeds City Council* [1977] relating to a social club, laid down four essentials before the rule can apply; the burden of proof being on the claimant. These are:

(1)  that the land is devoted to a purpose and would continue to be so devoted but for the compulsory acquisition

(2)  the purpose is one for which there is no general demand or market

(3)  there is a bona fide intention to reinstate and

(4)  if these conditions are satisfied the Tribunal's discretion should be exercised in the claimant's favour.

It should be noted that (2) above is not saying "there is no general market demand."

In *Zoar Independent Church Trustees* v *Rochester Corporation* (1974) rule 5 was held to apply in the case of a chapel, even though the congregation fell to only three or four by the date of entry, on the grounds that but for the compulsory acquisition, the purpose, ie the devotion by the congregation, would have continued where it was but was instead going to continue in its new location.

In *Nonentities Society (Trustees)* v *Kidderminster Borough Council* (1970) the rule was held to apply in the case of a theatre in Kidderminster, although the Tribunal opined that the position might have been different had the theatre been in London.

In *Harrison & Hetherington Ltd* v *Cumbria County Council* [1985] which case concerned a livestock market, it was held by the House of Lords that the fact that such a property might sell if it was put up for sale was not evidence of a general demand within the meaning of rule 5 and the fact that from time to time similar properties were marketed was not evidence of a general market. Rule 5 was therefore applicable. However *Wilkinson* v *Middlesborough Borough Council* [1982] CA concerned the acquisition of premises occupied by a veterinary surgeons' multi-principal practice, a type of property that perhaps also only comes to the market on rare occasions. In this case the Court of Appeal held that rule 5 did not apply as the claimants had not discharged the burden of proof of establishing that there was no general demand or market for property for carrying on veterinary surgeons' practices.

The case of *Festinog Railway Co* v *Central Electricity Generating Board* (1962) CA, is an example of the Tribunal's ability to exercise its discretion. In this case, although the Tribunal held that all the conditions specified in the rule were met, it was held that the Tribunal was not bound to assess compensation on a rule 5 basis, the Tribunal having decided that given the relationship between the cost of reinstatement and the value of the undertaking, it would not be reasonable to compensate on the basis of equivalent reinstatement. Having won the case, however, the CEGB made an *ex gratia* payment to enable the railway to reinstate.

Following the decision in *West Midland Baptist (Trust) Association (Inc)* v *Birmingham Corporation* [1968] the valuation date is the earliest date when reinstatement might reasonably have commenced. The *West Midland Baptist* case also confirmed that no deduction should be made for the age of the building, as if such an adjustment is made, the whole object of rule 5 will be defeated. Unless the authority meets the cost in full, the claimant will end up out of pocket and perhaps be unable to make up the shortfall. Compulsory acquisition could result in the extinction of such as charities through the application of a rule which was intended to enable them to survive. However in *Roman Catholic Diocese of Hexham & Newcastle* v *Sunderland CBC* (1964) it was held that the cost of essential repairs to the existing hall, and fees commensurate with those repairs, should be deducted from the compensation payable as this cost was already the claimant's liability. The decision is quite logical, a use will be reinstated without abatement of compensation whatever the age or obsolescence of the building provided that building is in repair, and similarly reinstated but with a deduction for the cost of repair if it is not.

Although rule 5 is directed at the cost of equivalent reinstatement, it is also considered that the rule might be applied in a case where instead of acquiring a site and building a new building, an acquiring authority might purchase another building and convert it to make it suitable for occupation for the purposes of the use which is to be reinstated.

However, the application of rule 5 is not sacrosanct in the case of specialist buildings. If the rule 2 basis of valuation exceeds the rule 5 figure, perhaps because the land has a higher value for redevelopment, the claimant is entitled to claim under rule 2.

Finally, statutory interest is payable in respect of compensation assessed under rule 5, even where an acquiring authority has met the various stage payments to be made under a construction contract such that the claimant has never been out of pocket; see *Halstead* v *Manchester City Council* [1997]. Interest is specifically recoverable under section 11 CPA 1965 until the final cost instalment is paid, which statute should not be interpreted as meaning that the right to interest should only run from entry until reinstatement takes place: see Study 2, below.

The final rule governing compensation is rule 6.

**Rule 6** "The provisions of rule (2) shall not affect the assessment of compensation for disturbance or any other matter not directly based on the value of the land"

Compensation for disturbance is a subject in itself. It is considered in detail later in this chapter.

# The other principle valuation requirements when assessing market value

## The scheme

In any case where compulsory purchase powers are used, or assumed to be used, the first task is to identify the scheme. This usually takes two forms; the statutory scheme provided for in section 6 and Schedule 1, LCA 1961 and the case law scheme.

Section 6 and Schedule 1 LCA 1961 provide that any increase or decrease in the value of land taken is to be ignored in so far as it is due to development or prospective development under the scheme, unless in the absence of the scheme it would have been likely to have occurred. Schedule 1 sets out four cases, two of which have been subdivided.

| | |
|---|---|
| Case 1. | Normal CPO cases |
| Case 2. | Comprehensive development areas and action areas |
| Case 3 and 3A. | New towns and extensions thereto |
| Case 4. | Town development plan cases |
| Case 4A. | Urban development area cases |
| Case 4B. | Housing action trust areas |

In each case the schedule sets out the development which is to be taken as part of the scheme, the effect on the value of which is to be disregarded, except in so far as it might reasonably have been expected to have taken place anyway, ie without the scheme. However, the statutory scheme of section 6 is often a very limited geographic area. It might only encompass one or two areas of land where the owners have resisted entering into any agreement to sell, believing themselves to have an advantageous bargaining position due to their now being located in an area that has (or is due to be) altered dramatically as a consequence of overall regeneration plans etc. So at this juncture, in steps the case law scheme. The case law scheme has the potential to widen the section 6 scheme.

Prior to the decision in *Waters v Welsh Development Agency* [2004] the classic exposition of the identification of the scheme was perhaps best illustrated in the speech of Lord Denning in *Myers v Milton Keynes Development Corporation* [1974]

> The valuer must cast aside his knowledge of what has in fact happened ... due to the scheme. He must ignore all future developments which will in all probability take place in the future ... owing to the scheme. Instead, he must let his imagination take flight to the clouds.

Milton Keynes was a 'new town' that had grown as a result of several phases. Being a new town, in general all the surrounding land had previously been fields. As a result of the new town designation, the Development Corporation had invested heavily in infrastructure such as roads, drainage etc. The subject land was identified in a compulsory purchase order relating to a later phase of the entire new town and consequently was surrounded by existing infrastructure. However, in valuing the subject land it was held that the existence of that infrastructure must be ignored. The scheme was the development of the entire new town, not just the area of the compulsory purchase order of which the subject land formed a part.

*Myers* followed a long line of cases establishing the principles relating to the scheme which, although that line had commenced many years beforehand, became known as the *Pointe Gourde* principle after the case of *Pointe Gourde Quarrying & Transport Co Ltd* v *Sub-Intendent of Crown Lands* [1947]. This case reiterated that, 'it is well settled that compensation on compulsory purchase of land cannot include an increase in value which is entirely due to the scheme underlying the acquisition' and proceeded to rule out a claimant's argument for additional loss of profit on account of the fact that the additional profit would not have been realised had the scheme works not taken place; on the face of it, a very logical conclusion. However, subsequent cases relating to the identification of the scheme, particularly in more recent years, led to increasingly more difficult mental gymnastics becoming necessary when seeking to ignore 'what has in fact happened' as expressed by Lord Denning.

Perhaps *Abbey Homesteads (Developments) Ltd* v *Northamptonshire County Council* [1991] illustrates the tortuous path of the scheme considerations. The case concerned land subject to a section 52 agreement reserving it for school purposes. The Court of Appeal had decided [1986] that a covenant in an agreement was a restrictive covenant running with the land and intended to be permanent, and declared that compensation for the land was to be determined on the basis that it was affected by the covenant. On remit, however, the Tribunal decided that the imposition of the restrictive covenant was part of the scheme underlying the acquisition and also that the agreement itself constituted an indication within section 9 LCA 1961. On appeal against this decision [1992] JPL 1133, the Court of Appeal stated that neither *Pointe Gourde* nor section 9 operated to remove the restriction nor require it to be disregarded. However, in the no-scheme world there would still have been a need for a school to be provided on an alternative site, in which case the section 52 agreement on the subject land could be discharged!

In *Waters*, the House of Lords reviewed in detail the case law relating to the scheme and concluded by identifying six points of general principle to be applied to its identification.

(1)  The *Pointe Gourde* principle should not be carried too far.
(2)  A result is not fair and reasonable where it requires an unreal or virtually impossible valuation exercise.
(3)  A valuation exercise that produces a wide variation between the value of the subject land and adjacent similar land that is not being acquired should be viewed with caution.
(4)  *Pointe Gourde* should be applied in an analogous manner to the statutory provisions of section 6 and Schedule 1.
(5)  Guidance as the extent of the scheme can be derived from the formal resolutions of the acquiring authority and the accompanying statutory documentation but these should not be thought of as conclusive.
(6)  If in doubt, a narrow rather than wide identification of the scheme should be adopted.

Following *Waters* perhaps less mental gymnastics are now needed but it remains to be seen.

If a premium value is entirely due to the scheme underlying the acquisition then it must be disregarded. However if it existed prior to the acquisition it should be taken into account because to ignore a pre-existent value would be to expropriate it without compensation, so contravening the fundamental principle of equivalence (*Horn* v *Sunderland Corporation* [1941]): see *Batchelor* v *Kent County Council* [1990] CA.

Altered circumstances affecting other land (not the claimant's) within the scheme land's boundary, which surrounds or adjoins the claimant's land, must be disregarded. In *Davy* v *Leeds Corporation* (1964) the question was whether the effects of the demolition of other property in the area had to be excluded from the valuation by virtue of section 6. It was held that there was no prospect of the land being cleared other than by the corporation, the claimant could only claim the value of his sites cleared for development but with the surroundings still encumbered with buildings.

## Set-off for betterment

Sections 7 and 8, LCA 1961 provide that where a claimant has land compulsorily acquired, if that claimant has an interest in other land contiguous or adjacent to the land taken which increases in value as a result of the scheme, then the compensation payable must be reduced by the amount of that increase in value. Thus, in *John* v *Rhymney Valley Water Board* (1964) the acquisition of land for a reservoir enhanced the value of the retained land, as it increased the prospect of planning permission being granted for housing, which increase in value was deducted from the compensation payable by way of set-off.

It should be noted that certain statutes make specific provision for set-off. For example, section 261 Highways Act 1980 provides special set-off provisions in the case of acquisitions of land for highway purposes and is not limited to contiguous or adjacent land. The betterment to be set-off in highway cases included increases in the value of any property owned by the claimant which fronts onto the same highway. If this land is itself later acquired, the amount deducted by way of set-off on the earlier acquisition must be repaid.

In *Leicester City Council* v *Leicester County Council* [1995] the Tribunal decided that it had no discretion to award the underlying existing use value of the land being acquired in a case where the agreed value that had to to be set-off entirely negated the compensation payable for the land being acquired.

In *Portsmouth Catholic Diocese Trustees* v *Hampshire County Council* [1980] the Tribunal decided that while the construction of a new road gave rise to the circumstances enabling planning permission for development to be granted, the grant of planning permission was not a direct consequence of the road construction, so section 7 did not apply.

At first sight, there might seem to be an inconsistency between the *Rhymney* and *Portsmouth* decisions but this is not so. In the former, the construction of the reservoir enhanced the prospect of planning permission being forthcoming, in the latter, the prospect existed, scheme or no scheme.

## Depreciation in value due to the compulsory acquisition

Under section 9 LCA 1961 any depreciation in the value of the land acquired by reason of designation or because of any other indication of an intention to acquire by compulsion is to be disregarded: see *Jelson Ltd* v *Blaby District Council* (1977) CA, where the reduction in value stemmed from a past indication of a likely acquisition. See also *Trocette Property Co Ltd* v *Greater London Council* (1974) and *London & Provincial Poster Group Ltd* v *Oldham Metropolitan Borough Council* [1991]. However, in *Thornton* v *Wakefield Metropolitan District Council* [1991] it was held that although a notification of the likelihood of acquisition in a planning permission to convert a former theatre fell within section 9, the claimant remained in full control and was not precluded in law from implementing his scheme prior to the deemed notice to treat.

It follows from section 9 that evidence of value derived from sales of comparables within the scheme area may well have to be ignored due to their being tainted by the prospect of the compulsory acquisition.

# Assumptions as to planning permission

Certain planning assumptions may be made in ascertaining the value of the relevant interest and the owner is free to choose whichever is the most valuable. A specific case may result in a number of such assumptions. Thus in *Co-operative Retail Services Ltd* v *Wycombe District Council* [1989] the existing use of a site was a coal yard; the town map showed it allocated for wholesale warehouse development, the acquiring authority required it for bus station development, and there was also an outstanding permission for light industrial purposes which, due to the introduction of the Town and Country Planning (Use Classes) Order 1987 (SI 1987 No 764) prior to the valuation date, could be translated as a consent for office development. This last was the basis adopted.

Section 14(2) LCA 1961 admits the value of any planning permission actually in force at the date of notice to treat.

Section 14(3) permits the payment of hope value if the evidence is that the market would allow it. Historically any development value would be appropriately discounted, *Co-operative Retail Services Ltd* v *Wycombe DC* (1989) however in the recently reported case of *Spirerose (in administration)* v *Transport for London* [2008] the Lands Tribunal determined compensation on the assumption that planning permission was in existence at the valuation date even though factually it was not. The decision is being appealed.

Section 64 P&CA 1991 added additional subsections (5) to (8) to section 14 LCA 1961 dealing with special planning assumptions for the construction of highway schemes, including alteration or improvement. If a planning determination for the purposes of assessing compensation is to be made under either section 16 or 17 it is to be assumed that if the subject land was not to be acquired, no highway would be constructed to meet the same, or substantially the same, need.

Section 15(1) assumes the grant of such permission as would permit development in accordance with the proposals of the acquiring authority (although rule 3 might apply to the valuation). There are no qualifying rules as to its likelihood in a no-scheme world and permission is to be assumed: see *Myers* v *Milton Keynes Development Corporation* [1974].

Under section 15(3) permission is to be assumed for any development of a class specified in either paragraph 1 or 2 of Schedule 3 T&CPA 1990, subject to Schedule 10 to that Act. Paragraph 1 allows the rebuilding of any building in existence on 1 July 1948 or any building in existence before that date but demolished or destroyed after 7 January 1937. Paragraph 2 allows the use as two or more separate buildings of a building used as a single dwelling house. Originally these provisions also provided for compensation to be paid in the event of a refusal of planning permission but this right was repealed by Schedule 19 P&CA 1991, the schedule only surviving as a planning assumption for the assessment of compensation on compulsory acquisition and for certain other purposes — please refer to chapter 10. However, the form of what can be rebuilt is also heavily restricted by case law. The effect of Schedule 3 is to preserve the existing use value of the land. It gives the notional right to rebuild and applies even in cases where the use is a non-conforming one for which real planning consent would not be granted. An unlawful use that is subject to enforcement proceedings is not so protected.

Section 15(5) states that regard should be had to any certificate of appropriate alternative development and to any conditions specified therein.

Under section 16(2) permission is to be assumed for development for which planning permission might reasonably be expected to be granted for the zoned use in the development plan, but in considering that question and also whether any other section 16 assumptions are applicable, regard shall be had to any contrary opinion expressed in a section 17 certificate. Para 15 of Schedule 15 P&CA 1991 substituted a new provision, section 14(3A) LCA 1961.

In considering planning assumptions under section 16 LCA 1961 regard shall be had to any contrary opinion to development expressed in a section 17 certificate (for details of which, see below). Also in connection with section 16, the Court of Appeal in *Provincial Properties (London) Ltd* v *Caterham and Warlingham Urban District Council* [1972] decided that the test of reasonableness could entail a planning permission being completely disregarded for the zoned use if that permission would not, in fact, have been granted.

## Section 17 Certificates of appropriate alternative development

Section 17, with its section 18 appeal procedure, provides a mechanism for determining what planning permission might be forthcoming where none of the foregoing provisions assist. Section 65 of the P&CA 1991 provides that an application for a section 17 certificate of appropriate alternative development may be made by either party in respect of any land in which an interest is to be acquired. Applications must be made to the local planning authority indicating what uses or development the applicant considers to be appropriate for the land in question, and reasons must be given to support the opinion expressed. In response, the planning authority must issue a certificate stating what, if any, planning permission would be forthcoming in the absence of the compulsory acquisition. These need not be the same as the uses proposed by the applicant. The certificate must also state that it is the authority's opinion that no other consents would be granted. Section 17(5) deals with any planning conditions in such certificates. If the applicant does not like the response, it can be appealed under section 18 in the same manner as any planning application decision. The authority is also entitled to seek a section 17 certificate, probably opining that no consents would be granted other than for its own proposals — the so-called nil certificate.

Section 65(2) P&CA 1991 modified the section 17 procedure so that either a positive or negative certificate must include the words "but would not be granted for any other development". These words, if attached to a nil certificate, will undermine any argument that the land has hope value, which, in the absence of a certificate, may have been arguable. Circular 48/59 states that the certificate system should be worked out on a broad common-sense basis; a certificate is not a planning permission but is to be used in ascertaining the fair market value of the land.

Section 65(2) P&CA 1991 also substitutes new paras (a) and (b) in section 17(4) LCA 1961 dealing with information to be included in all certificates and section 65(3) adds a new subsection (9A) to section 17, providing for expenses in connection with the issue of a certificate and for appeals determined in the applicant's favour to be recoverable.

The date at which the section 17 certificate is determined is at the relevant date under section 22(2) LCA 1961, which has been held to be the date when the proposed compulsory acquisition is first made public. It is to be assumed that on that date the proposed compulsory acquisition has been cancelled but otherwise the ordinary planning principles are then applied to the existing circumstances: see *Fletcher Estates (Harlescott) Ltd* v *Secretary of State for the Environment* [2000] and *Jelson Ltd* v *Ministry of Housing and Local Government* [1970].

In one case, a certificate was granted in respect of land within a green belt for "institutional use in large grounds". For valuation problems in these and similar circumstances see *Lamb's Executors* v *Cheshire County Council* (1970).

Any development specified in the Town and Country Planning (General Permitted Development) Order 1995 (SI 1995 No 418), must be taken into account as must be the Town and Country Planning (Use Classes) Order 1987 (SI 1987 No 764).

# Compensation when permission for additional development is granted after acquisition

Part IV LCA 1961 as amended by section 66 and Schedule 14, P&CA 1991 provides that where, within 10 years of completion of the acquisition, a planning decision is made granting permission for the carrying out of additional development of any of the land, compensation is to be reassessed on the assumption that permission for such additional development was available at the date of the original valuation, and any additional compensation must be paid to the claimant with interest from that date. The new section 23(3) LCA1961 excludes from these provisions, *inter alia*, acquisitions by urban development corporations and by highway authorities in connection with urban development areas and acquisitions made under the New Towns Act 1981. The provisions apply to sales completed after Schedule 14 came into force (25 September 1991).

### Study 1

*Compensation on the basis of open market value, with the benefit of alternative planning assumptions*
The freehold site (0.15 ha) of a factory built in 1890 but demolished after a fire eight years ago is to be acquired for housing purposes, for which purpose it is allocated in the development plan at a density of 200 persons per ha. It

has been occupied for the past five years by a car breaker. The present rent is £6,000 pa (exclusive) on a yearly tenancy, the rent having been fixed two years ago. The present full net rental value is £8,000 pa. Temporary planning permission for this use expires in three years' time. The rateable value is £5,000.

*Claim for freeholder*

The following matters have to be considered.

Section 5 (2) LCA 1961 — open market value.

Section 14 (3) LCA 1961 — it must not be assumed that planning consent would automatically be refused for uses other than industrial or residential which, in this case, are to be assumed under section 15(3) and (4) and section 16(1). If, therefore, further permission might be obtained for car breaking in the absence of the compulsory acquisition, this may be taken into account. A quick estimate would show, however, that this would produce a lower capital value.

Section 15(3) LCA 1961. Assume paragraph I, Schedule 3 to T&CP 1990 (subject to Schedule 10) — the right to rebuild factory plus 10% as the building demolished was an original building.

Section 16 (1). Residential development at a density of 200 persons per ha.

Valuations based on the following planning assumptions are allowable, the claimant being able to adopt whichever gives the highest value:

(a) car-breaking income for three years with reversion to the higher of either industrial or residential value (no Landlord and Tenant Act 1954 compensation is payable to the car breaker as he would serve a tenant's notice to quit if he could not use the site) or

(b) car-breaking income for, say, one year to allow time for possession after a landlord's notice to quit, with similar reversions but less compensation to the tenant under the Landlord and Tenant Act 1954 (Appropriate Multiplier) Order 1990 (SI 1990 No 363) of once or twice the RV, depending on the 14-year rule.

*Valuations required*:

(a) Site of 0.15 ha with planning consent to reinstate the original building plus 10% cube subject to limit of 10% increase in the gross floor space not being exceeded. A residual valuation might well be required, as comparables would allow for expected permitted plot ratio and more economic design and layout for modern factories,

Say, value for section 15(3) purposes                                                £250,000

(b) Site of 0.15 ha with planning consent for housing purposes at a density of 200 persons per ha. Gross site area, Say, 0.18 ha – 36 persons,

Say, 36 rooms at £6,000 per room (based on comparables)                              £216,000

Industrial site value is higher and therefore housing development can be ignored for valuation purposes. The question is whether possession would be sought as soon as possible for industrial purposes or whether it would be more profitable for the freeholder to enjoy the car-breaking income for the full remaining three years and then secure possession. The freeholder may claim on whichever basis gives the most compensation.

*Freehold value if possession is sought for redevelopment by service of a notice to quit at the first opportunity*

|  | £ pa | £ | £ |
|---|---|---|---|
| Term |  |  |  |
| Net income | 6,000 |  |  |
| YP at 1 year at 11% (allow one year to secure possession) | 0.95 |  | 5,400 |

| | £ pa | £ | £ |
|---|---|---|---|
| *Reversion* to industrial site value | | 250,000 | |
| *Less*: compensation to tenant, say 1 × RV | | 5,000 | |
| | | 245,000 | |
| PV £1 in 1 year at 8% (Note 1) | | 0.93 | |
| | | | 227,850 |
| | | | 233,250 |
| Say | | | £235,000 |

**Note**

(1)  The unsecured ground rent rate has been used, although the cost of capital rate could be argued for.

*Value if car-breaking income allowed to continue for full 3 years*

| | | £ pa | £ |
|---|---|---|---|
| *Term* | | | |
| Net Income | | 6,000 | |
| YP 1 year at 11% | | 0.9 | 5,400 |
| | | | |
| *Rent review*: to full rental value for car-breaking | | | |
| Net FRV | | 8,000 | |
| YP 2 years at 13% | 1.67 | | |
| PV £1 in 1 year at 13% | 0.88 | 1.47 | 11,760 |
| | | | |
| *Reversion*: to site value | | 250,000 | |
| PV £1 in 3 years at 8% | | 0.79 | 197,500 |
| | | | 214,660 |
| Value Say | | | £215,000 |

The compensation to the freeholder will be the higher figure, £235,000, plus fees and statutory interest. The yearly tenant is not entitled to receive a notice to treat. If notice of entry is served and possession taken, his compensation will be assessed under section 20 of the CPA 1965 having regard to section 47 LCA 1973. From the above figures, it appears that, in the absence of the scheme, the landlord could have been expected to secure possession as soon as possible for redevelopment. This will limit compensation for disturbance under section 20 under the principle set out in *Horn* v *Sunderland Development Corporation* [1941].

## Study 2

*Compensation on the basis of equivalent reinstatement*

A non-conformist chapel with a small congregation of about 30 is some 200 years old, in poor repair, and is sited in a shopping street. The adjoining shops are old and the area is allocated in the approved local plan for residential purposes, but with shopping frontage to the shopping street.

The chapel completely covers a site of 12 m frontage and 20 m depth which is to be compulsorily acquired. The chapel site with adjoining properties is to be used for shopping, each shop to have one floor for storage above. Notice to treat was served two years ago. Some one and a half years ago a site in a residential area was suggested for reinstatement, but the trustees have only recently decided to go ahead with its purchase at a cost of £80,000. The planning authority has made it a condition of the grant of planning permission for the new chapel that a car park be provided.

*Basis of valuation*

Two are available, open market value or equivalent reinstatement.

Open market value basis: section 5(2) LCA 1961. There are two possibilities here. One is the open market value of the building. The other is to take the site value having regard to the planning assumptions, less the cost of demolition. The planning assumptions in this instance are:

Section 14(3): Consent to use the building for storage purposes, consent for which would probably be available in the absence of compulsory purchase.

Section 15(1) and section 16 (2): Shopping site value less the cost of demolition.

Cost of equivalent reinstatement basis: section 5(5) LCA 1961. The date of valuation will be the date when redevelopment could reasonably have commenced. This could be a matter of dispute, owing to the trustee's delay in arriving at a decision. As to the car park, as there was not one with the original chapel, the cost of providing it could not be recovered unless the planning authority was insisting on its provision.

*Valuation*

Site value for shopping disregarding any increase in value due to the Scheme. A residual valuation may be required. For a good example of a residual valuation, see *St Pier Ltd* v *Lambeth Borough Council* [1976] where a higher plot ratio in the no-scheme world was the basis taken by the Lands Tribunal.

The Tribunal's views on the use of residual valuations in compulsory purchase references are well set out in *Clinker & Ash* v *Southern Gas Board* (1967) where it was said that the discipline of open market conditions achieves a balance so that the residual method can be a precision valuation instrument in valuing development sites. This discipline is absent in an arbitration case as there is in effect, a captive purchaser and a captive vendor. In *First Garden City Ltd* v *Letchworth Garden City Corporation* (1960) the Tribunal said; "once valuers are let loose upon residual valuations, however honest the valuers and reasoned their arguments, they can prove almost anything".

*Open market value basis*

| | | £ | £ |
|---|---|---:|---:|
| (a) | Site Value for shopping | | |
| | Say, cleared site value is | 150,000 | |
| | Less: cost of demolition | 3,000 | |
| | | 147,000 | 147,000 |

| | | £ pa | |
|---|---|---:|---:|
| (b) | Storage income from building | | |
| | Net income: 200 m² at £30 per m² | 6,000 | |
| | YP perp, Say | 9.0 | 54,000 |

*Equivalent reinstatement basis, rule (5) Compensation*

| | | £ |
|---|---|---:|
| i. | Purchase of alternative site | |
| | Purchase price, legal costs and other disbursements | |
| | in connection with the conveyance, plus surveyor's fees | 90,000 |
| ii. | Cost of new equivalent building | |
| | Construction cost, architect's and quantity surveyor's fees | 150,000 |
| | Plus Site works, costs and fees | 4,000 |
| iii | Removals and other disturbance items, say | 5,000 |
| iv. | Cost of providing car park on the new site | 15,000 |
| | Total | £264,000 |

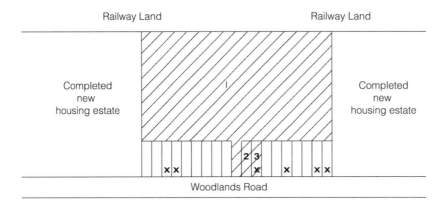

The rule (5) basis will therefore be adopted and compensation will be, say £265,000 (but subject to a possible deduction for repairs to the existing hall) plus legal and surveyor's fees in connection with the preparation and negotiation of the claim, and statutory interest on everything.

## Study 3

*The acquisition of back land and houses to provide access*
The backland shown hatched on the plan is allocated for educational use on the development plan as the site for a primary school. A compulsory purchase order has been made. This comprises:

Reference No 1: The school site, being freehold back land, previously allotments, and at present unused. It has an inadequate 4 m wide access to Woodlands Road.

Reference No 2: A freehold house adjoining the access referred to, subject to a regulated tenant and worth £130,000 as an investment in the open market.

Reference No 3: A freehold owner-occupied house with an open market value vacant possession value of £240,000.

The houses indicated on the plan are all detached, having frontages of 10 m, those marked X on the plan being owner-occupied. The remainder are let to statutory tenants.

*Compensation for reference No 1*
A certificate of appropriate alternative development should be applied for under section 17 of the LCA 1961. Assume that this specifies residential use by the erection of 40 three-bedroomed flats, subject to the provision of an access of a minimum width of 20 m to Woodlands Road.

| | £ |
|---|---|
| *Value of backland* | |
| 40 flats at £60,000 per flat, plot value | 2,400,000 |

*Less*: cost of procuring an access in the open market. No one here possesses the key to the development as any two houses would suffice, particularly the CPO houses as they adjoin the entrance. It is assumed the freehold interest in reference No. 2 could be purchased for £175,000 and reference No. 3 for £280,000. It is further assumed

that the tenant in the former would vacate for £30,000, a total of £485,000, plus
an allowance of £15,000 for demolition and clearance costs. Say, however,
£550,000 to allow a reasonable margin for risk                                   500,000

Open market value of the back land at present                                 £1,900,000

Note

If one owner possessed the key to the development the presumption, subject to rebuttal, might be that that owner would seek 50% of the back land value. Section 5, rule 3 (special suitability) does not operate to exclude the special value of references Nos. 2 and 3 as providers of access to the back land because that value does not exist solely because of a demand from authorities having compulsory powers — ordinary developers would pay extra in the no scheme world to secure an adequate access.

*Compensation for reference No. 2*
*Freeholder*:

Open market value of the freehold interest subject to the tenancy, but including any value the property has to the purchaser of the backland for development — see section 17 certificate regarding access requirements. Provided such development could be envisaged in the no-scheme world, the extra value is allowable, say £175,000. No home loss payment applies as the freeholder does not qualify for it (not in occupation for 12 months ending with displacement) but does qualify for a basic loss payment of 7.5% of the value of the interest, subject to a maximum of £75,000 — here, £13,125 — provided the freehold has been owned for more than a year before one of several specified dates (section 33A, LCA 1973). To this total must be added fees and statutory interest, including on the loss payment.

*Tenant*:

Possible home loss payment of £4,000, together with a disturbance claim under section 37 of the LCA 1973, plus fees and interest.

*Compensation for reference No 3*
*Freeholder*:

Freehold house with vacant possession including its value to the special purchaser. Section 50 of the LCA 1973 precludes any reduction in the compensation if the owner is rehoused by the acquiring authority. The claim is, therefore, £280,000 plus 10% home loss payment (maximum £40,000 as at February 2008) under sections 29 and 30 LCA 1973 as amended by sections 68 and 69 P&CA 1991. The total here is £308,000 plus disturbance, fees and interest.

It might be argued that a disturbance claim under section 5(6) of the LCA 1961 was inconsistent because some development value is included in the purchase price, but this is considered a weak argument — basically, the house is sold as a house for its existing use value, plus a premium to persuade the owner to part with it. It has not been valued at development site value for the construction of a block of flats.

Section 39 of the LCA 1973 is concerned with the duty to rehouse residential occupiers on compulsory purchase where suitable alternative residential accommodation on reasonable terms is not otherwise available. Trespassers and blight cases are excluded.

# Study 4

*Acquisition of farmland for new town purposes*

An owner-occupied farm of 150 ha is to be compulsorily acquired. It is in an area recently designated as part of the site of a new town. It is on the edge of a village about 2 miles from an existing motorway. The village has a railway station on the main line to London. When acquired, 135 ha of the farm will be used for housing, 12 ha for a school and 3 ha for road purposes.

Prior to the new town proposal the land was allocated for agricultural purposes, but owing to pressure of demand some land in the village was being released for housing each year.

The following preliminary matters have to be considered.

*Date of valuation*
See *West Midland Baptist (Trust) Association (Inc)* v *Birmingham Corporation* (1968) CA. See above for further details.

*Statutory basis of the compensation*: Section 5(2) LCA 1961. Open market value.

*Planning assumptions*:
Section 14 (3) LCA 1961. If the market would have paid more than agricultural value, then that should be taken into account: establishing this would be a matter of evidence.
Section 15(1) LCA 1961. Acquiring authority's proposed use — housing on 135 ha.
Section 16(2) LCA 1961. Development Plan allocated use. In this case it is the same as the authority's proposed use.
Section 15(5) LCA 1961. Section 17 certificate of appropriate alternative development on the 15 ha not allocated for housing — likely to be residential if not required for school or road purposes.

*Effect of section 6 and Schedule 1, LCA 1961*
Case 3 applies. Any increase in the value of the land taken due to the development or prospective development of the new town which would not have taken place but for the new town scheme is to be disregarded.

*Summary*
It is necessary to value the 150 ha with planning permission for housing in the no-scheme world, reflecting the rate of natural growth without any artificial influx which is due to the scheme, remembering that planning permission without demand does not create value: *Viscount Camrose* v *Basingstoke Corporation* (1966) CA.
The basic problem here is to assess what the demand for this land would have been in the no-scheme world, with planning permission. Suppose that it is agreed that there would have been a demand for 5 ha pa for five years but that thereafter any further demand would have been unlikely. If residential value, based on comparables, is £2,000,000 ha then the total sale price of the housing land over the next five years will be £50 million. This must be deferred for the average time for receipts — 2.5 years.

|  | £ | £ |
|---|---|---|
| Total proceeds from sale of housing land |  |  |
| 25 ha at £2 million | 50,000,000 |  |
| PV £1 in 2.5 years at 12%, Say (1)(2) | 0.75 | 37,500,000 |
|  |  |  |
| *Add*: Agricultural income from these 25 ha, also for | £ pa |  |
| an average period of 2.5 years (pending development). |  |  |
| 25 ha at say £240 (net) per ha pa | 6,000 |  |
| YP 2.5 years at 5% | 2.29 | 11,450 |
|  |  |  |
| *Plus*: Balance of 125 ha |  |  |
| 125 ha with planning permission but with the |  |  |
| remote possibility of demand (3). Say £20,000 ha | 2,500,000 |  |
|  |  |  |
| *Plus*: Basic Loss payment and occupier's loss payment (4) |  |  |
| Loss payment: 7.5% value of interest, up to maximum of 75,000 |  | 75,000 |
| Occupier's loss payment (5) up to a maximum of |  | 25,000 |
|  |  | 40,106,000 |

Compensation: £40.1 million plus fees and statutory interest.

Notes
(1)  12% reflects the risk of the estimated demand not being achieved.
(2)  An alternative approach giving the same result would be to capitalise the yearly income, £10 million by YP 5 years at 12%.
(3)  If it is unlikely that there would be a demand for any of the land, albeit with planning consent, in the no-scheme world, then only a small amount in respect of hope value over agricultural value would be justified.
(4)  Sections 33A and 33B, LCA 1973.
(5)  This will be the greater of 2.5% of the value of the interest, the land amount and the buildings amount, capped at a maximum of £25,000. See page 253 for details.

# Study 5

Acquisition of a factory within a town redevelopment project under section 36(4) of the T&CP 1990

An area predominantly comprising old warehouse and industrial buildings interspersed with two-storey Victorian cottages is to be redeveloped. A CPO has been made in respect of the whole area under section 226 T&CP 1990, the planned range of uses are shops, public buildings (including a library) offices, warehouses and flats.

You are acting both for the freeholder and the lessee of an old factory of 400 m² net sited within the area of the redevelopment scheme and allocated for use as part of a library site. The factory is subject to a lease now having eight years unexpired at a rent of £5,500 pa, without review, on tenant's internal repairing terms. The current net full rental value is £15,000 pa. Notices to treat were served on the owners of both interests two years ago and possession of the leasehold interest was taken six months ago, after the tenant had moved to a new factory on the outskirts of the town, the only reasonable alternative. That factory has an area of 300 m² net and the rent is £16,000 pa on full repairing and insuring terms.

*Preliminary considerations*
Section 5(2) LCA 1961. Open market value.
Section 15(1) LCA 1961. Development such as would permit the acquiring authority's proposals but using the development value approach will preclude making a disturbance claim: *Horn* v *Sunderland* [1941].
Section 15(3). Existing use value and a disturbance claim.
Section 16(4). Consider which of the planned range of users would have been appropriate in the circumstances of section 16(5). Probably warehouse use.
Section 17 certificate procedure: likely to result in an notional consent for warehouse or industrial use.
Section 6(1). Disregard so much of any increase or decrease in value in the circumstances of the second case in the first column of Schedule 1 as is attributable to development in column 2, subject to the qualifications of section 6(1)(a) and (b).

*Conclusion*
Valuation on the basis of the existing factory including Schedule 3 rights together with a proper disturbance claim will be the most profitable approach. The value in the no-scheme world must be used, ignoring any increase in value of the factory due to redevelopment in the area, because it is considered that, but for the scheme, no major redevelopment would have taken place.

*Date of valuation*
Freeholder's claim:  The date the valuation is being prepared.
Tenant's claim:  Date of entry, six months ago. It is assumed that values have not changed since then.

*Freeholder's interest*

|  | £ pa | £ pa | £ |
|---|---|---|---|
| Rent reserved |  | 11,000 |  |
| *Less*: External and structural repairs, Say, | 3,000 |  |  |
| Insurance (based on reinstatement costs) | 500 |  |  |
| Management (5% of rent passing) | 550 | 4,050 |  |
| Net income |  | 6,950 |  |
| YP 8 years at 11% |  | 5.15 | 35,793 |
| Reversion to net full rental value |  | 30,000 |  |
| YP perp deferred 8 years at 13% (freehold rack rent yield) |  | 2.89 | 86,700 |
|  |  |  | £122,493 |
| Say |  |  | £122,500 |

Compensation £122,500 plus Loss payment, fees and interest.

*Leaseholder's interest*

|  | £ pa |  |
|---|---|---|
| FRV (net, ie on FRI terms) | 30,000 |  |
| *Add*: External and structural repairs, management and insurance (as in freehold valuation) | 4,050 |  |
| Full rental value (internal repairing lease) | 34,050 |  |
| *Less*: Rent payable | 11,000 |  |
| Net profit rent | 23,050 |  |
| YP 8 years at 15% and 4% (Tax at 40%) | 3.12 |  |
| Value of leasehold interest |  | £71,916 |
| Say |  | £73,000 |

Compensation for land taken £73,000
*plus* Disturbance, basic and occupier's loss payments, fees and interest

# Leaseholder's disturbance claim

The basis of the claim under rule 6 is the actual loss suffered by the leaseholder. Rule 6 claims are examined in detail in part IV of this chapter.

It is assumed that the additional rent at the new premises is not compensatable in view of increased business efficiency. Items of claim may be expected to include the following: removals of plant and machinery, fixtures and fittings and stock, plus losses on forced sale where appropriate. Insurances against loss during the move. Duplicated expenditure, rent, rates, heating and lighting etc. from the date of renting the new accommodation until the date of possession by the acquiring authority. Crawley costs, legal and surveyor's fees in finding the new premises, negotiating the rent and agreeing the lease, and director's time and expenses in the search. The capitalised yearly increased distribution costs; permanent loss of profits if the level of production will be lower in the new premises as a result of the reduced floor area; temporary loss of profits as a result of the disruption of the business; and any other proper disturbance items.

# PART III: COMPENSATION WHERE PART ONLY OF THE LAND IS TAKEN

# Material detriment

The first consideration where an authority seeks to take only part of a property is whether section 8(1) of the CPA 1965 or section 53 of the LCA 1973 applies and whether these should be invoked by serving a counter-notice requiring the authority to take the whole property. The former applies to certain non-agricultural property while the latter provides an equivalent right in respect of agricultural property.

Section 8(1) applies if:

(a) the notice to treat is in respect of part of a house, building or manufactory or of part of a park or garden belonging to a house and

(b) the part cannot be taken without material detriment to the house, building or manufactory or, in the case of a park or garden, without materially affecting the amenity or convenience of the house, and section 58 LCA 1973 provides that in considering if there is material detriment, regard shall be had not only to the severance but also to the use of the whole of the works.

If the authority challenges the claimant's counternotice it will be for the Lands Tribunal to decide the question of material detriment. The burden of proof lies on the the authority which must show that the part can be taken without causing material detriment; the claimant does not have to show it cannot.

*Ravenscroft Properties Ltd* v *Hillingdon London Borough Council* (1968) clarified the test for material detriment. The case concerned the taking of the rear access, garage and garden of a property. The Tribunal determined that the test to be applied was whether or not the property became 'less useful or less valuable in some significant degree'. It concluded that the taking of the rear portion clearly transformed the property into something quite different from what it was before, the change being wholly for the bad, and so confirmed the section 8 notice.

It is important to note that in most cases, any notice seeking to operate section 8 must be served before the acquiring authority takes possession of the land; see *Glasshouse Properties Ltd* v *Department of Transport* [1994]. However it is perhaps even more important to note that often particular legislation will incorporate even more stringent time-limits on the service of a section 8 notice. Acquiring authorities such as Transport for London and London Underground Ltd regularly incorporate provisions that require any section 8 notice to be served within 21 days of the date of service of Notice to Treat.

Section 8 has been extended to apply to blight notices (section 166 T&CPA 1990). Thus in *Hurley* v *Cheshire County Council* (1976) the Tribunal held that while the inconvenience to the house through having part of the garden taken away could be minimised by landscaping, the amenity would be seriously affected by the road construction.

Section 8(2) CPA 1965 applies to land not in a town or built upon and deals with circumstances under which an owner can require the authority to acquire severed land of less than half an acre. Section 8(3) contains provisions whereby the acquiring authority may require the owner to sell severed land if the owner requires accommodation works that are uneconomic.

Assuming that a decision has been taken not to serve a section 8 notice and so accepted that the part only will be acquired, the first step is to consider the adequacy of any proposed accommodation works or, if none have been proposed, whether such works should be sought. Although the authority does not have to carry out accommodation works, it invariably does so because it will reduce the amount of compensation payable as the value of the works will be reflected in the after value of the retained land.

The Lands Tribunal cannot order accommodation works to be carried out by the acquiring authority but it may take their effect into account. It is important to remember though that cost does not equate to value. In *Stedman v Braintree District Council* [1990] the claimant spent £22,844 on a wall when part of his land, a private roadway, was taken and did so without discussing the works with the acquiring authority. He than sought to recover this sum, but the Lands Tribunal pointed out that there is no specific right to recover the cost of accommodation works. Furthermore, building the wall was not the only solution. He was awarded £5,128, being the reasonable cost of fencing and gates.

In a case where there is a likelihood that set-off (see later) could negate any compensation payable, accommodation works become extremely important as their value cannot be reclaimed.

## Severance and injurious affection

If part only is taken, the owner is entitled not only to the value of the land to be purchased but also to compensation for damage sustained as a result of severance, or injurious affection to other land held with it; section 7 CPA 1965. The case of *Cowper Essex v Acton Local Board* [1889] clarified the meaning of held with. It is enough to show that the land taken and the land alleged to be injuriously affected are so near together or so situated in relation to each other that the possession and control of each gave an enhanced value to the whole. Compensation is restricted to injury likely to arise from the proper exercise of statutory powers, so injury caused by unauthorised acts gives rise not to compensation but to an action for damages, or possibly an injunction. For example, if the contractor strays outside the area of land identified in the Notice to Treat and the retained land suffers injury, that injury will not be compensatable save through an action in tort.

Until 1973, a claimant could only claim compensation for injurious affection under section 7 CPA 1965 if (a) part of its land had been acquired and (b) the use of the scheme works constructed on that land acquired resulted in diminution in value to the land retained, the depreciation being measured only by reference to the use of so much of the works as were located on the land taken from the claimant; see *Edwards v Minister of Transport* [1964]. 1973 saw the enactment of section 44 of the LCA 1973 which until recently was understood to have widened the scope of section 7 CPA 1965 so as to provide for injurious affection to be paid in respect of all depreciation arising from the entirety of the works as they affect the land retained. However, in recent years doubt has been cast by the Lands Tribunal on this understanding of the effect of section 44 LCA 1973, to the extent that a number of cases where injurious affection was in issue have been settled prior to going to a hearing. This latest development has yet to be examined by a full hearing. The thrust of the argument is that before section 44 can be applied so as to widen the scope for the consideration of injurious affection, the taking of the land actually acquired has to be shown to cause diminution in value to the land retained. Thus, if a parcel of land is taken to form part of, say, a grass verge to a motorway (which of itself would be unlikely to result in diminution in value to the retained land due to its 'silent' use) then section 44 might not apply. This, if right, would seem likely to return a number of claims for compensation back to the pre-section 44 days.

A case having potentially wide implications is *Norman v Department of Transport* [1996] where in a reference to determine the validity of a blight notice, the Tribunal held that subsoil under the highway which passed immediately alongside the cottage was part of the cottage, occupation of the latter constituted occupation of the whole, including the land under the road. It was held that the blight notice was valid. In certain circumstances, therefore, if a road improvement is to be made which involves the taking of the subsoil, the adjoining owner might be brought within the compensation provisions of section 7 CPA 1965.

Two cases illustrating severance and injurious affection in the case of agricultural properties are *Cuthbert* v *Secretary of State for the Environment* (1979) at pp 1115 and 1176, and *Wilson* v *Minister of Transport* [1980]. The former is also of interest in relation to accommodation works.

## Study 6

*Part of claimant's land taken, but remainder subject to set-off for betterment*
Albert is the freehold occupier of a large stone house standing in grounds of 1 ha situated on the eastern outskirts of an industrial town. Permission to develop the whole for housing purposes has been refused on several occasions over the past 10 years on the grounds that any further extension of the town in the direction proposed would be inappropriate. The planning position, however, has recently been changed as a by-pass on the eastern side of the town has been approved which will entail the acquisition of an area of your client's land 15 m wide for the whole of its width of 80 m. The proposed road will be elevated along this section and will be, at its nearest, some 85 m from the house. In these changed circumstances planning permission would be forthcoming for residential development on the remainder of the land.

The following preliminary matters have to be considered: section 8 of the CPA 1965 and section 58 of the LCA 1973. A decision must first be reached as to whether to accept a notice to treat in respect of part only or whether to serve a counternotice to take the whole. If served, a counternotice would probably be upheld. However, on the assumption that the owner wishes to remain and is prepared to sell part only, the first step must be to agree the accommodation works which will be carried out by the acquiring authority. When agreed, a claim for compensation can then be prepared. This will consist of the value of land taken plus severance and injurious affection to land retained: (section 7 of the CPA 1965 and section 44 of the LCA 1973) less set-off for betterment of land retained (section 7 of the LCA 1961).

In addition a disturbance claim could be made if consistent with the basis of claim; that is if the land taken is acquired on the basis of existing use value. A claim for injurious affection must also be consistent with the basis of claim for land taken. For example, if part of a farm is taken and the basis of compensation is its development value for housing, injurious affection to the part of the farm retained due to the housing use on the land taken would not be proper.

A before less an after value gives a composite total for the compensation, but this must be broken down into its constituent parts to comply with section 4 of the LCA 1961 which deals with the submission of claims. Failure to comply may have an effect on costs. This before and after approach cannot be used in every case, however, and it was criticised by the Lands Tribunal in the circumstances of *Abbey Homesteads Group Ltd* v *Secretary of State for Transport* [1982] in favour of valuing the land taken and severance and injurious affection to the retained land separately. This concerned compensation for a strip of land for a road through land having residential development potential and meant that the land taken had to be valued as a strip only, as although severance depreciation in the claimant's retained land is payable this does not apply to severance depreciation in the value of the land taken from him. See also *Hoveringham Gravels Ltd* v *Chiltern District Council* (1977) CA, and *English Property Corporation plc* v *Kingston-upon-Thames Royal London Borough Council* [1998].

However, adopting the before and after approach:

*Assessment of compensation*

| | £ |
|---|---|
| *Before value* | |
| Value of house and garden prior to the scheme for the ring road. | |
| No reasonable prospect of planning permission | 550,000 |

*After value*
Either
Value of house with garden reduced by 1,200 m² in extent and having regard to the construction and use of the motorway and also to any works of mitigation carried out by the acquiring authority under part II of the LCA 1973 and to the value of agreed accommodation works,
Say                                                                                                                        320,000

or

0.88 ha housing land at, say, £1.5 million per ha, less cost of demolition, Say                 1,320,000

The after value, therefore, is £1.32 million as it will clearly be profitable to develop the retained land after the scheme. The betterment due to the scheme exceeds the value of the land taken and compensation will be nominal, say £5.

Notes
(1)   Where land is taken, there can be no claim under part I of the LCA 1973 (section 8(2)).
(2)   The freeholder will also be entitled to fees and interest. Whether set-off can negate these is not clear.

# Acquisition of part of a property subject to a lease — apportionment of the rent payable

If a part only of a property held on lease is taken, it will first be necessary to apportion the lease rent as to a rent for the part taken and the part retained; section 19 CPA 1965. Only the rental element in respect of the latter will continue to be payable under the lease. The Lands Tribunal has power to determine the rent apportionment in the absence of agreement. Compensation to both freeholder and leaseholder for land taken may then be arrived at on the basis of before and after valuations.

## Study 7

*Part of a tenanted property acquired, rent to be apportioned*
A shop has a net frontage of 6 m and a depth of 20 m. There is also a private forecourt 3 m deep and two floors of living accommodation set back above the shop. It is held on lease having five years unexpired at a net rent of £3,400 pa. The private forecourt and the front 3 m of the shop are to be acquired for road widening, but the upper part will not be affected. Accommodation works to reinstate the shop front have been agreed. Neither claimant wishes to compel the authority to take the whole under the material detriment rule. Comparables would indicate a present net full rental value of £200 per m2 for a zone A of 6 m depth on the basis of two 6 m zones and a remainder and halving back.
    Suggested apportionment of rent payable:

| Estimated full rental value: Before | £ pa |
|---|---:|
| Zone A 6 m × 6 m at £200 | 7,200 |
| Zone B 6 m × 6 m at £100 | 3,600 |
| Zone C 6 m × 8 m at £50 | 2,400 |
| Forecourt, Say, | 800 |
| Upper part | 2,000 |
| Full net rental value | £16,000 |

*Estimated full rental value After*
*(effectively loss of zone C and forecourt)* £ pa
Zone A 6 m × 6 m at £200 7,200
Zone B 6 m × 6 m at £100 3,600
Zone C 6 m × 5 m at £50 1,500
Upper part 2,000
Full net rental value £14,300

*Apportionment of the rent passing (£3,400 pa)*

Part retained: $\dfrac{14,300}{16,000} \times 3,400$ Say 3,040

Part taken: $\dfrac{1,700}{16,000} \times 3,400$ Say 360

£3,400

*Assessment of compensation*
*Freeholder's claim*

| | £ pa | £ | £ |
|---|---|---|---|
| Before value | | | |
| *Term* | | | |
| Net income | 3,400 | | |
| YP 5 years 7% | 4.1 | 13,940 | |
| | | | |
| *Reversion* to full net rental value | 16,000 | | |
| YP perp deferred 5 years 8% | 8.5 | 136,000 | |
| | | 149,940 | 149,940 |
| | | | |
| *Less*: After value | | | |
| Term | | | |
| Net income, apportioned for part retained | 3,040 | | |
| YP 5 years 7% | 4.1 | 12,464 | |
| | | | |
| *Reversion* to FRV, reduced area | 14,300 | | |
| YP perp deferred 5 years 8% | 8.5 | 121,550 | |
| | | 134,014 | 134,014 |
| Claim for freeholder for land taken (1) | | | £15,926 |

*Plus* Basic loss payment (2) fees and interest.

*Leaseholder's claim*

| | £ pa | £ |
|---|---|---|
| Before value | | |
| Net full rental value | 16,000 | |
| *Less*: rent payable | 3,400 | |
| Profit rent | 12,600 | |
| YP 5 years 9% and 4% (tax at 40%) | 2.5 | 31,500 |

| | | |
|---|---:|---:|
| *Less*: After value | | |
| Net full rental value, of the part retained | 14,300 | |
| *Less*: rent payable | 3,040 | |
| Profit rent | 11,260 | |
| YP 5 years 9% and 4% (tax at 40%) | 2.5 | 28,150 |
| Claim for leaseholder for loss of profit rent | | £3,350 |
| *Plus* Basic and occupier's loss payments, fees and interest. | | |

Notes
(1)  This will be 7.5% of £15,926, assuming the freeholder has owned the interest for the requisite period.
(2)  An alternative calculation would be to value the part taken on its own and to calculate the severance depreciation to the part left as a separate head of claim.
(3)  Compensation for the lessee for disturbance for, at the minimum, temporary disruption to trade during the carrying out of the accommodation works will also be recoverable.
(4)  Further, claims for severance and injurious affection for both parties might be justifiable under section 7 CPA 1965, reflected in lower After valuations.

# PART IV: COMPENSATION UNDER RULE 6

## Claims unrelated to the value of the land: rule 6 items, Disturbance and other matters

This is a complicated subject. The first thing to understand is that rule 6 covers all heads of claim that are not to do with the value of the land itself, of which disturbance is but one category. Damage to trade profits (goodwill) costs of attempted mitigation and of professional fees are three other examples. The second thing to get clear is that the rules are sometimes different depending upon whether the claimant is relocating or closing down a business.

The entitlement to claim compensation for disturbance is founded on rule 6, section 5(6) LCA 1961 which states:

Rule 6 "The provisions of rule (2) shall not affect the assessment of compensation for disturbance or any other matter not directly based on the value of the land".

Despite the vague nature of this wording, there is no doubt that it provides such an entitlement, and although the Acquisition of Land (Assessment of Compensation) Act 1919 (now section 5(2) LCA 1961) altered the basis of valuation for the land taken to open market value, section 5(6) LCA 1961 specifically leaves unaffected the basis of valuation for disturbance, which is value to owner.

However, compensation for disturbance is not a distinct and independent head of claim. It is one element in the assessment of the overall compensation to which a claimant is entitled. The value of the land to the claimant and the disturbance it incurs are inseparable parts of a single whole; *Hughes v Doncaster Metropolitan Borough Council* [1991] HL.

The general rule is that any loss suffered by a dispossessed owner which flows from the acquisition is compensatable provided that:

(a)  It is not too remote. *Bennett v Northampton Borough Council* (1989) concerned, inter alia, a claim for partial extinguishment of a business. A complex claim was made to substantiate a loss over a 12-year period but the Lands Tribunal held that the use of the Retail Price Index was too sophisticated and part of the loss claimed was too remote. Similarly in *Cawoods Aggregates (South*

*Eastern) Ltd* v *Southwark LB* [1982] LT, a claim for disturbance based on 10 year projections of profit was rejected on the basis that anything in excess of five years was too speculative.

b)  It is the natural and reasonable consequence of the dispossession: *Harvey* v *Crawley Development Corporation* (1957) CA.

A claimant is under a duty to mitigate its losses as far as possible; and must take all reasonable steps to minimise losses so that the claim does not exceed the natural and reasonable consequences of dispossession. If a claimant was faced with a straightforward move of office accommodation from one property to another, it would be expected to accept the lower of two comparable estimates. On the other hand, if the move involved, say, a Heidlberg printing press, it would be reasonable to employ Heidlberg engineers to effect the move, even though more expensive, particularly if doing so would obviate a claim for loss of profit due to a mishap during the removal. The duty is not to make the situation worse than it need be. This does not mean that the claimant must do himself down in order to minimise a claim — merely that he must act reasonably. See *K & B Metals Ltd* v *Birmingham City Council* [1976]. Costs incurred in an unsuccessful attempt at mitigation are recoverable — see *Pennine Raceway Ltd* v *Kirklees Metropolitan Council* (1981) LT; (1982) CA; [1989] LT; and [1989] CA. Fees incurred in attempting to mitigate are recoverable.

Disputes as to reasonableness and mitigation often relate to whether other premises should have been taken. Thus in *W C Jones & Co* v *Edmonton Corporation* (1957) a lease of premises of which the freehold was owned by the acquiring authority was reasonably rejected by the claimant; in *Rowley* v *Southampton Corporation* (1959) the rent of the alternative premises was higher than the claimant felt he could pay and his refusal to relocate was held to be justified; in *Bede Distributors Ltd* v *Newcastle upon Tyne Corporation* (1973) the claimants were held entitled to reject other suitable alternative accommodation owing to their financial position; in *Ind Coope (London)* v *Enfield London Borough Council* [1978] the claimants said that the projected trade at the new premises did not justify the rent and the net profits were likely to be reduced, which decision was held not to be unreasonable and one that they were, in their discretion, entitled to make; in *Knott Mill Carpets* v *Stretford Borough Council* (1973) LT the carpet company having rejected relocation in a precinct as they needed competition, were held justified in doing so as they were unlikely to increase turnover sufficiently to pay the higher rent and maintain profit.

However in *Bailey* v *Derby Corporation* [1965] the claimant, a builder, did not take alternative premises as he was too ill to move. The Court of Appeal supported the Tribunal and held that there was sufficient evidence to show that the ill-health was not due to the acquisition but was coincidental. It was therefore "an extraneous and independent matter which must be put on one side". This decision seems particularly hard and appears to conflict with the loss to the owner rule, furthermore the recent report of the Law Commission has recommended that the effect of this case law should be legislated against so if a similar case were to arise today it might be capable of challenge on human rights grounds. In *Hall* v *Horsham District Council* [1977] the claimant's failure to proceed with the relocation, coupled with his general conduct, was fatal to a claim for total extinguishment and in *Lamba Trading Company* v *Salford City Council* [2000] a claimant's disturbance claim was assessed on the basis of a notional relocation as it was held that its decision not to relocate was unreasonable in all the circumstances.

Losses incurred prior to the date of Notice to Treat are compensatable provided they arise as a direct consequence of the compulsory acquisition. In *Prasad* v *Wolverhampton Borough Council* [1983] the claimant, when threatened with inevitable displacement from land because of compulsory acquisition, acted reasonably in moving to other accommodation before notice to treat was served. It was held that he was displaced in consequence of the acquisition of the land and entitled to a disturbance payment.

This Court of Appeal decision followed earlier Scottish decisions, *Park Automobile Co Ltd* v *Strathclyde District Council* [1983] LTS and *Aberdeen City District Council* v *Sim* [1982].

In *Emslie & Simpson Ltd* v *Aberdeen City District Council* [1994] the Court of Sessions held that while in certain cases a loss of profit prior to dispossession might be recoverable, such as where it is due to steps taken to anticipate the dispossession, a loss due to the effect of blight on trading generally cannot be said to have been caused by the dispossession as it affected all traders whether their land was being taken or not. However, the same aspect was further considered in *Director of Buildings and Lands (Hong Kong)* v *Shun Fung Ironworks Ltd* [1995] a case involving property in Hong Kong at a time when English law prevailed, where the Judicial Committee of the Privy Council decided that compensation was payable for losses occurring before resumption [acquisition] of land by the Crown, the 'shadow period', provided that they arose in anticipation of resumption, because of the threat thereof, that they were not too remote and were not losses which a reasonable person would have avoided.

Section 46 of the LCA 1973 deals with persons who have attained the age of 60, who are displaced from trade or business premises and who may now elect for a total extinguishment basis subject to giving certain undertakings prescribed in section 46(3). The Court of Appeal held in *Sheffield Development Corporation* v *Glossop Sectional Buildings Ltd* [1994] that section 46 was satisfied where the claimant was trading when a compulsory purchase order was made and when a deemed notice to treat came into effect although he vacated 20 months before the acquiring authority took possession. He was "required to give up possession" as specified by section 46.

A disturbance claim should be for the actual loss suffered by the owner. Every foreseeable loss or expense which is likely to result from dispossession should be claimed, provided that it is the direct natural and reasonable consequence of being dispossessed. In addition a claimant may be entitled to recover such as fees for professional advice, loss payments and statutory interest but once a full and final settlement has been made, it is too late to claim further items, even though they would have been properly payable. A disturbance claim must be consistent within itself. If the claim for compensation is based on redevelopment value, no claim for disturbance is payable, *Horn* v *Sunderland Corporation* [1941].

A claimant cannot obtain more compensation under rule 6 than his actual costs or losses. However in *Palatine Graphic Arts Co Ltd* v *Liverpool City Council* [1986] CA, the claimant qualified for a 22% regional development grant in respect of certain disturbance items. The Court of Appeal held that the grant need not be deducted to arrive at the claimant's loss as it was not directly related to the compulsory purchase.

Cases sometimes arise, particularly in the case of family run businesses, where it is not apparent that an occupier has a sufficient interest in the land to entitle it to recover compensation. Cases of this nature gave rise to the expression 'piercing the corporate veil' as in *Smith, Stone & Knight Ltd* v *Birmingham Corporation* [1939] where it was held that a subsidiary company without a lease operated as the servant of a parent company owning the premises. In *DHN Food Distributors Ltd* v *Tower Hamlets Borough Council* [1976] CA, three companies, having no real separate identity, were treated as one and were able to claim compensation accordingly. However, the House of Lords in *Woolfson* v *Strathclyde Regional Council* [1978] concluded that on the facts of that case the strict view should prevail and the corporate veil was not to be pierced.

Sections 37 and 38, LCA 1973 now provide for disturbance payments for persons without compensatable interests comprising removal expenses and, if a trade or business, a disturbance claim so the incidence of corporate veil cases is likely to decline.

Where a claimant occupies land under a business tenancy protected by the Landlord and Tenant Act 1954, section 47 LCA 1973 requires the assumption to be made that a further tenancy within that Act would be granted. It should be noted however that this provision is only of relevance where the tenant's interest is acquired under the compulsory purchase process. If the circumstances are such that a developer is assembling land for, say, a town centre redevelopment with the assistance of compulsory purchase powers, then if that developer acquires the freehold interest it will, time permitting, be able to terminate the tenancy under its powers as landlord on grounds that the land is required for redevelopment. In this event, only the limited compensation available under the Landlord and Tenant Act 1954 will be payable (this payment being based on the rateable value) rather than the full disturbance entitlement under the compensation code.

# Basis of compensation under rule 6

The basis of valuation is value to owner. The acquisition may lead to the total extinguishment of a business, partial permanent loss of profits or only temporary disruption to trade, but what has to be considered is the loss to the owner. Thus where a business depends upon the personality of the owner reflected in higher profits, he will be entitled to claim on the basis of loss to himself. In *Sceneout Ltd* v *Central Manchester District Council* [1995] a total extinguishment case, loss to the owner of a laundry business exceeded market value and was the basis of the award.

Also fundamental to the basis of valuation is the principle of equivalence, established in *Horn* v *Sunderland*. A claimant is entitled to be put in the same position after the acquisition as it was prior to the acquisition, so far as money can do it.

Until quite recently, the perceived wisdom of some, but by no means all, valuers was that following the requirement that a claimant must mitigate its loss, a claimant could never be entitled to compensation on a relocation basis if this would exceed the amount of compensation payable on an extinguishment basis. So compensation for disturbance could never exceed the total extinguishment basis. However in *Director of Buildings and Lands (Hong Kong)* v *Shun Fung Ironworks Ltd* (1995) the Privy Council refused to accept the Crown's submission on this point and decided that the test should be how a reasonable businessman, using his own money, would behave in the circumstances. Although in the case of *Shung Fung* the claim based on relocation failed as it was held that the Lands Tribunal were entitled to conclude that the business planned for the alternative site was not the same business as that on the acquired site, a fundamental and arguably more equitable principle had been established.

So compensation under rule 6 will be assessed on one of two bases: total extinguishment or relocation.

# Total extinguishment claims

The main items of claim will be the value of the land taken (under rule 2 ) and, under rule 6, total loss of profits (regarding which please see later) redundancy payments and closing down costs. In the case of business properties there may also be a claim for loss on forced sale of plant and machinery, fixtures and fittings and stock. Stock is usually valued at the cost of purchase; however, in the case of the other items the correct approach is to compensate the owner for the loss it incurs on a forced sale and the claim will be as follows:

Value to the business of the plant and machinery
*Less*: Forced sale value (possibly only scrap value which the owner receives direct)
*Equals*: Loss on forced sale (payable as compensation by the acquiring authority).

In practice, many acquiring authorities will employ a specialist valuer to agree the value to the business directly with the claimant, which agreed figure is paid as compensation. This figure is paid in full and the acquiring authority takes possession of, and disposes of, the equipment etc. as it wishes.

In *Shevlin* v *Trafford Park Development Corporation* [1998] it was held that "value to the business" in para H 13.2 of the RICS's previous version of the Red Book (now the Valuation Standards) is consistent with value to the owner for compensation purposes, although in *Shevlin* there was a special extra value to the claimant to be allowed for.

No theoretical list of items claimable can possibly be comprehensive, as any reasonable loss or expense that is not too remote is compensatable. Thus in *Widden & Co Ltd* v *Kensington and Chelsea Royal Borough Council* (1970) the following were also properly recoverable:

* Redundancy payments and interest on a loan raised to pay them.
* Cost of cancellation of specific contracts.
* Holiday pay due on the termination of employment.
* Wages to staff to clear up after close-down.
* Advertising expenses in abortive search for new premises.
* A percentage of bad debts as these would be more difficult to recover once the business was wound up.

# Compensation for the loss of goodwill

## Total extinguishment cases

In the case of total extinguishment, the minimum that a claimant will lose will be the market value of goodwill. Traditionally a YP, which was based on market evidence, was applied to the adjusted net profit as previously determined. However, the Lands Tribunal in *W Clibbett Ltd* v *Avon County Council* [1976] drew attention to the lack of market transactions adduced as evidence, and until recently claims were often made on the basis of previous tribunal awards.

In the case of businesses in freehold or leasehold premises having Landlord and Tenant Act protected leases, YP's ranging from 2.5 to 4 became the norm; variations between the two having regard to the individual circumstances of the case. In other cases, the Tribunal adopted what became known as the 'robust' approach and simply determined a figure.

In *Payne* v *Kent County Council* [1986] the Tribunal adopted a more technical approach. The case concerned the acquisition of a petrol filling station ripe for demolition and rebuilding, the date of valuation being December 1982. The Tribunal took one-half of the net profit per gallon (7.1p) say 3.5p, as the rental element and applied this to a throughput of 1.5 million gallons giving £52,500 pa, capitalised at 6% to give say £875,000. £50,000 was added to reflect room for a small catering unit and £175,000 deducted as the cost of redevelopment. The award was £750,000.

In *Optical Express (Southern) Ltd* v *Birmingham City Council* [2005] the Tribunal went a stage further, heard evidence from forensic accountants and decided that the appropriate approach to determining compensation on total extinguishment of a business was the capitalisation of the estimated maintainable branch contribution before deduction of interest, tax, depreciation and amortisation (known as EBITDA). The appropriate multiple to be applied to the EBITDA was similarly derived from EBITDA evidence rather than P/E evidence due to the greater volume of evidence available in respect of EBITDA transactions.

The *Optical Express* decision equated to about 5+ YP on the traditional YP approach to the valuation of a business. While this decision is being argued, by those acting for acquiring authorities at least, as only appropriate to national retailers, it remains to be seen how long it is before the decision becomes accepted in the context of small/medium enterprises and how long before DCF and other valuation approaches are acknowledged by the Lands Tribunal.

# Damage to, or loss of, goodwill in relocation cases

On compulsory purchase, goodwill is not purchased by the acquiring authority; the owner is compensated for what he will lose by being displaced. Thus the issue is always to decide how much, if any, of the existing trade will be transferred. If none, then the minimum compensation will be the market value of the goodwill as, but for the acquisition, such a sum could have been obtained as part of the consideration on sale of the business. Loss to the owner might conceivably be a higher figure as, for example, in the case a small one-man business.

The value of goodwill is usually measured by calculating an average adjusted net profit and applying a YP. Conventional wisdom has it that the last three years' average profits from the actual accounts are usually considered but there is no foundation for this where there is a discernible profit trend, either upwards or downwards. In either of these cases, there is no reason not to adopt the final year as the level of maintainable profit. In any analysis, any year which is affected by the scheme should be disregarded, for example if profits have been reduced by nearby clearance, and any year in which there has been atypical expenditure should be adjusted to reflect more typical expenditure. A higher YP than normal will be adopted if the profits show a rising trend as past profits are only a guide as to what future profits might have been. Conversely, falling profits justify a lower YP than would be normal.

The following adjustments to the profit shown in the accounts must be considered when arriving at the figure which is to be used to assess the value of the goodwill.

## Adjustment for rent

If the claimant is the freeholder and in occupation, the full rental value of the premises must be deducted from the accounts to avoid double counting — it will already have been used to calculate the land value under rule 2. Mortgage or interest payments (deducted as costs in the trading accounts) are added back. This puts the valuation of the trade goodwill onto the same basis as a claimant trading from leasehold premises.

If the claimant is a leaseholder, the profit rent must be deducted (arrived at by adding back the rent paid and deducting the full rental value) again to avoid double counting — the profit rent will have been capitalised when calculating the value of the leasehold interest under rule 2. This puts the business accounts on the basis of occupying at FRV. It was decided in *Widden & Co Ltd v Kensington and Chelsea Royal Borough Council* (1970) that an increase in rent to full rental value could not merely be passed on by way of increased charges, and the profit rent had to be deducted.

## Deduction for interest on capital employed in the business (other than land and buildings)

This will comprise plant and machinery, stock, fixtures and fittings, vehicles etc., and also cash in hand and in a bank current account.

This was traditionally taken at 5% but see *Reed Employment Ltd* v *London Transport Executive* [1978] where 10% was taken although the claimants stated that their financial arrangements were such that the money was borrowed interest free: see also *RC Handley Ltd* v *Greenwich London Borough* (1970) and *Bostock Chater & Sons* v *Chelmsford Corporation* (1973) for the rationale in those cases for different rates of interest.

### Proprietor's remuneration in the case of a one-man business

Following *Perezic* v *Bristol Corporation* (1955) it is fairly well established that, on the principle of value to owner, no deduction need be made for the value of an owner's own services in such cases.

Deductions might have to be made for the value of a wife's services in the business if nothing or insufficient appears in the accounts, although this was not done in *Zarraga* v *Newcastle upon Tyne Corporation* (1968) and conversely, where a wife takes no active part in a business but receives a salary for tax reasons., her salary should be added back.

### Directors' remuneration

As far as companies are concerned, a reasonable sum should appear for the value of the work of each director: see *Shulman (Tailors)* v *Greater London Council* (1966). If, however, the company is virtually a one man business, the director's fees might be dealt with in the same way as the proprietor's remuneration in the case of a one-man business: see *Lewis's Executors and the Palladium Cinema (Brighton)* v *Brighton Corporation* (1956).

### Deduction from profits of a branch of a proportion of head office expenses

It is necessary to deduct from the branch profits the saving of head office expenses due to the closure: see *Reed Employment Ltd* v *London Transport Executive* [1978] — but only if there are any! If the branch closure does not result in any actual saving in the head office, no deduction should be made. It is wrong to say, for example, that if the branch contributed say 15% of the company's turnover, 15% of the cost of one member of head office staff should be deducted, because one cannot sack 15% of an employee, and furthermore while the amount of work at head office might be supposed to diminish, it does not follow that the various departments involved will be able to economise *pro rata* or at all — the office must still provide a full service to the rest of the firm.

# Other heads of claim in relocation cases

See above regarding damage to goodwill. In addition, the following are commonly encountered. However, because any reasonable loss suffered by the owners is compensatable, no comprehensive list of items under this heading can be made.

### Fees incurred in obtaining new premises

Reasonable fees incurred in obtaining new premises of a comparable kind were allowed in *Harvey* v *Crawley Development Corporation* (1957) CA. A claim may extend to reasonable abortive expenditure in seeking alternative accommodation (including survey fees) the costs of terminating a mortgage and taking out a new one, travelling expenses and the loss of earnings during time reasonably spent in searching. These costs are not restricted to those in connection with the property eventually chosen. Where, for example, one or more possible properties are rejected on survey, their costs are equally allowable. This principle would also extend to the costs incurred in acquiring a freehold interest to replace a leasehold one where no suitable leasehold relocation opportunity was available.

Interest on a bridging loan for a new property would be a proper item and also financial losses involved in having to terminate a low-interest fixed-rate mortgage and taking a new one at a higher rate. Moving might necessitate the purchase of new school uniforms, which would be payable. However claims must be reasonable. Thus in *Mogridge (WJ) (Bristol 1937) Ltd v Bristol Corporation* (1956) costs of drawings and preparation of quantities were disallowed, being unnecessary to arrive at a decision whether to build.

Schedule 15, para 2 P&CA 1991 provides, as section 10A LCA 1961, that where the interest of a person not in occupation is acquired, charges and expenses of re-investment in other land incurred within one year after date of entry are claimable.

## Removal costs

Removal costs including the cost of dismantling, adapting and reinstalling plant and machinery or of fixtures and fittings, and any loss on the forced sale of those items which cannot be taken. A claim may extend to extra temporary supervision after reinstallation of machinery. Minor items would include notification of the new address, reprinting stationery and telephone removal. Mitigation is often an important aspect of this element of a claim, thus the usual expectation is that competitive estimates of cost will be obtained and the lowest taken unless there are exceptional and demonstrable reasons for taking the higher estimate.

In *Succamore v Newham London Borough Council* [1978] removal 31 miles away was held to be reasonable and properly compensatable although other premises were available within a shorter distance.

## Duplicated expenditure: double overheads

These claims arise where it is necessary to operate two premises for an overlap period in order to mitigate the claim. Examples of such circumstances might be where no rent free period could be negotiated in respect of the fit-out period for relocation premises, or where a production line was moved in phases in order to maintain production and so obviate a loss of profit claim. The claim would extend to double rent, rates and additional wages, heating, lighting and telephone. If the new premises are not rented, but purchased, then loss of interest on capital could be included over a negotiated period.. The claim should run until the date of completion of purchase or, if earlier, the date the acquiring authority takes possession, as it will pay interest on the purchase price from that date and the claimant should have no financial liability in respect of the old property after the acquiring authority has taken possession of it.

## Repairs and cost of adapting new premises

Where the new premises are in need of repair, the purchase price or lease rent should have been reduced to take this into account. If it has been, the cost of the work would not be recoverable from the authority because the claimant would, in doing it, be receiving value for money.

However, necessary adaptations in order to maintain the business operation would be recoverable. Normally these would not be expected to increase the value of the new premises, for example the creation of pits in the floor in which to install machinery. Other examples can arise where, notwithstanding that the alteration increases the value of the new premises, the cost of alteration is recoverable because it maintains the operation of the business, for example the installation of a three phase electricity supply in a building lacking one where such a supply is necessary for the operation of the machinery of the relocated business: see *M&B Precision Engineers Ltd* v *Ealing London Borough Council* (1972) and *Smith* v *Birmingham City Council* (1974).

## Cost of new premises and increased overheads

A higher price for alternative accommodation will not normally be compensatable as the owner will be presumed to have obtained value for money, but this a principle capable of rebuttal, for example where it might be argued that a claimant had no alternative but to relocate to more expensive premises. So in *Metropolitan and District Railway Co* v *Burrow* [1884] HL, and *Mogridge (WJ) (Bristol 1937) Ltd* v *Bristol Corporation* (1956) circumstances arose where increased overheads were allowed The decisions in *Greenberg* v *Grimsby Corporation* (1961) concerning a new shop and *J Bibby & Sons Ltd* v *Merseyside County Council* [1979] dealing with new and larger replacement offices, applied the principle that greater efficiency of the new premises should be set off against increased operating costs although the latter decision did not dismiss the prospect of the recovery of increased overheads in appropriate cases.

In *Service Welding* v *Tyne & Wear County Council* [1979] CA, bank interest and loan charges to build an alternative factory were held not to be compensatable as the claimants had obtained value for money.

## Relocation of equipment

*Tamplins Brewery Ltd* v *Brighton County Borough Council* (1971) addressed the proper method of ascertaining loss where new plant and machinery were installed in alternative premises even though the old plant and machinery had a useful life. The relocation scheme adopted, with the concurrence and assistance of the corporation, was the best and cheapest way of mitigating the claimant's loss. It was held that the loss could not be measured by the open market value of the existing plant and machinery and that the cost of the new plant and machinery should be taken as the first step less the scrap value of the old and the value of the new plant after 10 years (the estimated life of the old plant) less also capitalised savings on running costs.

### Study 8

*Leasehold disturbance claim involving adjustment of accounts*
A small corner back street shop with a residential upper part on two floors is to be acquired compulsorily for public open space purposes. The property has been occupied for 10 years by the lessee, who lives in the upper part and carries on a small grocery business from the shop. The rent payable for the whole is £3,000 pa exclusive on an internal repairing lease having 10 years unexpired without review. The full rental value of the property is £12,000 pa net of which £10,000 is attributable to the shop.

The accounts show that the outgoings for the upper part have been deducted as well as the rent of £3,000 for the whole premises.

Other relevant details are as follows:

The net profit after deducting mortgage interest of £800, repairs £720, and business rates of £1,840 is £44,000. The figures for repairs relate to the whole structure and rates to the business part only No deduction appears for the owner's or his wife's services. She works about 10 hours a week in the shop and he is virtually full time. There are no suitable alternative shop premises available and the acquiring authority agree that this is a total extinguishment case.

Prepare a claim for compensation for the lessee.

*Leasehold claim on basis of the existing use value plus disturbance basis section 5(2) and (6) LCA 1961*

| | £ pa | £ |
|---|---|---|
| Valuation of the leasehold interest in the land taken | | |
| Full rental value of whole premises (net) | 12,000 | |
| *Add*: Landlord's outgoings (external repairs, insurance and management) Say | 1,200 | |
| Full rental value, internal repairing lease terms | 13,200 | |
| *Less*: Rent payable under lease on the same basis | 3,000 | |
| Net profit rent | 10,200 | |
| YP 10 years 12% and 4% (Tax at 40%) | 3.86 | |
| | | 39,372 |
| Value of land taken (leasehold) Say | | £39,500 |

*Disturbance claim*

The accounts include outgoings for the residential upper part. The profit for the shop is isolated by adding back the rent paid and internal repairs (both relating to the whole property) and deducting figures applicable to the shop only.

| | | £ |
|---|---|---|
| Unadjusted net profit | | 44,000 |
| *Add back* | | |
| Mortgage interest | 800 | |
| Rent payable | 3,000 | |
| Internal repairs | 720 | |
| | | 4,520 |
| Profit if no deductions for rent or repairs | | 48,520 |
| *Less*: Full rental value of the shop on internal repairing terms: Say | 11,000 | |
| *Add*: Tenant's internal repairs on shop | 600 | 11,600 |
| | | 36,920 |
| *Less*: Interest on tenant's capital 5% of £10,000 | | 1,000 |
| | | 35,920 |
| No deduction made for proprietor's remuneration but, say, £100 per week for wife however see *Zarraga* v *Newcastle upon Tyne Corporation* (1968)) | | 5,200 |
| Adjusted net profit | | 30,720 |
| YP based on comparables (back street shop) Say | | 2 |
| Compensation for total loss of goodwill | | 61,440 |
| *Plus*: Loss on forced sale of tenant's fixtures and fittings, Say 80% of £6,000 | | 4,800 |
| Loss on forced sale of stock, Say 50% of £6,000 | | 3,000 |
| Abortive costs of seeking alternative premises, Say | | 800 |

Home loss payment for residential upper part. The lessee qualifies under the substituted section 29(2) LCA 1973 and possesses an owner's interest under the substituted section 30. A 10% home loss payment based on the apportioned value of the lease of the residential upper part would only be about £658 (10% of approx. 1/6th of £39,500).

| | |
|---|---:|
| Therefore the minimum of £4,000 applies. | 4,000 |
| *Crawley* costs and removals for the upper part, Say, | 600 |
| Total disturbance claim | 76,140 |
| Value of land taken | 39,500 |
| | £115,640 |

Total claim, Say £115,650

*Plus* Apportioned basic and occupier's loss payments with regard to the shop, fees and statutory interest.

## Study 9

*A total extinguishment claim in a blight notice case*

You have been asked to advise a client, a grocer. He is now 70 years old and wishes to retire but his efforts to sell his business have been frustrated as the property is allocated as a future library site, a proposal which is to be implemented in about 10 year's time. His shop and upper part are in a poor trading position. He has a lease having nine years unexpired at £2,000 pa without review on a tenant's internal repairing lease. The rateable value of the business part is £9,400 and the present full net rental value of the whole is £10,700 pa.

He has been trading as a grocer for the past 10 years and his net profits for the last three years were:

£31,000
£34,000
£32,400 (most recent)

Both he and his wife work in the business but no figure for remuneration appears in the accounts for either of them. The wages of a part-time help and delivery-boy have been deducted in the accounts.

*The following preliminary matters have to be considered*:

The blight notice provisions — sections 149–171 of, and Schedule 13 to, T&CP 1990. These enable qualifying owner-occupiers affected by planning proposals to require purchase of their interests. The categories of qualifying blight are specified in Schedule 13.

If part only is affected by a scheme, the notice must be served in respect of the whole property (section 150(3)) but if the authority serve a counter notice that they require part only (section 151(4)(c)) the Lands Tribunal can decide whether the blight notice will apply to the whole or part (section 153(6)). In *Smith* v *Kent County Council* [1995] a counter notice was not upheld when a small part of a front garden was required for road widening.

The date for determining whether the land is blighted is the date of the counternotice. See *Carrell* v *London Underground Ltd* [1996] and *Sinclair* v *Secretary of State for Transport* [1997] a change of plan after that date being irrelevant; *Entwistle Pearson (Manchester) Ltd* v *Chorley Borough Council* (1993).

Further details will be found in Chapter 10.

For the purposes of this Study, it is assumed that the blight notice has been accepted.

*Basis of claim*

A certificate of appropriate alternative development could be applied for, but a claim on the basis of existing use value plus disturbance is likely to be the most favourable.

Section 46 of the LCA 1973. As the claimant who is carrying on the trade or business has attained the age of 60, compensation for disturbance may, at his option, be assessed on the basis of total extinguishment, subject to his giving certain undertakings.

The claimant is entitled to a home loss payment regarding the residential part, and to both the basis and occupier's loss payments regarding the remainder.

| *Open market value of leasehold interest* | £ pa | £ |
|---|---|---|
| Full rental value of whole premises, net (FRI) basis | 10,700 | |
| Add: Landlord's external repairs, insurance and management | 1,100 | |
| Full rental value (internal repairing lease basis) | 11,800 | |
| *Less*: Rent payable on the same basis | 2,000 | |
| Profit rent | 9,800 | |
| YP 9 years at 10% and 4% (Tax at 40%) | 3.88 | |
| Value of land taken | | 38,024 |

| *Disturbance claim for shop part* | £ pa | £ pa | £ |
|---|---|---|---|
| Average unadjusted net profit | 32,466 | | |
| Assuming that the rent of £2,000 pa is for the whole but repairs for the shop alone have been allowed for in the accounts, it is only necessary to make an adjustment for rental value, therefore | | | |
| *Add back*: rent paid | 2,000 | | |
| | 34,466 | | |
| *Deduct*: full rental value of shop only on tenant's internal repairing lease, Say | 8,400 | | |
| Profit after making a deduction for the full rental of the shop | | 26,066 | |
| *Less*: Interest on tenant's capital 5% of £20,000 | 1,000 | | |
| Wife's wages: Say £160 per week (but see *Zarraga* v *Newcastle upon Tyne Corp.* (1968)) See Note (1) below | 8,320 | 9,320 | |
| Adjusted net profit | | 16,746 | |
| YP for total extinguishment based on comparables for this type of business | | 3 | |
| Value of goodwill on extinguishment basis | | | 50,238 |
| *Add*: Loss on forced sale of fixtures and fittings. Say 80% of £5,000 | | | 4,000 |
| Loss on forced sale of stock,(mainly tinned food which could be disposed of easily). Say 30% of £6,000 | | | 1,800 |
| | | | £94,062 |

Claim for land taken and disturbance claim on a total extinguishment basis say £94,000
*Plus* Basic and occupier's loss payments on the business premises, fees and interest, and in addition the items listed below with regard to the residential upper part.

*Disturbance claim for the residential upper part*
This would include:
- *Crawley* costs based on actual and reasonable costs incurred.
- Removal costs (including telephone) and expenses incidental thereto similarly assessed.
- Depreciation of fixtures and fittings due to removal or loss on forced sale, as appropriate.
- Home loss payment. As the minimum home loss payment of £4,000 will clearly exceed 10% of the value of the lease of the residential upper part the payment will be £4,000.

Note (1)  If the claimant's wife had been an employee, he would have had to make her redundant and her redundancy payment would have been a valid item in the disturbance claim.

## Study 10

In which the reader is invited to examine a rule 6 claim in detail.

*Background*

The CPO notices were served in April with entry in September.

Smudgeprint Ltd is a small but expanding printing company occupying Oldprop comprising five individual leasehold industrial units on a 1950s industrial estate near the centre of Bigtown. It is owned and operated by Mr and Mrs Script who each draw £60,000 pa as directors of the company. Profits have built up steadily and last year Smudgeprint showed a net profit of £500,000 equating to approximately 15% of turnover.

The client base is predominantly local but the company's growing reputation for quality work using the latest technology has resulted in it being appointed the sole printer for 'outers' for a prepared-food catering company, Nosh It, with national distribution. This one contract is worth £100,000 pa in terms of net profit to Smudgeprint. Nosh It operates on a 'just-in-time' basis where its requirements are faxed to Smudgeprint 48 hours before delivery is required. Smudgeprint will incur contractual penalties if delivery is not made on time. Smudgeprint do not have space at Oldprop to build up 'outers' stock and in any event have no way of knowing what design of 'outers' will be called for during the move period.

Smudgeprint also provides an update service for a medical journal. This requires Smudgeprint to maintain a minimum stock of all past editions ready for supply to subscribers and to reprint used stock as necessary.

Building A comprises a terrace of three adjoining units and is where the actual printing takes place using two Heidlberg printing presses. The presses will be down for three weeks while they are moved from Oldprop to Newworks. A third press was to be installed at Oldprop and had been ordered in January, with a deposit paid, by the Scripts. However, immediately following the placing of the order in February the CPO was confirmed. With the agreement of Heidlberg, installation of the third machine was delayed until Smudgeprint had relocated to Newworks.

Building B, the fourth unit, is the warehouse for raw product, ie paper, inks etc. Building C, the fifth unit, is the warehouse for finished product. All product stored in Buildings B and C is shrink-wrapped but only because these units are not 100% weatherproof.

All five units are alarmed (both for fire and intruders) and connected by data and telephone links. The telephone system was newly installed three years ago, designed especially to link the separate units. The computer system provides 5 terminals in Building A with one each in Buildings B and C. Building A is fitted out with partitioned offices, two dark rooms, a chemical store, an image inspection room with light boxes to enable close-up viewing of negative film, staff mess room, showers and toilets.

The premises are held on three separate leases, the overall rent payable is £100,000 pa and there is no profit rent.

*Relocation facts*

In the time available following service of notices Smudgeprint has been unable to identify a suitable size leasehold unit, or units, into which to relocate in the immediate vicinity in Bigtown. It has therefore purchased a stand-alone freehold property, Newworks, in Little Hamlet, a village about five miles to the north of Bigtown. Some repair work is required to this building as it has been vacant for some time. The shell is sound but it has no internal fit-out whatsoever apart from minimal toilet facilities, and has been vandalised.

This property has a *sui generis* planning permission for its previous use (an agricultural supply cash and carry warehouse) but the planning authority has confirmed that an application for change of use to B1 will be granted consent, subject to a condition restricting lorry movements to between 8.00 a.m. and 7.00 p.m. on weekdays only.

The freehold property, Newworks, in Little Hamlet cost £1.25m and has been purchased by Mr and Mrs Script with the aid of a mortgage. At present it is too large for the company's immediate needs. Mr and Mrs Script in turn have let the property to Smudgeprint on a 20 year, FRI lease within the LTA 1954 at full open market rental value of £125,000 pa. It has been done in this manner so as to set up a pension scheme for Mr and Mrs Script on the advice of their accountant. An initial three month's rent free period was granted to allow for the fitting out of Newworks.

Two staff will become redundant as a result of the move as they have no means of getting to Little Hamlet and it is not well served by public transport. Smudgeprint has its own redundancy scheme based on length of service which is written into its employee's contracts. Under the Smudgeprint scheme the two will be entitled to £3,000 each; under the statutory scheme they are only entitled to £2,000 each. A third employee, an experienced print industry engineer responsible for the maintenance and operation of the Heidlbergs is in a similar position with no transport, but because of his importance the company has agreed to pay his taxi fares until he retires in five years time, during which period he will recruit and train up a successor, a school leaver who was appointed a week before the move to Newworks. The cost of this over the full five year period is estimated to be £5,000.

The following table sets out the itemized claim. Alongside the table is a column for your answer. At the end of this chapter is another table with the suggested answers.

*The claim*

| Item | Description | £ Cost | Yes/No/ Perhaps etc. |
|---|---|---|---|
| 1 | Purchase price of Newworks | 1,250,000 | |
| 2 | Legal costs of acquisition | 5,000 | |
| 3 | Stamp Duty on purchase of Newworks | 50,000 | |
| 4 | Commercial agents' fees — finding/ negotiating price of Newworks | 12,500 | |
| 5 | Stamp Duty Lease Tax | 5,000 | |
| 6 | Mortgage arrangement fee | 500 | |
| 7 | Structural survey fees of two leasehold properties considered for occupation | 800 | |
| 8 | Accountant's fee in connection with the acquisition | 500 | |
| 9 | Planning fees | 1,000 | |
| 10 | Space planner's fee for devising internal layout of Newworks | 500 | |
| 11 | Repairs to roof of Newworks | 20,000 | |
| 12 | Cost of increasing roof height to accommodate one relocated press | 5,000 | |
| 13 | Provision of three phase electric supply from road into Newworks | 15,000 | |
| 14 | Provision of internal three phase electrics, distribution boards etc to presses | 10,000 | |
| 15 | Provision of additional power sockets for normal power consumption | 5,000 | |
| 16 | Reconfiguration of lighting system around presses | 3,000 | |
| 17 | General up-grading of lighting system elsewhere | 2,000 | |
| 18 | Provision of partitioned offices | 25,000 | |
| 19 | Provision of dark rooms | 5,000 | |
| 20 | Provision of image inspection room and light boxes | 2,500 | |
| 21 | Provision of constructing chemical store in stand alone building in yard | 5,000 | |
| 22 | Provision of showers | 1,500 | |
| 23 | Up-grading of toilet facilities | 1,500 | |
| 24 | Provision of toilet facilities for disadvantaged persons | 500 | |
| 25 | Provision of staff mess room | 500 | |
| 26 | Provision of foundations for three presses | 9,000 | |
| 27 | Arrangement fee for short term loan for fit out costs | 250 | |
| 28 | Cost of interest on fit out loan for five year term of loan | 5,000 | |

| Item | Description | £ Cost | Yes/No/ Perhaps etc. |
|---|---|---|---|
| 29 | Cost of materials and labour to replicate at Newworks bespoke fitted storage racking in Building C at Oldprop | 2,750 | |
| 30 | Removal from Building B at Oldprop and re-erection at Newworks of Dexion type racking (£1,500) + purchase of additional racking to complete the 'run' at Newworks (£500) | 2,000 | |
| 31 | General removal costs; two estimates — £3,000 and £4,000 | 4,000 | |
| 32 | Removal and re-installation of two presses; two estimates — Heidlberg £10,000, Lift 'n Drop £3,000 | 10,000 | |
| 33 | Installation of press where purchase was delayed | 5,000 | |
| 34 | New telephone system | 4,000 | |
| 35 | New fire and intruder alarm system | 2,500 | |
| 36 | Security shutters to deter local vandals (Little Hamlet is known for vandalism) | 1,500 | |
| 37 | New data network to provide 10 computer terminals | 2,500 | |
| 38 | Printing of 20,000 letterheads, 10,000 continuation sheets and printed envelopes | 1,250 | |
| 39 | Printing of business cards | 500 | |
| 40 | Artwork for new company brochure | 1,000 | |
| 41 | Printing of 2,500 new brochures | 5,000 | |
| 42 | Double overhead — three months rent of Oldprop during fit out of Newworks | 20,000 | |
| 43 | Double overhead — buildings insurance of Newworks for fit out period | 500 | |
| 44 | Double overhead — business rates of Oldprop during fit out of Newworks | 10,000 | |
| 45 | Cost of shrinkwrap to shroud all paper products during move | 500 | |
| 46 | Cash payment to A N Other to clear rubbish from Oldprop after move | 250 | |
| 47 | Redundancies | 6,000 | |
| 48 | Taxi fares for essential employee | 5,000 | |
| 49 | Extra travelling costs claimed by two other employees | 8,000 | |
| 50 | Cost of out-sourcing printing of outers for Nosh It during the move period. A local firm has quoted £22,000. A firm further away has quoted £30,000 and is thought to be more reliable although there is no real evidence to cast doubt on the ability of the lower quoting firm | 30,000 | |
| 51 | Staff time incurred in preparing for the move | 15,000 | |
| 52 | Directors' time in looking for a relocation property following the cpo announcement | 10,000 | |
| 53 | Directors' time incurred in the move | 10,000 | |
| 54 | Loss of profit during press down-time while reinstallation takes place | 30,000 | |
| 55 | Loss of profit on out-sourced Nosh-It work | 4,500 | |
| 56 | Surveyor's fees | 20,000 | |
| 57 | Loss of profit due to extra rent payable £25,000 pa × 4 YP | 100,000 | |

*Compulsory Purchase*

# PART V: MINOR INTERESTS AND OTHER SPECIAL CASES

This part deals with:

- short tenancies
- land subject to agricultural tenancies (see also Chapter 1)
- persons without compensatable interests
- unfit housing
- Leasehold Reform Act 1967 and compulsory purchase
- tax on compensation.

## Short tenancies — section 20 CPA 1965

The right of a lessee to apply for a new lease of business premises under Part II, Landlord and Tenant Act 1954, is taken into account and reflected in the YP for loss of goodwill. However, yearly tenants or less are not entitled to a notice to treat and may be dealt with in one of two ways.

1. By the purchase of the superior interest and the service of a notice to quit under the Landlord and Tenant legislation. Where notice to quit is served, sections 37 and 38 of the LCA 1973 as amended by Schedule 13 to the Housing Act 1974 apply.
2. By the service of notice of entry and the taking of possession under compulsory purchase powers. Compensation will be assessed under section 20 CPA 1965. Section 47(2) LCA 1973 provides that compensation under section 20 will now reflect any right to continue in occupation.

However, in *Bishopsgate Space Management Ltd and Teamworks Karting Ltd v London Underground Ltd* LT it was held that as the contracted-out lease under which Teamworks occupied contained a landlord's break clause, it must be assumed that notice to break the lease would be served as soon as possible after possession of the land had been taken. The value of Teamworks' interest was therefore limited to the profit it would have been likely to earn between the date of possession and the date some five months later when the break notice would have taken effect, notwithstanding that the evidence was that in the no-scheme world it was almost certain that Teamworks would have been granted further occupational leases for a number of years to come.

Under P&CA 1991 Schedule 15 para 4, tenants may now claim for losses due to any severance of the tenanted land from other land which they occupy or own. Section 20 (2) CPA 1965 related only to land held on the same tenancy.

## Land subject to agricultural tenancies

Please also refer to Chapter 1.

In valuing either the freehold interest subject to a tenancy or compensation to tenants of agricultural holdings dispossessed by an acquiring authority, the issue of whether possession could be obtained under the Agricultural Holdings Act on the grounds that planning permission has been given for a non-agricultural purpose is important: see case B, Schedule 3 to the Agricultural Holdings Act 1986. The acquiring authority's scheme is normally such a ground and in *Rugby Joint Water Board v Foottit* (1972) and *Minister of Transport v Pettit* (1968) it was held that the interests had to be valued allowing for possession which increased the freeholder's compensation at the expense of the tenant. Section 48 of the LCA 1973 now provides, in effect, that possession shall be considered only if it would have been obtainable apart from the authority's scheme. It is necessary to determine how long the occupancy might have continued but for the compulsory purchase. The Lands Tribunal's approach to the uncertainties involved in valuing, in effect, a life expectancy of profit was illustrated in *Wakerley v St Edmundsbury Borough Council* (1977). The rights to succession on death or on retirement of the tenant are now covered by part IV of the Agricultural Holdings Act 1986 which consolidated and amended certain enactments relating to agricultural holdings. Schedule 3, cases A to H list the circumstances in which the consent of the Agricultural Land Tribunal to the operation of notices to quit is not required.

After 1 September 1995 new tenancies cannot be granted under the Agricultural Holdings Act 1986, other than succession tenancies. The parties are free to fix the terms of the tenancy and an open market rent applies on review, unless agreed otherwise. The Agricultural Tenancies Act, 1995 also provides that to obtain possession, for a tenancy which is either from year to year or a fixed term of greater that two years, the notice requirement is 12 to 24 months expiring on the term date. For terms of two years or less, no notice is required.

## Persons without compensatable interests

Sections 37 and 38 LCA 1973 as amended by Schedule 13 para 39 Housing Act 1974 apply to persons without compensatable interests displaced in consequence of compulsory acquisition provided they are in lawful possession. Disturbance payments must now be made to such persons assessed under section 38, comprising expenses in removing and, in addition, in the case of a business, disturbance having regard to the period the business might have continued. Section 37(4) restricts compensation to either section 38 compensation or Landlord and Tenant Act 1954 compensation, but not both.

In the context of the statutory requirement to have regard to the likely period of continuing occupation, is interesting to contrast the compensation position of a claimant occupying under an arrangement qualifying for compensation under section 37 with that of a claimant occupying under a formal but contracted-out lease with a landlord's break clause as was the situation in the *Teamworks* case referred to in relation to section 20 above.

In *Wrexham Maelor Borough Council v MacDougall* [1993] the Court of Appeal held that MacDougall, who was the lessee, was in possession for the purposes of section 37 and a company, from which earned his living, had exclusive occupation from him. The company was therefore in lawful occupation under section 37.

Compensation paid to a tenant under section 37 Landlord and Tenant Act 1954 is not liable to capital gains tax: *Drummond (Inspector of Taxes) v Austin Brown* [1985] nor is compensation for disturbance under section 34 (1) of the Agricultural Holdings Act 1948: *Davis (Inspector of Taxes) v Powell* [1977]. In neither case did the sum payable derive from the asset.

## Unfit housing

Schedule 9 to the Local Government and Housing Act 1989 made important changes to the Housing Act 1985 and the law relating to unfit housing and compensation. In particular para 83 of Schedule 9 substituted a new section 604 into the Housing Act 1985 containing a revised fitness for habitation standard.

The site value compensation provisions in sections 585 to 595 of the Housing Act 1985 ceased to have effect (para 76). Consequently section 10 of, and Schedule 2 to, LCA 1961 dealing with ceiling values, owner-occupier supplements and well-maintained payments were no longer necessary and were repealed by Schedule 12. The normal open market value basis applies and this will reflect any lack of fitness and amenities and disrepair.

Section 584A provides for compensation for properties subject to closing or demolition orders based upon the diminution in the value of the owner's interest as a result.

# Listed buildings in poor repair

If the building is being acquired under the minimum compensation rules of section 50 T&CP Act 1990, which apply where the building has been deliberately allowed to fall into disrepair, the planning assumptions are that planning permission would not be granted for any development or redevelopment and that the only listed building consent available would be for the restoration and repair of the building.

# Leasehold Reform Act 1967

If a valid notice to enfranchise is served before a notice to treat, the notice to enfranchise will cease to have effect but the compulsory purchase compensation will have regard to it (section 5(6)(b)). If, however, the notice to treat precedes a notice to enfranchise, section 5(6)(a) states that the latter notice will be of no effect. For the effect on valuations: see *Boaks* v *Greater London Council* (1978). Section 5(6)(a) will be of particular importance where a lease has less than three years unexpired on the notional open market sale date, so that a prospective purchaser can never qualify to enfranchise. The value of the leasehold interest will therefore be considerably less, reflecting Landlord and Tenant 1954 part I rights only and the freehold interest correspondingly more. *Sharif* v *Birmingham City Council* (1978) concerned a leasehold interest with the benefit of a valid notice to enfranchise; the award was on the basis of the freehold value less the enfranchisement price and surveyor's and legal fees. No end deduction for contingency costs and time and trouble was made.

# Tax on compulsory purchase compensation

If land is disposed of under a compulsory purchase procedure, or in the face of a threat thereof, the total compensation can be apportioned for tax purposes between its constituent parts (section 52(4) TCGA, 1992). Compensation for loss of profits or on stock and reimbursement of removal expenses are assessable under case I or II of Schedule D. (Statement of practice SP8/79). Compensation for the land itself will be subject to capital gains tax and any amount apportioned as injurious affection or severance compensation is treated as a part disposal of the land retained for capital gains tax: see *Stoke-on-Trent Council* v *Wood Mitchell & Co Ltd* [1978].

Under the TCGA 1992, sections 247 and 248 a landowner who realises a capital gain on the disposal of land to an authority having compulsory purchase powers may be able to claim roll-over relief if he applies the proceeds of the disposal in the acquisition of new land which meets certain conditions. He must, however, not have advertised the land for sale or taken any other steps to make known to the authority, or others, that he was willing to sell the land.

# PART VI: LOSS AND OTHER PAYMENTS
## Introduction to loss payments

There are two sorts of supplementary payment to which claimants with qualifying interests are entitled. They apply to all cases of compulsory purchase, including where the acquisition has been triggered by the service of either a blight notice or a purchase notice. Statutory interest is payable on these supplements.

Home loss payments are to be made to resident occupiers displaced by the acquisition; Loss payments are to be made where a home loss payment is not available to the claimant. This will include the landlords of residential property.

It could be said that the LCA 1973 undid rule of 1 section 5 LCA 1961 when it introduced home loss and farm loss payments; these being payments made to those entitled solely on account of the acquisition of their interests by compulsion. However, as their names suggest, they were payable only in the case of residential or agricultural property; the occupiers and/or owners of business property interests were not entitled to anything similar.

This lack was remedied by the PCPA 2004 which added sections 33A–K to the LCA 1973 and, with effect from 31 October 2004, introduced basic loss and occupier loss payments, which are payable where no home loss is payable. If an interest comprises both residential and non-residential property, the value of the non-residential property is deducted from the total value before calculating either the basic or occupier loss payment. The PCPA 2004 also laid down a new basis for the calculation of farm loss payments, re-naming them agricultural loss payments in the process.

## Home loss payments

These are provided for by sections 29 and 30 LCA 1973 (as amended by sections 68 and 69 P&CA 1991). The current position is that a person is now entitled to a home loss payment if the following conditions are satisfied throughout the period of one year (previously five years) ending with the date of displacement:

(a)   The claimant has been in occupation of the dwelling, or a substantial part of it, as his only or main residence and

(b)   has been in such occupation by virtue of a qualifying interest which includes statutory tenancies under the Rent Acts (section 29(4)).

The amended section 30 provides that, if the interest of a person who so qualifies is that of an owner, as defined in section 7 ALA 1981, the amount of the home loss payment is 10% of the market value of the owner's interest in the dwelling, subject to a maximum of £38,000 and a minimum of £3,800. If the qualifying interest is other than that of an owner, the home loss payment is £3,800.

These payments also apply in the case of blight and purchase notices (section 68(2) P&CA 1991) and in a number of other instances — see section 29(1).

Home loss payments are also available to qualifying caravan dwellers.

Discretionary payments may also be made if (a) and (b) above are satisfied but the occupation has been for less than one year (section 29(2)).

# Loss payments

There are two types of loss payment. The basic loss payment is available to qualifying freeholders and leaseholders; the occupier's loss payment is available to those in occupation. An owner occupier, or an occupying leaseholder, is entitled to both. There are some exclusions — see section 33D, LCA 1973.

## Basic loss payments

These are payable in the case of a claimant who has a qualifying interest; this being either a freehold or leasehold interest subsisting for more than one year as at, essentially, the valuation date. The payment is equivalent to 7.5% of the compensation paid for the value of the interest in land up to a maximum of £75,000. This carries interest from the date of entry.

## Occupier's loss payments

Different methods of calculation apply depending upon whether the interest in land relates to non-agricultural or agricultural land. A claimant is entitled to an occupier's loss payment if it has a qualifying interest for the purpose of a basic loss payment and has been in occupation for a period of one year ending with again, essentially, the valuation date. These payments carry interest from the date of displacement.

## Non-agricultural land

The the claimant is entitled to adopt whichever one of three basis produces the highest figure, subject to an upper limit of £25,000. The old farm loss payments have been phased out. The bases are:

1.  25% of the value of the interest in land
2.  the land amount: which is the greater of
    i.   £2,500 if the whole of the claimant's interest is acquired or
    ii.  £300 if only part is acquired, and, in either case
    iii. £2.50 per m² of the area of the land acquired
3.  the building amount: £25 per m² of the gross floor space of any buildings on the land, measured externally.

## Agricultural land

As for non-agricultural land, there are three alternative basis, subject to an upper limit of £25,000. Bases 1 and 3 are calculated as described above. In the case of base 3 however, the land amount is the higher of £300 or:

1.  Where the land taken does not exceed 100 ha, £100 per hectare,
2.  Where the land taken exceeds 100 ha, £100 per hectare for the first 100 ha and £50 per hectare thereafter, up to a limit of 300 ha.

# Advance payment of compensation

Section 52 LCA 1973, as amended by section 63 P&CA 1991, provides that if an acquiring authority has taken possession it must, if requested, make an advance payment on account of compensation payable, of 90% of the agreed amount, or, if not yet agreed, of the compensation estimated by the authority.

An additional section 52A provides for the advance payment to include accrued interest from the date of entry. To avoid double counting, section 52A(9) provides that in such a case the right to interest under section 11(1) CPA 1965 does not apply.

# Interest on compensation

Section 80 P&CA 1991 provides for the payment of interest on compensation at the rate prescribed under regulations made under section 32 LCA 1961 in cases listed in part 1 of Schedule 18, which lists circumstances where legislation did not already provide for interest, for example section 31 LCA 1961 dealing with compensation for withdrawal of notice to treat. Interest is payable from the date specified in the schedule until the date the compensation is paid. In compulsory purchase cases, it is payable on the entire claim. Where an advance payment is made, interest ceases to run on the amount advanced, but the interest is due on the full amount, not on the 90% advanced.

Part 2 of the schedule lists provisions that already provided for interest, *inter alia*:

Section 11(1) CPA 1965 (entry on land)

Section 18 LCA 1973 (claims under part 1 of that Act)

Section 63(1) LCA 1973 (McCarthy claims where no land taken)

Section 11(1) CPA 1965 was at issue in *Halstead & Others (Members of the Whalley Range Methodist Church Council and for the Custodian Trustees for Methodist Church Purposes) v Manchester City Council* [1997] it was held that statutory interest is payable in respect of compensation assessed under rule 5, even where an acquiring authority has met the various stage payments to be made under a construction contract such that the claimant has never been out of pocket; interest being specifically recoverable, until the final cost instalment is paid, under section 11 CPA 1965.

In *Mallich v Liverpool City Council* (1999) CA, a claim for loss of investment income as a result of a very low advance payment was rejected as the claimant had been compensated on a total extinguishment basis, and any delay in full payment could only be compensated by way of interest at the prescribed rate.

If, on the annual anniversary of the first advance payment, the amount of accrued interest exceeds £1,000, the authority must pay this — unprompted. The claimant does not have to ask for it. The Act says "... the authority shall ..." If the amount is less than £1,000 it is carried forward to the next annual anniversary.

Interest and advance payment calculations are not easy. Detailed guidance and explanation will be found in the *Handbook of Land Compensation*.

# PART VII: COMPENSATION WHERE NO LAND IS TAKEN

Where no land is taken, compensation may be recoverable for depreciation in the value of the claimant's interest in one (or both) of two instances.

# Compensation under section 10 CPA 1965

Prior to the LCA 1973 the only legislation providing for compensation where no land is taken was section 10 CPA 1965. The effect of this section was clarified by the four rules in *Metropolitan Board of Works* v *McCarthy* (1874). To obtain compensation, depreciation in the value of land has to be caused by:

1. the lawful execution of the works
2. which would have been actionable but for the statutory powers to proceed with the works
3. which physically interfere with a right, private or public, which adds value to the land of the claimant and
4. that depreciation must arise as a result of the carrying out of those works, not from the subsequent use of them.

In *Argyle Motors (Birkenhead) Ltd* v *Birkenhead Corporation* [1974] HL, it was confirmed that section 10 will not allow compensation to be paid for mere loss of profit. However that case indicated that if the value of the interest in land was derived from its profit-making potential then section 10 would apply. *Wilson's Brewery Ltd* v *West Yorkshire Metropolitan County Council* [1977] was a claim for loss of support for a gable end wall by the demolition of an adjoining property. The Tribunal decided that a mutual right of support had not been abandoned, nor were any rights subsequently obtained "clam" (that is acquired secretly). A claim under section 10 was therefore proper. A case involving the over-riding of a restrictive covenant was *Wrotham Park Settled Estates* v *Hertsmere Borough Council* [1993]. The Court of Appeal held that it was not inappropriate for compensation to be assessed by reference to the diminution in value of the benefited property consequent on the carrying out of the authorised works and not a sum representing a price for relaxing the covenant.

Until recently it was thought that section 10 would not enable a claim for temporary loss of profit as a result of the execution of the works since, the valuation date being the date when the works ended, any temporary loss would arguably have ceased by then. *Wildtree Hotels Ltd* v *Harrow London Borough Council* (2000) HL, decided that section 10 can apply to a reduction in the letting value during the execution of works. However this case went on to hold that it would be almost impossible for damage caused by noise, dust or ventilation to satisfy all the rules. Similarly, hoardings erected to prevent the public from suffering injury from building works which caused interference to businesses had for many years been held not to give rise to a claim under section 10, see *Ricket* v *Metropolitan Railway Company* [1867] HL. However, the Court of Appeal in *Ocean Leisure Ltd* v *Westminster City Council* [2004] has made it clear that a claimant will be entitled to compensation should such hoardings cause a temporary reduction in value and it will be no defence to argue that the hoardings were only *in situ* for a reasonable period of time.

## Study 11

The land A B C D has been compulsorily acquired from a third party and a primary school is being built close to Blackacre. A restrictive covenant in favour of Blackacre limits the use of the site to agricultural use. A vehicular right of way to Greenacre crosses its northern part and will be extinguished, as the land is essential for inclusion in the primary school site.

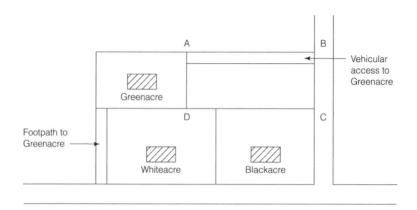

*Claim for freeholder of Greenacre under section 10 CPA 1965 for loss of right of access*

| | £ |
|---|---|
| Before value: | |
| Value with both vehicular and footpath access | 240,000 |
| *Less*: After value: footpath access adjoining Whiteacre only | 220,000 |
| Claim for depreciation | £20,000 |
| Plus fees and interest. | |

*Claim for freeholder of Blackacre under section 10 of the CPA 1965 for overriding a restrictive covenant*

| | £ |
|---|---|
| Before value: | |
| Open market value with the benefit of the restrictive covenant | 300,000 |
| *Less*: After value: value with primary school adjoining | 260,000 |
| Claim for depreciation | £40,000 |
| Plus fees and interest. | |

The owner of Whiteacre has no claim under section 10 of the CPA 1965 as he does not fall within the *McCarthy* rules.

# Compensation under part 1 of the LCA 1973

Part 1 of the LCA 1973 greatly increased the scope for compensation for injurious affection where no land was taken by providing that, in addition to any rights under section 10 of the CPA 1965, compensation is payable for depreciation by the use of certain public works, particularly highways, but only for depreciation caused by the physical factors specified in section 1(2) — noise, vibration, smell, fumes, smoke, artificial lighting and the discharge on to the land of any solid or liquid substance. Public works are defined as any highway, aerodrome or other works on land provided, or used in the exercise of, statutory powers. Section 9 deals with alterations to public works, both carriageway alterations (section 9(5)) and runway aprons (section 9(6)) but mere intensification of use of public works does not qualify.

Section 1(6) prescribes that compensation under part 1 does not apply except in highway cases unless immunity from nuisance in respect of the use other than a highway is conferred by the relevant statute. *Vickers* v *Dover District Council* [1993] concerned the provision of a car park. It was held that the Tribunal had no power to award compensation as section 32 of the Road Traffic Regulation Act 1984 does not either expressly or by implication, confer immunity from actions for nuisance. Also see *Marsh* v *Powys County Council* [1997] where a similar decision was reached concerning the use of a primary school.

Section 2 sets out interests which qualify for compensation; an owner or owner-occupier in the case of residential property, an owner-occupier of a farm or an owner-occupier of other premises with an annual value of £29,200 or below. Section 171 of the T&CPA 1990 defines annual value as the rateable value if the property is non-domestic and no part is exempt from rating. Special rules apply to hereditaments which include domestic or exempt property or which are entirely exempt from rates. An owner includes a leaseholder with not less than three years unexpired at the date of the notice of claim. The interest must have been acquired before the relevant date, which is the date on which the works were first used (see section 11 for the rules for inherited interests).

Section 3 as amended by section 112 of the Local Government, Planning and Land Act, 1980 deals with the submission of claims, which must be made within the six years provided for by the Limitation Act, 1980.

Section 4(1) specifies that compensation will be based on prices current on the first day of the claim period, which is 12 months after the works began to be used. Account is to be taken of the use of the works at that date and of any intensification that could then have reasonably been expected.

Section 5 deals with the assumptions as to planning permission on the relevant land. Planning permission is not to be assumed except for Schedule 3, T&CP 1990 rights. Thus depreciation to development value is not covered. Existing planning permissions that have not been implemented are also to be disregarded.

Section 6 provides for a reduction of compensation for any betterment conferred by the works, including to contiguous and adjacent land, and any works undertaken by the authority to mitigate damage. So, for example, in *Hillard* v *Gwent County Council* [1992] the Tribunal took depreciation at 6% of the unaffected value, but reduced the award to 4% to reflect 2% betterment.

*Shepherd* v *Lancashire County Council* (1976) concerned a council tip: the Tribunal found that the property had been depreciated by the works but this was due to the proximity of the tip and not the use and the claim therefore failed. The Tribunal accepted, however, that depreciation need not be permanent to be compensatable.

It was held by the Lands Tribunal in *Blower* v *Suffolk County Council* [1994] that depreciation caused by the view from the house of the light emanating from distant street lighting was compensatable.

In *Hallows* v *Welsh Office* [1995] the Tribunal said that the depreciation in value may be calculated in many ways such as by a percentage of the pre-works market value (ie unaffected value) by taking part of the total reduction in value caused by the works as due to the physical factors or even as a spot figure based on the valuer's experience of the locality. The Tribunal may also be able to assess the depreciation from agreed settlements.

*Wakely* v *London Fire & Civil Defence Authority* [1996] included a claim in respect of depreciation from the use of a new fire station on the freehold interest in a house let on a regulated tenancy, the award being 6.25% of the unaffected value subject to the tenancy, while *Clwyd Alyn Housing Association Ltd* v *Welsh Office* [1996] was concerned with a claim for the owner of sheltered housing let on secure tenancies. It was held that the evidence of settlements for owner occupied houses could not be applied to sheltered housing and the claim failed through lack of evidence of any depreciating effect of a by-pass on rents.

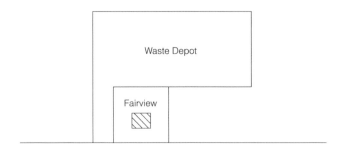

See *Valuation: Special Properties and Purposes,* Edited by Phil Askham (EG Books 2003) for further details.

## Study 12

Fairview is a house on the edge of a town which had open country at the rear. Some 2 ha of this land immediately behind the house has been purchased by the local council from a farmer for a waste depot. There are two cranes in use for some 10 hours a day. The site is also a disused car disposal point and is also used for sorting glass and waste paper. About 50 vehicles a day use the road alongside Fairview to obtain access to the site. The value of the house has been depreciated by the loss of privacy and view as well as by the operations in the area, particularly noise.

Compensation is limited to the physical factors of section 1(2) and will not cover loss of view or privacy. Unless the road leading to the site is new, or has been altered, no compensation is recoverable for traffic noise emanating from vehicles visiting the site, but will be recoverable for any depreciation attributable to their noise once they are on it. A before and after approach can be adopted, as below:

| | £ |
|---|---:|
| Before value | |
| Open market value assuming that the works are in existence but without the physical factors (to confine compensation due to use). This is obviously a valuation that cannot be substantiated by comparables | 240,000 |
| *Less*: After value | |
| Open market value allowing for the physical factors and any betterment | 200,000 |
| Compensation for depreciation | £40,000 |
| Plus interest (section 18(1)) and fees | |

Alternative approaches to the claim are to take say 12% of the unaffected (pre-scheme) value of £320,000 (£40,000) or 33% of the total reduction in value of £120,000 (£260,000 *less* £200,000) (£40,000) again plus fees and interest.

Had the land been taken for a modern sewage works capable of operating without noise or smell, but nevertheless the market value of the house had been substantially reduced because of public prejudice against its close proximity, a claim under part 1 of the LCA 1973 would fail as the depreciation would not be caused by the physical factors specified in section 1(2).

If any part of Fairview had been taken, even a small part of the garden to improve access, then compensation would not have been limited as under part 1 of the LCA 1973. Instead the way would have been opened for a full compensation claim as described in part III of this chapter.

# Appendix

Suggested approach to compensation payable on the disturbance claim of Smudgeprint

| Item | Description | Comment |
|---|---|---|
| 1 | Purchase price of Newworks | In most cases, the presumption will be that this is 'value for money' and so not compensatable but this is a rebuttal presumption should circumstances dictate — *Powner & Powner* v *Leeds Corporation* [1953] |
| 2 | Legal costs of acquisition | Despite 1 above being value for money, if the only alternative to acquiring a freehold is extinguishment of a viable business then this cost is compensatable |
| 3 | Stamp Duty on purchase | Will follow 2 above |
| 4 | Commercial agents' fees | Will follow 2 above |
| 5 | Stamp Duty Lease Tax | What has to be considered is whether or not the lease arrangement was justified in the circumstances |
| 6 | Mortgage arrangement fee | Will follow 5 above |
| 7 | Structural survey fees | As none of the leasehold properties were of the right size, no survey fees should be paid — why commission surveys for unsuitable properties? |
| 8 | Accountant's fee | Will follow 5 above |
| 9 | Planning fees | Value for money |
| 10 | Space planner's fees | As the relocation property is much larger, the need for space planning has to be fully justified |
| 11 | Repairs to roof of Newworks | Not compensatable — value for money |
| 12 | Cost of increasing roof height | Compensatable in full — necessary adaptation |
| 13 | Provision of three phase electrics | Compensatable in full — necessary adaptation |
| 14 | Internal three phase electrics etc | Compensatable in full — necessary adaptation |
| 15 | Additional power sockets | Not compensatable — value for money |
| 16 | Reconfiguration of press lighting | Compensatable in full — necessary adaptation |
| 17 | General up-grading of lighting | Not compensatable — value for money |
| 18 | Provision of partitioned offices | Compensatable in part. The production space needs to be separated from the offices but any further sub-division of the office area will be value for money. |
| 19 | Provision of dark rooms | Compensatable in full — necessary adaptation |
| 20 | Provision of image inspection etc | Compensatable in full — necessary adaptation |
| 21 | Constructing chemical store in yard | Compensatable in full — HSE requirement |
| 22 | Provision of showers | Compensatable in full — HSE requirement |
| 23 | Up-grading of toilet facilities | Not compensatable — value for money |
| 24 | Provision of DDA toilet facilities | Not compensatable — value for money |

| Item | Description | Comment |
|------|-------------|---------|
| 25 | Provision of staff mess room | Not compensatable — value for money |
| 26 | Foundations for three presses | Compensatable only in respect of the two existing |
| 27 | Arrangement fee for fit out loan | Strictly compensatable only in respect of those items attracting compensation but in practice, probably a fixed fee |
| 28 | Cost of interest on fit out loan | As for 27 |
| 29 | Cost of 'C' replacement racking | Compensatable in full |
| 30 | Removal of 'B' racking + additional | Compensatable in part — only the cost of removal and re-erection of existing |
| 31 | General removal costs; two estimates | The lower estimate would usually apply |
| 32 | Removal and re-installation presses | It would be usual to only compensate in respect of the lower estimate but there could be a case for accepting the higher one. In the case of printing presses, it could be the case that a specialist remover (eg Heidlberg) would guarantee the presses working following removal thus avoiding a potentially more costly temporary loss of profit claim |
| 33 | Installation of additional press | Not compensatable as not installed prior to relocation |
| 34 | New telephone system | If the existing is so old that it will not relocate then perhaps a nominal contribution (10%) to the cost of the new. If the existing is virtually new but will nevertheless not relocate for whatever reason then a significant contribution to the cost of the new is justified |
| 35 | New fire and intruder alarm system | Compensatable in full — no self-respecting supplier of such equipment relocates existing equipment |
| 36 | Security shutters | Not compensatable — value for money |
| 37 | New data network | Strictly compensatable only in respect of what exists, ie 7 terminals but practically probably little difference to full cost |
| 38 | Printing of stationary | Strictly compensatable only in respect of what exists as at date of removal |
| 39 | Printing of business cards | As for 38 |
| 40 | Artwork for new company brochure | Compensatable in full |
| 41 | Printing of 2,500 new brochures | As for 38 |
| 42 | Double overhead — rent | In this case — not compensatable. There was a 3 month rent free period |
| 43 | Double overhead — buildings insurance | Compensatable in full |
| 44 | Double overhead — business rates | Business rates would not normally be charged on vacant industrial property during such a fit-out but to the extent that they are charged then compensatable in full |

| Item | Description | Comment |
|------|-------------|---------|
| 45 | Shrink-wrapping protection | Compensatable in full as it avoids a claim for damage to goods during the move |
| 46 | Cash payment to A N Other | Strictly compensation is only payable against proven losses but some commonsense should be shown in respect of such items; a 'formal' job would probably have cost more! |
| 47 | Redundancies | Only compensatable to the extent that they equate to the statutory payments. If a benevolent employer decides to make a higher payment then the excess is down to him |
| 48 | Taxi fares for essential employee | There is case law to support a claim for payment of increased travel expenses to an employee, *Rutter* v *Manchester Corp*. Whether or not this payment should be made will depend upon such factors as the duration of the payment, how long it would take and the cost of training a replacement all measured against the likely level of a loss of profit claim. |
| 49 | Extra employee travel costs | As for 48 |
| 50 | Out-sourcing printing of outers | Compensatable in principle having regard to the level of any potential loss of profit claim. Enquiries would have to be made to justify paying the higher of the two quotes |
| 51 | Staff time incurred in the move | Compensatable in principle provided there is no duplication of this cost in any subsequent loss of profit claim |
| 52 | Directors' time — relocation property | Compensatable in principle provided that there is no duplication of involvement with professional advisers |
| 53 | Directors' time incurred in the move | As for 51 |
| 54 | Loss of profit during press down-time | Compensatable in full |
| 55 | Loss of profit on out-sourced work | Compensatable in principle depending upon the exact terms of the out-sourcing agreement; eg if materials provided to the out-source company. |
| 56 | Surveyor's fees | Compensatable in full provided agreed with the acquiring authority beforehand (see RICS guidance) |
| 57 | Loss of profit due to extra rent payable | Increased rent is generally presumed to represent value for money but this is a presumption capable of rebuttal. If there is no alternative relocation property to which the claimant can relocate, then increased rent could be payable as temporary loss of profit |

# Further reading

*Encyclopaedia of Compulsory Purchase Compensation*, Clive Brand — Sweet and Maxwell.
*Handbook of Land Compensation*, Richard Hayward — Sweet and Maxwell.
*Compulsory Purchase and Compensation* (8th Ed.) — Barry Denyer-Green, Estates Gazette.

# Compensation for Planning and other Restrictions on the Use of Land

All references in this chapter are to the Town and Country Planning Act 1990, as amended, unless otherwise stated. The following abbreviations are used:

| | |
|---|---|
| 1949 Act | National Parks and Access to the Countryside Act 1949 |
| 1961 Act | Land Compensation Act 1961 |
| 1973 Act | Land Compensation Act 1973 |
| 1991 Act | Planning and Compensation Act 1991 |
| 2004 Act | Planning and Compulsory Purchase Act 2004 |
| T&CP Act | Town and Country Planning Act 1990 |
| GPDO | The Town & Country Planning (General Permitted Development) Order (presently 1995), as amended |
| HS Act | Planning (Hazardous Substances) Act 1990 |
| HSA | Hazardous Substances Authority |
| HS Consent | Hazardous Substances Consent |
| LB Act | Planning (Listed Buildings and Conservation Areas) Act 1990 |
| LB Consent | Listed Buildings Consent |
| LPA | Local Planning Authority |
| SM Consent | Scheduled Monument Consent |
| T&CP | Town and Country Planning |

## Introduction

This chapter deals with a wide range of material, but even so is not exhaustive. Readers who require a detailed examination of the subject will find one in the author's *Handbook of Land Compensation*. Claims under Part I of the 1973 Act are not dealt with here: please refer to Chapter 9, p. 256. For more detail, see: *Valuation: Special Properties and Purposes*[1] or the *Handbook*.

---

1      P Askham (ed), *Estates Gazette* 2003.

So far as the RICS *Valuation Standards* (the Red Book) is concerned, all valuers have to be mindful of its requirements. It does not, however, govern statutory compensation, for which any definitions laid down in the statutes or regulations have to be used. If nothing is said, then regard should be paid to the Red Book.

The question of responsibility for site decontamination would in practice arise in some of the scenarios used in this chapter. Decontamination issues will be site-specific, and nothing would be gained — in terms of explaining statutory compensation — by making the sudies more complicated by raising questions of this sort. Readers who insist on worrying about this aspect should understand that all these imaginary properties stand on deep layers of impermeable clay!

Last, a word about statutory instruments. The production rate for these has significantly increased during the last few years, as has the tendency to follow up with amendments. Devolution has ensured that England and Wales now often — but not always — have their own individual variants, sometimes with subtle differences, while Scotland has her own. It is important, therefore, where regulations govern claims, to be careful to use the set appropriate to the claim — different regulations apply, for example, to tree preservation orders made before and after 2 August 1999. References to specific statutory instruments have, therefore, been kept to the minimum, if only because one cannot be sure of their life expectancy.

## General principles governing statutory compensation

There is no general right to compensation for adverse planning decisions: the compensation code is statutory and claims must meet the statutory requirements if they are to succeed. The events which may give rise to a claim, those who may be entitled to claim, the type of loss, the structure of the claim and the method of claiming compensation are all prescribed, either by the statutes or by regulations. The rules are different for the different sources of compensation.

As a general principle, it is likely that a claim will be possible if an existing right is cut back or interfered with, but not if a planning consent is refused from the outset. For example, there is no compensation because planning permission is refused, but there is if it is revoked or modified. However, there are exceptions: tree and scheduled monuments providing two examples where an outright refusal may lead to a compensation claim.

Eligible claimants are sometimes defined as those having an interest in the land or being interested in it; in other cases the net is cast more widely to include anyone who has suffered a direct, consequential, loss. Eligibility is often strictly defined, sometimes involving the length of occupancy — blight notice cases provide an example.

The measure of compensation varies between the types of claim. It may be any direct loss or the depreciation in the value of an interest in the land. The former category embraces the latter. The loss may be calculated by reference to land values or, on occasion, by reference to a loss of profits; the claim may include abortive expenditure and additional running costs. In some instances, notably discontinuance, a full disturbance claim may arise. In most cases, professional fees incurred in the preparation of the claim are recoverable.

Claims must be correctly submitted, usually within six months, and are governed by regulations[2]. Disputes go to the Lands Tribunal. There are some instances where there is no prescribed form for the

---

2    Town and Country Planning General Regulations 1992; Planning (Listed Buildings & Conservation Areas) Regulations 1990 — both as amended; and others.

claim, but where an Act states that particulars must be given it is necessary to specify the amount sought. It can prove fatal to leave the matter imprecise. "An amount to be agreed above £50" was insufficient to satisfy the requirement of Part I of the Land Compensation Act 1973: *Fennessy* v *London City Airport* [1995]. This may be contrasted with the position under section 186 of the T&CP Act (stop notices), where there is no prescribed form.

Not all activities require the approval of the local planning authority. Section 55 defines development and lists several things which do not constitute development at all. The GPDO gives planning permission for a lot, and compensation is recoverable if these consents are cut back (see below, Revocation). The reader is referred to the order for details.

# Valuation principles
## *General principles affecting claims*

There are some principles of general application. The first concerns the definition of Schedule 3 value, formerly existing use value, which is a frequent ingredient in the after value used when calculating depreciation. It is complicated, and is dealt with separately further on in the chapter — the other principles are easier and are discussed here.

The statute will specify which, if any, of the six valuation rules of section 5 of the 1961 Act are to be applied "modified as necessary". Commonly, one is required to value for sale in the open market, assuming willing parties. Where lost development value or lost developer's profit is at issue, one is not obliged to hypothesise a sale to a speculator: the claimant is in the market and may carry out the development: *Richmond Gateways Ltd* v *Richmond upon Thames London Borough* [1989] CA. This was a claim for lost Schedule 3 rights — a type of development that would not attract the interest of outsiders. In other cases, also involving small-scale development — deductions for profit and risk have been made: see, for instance, *Newham LBC* v *Hussain* [2006] LT; *Watford Borough Council (No respondent)* [2003] LT; *Corton Caravans & Chalets Ltd* v *Anglian Water Services Ltd* [2003]; *Smith* v *South Staffordshire District Council* [2006] LT. This is discussed in more detail in note (5) to Study 3.

Claimants are duty-bound to minimise their losses. A compensation case is not an invitation to raid the resources of the compensating authority. Losses incurred in a genuine attempt to reduce losses are recoverable: *Pennine Raceway Ltd* v *Kirklees Metropolitan Borough Council* (1984) LT.

Valuation evidence should be based on the analysis of market comparables. Where evidence is derived from the subject property, care must be taken to ensure that the transactions were genuine and at arm's length. The residual method is suspect unless the calculations were prepared for purposes other than the claim: they are much more credible if prepared in advance of the adverse planning decision — as part of the pre-purchase appraisal, for example.

A claim must be internally consistent. As an example, a valuation that was based on the development value of a piece of land would be inconsistent if the claim also included an amount for injurious affection to the remainder of the estate if this was caused by the very development that generated the development value. The reason is that in the open market the owner would have to accept the depreciation in order to secure the development value. If the depreciation was unacceptable, the land would not be sold for development and would be valued on the basis of its existing use. Those who are expert in compulsory purchase will appreciate that injurious affection caused by the wider scheme would nevertheless be recoverable.

Great care must be taken to avoid double counting — for instance, a claim for lost development value or profit must reflect the development costs without which that value could not be realised. This can require particular clarity of thought when claiming for abortive expenditure where planning permission has been modified. It is also another example of what is meant by ensuring a valuation is internally consistent.

The costs of compliance with other legislation must be included in the development costs — one is attempting to shadow what would have happened had there been no interference by the planning authority: *Church Cottage Investments Ltd* v *Hillingdon London Borough* [1990] CA.

Each case must be considered and valued according to its own facts. It is both dangerous and wrong to rely on earlier Tribunal decisions as precedent other than as giving an indication of possible methods and of how the law should be applied. It is invariably a mistake to take percentage deductions for risk, or other valuation data, from one case to another. Each case is unique so far as its facts and valuation evidence are concerned.

## Depreciation in the value of the interest

The before-and-after approach is recommended. It is far more reliable to value the entire holding rather than just that part affected by the adverse decision, as this helps one to avoid mistakes such as double counting or the inadvertent omission of consequential losses such as severance.

The before value is that obtainable on the open market assuming that there had been no adverse planning decision. In a revocation case, for example, one assumes consent for the planning permission in hand prior to the making of the order.

The after value is the value subject to the adverse decision, with the benefit of any planning consents in hand, but the statutory provisions must be referred to in each case to discover which planning assumptions are to be made. In revocation cases, for example, one assumes consent for Schedule 3 development (as defined below).

The before-and-after approach has been criticised in connection with the calculation of injurious affection and severance, but was approved by the House of Lords in *Duke of Buccleuch* v *Metropolitan Board of Works* (1872). Where two possible approaches vie for attention, it is sensible to use both and to worry if the results are significantly different.

## Other direct losses

Where recoverable, these are what they say and may include losses under rule 6[3], including disturbance. Not all categories of claim specifically mention rule 6. However, if the Act uses the words "direct loss or damage" it means exactly that: everything that can be said to flow directly from the interference or restriction which has triggered the claim. The loss must be a consequence of the adverse decision: see *Loromah Estates Ltd* v *Haringey London Borough Council* (1978), LT.

## Depreciation and disturbance in the same claim

In *Horn* v *Sunderland Corporation* [1941] a compulsory purchase case, it was held that one may claim so as to receive the greater of development value on its own or existing use value plus disturbance. One cannot have both development value and disturbance.

---

3     1961 Act, section 5(6).

In places, the planning compensation rules use the word "or" — sometimes in ways which imply "and/or". One must not forget the Golden Rule: the purpose of compensation is to put the claimant — so far as money can do it — in the position of not having been interfered with.

In planning cases, one must distinguish carefully between appreciation and depreciation in the value of an interest. In revocation, the original planning consent appreciated the land value, the revocation then depreciated it. If the claim is based on the recovery of the development value which came and went, one should not also claim for disturbance as other losses — *Horn* would prevent it.

Discontinuance, however, presents other variants and a more difficult picture. The order may require either a total cessation or merely a diminution of the established use. If the order results in a decrease in the land value, in the loss of business profits and in various expenses, clearly one will be entitled to claim for everything, including disturbance. However, an alternative planning consent may be forthcoming, either as part of the order itself or subsequently (following the service of a purchase notice) and its value must be brought into the reckoning.

Where appropriate, similar considerations may apply to other categories of compensation claim.

## Loss of developer's profit

This can be an alternative to a claim for loss of development value, typically in a revocation case — one may opt for either, but not both. It is not to be confused with the loss of business profits or damage to goodwill.

Where a claim for loss of the developer's profit is made, one must recognise that such profits would have been speculative upon the successful completion and sale of the project which has been frustrated, and a deduction must be made for the risk and for the length of time it would take to secure the profit. The amount of any deduction for risk will vary according to the facts of the case; the better prepared the scheme, and the more advanced the work, the lower the risk. Another factor will be the amount retained (if any) against contingencies over and above that identified as being available for risk and profit. Thus, a carefully specified scheme for which tenders have been sought will contain less inherent risk than one which is only at an early stage of detailed preparatory work, while one for which a tender has been accepted will require even less.

Cases in which loss of profit was discussed or allowed include:

- *Excel (Markets) Ltd* v *Gravesend Borough Council* (1968) LT, loss of developer's profit following revocation of consent
- *Hobbs (Quarries) Ltd* v *Somerset County Council* (1975) LT, damage caused by loss of profit from working the minerals rather than to the value of the minerals
- *Burlin* v *Manchester City Council* (1976) LT — despite the appearance of a claim for lost profit, the claim was really for lost interest on the compensation, and that part was disallowed
- *Cawoods Aggregates (South Eastern) Ltd* v *Southwark London Borough Council* (1982) LT, damage to business following revocation — both depreciation in the existing use value of the land and loss of trading profit were allowed
- *Richmond Gateways Ltd* v *Richmond upon Thames London Borough Council* [1989], no deduction for developer's profit need be made, since the claimant was a developer (but see *Newham London Borough Council* and other cases, *supra*, in which these deductions were made, and note (5) to Study 3

- *Pennine Raceway Ltd* v *Kirklees Metropolitan Borough Council* (1981) LT; (1983) CA: (i) licensee may claim CA (1982), (ii) loss of projected trading profit for each of five years allowed, plus other matters including losses on an abortive attempt to mitigate loss by opening elsewhere, (iii) compensation not to be reduced for tax[4].

## Interest on the claim

In section 80 of Schedule 18 to the 1991 Act the events from which interest will run are set out. These are not restricted to planning compensation cases. The Tribunal has power to award interest. Rule 32 of the Lands Tribunal Rules 1996 (SI 1996 No 1022, as amended) governs the position and applies section 49 of the Arbitration Act 1996 to all proceedings.

## Losses arising on other land

Provided the loss is a direct result of the adverse planning decision, a claim may be considered. A reduction in business profits is one example: see *Cawoods Aggregates (South Eastern) Ltd* v *Southwark London Borough Council* (1982) LT. Depreciation in the value of other land would be much harder to establish, although a revocation might have an adverse effect on the value of contiguous land. The key issues would be the physical relationship of the sites, the interest of the claimant in each of them, and the precise wording of the statute. The rules against remoteness would have to be met.

## Professional fees

Valuer's and solicitor's fees are usually allowable. Note the singular!

## Tax on compensation

Following *Pennine Raceway Ltd* v *Kirklees Metropolitan Borough Council* [1989] CA, it is now clearly established that compensating authorities must not reduce the compensation for tax just because they do not think the claimant will actually be taxed: tax is a matter between the Revenue and the claimant. Compensation may be so reduced only where it is clear "beyond peradventure" that the Revenue will not be levying a tax on the compensation and that the claimant could not have made the money ordinarily without paying tax, otherwise compensation is to be paid gross of tax and the tax accounted for to the Inland Revenue by the claimant. The question of tax is ignored in all the examples which follow.

Some compensating authorities deduct tax from interest payments unless the claimant produces a tax exemption certificate.

If the interference with the claimant's rights causes the loss of an exemption from tax, as happened in *Colley* v *Canterbury City Council* (1990) LT, it would be necessary to gross up the compensation to leave the claimant in the same position after paying the tax on the compensation as would have been enjoyed had the compensation been received without such liability for tax. In *Colley*, the revocation of

---

4    LT 1988 (1984 reissued consequential upon CA ruling re tax); CA [1989] 1 EGLR 30.

consent for rebuilding a house which had already been demolished resulted in a capital gains tax charge incurred after the sale of what would otherwise have been the claimant's main residence.

In *Loromah Estates Ltd* v *Haringey London Borough Council* (1978) the claimant, among other things, sought extra compensation to reflect a liability to pay development land tax (DLT). This liability arose for the first time in the period between the confirmation of the revocation order (May 1973) and the payment of the compensation (1978) — the appointed day for the Development Land Tax Act 1976 being 1 August 1976. If the compensation had been paid before August 1976 DLT would not have been payable. This part of the claim was dismissed on the ground that the liability for DLT was a consequence of the passing of the DLT Act and not a consequence of the revocation order — in other words, the loss was too remote. It is more important to note that the local planning authority paid compensation in accordance with the valuation and not that sum minus the new tax liability.

# Schedule 3

At its simplest, the schedule preserves an owner's right to carry on with his legitimate use of the land, rebuilding as and when necessary. The schedule does not give planning permission for anything — it provides a notional consent that underpins, and assures, the continuation of what is going on. For example, if a factory that is a non-conforming but lawful user is destroyed, it is fair to say planning permission for its reconstruction would not be given; the schedule, however, allows the value of a notional right to rebuild to be included in any claim for compensation, and, if a purchase notice is served, that Schedule 3 deemed consent provides the base value underpinning the claim.

## *Abandonment or loss of Schedule 3 rights*

Whether called Third Schedule, or Schedule 3 makes no difference. The rights contained in the schedule cannot be abandoned, and they survive even if the use itself is abandoned, the land left idle and the building demolished. In *Dutton and Black* v *Blaby DC* [2005] CA, the use had been abandoned for 43 years, but the claimant was held to be right in arguing that planning consent to rebuild could still be assumed under Schedule 3 for the two cottages that had been standing on 7 January 1937 and 1 July 1948. The council also failed in its argument that the value of this right would have to be ignored by virtue of rule 4 (illegal user)[5], it being held that in a hypothetical world where planning consent for rebuilding was to be assumed, the use of the new property would not be contrary to law. See also *Old England Properties Ltd* v *Telford and Wrekin Council* (1999) in which the same conclusion was reached.

Schedule 3 rights will only expire if there is an intervening change of use or a change to the character of the land — see *Ivens and Sons (Timber Merchants) Ltd* v *Daventry District Council* (1976)

## *Schedule 3 development and the concept of existing use value*

Any development will either fit within the descriptions set out in Schedule 3 or it will not. The distinction between these two groups has been central to the concept of existing use value ever since the Act of 1947, and still is. The old Third Schedule development formed part of the existing use value.

---

5    1961 Act, section 5(4).

Its descendant, Schedule 3, is a valuation schedule. It does not give planning consent but lists development for which consent may be assumed when valuing for compensation.

Development lying beyond the boundaries of Schedule 3 used to be called "new development", but the 1991 Act wiped the term new development from the legislation while preserving the distinction between Schedule 3 development and the rest.

The scope of Schedule 3 has also been amended. Part I contains two paragraphs. The first deals with rebuilding (plus an increase in the gross volume of 10%). This is subject to the additional restrictions imposed by Schedule 10, of which more later. The second paragraph deals with the conversion of any building used as a single dwellinghouse into two or more separate dwellinghouses and is not subject to Schedule 10.

Part II was repealed by the 1991 Act.

The term "existing use value" has been replaced by "Schedule 3 value".

## Existing use value replaced by Schedule 3 value

The concept of an existing use value which includes certain notional development rights remains. It is the measure of those rights which is altered, and it is this change which is reflected by the change of name.

Schedule 3 value is of significance in compulsory purchase, in purchase notice cases and as the basis for calculating the depreciated value of an interest in some planning compensation cases. It is the value of the interest assuming consent for any development described in Part I of the schedule.

*Para 1*: Rebuilding, subject to a maximum increase of 10% in the gross volume and to the restrictions of Schedule 10. The latter imposes a 10% limit on any increase in the gross floor space, and no floor space increase is permitted for replacement buildings.

*Para 2*: The use as two or more dwellinghouses of a building used as a single dwellinghouse.

So far as rebuilding is concerned, the replacement must be a replacement and not a wholly new creature bearing little resemblance to what went before, although new construction methods and some internal rearrangement — such as the omission of internal walls — could perhaps be claimed as falling within Part I.

The rebuilding work may incorporate an enlargement of not more than 10% in the gross volume, except for dwellings, where the maximum is the greater of 1,750 cu ft or 10% of the volume.

## Effect of Schedule 10 on Schedule 3 value

An original building can be regarded as the first building on the site, regardless of whether it was erected before or after 1 July 1948. If and when it is rebuilt (after that date), it is the original which remains original, for the purposes of Schedules 3 and 10, and any increases in the volume or gross floor space are measured by reference to that building and not to the replacement. It is important, therefore, to keep records of the original, as its ghost will survive into the future: Schedule 3, paragraph 13, and Schedule 10, paragraph 5.

When an original building is first replaced, no matter whether it was erected before or after 1 July 1948, planning permission for an increase of not more than 10% in the gross volume and gross floor space may be assumed for valuation purposes: Schedule 3, paragraphs 1 and 10, with Schedule 10, paragraph 1.

When a post 1 July 1948 replacement building has itself to be replaced or altered, Schedule 10 limits the amount of gross floor space which may be used for any purpose in the new or altered building to the amount which was last used for that purpose in the replacement building immediately before it was replaced. This blocks second helpings of enlargement in terms of floor space: Schedule 10, paragraphs 2 and 5. Schedule 3, paragraphs 1 and 13, block any increase in volume for such a second replacement.

Volume and gross floor space are measured externally. Increases in the floor space are measured by reference to the areas used in the original building. Where floor space is used for more than one purpose, it is apportioned rateably: Schedule 3, paragraph 10 and Schedule 10, paragraph 4.

## Schedule 3 value and permitted development value

It is important not to confuse permitted development value (PDV) with the classes of permitted development contained within Schedule 2 to the GPDO, and equally important to avoid confusion between these classes and the development described in Schedule 3.

There are times when the measure of compensation is the difference between PDV and Schedule 3 value — for example if the Secretary of State gives a direction as to the availability of some other consent in a purchase notice case. If this consent is less valuable than the Schedule 3 value, compensation payable for the difference and the purchase notice does not take effect.

Schedule 3 value is the value of an interest with the benefit of permission for the development specified in Schedule 3, Part I, subject to the restriction of Schedule 10 so far as rebuilding is concerned.

PDV is the value of an interest on the basis that planning consent will be given for development specified by the Secretary of State, but for nothing else, and that no other consents will be forthcoming.

Both Schedule 3 value and PDV are defined in section 144.

# Refusal of consent for development lying within Schedule 3 Part I

This does not rank directly for compensation, but section 137 allows the owner to serve a purchase notice.

### Study 1

*Purchase notice following refusal of Schedule 3 development*
In this case the land is rendered incapable of reasonably beneficial use.

A small freehold factory, built in 1930 and occupied by the freeholder, has been burned down. The local planning authority has refused consent for rebuilding because the site lies on the line of a proposed by-pass. The owner's purchase notice, served on the district council under section 137, has been accepted by the highway authority under section 139. The purchase price is to be assessed.

The site measures 18 m frontage by 45 m deep; the building had been of 600 m$^2$ gross floor space and 2,200 m$^3$ gross volume. Industrial land is worth £1,240,000 per ha.

Site value assuming planning permission for a factory of 2,420 m$^3$: (1)
Schedule 3 value 810 m$^2$ at £124 per m$^2$ industrial site value (2) Say                    £100,000
Purchase price equals value of site: £100,000

Notes

(1) When considering whether land has become incapable of reasonably beneficial use, one must disregard any unauthorised prospective use. This is defined in section 138(2) as anything involving the carrying out of development except that specified in Schedule 3, paragraph 1 (subject also to Schedule 10) and paragraph 2, but see the section about purchase notices at the end of this chapter.

(2) The land value should be found from the analysis of sales of comparable sites, reduced to some appropriate unit of comparison, and taking into account differences between sites. Only if such evidence is scant should a residual approach be used, working from the value of the completed development. It must be remembered that any valuations of comparable sites will have taken account of the plot ratios applicable to those sites, whereas all that can be claimed for the subject land is the right to rebuild the old factory plus 10%. The effect of this restriction should be reflected in the valuation.

(3) The purchase notice having been accepted, the valuation is as for compulsory purchase: see Chapter 9.

(4) The loss of the building itself ought to be an insurance matter.

(5) Paragraph 1 of Schedule 3 deals with an owner's right to rebuild, which is not the same as putting up a different type of structure by way of improvement. A distinction must be made between original and non-original buildings. An original building is the first building put up on the site. A replacement building is a replacement structure put up after 1 July 1948.

   If rebuilding an original building (whether it is a pre-July 1948 structure or one first erected after that date), consent may be assumed for the replacement to include an enlargement of not more than 10% gross volume and 10% gross floor space.

   Should it ever come about that the replacement building has itself to be replaced, it will be the measurements of the old original building that remain relevant: Schedule 3, paragraph 13 and Schedule 10, paragraph 5. Only one 10% increase may be assumed, no matter how often the building is rebuilt.

(6) The compulsory purchase rules apply to valuations for purchase notices — one is not restricted to Schedule 3 as the only source of planning assumptions. However, the reality of the situation where a purchase notice has had to be used may well be that Schedule 3 is the only (or best) planning assumption available — if something better is truly available, how come the land can be found to be devoid of beneficial use? The answer is that when considering whether the land has been rendered incapable of beneficial use, the planning assumptions are restricted to Schedule 3 development subject to the limitations of Schedule 10: see section 138.

## Study 2

*Purchase Notice and PDV*

This illustrates the calculation to be made when some worthwhile planning consent is available. In these circumstances the owner retains the site and is compensated for its loss in value.

   The same factory as in Study 1, but in this case sited in a residential area. Planning permission for its reconstruction has been refused on the ground that a factory is a non-conforming and unacceptable user.

   The owner's purchase notice has been sent to the Secretary of State under section 139 and he has directed in section 138 that planning permission for two houses is to be granted if applied for.

   Industrial land is worth £900,000 per ha and residential building plots are worth £45,000 each.

   The compensation to be paid would be assessed as follows:

| | £ |
|---|---|
| *Schedule 3 Value:* (1) | |
| Industrial site, 810 m² at £124 per m² site value. | |
| Assume consent for factory of 2,420 m² | 100,500 |
| *Less:* PDV (2) | |
| 2 residential building plots at £45,000 | 90,000 |
| Compensation (3), under section 144(2) | £10,500 |

Notes

(1) The Schedule 3 value is calculated by reference to an assumed consent for Schedule 3 development, paragraph 1 (rebuilding), subject to the restrictions of Schedule 10. The factory being an original building, one has the right to assume permission for a 10% increase in the volume and gross floor space.

(2) The PDV is calculated on the assumption that no consent would be given other than in accordance with the Secretary of State's direction: see section 144(6).

(3) The claimant need not be the freeholder. Owner is defined in section 336 as the person entitled to receive the rack-rent: section 138.

## Study 3

*Schedule 3: Replacement building*

This shows the position where a pre-1948 structure which has already been replaced once after July 1948 has again to be rebuilt.

In this case the 1930 factory was of 2,200 m$^3$ (600 m$^2$) gross, but it was replaced in 1950 with a similar structure of 2,420 m$^3$ (660 m$^2$) designed to support overhead gantry-cranes. Planning policy has since been reversed and the area is now zoned for residential use. Two years ago the replacement factory was burned down and planning consent for rebuilding was refused. The subsequent history is the same as in Study 2.

The site is of 810 m$^2$, but there is no comparable evidence to determine industrial site value. Residential site value is £60,000 per plot and permission for two houses would be granted.

The compensation would be assessed as follows:

|  | £pa | £ |
|---|---|---|
| Schedule 3 value: (1) | | |
| Gross development value (GDV) | | |
| 660 m$^2$ gross, say 630 m$^2$ at £80 pa per m$^2$ (full rental value) | | |
| Net income | 50,400 | |
| YP perp at 7.5% | 13.33 | 672,000 |
| | | |
| *Less*: Costs of development: | | |
| Construction 660 m$^2$ at £535 per m$^2$, Say | 353,100 | |
| Site clearance, Say | 20,500 | |
| | 373,600 | |
| Professional fees at 12% | 44,832 | |
| | 418,432 | |
| Interest on £209,216, 1 year at 12% | 25,106 | |
| Legal and agents' fees at 3% of GDV | 20,160 | |
| Developer's profit at 15% of GDV (4 & 5) | 100,800 | |
| | 564,444 | 564,444 |
| | | 107,556 |
| PV £1 in 1 year at 12% (6) | | 0.8929 |
| Schedule 3 value, industrial site value, Say | | 96,037 |
| *Less*: PDV | | |
| PDV, 2 buildings plots at £45,000 | | 90,000 |
| Compensation equal to depreciation | | £ 6,037 |

Notes

(1)  Schedule 3 value and PDV are defined in section 144(6). Schedule 3 value takes account of development within Schedule 3, subject to the provisions of paragraph 13. This makes the 1930 building the original building, not the 1950 replacement, and so bars any attempt to increase the volume to 2,662 m$^3$ (with a corresponding increase in the floor space).

(2)  If the factory had been erected for the first time after 1 July 1948, it would still be permitted to assume a 10% increase when calculating Schedule 3 value because section 144(6) makes no distinction between buildings first erected before or after 1 July 1948.

(3)  Valuations which rely on the residual method are not readily accepted by the Lands Tribunal. The lack of hard market evidence of site value in this study is artificial, but see note (5) below.

(4)  An owner who could show that, if given permission to rebuild, the work would be put in hand for owner-occupation, could well avoid the item: seen note (5).

(5)  The question of whether a deduction for developer's profit need be made will be answered by the facts of the case. Some valuers may argue that it ought always to be included as a cost when the average open market site value is being sought. Others may prefer to say (where they can justify it) that the owner would rebuild without reference to a speculative developer and accordingly need not deduct the latter's allowance for risk and profit. The omission of this deduction would, of course, increase the estimated depreciation, which is the measure of the compensation in these cases. In *Richmond Gateways Ltd* v *Richmond upon Thames London Borough Council* [1989] CA, a Part II case, the claimant property company successfully maintained that it would have carried out the development itself, and the compensating authority lost its argument for the inclusion of developer's profit as a cost. Much will depend on the facts of the case. A further consideration will be the measure of the compensation. In some instances, the measure is the loss suffered by the claimant and one must then take the claimant as he or she is found. In other cases, as here, the compensation is the depreciation in the value of the interest, and in such cases an open market approach would be applicable. What would the market have paid for the claimant's interest on each of two sets of planning assumptions? The difference is the compensation. The general market would have regard to the need to allow for risk and profit, but under the six rules of compulsory purchase valuation (which apply by virtue of section 117: see section 144(5)), the claimant is deemed to be in the market as a prospective purchaser and may therefore be able to refute the argument as to retentions for risk and profit, particularly as here where the rebuilding would be an insurance matter in which neither retention would feature.

(6)  This represents an allowance for the interest on the purchase of the land. Strictly, the other costs of purchase (stamp duty and legal fees) have also to be deducted.

# Refusal of development not coming within Schedule 3

There is no right to compensation of what used to be known as new development, but if, following a planning refusal or the conditional grant of permission, land is found to be incapable of reasonably beneficial use, a purchase notice may be served. This topic is discussed at the end of the chapter.

Rather than confirm a purchase notice the Secretary of State may instead indicate the availability of some alternative consent. If this happens, and the alternative development is less valuable than the Schedule 3 value of the land, compensation for the difference becomes payable.

## Study 4

*Alternative consent following a purchase notice*
Land is rendered incapable of reasonably beneficial use, but the Secretary of State promises an alternative consent instead of confirming the purchase notice.

The law is contained in sections 137–144.

No 7 is an old, dilapidated, detached cottage adjoining some shops but protruding beyond the rear line of the pavement. The owner has repeatedly tried to sell it in its present condition but has been unable to do so because the council is threatening to make a demolition order.

No 10, a comparable cottage, was renovated by its owner at a cost of £90,000 and later sold for £250,000, but it was never in so serious a state of disrepair as No 7. Further down the road a vacant building plot with consent for a small house fetched £85,000 at a recent auction.

Following the refusal of planning consent for material changes to the external appearance of No 7, the owner served a purchase notice on the district council under section 137 which the council refused to accept. The Secretary of State has directed that consent be given for a modern house set back to the building line. The amount of compensation is to be assessed.

Compensation would be assessed as follows:

| | £ | £ |
|---|---|---|
| Schedule 3 value: (1) | | |
| Value with the right to rebuild and/or extend the cottage by up to the greater of 1/10th of its cubic content or 1,750 cu ft (2) | | 270,000 |
| *Less*: Costs of renovation and extension (3) | | 125,000 |
| Schedule 3 value | | 145,000 |
| *Less*: PDV: (1) | | |
| Single vacant building plot, value based on comparable | 85,000 | |
| *Less*: Cost of demolition and clearance net of salvage | 12,000 | |
| Permitted development value | 73,000 | 73,000 |
| Compensation equals depreciation plus fees and statutory interest | | £72,000 |

Notes
(1) See section 144(6) for the definitions of Schedule 3 value and PDV.
(2) Value based on the sale price of No 10 with an allowance for the increase in size allowed to No 7.
(3) In practice these should be accurately costed, and detailed. It includes an allowance for risk and profit.
(4) The owner must claim his compensation from the local planning authority: see section 144(2).
(5) Had the Secretary of State confirmed the purchase notice then the ordinary compulsory purchase valuation rules would apply and the owner would receive £145,000 for his property: see Chapter 9. As it is, he receives £72,000 for the depreciation and keeps the property.

# Compensation for the revocation or modification of a planning consent

Sections 97–100 deal with the compensation. Sections 137–148 have relevance if, as a result of the authority's section 97 order, a purchase notice is served.

Planning consent for building or other operations may be revoked at any time before they have been completed, but the order cannot affect so much of the work as has already been done. Consent for a change of use may be revoked at any time before the change is made.

Where consent derives from an order such as the GPDO, but a direction requires that specific consent be obtained, compensation is available under section 107 if that consent is refused or is granted subject to conditions. The compensation is calculated in the same way as for the revocation of a specific planning permission.

Section 117 applies to cases coming within Part IV. This attaches the six valuation rules of section 5 of the 1961 Act, "modified as necessary", and deals with cases where an interest is mortgaged. Part IV encompasses revocation and discontinuance (including mineral consents) and the apportionment and repayment of that compensation.

Schedule 18 to the 1991 Act provides for the payment of interest from the date of the section 97 order for revocation and modification cases and from the date of the refusal of planning permission for cases where permitted development rights have been reduced or taken away following the making of an article 4 direction.

The claimant need only be a "person interested in the land". It is a nice point whether — for compensation purposes under section 107 — there is any difference between a person "having an interest in land", and one who is "interested in the land". The last is the form of words used in section 107: see *Pennine Raceway Ltd* v *Kirklees Metropolitan Borough Council* (1982) CA. In this case the owner of a licence with an enforceable right against the freeholder was able to claim compensation. A prospective purchaser who owned an enforceable option to purchase would be entitled to claim. A contractor employed by a person entitled to claim may have to proceed against the latter for damages for breach of contract rather than against the planning authority concerned, but in this event the damages would be added to the compensation claim itself as being "loss or damage which is directly attributable to the order".

Compensation is to be paid for any expenditure on works which have been rendered abortive, and for any other loss or damage which is directly attributable to the order. Compensation is not available for works undertaken before the planning permission was actually granted, except that architect's and other professional fees are always recoverable. This includes the costs incurred in getting the consent in the first place. Other loss or damage includes depreciation in the value of the interest, and, as an alternative, lost development profit may be substituted if this is greater. Subject to the rules against remoteness of damage, the loss does not have to arise on the land subject to the order: see *Cawoods Aggregates (South Eastern) Ltd* v *Southwark London Borough* (1982).

The most reliable method of calculating depreciation is to value the interest before and after the revocation or modification of consent. If consent has been revoked, the after value assumes consent for Schedule 3 development only: see section 107(4) and *Burlin* v *Manchester City Council* (1976) LT. If the revoked consent had itself been for Schedule 3 development, the after value is nevertheless to be prepared on the assumption that consent is available for Schedule 3 — see *Colley* v *Canterbury City Council* [1993] CA — even though this strict interpretation of section 107(4) eliminates the claim for depreciation. Other losses and fees can, however, still be recovered. The remedy here is to follow up the revocation claim with a purchase notice: see Study 7 and note (2) to Study 5.

Compensation for the revocation, modification and discontinuance of mineral working is complex, but as a generality follows the same rules as for the revocation of planning permission. One difference is that those interested in the minerals have the same rights to compensation, even if they do not own an interest in the land: see also Chapter 11.

## Study 5

*Revocation of planning permission*
A straightforward case where planning permission is revoked before the land is sold.

The owner of a house and orchard employed a surveyor to obtain planning consent for a house on part of the orchard plot and to sell that land with the benefit of the consent.

The surveyor obtained planning permission and received a genuine and fair offer of £100,000 for the plot, but before this could be accepted the consent was revoked and the offer withdrawn. The surveyor charged £900 for obtaining the consent and £2,500 in connection with the attempted sale.

| The freeholder's claim would be as follows: | £ | £ |
|---|---|---|
| 1. Depreciation in value because of the revocation of consent: (1) | | |
| Site value with permission for one house | 100,000 | |
| *Less*: Value subject to section 97 order (2), Say | 12,000 | |
| Depreciation | 88,000 | 88,000 |
| | | |
| 2. *Plus*: Abortive expenditure | | |
| Cost of obtaining consent | 900 | 900 |
| Agent's fee for the attempted sale | | 2,500 |
| Total claim | | £91,400 |
| Plus: Professional fees for preparing the claim | | |

Notes

(1) See section 107 for heads of claim.

(2) The value subject to the order is to be calculated on the assumption that consent would be given for Schedule 3 development, subject to the provisions of Schedule 10. This is the only consent to be assumed in this case. If some other consent remained alive, it is arguable that its value should be brought in as reflecting the true after value. One would do so in a modification case if the surviving consent was the remnant of the original. A truly independent consent is another matter. If it and the revoked consent were not for mutually exclusive developments, then its value would also be present in the before valuation and might be left out of both the before and after calculations, provided there was no financial or physical linkage between the two schemes. If, however, the alternative consent is a true alternative rather than a companion, to include its value in the after valuation would seem contrary to section 107(4) and would reduce the depreciation below that actually attributable to the revocation.

(3) Any compensation received for the depreciation in the value of an interest is registered as a compensation notice and may be repayable on the grant of a subsequent planning permission: see sections 107 to 111.

(4) The costs of all professional fees for plans and obtaining planning consent are recoverable under section 107(2) even though these may precede the grant of the planning consent: see *Burlin* v *Manchester* (above) and earlier cases, in particular *Holmes* v *Bradfield Rural District Council* [1949].

(5) Had the owner of the plot, instead of attempting to sell the land, set about having the house built for his own use, then upon revocation he would be entitled to claim such direct losses as compensation to a builder for breach of contract. This would be in addition to compensation for the depreciation in the value of his interest.

(6) Losses claimed, including depreciation and/or loss of profits, need not themselves have arisen on the site that is the subject of the revocation or modification order, provided they were directly caused by the order. The test is that the loss must not be too remote: there must be causation: see *Cawoods Aggregates (South Eastern) Ltd* v *Southwark London Borough Council* [1982] in which the Member said, "... There seems ... to be nothing in section 164 of the 1971 Act [now section 107 of the T&CP Act 1990] which limits loss to the land which is the subject of a revocation order".

# Study 6

*Modification of planning consent*

This shows the calculation of a claim for a developer who had started work but whose planning permission was modified before the development had been completed. Loss of profit is taken into account.

The developer bought 1 ha of land with outline consent for 28 houses some years ago for £370,000. He later obtained full approval and started work. An estate road was built, a sewer laid and construction started on 10 houses. The local planning authority, responding to intense local pressure, revoked the consent on 0.25 ha where no work had been done. The district council has offered to buy this part of the site for £15,000 to add to the village green, otherwise the land must remain undeveloped as amenity land.

The freeholder's claim would be as follows:

|  | £ | £ |
|---|---|---|
| *Value before revocation order*: | | |
| GDV: 0.25 ha with consent for 7 houses (1), | | |
| 7 houses at £250,000 | | 1,750,000 |
| *Less*: Costs of development: | | |
| Construction, 7 houses at £125,000 | 875,000 | |
| Site works, sewer, road etc say | 70,000 | |
| | 945,000 | |
| Architect's and quantity surveyor's fees | 118,125 | |
| | 1,063,125 | |
| Interest on £532,000, 2 years at 8% | 88,525 | |
| Legal and agent's fees at 3% of GDV | 52,500 | |
| Developer's risk and profit, 5% of GDV (2) | 87,500 | |
| | 1,291,650 | 1,291,650 |
| | | 458,350 |
| PV £1 in 2 years at 8% | | .8573 |
| Value before order, Say | | 392,943 |
| *Less*: Value subject to the order | | 5,000 |
| Direct loss | | £377,943 |

*Plus*: Abortive expenditure (see note (3)), professional fees in connection with plans (note (4)) and professional fees in connection with the preparation of the claim, and statutory interest

Notes
(1)   The 0.25 ha could be valued either as here or by valuing the 1 ha in its totality both before and after the order. A problem could arise if there is some abortive expenditure to identify, as care must be taken to avoid double counting: see note (2), Study 7.
(2)   The compensation under section 107 is to be full compensation for loss or damage. Subject to the rules against remoteness and double-counting, this can include profits, provided they are reasonably close at hand and certain. Generally the development value of land reflects the possibility of profit from a purchaser's point of view. The profits in this example are close at hand, and this is allowed for by deducting only 5% for developer's risk instead of the more usual 15%.
(3)   The sewer laid on the 0.75 ha part of the site may now be larger than is strictly necessary. Whether or not this extra cost could be recovered is arguable. It probably could be if none of the developer's profit came into the compensation, but since the profit on the 0.25 ha could only be made by providing the services for it, and as that profit has been included in the compensation, it is not thought reasonable to claim for the excess capacity in the services as well: see section 107(3). There would be a valid claim for abortive expenditure if the sewer had to be relaid.
(4)   If the 0.25 ha on its own was no longer capable of agricultural or other beneficial use, and had the council not wanted to buy it, a purchase notice could be served.

## Compensation for the refusal of permitted development

Section 108(2) imposes a 12-month time-limit for claiming compensation following the revocation or modification of a planning consent originally granted by a development order. Subsection (4) allows regulations to exclude permitted development for demolition from the requirements of section 108.

A planning authority may, with the consent of the Secretary of State, make a direction under article 4 of the GPDO to strike out some or all classes of permitted development. The direction may be made applicable to an entire district or to a single property. Similar powers may be attached to special development orders. The effect is to remove the relevant consent(s), so requiring those whose properties have been affected to seek specific planning permission. If this is refused, or is granted subject to conditions different from those imposed by the development order, the applicant has been disadvantaged by comparison with people whose properties do not come within the scope of the direction. Compensation may, therefore, be claimed under section 107 (modification or revocation of planning consent). Compensation is calculated in exactly the same way as for any revocation or modification case, as shown in Studies 5 to 7. This basis is the same regardless of whether the lost consent had been for development within or beyond Schedule 3: the development order originated a consent which has been revoked or reduced.

Schedule 18 to the 1991 Act provides for the payment of interest from the date of the adverse planning decision which denied or reduced the permitted development.

A more recent case is *Land & Property Ltd v Restormel Borough Council* [2004] LT, in which permitted development for the retail use of a ground floor was removed. The report deals with the basis of valuation including depreciation.

## Study 7

*Compensation where a purchase notice follows a revocation order*
The revocation of planning permission, the land being rendered incapable of reasonably beneficial use.

The owner of a piece of scrubland sought planning permission for one house. This was granted, but, before offering the land for sale, the owner, in the hope of receiving a better price, spent £11,000 on demolishing a derelict shed and laying a drain. The consent was, however, revoked before the drain was finished.

Comparable unimproved building plots are worth £90,000. The owner paid £800 to the land agent who obtained the consent. Deprived of its consent the land is worthless, save perhaps as a car park for the parish hall.

A purchase notice has been served on the local council and accepted. The purchase price is agreed at £10,000. The claim is to be prepared.

| Compensation under section 107(1) (see note 1): | £ | £ |
|---|---|---|
| | | |
| Value with consent for one house (2): Unimproved site value, Say | 90,000 | |
| *Less*: Value subject to the revocation order (3) | 10,000 | |
| Depreciation | 80,000 | |
| *Plus*: Compensation under section 107(2), fees | 800 | |
| Section 107(1) and (3), expenditure on work rendered abortive | 11,000 | |
| To be received from the local planning authority | 91,800 | 91,800 |
| | | |
| Compensation under section 137 (see note 4) | | |
| Purchase price (4) | 10,000 | |
| *Less*: If compensation is paid for expenditure on abortive work, the value of that work (5), Say | 6,000 | |
| To be received from the district council | £4,000 | 4,000 |
| Total claim | | £95,800 |

*Plus* the costs of preparing the claim and statutory interest

Notes

(1) See sections 107, 137 and 144.

(2) This is unimproved site value. The use of the improved site value here would lead to double counting if the cost of the works (£11,000) is also claimed. The work would not be abortive if it raised the site value from unimproved to improved value, but it does not.

   In practice, evidence of value based on comparables would have to be carefully scrutinised to distinguish between improved and unimproved sites.

   It would be open to the claimant to opt for whichever calculation gave the most compensation. The choice here is between unimproved site value plus abortive expenditure on one hand and improved site value only on the other. While it might be expected that, when completed, the site works would add more than their cost to the value of the site, it is more likely that in a partially completed state the cost incurred would exceed the value added.

(3) See section 107(4). In addition to any real planning permission in hand, planning consent for Schedule 3 development (and only Schedule 3) is to be assumed. In this study the assumption of consent for Schedule 3 does not help. The existing use value is established at £10,000.

(4) A purchase notice valuation follows the ordinary compulsory purchase rules, modified as directed by the Act. Section 144 is the starting point for the valuation rules. This brings in section 117 if an alternative consent is given: see subsections (2) and (5). See also Chapter 9 for details of compulsory purchase valuations.

(5) Where compensation is paid for expenditure on work under section 107, compensation on acquisition is reduced by the value of the works in respect of which compensation was paid under section 107. Section 107 pays compensation based on the expenditure, whereas section 144(1) deducts the value of the works. There is no reason to suppose that cost and value are equal — see note (2) above.

   Whether fees paid for under section 107(2) are to be deducted from the purchase price is not clear from section 144(1).

   The key question may be whether the "value with consent" used in the calculation of depreciation under section 107 takes account of the value of the work, for if it does it may be possible to argue that the work was not abortive. In many cases, however, it is likely that the cost of preliminary work will not be met by an increase in value. Carelessness here in the preparation of the claim could leave the claimant out of pocket when the total received is compared with the apparent intention of section 107, which is to pay full compensation. Perhaps the intention of section 144(1) is merely to avoid double payment.

# Compensation for discontinuance orders, including those requiring some reduction in the intensity of a use

Section 102 authorises the making of a discontinuance order, sections 115 to 117 deal with the compensation, and section 137 with the possibility of serving a purchase notice. Schedule 18 to the 1991 Act provides for the payment of interest from the date the damage is suffered or the expense is incurred.

If a purchase notice is resorted to, section 144(7) bars the recovery of section 115 compensation for the discontinuance order if the notice results in the purchase of the claimant's interest. It is also barred if the availability of an alternative consent is indicated and compensation is paid under section 144(5) and 117. A purchase notice cannot be used if a valuable consent is attached to the discontinuance order: see under purchase notices below.

Where a planning permission is included with the discontinuance order, a distinction may have to be made between a consent which authorises the existing use but in a restricted way and one which is for some development that is entirely different. If the value of a permission attached to the order is less than the value prior to the discontinuance, there is no difficulty: one has suffered depreciation and the claim would allow both that and disturbance. If, however, the alternative consent produces a higher land value than the established (pre-discontinuance) use, then there is an appreciation in the development value. Whether this could bar disturbance in a case involving a total cessation of the business and its forced removal elsewhere is open to doubt, bearing in mind that discontinuance is one occasion when a landowner may have a planning permission thrust at him without having made an application. One relevant consideration may be whether the enhancement in land value actually covers the costs of disturbance: one goes back to the basic principle that compensation is to put the claimant in the position, so far as money can do it, of not having been interfered with. One thing is clear: the totality of the compensation is a single entity, no matter how that total is arrived at, and this suggests that disturbance is not immune from downward adjustments.

In *Andrew Williamson v Mid Suffolk District Council* [2006] LT, a discontinuance order wiped out a planning permission. The council was estopped from asserting that the conditions which had belonged to that consent then moved across to attach themselves to an earlier, less restrictive, permission that was still in existence.

There are special rules for the assessment of compensation in the case of mineral workings which are the subject of a discontinuance order.

The claimant can be anyone who has suffered damage as a result of the order, provided the loss is by way of depreciation in the value of an interest or by disturbance. There is no need for the claimant to be interested in the land, let alone have an interest in it. Any person who has carried out work in compliance with the order may claim any reasonable expenses incurred.

## Study 8

*Discontinuance Order requiring a reduction in the use*
In which a discontinuance order requires a reduction in the intensity of use which in turn depreciates the value of both the freehold and leasehold interests.

The tenant and the freeholder of a factory near a residential area have been served with a section 102 order requiring a reduction in the factory's working from 24 to 15 hours per day. The tenant is willing to comply, but the loss of night production will necessitate alterations to some of the plant at a cost of £340,200 and profits will fall by 25% from their present average of £1,386,000 pa before tax.

The tenant holds on a lease having five years to run at a net rent of £250,000 pa. The current net full rental value is £420,000 but with reduced operating time this will fall to £350,000 pa.

The compensation claims for freeholder and tenant are:

| *Freeholder's Claim* | £pa | £ | £ |
|---|---|---|---|
| Valuation of freehold interest prior to the order | | | |
| Net income | 250,000 | | |
| YP 5 years at 8% (1) | 3.99 | 997,500 | |
| Reversion to full net rental value | 420,000 | | |
| YP perp deferred 5 years at 9% | 7.22 | 3,032,400 | 4,029,900 |
| *Less*: Value of freehold interest after order (2) | | | |
| Net income | 250,000 | | |
| YP 5 years at 8% (1) | 3.99 | 997,500 | |
| Reversion to full net rental value | 350,000 | | |
| YP perp deferred 5 years at 9% | 7.22 | 2,527,000 | 3,524,500 |
| Freeholder's total claim | | | £505,400 |

*Plus*: Professional fees for preparing the claim and statutory interest

*Tenant's Claim*

| | £ pa | £ |
|---|---|---|
| Value of leasehold interest prior to the order | | |
| Rent received, full rental value | 420,000 | |
| *Less*: Rent paid | 250,000 | |
| Net profit rent | 170,000 | |
| YP 5 years at 11% and 4% (Tax at 40%) | 3.394 | 576,980 |
| Thereafter Landlord and Tenant Act lease at FRV | | |
| *Less*: Value after the order | | |
| Rent received, full rental value | 350,000 | |
| *Less*: Rent paid | 250,000 | |
| Net Profit rent | 100,000 | |
| YP 5 years at 11% and 4% (Tax at 40%) | 3.394 | 339,400 |
| | | |
| Thereafter Landlord and Tenant Act lease at FRV | | |
| Depreciation | | 237,580 |
| *Plus*: | | |
| 1.   Loss of profit (3) | 346,500 | |
|     YP Say | 3 | 1,039,500 |
| 2.   Cost of complying with the order: | | |
|     alteration to plant | | 340,200 |
| Tenant's total claim | | 1,617,280 |

*Plus*: Professional fees for preparing the claim and statutory interest

Notes
(1) A lower rate is used because the income is secure.
(2) If, as a result of the order, the investment was sufficiently changed to justify a change in the yields, this should be done. The term here is still secure, although secured on a reduced profit rent.
(3) Any other losses that the claimant could show would also be recoverable. An example might be interest on the excess capacity of plant and machinery now used less intensively.

# Study 9

*Discontinuance order with alternative consent*
In which a discontinuance order extinguishes a use, requires work to be done and grants planning permission for some alternative development on the site.

A builder's merchant's yard adjoining a residential area. It has been in use for many years and is operated by the freeholder. Part of the work done in the yard includes the mechanical grading and bagging of aggregates.

A discontinuance order has been confirmed, requiring the cessation of the business on the site and the removal of the old machinery and stacks of aggregate and other materials. It also gives planning permission for four houses on the land.

In its present use the site is worth £450,000. If cleared for housing it is worth £320,000. A suitable alternative site for the business could be bought for £475,000. This site is two miles away.

The freeholder's compensation claim is as follows:

| | £ | £ |
|---|---|---|
| Depreciation (1) | | |
| Value prior to the order | 450,000 | |
| *Less*: Value subject to the order (2) | 320,000 | |
| Depreciation | 130,000 | 130,000 |

| | | |
|---|---:|---:|
| *Plus*: Cost of complying with the order (3) | | |
| Demolition of buildings, grading machinery etc | 56,000 | |
| Removal of bulk aggregate and other stock | 9,500 | |
| | 65,500 | 65,500 |
| | | 195,500 |
| *Less*: Salvage value received: (4) | | |
| Hardcore etc from demolition of buildings | | 3,500 |
| | | 192,000 |
| *Plus*: Disturbance: (1) (5) | | |
| (a)   Loss on forced sale of plant and machinery: (5) | | |
|        Value to incoming tenant | 36,000 | |
|        *Less*: Scrap value | 5,400 | |
| | 30,600 | 30,600 |
| (b)   Other Losses, including temporary and/or permanent loss of profit, any reasonable increase in distribution costs (capitalised), duplicated expenditure, *Crawley* costs, etc (6) (7) (8) Say | | 72,000 |
| Total compensation claim | | £294,600 |

*Plus*: Professional charges incurred because of the order, plus statutory interest

Notes
(1) See section 115(2).
(2) Cleared site value with planning permission for four houses.
(3) See section 115(3).
(4) See section 115(4). Note that the salvage is to be deducted from the total claim and not just from the costs of compliance.
(5) See Chapter 9 for details of disturbance claims.
    So far as the plant and machinery is concerned, the loss is the difference between the scrap value achieved and what would have been paid for it by an entrepreneur coming in to operate the business on the site. It has been assumed that it would not be economic to remove the machinery to the new site. A claimant must always minimise his loss.
(6) For these items of claim, only an indication of some likely ones has been given and a spot figure assumed. It is important to avoid double counting. The cost of removing the stock, required by the order, has already been claimed. Any other removal expenses, however, should be claimed here.
    It is also important to exclude items which are not allowable. The £25,000 difference in site values between the two yards would not, for instance, be recoverable. The reason for this is that the claimant will be deemed to have received value for money in the purchase of the new yard — it is up to him if he buys a better one.
(7) See *K&B Metals Ltd* v *Birmingham City Council* (1976) regarding compensation for activities between the making and the confirmation of the order. A section 102 order does not take effect until confirmed.
(8) For a further example of a valuation following a section 102 order, in which one use was extinguished and an alternative consent given: see *Raddy* v *Cornwall County Council* (1967).

# Study 10

*Compensation where a purchase notice follows a discontinuance order*
In which a use is extinguished and the land is rendered incapable of reasonably beneficial use.
    0.25 ha of land on the edge of a heathland nature reserve has been used since before 1964 as a dump for scrap metal (see note 1), a dealer paying the freeholder the full rental value of £8,000 pa for the use of the land.

A discontinuance order requiring the cessation of the use and the removal of the scrap has been confirmed. N~~ alternative consent has been given. Neither the freeholder nor the dealer has other land in the vicinity. As bar~ heathland the site is worth £2,000.

The freeholder's claim is as follows, but see note (2) below:

| | £ pa | £ |
|---|---|---|
| Existing use value | | |
| Open storage: (4) | | |
| Net income | 8,000 | |
| YP in perp at 11% | 9.09 | 72,700 |
| Purchase price if the purchase notice is upheld | | £72,700 |
| *Plus* professional fees plus statutory interest | | |

Notes

(1)   The use is established regardless of whether planning permission was obtained or not: see section 191. If a limited consent had been granted, the effect of the limitation would have to be taken into account.

(2)   The freeholder has the choice of whether to claim under section 115 or to serve a purchase notice under section 137. The latter appears to be the better choice here. This would preclude a claim by him under section 115, bu~ the scrap dealer's rights to claim under section 115 would not be affected. Note the word's "of that order" in section 144(7).

(3)   The scrap dealer may claim under section 115 regardless of whether his interest is sufficient for him to satisfy section 137(2)(b) or not. Section 115 allows a claim to be made by any person who has suffered damage as a consequence of the order; section 137(1)(c) and (2)(b) allow a purchase notice to be served by any person entitled to an interest in the land.

(4)   The valuation rules under section 137 are those of the Land Compensation Act 1961, sections 5, 14 to 16 etc. One of the values thereby allowed is Schedule 3 value, plus disturbance if applicable.

In this study a "nil" section 17 certificate of appropriate alternative development is assumed: see Chapter 9.

(5)   If the freeholder had been running the business on the site, he would have the same choice between section 115 and section 137. If he chose the latter he could not recover the cost of complying with the order, but would sell as if his interest were being compulsorily acquired. The disturbance compensation would depend upon the availability or otherwise of alternative accommodation. The value of the site as a scrap dump would form part of the total compensation.

# Compensation for stop notices

Sections 183 and 184 deal with the power of a local planning authority to serve a stop notice and section~ 186 with a claimant's right to compensation. Sections 185 and 187 deal with service and penalties. Schedule 18 to the 1991 Act provides for the payment of interest from the date the stop notice is served.

The 2004 Act made provision for temporary stop notices. These are discussed immediately after Study 11.

There is no prescribed form for a claim under section 186, but the claimant's letter must, if it is to~ count as a claim, make it clear that a claim is being made, not that one will be made at some future~ date. Neither the amount of compensation nor other details need be given at this stage: see *Texas Homecare Ltd* v *Lewes District Council* [1986].

Compensation becomes available only in the following circumstances.

1.   The withdrawal of the stop notice, for whatever reason.
2.   If the supporting enforcement notice is quashed on the ground that planning permission ought to have been granted.

3.   If the enforcement notice is varied or withdrawn in similar circumstances.

There is, therefore, no compensation if planning consent is subsequently given or if planning conditions are lifted. Compensation is also excluded for the prohibition of an activity which was, or contributed to, a breach of planning control at the time the notice was in force. Neither is it available for loss or damage which could have been avoided had the claimant responded properly to the local planning authority's notices. Where this happens compensation may be reduced to the extent that the damage was avoidable.

Where entitlement to compensation is established, any person having an interest in, or occupying, the land at the time the notice was first served may claim for:

1.   any loss or damage directly attributable to the prohibition contained in the notice or
2.   if the enforcement notice is varied, for loss or damage due to the prohibition of any activity which ceases to be prohibited on the variation of the enforcement notice — but subject to the exclusions described above.

Claims may include sums paid for any breach of contract caused by the need to comply with the stop notice's prohibitions.

See *Robert Barnes & Co Ltd v Malvern Hills District Council* [1985] for a case which raised a number of heads of claim, including interest charges. Interest on the purchase of the affected site was allowed. A claim for interest on the postponed receipt of development profits was disallowed because the claimant failed to show that the profits were any less than they would have been without the stop notice. It was the claimant's failure to establish the extent of the loss, not the concept of claiming interest on postponed profits, which was defective.

See *J Sample (Warkworth) Ltd v Alnwick District Council* [1984] for a review of the question of remoteness of loss. The meaning of "directly attributable" is not qualified by the concept of "reasonable foreseeability".

See *Graysmark v South Hams District Council* [1989]. In this case the Lands Tribunal (upheld by the Court of Appeal) allowed claims for interest on the cost of the development which had been stopped and for the deferment of the profit which would have been made had there been no stoppage. Claims for worry, loss of repute with bankers and the like were held to be too remote.

## Study 11

*Stop notice compensation*
Compensation for a stop notice attached to an enforcement notice which is quashed on appeal.

A householder enlarged his house to the full extent permitted by Schedule 2 Part 1 class A of the GPDO. Some time later, and without planning permission, he contracted with a local builder for the erection of a brick-built hobbies room. This was partly constructed when an enforcement notice, coupled with a stop notice, was served on him. He appealed against the former under section 174(2)(c) on the ground that the room came within Part 1 class E. Two years later he won his appeal but the cost of the work had risen from the original £16,000 to £19,750. In addition he spent £600 protecting the part-built structure.

| Assessment of the compensation: | |
|---|---|
| Direct loss or damage (1) | £ |
| Increase in the cost of construction | 3,750 |
| Expenditure on protective work | 600 |
| Compensation | £4,350 |

Plus: Interest from the service of the notice (2) and professional fees for the preparation of the claim, but not of the appeal (3).

Notes
(1)   Section 186(2) specifies that compensation is for "any loss or damage directly attributable to the prohibition ...". Subsection (4) specifically allows the recovery of payments made in respect of breaches of contract caused by the obligation to comply with the stop notice.
(2)   Schedule 18 to the 1991 Act provides for the payment of interest from the date on which the stop notice is served.
(3)   The costs of the appeal against the enforcement notice are not recoverable, since that notice does not stem from the stop notice but precedes it: see *Barnes* v *Malvern* and *Sample* v *Alnwick*, above.
(4)   It is arguable that interest should be claimed for the loss of any money spent on the work prior to the stop notice, since this has effectively been tied up without either interest or use for the length of time taken by the appeal.

Sections 171E to 171H, introduced by the 2004 Act, deal with the power of the LPA to serve a temporary stop notice. Compensation is recoverable for any direct loss or damage that is attributable to the notice, but only if the conditions set out below are met, and is calculated on the same basis as for a full — but wrong — stop notice. It is payable to any person who had an interest in the land at the time the notice was served, but it does not have to have been served on that person.
    Compensation is only payable if:

1.   the activity was authorised by a planning permission, development order or local development order before the date on which the notice was first displayed or
2.   there was a certificate in respect of the activity issued under section 191 or granted under section 195 or
3.   the local planning authority withdraws the temporary stop notice for reasons other than it has since granted planning consent for the activity.

## Compensation for listed buildings

The owner of a listed building has two hurdles to jump before work may go ahead: to obtain listed building consent, and to get planning permission if the work amounts to development. There is no compensation for the refusal of listed building consent, nor for the refusal of planning permission. Compensation is recoverable for the revocation or modification of listed building consent, but no claim may be made for abortive expenditure on work undertaken in advance of the consent. It is also available if a building preservation notice is served but the building is not subsequently listed. None of this affects the availability of compensation for the revocation or modification of planning consent.
    Schedule 18 to the 1991 Act provides for the payment of interest from the date of the section 23 revocation order under the LB Act or from the date a building preservation notice is served.

The LB Act provisions apply to conservation areas, but much is then deleted by the LB Act section 75 and by regulations. The overall result is that demolition in conservation areas requires conservation area consent, and compensation is available if that consent is subsequently revoked or modified (unless the revocation is unopposed).

Any building or structure within the curtilage of a listed building is treated as part of the listed building provided it is attached to the land and has been so since before 1 July 1948: see LB Act section 1(5). Listed building consent is, therefore, needed for work to these buildings. The distinction between "building" and "listed building" was before the House of Lords in *Shimizu (UK) Ltd* v *Westminster City Council* [1997]. There cases on the meaning and extent of curtilage crop up from time to time: see, for example, *Lowe* v *First Secretary of State* [2003].

The Coal Mining Subsidence Act 1991 makes provision for listed buildings affected by mining subsidence.

If a building is listed, section 7 of the LB Act requires that specific consent be obtained for any work of demolition, alteration or extension which would affect its character as a building of special architectural or historic interest. This is all-embracing and wider than the definition of development contained in section 55. It means that specific consent must be obtained even if the proposed work fits within section 55(2)(a) or if consent might already appear to have been given by Schedule 2 to the GPDO (permitted development).

So far as dwellings are concerned, the development permitted by Schedule 2, Part 1, class A, of the GPDO is excluded if it would consist of or include the erection of a building within the curtilage of a listed building: in these cases the right to erect a building within the curtilage would derive from class E, which limits the volume to no more than 10 m$^3$.

Apart from the above restrictions, there is no general exclusion of permitted development rights for the enlargement, alteration or improvement of a dwellinghouse which is a listed building. Those works still require specific listed building consent under the LB Act, section 8, and permitted development rights cannot be exercised in its absence without giving rise to an offence under section 9.

If the specific refusal of permitted development rights (via the making of an article 4 direction) ran with a refusal of listed building consent, some scope for argument might exist about remoteness and the true cause of the loss.

## Compensation for the revocation or modification of listed building consent

The provisions of LB Act, section 28 are similar to those of section 107 except that there is no equivalent to section 107(4).

There is no compensation if LB consent is revoked or modified by agreement. Such modification is commonplace where it becomes necessary to change the terms of an original consent in the light of fresh discoveries about the building made during the course of the works: what was originally envisaged and approved may turn out to be impossible or too damaging, so that an alternative solution has to be agreed. The best advice is to foster close co-operation between the owner, the builder and the local planning authority.

Schedule 18 to the 1991 Act provides for the payment of interest from the day of the order.

## Study 12

*The modification of listed building consent*

Planning permission was obtained for the division of a listed building into three houses and for the construction of three garages. The conversion work involved changes to the façade of the building. The consent was modified, against the owner's wishes, to allow the internal division of the building into two units with a common front door.

The freeholder's claim for compensation:

| | £ | £ | £ |
|---|---|---|---|
| Value with the original listed building consent: (1) | | | |
| 3 houses at, say, £200,000 | | 600,000 | |
| *Less:* | | | |
| Costs of conversion | 250,000 | | |
| Cost of 3 garages | 24,000 | | |
| Architect's fees | 36,000 | | |
| | 310,000 | 310,000 | |
| Value | | 290,000 | 290,000 |
| *Less:* Value with consent as modified: (2) | | | |
| 2 houses at say, £250,000 | | 500,000 | |
| *Less:* | | | |
| Costs of conversion | 200,000 | | |
| Cost of 3 garages | 24,000 | | |
| Architect's fee | 32,000 | | |
| | 256,000 | 256,000 | |
| Value | | 244,000 | 244,000 |
| Depreciation | | | 46,000 |
| *Plus:* Professional fees rendered abortive, calculated by apportionment of the fees actually incurred (3): Abortive element, Say | | | 3,200 |
| Total claim | | | £49,200 |

*Plus*: Professional fees for preparing the claim and statutory interest

Notes

(1) It is irrelevant whether the consent is a listed building consent or otherwise. In either case a planning consent has been revoked or modified, and the consequential loss is recoverable. Compare section 107(1) with LB Act, section 28(1).

(2) Note there is no mention of Schedule 3 in LB Act, section 28. The after value will be the value of the interest on the open market subject to the revocation or modification of the listed building consent — which may or may not have involved the removal of development lying within Schedule 3.

(3) The modification would involve further work by the architect, but not all his preliminary work would be wasted. Compensation is due for the extra fees involved if this is the true measure of the cost of the modification to the owner.

(4) For simplicity, interest charges, developer's profit etc have been omitted. These issues have been discussed in earlier studies to do with the revocation of planning consent.

# Compensation for the imposition of a building preservation notice

A building preservation notice under the LB Act, section 3 has the effect of temporarily listing a building for six months, after which it will lapse. It also lapses when and if the building is listed or the Secretary of State notifies the local planning authority that he does not intend to list it.

If the building is not subsequently listed, compensation under section 29 becomes payable for all direct loss or damage. This includes any compensation paid by the claimant to a third party for breach of contract caused by having to discontinue or countermand works: see section 29(2) and (3).

If the building is listed, there is no compensation under section 29. Listed building consent would have to be sought in the normal way. The owner of an unlisted building who, acting properly, started to carry out permitted development under the GPDO, but then received a building preservation notice would have to stop work and seek listed building consent — or wait. If the building is subsequently listed, the owner would suffer a loss — perhaps for breach of contract, perhaps for depreciation, or both. Had the loss resulted from a planning refusal following the making of an article 4 direction, compensation would have been recoverable as if consent had been revoked. However, sections 29 and 28 do not appear to provide for compensation under these circumstances.

Schedule 18 to the 1991 Act provides for the payment of interest from the date the building preservation notice is served.

Claims must be made in accordance with the Planning (Listed Buildings and Conservation Areas) Regulations 1990 SI 1990 No 1519, as amended, within six months from the date of the decision — presumably the lapse, not the imposition, of the notice since the latter has a life of six months and no claim is possible until it has lapsed.

## Study 13

*The imposition of a building preservation notice*
In which a building preservation notice is imposed but the building is not subsequently listed.

The owner of an old house obtained approval under the building regulations for extensive internal alterations at an estimated cost of £30,000. He was served with a building preservation notice forbidding any alteration, but the building was subsequently not listed. By the time the building preservation notice ceased to have effect the builder's estimate had risen by 15% and house values had fallen by 3%.

The claim may be contentious if the parties cannot agree what is the direct and reasonable consequence of the order. The most basic claim may look like this:

|  | £ |
| --- | --- |
| Cost of work at time notice ceased to have effect | 34,500 |
| *Less* Cost of work at time of revocation | 30,000 |
| Direct loss | £4,500 |
| *Add* Other direct loss or damage | |
| Losses attributable to delay, Fees etc | |

*Compensation*: The additional cost of the work, £4,500 (15% of £30,000) plus other direct loss or damage, including fees, plus statutory interest

Notes
(1)  If the claimant sought compensation for the fall in house prices, the compensating authority would be right to argue that the fall in values is in no way attributable to the imposition of the building preservation notice. LB Act section 29(2) awards compensation for any loss or damage directly attributable to the effect of the notice.

See *C&J Seymour (Investments) Ltd* v *Lewes District Council* [1992]: held that the claimant buys and sells in the same market and is disadvantaged only if deemed to sell in a lower-priced market than that in which the replacement or reinvestment is purchased — as happened in *West Midland Baptist (Trust) Association (Inc)* v *Birmingham Corporation* [1970].

(2)  A contrary, but weak, view may be that because the compensation is for all direct loss or damage, the fall in the value might be contended for — but the market change is not a directly attributable consequence.

(3)  If the alterations reduced the running costs of the building, then the additional running costs incurred as a result of the enforced postponement would come within section 29(2) provided they were not too remote. Losses included in the claim would have to pass the test of being a direct consequence of, or not too remote from, the service of the preservation notice.

## Purchase notices and listed buildings

See the LB Act, sections 32 to 37. The general rules are similar to other purchase notice cases: see Studies 7 and 10. The test is that the building and land have become incapable of reasonably beneficial use in their existing state and cannot be rendered capable by the execution of works for which listed building consent has been granted or promised and that the use of the land is subsequently inseparable from the building (so that the two should be treated as a single holding). When considering what might be regarded as a reasonably beneficial use, prospective development outside Schedule 3 is to be disregarded, and so is any work requiring listed building consent if that consent has not been promised by either the local planning authority or the Secretary of State.

The LB Act, section 48, restricts the service of a listed building purchase notice if a repairs notice has been served, but does not utterly prevent it.

Sections 49 and 50 set out the valuation rules. Under normal circumstances, listed building consent may be assumed for the alteration of the building or for its demolition in connection with Schedule 3 development, except in so far as compensation may have already been paid under section 27 for any refusal. (The 1991 Act repealed, section 27.)

If the building is being acquired under the minimum compensation rules of section 50, which apply where the building has been deliberately allowed to fall into disrepair, the planning assumptions are that planning permission would not be granted for any development or redevelopment and that the only listed building consent available would be for the restoration and repair of the building.

## Conservation area consents

The listed building provisions are transferred to apply to conservation areas, but many have been deleted by LB Act, section 75, and by regulations.

The result is that demolition in conservation areas requires conservation area consent, and compensation is available if this is subsequently revoked or modified, unless that change is unopposed. The basis of compensation is the same as for the revocation or modification of a listed building consent. This is the only conservation area compensation.

## Conservation area purchase notices

The rules are the same as for listed building purchase notices where the incapacity is caused by the refusal, revocation or modification of consent in prescribed circumstances.

# Compensation for the preservation of ancient monuments

The Ancient Monuments and Archaeological Areas Act 1979 has been considerably amended by the National Heritage Act 1983. The 1979 Act deals with the protection of ancient monuments, guardianship agreements, the acquisition of monuments (including by compulsory purchase), adjoining land and easements, and with compensation. A scheduled monument may not be listed, but a listed building may subsequently be scheduled.

The Secretary of State has issued class consents for six categories of work — see the Ancient Monuments (Class Consents) Order 1994 SI 1994 No 1381. If these are withdrawn, whether generally or with reference to a specific monument, compensation is recoverable in the same way as for an express consent, for details of which see below.

A compensation claim may arise in one of four ways.

1. *Following the refusal of scheduled monument consent or its grant subject to conditions.*
   This is dealt with in the 1979 Act, sections 4, 7, 27 and 47. To give rise to a claim the loss must stem from the refusal of scheduled monument consent for work within one of three categories.

   (a)  Work in connection with development for which express planning permission was given (ie not by way of a development order) prior to the scheduling of the monument. This permission must itself be effective when the scheduled monument consent is applied for. A permission that has time-expired will not do.

   (b)  Work which is not development at all or which is permitted development under a development order: see section 55, 1990 Act. Even so, no compensation is payable if the work would result in even the partial destruction or demolition of the monument, unless that work is incidental to agriculture or forestry: see section 7(2) and (4).

   (c)  Work that is necessary for the continuation of the existing use(s) to which the monument was being put immediately before the application for scheduled monument consent. Such use(s) must not be in breach of any legal restrictions.

   Where an express planning consent is frustrated by the refusal of scheduled monument consent (or its grant subject to conditions), compensation is not recoverable if the planning consent was granted after the monument was scheduled. A planning authority may, of course, refuse planning consent in order to protect a monument.

   So far as compensation is concerned, the link between monument protection and planning control is akin to that between listed building consent and planning permission in general. In *Hoveringham Gravels Ltd* v *Secretary of State for the Environment* (1975) it was held that the value of the claimant's interest had been damaged chiefly by the failure to obtain planning permission to work minerals rather than by the imposition of a preservation order which merely frustrated agricultural permitted development under the GPDO. The compensation was, therefore, severely limited. This position has now been reinforced, in that compensation is not payable if the implementation of a planning permission is frustrated by the refusal of scheduled monument consent if that planning permission was granted after the monument was scheduled.

   The measure of any compensation under section 7 is the amount of expenditure or other loss or damage suffered by a person having an interest in all or part of the monument. It will include the depreciation in the value of the interest — a before-and-after valuation in accordance with section 5 of the 1961 Act, for which see Chapter 9, and professional fees.

Section 8 of the 1979 Act deals with the recovery of compensation if consent is subsequently granted.

2.  *Compensation following the cessation of authorisation of works affecting a scheduled monument following the modification or revocation of scheduled monument consent*
    Sections 9, 27 and 47 of the 1979 Act deal with this. The compensation being for expenditure on works rendered abortive because further work has ceased to be authorised, and for any other loss or damage directly due to this cessation, including the depreciation in the value of an interest in all or part of the monument. The entitlement to compensation is based on sections 107 and 108 of the 1990 Act (revocation orders). Professional fees would be recoverable.

3.  *Compensation for damage caused by the exercise of certain of the powers contained within the 1979 Act*
    A claim may be made if the exercise of the powers of entry and investigation contained within sections 6, 26, 38, 39, 40 and 43 results in damage to land or to any chattels thereon: section 46.

4.  *Compensation in connection with rescue archaeology*
    Provision is made by sections 33-41, 43 and 47 for the designation of land as being an area of archaeological importance. Within such areas, six weeks notice must be given before any operations that are likely to disturb the ground may be undertaken. This is all to do with rescue archaeology and gives the investigating authority six weeks to explore the site. If notice of intention to excavate is served, development will be held up for six months. Any damage caused to the land or chattels by the excavations is compensatable. If the investigations result in the site becoming scheduled, compensation under section 7 is likely to be available. An essential ingredient is that planning permission for the operations which caused the six weeks' notice to be served must have preceded scheduling, and in these circumstances that will be the case: see *Hoveringham*, above.
    Sections 10, 11 and 15 provide for the acquisition of ancient monuments by compulsion and by agreement, and section 16 with the acquisition of easements and similar rights over land in the vicinity of an ancient monument.
    Compensation claims must be submitted within six months of the adverse event and are governed by the Ancient Monuments (Claims for Compensation) Regulations 1991, SI 1991 No 2512 in England, and SI 1992 No 2647 in Wales. Interest is recoverable: see 1991 Act, Schedule 18.

# Compensation derived from tree preservation orders

Trees are separately protected under the Forestry Act 1967 and under the 1990 Act. There are complex interactions between the two. Both carry compensation rights. This part of the chapter only deals with the provisions of the latter, but the whole topic is a difficult one containing much small print.

The relevant T&CP 1990 Act provisions are in Part VIII. The powers and duties of the local planning authority to preserve trees and woodlands in the interest of amenity are contained in sections 197 to 202. Planting requirements, enforcement, and penalties are set out in Sections 206 to 210; injunctions in section 214A and rights of entry in sections 214B to 214D. Where compensation becomes payable, interest runs from the date of the refusal or conditional grant of permission: see Schedule 18 to the 1991 Act.

Unless it is dead, dying or dangerous, specific consent is required, with minor exceptions, to lop, top or fell any tree that is the subject of a tree preservation order. A deemed refusal of consent to fell is a refusal on which a claim may be made. The articles of a tree preservation order are freestanding and are not governed by the 1990 Act: see *Duncan & Prudential Assurance Co Ltd* v *Epping Forest District Council* [2004] LT. There appears to be no mechanism whereby a tree preservation order, once made, can be revoked — it may outlive the trees themselves.

Trees in conservation areas are similarly protected, in effect by a blanket tree preservation order, but specific consent need not be obtained from the authority before a prohibited act is undertaken — one must merely give the authority six weeks' notice of intent. The authority may then either consent, impose a specific tree preservation order, or say nothing. In the latter event, protection will lapse for two years. Exceptions are made to this protection by the Town and Country Planning (Tree Preservation Order) (Amendment) and (Trees in Conservation Areas) (Exempted Cases) Regulations 1975 (SI 1975 No 148).

A tree preservation order does not survive the grant of planning permission for development if work to the tree is immediately required for the purpose of carrying out that development. There are other exceptions: see sections 198 and 200 (forestry operations on land subject to a forestry dedication agreement) and the model order.

The refusal, or conditional grant, of a tree preservation order consent gives rise to compensation rights, and interest runs from the date of the adverse decision. Compensation is governed by regulations, of which there are two different sets — the old and the new. The latter give far less compensation.

The regulations are:

- *The Old Regulations*: Town & Country Planning (Tree Preservation Order) Regulations 1969. These apply to all tree preservation orders made before 8 August 1999. The Town & Country Planning (Tree Preservation Order)(Amendment) and (Trees in Conservation Areas) (Exempted Cases) Regulations 1975 also apply: SI 1969 No 17, much amended, and SI 1975 No 148 respectively
- *The New Regulations*: Town & Country Planning (Trees) Regulations 1999 SI 1999 No 1892. These apply to orders made on or after 2 August 1999. They distinguish between forestry cases (compensation for replanting directions) and everything else.

## Refusal of consent to fell, or the grant of conditional consent: section 203
### Compensation under the 1969 regulations (amended)

Compensation is recoverable under section 203 by anyone who suffers loss or damage caused by the adverse decision (which includes conditional approvals), including compensation for any depreciation in the value of an interest. Where land value is affected, a before-and-after valuation is a good starting point, but losses and costs unrelated to land value must be assessed separately. Where a claim is for depreciation, one must be careful to ensure the loss is due to the refusal of the tree preservation order consent and not to the refusal of planning permission — which can be difficult if one of the reasons for the refusal of planning consent is to protect the trees. It is therefore sensible to apply for the order consent before applying for planning permission itself. Compensation is also recoverable in connection with replanting directions — for which see later. Losses, such as repair costs, incurred before the refusal are not recoverable. The revocation or modification of consent is also compensatable.

In *Bell* v *Canterbury City Council* [1988] the Court of Appeal ruled that loss or damage included damage to the value of the claimant's interest in the land and that the loss arose at the moment the consent was refused rather than at the time the order was made. In *Bell* the loss was the difference between the value of the land as woodland and as grazing. The council also argued, in the context of what is now section 203, that if they had granted consent they would also have imposed a replanting requirement. This would not, however, affect the overall compensation, since section 204 also carries compensation rights.

In *Fletcher* v *Chelmsford Borough Council* [1991] LT a claim followed the refusal of consent to fell a mature tree which was considered to be endangering the foundations and stability of a nearby house. The opinions of an aboricultural consultant and the planning authority were in conflict, so the owner commissioned a more detailed report which involved measuring the movement of a wall for a year. Although insurers paid for the subsequent underpinning, the authority denied liability for the cost of the report on the ground that it was too remote, because the report had been deliberately commissioned, and months after the refusal of consent to fell at that. The Lands Tribunal held that the cost of the report was properly incurred as part of the process of counteracting the effects of a known hazard on the land.

In *Deane* v *Bromley London Borough Council* [1992], the claimant was given permission to reduce the height of 26 mature boundary trees after a large limb had fallen from one on to adjoining land. The council imposed a condition which required that the work be carried out by an approved contractor, which was done at a cost of £1,437.50. The claimant sought compensation on the basis that he and his wife maintained their garden and, but for the condition, would have done the work for nothing. The council argued that the claimant benefited from the work, being under an obligation to see that the trees were safe, and denied that there was any loss. The Tribunal awarded £977.50, being the total cost less what the claimant would have spent hiring additional equipment and the wear and tear thereon (£499 plus VAT).

In *Bollans* v *Surrey County Council* (1968), the Lands Tribunal discussed several important issues and concluded that:

1. damage or expenditure consequential on a refusal under a tree preservation order is analogous to loss or damage directly attributable to a revocation of planning permission
2. loss of profit may be too remote if the use for the profitable purpose had still to be started at the time of the revocation or order
3. it would not be double counting to allow a claim which included both the value of the tree as timber and the profit from its subsequent disposal if the claimant operated both the growing and retailing businesses from which those profits came.

## Study 14

*Refusal of consent to fell (old regulations)*
In which consent is refused for the conversion of woodland, the subject of a tree preservation order, to pasture.

The freeholder of a farm sought consent for the felling of six acres of woodland, the subject of a tree preservation order made under section 198, in order to convert the land to pasture. A ministry grant would be forthcoming. The local planning authority have refused consent. The land would be worth £3,500 per acre as pasture or £1,500 per acre as woodland, inclusive of the value of the trees. Compensation is payable under section 203.

| | £ | £ |
|---|---|---|
| Value had consent been given (1) | | |
| 6 acres pasture at £3,500 | 21,000 | |
| PV £1 for 3 years at 4% (2) | 0.8889 | |
| | 18,667 | 18,667 |

| | | |
|---|---:|---:|
| *Less*: Cost of reclamation | 8,700 | |
| PV £1 for 1 year at 4% (3) | 0.962 | |
| | 8,369 | |
| *Less*: Proceeds from sale of trees | 2,600 | |
| | 5,770 | 5,770 |
| Value with consent for reclamation | | 12,897 |
| *Less*: Value subject to refusal of consent | | |
| 6 acres woodland at £1,500 | | 9,000 |
| Depreciation in value | | 3,987 |
| *Plus*: Consequential losses and disturbance | | |
| Value of any grant towards cost of reclamation (4), Say | 1,000 | |
| PV £1 for 1 year at 4% (5) | 0.962 | |
| | 962 | 962 |
| Total claim | | £4,859 |
| *Plus*: Professional fees for preparing the claim and statutory interest. | | |

Notes

(1) See *Bell* v *Canterbury City Council* (1986) for a case in which many of the principles are discussed and on which this study is based.

(2) The deferral reflects the time which must elapse following grubbing out before the pasture is fully established.

(3) The Lands Tribunal considered that the reclamation costs should be deferred for one year in *Bell*'s case but did not defer the sale of timber. The report does not discuss the reasoning, but presumably the timber would be sold in advance of reclamation.

(4) The cost of reclamation used (£8,700) is the full cost. If a grant towards the cost of the work is likely to be forthcoming — for example, from the Ministry — then its amount should be brought into the calculation, either as here or by using the net-of-grant cost of reclamation (£7,700) in place of the £8,700 in the earlier part of the valuation.

(5) The Tribunal did not show deferral of grant for one year in *Bell*'s case, but added it directly as a consequential loss, perhaps because the grant figures were not disputed. It would, however, seem consistent to discount this for the same period as the reclamation costs, since the grant would not be paid until the work had been carried out. Failure to discount would result in different final totals between the two approaches.

## Refusals under the new regulations (1999)

These regulations distinguish between refusals in forestry cases, and everything else. They are much less generous than the old.

## Forestry cases

Only the owner of the land may claim, but this is not restricted to the freeholder. Compensation is limited to the depreciation in the value of the trees which is attributable to the deterioration in the quality of the timber because of the refusal of consent to fell: see the model order article 9(3) and (5), plus the Forestry Act 1967 section 11(3) to (5). Section 11 applies to the assessment of compensation as it does to the refusal of felling licences under section 10.

## All other cases

The claim must exceed £500.

No compensation is available for any loss of development value. This is defined as any increase in value which is due to the prospect of development, and development here includes simple land clearance. So compensation has ceased to be available for the loss of any increase in land value by converting woodland to agricultural use, as well as for the more obvious losses when the refusal of tree preservation order consent impacts on the density of development. *Bell* (see above) would not receive compensation under the new regulations.

No claim may be made for any other depreciation in the value of the land.

These two provisions reverse a number of Lands Tribunal decisions, so cases decided under the old regulations must now be treated with caution.

No claim may be made for any foreseeable loss or damage where the claimant fails to take reasonable steps to mitigate.

The claimant must include a statement of reasons for making the application for consent when the application is submitted to the local planning authority.

Where the refusal of consent, or the conditional grant, results in a loss which was not reasonably foreseeable at the time of the decision (having regard to the statement of reasons and any evidence put forward to support it), no claim may be made.

## *Replanting directions: section 204*

If the local planning authority directs that land cleared in the course of forestry operations be replanted, and the Forestry Commission decides not to make a grant or loan under section 1 of the Forestry Act 1979 because the direction frustrates the use of the land for commercial forestry, compensation is payable for any resultant loss or damage. Interest on the compensation runs from the date the direction is given. If the claim arises under the old regulations, the *quantum* is the full amount of the loss or damage; under the new it is far less generous.

Exceptions to the availability of compensation are where the authority certified that:

- its decision was made in the interest of good forestry or
- the tree(s) have a special or outstanding amenity value.

If the replanting direction is made under the new regulations, the claim must be for £500 or more, and no compensation is available for any diminution in the value of the land.

There is, of course, no right to compensation attached to the automatic obligation, contained in section 206, to replant trees which were removed in contravention of a tree preservation order.

## *Trees and purchase notices under section 198*

Where a tree preservation order is in force and refusal of consent concerning the trees renders the land incapable of reasonably beneficial use, a purchase notice may be served: see section 198(3)(c) and (4)(b).

Tree felling does not come within the definition development in section 55. When considering whether land has beneficial use, one must disregard the prospect of any unauthorised prospective use of the land as defined in section 138(2). Once it has been established that following the refusal of consent the land has become incapable of reasonably beneficial use, the normal compulsory purchase code applies.

Purchase notices are considered later in this chapter.

# Compensation for losses under the Wildlife and Countryside Act 1981

There are various sources from which payments may be available in connection with the countryside which are not compensation payments resulting from interference. Compensation is not available for access by the public to open country under Part I of the Countryside and Rights of Way Act 2000, but it is for access orders made under sections 70–73 and 107 of the 1949 Act, where the measure is the depreciation in the value of the interest. The various provisos and exclusions that apply are beyond the scope of this chapter.

## Areas of Special Scientific Interest (SSSI)

Section 28 of this Act provides for the protection of areas of special scientific interest, notice being served on the owner or occupier. The effect is that any operations specified in the notice must not be carried out without the Nature Conservancy Council having been given prior notice — the latter may then enter into an agreement, which may involve making payments to the owner or occupier. There is no right to compensation under section 28.

## Nationally important SSSIs: nature conservation orders

Section 29 provides additional protection for sites of special scientific interest which are of particular national or international importance, designation being by way of a nature conservation order prohibiting specified operations. Compensation is available under section 30(2) to any person having an interest in land comprised in an agricultural unit for depreciation in the value of that interest. The compensation is calculated by reference to the value of the interest subject to, and free from, the order: see *Cameron v Nature Conservancy Council* [1992] — the first reference under the Wildlife and Countryside Act 1981.

Section 29 permits the council to prevent any specified operations, so that the owner or occupier seeking to execute them must notify the council. Subsections (5)–(7) extend the period for making an agreement from three to 12 months and permit compulsory acquisition. Compensation is available under section 30(3) to anyone with an interest in the land at the time the notice was given who has, as a result of the barring of specified operations under subsection (5), as modified by subsections (6) or (7), either:

1. reasonably incurred expenditure which has been rendered abortive, or spent money on work which has been rendered abortive or
2. has suffered direct loss or damage (excluding damage to the value of an interest in land).

Interest is recoverable from the date of the claim until payment. The relevant regulations are the Wildlife and Countryside (Claims for Compensation under section 30) Regulations 1982 (SI 1982 No 1346).

## Management schemes and management notices

These were introduced from 20 January 2001, by sections 28J–28M of the Countryside and Rights of Way Act 2000. Conservation bodies may devise schemes for the whole or part of an SSSI to conserve or restore flora, fauna or its physical characteristics. Prior consultation with owners and occupiers is required. Discretionary payments may be made, the amount of which is subject to ministerial guidance. If no agreement is possible on reasonable terms, or if proper effect is not being given to a scheme, a management notice may be served stating what must be done. If not complied with, the conservation body may enter, do the work and charge the cost. Compensation is recoverable for losses suffered because a management agreement is cancelled.

# Compensation for the closure of a highway to vehicles, and similar events

The Highways Act 1980 contains a number of provisions which entitled those who are adversely affected to claim compensation.

A local planning authority may apply to the Secretary of State for an order banning vehicles from a highway (other than a principal or trunk road) so as to improve the amenity of the area: section 249. The status of the highway is then reduced to that of bridleway or footpath, as happens when a street is turned into a pedestrian precinct. Subsections (3) and (4) allow for exceptions to the ban, in terms of vehicle type, particular persons or time of day.

Section 250 deals with compensation for anyone having an interest in land which has lawful access to the highway.

The measure of compensation is the depreciation in the value of the interest directly attributable to the order, plus any other direct loss or damage. This would include the costs of making the claim.

Schedule 18 to the 1991 Act provides for the payment of interest from the date of the section 249 order.

A simple before-and-after valuation would determine the depreciation. Direct loss or damage could include increased operating costs, loss of profits (if any), and so on, provided these were directly attributable. If the scheme improved the amenity of the area, any claim for losses would need careful preparation. There appears to be no provision for set-off, or for the repayment of compensation if the scheme is subsequently done away with.

In *Ward* v *Wychavon District Council* [1986] LT, no compensation was awarded, the new access arrangements being found, as a matter of fact, better than those which were stopped up. In *Saleem* v *Bradford City Metropolitan Council* [1984], the claimant argued that falling profits were due to the closure of the highway to vehicles, but the valuation evidence was poorly assembled and inconsistent and the Lands Tribunal found that the failure of the business was due to inexperience and excessive borrowing. See also *Leonidis* v *Thames Water Authority* [1979] for a successful claim for damage to a business following the closure of one end of a road for a year (albeit under the Public Health Act).

The Highways Act 1980, sections 116–120 gives powers for the stopping up and diversion of all types of highway. Compensation is only available if public footpaths or bridleways are stopped up or diverted by the making of public path extinguishment or diversion orders. Claims may be made regarding both the land which enjoyed the old right of way and land affected by the diversion.

Sections 124–126 of the 1980 Act and section 248 of the 1990 Act provide arrangements whereby a private access to a highway may be closed, and for the provision of an alternative access to the premises. Compensation for depreciation and disturbance are recoverable under section 126(2), and this will include compensation for damage to business profits. Interest runs from the date of the claim: 1991 Act, section 80 and Schedule 18.

Any person who suffers damage because of work undertaken for the safety of pedestrians, for example by the provision of kerbs, fencing or pavements under sections 66 and 70 of the 1980 Act is entitled to compensation.

Compensation is available under section 28 of the 1980 Act for those whose interests in land are depreciated by, or who suffer disturbance by reason of, the creation of new footpaths and bridleways. Such creation usually involves compulsory acquisition.

The 1980 Act also gives various powers to highways authorities to alter levels, improve corners and require the removal of projections from buildings. Compensation is available for damage suffered or expense incurred. Similarly, building and improvement lines may be defined, works in front of which require the approval of the highways authority. If injurious affection is caused, compensation may be claimed, but there are time limits for this. If planning consent is refused to safeguard future road improvements and no such line has been defined, compensation is not available; but if the refusal is because a line has been defined, then compensation must be paid: 1980 Act, sections 73 and 74.

There are numerous other provisions to do with highways and rights over land, by no means all of which derive from the Highways Act.

Claims may also be made under section 10 of the Compulsory Purchase Act 1965 — injurious affection where no land is taken: see Chapter 9.

# Compensation for restrictions on advertising

Regulations are made under sections 220 and 221. Section 222 gives certain planning permissions, section 224 deals with enforcement, section 223 with compensation for advertisements in place before 1 August 1948, and section 220(2)(c) and (3)(b) with purchase notices. The regulations in England are the Town and Country Planning (Control of Advertisements)(England) Regulations 2007 (SI 2007 No 783). They come with a new circular, DCLG 03/2007. At the time of writing, corresponding regulations for Wales have not been published.

Control over advertisements is exercised by the regulations. Advertisements which satisfy their requirements do not need planning permission, but the regulations may specify classes of advertisements which do require consent. There is no compensation for the outright refusal of consent.

Compensation is available in only two cases, as set out below.

1. *The revocation or modification of an express consent for an advertisement*
   Express consent may be revoked or modified under regulation 16 and is compensatable under regulation 17. An express consent is one which has been granted by the local planning authority or the Secretary of State, as distinct from one which is granted by the regulations.

   Compensation is available to any person suffering loss or damage. The measure is the amount of expenditure rendered abortive (including expenditure on the preparation of plans, etc), plus any other direct loss or damage, but excluding depreciation in the value of any interest in land.

   The claim must be made within six months of the date on which the order is approved.

2.  *The removal of a prohibited advertisement or the discontinuance of an advertising station*
    Regulation 8 allows the local planning authority to serve a discontinuance notice affecting advertisements which have a deemed consent.

    Under section 223 and regulation 20, the forced removal of an advertisement which was on display on 1 August 1948, or the discontinuance of the use of land as an advertising station if so used since that date, is compensatable.

    Compensation is limited to the expense of removing the advertisement or of discontinuing the use of the land. It is available to any person carrying out works to comply with the regulations. It must be claimed within six months of completing the work. Schedule 18 to the 1991 Act provides for the payment of interests from the date the expenses are incurred.

# Compensation for restrictions on mineral workings

The 1991 Act, the T&CP (Minerals) Regulations 1995 (SI 1995 No 2863) and the Town and Country Planning (Compensation for Restrictions on Mineral Working and Mineral Waste Depositing) Regulations 1997 (SI 1997 No 1111) have made significant changes. Matters are complicated: see also Chapter 11.

# Compensation in connection with hazardous substances

The HS Act 1990 has been amended by the Environmental Protection Act 1990 and by the 1991 Act.

Compensation may arise out of either of two events: the revocation or modification of any hazardous substances consent under section 14(1) and, in defined circumstances, the revocation or modification of consent following an application for its continuation under sections 17 to 19.

The HS Act admits other powers of revocation, but these do not carry any entitlement to compensation: see HS Act, sections 13 and 24.

## Revocation or modification of hazardous substances consent by order under section 14(1)

Section 14 allows a hazardous substances authority to revoke or modify consent. The reason for so doing must lie within either 14(1) — that, having regard to any material consideration, it is expedient — or 14(2), which specifies four cases that may be summarised as change of use or abandonment. There is no compensation if section 14(2) is used.

The use of the expediency provision of section 14(1) opens the way for a claim under section 16. There are three possible components: depreciation, disturbance and the expense of works necessary to comply with the order. The calculation of any claim will be under section 117 and 118 of the T&CP Act as if the compensation was being calculated for the purposes of a discontinuance order under section 115.

Any person who has suffered damage as a consequence of the order may claim for either or both:

1.  the depreciation in the value of an interest
2.  disturbance.

Any person who carries out works of compliance may recover the reasonable expenses so incurred from the authority.

Schedule 18 to the 1991 Act provides for the payment of interest from the date of the section 14(1) order.

## *The revocation or modification of hazardous substances consent following an application for its continuance under section 17*

Sections 17 to 19 refer. The revocation power is contained within section 18, and compensation is payable under section 19. Schedule 18 to the 1991 Act provides for the payment of interest from the date of the revocation or modification.

There are no specific rules for claiming compensation — as to either time-limits or method.

Section 6(2) states that a hazardous substances consent enures for the benefit of the land to which it relates unless the terms of the consent provide otherwise. Such consent is, however, automatically revoked if there is a change in the person in control of part of the land, unless prior application is made by him to the hazardous substances authority for the continuation of that consent after the change. If such an application is made, the authority — in addition to agreeing to the change — have the right to modify or revoke the consent, but if they do other than agree they must then pay to the person who was in control of the whole of the land before the change, compensation for any loss or damage sustained by him which is directly attributable to the modification or revocation of the consent. If the person who is in control of the land omits to apply, in advance, for the hazardous substances consent to continue after the change of control, then no compensation will be payable for the automatic revocation of consent which will follow that change of control.

# Compensation for statutory undertakers

Sections 262 to 283 provide for compensation to be paid to statutory undertakers on a special basis under a variety of circumstances which include the refusal, conditional grant, revocation or modification of planning permission on operational land, the imposition of a requirement to remove or re-site apparatus, the extinguishment of a right of way, and compulsory purchase.

In the case of compulsory purchase, a statutory undertaker may elect to waive the special basis of section 280 and instead to invoke section 281 and have matters dealt with in the ordinary way.

Not all planning refusals are compensatable.

A distinction is made between operational and non-operational land. The list of those deemed to be statutory undertakers varies: it depends upon the circumstances as to whether one is "in" or "out". For example, telecommunications operators are not statutory undertakers but are intermittently included as such.

The compensation and valuation rules are set out in sections 279 to 282.

For claims coming within section 279(1), Schedule 18 to the 1991 Act provides for the payment of interest from the date of the decision under section 266 (planning refusals) or of the order under section 97 (revocation or modification of consent). For claims under section 197(2) (extinction of a right or imposition of a requirement under section 271) interest is payable from the date the right is extinguished or the requirement imposed.

For the refusal, or conditional grant, of planning consent to be compensatable, it must have been for what would have been permitted development but for the making of a direction and is not development which has received specific parliamentary approval within the meaning of section 264(6).

The right to compensation is established by section 279 and the special basis for calculating it is in section 280. The basis is that compensation is recoverable for:

1. the reasonable expenses of any acquisition made necessary
2. the same, of works made necessary
3. the loss of profits, calculated in accordance with subsection (3)
4. the reasonable expenses of removing apparatus as required by an order under section 279(2) or (3) — extinguishment of a right vested in a statutory undertaker or the imposition of a requirement. Subsection (2) deals with statutory undertakers and flows from section 271. Subsection (3) applies to telecommunications operators and flows from section 272. Claims for interest appear to be excluded so far as the telecommunications operators are concerned, there being no mention of section 279(3) in Schedule 18 to the 1991 Act.

# Compensation for planning blight

The law is to be found in sections 149 to 171 and Schedule 13.

Where property is adversely affected by the prospect of an impending acquisition for public works, the owner may serve a blight notice on the appropriate authority requiring it to acquire early. Generally, it must be shown that no sale could be achieved on the open market except at a substantially reduced price. The events which permit the use of a blight notice are specified in Schedule 13 to the 1990 Act.

The recipient authority may serve a counternotice challenging the validity of the blight notice for one or more specified reasons. The terms of the authority's counternotice are definitive and the date of that notice is the material date. Thus, in *Burn v North Yorkshire County Council* (1991), the authority could not avoid a blight notice where the scheme came to be abandoned after the date of the counternotice.

If a counternotice is served the matter will — unless resolved — go to the Lands Tribunal. If the Tribunal upholds the blight notice, the valuation rules are as if the relevant compulsory purchase procedure had arrived at the notice to treat stage. The reader is referred to Chapter 9.

The claimant must have a qualifying interest. These are defined as:

1. resident owner-occupier (freeholder or tenant with three years' unexpired term) who has been in occupation for not less than six months
2. owner-occupier of any property with a net annual value of not more than £29,200 (The limit was raised from £24,600 on 1 April 2005: SI 2005 No 406 for England; SI 2005 No 367 for Wales.)
3. owner-occupier of an agricultural unit with six months' occupation of the whole.

The six months' rule can also be satisfied if property has been unoccupied for not more than 12 months, and six months' occupation occurred prior to the property becoming vacant.

With two exceptions, the property must first be tested on the open market by attempted sale. No sale need be attempted where a compulsory purchase order is in force, or if compulsory acquisition is authorised by a special enactment.

Blight notices do not bar home loss or loss payments. A valid blight notice is a deemed notice to treat and attracts the full compensation provisions of the appropriate acts: see Chapter 9. A disturbance claim under section 5(6) of the 1961 Act is not excluded. The argument sometimes put forward, that the claimant brought the disturbance on himself and should not be compensated, is not accepted. The blight was brought on the claimant by the inclusion of the property in the specified descriptions and became the source of loss to hand when the owner decided to sell — whether selling by choice or of necessity. The statute does not provide for the exclusion of disturbance, and accordingly the vendor under a blight notice is in no different a position than one who sells at the instigation of the authority itself.

Section 157 provides for compensation to be reduced for listed buildings in disrepair and for slum clearance cases. This stops blight notices being used as a way round these provisions.

# Purchase notices and "reasonably beneficial use"

Section 137 enables an owner to serve a purchase notice on the district council if, following either the refusal of planning permission or its grant subject to conditions, his land is "rendered incapable of reasonably beneficial use in its existing state".

Purchase notices may also be served following revocation or discontinuance orders and in several other situations, including in connection with listed buildings[6]. In discontinuance cases, the claimant need only be entitled to an interest as distinct from being the owner.

Once a purchase notice has been served and not contested, or confirmed by the Secretary of State, the procedure is the same as if a notice to treat has been served and compensation follows the compulsory purchase rules: see Chapter 9. Home loss, loss and disturbance claims are not excluded.

The refusal of consent does not have to reduce the value of the land, indeed an outright refusal can never do this, since planning permission need never be sought merely to continue what is already lawfully established. (An exception is if planning permission was originally given for a limited time and that time has expired. Apart from this one case the only way in which a lawful or established use can be upset is by the service of a discontinuance order.) It follows, therefore that since the refusal of consent cannot reduce value, neither can the refusal itself render the land incapable of reasonably beneficial use. The worst that a refusal can do is to prevent an owner putting his land to some more advantageous use.

So far as entitling an owner to serve a purchase notice is concerned, it must be because events other than the planning refusal have so upset the status quo that the current use has ceased to be of reasonable benefit, thus forcing the owner to seek permission to carry out some form of development. If this permission is refused, or is granted subject to conditions, so that the owner is pinned down and unable to shift to a beneficial use, then he may serve a purchase notice. This in effect says to the council: "If you like my property so much the way it is, then buy it from me or find someone else to do so, for it is useless to me". However, a purchase notice is not intended to provide a remedy merely because an owner is refused permission to release development value. It is to deal with the case where an owner is trapped with land that has become incapable of reasonably beneficial use.

It has, by some, been considered a nice point as to whether a purchase notice could be used to help an owner whose land has always been useless. That it may be so used was decided in the case of *Purbeck District Council* v *Secretary of State for the Environment* [1982], where it was held that it is only necessary to look at the situation as it has become following the planning decision, there being no need to look at the history of the land, nor to be concerned with what brought the present situation about: section 137 refers only to the land's existing state, not its former condition. This applies even if it is the owner's or occupier's activities which have rendered the land useless: there is nothing in the legislation which states that the cause has to be involuntary. It should, however, also be noted that in this case the Secretary of State's decision to uphold the purchase notice was quashed because the

---

6       For details of this, see earlier in this chapter.

occupier, the claimant's tenant, had rendered the land useless by persistent defiance of conditions attached to a planning consent, despite the council's best efforts to force compliance. It was held that a claimant should not be allowed to take advantage of his own wrongdoing to foist the land on an unwilling local authority, and that section 137 excludes situations where the land has become incapable of reasonably beneficial use because of (in planning terms) unlawful activities on the land: 'It would be monstrous if a local planning authority were not entitled to say "We should not have to buy because the reason the land is incapable of reasonably beneficial use is because conditions we imposed in giving planning permission were not complied with".' A breach of planning control is unlawful: see *LTSS Print and Supply Services Ltd* v *Hackney London Borough Council* [1975] CA.

This decision follows that in *Adams & Wade Ltd* v *Minister of Housing and Local Government* (1965) 18 P&CR 60 at p67:

> The purpose of the section is to enable an owner whose use of his land has been frustrated by a planning refusal, to require the local authority to take the land off his hands. The reference to beneficial use must therefore be a reference to a use which can benefit the owner (or prospective owner) and the fact that the land in its existing state confers no benefit or value upon the public at large would be no bar the service of a purchase notice.

If part of a parcel of land is rendered incapable of reasonably beneficial use and part is not, the owner can oblige the authority concerned to purchase only the part which is not capable of such use: see the Court of Appeal decision in *Wain* v *Secretary of State for the Environment* [1982].

When deciding whether land is capable of reasonably beneficial use, section 138 states that no account is to be taken of any unauthorised prospective use. Note the word "unauthorised", which here means carrying out development other than that specified in Part 1 of Schedule 3. If the notice is served following the refusal (or conditional grant) of planning permission, paragraph 1 of the schedule (rebuilding) is subject to the limitations of Schedule 10 (restrictions on increases in floor space).

Authorised prospective uses, which should be taken into account, include activities and changes which do not amount to development at all — section 55 refers.

The claimant must also satisfy section 137(3) or (4) by showing that the land cannot be rendered capable of reasonably beneficial use by carrying out development for which planning permission is already in existence, whether by a specific grant or by way of a development order consent. In cases deriving from the refusal of permission or the conditional grant of permission, or from revocation and modification orders, one must also show that the land cannot be rendered capable by carrying out development for which consent has been promised by either the Secretary of State or the local planning authority. Such promises need not be considered in discontinuance cases: compare section 137(3) with section 137(4) — the latter has no equivalent to subsection (3)(c).

Section 138(1) is clearer than its predecessor, section 180(2) of the 1971 Act. The key word in section 138 is "unauthorised". Section 180 had it that prospective new development must be disregarded, which was in conflict with section 180(1) so far as available planning consents were concerned, a conflict which was resolved in *Gavaghan* v *Secretary of State for the Environment* [1989] where it was held that such consents should be considered.

There was also some inherent difficulty when the refusal of consent was for what is now Schedule 3 development itself — did one have to assume its continuing availability when considering whether the land had reasonably beneficial use? Clearly, to do so would make a nonsense of the intention behind the purchase notice provisions. Again, the word "unauthorised" resolves the problem: if Schedule 3 development has been refused it can no longer be an authorised prospective use and, being unauthorised, would have to be disregarded.

The Secretary of State has gone so far as to say that Schedule 3 is relevant as a guide to compensation once a purchase notice has been accepted, but has no relevance to the question of whether or not the land has reasonably beneficial use.

In summary, when deciding whether a purchase notice can be served, the existing use must be looked at. If the land has become of no reasonable benefit and if it cannot be made beneficial without carrying out development which requires planning permission (and this has not been promised), then a purchase notice can be served. It is essential to realise that "incapable of reasonably beneficial use" is not just another way of saying "less valuable" or even "less useful". So far as agricultural land is concerned, it will be incapable of reasonable beneficial use if its size, shape or location is such that farming is not practicable and any other use requires development and, hence, planning consent.

It is wrong to say that a purchase notice can only be served once permission for Schedule 3 development has been refused. One can be served whenever land is left incapable of reasonably beneficial use, including after the refusal of Schedule 3 development. Once a purchase notice has been served and accepted, the valuation rules are those of the Land Compensation Act 1961. The planning assumptions to be made are those of sections 14 to 17 of that Act: see Chapter 9.

There are two distinct stages to be thought about. The first is whether the land has been rendered incapable of reasonable or beneficial use, and here one ignores the possibility of any unauthorised prospective use of the land. Once that stage has been passed, and a valid purchase notice resorted to, one moves on to the second stage, which is to prepare the valuation for compensation. At this stage one takes account of the planning assumptions, which may bring in the value of development lying both within and beyond Schedule 3.

The following is a summary of paragraphs 12 to 19 of Circular 13/83, which relate to reasonably beneficial use.

1.  The question in each case is whether the land in its existing state, and taking account of operations and uses for which planning permission or listed building consent is not needed, is incapable of reasonably beneficial use.

2.  No account shall be taken of:
    (a) the prospect of any unauthorised use
    (b) works for which listed building consent would be needed, unless such consent has been promised by the local planning authority or the Secretary of State
    (c) a use which would be beneficial to someone other than the owner or prospective purchaser of the land. For the latter there must be a reasonably firm indication that such a person really exists.

3.  Relevant factors when considering the land's capacity for use are:
    (a) the physical state of the land
    (b) its size, shape and surroundings
    (c) the general pattern of uses in the area
    (d) a use of relatively low value may be beneficial if such a use is common for similar land in the vicinity
    (e) it may be possible to render a small piece of land capable of reasonably beneficial use by using it in conjunction with some larger parcel, provided, in most cases, that the latter is owned by the owner or prospective purchaser of the small piece.

4.  Valuation evidence:
    (a) profit from the land may be a useful test, but to point to an absence of profit is not conclusive evidence that the land has no beneficial use. The notion of reasonably beneficial use is not specifically identifiable with profit
    (b) when considering whether the land has become incapable of reasonably beneficial use, a relevant test — in appropriate cases — is to consider the difference between the annual value of the land in its existing state and the annual value if Schedule 3 Part 1 is disregarded.

5.  "Land cannot be rendered capable of reasonably beneficial use by carrying out development for which planning permission has been granted or promised": section 137(3) and (4).
    Such permission or promise must have been given before the service of the purchase notice. If neither permission nor promise has been given, but the local planning authority thinks that some type of development not sought in the application ought to be allowed, and that this would render the land capable of reasonably beneficial use, the authority should ask the Secretary of State to direct that such a permission would be granted if asked for. Where such a direction is made, the Secretary of State will not confirm the purchase notice.

6.  The Secretary of State would normally expect to see evidence that the claimant had tried to dispose of the interest, by sale or letting, before being satisfied the land had become incapable of reasonably beneficial use.

7.  The whole of the land which is the subject of the purchase notice must be incapable of reasonably beneficial use — if part is capable of such use, the notice will fail.

The following cases are of particular relevance:

*   *R v MHLG, ex parte Chichester Rural District Council* [1960]
*   *General Estates Co Ltd v MHLG* (1965)
*   *Trocette Property Co Ltd v Greater London Council* (1974).

## Compensation where an alternative consent is given

Under section 141(3) the Secretary of State may, instead of confirming the notice, direct that some specified planning consent be given if sought. If the value of this permitted development is less than the Schedule 3 value, compensation is available for the difference: see section 144(1) and (6), and Studies 2 and 4.

## Studies which involve purchase notices

Studies 1, 2, 3, 4, 7, and 10 involve purchase notices.

# Further Reading

Brand, Clive *Encyclopedia of Compulsory Purchase & Compensation* Sweet and Maxwell *

Davies, Keith *Law of Compulsory Purchase and Compensation* Tolley

Denyer-Green, Barry *Compulsory Purchase and Compensation* Estates Gazette

Grant, Malcolm *Encyclopedia of Planning Law and Practice* Sweet and Maxwell *

Hayward, Richard *The Handbook of Land Compensation* Sweet and Maxwell*

* Looseleaf with update service

# Minerals

## Introduction

There are a number of factors to be considered in the valuation of mineral properties which distinguish them from valuations of other types of property.

One of these factors is that the mineral to be valued is hidden from view and various techniques must be employed to elucidate its extent and quality and this will have a bearing on the value of the mineral.

In early mining and quarrying times, the extent and quality of a mineral was very much established by trial and error until a local knowledge was built up giving indications of its extent, quality and nature of disposition. With technological advancement, techniques for evaluating a mineral reserve have now reached a very advanced stage. For example, the evaluation of oil reserves will employ a detailed geological and geophysical investigation prior to drilling. The use of computers has greatly assisted the interpretation of geophysical information and therefore increased the reliability of these non-intrusive investigation techniques. Even so the majority of oil wells drilled are still unproductive. Even after extensive exploration work the detailed nature of some mineral deposits remains largely unknown until they are worked. Certain mineral deposits are relatively easily understood while others are extremely complex. With underground mines it is more difficult, and hence more expensive, to undertake exploratory work and the points of geological reference, for economic reasons, become more widely spaced than with surface deposits. The geological interpretation therefore becomes less reliable and underground mining is particularly prone to encounter unforeseen geological and mining problems. Modern mining is very capital intensive and the need for thorough prior investigation has never been greater. The valuer must therefore have a good understanding of the geology and its effect on the viability of the mining/quarrying operations. A knowledge of the methods of working the mineral is also essential.

The mineral must be capable of being economically extracted and processed to a state to meet market requirements. The quality and specification requirements are now very important and British, European and International standards have to be considered, as well as customer requirements. The valuer should have a good knowledge of mineral processing techniques and the particular market conditions.

The market determines the price of minerals and any individual mineral working must of essence be capable of economic operation, ie unit working plus investment costs must be less than selling price. The selling price is determined by what the customer is prepared pay at the point of use. Therefore, a mineral low value bulk minerals such as aggregates the delivery costs can often exceed the mineral's ex-pit selling price. With higher value minerals the selling price is usually determined by the internationally traded price.

Mineral deposits vary considerably in their size, quality, geological complexity and ease of working and there can, therefore, be wide variations in operating costs and profitability from one site to another. The valuer should have a good understanding of the economics of mineral working in order to be capable of independently assessing, in the broadest terms, the viability of individual operations.

Another important consideration in the valuation of minerals within the United Kingdom is the Town and Country Planning legislation. The development of new mineral reserves is now controlled by the broad umbrella of the Regional Spatial Strategy and in detail by local development schemes. This planning framework is set out in the Planning and Compulsory Purchase Act 2004 and supporting Town and Country Planning legislation. If a mineral deposit does not have planning permission then no value other than hope value can be attached to it. Again it is essential that the valuer has a sound knowledge of planning law and its effect on mineral working.

Additionally, mineral extraction is now also impacted on by environmental legislation and there is other specific legislation such as the Mines (Working Facilities and Support) Act 1923 as amended by the Mines (Working Facilities and Support) Act 1966 and 1974 which may need to be considered. The list of specific legislation affecting minerals is too great to detail here and valuers will have to refer to text books on law and mineral land management. The valuer should have good knowledge of environmental and other legislation so far as it affects mineral workings.

Surface mining in the past was often very destructive, the surface of the land suffering high levels of dereliction. Modern restoration techniques now offer beneficial after use and in some cases a greater development potential can be achieved. Valuation of a mineral working may now require input from a number of specialists to comment on the problems of each development phase, ie the mineral extraction, the landfilling and the afteruse development potential. The valuation of land containing a mineral reserve may therefore involve a multi-disciplinary approach.

The ownership of minerals is far from simple and while many minerals are still vested in the surface owner it is not uncommon for the surface and mineral ownerships to be severed. Several minerals have been taken into public ownership over the past 60 years or so and the ownership of mines of gold and silver are vested in the Crown. With former copyhold land it is common for the minerals to be severed and vested in the Lord of the Manor. There are still traditional rights to work lead in the Peak District of Derbyshire and coal and iron in the Forest of Dean. The valuer needs to be familiar with the law of mineral ownership while not overlooking that legal guidance will have to be sought where necessary.

While underground mining can take place without destroying the surface of the land, open pit mining or quarrying by necessity remove the surface of the land, thus precluding the use of the land for any other purpose while mining takes place. The mineral working is a temporary use of the land but the character of the land is changed forever. The same minerals cannot be worked twice and as a result the working of minerals is often referred to as a 'wasting asset', rather similar to the reducing term of a leasehold interest.

As in any valuation, it is essential to establish the purpose of the valuation and the basis on which it is to be prepared.

Valuations fall into two broad categories.

- *Valuations statutory for purposes*
    - Compulsory purchase
    - Compensation for planning decisions

- – Estate duty
- – Inheritance tax
- – Rating
- – Capital gains tax
- – Mining code

- *Market valuations*
  - – Valuations for loan security purposes
  - – Valuations for sale
  - – Valuations for purchase
  - – Asset valuations.

The valuer will need to look at the basis of valuation. The basis for statutory purposes is usually defined.

Non-statutory valuations need a wholly different approach and the basis of valuation and the methods to be adopted in arriving at an opinion of value are usually in accordance with the *Valuation Standards* (The Red Book) issued by the Royal Institution of Chartered Surveyors. The Red Book defines a variety of valuation bases: market value and existing use value. UK Practice Statement 1.10 and Guidance Note 4 of the Red Book are concerned with minerals and set out the considerations which have to be taken into account. Market value is defined in the Red Book at PS 3.2 along with commentary as to the interpretation of the definition.

> Market value is defined as: The estimated amount for which a property should exchange on the date of valuation between a willing buyer and a willing seller in an arm's-length transaction after proper marketing wherein the parties had each acted knowledgeably, prudently and without compulsion.

Whatever valuation basis is required the valuer must have regard to the fact that the value of mineral properties is reduced with each tonne of mineral extracted, processed and removed from the land. This reduction however may not be significant for a long period where the property concerned has large mineral reserves that will support the level of operations for many years.

The essence of any valuation is that where possible it should be achieved by way of comparison with similar recent transactions in similar circumstances. This simple principle becomes more complicated when, as in the case of minerals, the assets being valued are of a type which come on to the market relatively infrequently and comparable transactions may be scarce or non-existent. Mineral valuers attempt to overcome the problem by looking for comparable evidence such as rents, royalties and profits, etc. When valuing mineral properties the traditional approach is to capitalise the rent or royalty. An addition may then be applied to take into account any plant which might exist and the benefit of any trading agreements or market share which the quarry might enjoy. Such an approach can give a good indication of the value which might be considered prudent for the purposes of security for a loan or for an asset valuation but such a valuation is likely to be well short of the selling price which an attractive mineral property might achieve if sold on the open market. It is therefore considered necessary in many cases to take into account the income that is likely to be generated from working the minerals.

There are more lease transactions than sales of mineral properties and these are sources of evidence for comparison when preparing mineral valuations. Landlords and tenants will, within a mineral lease, have agreed the appropriate royalty for the right to extract, process and sell the mineral and this will have taken market conditions and other factors into account. There may well be other payments within the lease and these will need consideration in order to arrive at the total amount agreed for the right to extract the mineral. It must however be borne in mind that the royalties passing are only an indication and the key factors in arriving at the value of the property are likely to be:

(i) the annual income that can generated and
(ii) the period over which this income can be enjoyed.

These two key factors can only be arrived at after a thorough analysis of the mineral property and the commercial circumstances of the mineral involved.

Mineral/mining leases usually provide for three kinds of rent.

1 A minimum rent (sometimes referred to as a dead or certain rent) is the rent payable regardless of the quantity of minerals extracted. This often merges with the royalty rent whereby minerals to the value of the minimum rent can be worked before royalty payments become due.
2 A royalty rent, which is payable for each tonne (or some other measure such as volume) of mineral extracted, processed and sold.
3 A surface rent in respect of surface occupied. This can vary in the case of surface mineral extraction according to the extent of land occupied.

## Study 1

This shows an example of the rents payable to a landlord under a mining lease incorporating the usual rent elements of such a lease. Although the valuation of minerals require special consideration they are not divorced from the general economic considerations and, as with most commodities in the 21st century, require consideration not only of national but also European and world markets.

The freeholder of an area of land containing limestone has been approached by the lessee, who wishes to purchase the freehold interest. The lessee is the operator of a quarry that is capable of producing 400,000 tonnes pa and has proven reserves of 5 million tonnes with possible further reserves of 3 million tonnes. The expected output over the next four years will be 100,000; 200,000; 250,000 and 350,000 tonnes respectively, building up to full production of 400,000 tonnes in the fifth year. Approximately 2 ha of surface land are occupied for the purpose of processing limestone. The use of royalty value is considered appropriate in this case because the value to the lessor is being established.

The terms of the lease are:

| | |
|---|---|
| Total area included in lease | 50 ha |
| Rent for surface land occupied | £250 per ha/pa |
| Minimum rent | £50,000 pa |
| Shortworkings[1] clause | |
| Royalties: | first 200,000 tonnes pa at 40p per tonne. |
| | Remainder at 35p per tonne. |

Term: 25 years unexpired with break clause if minerals become exhausted.
Restoration: area to be restored by the lessee to agricultural use.

---

1 Shortworkings are where the royalties are less than the minimum rent due for any year and some leases allow this shortfall (shortworkings) to be recovered from 'overworkings' in subsequent years. Overworkings are the royalties in excess of the minimum rent.

Assuming there are shortworkings to date of £20,000, advise the lessor as to the value of his freehold interest in the 50 ha.

## Table 1

| Life of Quarry | tonnes | tonnes |
|---|---|---|
| Proven reserves | | 5,000,000 |
| Deduct next 4 years workings | | |
| Year 1 | 100,000 | |
| Year 2 | 200,000 | |
| Year 3 | 250,000 | |
| Year 4 | 350,000 | |
| Remaining reserves when full production commences | | 4,100,000 |
| Divided by future annual output | 400,000 | |
| Life at full production: | 10.25 years | |
| Possible additional reserves | 3,000,000 | |
| Divided by annual output | 400,000 | |
| Possible further quarry life: | 7.5 years | |

*Note*: The numbers in parenthesis, thus (1), (2), in the following Tables refer to the numbering of the subsequent explanatory notes.

## Table 2

| Royalty values | Output Tonnes | Royalty £ per tonne | Amount Due £ | Total Royalties £ |
|---|---|---|---|---|
| Year 1 | 100,000 | 0.40 | 40,000 | |
| Year 2 | 200,000 | 0.40 | 80,000 | |
| Year 3 | 200,000 | 0.40 | 80,000 | |
| | 50,000 | 0.35 | 17,500 | 97,500 |
| Year 4 | 200,000 | 0.40 | 80,000 | |
| | 150,000 | 0.35 | 52,500 | 132,500 |
| Years 5–21 | 200,000 | 0.40 | 80,000 | |
| | 200,000 | 0.35 | 70,000 | 150,000 |
| Year 22 (1) | 200,000 | 0.40 | 80,000 | |
| | 100,000 | 0.35 | 35,000 | 115,000 |

## Table 3  Minimal rent, royalties and shortworkings reconciliation

| | Royalty Value £ | Min. Rent £ pa | Shortworkings £ | Overworkings £ | Accumulated shortworking £ | Amount Due £ |
|---|---|---|---|---|---|---|
| Year 0 | | | | | 20,000 | |
| Year 1 | 40,000 | 50,000 | 10,000 | | 30,000 | 50,000 |
| Year 2 (2) | 80,000 | 50,000 | | 30,000 | | 50,000 |
| Year 3 | 97,500 | 50,000 | | 47,500 | | 97,500 |
| Year 4 | 132,500 | 50,000 | | 82,500 | | 132,500 |
| Year 5 | 150,000 | 50,000 | | 100,000 | | 150,000 |
| Year 6 to 20 | 150,000 | 50,000 | | 100,000 | | 150,000 |
| Year 21 | 150,000 | 50,000 | | 100,000 | | 150,000 |
| Year 22 | 115,000 | 50,000 | | 65,000 | | 115,000 |

**Table 4  Lessor's interest**

| | | Minimum rent<br>£ pa | Present value<br>£ |
|---|---|---|---|
| Minimum rent | | 50,000 | |
| Years' Purchase 22 years @ 7% (3) | | 11.061 | 553,050 |
| *Add* Royalties: | | | |
| *Year 3* (royalty in excess of minimum rent) | 30,000 | | |
| YP for 1 year @ 8% (4) | 0.9259 | | |
| Deferred 2 years @ 8% | 0.8573 | 0.7938 | 23,814 |
| *Years 4–22* (royalty in excess of | | | |
| minimum rent on first 200,000 tonnes pa | | 30,000 | |
| YP for 19 years @ 10% | 8.3649 | | |
| Deferred 3 years @ 10% (5) | 0.7513 | 6.2845 | 188,536 |
| *Year 3* (royalty in excess of 200,000 tonnes pa) (6) | | 17,500 | |
| YP for 1 year @ 11% | 0.9009 | | |
| Deferred 2 years @ 11% | 0.8116 | 0.7312 | 12,795 |
| *Year 4* (royalty in excess of 200,000 tonnes pa) (6) | | 52,500 | |
| YP for 1 year @ 12% | 0.8929 | | |
| Deferred 3 years @ 12% | 0.7118 | 0.6356 | 33,367 |
| *Years 5–21* (royalty in excess of 200,000 tonnes pa) | | 70,000 | |
| YP for 17 years @ 12% | 7.119 | | |
| Deferred 4 years at 12% (7) | 0.6355 | 4.5241 | 316,689 |
| *Year 22* | | 35,000 | |
| YP for 1 year @ 12% | 0.8929 | | |
| Deferred 21 years @ 12% | 0.0926 | 0.0827 | 2,894 |
| *Add* rent for surface land | | | |
| 50 ha @ £250 pa | | 12,500 | |
| YP for 23 years @ 10% (8) | | 8.8832 | 111,040 |
| *Add* reversion of land to agricultural use | | | |
| 50 ha @ £200 pa | 10,000 | | |
| YP in perp @ 5% | 20 | | |
| Deferred 23 years @ 5% (9) | 0.3256 | 6.5120 | 65,120 |
| Total | | | 1,307,304 |

Present Capital Value, Say £1,300,000

*Notes*

| | | |
|---|---|---|
| (1) | Reserves when full production commences | 4,100,000 tonnes |
| | Possible reserves | 3,000,000 tonnes |
| | Total | 7,100,000 tonnes |
| | *Less* | |
| | 17 years working @ 400,000 pa | 6,800,000 tonnes |
| | Final year | 300,000 tonnes |

(2)  Amount due, being minimum rent plus the difference between excess workings and short workings to date, ie £50,000 plus £30,000 minus £30,000.

(3)  Minimum rent considered at risk rate of 7% as less than 50% of expected total income and is considered to be more assured than royalty rent.

(4)  Royalties over and above the minimum rent but below expected full production.

(5)   The minimum rent will be paid whether the quarry is producing or not, but royalties are paid on output, which from the lessor's viewpoint is not so secure as the minimum rent; therefore the royalties are given a higher risk rate.

(6)   The difference between the amount due and the minimum rent plus royalties paid at 40p when merged, ie £97,500 less £80,000.

(7)   The royalties at 35p are not as certain as those at 40p and therefore given a higher risk rate.

(8)   Surface rent for land occupied.

(9)   Reclaimed agricultural land.

# Rating

The purpose of this study is to provide advice to a client on the expected rating assessment for a new quarrying operation. Advice will be in the form of an estimate based on the planned quarrying operation and mineral output from the quarry. It is not overlooked that the ratepayer could seek the advice of the Valuation Office Agency. However, there will be occasions in the early stages of project planning when the company may want to restrict the details of the scheme to company personnel.

In order to arrive at the estimated rating assessment it is necessary to find the correct rateable value for the 'hereditament' according to law.

The rateable value is defined by paragraph 2(1–1B) of Schedule 6 to the Local Government Finance Act 1988 (as amended)[2]:

> The rateable value of a non-domestic hereditament none of which consists of domestic property and none of which is exempt from local non-domestic rating shall be taken to be an amount equal to the rent at which it is estimated the hereditament might reasonably be expected to let from year to year on these three assumptions-
>
> (a)   the first assumption is that the tenancy begins on the day by reference to which the determination is to be made;
>
> (b)   the second assumption is that immediately before the tenancy begins the hereditament is in a state of reasonable repair, but excluding from this assumption any repairs which a reasonable landlord would consider uneconomic;
>
> (c)   the third assumption is that the tenant undertakes to pay all usual tenant's rates and taxes and to bear the cost of the repairs and insurance and the other expenses (if any) necessary to maintain the hereditament in a state to command the rent mentioned above.

For the purposes of the above, the state of repair of a hereditament at any time relevant for the purposes of a list shall be assumed to be the state of repair in which, under sub-paragraph (1) above, it is assumed to be immediately before the assumed tenancy begins.[3]

The generally accepted method of valuing a quarry for rating purposes is to adopt a royalty per tonne which a tenant would be prepared to pay for the right to work the mineral. The right to work the mineral often, but not always, includes ancillary rights such as the right to process and make merchantable without the payment of additional rent. The royalty is by convention multiplied by the tonnage of mineral extracted from the quarry during the preceding year. To this sum is added the rent of the surface or other land in the hereditament and also an appropriate percentage of the capital value of the buildings and rateable plant and machinery. Too often this is carried out as a mathematical

---

2   By the Rating (Valuation) Act 1999: see Schedule 6 to the Local Government Finance Act 1988 (non-domestic rating: valuation).

3   Schedule 6 to the Local Government Finance Act 1988 (paragraph 2(8A).

exercise. It must be remembered that the process is merely one method of arriving at the rent which the yearly tenant might be expected to pay. The important stage often overlooked is a 'stand back and look' where all the peculiarities of the hereditament which might affect the bid of the hypothetical tenant are considered.

## Study 2

*Rating*

This study is to advise Sherwood Quarrying Company on the probable rateable value for the proposed Sherwood Farm Quarry

Sherwood Farm lies adjacent to a main road 12 miles from Bigoak, a major commercial centre. The farm which extends to 100 ha is owned by Sherwood Quarrying Company Ltd. The quarry was purchased by means of a capital sum many years ago. A sand and gravel deposit about 4 m thick underlies an average of 1.5 m of overburden. The base of the sand and gravel is above the water table. The underlying solid strata is a sandstone which is a local aquifer providing water for a nearby brewery. The sand and gravel contains approximately 60% sand and 40% gravel. Planning permission has recently been granted, and a contract has been signed, for the construction of a sand and gravel processing plant capable of producing 90 tonnes per hour. Anticipated quarry output is expected to be in the order of 90% of plant capacity. Restoration of the quarry is to low level agriculture.

Much Aggregates Ltd have a quarry which they hold under a lease. A haul road 50 m long has been constructed to give access to a main road. The quarry is 5 miles from Sherwood Farm and 11 miles from Bigoak. The geology of the Much Aggregates Quarry is similar to Sherwood Farm, the sand and gravel being some 4.5 m thick which underlies overburden some 1.2 m thick. The silt content of the sand and gravel deposit is slightly higher. The quarry is subject to a lease which was signed in October, two and a half years prior to the new Rating List coming into force. The term is for 21 years, the certain rent is based on an output of 100,000 tonnes pa. The opening royalty was £1.40 per tonne. The royalty is reviewed annually to the Retail Price Index (RPI) and there is an open market review every three years. There is a termination clause on mineral exhaustion. The quarry output is around 200,000 tonnes pa. Restoration of the quarry is to low level agriculture.

Big John Quarrying Co have a sand and gravel quarry which is situated 22 miles from Bigoak and is reached by narrow country lanes. The sand and gravel is some 5 m thick underlying overburden which is approximately 1 m thick. The geology differs from Sherwood Farm in that the bottom 4 m of sand and gravel are below the water table. The material is processed at a nearby, but not adjacent, plant site. There is a planning restriction which prohibits the erection of a mineral processing plant within the mineral deposit area. The plant site is leased at an annual rent of £20,000 pa. The sand and gravel is held on lease which was signed two years before the Rating List was due to come into force. The term is for 18 years with annual reviews to RPI and open market reviews every three years. The opening royalty was £1.10 per tonne. The output of the quarry is of the order of 200,000 tonnes pa. The landlord is keen to develop water based activities such as fishing after the quarrying is complete.

*Note* All the numbered notes referred to in this study will be found at the very end, following the valuation which forms Table 6.

*Analysis of Information — Assumptions*

The valuation date for the Rating List is two years before the list comes into force.[4] This is called the antecedent valuation date.

Inflation is currently running at less than 3% pa.

---

4    Local Government Finance Act 1988, Schedule 6(3)(b).

*Big John Quarrying Co — Analysis*
Lease details are compatible with the statutory definition of rateable value. The royalty was agreed on the antecedent valuation date and consequently no adjustment is necessary.

| | |
|---|---|
| Royalty payable (1) | £ 1.10 per tonne |
| Adjust for remote plant site — transport (1) | + 0.14 |
| Adjust for distance from market (2) | + 0.08 |
| Adjust for part of deposit being below water table (3) | + 0.06 |
| Equivalent royalty for Sherwood Farm say | £1.38 per tonne |

*Much Aggregates Ltd — Analysis*
The lease is compatible with the statutory definition of rateable value but adjustment to the antecedent valuation date is required.

| | |
|---|---|
| Royalty payable | £1.40 per tonne |
| Adjust to the valuation date (4) | + 0.016 |
| Equivalent royalty for Sherwood Farm (5) | £1.416 per tonne |

*Valuation of Sherwood Farm*
In terms of tone of the list royalty, see note (6), the evidence from Much Aggregates is preferred. The reasons for this are:

1. it is closer to Sherwood Farm than Big John Quarry, and a similar distance to the market
2. there are fewer adjustments to be made to the royalty to make the evidence comparable to the situation at Sherwood Farm
3. although the evidence is six months prior to the valuation date, it is not so remote that adjustment would make it unsafe as comparable evidence.

An appropriate tone of list royalty for Sherwood Farm would therefore be, say £1.42 (rounding of the actual figure of £1.416) and the evidence from Big John Quarry, at an adjusted royalty of £1.38 would give a general level of support for this tone of list royalty.

*Valuation of buildings, plant and machinery*
Buildings, rateable plant and machinery situated on mines and quarries are valued using the contractor's test method of valuation — see note (7). This method is adopted because it is rare to find evidence of rental value for buildings plant and machinery on mineral producing hereditaments. The fundamental theory behind the contractor's test method is that the hypothetical tenant would be prepared to pay as a rent a figure which represents a reasonable percentage of the effective capital value of the buildings, plant and machinery. Effective capital value is determined having regard to the cost of construction (see note 8), deductions being made as necessary for age, functional and technical obsolescence.

Legislation prescribes the decapitalisation rates to be used in contractor's test valuations. For the 1990 and 1995 Rating Lists these were 6% and 5.5% respectively. For the 2000 Revaluation, the rates were 3.67% for defence, educational and healthcare hereditaments, and 5.5% in all other cases. The figures for the 2005 Rating List are 3.33% for defence, educational and healthcare hereditaments (3.3% Wales) and 5% in all other cases — see note (10).

## Table 5

| Ref | Item | Notes N/R, unit, etc. | Size | Unit cost £ | Effective capital value/ allowance | Adjusted capital value £ |
|---|---|---|---|---|---|---|
| | *Buildings*: | | | | | |
| B1 | Weigh Office/Office | m² | 16.10 | 300 | 1.0 | 4,830 |
| B2 | Mess Room (in poor condition) | m² | 15.50 | 275 | 0.9 | 3,836 |
| B3 | Stores | m² | 20.00 | 250 | 1.0 | 5,000 |
| B4 | Workshop | m² | 300.00 | 220 | 1.0 | 66,000 |
| B5 | Control Cabin (in poor condition) | m² | 13.30 | 275 | 0.9 | 3,292 |
| | Supports | m³ | 15.96 | 40 | 1.0 | 638 |
| B6 | Steel Security Store | m² | 14.64 | 100 | 1.0 | 1,464 |
| B7 | Steel Security Store | m² | 14.64 | 100 | 1.0 | 1,464 |
| | *Plant & Machinery*: | | | | | |
| P1 | Primary Feed Hopper supports/grizzly | m³ | 40.00 | 80 | 1.0 | 3,200 |
| | Foundations | | 90.00 | 275 | 1.0 | 24,750 |
| P2 | Conveyor (800mm) and walkway | m | 20.00 | 280 | 1.0 | 5,600 |
| P3 | Conveyor (700mm) and walkway | m | 14.40 | 250 | 1.0 | 3,600 |
| P4 | Barrel wash supports | m³ | 280.00 | 80 | 1.0 | 22,400 |
| | Foundations | m³ | 60.00 | 275 | 1.0 | 16,500 |
| P5 | Conveyor (800mm) and walkway | m | 25.00 | 280 | 1.0 | 7,000 |
| P6 | Surge Bin | N/R | | | 1.0 | Nil |
| | Foundations/feed ramp | | 3.60 | 250 | 1.0 | 900 |
| P7 | Conveyor (700mm) and walkway | m | 25.00 | 250 | 1.0 | 6,250 |
| | Trestle | m | 6.00 | 190 | 1.0 | 1,140 |
| P8 | Bins | N/R | | | 1.0 | Nil |
| P8 | Supports | tonnes | 14.00 | 1,800 | 1.0 | 25,200 |
| | Foundations | m³ | 65.00 | 275 | 1.0 | 17,875 |
| P9 | O/size conveyor | m | 5.00 | 150 | 1.0 | 750 |
| P10 | Crusher Foundations | m³ | 7.50 | 275 | 1.0 | 2,063 |
| | Crusher supports | tonnes | 11.00 | 1,800 | 1.0 | 19,800 |
| P11 | Return conveyor | m | 25.00 | 180 | 1.0 | 4,500 |
| | Supports | tonnes | 3.50 | 1,800 | 1.0 | 6,300 |
| P12 | Radial conveyor | m | 12.00 | 120 | 1.0 | 1,440 |
| P13 | Radial conveyor | m | 12.00 | 120 | 1.0 | 1,440 |
| P14 | Radial conveyor | m | 12.00 | 120 | 1.0 | 1,440 |
| P15 | Radial conveyor | m | 12.00 | 120 | 1.0 | 1,440 |
| P16 | Bund Wall to oil tanks | m² | 14.00 | 55 | 1.0 | 770 |
| P17 | Bund Wall to diesel tanks | m² | 14.50 | 55 | 1.0 | 798 |
| P18 | Aggregate Bays | | 85.00 | 160 | 1.0 | 13,600 |
| P17 | Concrete | | 1,550.00 | 22 | 1.0 | 34,100 |
| P18 | Electrics | | 1.00 | 18,000 | 1.0 | 18,000 |
| | Services | | 1.00 | 20,000 | 1.0 | 20,000 |
| | Sub Total | | | | | 347,379 |
| | Location factor (9) | | | | | × 0.92 |
| | | | | | | 319,589 |
| | Contract size adjustment 5% (9) | | | | | × 1.05 |
| | | | | | | 335,568 |
| | Contract fees @ 8% (9) | | | | | × 1.08 |
| | Total: Buildings, rateable plant and machinery | | | | | £362,414 |

| | |
|---|---|
| Total Buildings rateable plant and machinery | £362,414 |
| Capitalise @ 5.0% | |
| Rateable value | £18,121 |

## Table 6  Valuation of Sherwood Farm Quarry

| | Unadjusted Rateable Value (11) £ | Rateable Value £ |
|---|---:|---:|
| 180,000 tonnes /annum (say) @ £1.42 | 255,600 | 127,800 |
| Buildings, rateable plant and machinery | 18,121 | 18,121 |
| Totals | £273,721 | £145,921 |

It would then be appropriate to advise Sherwood Quarrying Co Ltd that:

Rateable Value × Unified Business Rate = Rates payable

The rates would be payable on a pro rata basis for the likely period of occupation during the rate year.

*Notes*

(1) Many mining leases as well as granting rights to extract the mineral give the right to erect a processing plant. It would have to be assumed that the royalty agreed reflected the fact that the tenant also had to pay an additional rent for the remote plant site and meet the cost of transporting the unprocessed sand and gravel to it .

(2) The analysis aims to arrive at the rent which would be payable at Sherwood Farm which is closer to the main market. It is therefore necessary to make an adjustment to the rent at Big John Quarry to compensate for the difference in distance to the main market.

(3) There is a further disability because part of the deposit is below the water table, with increased working difficulty, and the same principle will apply as in (2) above.

(4) In the absence of any further information it has to be assumed that sand and gravel royalties are moving in line with inflation. The adjustment of royalties by an appropriate index, in the absence of market evidence, has been accepted by the Lands Tribunal to be an appropriate method for adjusting mineral royalties to the valuation date: see *Hodgkinson (VO) v ARC Ltd* (RA 342–343/93). The appropriate adjustment is, therefore, the change in the RPI between October and the following April.

(5) In terms of a mineral royalty, the construction of a 50 m road for access to the public highway and the slight variation in silt content are not considered sufficient disabilities to require further adjustments to the market royalty for Sherwood Farm Quarry.

(6) 'Tone of List' is not a term recognised in rating law, as the law requires the valuation of each hereditament (see *Ladies Hosiery and Underwear Ltd v West Middlesex Assessment Committee* [1932]. However, tone of the list is accepted by rating practitioners as a means of comparing similar properties by adopting similar unit rates to arrive at rateable value.

(7) Buildings, rateable plant and machinery in mines and quarries are valued using the contractor's test because rental evidence for them is rare. The method is detailed in *Ryde on Rating*, is included in RICS Guidance Notes, and has often been considered by the Lands Tribunal: see *Dawkins (VO) v Leamington Spa Corporation and Warwickshire County Council* [1961]. The theory is that the tenant would be willing to pay a reasonable percentage of the effective capital value of the buildings, plant and machinery as rent.

(8) Capital value is found from the cost of construction, with deductions for age, functional and technical obsolescence and suchlike: see *Monsanto plc (now Solutia UK Ltd) v Farria (VO)* [1998], and *Shell Exploration and Production Ltd v Assessor for Grampian Valuation Joint Board* LTS/VA/1998/47. The state of repair should also be considered: see explanatory notes to the Rating (Valuation) Act 1999. The changes to the law including changes to the statutory definition of rateable value were brought about by the government's response to the Lands Tribunal decision in *Benjamin (VO) v Anston Properties Ltd* [1998].

(9)   If appropriate, the valuer should consider professional fees, location factor and scale of contract, either within the unit rate or as additional items, to arrive at the total adjusted capital value for the buildings, rateable plant and machinery.

(10)  The decapitalisation rates to be adopted for the 2005 Rating List are set out in the Non-Domestic Rating (Miscellaneous Provisions) (Amendment) (England) Regulations 2004 (SI 2004 No 1494): see also corresponding legislation for the Welsh Assembly. Practitioners should note that different rates apply for the 1990, 1995 and 2000 Rating Lists.

(11)  There is a special provision regarding mines and quarries which allows for the capital element which is part of the royalty payment. Regulation 5 of the Non-Domestic Rating (Miscellaneous Provisions) Regulations 1989 (SI 1989 No 1060) as amended applies to any hereditament which consists of or includes a mine or a quarry. The regulation provides that for the purposes of assessing the rateable value of a mine or quarry, no account shall be taken of sums payable in respect of the extraction of minerals from such land in so far as such sums are attributable to the capital value of the minerals extracted. The allowance also applies to land occupied for the purposes of winning, working, grading, grinding and crushing of minerals. There is no statutory term for the application of the 50% reduction, but it is generally known as the unadjusted rateable value.

## Study 3

*Compensation for Compulsory Purchase of Mineral Bearing Land*

A mineral operator owns an area of land with the underlying gypsum some 80 ha in extent. Approximately 16.5 ha have been worked out by surface extraction methods and 4 ha are used for processing the gypsum, leaving 59.5 ha to be worked.

The gypsum is 4.5 m thick and it is envisaged that 5 ha of land will be excavated each year.

A motorway is to be constructed over the unworked portion of land and a compulsory purchase order has been made in respect of 5 ha of land.

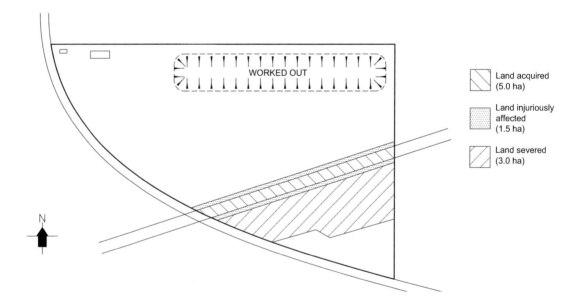

The sketch above shows the details of the scheme. What is the owner's claim for compensation?

**Table 7**

| Mineral Resource Sterilised | | | | | Tonnage |
|---|---|---|---|---|---|
| Seam Thickness (m) | Bulk Density (t/m³) | Tonnes/m² | Tonnes per ha | Total Area | of gypsum sterilised |
| 4.5 | 2.3 | 10.35 | 103,500 | 9.5 | 983,250 |

| Land Acquired | | | | | |
|---|---|---|---|---|---|
| Area (ha) | Rate: £ per ha(1) | Value £ | Less Agri-cultural Value | | Compensation £ |
| 5.0 | 150,000 | 750,000 | n/a | 750,000 | |

| Land Injuriously affected (2) (3) | | | | | |
|---|---|---|---|---|---|
| 1.5 | 150,000 | 225,000 | 15,000 | 210,000 | |

| Severance(4) | | | | | |
|---|---|---|---|---|---|
| 3.0 | 150,000 | 450,000 | 30,000 | 420,000 | 1,380,000 |

| Disturbance (5) | | | | | |
|---|---|---|---|---|---|
| Annual value of plant | YP 2 years @ 16% and 2.5% Tax 40% | Present Value £ | Deferred 10 years @ 16% | Compensation £ | |
| 120,000 | 1.02 | 122,400 | 0.227 | 27,785 | 27,785 |

| Break up value at closure at 10 years(5) | | | | |
|---|---|---|---|---|
| Value £ | | PV £1 in 10 years @ 16% | Discounted value £ | |
| 80,000 | | 0.227 | 18,160 | |

| Break up value at closure at 12 years | | | | |
|---|---|---|---|---|
| Value £ | | PV £1 in 12 years @ 16% | Discounted value £ | |
| 80,000 | | 0.168 | 13,440 | |

| | |
|---|---|
| Difference in break up value | –4,720 |
| Total Claim | £1,403,065 |

*Notes*
(1) Comparable transactions have been analysed and adjustments made for differences between them and the land being considered. A figure of £150,000 per ha is indicated, including the minerals.
(2) The motorway must be supported on either side, which means that a strip of minerals either side cannot be worked. These minerals are not acquired but their value is injuriously affected by the acquisition to the extent that it is totally eliminated. These areas are shown stippled on the drawing.
(3) On the south side of the motorway the land injuriously affected and the land severed (see note 4) will continue to be used for agricultural purposes. The loss suffered, therefore, for those portions is the mineral value less the agricultural value.
(4) The piece of land to the south of the motorway is severed. Neither the planning authority nor the economics of extraction will permit the working of this small area in isolation and there are no prospects of it being worked with other adjacent land. The cost of providing a bridge over/under the motorway is in excess of £650,000.

(5)  The loss of 9.5 ha of land means that the processing plant will reach the end of its useful life two years earlier than it would have done but for the motorway. That is, at the end of the 10th year instead of at the end of the 12th year. The acquisition is, therefore, the direct cause of the loss of the annual value of the plant for a two year period. However, the break-up value of the plant will be realised two years earlier, so the actual loss suffered for disturbance will be reduced by the gain made on the earlier realisation of the break up value of the plant.

(6)  Present annual value of the plant for years 11 and 12 less the difference between the present break up value in 10 years and 12 years time respectively.

# Compulsory Purchase and Compensation under the Mining Code

The Railway Clauses Consolidation Act 1845 sections 77 to 85, known as the Mining Code, introduced a code of practice regulating the right of support for railways from the underlying minerals.

The mining code was amended by Part II of the Mines (Working Facilities and Support) Act 1923, so far as it applies to railways, providing for the mine owner to serve the notice of approach and determines an area of protection, being the width of the protected works plus a distance of 40 yds or half the depth of the seam, whichever is the greater. The area of protection is divided into an inner area and an outer area, the inner area being the width of the protected works plus 40 yards, the remainder being the outer area. The compensation is to be assessed separately for the mine owner and the royalty owner and is specified at a rate per tonne for the total area and payable at 100% for the inner area and 33.333% (one third) for the outer area.

The mining code is incorporated in other enactments such as the Waterworks Clauses Act 1948, Public Health Act 1875, Support of Sewers (Amendment) Act 1883, Public Health Act 1936, Pipelines Act 1962, Water Act 1989 and Water Resources Act 1990. There is statutory provision for the mining code to apply to land and buildings that are included in compulsory purchase orders. The 40 yds has been replaced by 37 m in the more recent legislation.

Briefly, the code requires a mine owner, who can be the owner, lessee or occupier of the minerals, to serve a notice of approach to a railway or other protected utility when any workings reach 40 yds distance from any railway, buildings or works. If the railway company or utility operator requires the minerals to remain unworked to support the railway or protected structure then they must serve a counternotice requiring support, and pay compensation to the mine owner for the minerals unworked. The amount of compensation is the loss suffered by the owner by virtue of leaving the minerals unworked, plus any additional expense incurred. The extent of compensation payable has recently been examined in the High Court case of *National Grid Gas plc v Lafarge Aggregates Ltd* [2006] and part of the summary succinctly sets out the position as follows:

> ... in the light of the House of Lords decision in the BA Collieries[5] and its endorsement of the approach in Bwllfa[6] there is scope for compensation for heads of loss which relate not just to the minerals themselves which cannot be worked, but also to other consequences which may follow from the prevention of such working.

It should be noted, however, that if the mine owner has a right to work and withdraw support then all the minerals are deemed to be within the inner area.

---

5    *BA Collieries v LNER* [1945].
6    *Bwlfa and Merthyr Dare Steam Colleries (1891) v Pontypridd Water Works Co* [1903].

## Study 4

*Mining code*

A gas company wishes to obtain a pillar of support for a length of pipeline and has served a counter notice to a notice of approach served under the Pipelines Act 1962 by the operator of a fluorspar mine. The lessee is a substantial operator making a net profit of £10 per tonne. The order authorising the pipeline construction designated the distances prescribed in the Mines (Working Facilities and Support) Act 1923. The relevant terms of the lease are — royalty £0.75 per tonne of crude ore; minimum rent £4,000 pa; there is an average clause; prior to the acquisition of the land by the gas company there was a right to work the minerals but not to let down the surface.

The mineral vein is restricted to the carboniferous limestone strata which in this area is about 60 m thick and this is overlain by about 100 metres of non-productive shales. The vein varies considerably in width but averages 2.5 m. The vein is for all intents and purposes vertical. Fluorspar is the major constituent of the vein, with a grade of about 50%, the remaining constituents are mainly barytes, calcite, silica and galena. The fixed royalty of £0.75 a tonne is paid irrespective of variations in grade. It has been agreed that the loss of minerals over the next four years will be as follows:

Year 1  13,560 tonnes within the inner area, and 3,130 tonnes within the outer area of the proposed pillar of support.
Year 2  16,000 tonnes within the inner area, and 8,000 tonnes within the outer area
Year 3  8,000 tonnes within the inner area, and 7,040 tonnes within the outer area
Year 4  8,000 tonnes within the inner area, and 7,040 tonnes within the outer area of the proposed pillar of support.

In addition, it has been agreed that there will be 5,480 tonnes of unworkable fluorspar of which 2,740 tonnes would have been worked during the second year and 1,370 tonnes in the third and fourth years.

Leaving the pillar of support unworked will interrupt production from the mine while access roads are driven in the country rock to reach the vein beyond the pillar.

Advise the freeholder and the operator of the fluorspar of their claims for compensation from the gas company for the proposed sterilisation of the fluorspar.

*Lessee's Claim*

| Tonnage | Unit: Income per tonne £ | Income £ | PV £1 in 1 year @ 8% | Value £ | Compensation £ |
|---|---|---|---|---|---|
| **Year 1** | | | | | |
| 13,560 | 10 (1) | 135,600 | 0.926 | 125,565 | |
| 3,130 | 3.33 (2) | 10,423 | 0.926 | 9,652 | 135,217 |
| **Year 2** | | | | | |
| 16,000 | 10 (1) | 160,000 | 0.857 | 137,120 | |
| 8,000 | 3.33 (2) | 26,640 | 0.857 | 22,830 | |
| 2,740 | 3.33 (3) | 9,124 | 0.857 | 7,819 | 167,770 |
| **Year 3** | | | | | |
| 8,000 | 10 (1) | 80,000 | 0.794 | 63,520 | |
| 7,040 | 3.33 (2) | 23,443 | 0.794 | 18,614 | |
| 1,370 | 3.33 (3) | 4,562 | 0.794 | 3,622 | 85,756 |
| **Year 4** | | | | | |
| 8,000 | 10 (1) | 80,000 | 0.735 | 58,800 | |
| 7,040 | 3.33 (2) | 23,443 | 0.735 | 17,231 | |
| 1,370 | 3.33 (3) | 4,562 | 0.735 | 3,353 | 79,384 |

Compensation to Lessee for Loss of Mineral Income                                                     £468,127

*Compensation for Premature Development Work*
Leaving the pillar of support unworked has meant that development work has had to be brought forward so that the vein beyond the pillar can be accessed. The cost of this work is estimated to be £210,000 and this needs to be expended immediately instead of over a period of three years. It is assumed that the development work would have been spread equally over the three-year period and that the cost would have been £70,000 pa.

*Lessee's claim for compensation for premature development work*

| Immediate development cost 210,000 | Previous annual development costs £ | YP 3 years @ 10% | Present value | Compensation |
|---|---|---|---|---|
| *Less* | 70,000 | 2.487 | 174,090 | |
| Compensation for Premature Development Work (4) | | | 35,910 | |
| Lessee's Total Claim | | | | £504,037 |

*Freeholder's claim*

| Tonnage | Unit: Income per tonne £ | Income £ | PV £1 in 1 year @ 6% (5) | Value | Total for year |
|---|---|---|---|---|---|
| Year 1 | | | | | |
| 13,560 | 0.75 (1) | 10,170 | 0.943 | 9,590 | |
| 3,130 | 0.250 (2) | 783 | 0.943 | 738 | 10,328 |
| | | | | | |
| Year 2 | | | | | |
| 16,000 | 0.75 (1) | 12,000 | 0.89 | 10,680 | |
| 8,000 | 0.250 (2) | 2,000 | 0.89 | 1,780 | |
| 2,740 | 0.250 (3) | 685 | 0.89 | 610 | 13,070 |
| | | | | | |
| Year 3 | | | | | |
| 8,000 | 0.75 (1) | 6,000 | 0.84 | 5,040 | |
| 7,040 | 0.250 (2) | 1,760 | 0.84 | 1,478 | |
| 1,370 | 0.250 (3) | 343 | 0.84 | 288 | 6,806 |
| | | | | | |
| Year 4 | | | | | |
| 8,000 | 0.75 (1) | 6,000 | 0.792 | 4,752 | |
| 7,040 | 0.250 (2) | 1,760 | 0.792 | 1,394 | |
| 1,370 | 0.250 (3) | 343 | 0.792 | 271 | 6,417 |
| Compensation to freeholder for sterilisation of minerals | | | | | £36,621 |

Notes
(1) The inner area at full value.
(2) The outer area at one third value.
(3) Additional mineral made unworkable.
(4) Additional compensation under the mining code.
(5) The risk is not as high for the royalty owner; he would receive a minimum rent irrespective of whether the minerals were worked.

# Study 5

*Computer Spreadsheets and Discounted Cash Flow*

Computers and spreadsheets are commonly used for valuation purposes, and a number of specialist programmes are available commercially. It is not difficult to format a spreadsheet to prepare either valuations using the YP or DCF approach.

Discounted cash flow comes into its own when cash flows vary and when sensitivity analysis has to be undertaken. However, since the YP is merely the summation of the present values over the period under consideration, the YP will arrive at a capital value with less fuss provided the income remains unchanged — or does not change too often. If the discount rate and income remain unchanged, the YP and DCF methods will produce identical results.

An explanation of the construction and use of DCF spreadsheets will be found in most introductory valuation textbooks, and of the mathematics of the valuation tables in *Parry's Valuation Tables*. However, a few words are offered here, and Tables 8, 8a, 8b and 8c show a spreadsheet calculation using the same basic data from study 1, calculating the value by use of DCF as opposed to the YP approach used in that study.

Spreadsheets really come into their own when the risk rate varies, or when there are variations in the income, or in expenditure, during the period involved. The more frequent such variations, the greater the justification for using DCF. It is also useful when the discount periods are shorter than annual, since it is easy to calculate the PV of £1 on a monthly or quarterly basis. In the absence of variations, the YP approach is quicker and simpler. In YP calculations, the percentage used to calculate the YP, the yield, is the remunerative rate of return applicable to the valuation of the income flow. In a DCF calculation, the discount rate may, of course, be the remunerative rate, in which case the calculation will be on the same basis as a conventional valuation. It is, however, more usual for it to be either the company's borrowing rate, or its internal cost of capital. Indeed, DCF really comes into its own not for valuation, but to assist when deciding whether a proposed investment is worthwhile or not. In these cases, one is undertaking appraisal rather than valuation, experimenting by varying the discount rate until the net present value of the cash flow (the figure at the end of a DCF calculation) is £ Nil. This will occur when the discounted receipts and expenditures exactly cancel each other out, and the discount rate at which this happens is known as the Internal Rate of Return. It is the rate at which the cash flows balance, and if it is above the company's cost of capital it is a *prima facie* indication that the project will result in a loss. There may, of course, be other considerations that override this, but these are beyond the scope of this chapter. Table 9 shows a simple DCF spreadsheet for a cash flow running for 11 years. There is an initial receipt of £12,000, which, of course, does not have to be discounted. This occurs in Year 0, ie immediately. For no better reason than to show how such things are handled, there is an expenditure of £6,000 in the sixth year. For illustrative purposes the royalty per tonne increases every three years, but in a valuation only predetermined increases should be taken into account and future inflation increases ignored. The output is variable to show the flexibility of the DCF approach. As can be seen, the current value of this cash flow is £20,850, but say £20,800 assuming a discount rate, or yield, of 10%.

# Study 6

*Valuation matrices*

Where there is a large number of variables it may be appropriate to construct a valuation matrix. This will assist the valuer in arriving at the valuation figure and will demonstrate to the client the volatile nature of the parameters and their effect on value.

An example of where a matrix may be appropriate is with underground coal mines. Modern capital intensive longwall mining can be highly productive but requires relatively large areas which are free from geological faulting, even minor faulting can severely disrupt production. Relatively small shortfalls in budgeted output can turn a profit into a loss. Many mines operate with only a broad outline knowledge of the geology in front of the workings and the detailed geology is proved by the driving of development roadways. There may only be a year or so of 'measured' coal reserves but 'indicated' coal reserves may be sufficient to last several decades. The mine could therefore theoretically work profitably for the foreseeable future but it could hit geological problems which could close it in, say, less than two years.

# Table 8

Capitalisation Using Discounted Cash Flow
Study 1

## Table 8a

| | Minimum Rent Capitalisation | | | |
|---|---|---|---|---|
| Year | Min Rent | Risk Rate | PV £1 | Present Value |
| 1 | £50,000 | 7% | 0.934579 | £46,729 |
| 2 | £50,000 | 7% | 0.873439 | £43,672 |
| 3 | £50,000 | 7% | 0.816298 | £40,815 |
| 4 | £50,000 | 7% | 0.762895 | £38,145 |
| 5 | £50,000 | 7% | 0.712986 | £35,649 |
| 6 | £50,000 | 7% | 0.666342 | £33,317 |
| 7 | £50,000 | 7% | 0.62275 | £31,137 |
| 8 | £50,000 | 7% | 0.582009 | £29,100 |
| 9 | £50,000 | 7% | 0.543934 | £27,197 |
| 10 | £50,000 | 7% | 0.508349 | £25,417 |
| 11 | £50,000 | 7% | 0.475093 | £23,755 |
| 12 | £50,000 | 7% | 0.444012 | £22,201 |
| 13 | £50,000 | 7% | 0.414964 | £20,748 |
| 14 | £50,000 | 7% | 0.387817 | £19,391 |
| 15 | £50,000 | 7% | 0.362446 | £18,122 |
| 16 | £50,000 | 7% | 0.338735 | £16,937 |
| 17 | £50,000 | 7% | 0.316574 | £15,829 |
| 18 | £50,000 | 7% | 0.295864 | £14,793 |
| 19 | £50,000 | 7% | 0.276508 | £13,825 |
| 20 | £50,000 | 7% | 0.258419 | £12,921 |
| 21 | £50,000 | 7% | 0.241513 | £12,076 |
| 22 | £50,000 | 7% | 0.225713 | £11,286 |
| | | | | £553,062 |
| | | | | (i) |

## Table 8b

| <200Ktpa Royalty Capitalisation | | | |
|---|---|---|---|
| Royalty Rent | Risk Rate | PV £1 | Present Value |
| | | | |
| | | | |
| £30,000 | 8% | 0.793832 | £23,815 |
| £30,000 | 10% | 0.683013 | £20,490 |
| £30,000 | 10% | 0.620921 | £18,628 |
| £30,000 | 10% | 0.564474 | £16,934 |
| £30,000 | 10% | 0.513158 | £15,395 |
| £30,000 | 10% | 0.466507 | £13,995 |
| £30,000 | 10% | 0.424098 | £12,723 |
| £30,000 | 10% | 0.385543 | £11,566 |
| £30,000 | 10% | 0.350494 | £10,515 |
| £30,000 | 10% | 0.318631 | £9,559 |
| £30,000 | 10% | 0.289664 | £8,690 |
| £30,000 | 10% | 0.263331 | £7,900 |
| £30,000 | 10% | 0.239392 | £7,182 |
| £30,000 | 10% | 0.217629 | £6,529 |
| £30,000 | 10% | 0.197845 | £5,935 |
| £30,000 | 10% | 0.179859 | £5,396 |
| £30,000 | 10% | 0.163508 | £4,905 |
| £30,000 | 10% | 0.148644 | £4,459 |
| £30,000 | 10% | 0.135131 | £4,054 |
| £30,000 | 10% | 0.122846 | £3,685 |
| | | | £212,356 |
| | | | (ii) |

## Table 8c

| >200Ktpa Royalty Capitalisation | | | |
|---|---|---|---|
| Royalty Rent | Risk Rate | PV £1 | Present Value |
| | | | £0 |
| | | | £0 |
| £17,500 | 11% | 0.731191 | £12,796 |
| £52,500 | 12% | 0.635518 | £33,365 |
| £70,000 | 12% | 0.567427 | £39,720 |
| £70,000 | 12% | 0.506631 | £35,464 |
| £70,000 | 12% | 0.452349 | £31,664 |
| £70,000 | 12% | 0.403883 | £28,272 |
| £70,000 | 12% | 0.36061 | £25,243 |
| £70,000 | 12% | 0.321973 | £22,538 |
| £70,000 | 12% | 0.287476 | £20,123 |
| £70,000 | 12% | 0.256675 | £17,967 |
| £70,000 | 12% | 0.229174 | £16,042 |
| £70,000 | 12% | 0.20462 | £14,323 |
| £70,000 | 12% | 0.182696 | £12,789 |
| £70,000 | 12% | 0.163122 | £11,419 |
| £70,000 | 12% | 0.145644 | £10,195 |
| £70,000 | 12% | 0.1300 | £9,103 |
| £70,000 | 12% | 0.116107 | £8,127 |
| £70,000 | 12% | 0.103667 | £7,257 |
| £70,000 | 12% | 0.09256 | £6,479 |
| £35,000 | 12% | 0.082643 | £2,892 |
| | | | £365,779 |
| | | | (iii) |

## Table 8d

| Agricultural Land Capitalisation | | | |
|---|---|---|---|
| Ag Rent | Risk Rate | PV £1 | Present Value |
| £12,500 | 10% | 0.909091 | £11,364 |
| £12,500 | 10% | 0.826446 | £10,331 |
| £12,500 | 10% | 0.751315 | £9,391 |
| £12,500 | 10% | 0.683013 | £8,538 |
| £12,500 | 10% | 0.620921 | £7,762 |
| £12,500 | 10% | 0.564474 | £7,056 |
| £12,500 | 10% | 0.513158 | £6,414 |
| £12,500 | 10% | 0.466507 | £5,831 |
| £12,500 | 10% | 0.424098 | £5,301 |
| £12,500 | 10% | 0.385543 | £4,819 |
| £12,500 | 10% | 0.350494 | £4,381 |
| £12,500 | 10% | 0.318631 | £3,983 |
| £12,500 | 10% | 0.289664 | £3,621 |
| £12,500 | 10% | 0.263331 | £3,292 |
| £12,500 | 10% | 0.239392 | £2,992 |
| £12,500 | 10% | 0.217629 | £2,720 |
| £12,500 | 10% | 0.197845 | £2,473 |
| £12,500 | 10% | 0.179859 | £2,248 |
| £12,500 | 10% | 0.163508 | £2,044 |
| £12,500 | 10% | 0.148644 | £1,858 |
| £12,500 | 10% | 0.135131 | £1,689 |
| £12,500 | 10% | 0.122846 | £1,536 |
| £12,500 | 10% | 0.111678 | £1,396 |
| | | | £111,040 |
| | | | (iv) |

Reversion in perpetuity

| | |
|---|---|
| £10,000 | Annual Rent |
| 20 | YP in perpetuity @ 5% |
| £200,000 | |
| 0.3256 | deferred 23 years @ 5% |
| £65,114 | Present Value |
| | (v) |

| Total Value | £1,307,351 | *ie* Sum (i), (ii), (iii), (iv), (v) |
|---|---|---|

**Note**

There is a slight variation in the figures arrived at by the two methods due to the DCF being calculated to 18 decimal points and the YP to four decimal points

**Table 9  Discounted Cash Flow**

| A Year | B Output tonnes | C Royalty £ per tonne | D Receipts £ | E Expenditures £ | F Cash Flow £ | G Discount or risk rate % | H Present Value of £1 | I Discounted cash flow | J Running total or NPV |
|---|---|---|---|---|---|---|---|---|---|
| 0 | | | 12,000 | | 12,000 | 10 | 1.0 | 12,000 | 12,000 |
| 1 | 1,000 | 1 | 1,000 | | 1,000 | 10 | 0.9091 | 909 | 12,909 |
| 3 | 1,000 | 1 | 1,000 | | 1,000 | 10 | 0.8265 | 826 | 13,735 |
| 4 | 1,500 | 1.25 | 1,875 | | 1,875 | 10 | 0.7513 | 1,408 | 15,144 |
| 5 | 1,500 | 1.25 | 1,875 | | 1,875 | 10 | 0.6830 | 1,280 | 16,424 |
| 6 | 1,750 | 1.25 | 2,187 | | 2,187 | 10 | 0.6209 | 1,358 | 17,783 |
| 7 | 2,000 | 1.5 | 3,000 | 6,000 | -3,000 | 10 | 0.5645 | 1,693 | 16,089 |
| 8 | 2,000 | 1.5 | 3,000 | | 3,000 | 10 | 0.5132 | 1,539 | 17,629 |
| 9 | 2,000 | 1.5 | 3,000 | | 3,000 | 10 | 0.4665 | 1,399 | 19,028 |
| 10 | 2,000 | 1.75 | 3,500 | | 3,500 | 10 | 0.4241 | 1,484 | 20,513 |
| 11 | 500 | 1.75 | 875 | | 875 | 10 | 0.3855 | 337 | 20,850 |

Net Present Value £20,850

The matrix shown in Table 10 is prepared on the assumption that there is sufficient proved coal to last about two years and possibly sufficient coal to last at least a further 10 years where detailed geology has yet to be proved. The financial risks associated with underground coal mining are high and we have accordingly used an interest rate of 18% for the first two years and interest rates varying between 25% and 40% for the remaining 10 years.

It will be seen from the matrix that similar present values are arrived at using lower outputs and lower interest rates to that using higher output and higher interest rates. The matrix shows a range of values between £2.1 m and £15.4 m but for all practical purposes the range is from £2.9 m to £11.2 m since one would not combine a low risk rate with high output levels. The valuer would have to carry out a detailed analysis of the circumstances at the mine before deciding on the appropriate value.

**Table 10  Valuation matrices**

| Annual Production Tonnes | Estimated Profit £ per tonne | Estimated Annual Profit | Interest Rates 15% & 25% £ | Interest Rates 18% & 30% £ | Interest Rates 18% & 35% £ | Interest Rates 18% & 40% £ |
|---|---|---|---|---|---|---|
| 750,000 | 1 | 750,000 | 2,888,000 | 2,546,000 | 2,292,000 | 2,098,000 |
| 800,000 | 2 | 1,600,000 | 6,161,000 | 5,412,000 | 4,889,000 | 4,475,000 |
| 850,000 | 2.5 | 2,125,000 | 8,183,000 | 7,214,000 | 6,493,000 | 5,944,000 |
| 900,000 | 3 | 2,700,000 | 10,397,000 | 9,166,000 | 8,250,000 | 7,552,000 |
| 950,000 | 3.5 | 3,325,000 | 12,804,000 | 11,288,000 | 10,159,000 | 9,300,000 |
| 1,000,000 | 4 | 4,000,000 | 15,403,000 | 13,580,000 | 12,222,000 | 11,188,000 |

# Compensation for mineral planning decisions

The current provision for compensation for mineral planning decisions can be found in the following legislation.

1. The Town and Country Planning Act 1990.
2. Planning and Compensation Act 1991, guidance can be found within Mineral Planning Guidance (MPG) 9.
3. The Environment Act 1995: Review of Mineral Planning Permissions. Guidance can be found within MPG14.
4. The Town and Country Planning (Compensation for Restrictions on Mineral Workings and Mineral Waste Depositing) Regulations 1997 (SI 1997 No. 1111), which came into force on March 25 1997. Guidance can be found within MPG4.

There are three broad areas which can give rise to compensation following mineral planning decisions.

(i) The Environment Act 1995 (the 1995 Act) provides for the initial and subsequent periodic reviews of mineral planning permissions to keep them in line with changing environmental practices and standards. The circumstances relevant to initial and periodic reviews are set out in MPG14. The majority, if not all, initial reviews will by now have been carried out and future reviews of mineral planning permissions will mostly be concerned either with periodic reviews or cases where dormant sites are to become active again. Compensation may be payable if new conditions, other than restoration and aftercare conditions, restrict working rights. Compensation can either stem from the Mineral Planning Authority (MPA) serving notice and including within the notice a statement that the new conditions would adversely affect, to an unreasonable degree, either the economic viability or the asset value of the site. Alternatively, the land or mineral owner may be of the view that the new conditions will affect the viability of the site. In these circumstances Parts IV and XI of the Town and Country Planning Act 1990 (the 1990 Act) have effect as if a modification order had been made and confirmed under sections 97 and 98 of the 1990 Act. Persons having an interest in the land or the minerals are able to claim compensation, eg the mineral and/or land owner and the mineral operator.

(ii) The Planning and Compensation Act 1991 (the 1991 Act) introduced new procedures for dealing with permissions for the winning and working of minerals or the depositing of mineral waste, originally granted under Interim Development Orders (IDOs). The 1991 Act refers to the IDOs as old mining permissions. The 1991 Act required certain actions to be taken in respect of the conditions attached to old mining permissions if they were to continue to be valid. In the case where a person with an interest in the land or minerals has submitted a scheme of working including restoration conditions and the MPA impose conditions different from those submitted in the application, then there is a right of appeal to the appropriate Secretary of State. The government, when introducing the 1991 Act, made it clear that they only envisaged conditions applying to environmental and amenity aspects of mineral working sites and it did not anticipate that any new conditions applied would affect the asset value or economic structure of the operation. The guidance notes indicate that conditions which would significantly affect the asset value are more appropriate for MPA reviews introduced under the Town and Country Planning (Minerals) Act 1981. Compensation does not seem to be a concept envisaged under the IDO registration and agreement of conditions. It is assumed that compensation would arise from a review of planning permissions under the Environment Act 1995 or by the MPA serving an order under the 1990 Act as detailed in the following paragraph.

(iii) MPAs still retain their powers to make orders revoking, modifying, discontinuing, prohibiting or suspending mineral workings. In summary the following may attract compensation from the MPA if they are confirmed by the Secretary of State:

(a) revocation and modification orders (section 97 and Part II of Schedule 5, 1990 Act)

(b) discontinuance orders (section 102 and paragraphs 1 to 3 of Schedule 9, 1990 Act)

(c) prohibition orders (paragraphs 3 and 4 of Schedule 9, 1990 Act) and

(d) suspension orders and supplementary suspension orders (paragraphs 5 to 9 of Schedule 9, 1990 Act).

A claim must be made under the appropriate provisions of the 1990 Act. The provisions are section 107 (or 279 as appropriate) in the case of revocation and modification orders, or section 115 (or 280 as appropriate) in the case of other orders. It should be noted that the Town and Country Planning (Compensation for Restrictions on Mineral Workings and Waste Depositing) Regulations (the 1997 Regulations) define the circumstances in which compensation is not to be payable following the making of a modification or discontinuance order. The 1997 Regulations also modify section 115 of the 1990 Act in its application for claims for compensation following the making of prohibition, suspension or supplementary orders. The 1997 Regulations revoked the Town and Country Planning

(Compensation for Restrictions on Mineral Working) Regulations 1985 (SI 1985 No 698) and the Town and Country Planning (Compensation for Restrictions on Mineral Working) (Amendment) Regulations 1990 (SI 1990 No 803).

# Compensation

MPG14 states the government's intention to bring the compensation entitlement following a revocation, modification or discontinuance order into line with the compensation provisions for periodic reviews, ie there would be no compensation for orders which imposed restoration or aftercare conditions, nor for orders which imposed conditions which did not affect working rights. Working rights are affected if any of the following are reduced or restricted:

(a)  the size of area which may be used for the winning and working of minerals or the depositing of mineral waste
(b)  the depth to which the operations for the winning and working of minerals may extend
(c)  the height of any deposit of mineral waste
(d)  the rate at which any particular mineral may be extracted
(e)  the rate at which any particular mineral waste may be deposited
(f)  the period at the expiry of which any winning and working of minerals or depositing of mineral waste is to cease or
(g)  the total quantity of minerals which may be extracted from, or of mineral waste which may be deposited on site.

The circumstances where *no* compensation is payable following the making of orders are set out in the 1997 Regulations and summarised in MPG4.

It is not proposed to repeat all the circumstances here as the conditions where compensation will arise and the amount of compensation due varies under the orders. Any valuer dealing with a claim for compensation will need to refer to the 1990 Act and the 1997 Regulations (guidance from MPG4). As an example it may be useful to consider the question of compensation where a review is made under the Environment Act 1995.

The scenario being that revised conditions have been submitted for an active Phase 1 site under the provisions of the 1995 Act (Phase 1 sites are detailed under the 1995 Act but in brief refer to active sites where the predominant planning permission/s was/were granted after June 30 1948 and before April 1 1969.) The MPA has not accepted the conditions submitted by the applicant but has substituted amended ones. Some of the conditions will affect the economic viability of the site. The MPA has served a notice (see Schedule 13 paragraph 15, to the Environment Act 1995) stating that it is their opinion the revised conditions could affect the economic viability of the site. The same scenario would arise where the applicant has made a successful appeal (see Schedule 13, paragraph 11 to the Environment Act 1995). Parts IV and XI of the 1990 Act will have effect as if a modification order had been made and confirmed under sections 97 and 98 of the 1990 Act. The applicant will be able to make a claim for compensation under section 107 of the 1990 Act, as modified by the Town and Country Planning (Compensation for Restrictions on Mineral Working and Mineral Waste Depositing) Regulations 1997 (SI 1997 No 1111).

First, dealing with the modification under the 1997 Regulations. Regulation 3 provides that no compensation is payable where the following conditions are satisfied:

(a) the order does not impose any restriction on working rights other than restoration or aftercare conditions (see paragraphs 2 (a)(i) & (ii))

(b) that either the permission was granted not less than five years before the grant of the order, or the planning permission was granted before February 22 1982 (see paragraphs 2(b)(i) & (ii)) and

(c) the order was made more than five years after any previous order or orders in respect of the same land, and more than five years after an application for determination of conditions under Schedule 2 to the Planning and Compensation Act 1991 or under Schedule 13 or 14 to the 1995 Act was finally determined (see paragraphs 2(c)(i) & (ii)).

If these conditions are not satisfied then section 107 of the 1990 Act applies and unmodified and unabated compensation is payable.

Section 107 of the 1990 Act sets out that a person with an interest in the land or minerals can make a claim, within the prescribed time and prescribed manner, to the local planning authority if it can be shown that they have:

(a) incurred expenditure in carrying out work which is rendered abortive by the order

(b) otherwise sustained loss or damage which is directly attributable to the order.

## Study 7

*Compensation for the modification of mineral planning permission*

A quarry operator has an active phase 1 site and has submitted an application for determination of new conditions (1). All the correct procedures have been adopted but both parties agreed to a time extension to determine conditions. After lengthy discussions the MPA has determined conditions different from those submitted by the applicant.

The revised conditions relate to the construction of a new access (which avoids residential dwellings and a school). The new access includes a screened entrance but will sterilise 180,000 tonnes of limestone. The limestone would have been worked in five years time as shown on working plans previously submitted to the MPA. The current profits on sales are £1.80 per tonne. Some £15,000 was spent on a new weighbridge and wheelwash at the existing access some two years ago. It will be possible to relocate these items to the new access at a cost of £12,000. The cost of the new access will be £120,000. Consultants' fees for revising the restoration plans and construction details for the new access including negotiations with the Highways Authority and supervision of the works will be £3,750. The quarry production has averaged 90,000 tonnes over the last eight years and this is supported by statutory returns.

Stripping costs of adjacent land containing reserves will now take place in four years time instead of six years time. Some 10,500 $m^3$ of soils and 9,000 $m^3$ of overburden is involved. The costs of moving the soils is £0.70/$m^3$ and the cost of moving the overburden is £0.60/$m^3$.

There is no opportunity for landfill. The proposed restoration of the whole site is to low level restoration with afteruse for agriculture, with some amenity woodland planting. The restoration plan along with a five-year aftercare scheme has been agreed with the MPA. The conditions within the original planning permission in respect of restoration required all plant and machinery to be removed from the site at the cessation of mineral extraction. The soils stripped from the site will be re-spread over the quarry floor and the land returned to agricultural use. The quarry operator has estimated that on current costs the agreed restoration plan and aftercare scheme will cost a further £15,000 above the cost envisaged for the proposed restoration under the original planning permission — see note (2) below.

Advise the quarry owner on the amount of compensation to be submitted in a claim to the MPA.

**Table 11**

| | £ | £ | Compensation £ |
|---|---|---|---|
| Cost of constructing new access | | | 120,000 |
| Relocation of wheelwash and weighbridge | | | 12,000 |
| | | | |
| Loss of limestone Year 5 | | | |
| — 90,000 tonnes at £1.80 | 162,000 | | |
| PV of £1 in 5 years @ 8% | 0.6806 | 110,257 | 110,257 |
| | | | |
| Loss of limestone Year 6 | | | |
| — 90,000 tonnes at £1.80 | 162,000 | | |
| PV of £1 in 6 years @ 8% | 0.6302 | 102,092 | 102,092 |
| | | | |
| Fees incurred in plan preparation etc | | | 3,750 |
| Cost of moving soils in 4 years time | | | |
| 10,500m³ at £0.70 | 7,350 | | |
| 9,000m³ at £0.60 | 5,400 | | |
| | 12,750 | | |
| PV of £1 in 4 years @ 8% | 0.735 | 9,371 | |
| | | | |
| *Less* | | | |
| Cost of moving soils in 6 years time | | | |
| 10,500m³ at £0.70 | 7,350 | | |
| 9,000m³ at £0.60 | 5,400 | | |
| | 12,750 | | |
| PV of £1 in 6 years @ 8% | 0.6302 | 8,035 | |
| Additional cost of moving soils | | 1,336 | 1,336 |
| Total claim | | | £349,435 |

*Notes*
(1)  See Planning and Compensation Act 1991 and MPG8 for procedures and time scale.
(2)  The claim for compensation for additional restoration and aftercare costs is excluded under the Town and Country Planning (Compensation for Restrictions on Mineral Workings and Mineral Waste Depositing) Regulations 1997 (SI 1997 No 1111): see Regulation 3.

© PR Deakin and MC Jervis 2008

# Rating

Rating continues to be a reliable and relatively efficient system of raising revenue for the government. Many adjustments, alterations and fine tunings have been undertaken over the years to try and ensure equity and minimise the political dimensions. At the time of writing this chapter the council tax liabilities have been reassessed for Wales and this has necessitated creating a new tax band. However, the revaluation for council tax for England has been postponed. It may be that with the recent history of sustained growth in residential capital values, issues of equity and liability may again be questioned.

This chapter deals with the principles of valuation practice as applied to rating. Although rating valuations are influenced by and impinge upon particular legal principles, underpinning statute and case law, it is not the purpose of this chapter to consider these other than where it is felt particularly appropriate.

## Chargeable dwellings

The definition of a dwelling is contained in section 3 of the Local Government Finance Act 1992. Basically a chargeable dwelling is a 'hereditament' that certifies certain conditions set out in the Act. There is one bill per dwelling, based on bands related to capital value, as at 1 April 1991. The tax has both property and personal elements. The personal element is based on an assumption that the dwelling is occupied by two adult residents and the bill is discounted by 25% where there is only one. The basis of valuation is the amount that the property would have sold for on the 'open market' by a willing vendor on 1 April 1991 on assumptions contained in regulation 6 of the Council Tax (Situation and Valuation of Dwellings) Regulations 1992 SI 1992 No 550.

The main valuation assumptions are:

- that the sale was with vacant possession
- that the interest sold was freehold except for flats when a lease for 99 years at a nominal rent is to be assumed (the actual lease term is ignored)
- that the dwelling had no potential for any building work or other development requiring planning permission
- that the size, layout and character of the dwelling and the physical state of its locality which existed on 1 April 1991 were the same as those which actually existed on 1 April 1993 or a later date depending on why the alteration is being made

- that the dwelling (and any common parts such as an entrance or hallway shared with other dwellings) was in a 'state of reasonable repair'.

## Size, layout, character and locality

While the dwelling is valued at 1 April 1991, the size, layout character and locality of the dwelling are required to be considered as they actually existed at 1 April 1993 or, at a later date, depending on the facts.

If there needs to be a correction of an error in the valuation list when the council tax came into effect, the date would be when the list became inaccurate. The dwelling is valued as if all the physical factors existed when the valuation first came into force.

If there is a division or merger of dwelling(s), the date would be the date of the split or merger. Dwellings are valued having regard to all the physical factors as they existed at the date of the completion of the works.

If the dwelling was improved, the date would be the date when the building is subsequently sold as a dwelling cannot be re-banded due to increases arising from improvements until it is next sold. The dwelling is revalued having regard to all the physical factors as they existed at the date of sale.

If there is a change in the balance of the business/domestic use then the date would be that when the change occurred, with the whole dwelling valued having regard to the physical factors as they existed at the date when the change of use occurred.

If part of a dwelling is demolished with no intention of rebuilding, then the date is the date of demolition. If however the demolition is part of planned future works then no alteration may be made. Regard is given to the physical factors as they existed at the date of demolition.

If the value is reduced due to alterations for making them suitable for a physically disabled person, then the date would be the date of adaptation with the physical factors considered as at that date.

Where there is a change in the physical state of the dwelling's locality, the date would be the later of:

(a) date of previous alteration
(b) date of any previous sale giving rise to an alteration
(c) 1 April 1993.

Note that if a reduction in value has been caused by external factors, these cannot be offset by improvements executed by the current taxpayer. The state of the locality would be taken as that when the changes to the locality took place.

# Non-domestic hereditaments

For non-domestic rating the basic essentials of rateability remain unchanged and the term hereditament has the same meaning that it always had. To be included in a local rating list a hereditament has to be either used wholly for non-domestic and non-exempt purposes or there has to be at least some part of it which is neither domestic property nor which is exempt. Details of exemptions are to be found in Schedule 5 of the Local Government Finance Act 1988 and include among other exemptions, agricultural premises, places of religious worship, parks and property used for the disabled. The legislation provides that property is domestic if it is used wholly for the purpose of living accommodation. This may be taken to be accommodation or property used for sleeping, eating and associated purposes. Included within the specific definition of domestic property are private garages used wholly or mainly for the accommodation of a private motor vehicle and private storage premises used wholly or mainly for the storage of articles of domestic use. Excluded from the definition of domestic property is certain 'short-stay' accommodation; 'property' is not domestic if it is wholly or mainly used for the purposes of a business for the provision of 'short-stay' accommodation ie (*inter alia*) accommodation which is provided for short periods to individuals whose sole or main residence is elsewhere. It is however generally considered that a minor non-domestic use of an otherwise wholly domestic hereditament will not give rise to liability to non-domestic rates, if that use is *de minimus* and does not materially detract from the domestic use. The boundary between domestic and non-domestic use is a grey area. All domestic property comprises, or is ancillary to, living accommodation, but some living accommodation comes within the definition of rateable non-domestic property because it is used for the purposes of a business for the provision of short stay accommodation. Hotels for example comprise living accommodation but are rateable because they provide short stay accommodation for persons whose sole or main residence is elsewhere. Rooms occupied by the long stay residents and accommodation occupied by the proprietors or staff, as their sole or main residence, do come within the definition of domestic property. The value attributable to the use of these parts has to be excluded from the rating valuation — see composite hereditaments below. Single units of self-catering holiday accommodation are rateable as are second homes made available for letting commercially for more than 140 days in a year. A particular difficulty is the seasonal hotel which reverts to wholly domestic use out of season. A detailed discussion of these problems is beyond the scope of this chapter.

# Composite hereditaments

A composite hereditament is defined in section 64(9) of the Local Government Finance Act 1988 which states that, 'A hereditament is composite if part only of it consists of domestic property'. Composite properties mainly comprise those which include a non-domestic element which appears (or will appear) in the non-domestic rating list. However, there is another type that exists, that which includes a non-domestic element which is exempt from rating, and for that reason does not appear in the non-domestic rating list.

The valuation basis for composite hereditaments is provided in Schedule 6 paragraphs 1A and 1B as inserted by Schedule 5 paragraph 38(4) of the Local Government and Housing Act 1989. Hereditaments which comprise both rateable non-domestic property and domestic property are known as composite hereditaments. For such hereditaments it is only the non-domestic use which is to be valued, but for rating the valuation must reflect the benefit of the presence of the domestic accommodation within the larger hereditament. The extent of the non-domestic use which is to be valued is either the actual physical extent of that use, or possibly a notional part of the hereditament consistent with the prevailing pattern of domestic/non-domestic use in the locality. This concept of 'notionality' is intended to overcome the bizarre and atypical distribution of uses which are sometimes encountered. Notionality applies only to distributions of uses within composite hereditaments. It cannot be used to bring into rating premises built as business premises, but which are used wholly as domestic accommodation, against the prevailing local pattern of uses.

# Valuation

Rating valuation follows usual valuation methodology. The main differences really emerge from the rating hypothesis implied by the statutory framework, such as valuing to the antecedent valuation date and the hypothetical tenancy. These are covered later. It could even be argued that, over the years, rating has helped develop valuation methods due to the standardising of the underlying assumptions, a common date of valuation and a process of resolving disputes through valuation courts and the Lands Tribunal to determine the most appropriate approach.

Originally hereditaments were valued to gross value or net annual value as defined by section 19 of the General Rate Act 1967. Now, all non-domestic hereditaments are to be valued to rateable value by virtue of the Local Government Finance Act 1988 Schedule 6, paragraph 2(1) as amended by the Rating (Valuation) Act 1999. This definition is identical to the old section 19(1) and (3) net annual value of the General Rate Act 1967 and is effectively the annual rent with the tenant carrying out all repairs, insurance and paying all outgoing necessary to keep the hereditament in a reasonable standard of repairs. Hence rateable value is essentially a full repairing and insuring rental, and follows current market practice. This enables a more direct comparison with market data than had the original valuation to gross value been maintained. The valuation date is defined by Schedule 6, paragraph 2(3), and for the purposes of a new rating list is the day on which the list must be compiled, or such day preceding that day as may be specified by the Secretary of State. For the 2005 list the date was 1 April 2003, and for the 2010 list will be 1 April 2008. However, Schedule 6, paragraph 2(6) states that the state of the hereditament and the environment are to be taken as at the date the list was compiled. This in effect means that for 1 April 2003, the assessment of a hereditament was based on 1 April 1998 values, modified to take account of physical circumstances as they were expected to be on 1 April 2000. Essentially, this means that a property is valued at the date of the valuation, but that values which existed two years prior to that date are applied. This is known as the antecedent valuation date. While this may not be the most equitable approach in that the values applied are effectively out of date, it is more certain. The values that are applied to the list have been well established, are capable of being readily defended and reduce uncertainty and potential grounds for appeals and unfair assessments. All valuations are now to the 'tone of the list'. For the 2005 list 'absolute' falls in value after 1 April 2003 did not give rise to reductions in rateable value between revaluations. The matters to be considered at the date the list was compiled are set out in Schedule 6, paragraph 2(7) and are:

(a)  matters affecting the physical state or physical enjoyment of the hereditament
(b)  the mode or category of occupation of the hereditament
(c)  the quantity of minerals or other substances in or extracted from the hereditament
(cc) the quantity of refuse or waste material which is brought onto and permanently deposited on the hereditament
(d)  matters affecting the physical state of the locality in which the hereditament is situated or which, though not affecting the physical state of the locality, are nonetheless physically manifest there and
(e)  the use or occupation of other premises situated in the locality of the hereditament.

Where the rateable value is required in order to make an alteration to a list which has already been compiled, these matters are to be taken as they existed at the date of the proposal: Schedule 6, paragraph 2(6).

# The hypothetical tenancy

All valuations are, of course, in essence the valuer's opinion of the outcome of imagined negotiation between a purchaser/lessee and a vendor/lessor. Valuations for different purposes require a variety of scenarios for the hypothetical bargain. Rating valuation requires the valuer to determine a rent, which is the result of a negotiation for a hypothetical tenancy. The terms of the tenancy are based in the legislation as interpreted by decisions of the Lands Tribunal and higher courts. In considering the likely rental bids all possible occupiers, including the actual occupier and owner, must be taken into account (*R. v London School Board* [1886]). However, where there is only one possible tenant, this would place that tenant in a strong negotiating position which must be taken into account in determining the assessment (*Tomlinson (VO) v Plymouth Argyle Football Club Ltd* [1960]). In these circumstances the ability of the actual tenant to pay the rent is to be considered. Usually there is more than one potential tenant and the hypothetical tenant is to be taken as being neither an exceptionally good nor an exceptionally bad business person, but rather someone of more middle-of-the-road commercial capabilities.

The landlord must not necessarily be assumed to be a local landlord with special interests (*Coppin (VO) v East Midlands Joint Airport Committee* [1971]), but will generally let to the highest bidder. Both parties are to be considered to be reasonably minded but commercially prudent. Although the letting is assumed to be from year to year it would normally have a reasonable prospect of continuance (*R v South Staffordshire Waterworks Co* [1885]). The state of repairs to be envisaged for the valuation is not necessarily the existing state of repair at the valuation date. Following the decision in the case of *Wexter v Playle (VO)* [1960] 2 WLR 187 before the Court of Appeal, the party having the liability to maintain premises in repair or to keep premises in repair, also has the obligation to put the premises in repair. This assumption has been put beyond doubt by the Rating (Valuation) Act 1999, reversing the Lands Tribunal decision of *Benjamin (VO) v Anston Properties Ltd* [1998]. The rent required by the rating valuations is that which the tenant would bid to reflect the obligation to first put the hereditament into, and then to maintain it in an appropriate standard of repair. The standard of repair to be assumed in determining a rating assessment was considered in the Lands Tribunal case of *Brighton Marine Palace and Pier Co v Rees (VO)* (1961) and will differ from hereditament to hereditament depending upon such factors as the age of the building, the quality of the locality and the type of tenant. As is the case elsewhere in the rating hypothesis the landlord is to be taken to be a reasonably minded person. The facets above are considered to provide an underlying philosophy behind the hypothetical tenancy concept. There are many more aspects which have been explored by the Lands Tribunal.

# Valuation methods

The valuation methods of the rating valuer remain unchanged with rental comparison being the preferred approach wherever possible. Whichever method is used, the end result should theoretically be the same when correctly applied, but the least appropriate methods require unacceptable degrees of adjustment. All buildings should be measured in accordance with RICS guidelines.

## Rental comparison

Market rental evidence when it is available, is always to be preferred, as being the most direct evidence and hence rental comparison is for the majority of hereditaments the usual valuation method. Ideally the letting should be grants of fresh tenancies entered into on rateable value terms on the antecedent valuation date. This was the 1 April 2003 for the 2005 list. In reality such evidence will rarely be available and the rents will need considerable adjustment to be 'reconciled' with the statutory definition. This makes demands upon the valuer's skill and is greatly aided by the backup of an in house research department.

The rental adjustment process involves considering the actual bargain at the date it was entered into to establish the rent which would have been paid at that time had the letting been in the terms of the hypothetical tenancy. Assisted by the evidence, a view is then taken of the rent which would have been paid on those terms at the valuation date. The general process of adjustment of rents to terms of rateable value is well established. However, it must always be remembered that the greater the extent of the adjustments which have to be made the less reliable is the rent as evidence.

Tenants frequently execute works of improvement and extension to their property during the tenancy. The total cost to the tenant of the occupation of the property, the virtual rent, includes both the rent reserved and the cost of those works. The valuer's task is to consider the actual economic decisions of the tenant to determine the rent that would have been paid for the improved property, in terms of rateable value, had it been so improved or extended at the date the current rent was fixed. These improvements will potentially have a value to the tenant only for so long as they are not reflected in the rent he pays to his landlord. Thus the valuer will have to consider the Landlord and Tenant Acts and the terms of the tenancy to establish when the value of the improvements may become included in the rent. This represents the maximum period for which the improvements will have any value to the tenant. The period may be shorter if the 'life' of the improvements, either physical or economic, comes to an end earlier: see *Edma (Jewellers) Ltd* v *Moore (VO)* [1975].

If the rent is not on full repairing and insuring terms the adjustment for these items must be carried out at the final stage. It is customary for a percentage adjustment to be used for this purpose but a more refined approach, based upon the actual costs, would be preferable. The rents so adjusted to terms of rateable value, will require analysis to unit prices for the comparison purposes. This may entail the use of end allowances which are discussed later. Review periods greater than one year may require adjustment to take account of overage (the amount by which the rent fixed for several years may differ from that fixed on an annual basis), although, in practice, such annual lettings may not always be reviewed on a yearly basis. Various formulae exist which attempt to quantify this figure. However, it may well be that evidence does not exist in the market in a suitable form to substantiate such an approach for rating assessments.

It is to be noted that not all evidence of lettings is necessarily of equal value. Rent reviews during the course of a lease or renewal of a lease determined under the Landlord and Tenant Act 1954, may require a market rental. This does not mean that a market rent is determined. Many factors may influence the eventual figure, not least the relationship of the actual landlord and tenant. The only true market rent is that achieved when a property, with the benefit of vacant possession, is freely and efficiently exposed to the market with adequate time allowed for negotiation.

## Profits or accounts (receipts and expenditure) method.

This method is based on the premise that the rental value will depend upon the profit earning capability of a hereditament. The greater the estimated profitability of a business conducted at the premises the more the tenant can afford to pay. Certain hereditaments are valued by this method rather than rental comparison. This is usually because rental evidence does not exist in a form that allows for direct comparison. This would typically be where a hereditament enjoys some form of uniqueness, normally resulting from a factual or legal monopoly. A hotel opposite the only railway station in a provincial town with few other hotels, would be a typical example of a factual monopoly. The license required by a public house to sell intoxicating liquor, would similarly be representative of a legal monopoly.

The method entails an analysis of the trading figures to determine the level of profit the hypothetical tenant might expect from the business conducted at the premises and from that, the amount of rent that he could reasonably afford to bid. This would naturally be on rateable value terms with the occupier assumed to be liable for all repairs and insurance. The gross income is estimated having regard to the actual accounts. From this are deducted the purchases and working expenses (excluding rent but including rates) to give the divisible balance. This figure is then divided between the tenant's share and the amount available for payment of rent. It is to be noted that the tenant's share is a most important constituent and must be sufficient to induce him to run the business. An adequate return on the capital tied up in the business must be allowed before considering the sum available for rent and rates.

This method calls greatly upon the skill of the valuer to interpret the accounts which will not have been prepared for the purposes of a rating valuation. The trading accounts to be considered are those for the (three) years preceeding the valuation date as these are deemed to be available to the hypothetical tenant.

## Contractor's method

This is often described as the 'method of last resort'. It is certainly true to say that it is only used where the previous methods are not applicable. It is based upon the premise that cost approximately equates to value. If a hereditament were not available to rent then a prospective tenant must construct one. The land would need to be acquired and the building constructed. The occupier would have to borrow the capital in order to construct the building and pay interest on this loan. The amount of the annual interest payable would be equivalent to the maximum rent he might pay. Briefly the process is as follows. The building costs as at the valuation date are determined. They are then adjusted to take account of any age and/or obsolescence disabilities the actual building may suffer by comparison with a new building, or its equivalent, from which the costs will have been derived. The resulting adjusted replacement cost of the building(s) plus the value of the site works and fees is then 'rentalised' or decapitalised. The percentage rate at which the total cost is decapitalised represents the actual costs of making the money available, adjusted to reflect the essential differences between the benefits to be derived from owning the property rather than renting it, on the terms of the hypothetical tenancy. While the philosophy of the decapitalisation rate is interesting it is somewhat academic, as the percentage rates are now prescribed by law. For the 2005 rating list they were 3.33% for hospitals and educational establishments and 5% for other hereditaments: The Non-Domestic Rating (Miscellaneous Provisions) (Amendment) (England) Regulations 2004 SI 1994 No 1494.

The fifth stage of the approach, and possibly the most important, requires the valuer to stand back and check the results of the calculation against known values of properties as similar in type as possible. Adjustments at this fifth stage may be either downwards or upwards and are aimed at ensuring that the rental determined is realistically representative of what the market might have paid, were rental evidence to exist.

The contractor's method was scrutinised in *Imperial College of Science & Technology* v *Ebdon (VO)* [1985], LT and [1987], CA, and more recently in *Monsanto plc* v *Farris (VO)* [1998].

## Phasing or transitional arrangements

For the 1990 revaluation, the government enacted legislation to ease the burden on businesses facing high increases in payments of local taxation. They placed limits on proposed increases and these were funded by similar limits placed on reductions. These were known as transitional arrangements (also known as phasing) and they have been retained and developed in the last revaluation, and amended since. Transitional arrangements are not part of a valuation and consequently are not considered in detail here. However, their implications for the actual liability of a ratepayer must be taken into account. Transitional arrangements were maintained for the 2005 revaluation, and in December 2004 a transition scheme was confirmed for a four year period ending in 2010. This set limits by which bills could increase or decrease until they reached their true rate liability in the fifth year. Transitional relief is automatically calculated by the rating authority and taken into account in the bill.

## Reconciling evidence

The rating hypothesis effectively requires a rental to be determined on the terms of the hypothetical tenancy. Rental evidence is commonly not directly comparable. There may be many reasons for this. For example, the repairing and insuring terms may be different from those envisaged by the hypothetical tenancy, the date of the transaction may be different from the antecedent valuation date or the rental agreed may not be on annual terms. Added to this are commonplace market practices when letting properties, such as the payment of premiums, reverse premiums and rent-free periods. These will need to be considered when analysing the comparable evidence and reconciled to the requirements of the hypothetical tenancy.

### Study 1

An office recently let at £15,500 on internal repairing terms only. As the rating hypothesis requires a valuation to what effectively is full repairing and insuring terms, external repairs and insurance must be deducted.

|  |  | £ pa | £ pa |
|---|---|---|---|
| Rent passing | | | 15,500 |
| *Less* | External repairs 5%(1) | 775 | |
| | Insurance 3.5% (1) | 542 | 1,317 |
| | Rental in rateable value terms | | 14,183  pa |

Note
(1)   These rates will normally have been established and the valuer will usually expect to follow custom and practice, especially when reconciling assessments with the tone of the list.

The 1995 revaluation created particular problems due to the nature of the property market, particularly in the south of the country where rental values were falling, this is in contrast to the 1990 revaluation where values had increased. Recent years have seen a return to increases since the last edition of this chapter was published. Changes in values are not always reflected in the headline rent for a number of reasons. For instance, there is a general reluctance to reduce the actual rent paid as this could have a knock-on effect when these transactions were used as comparables. Where this might happen, reverse premiums have sometimes been paid by landlords to tenants to maintain the rent apparently passing. Similarly, rent free periods, common in normal market conditions have been extended, effectively reducing the rental value while keeping the actual rent paid artificially high. Sometimes landlords have contributed to the costs of fitting out premises, to the same effect.

The previous paragraph refers to particular circumstances arising from the state of the property market at that time. It is important to appreciate that when such market conditions do occur, these factors may need to be taken into account when analysing rents for rating purposes. The following examples are given to illustrate the various concepts and suggest one of the various approaches to solving the problem.

## Study 2

The letting of a shop unit is agreed at a rent of £115,000 pa on FRI terms for the next 5 years. The landlord paid a reverse premium to the tenant of £50,000.

As a reverse premium was paid to the tenant, the true rental value would be lower than the agreed rent passing of £115,000. To adjust the rent, an annual equivalent of the reverse premium would be deducted.

|  |  | £ | £ |
|---|---|---|---|
| Rent passing |  |  | 115,000 |
| *Less* | Reverse premium | 50,000 |  |
|  | *Decapitalise* at YP 5 years 8% (1) | 3.99 | 1,253 |
|  |  |  | 113,747 |
|  | Actual rental value say |  | £113,750 pa |

Note
(1)  This and subsequent decapitalisations have all been undertaken using single rate figures, ie from a freehold landlord's position. It could be argued that the figures should also be analysed from the tenant's perspective, using dual rate tax adjusted figures and that the actual rent paid would reflect these two valuations.

The concept is similar to that commonly used for valuations in connection with surrender and renewals.

## Study 3

A letting has been agreed of a shop unit at a rent of £145,000. The landlord has contributed £25,000 towards the costs of fitting out the unit. There will be a review to full rental value in five years time.

This is very similar to the above example. The actual rent paid reflects the fact that the landlord is contributing towards what would normally be a tenant's expenditure. Consequently the true rental value would be less than the £145,000 pa agreed.

|  |  |  | £ pa |
|---|---|---|---|
| Rent passing |  |  | 145,000 |
| *Less* | Reverse premium | 25,000 |  |
|  | *Decapitalise* at YP 5 years 8% ((1) above) | 3.99 | 6,266 |
|  |  |  | 138,734 |

Actual rental value say £138,750 pa

## Study 4

The rent of a shop is agreed at £175,000 pa until it is reviewed to the open market rental in five years time. A rent-free period of two years has been granted.

Again the actual rent agreed does not reflect the actual rental value because a rent-free period has been agreed and the real rental value will be less. In order to calculate the real rental value, the value of the rent-free period must be taken into account. On the basis that the landlord is not losing anything of value, the capital value of the term before review, must be the same for the actual agreed terms as they would be if the actual market rent was being paid, ie $Vtar = Vtomr$ where $Vtar$ = the value of the term on the actual rental and $Vtmor$ = the value of the term at an open market rental.

Value of Term on the actual rental ($Vtar$)

|  |  | £ pa |
|---|---|---|
| Rent passing |  | 175,000 |
| YP for 3 years at 8% (1, above) | 2.577 |  |
| *Deferred* 2 years at 8% | 0.857 | 2.21 |
|  |  | 386,750 |

Capital Value of term £386,400

As this must equal the value of the term at the open market rental ($Vtar = Vtom$), we can decapitalise it to find what that rental should have been:

|  |  |  |
|---|---|---|
| Capital Value of term | 386,750 |  |
| *Decapitalise* YP 5 years at 8% (1, above) | 3.99 |  |
|  |  | 96,929 |

True open market rental, say £96,850 pa

Although studies 2 and 3 deal with a falling market, the principles used to undertake analysis of the data are relatively straight forward. They would be used to analyse expenditures undertaken by the tenant, for example where the tenant has undertaken some expenditure as a condition of entry. This is commonly experienced, for example, where improvements are undertaken upon entry or a premium or key money are paid upon entry. The analysis of premiums and key money is not necessarily straightforward in that a decision has to be made regarding what the premium or key money is paid for. To be analysed and added to the rental, the key money or premium would have to have been paid for factors which would genuinely have added to the rental value. For example, payments may have been made to reflect the additional value attributable to goodwill, fixtures and fittings, improvements or in respect of a reduction in rent. Goodwill and fixtures and improvements should not be included as these would not contribute to the rental paid by the hypothetical tenant. Improvements may be personal and, therefore, would similarly not count towards the rent paid under the hypothetical tenancy.

## Study 5

A shop is let on a 15 year lease on full repairing and insuring terms at a rent of £70,000 pa exclusive subject to five year rent reviews to full rental value. A premium of £35,000 was paid upon entry.

|  |  | £ pa |
|---|---|---|
| Rent passing |  | 70,000 |
| *Add* |  |  |
| Premium paid | 35,000 |  |
| YP 5 years at 8% (1) | 3.99 |  |
| Annual equivalent |  | 8,772 |
| Full rental value |  | 78,772 |

Note: (1) The premium is taken to account for a reduction in rent paid until the first rent review, when the rental would return to full rental value.

Similar adjustments would have to be made for tenant's improvements where they added value to the rent paid for a hereditament. Care would need to be exercised that any improvements did actually add value, and that any additional rental element came within the terms of the hypothetical tenancy. In the previous example, the term of years for decapitalisation is clear, ie it reflects the reduction in rent paid until full rental value is reached. With improvements, the period may not be that straightforward. It could be the time till the next rent review, the end of the lease or the economic life span of the improvement.

The following case studies relate to England and Wales only.

# Shops*

These are normally valued by direct rental comparison as comparable rental information is reasonably common. The method most often used to compare one shop with another is that of Zoning. This has not always been the case and the method has a chequered history. It is probable that the method arose from the valuation of 'parlour' shops in the earlier part of the 20th century, when shops commonly comprised the front room of a converted house, for sales space, with the rear room for storage. The rear room (storage) was often valued at half the rate of the front room (sales), the depth of the rooms determining the zone (commonly 15 or 20 ft). Over the years the dividing walls were removed but the process of using 15/20 ft zones and halving back continued.

The method is not without its problems. In some circumstances great sophistication is required in analysis and valuation. Slavish adherence to the principle of halving back is not to be recommended. 'Zoning should be subservient to the valuation and not valuation subservient to zoning.' Whatever the valuation entails, it should be remembered that 'as you analyse so must you value.'

Essentially a shop is zoned back from the main frontage in strips. These used to be 20 feet but for rating purposes are now usually 6.1 m. However, it is possible to use natural zones where the shops

---

\*     See also Chapter 7.

used in analysis and valuation have some obvious common depth. This could be the case in a modern shopping development. There are many advantages in using a natural zone for a specific valuation. The main disadvantage is that by definition, the natural zone will not be common to many other shops. Consequently any analysis will be of limited use. However, if the more conventional zone pattern is adopted, the analysis can be used for any valuation for which the comparable is itself valid.

The first zone is considered the most valuable, with value decreasing with distance from the main frontage. It is established valuation convention to halve back. Hence the first zone (zone A) will have a value of £X. The next zone (zone B) will have a value of £0.5X and so on. However, the valuer must be watchful for local market rents which may show this to be inappropriate. The reduction used should follow the local market evidence. The other areas of the shop which have value are similarly assessed 'in terms of zone A, (ITZA)' in order to obtain a total picture as to the likely value.

The resulting fractions, or 'X' factors as they have sometimes been known, may seem very odd if not interpreted in monetary terms (*Lewis's Ltd* v *Grice (VO)* (1982)). Hence market analysis might show the zone A to be £250 per m2, and some other part of the shop to be worth £16.00 per m². The X factor would therefore be 16/250 or 0.064. Where a significant differences in area occurs between rented shops and those to be valued, the figures may be reconciled using an adjustment known as a quantity allowance. Where the quality of shops is significantly different, a quality allowance may be used. Where the layout of a shop has significant disadvantages over another a disability allowance may be used.

Any rateable plant and machinery may be valued by an annual equivalent on the depreciated replacement cost (Contractor's Method) if not more conveniently incorporated into the value of the building directly.

While most shops can reasonably be analysed using zoning methods, it may sometimes be appropriate to use an overall basis. This depends not only on the subject property but also the nature of the comparables. In *Lewis's Ltd* v *Grice (VO)*, cited above, a very large department store in Liverpool comprising over 37,000 m² over seven floors was valued in terms of zone A due to the nature of the comparables. However in *Argyll Stores (Properties) Ltd* v *Edmonson (VO)* [1989], LT the Lands Tribunal accepted a valuation of a supermarket on an overall basis, as there were acceptable comparable supermarket style shops capable of comparison on this basis. It should also be remembered that the needs of the hypothetical tenant of a large store may be quite different from the hypothetical tenant of small lock-up shop.

When analysing rents of shops, it is useful wherever possible, to start with a simple unit free from disabilities, return frontages and similar peculiarities. This should enable the differences to be highlighted in order to construct an overall picture.

The valuations of certain types of shop are considered to follow a specialised approach reflecting the particularities of their trades, for instance betting shops, department stores, supermarkets and hypermarkets, large shops, kiosks, off licences and post offices.

## Study 6

Analyse 1, The High Street, a city centre lock-up shop 5 m wide by 12.2 m deep, situated in the Midlands. Use the information to assess a similar shop 5, The High Street, which is 6 metres wide by 12.2 metres deep. No 1 was let on FRI terms in March 2003 on a 15 year lease with 5 yearly rent reviews at £71,000 pa.

*Analysis*

Taking 6 metre zones and halving back:

|  | Size | Area m² | 'X' factor | Area m² (ITZA) |
|---|---|---|---|---|
| Zone A | 5 m × 6.1 m | 30.5 m² | A | 30.50 |
| Zone B | 5 m × 6.1 m | 30.5 m² | A/2 | 15.25 |
| Total area ITZA |  |  |  | 45.75 m² |

£71,000/45.75 = Zone A rent of £1,552 per m² ITZA in RV terms (1)

*Valuation (i)*

|  | Size | Area | 'x' factor | Area (ITZA) | RV at £1,552 |
|---|---|---|---|---|---|
| Zone A | 6 m × 6.1 m | 36.6 m² | A | 36.6 m² |  |
| Zone B | 6 m × 6.1 m | 36.6 m² | A/2 | 18.3 m² |  |
| Total area ITZA |  |  |  | 54.9 m² | £85,204 |

RV = £85,200

or

*Valuation (ii)*

|  | Size | Area m² | at £/m | RV £ |
|---|---|---|---|---|
| Zone A | 6 m × 6.1 m | 36.6 m² | 1,552 (A) | 56,803 |
| Zone B | 6 m × 6.1 m | 36.6 m² | 776 (A/2) | 28,401 |
|  |  |  |  | 85,204 |

RV = £85,200

Notes
(1) Although there are five yearly rent reviews, in this case there is no market evidence to support a deduction for overage.
(2) There are staff toilets on the premises, but the floor area of staff toilets is excluded as the value of the toilets has been taken into account in the Zone A rent when it was agreed.

# Study 7

Using the comparable evidence above, value 7, High Street, which is 6 metres wide by 17.2 metres deep. To the rear there is a store of 30 m².

*Valuation*

|  | Size | Area m² | at £/m | £ |
|---|---|---|---|---|
| Zone A | 6 m × 6.1 m | 36.6 | 1,552 (A) | 56,803 |
| Zone B | 6 m × 6.1 m | 36.6 | 776 (A/2) | 28,402 |
| Remainder | 6 m × 5 m | 30.0 | 388 (A/4) | 11,640 |
| Store |  | 30.0 | 155 (A/10) | 4,650 |
|  |  |  |  | 101,315 |

RV = £101,300

Note: (1) 'x' factor from analysis of rents used to derive price per m².

## Study 8

If a similar shop, 9, High Street had air conditioning on the ground floor, then the valuation would have to include an increase in rental value to reflect the air conditioned space.

*Valuation*

| | Size | Area m² | at £/ m² | £ pa |
|---|---|---|---|---|
| Zone A | 6 m × 6.1 m | 36.6 | 1,552 (A) | 56,803 |
| Zone B | 6 m × 6.1 m | 36.6 | 776 (A/2) | 28,402 |
| Remainder | 6 m × 5 m | 30.0 | 388 (A/4) | 11,640 |
| Store | | 30.0 | 155 (A/10) | 4,650 |
| *Additional item* | | | | |
| Air conditioning (1) | | 102.0 | 7 | 714 |
| | | | | 102,209 |

RV = £102,200

Note: (1) Only the ground floor retail space is air conditioned.

## Study 9

Using the above comparable in study 7, value 11, High Street which has a return frontage. The unit is 6 m wide by 20 m deep. You consider the basic layout to have an inherent trading disability compared with comparable properties. In this example, the disability result from an odd shape and slope on the internal wall.

*Valuation*

| | Size | Area | at £/ m² | £ pa |
|---|---|---|---|---|
| Zone A | 6 m × 6.1 m | 36.6 | 1,552 (A) | 57,803 |
| Zone B | 6 m × 6.1 m | 36.6 | 776 (A/2) | 28,402 |
| Remainder (1) | 6 m × 7.8 m | 46.8 | 388 (A/4) | 18,158 |
| | | | | 104,363 |
| *Add*: | | | | |
| Return frontage 5% (2) | | | | 5,218 |
| | | | | 109,581 |
| *Deduct*: | | | | |
| Disability 2% (3) | | | | 2,087 |
| | | | | 107,494 |

RV, say £107,500

Notes
(1) In this example two zones and a remainder have been used. In rating this is the most common approach, although three zones plus a remainder may be encountered.
(2) From an analysis of similar shop units with return frontages. There are a number of ways that return frontages can be analysed. A figure per metre run is the most common. A spot figure or a percentage addition to the value of the ground floor can also be used. The most appropriate method will be determined by the nature of subject property and the comparables. This study uses an increase in the value of the ground floor sales area of 10%. This would suggest that return frontage is of good quality along its length and that the street is a good retail trading street.
(3) To allow for the disability and enable the assessment to be reconciled with similar properties excluding the disability. The percentage should again come from market analysis. Other end allowances such as quantity allowances would be deducted at this stage of the valuation as appropriate.

# Study 10

## Shops

Shops frequently cover more than the ground floor. The following example is a small shop in a small North Midlands town. The Zone A rental has come from comparison with similar shops in the area.

| Ground floor | Area m² | at £/m² | £ pa |
|---|---|---|---|
| Zone A | 40 | 660 (A) | 26,400 |
| Zone B | 40 | 330 (A/2) | 13,200 |
| Zone C | 40 | 165 (A/4) | 6,600 |
| Remainder (1) | 25 | 82.50 (A/8) | 2,062 |
| | | | |
| *First floor* | | | |
| Internal storage | 45 | 33.00 (A/20) | 1,485 |
| Kitchen | 3 | 33.00 (A/20) | 99 |
| | | | |
| *Second floor* | | | |
| Internal storage | 50 | 26.40 (A/25) | 1,320 |
| | | | 51,166 |

Note: (1) This shop is in a location where the custom and practice has been to use three zones and a remainder, halving back for each zone. Rental values for the first floor and second accommodation have come from analysis of similar properties in the same area. Note that in some areas, Zone C is calculated at A/10.

# Study 11

Value a supermarket which is situated on its own site with its own car parking area and petrol filling station.

*Valuation*

| Description | Area in m² | Price/m² | £ pa |
|---|---|---|---|
| *Ground floor* | | | |
| Entrance (1) | 90 | 168.00 | 15,120 |
| Sales area (2) | 10,000 | 210.00 | 2,100,000 |
| Preparation (3) | 20 | 112.00 | 2,240 |
| Warehouse (3) | 280 | 105.00 | 29,400 |
| External storage (3) | 25 | 52.50 | 1,312 |
| | | | 2,148,072 |

RV, say £2,148,000

Notes
(1)  The entrance space is weighted at a percentage of the sales space, based on other supermarkets in the area. Alternatively a gross figure per m² might be used. 10,390 m² at £206.62 = £2,146,781. Add external storage at £1,312 = £2,148,093, say £2,148,00. Note that comparables may be located across a much larger geographic area than would be expected with normal shops.
(2)  An overall treatment for the sales space, rather than zoning is used, as this is a large unit with good comparable evidence of similar sizes and there is no direct relationship with small 'zoned' shops.
(3)  Weighted at a percentage of the sales space, based on other supermarkets in the area.

## Study 12

In the following example a large versatile space is used as a retail warehouse. This type of accommodation has become very common since the last edition of this chapter and consequently an adequate number of comparable properties may well exist. Analysis and valuation may take the basis of net internal area or gross internal area, depending upon the circumstances.

Value a retail warehouse which is situated in a retail park on the edge of town. The accommodation comprises retail space, warehousing, a garden centre and car parking facilities shared with other retail units on the site.

*Valuation*

| Description | Area in m² | Price/m² | £ pa |
|---|---|---|---|
| Ground floor sales | 3,030 | 85.00 | 257,550 |
| Ground floor warehouse | 225 | 85.00 | 19,125 |
| Ground floor garden centre | 900 | 12.75 | 11,475 |
| | | | 288,150 |

RV, say £287,000

Notes
(1)  The car parking is reflected in the rental applied to the retailing space.
(2)  Where there are first or mezzanine floors, these will have often been added by the occupier and will not be reflected in the rental evidence. Consequently there is often a shortage of good quality rental evidence. In such circumstances it may be acceptable to calculate a rental by taking an annual equivalent of the costs of providing the first or mezzanine floor. If there is a first or mezzanine floor, good comparable information is to be preferred.

# Offices*

Many types of offices exist but they should all be capable of assessment by rental comparison. The valuation may require assessment of a single office, suite of offices or a whole office block, depending on the extent of the rateable occupation. In any event the basic principles are similar. Modern offices are measured to net internal area (NIA), which excludes common areas such as staircases, toilets, fire corridors etc. In a multiple occupation building, reception areas, high quality atria etc., which benefit all tenants, will generally be reflected in the value of the separate occupations. Occasionally, some residual value may remain in the hands of the landlord as a separate assessment.

The nature of the demountable partitioning needs to be considered to determine whether it is part of the setting within which the business is carried out (*British Bakeries Ltd* v *Gudgion (VO)* [1969]). If this is the case then the partitioning is rateable and must be reflected in the assessment. This is usually done by increasing the unit price or taking a percentage of the depreciated replacement cost of the partitioning. If, on the other hand, the partitioning is necessary for the purposes of the particular business it is considered to be plant, and so long as demountable partitioning remains unlisted in the relevant statutes, will not be rateable. Older office buildings, frequently the product of the conversion of residential accommodation, will often be measured and valued on an effective floor area basis (EFA).

---

*     See also Chapter 6.

## Study 13

This is a valuation of an office on a self contained plot with dedicated car parking. The office space occupies the ground, first and second floors, and a basement. There are no landlord's services.

| Description | Area in m² | Price/m² | £ pa |
|---|---|---|---|
| Ground floor | 700 | 72.00 | 50,400 |
| 1st floor (1) | 11,000 | 72.00 | 79,200 |
| 2nd floor | 220 | 72.00 | 15,840 |
| Basement | 100 | 35.00 | 3,500 |
| Car parking 50 spaces at £350 | | | 17,500 |
| | | | 166,440 |

RV, say £166,400

Note: (1) The accommodation has good access to all floors serviced by a modern lift.

# Factories and warehouses*

Factories have always been valued to net annual value so there is virtually no change in the valuation process. Many factories are similar to each other in the basic facilities that they provide. These can normally be assessed using comparable rental assessments. However, certain factories may be specialised with much plant and machinery, as in the case of oil refineries. These will need special treatment which requires specialist knowledge which is outside the scope of this chapter.

Warehouses were previously valued to gross value and must now be assessed to rateable value. Modern warehouses are more than just boxes, being designed for efficient storage and retrieval. Regard must be paid to the working height and units with a height less than 7.6 m are not likely to be so well suited to this purpose. The unit assessment must be reduced accordingly. Where good comparable evidence of a similar style exists locally, it may be that valuation can take place on an overall or gross internal area basis. In other cases a more sensitive approach may be needed. Some parts of the country have their own custom and practice, but whichever method is adopted the RICS measuring codes are to be preferred.

## Study 14

Value the Enterprise factory and warehouse which is situated in the North of England on its own site and with its own loading facilities. It comprises several buildings with different uses. You have a comparable, factory and warehouse of similar construction which was let in March 2003 on FRI terms at £90,700 pa.

*Analysis*

| Description | Area m² | 'x' factor | Total |
|---|---|---|---|
| Ground Floor | | | |
| Office | 310 | 1.1 | 341 |
| Workshop | 3,200 | 1.0 | 3,200 |
| Warehouse | 1,700 | 0.95 | 1,615 |
| | | | 5,156 |

\*   See also Chapter 5.

£90,700 ÷ 5,156 = £17.59 m$^2$ main space in RV terms

*Valuation of Enterprise Factory and Warehouse*

| Description | Area m$^2$ | 'x' factor | £/m$^2$ | Total £ pa |
|---|---|---|---|---|
| *Ground Floor* | | | | |
| Office | 390 | 1.10 | 19.35 | 7,547 |
| Workshop | 4,590 | 1.00 | 17.59 | 80,738 |
| Warehouse | 2,640 | 0.95 | 16.71 | 44,114 |
| Store | 240 | 0.95 | 16.71 | 4,010 |
| Internal storage | 55 | 0.95 | 16.71 | 919 |
| Canopy | 300 | 0.15(2) | 2.64 | 792 |
| External storage | 120 | 0.75 | 13.19 | 1,583 |
| | | | | |
| *Mezzanine* | | | | |
| Workshop | 400 | 0.2 | 3.52 | 1,408 |
| | | | | 141,111 |

RV, say £141,000

Notes
(1)   The work area determines the value of the unit and other values are related to it. Hence the weighting of 1.00.
(2)   The 'x' factor applied to this space and the mezzanine space comes from an analysis of other similar comparables, which have this space.

# Hotels*

The term hotel is quite broad and covers a variety of forms. What they all have in common is some form of monopoly eg by virtue of their position. Lettings are not that frequent and so good direct comparable evidence is rarely available. An exception to this may be found in seaside and other tourist areas where many hotels of a similar nature may exist side by side. Perhaps the best example might be the small hotel found in Blackpool, where row upon row of houses have been converted into privately run hotels. At the other end of the spectrum are five-star hotels found in city centres, through to commercial and budget hotels. These hotels are generally assessed using the profits, or receipts and expenditure method. In some cases a sufficient number of comparables enable a rating assessment to be made using comparable rental evidence. This would be the case with a number of hotels in a seaside resort such as Blackpool. Normally the receipts and expenditures prior to and closest to the antecedent valuation date will be used. If none existed, then those receipts and expenditures which do exist will have to be adjusted accordingly. Although the valuation which follows is mathematical in style and based upon the trading accounts, it must be remembered that it is actually the hereditament subject to the business which is being valued. The 'numbers' used should not cloud the valuer's judgment as to what is actually being valued: the property, and location let on a hypothetical tenancy to a hypothetical and 'reasonable' tenant. Appropriate licences should be in place and the property should be being maintained and managed properly. The occupier is not supposed to be a particular expert in the business, nor is he allowed — for rating purposes — to be incompetent, but is expected to be able to generate and maintain

---

*      See also Chapter 18.

a fair maintainable level of trade. Whether this is the case can be judged by reference to similar hotels in and around the locality. A useful guide to whether a hotel is trading well is the average occupancy rate. Depending upon the hotel type, this might be expected to be in excess of 75%. The fair maintainable trade can be judged by a consideration of the fair maintainable receipts. These would be those which existed at 1, April 2003 and would take into account trends, risks and any uncertainties.

In the past valuations would involve an analysis of valuation of the trade using the actual trading accounts. In recent years there has been a move towards a more simplified valuation based directly on gross receipts. This is also a reflection of the wider valuation world outside rating. Discussions between the VOA and agents representing the British Hospitality Association have resulted in an agreed valuation scheme. This chapter does not examine that scheme, but considers the two main options where this would not be the case.

## Study 15

A good class hotel with excellent all the year round trade. Three years' accounts have been examined and show that the trading pattern is reasonably consistent. The following figures are based on the accounts for the year ending 31st December 1992.

| | £ pa |
|---|---|
| *Gross Receipts* | |
| Bedroom letting | 450,000 |
| Restaurant | 550,000 |
| Bar | 235,000 |
| Total Income | 1,235,000 |
| *Less:* | |
| Purchases | 116,000 |
| Gross Profit | 1,119,000 |
| Working Expenses: *viz* Salaries and wages(including National Insurance), electricity, gas, water(metered), stationery, printing, postage, telephone and advertising, laundry and cleaning materials, repairs and renewals of furnishing and equipment (1) | |
| Repairs to structure (2) | |
| Insurance: structure and third party, rates payable and sundry expenses | 828,000 |
| Net Trading profit | 291,000 |
| *Less*: | |
| Interest on capital (3): | |
| Chattels | |
| Cash | |
| Stock £450,000 at 8% | 36,000 |
| Divisible Balance | 255,000 |
| *Less* | |
| Tenants share at 40% (4) | 102,000 |
| Amount left for rent | 153,000 |

RV £153,000

Notes
(1)   This figure should represent the annual amount necessary to maintain the trade at a consistent level.
(2)   Repairs and insurance of property are required to bring the rent to RV terms.

(3) In the rating hypothesis, the tenants have to provide these items from their own capital. They are therefore entitled to the interest on the opportunity cost of the capital.

(4) This share should be sufficient to induce the tenant to trade. This could be considered as being made up from a salary or remuneration and a reward for the risk associated with running the business. It is the amount that would be required for running the business. It is common for a percentage to be used.

The RV as determined by the profits method and can be analysed and valued using a shortcut valuation, known as the Double Bed Unit or DBU basis. A standard unit basis is adopted taking the double bedroom as 100% or a factor of 1.00. A rate is then applied to other bedrooms to bring them into terms of the DBU. This can be considered a similar practice to that of using the 'x' factor or zoning in the previous studies.

## Study 16

Analyse The Lodge, an average quality hotel, typical of its type, in a large sea side town with good comparable evidence, having lift access to all floors. A rent of £185,000 was agreed at 31 March 2003 on RV terms. Use the information to value the nearby Cartref hotel.

### Analysis: The Lodge

*Accommodation*

| | |
|---|---|
| *Ground floor:* | Lounge, bar (full on-licence), residents' dining room. |
| *First floor:* | 10 double, 3 single bedrooms all with sea views. |
| *Second floor:* | 8 double bedrooms and 4 family rooms. |
| *Third floor:* | 5 double bedrooms and 4 suites. |

| Floor | Bedrooms | | DBUs |
|---|---|---|---|
| First floor | 10 | double at 1.00 | 10.00 |
| | 3 | single at 0.75 | 2.25 |
| Second floor | 8 | double at 1.00 | 8.00 |
| | 4 | family at 1.25 | 6.00 |
| Third floor | 5 | double at 1.00 | 5.00 |
| | 4 | suites at 1.50 | 6.00 |
| | | | 37.25 |

£185,000 ÷ 37.25 = £4,996 per DBU.

### Valuation: Cartref hotel

*Accommodation*

| | |
|---|---|
| *Ground floor:* | Lounge, bar (full on-licence), residents' dining room. |
| *First floor:* | 8 double, 5 single bedrooms, all with sea views. |
| *Second floor:* | 7 double bedrooms and 5 family rooms. |
| *Third floor:* | 4 double bedrooms and 5 suites. |

| Floor | Bedroom | DBUs | |
|---|---|---|---|
| First floor | | 8 double at 1.00 | 8.00 |
| | | 5 single at 0.75 | 3.75 |
| | | | |
| Second floor | | 7 double at 1.00 | 7.00 |
| | | 5 family at 1.25 | 6.25 |
| | | | |
| Third floor | | 4 double a 1.00 | 4.00 |
| | | 5 suites at 1.50 | 7.50 |
| | | | 36.75 |

36.75 DBUS at £4,996 = £182,354

RV, say £182,350

Note: (1) This method relates to direct evidence and has particular value for this reason.

# Schools and nurseries

Most schools, both independent and those occupied by Local Education Authorities, are now assessed on a contractors' basis. However, some independent schools may have sufficient comparable rental evidence to be valued by rental comparison. This may be by comparison with other schools or with similar premises which are let for other uses. Hence a large Victorian house might be let for use as an office or as a private school, at similar rentals, if planning permission were available.

Particular problems will be experienced if the buildings are old, of great historical and architectural merit but functionally obsolete. In such circumstances a valuer would be well advised to have regard to *Eton College* v *Lane (VO)* [1971], which is considered later.

# Public buildings

Public buildings may be valued using normal comparison methods. For example, office accommodation, workshops and storage may be rented on commercial leases. Even where this is not the case, were they to be available to let on the open market, direct comparison methods may well be the appropriate valuation methodology. However, there will be certain public buildings, such as public museums or libraries where it would be unlikely that suitable rental evidence exists. In these cases it will usually be appropriate to use the contractors' method. This might also be the case with municipal sports facilities which at first sight may seem suitable candidates for the profits method. However, these facilities may not be run on a commercial basis, but rather to provide the facilities to the local community, and as such the trading accounts would prove unsuitable. (At the time of writing, early in 2007, the valuation approach for these types of property is the subject of dispute.)

## Study 17

A modern dry ski slope owned and operated by a local district council in an area of relatively low land values. The unit comprises a dry ski slope, mechanical lift, car parking and ancillary buildings.

*Valuation*

|  | £ |
|---|---|
| Site works including fencing, services, floodlighting, car parking and mechanical lift (1) | 390,000 |
| Building costs including all fees | 419,000 |
| Total building costs | 809,000 |
| *Add*: |  |
| Land costs: 9 acres at £140,000 | 1,260,000 |
| Total effective capital value of land and buildings (2) | 2,069,000 |
| Decapitalise at 5.0% (3)   (= £2,069,000 ÷ 20) | 103,450 |

RV, say £103,500

Notes
(1)   The extent to which certain equipment will be considered part of the rateable hereditament will be governed by the Plant and Machinery Regulations as disused at the end of this chapter.
(2)   As the ski slope is of modern construction obsolescence is minimal.
(3)   Statutory decapitalisation rate, as provided for by the Non-Domestic Rating (Miscellaneous Provisions) (Amendment) (England) Regulations 2004 (SI 2004 No 1494).

## Study 18

A modern crematorium built by a district council. The premises are working at near full capacity.

*Valuation*

|  | £ |
|---|---|
| Estimated buildings costs |  |
| Crematorium | 970,000 |
| Cremators (furnaces) | 231,000 |
| Site works | 150,000 |
|  | 1,351,000 |
| *Add*: |  |
| Land | 330,000 |
| Effective capital value (2) | 1,681,000 |
| Decapitalise at 5.0%   (= 1,681,000 ÷ 20) | 84,050 |

RV, say £84,000

Notes
(1)   Such items are rateable as plant even though they do not occur in an industrial building.
(2)   If the crematoria had surplus capacity, there would be a valid reason to reduce the assessment. For example, if there were 25% surplus capacity, the effective capital value might be reduced by 25%.

## Study 19

A leading case regarding an independent school is *Eton College* v *Lane (VO) and Eton Urban District Council* [1971] RA 186 and this study is based upon the valuation principles approved by the Lands Tribunal. Eton College comprises buildings that are old, of great intrinsic merit and interest but which are less than ideal for use as a large boarding school. The valuation approach can interestingly be compared with the Scottish example of *Aberdeen University* v *Grampian Regional Assessor* [1990], LTS, a rather detailed consideration of the problems encountered in the valuation of a university.

*Valuation*

| | £ | £ |
|---|---:|---:|
| *Buildings* | | |
| Estimated replacement costs | 16,000,000 | |
| *Less:* | | |
| Allowances for age and obsolescence factors, say 63% | 10,080,000 | |
| Effective capital value | 5,920,000 | 5,920,000 |
| *Site* | | |
| 22.67 ha at £40,000 per ha | 906,800 | |
| *Less:* 63% | 571,284 | |
| | | 335,516 |
| Effective capital value of land and buildings | | 6,255,516 |
| *Playing fields* | | |
| 80 ha at £5,000 per ha | | 400,000 |
| Total effective capital value | | 6,655,516 |
| Decapitalise at 3.67% (NB see Note 1) | | 244,257 |
| *Deduct*: end allowance 10% | | 24,426 |
| | | 219,831 |
| RV, say £219,830 | | |

*Note*

(1) This was the decapitalisation rate used at the time, not that provided by the Non-Domestic Rating (Miscellaneous Provisions) (Amendment) (England) Regulations 2004) (SI 2004 No 1494).

# Cinemas

The birth of the multiplex cinema heralded a new age for cinemas which had been in decline. This type of hereditament is specialised and it is unlikely that the average valuer will be required to undertake such a rating valuation. The VOA agreed a valuation scheme having regard to the fair, maintainable Gross Receipts, from all sources (exclusive of VAT) at the antecedent valuation date. A scale giving the RV per seat was agreed. All seats, including disabled spaces are to be calculated.

## Study 20

Value a multiplex cinema built near the Home Counties with 8 screens.

|  | Seats |
|---|---|
| Screen 1 | 150 |
| Screen 2 | 175 |
| Screen 3 | 210 |
| Screen 4 | 350 |
| Screen 5 | 210 |
| Screen 6 | 210 |
| Screen 7 | 350 |
| Screen 8 | 175 |
|  | 1,830 |

1830 seats at £275 (1) = £503,250

RV, say £503,000

Note:
(1)    From table agreed between VOA and industry representatives.


# Petrol filling stations and garages*

These valuations are, again, unlikely to be encountered by most valuers. Petrol filling stations have experienced a substantial change over recent years. Environmental controls have increased running costs, and supermarkets have further increased their share of the market. Large numbers of stations have closed as a result and some partnerships have emerged between the petrol companies and supermarkets. The modern petrol filling station is usually associated with a supermarket, or a convenience store is associated with the petrol filling station. Although throughput is fundamentally important, the rental will be expected to comprise the elements present in the hereditament namely: petrol sales, car wash, shop sales and any other uses such as associated buildings.

The petrol sales element is based on a consideration of the maintainable throughput. This would be the total annual sales of all fuels, without distinguishing between different types of unleaded petrol or diesel.

The gross profit to the occupier produced by the petrol sales may be calculated by multiplying the throughput by the dealer's margins. In such circumstances it is worth noting that the profit per litre made from say super unleaded, unleaded petrol and derv may be different. Care should accordingly be exercised when comparing stations with broadly similar throughputs, as the make up of those throughputs may affect the profitability.

Car wash plant is not rateable and is notionally cleared from the site. It is the potential of the site for car wash use which is valued by reference to potential turnover.

---

*    See also Chapter 14.

## Study 21

A filling station on a main road without apparent competition. The opening hours are 06:00 to 22:00 hours each day of the except Christmas Day.

| | £ | £ |
|---|---|---|
| Petrol (Forecourt) sales | | |
| 3,525,000 litres at £2.80 per 1,000 litres (1) | | 9,870 |
| Shop | | |
| Turnover at the antecedent valuation date (2) | £570,000 | |
| Rental element at 4% | | 22,800 |
| Car wash | | |
| Turnover (3) | 22,000 | 3,000 |
| | | 35,670 |

RV, say £35,600

Notes
(1) This throughput is considered to be the fair maintainable throughput. This is the figure that would be achieved by a competent operator and would have regard to competition and other factors such as traffic flow and road proposals. Annual throughputs of less than 2,500,000 litres are substantially less profitable. Ideally one might expect a throughput in excess of 4,000,000 to be certain of it being maintainable.
(2) This would also be based on a consideration of the fair maintainable trade that would be expected in the hands of a competent operator. It is interesting to note that, although at first sight customers might all be expected to be motorists, in reality a substantial element of the trade can come from pedestrians, nearby residents and young people.
(3) This figure would again be based on the fair maintainable turnover at the antecedent valuation date, which would be based on an analysis of the preceding years.

# Plant and machinery

Plant and machinery is, again, a specialist valuation area of rating. Plant and machinery may form part of the rateable hereditament in many types of property, not only industrial hereditaments. It is only rateable if its inclusion is in accordance with the relevant Plant and Machinery Order in existence at that time. Certain plant is valued in so far as it adds value to the hereditament. This would be the case with plant and machinery falling within Classes 1 to 3, for example heating. As a result, such valuations should normally be within the capability of a reasonably competent valuer. However, plant and machinery falling within Class 4 would normally be valued as an addition in its own right. Such valuations would be on a cost basis, contractors' valuation and would require a specialist knowledge of the machinery and costs.

For the 2005 valuation, the appropriate legislation was the Valuation for Rating (Plant and Machinery) ( England ) Regulations 2000 (SI 2000 No 540). It is interesting to note that windmills are still in the lists. When the first edition of this text was published they were in decline. With the drive for renewable energy they are enjoying a resurgence. This is reflected in the Order now including aero-generators, wind turbines, solar cells and solar panels. Detailed tables of the plant and machinery to be considered are included within these regulations and consequently are not stated here. However, the main elements are:

- Class 1: Power plant
  Eg Steam boilers, dynamos, storage batteries, transformers, cables, water wheels, compressors, windmills, wind turbines, solar cells.

- Class 2: Services
  Eg  Heating cooling and ventilation, lighting, draining, supplying water, protection from hazards.

- Class 3: Conveyors and pipelines
  Eg railways, lifts, cables, poles and pylons, towers, pipe lines)

- Class 4: Plant which is in the nature of a structure
  Class 4 contains two tables of plant and machinery Table 3 and Table 4. Plant and machinery in Table 3 is automatically rateable. Plant and machinery in Table 4 is not included if its

  total cubic capacity (measured externally and excluding foundations, settings, supports and anything which is not an integral part of the item) does not exceed four hundred cubic metres and which is readily capable of being moved from one site and re-erected in its original state on another without substantial demolition of any surrounding structure.

  Four hundred cubic metres is about the size of a modest two storey house.

Plant and machinery in class 4 must be considered to be in the nature of a structure, and reference should be made to the relevant case law. In all other cases the plant and machinery is automatically rateable by virtue of inclusion in this order.

The value of plant and machinery may be included by an increase in the unit price, as with central heating. In other cases it may be included by taking an annual equivalent of the depreciated replacement cost. This usually requires specialist knowledge. The current decapitalisation rate is 5%, but this is subject to amendment.

© MR Jayne 2008

# Development Properties

13

Development property is that which has the potential for a more valuable use. This may be by a complete redevelopment of a site or a partial redevelopment and refurbishment, or the refurbishment of an existing building. When valuing an existing building by the investment method, it is conventional to capitalise the rent payable by the tenant(s) into perpetuity. It is increasingly apparent that the widely held assumption that property can be valued in its existing use into perpetuity is incorrect. A profitable redevelopment scheme is frequently possible after as little as 25 years into the life of a building. Redevelopment or refurbishment requires the investment of capital to show a return appropriate to the risk accepted and the growth potential. It may be prompted by a variety of market conditions and some of these are considered in the studies below where reference is made to the terms feasibility and viability. Sometimes these terms are used synonymously but it is correct to distinguish between them.

Feasibility implies that it is possible to complete a redevelopment scheme whereas viability suggests that it can be completed profitably. The question remains, what is an acceptable level of profit for a development scheme? This will depend upon the risk involved in undertaking the project from the viewpoint of the party involved. A long-term funder such as a Life Assurance fund may take a different view of viability from that of a developer primarily concerned with realising a short-term profit in contrast to the fund's long-term interest. This will be further considered in the valuation studies below, but an important point is that no measure of return can be assessed in isolation. Measures of return are only meaningful when compared to returns from other, competing, media all of which have different risk factors. Property development is risky in varying degrees for all the parties involved, and the returns must compensate for the risk accepted.

## Planning policy and legislation

Policy changes such as permitting changes between uses may make a refurbishment scheme feasible. The Town and Country Planning (Use Classes) Order 1987 and the Town and Country Planning (General Permitted Development) Order 1995 have both undergone amendment over the years. One of the results of this legislation is that buildings in General Industrial Use B2 can change to Business Use B1, and, where office use is more valuable, refurbishment or redevelopment may be initiated. The more recent revision to Planning Policy Guidance Note 3 (PPG 3) (now replaced by Planning Policy Statement 3) has resulted in developers promoting high density residential schemes in areas where consent would previously not have been forthcoming. The proposed loosening of planning control in respect of out-of-town retail development proposed in the second Barker report may have similar effects for sites on the periphery of major settlements if its recommendations are followed up in the promised 2007 White Paper.

## Tenants' requirements

The value of any property depends upon its usefulness to the end user and as tenants' requirements change refurbishment or redevelopment may become feasible. Multiple retail tenants, for example, often require units which are far larger than the standard shops that they once were happy to occupy. Shopping centre refurbishments have been prompted by such market changes.

## The institutions

Life Assurance Funds and Pension Funds dominate ownership of stock market quoted shares and define what is to be regarded as a "prime" property. They are often the providers of both short and long term funds for property development, and as such have a major influence over matters such as specification, location and tenant's covenants.

## Macro economic conditions

Property development is very sensitive to changes in economic conditions such as Minimum Lending Rate (MLR), economic growth, inflation rate and the rate of change of incomes. The major factor in causing house prices to rise is an increase in national income and this will result from higher economic growth. Property developers have been described as 'price takers' in the sense that they can have no influence on rents and yields but can only respond to economic signals to build or refurbish when it appears viable. This is not exactly accurate if a developer has a large land bank and can, to some extent, control the rate of development and the price (or rent) of the product. One problem for developers is that the product they supply is price inelastic whereas economic conditions can change very quickly in the short run. The result may be that a developer who undertakes a scheme late in an economic boom may be caught in changing economic conditions of rising inflation and interest rates. This in turn may lead to the retail tenant market softening or house prices falling for a time. There were many examples of developers being surprised in this way in the boom and subsequent recession of the late 1980s and early 1990s.

## Infrastructure improvements

Factors such as motorway extensions and new roads can influence the timing of development. Similar effects can be seen in the establishment of a new rail head or an airport extension. The extension of the M3 motorway to Sunbury resulted in development in the area becoming viable as the potential value of new development rose on the back of increasing tenant demand. Without the expansion of Heathrow airport the redevelopment of the 600 acre Slough industrial estate (owned by Slough Estates plc) would not have proceeded with the speed that it has.

# Valuation purposes

Valuations of land with development potential are carried out for a variety of reasons depending on the incentives of each party involved with a particular site. The main purposes of valuations of this type are:

(i)   to establish the open market value of a site with development potential
(ii)  to predict the developer's profit where a site has been acquired
(iii) to arrive at a ground rent and rental shares where a partnership scheme is envisaged
(iv)  to arrive at a maximum build cost where capital values can be predicted with some certainty.

Development valuations are also used to appraise complex development situations where, for example, part of a site is purchased for development with the remainder being initially retained and then sold in its existing use.

There is a fundamental difference between a valuation and a financial appraisal, although both are associated with development land. An open market valuation is carried out with no particular purchaser in mind. The eventual purchaser of the land is hypothetical and there are no restrictions as to who might bid. A development appraisal, in contrast, is prepared for a particular client who may well have already acquired the site or may have secured it in some way, perhaps through an option agreement or a conditional contract to purchase. It is therefore not a valuation but a calculation of worth. In the purposes outlined above (i) is a valuation and (ii) is an appraisal whereas (iii) and (iv) may be valuations or appraisals depending on whether the site has been secured by a client and that client's particular circumstances are being taken into account. An appraisal, for example, will reflect a client's funding arrangements, market and investment sentiment, and any design or planning work that has been completed. The level of profit predicted by a development appraisal is nevertheless a residual figure as it is arrived at by subtracting development cost from a net development value. For this reason the calculation is sometimes referred to as a "residual appraisal" although for the sake of clarity it should be called a development appraisal. Advice provided by the Royal Institution of Chartered Surveyors (RICS) (see below) draws a clear distinction between valuation and development appraisal.

# The valuation of land with development potential

Land with development potential is valued in two ways; the comparative method and the residual method. The former is problematical as no site is likely to be helpfully comparable with another. There are likely to be significant differences in the type of development, size of site, planning, infrastructure requirements, market differences and timing of development. For these reasons the method usually adopted is the residual. This requires the valuer to envisage the most valuable building(s) that could be developed on the site and to value this scheme as though it existed. From this value is deducted all the costs of development, such as construction cost, letting fees planning fees. Also deducted are developer's profit and the amount incurred in borrowing all the development monies for the duration of the project. The resulting value, when adjusted for the charge to interest arising from buying and holding the site for the period of the project, is the land value. This land value does, of course, reflect the assumptions that the valuer has made about the size and use of, and the rental income from, the building to be developed.

# Professional guidance

In addition to the material in the RICS Red Book — *Valuation Standards* — guidance is to be found in the RICS's Valuation Information Paper 12: Valuation of Development Land, published in March 2008. This deals with the subject under the following heads:

- Introduction
- Establishing the facts
- Assessing the development potential
- Valuing by the comparison method
- Valuding by the residual method
- Assessing the land value
- Reporting the valuation
- Conclusion.

The valuer is recommended to retain records to support the assumptions made in case it is necessary to justify the valuation at a later date. The advice contained in the guidance note is followed in the residual valuation used in Study 1 below.

# Preliminary investigations

Before preparing a residual valuation the valuer will carry out preliminary investigations and make some overall assumptions. These will be stated in the valuation report to the client.

# Site investigations

It is axiomatic that the valuer should inspect the site as part of the valuation process. However, it is common for development surveyors employed in-house by property companies to carry out valuations on very thin site knowledge to discover if current rent and yields in the location justify any further interest in the proposition. It will be appreciated that rent and yield must be at a sufficient level if any profit is to be produced from the scheme.

If the site is located in a location which appears viable, then it is worthwhile carrying out initial investigations. The word "viable" in this context is synonymous with "profitable", whereas "feasible" can be taken to mean "practicable". However, as with many terms used in the development industry "viable" and "feasible" are often regarded as synonymous.

Site boundaries and areas should be rigorously checked by reference to the Ordnance Survey map at an appropriate scale. It is probably not necessary to physically pace out the site (an activity favoured by a property company chairman and previous employer of the author) but agents' particulars can be inaccurate. Boundaries can be indistinct or unmarked, and encroachment or any evidence of a right of way must be noted. Any restrictive covenants which encumber the title should be investigated and the risk discussed with the purchaser's solicitor. If a site is acquired and is encumbered with a restrictive covenant it may be possible to remove it on application to the Lands' Tribunal. In other circumstances it may be necessary to buy out the covenant from the holder, and sometimes a covenant may safely be ignored. Boundaries often pose a particular problem whether the site is registered or unregistered. The Land Registry office copy plan may well be vague and the property deeds, in the case of unregistered and, similarly so. Inspection may show such features as a hedge, ditch or bank between the site boundary and an adopted road. Again this must be noted, as the valuer will assume that access is available to an adopted highway. If the site is eventually purchased, this problem might be overcome by a vendor's indemnity (if the vendor's covenant is of sufficient strength), and evidence such as a statutory declaration by a long standing resident may be enough to enable the purchaser to secure insurance.

Before purchase, the purchaser will arrange for trial pits and bore hole surveys to be undertaken, but the valuer should note any early evidence of subsoil problems. The presence of a previous industrial use, or irregular topography, may indicate the presence of contamination or recent fill. If the site slopes away from an adopted highway it is likely that foul and surface water drainage will have to be pumped to the sewers in the highway. The valuer should allow for this cost in the valuation and also the effect it will have on site density.

For an open market valuation, such as a residual, the valuer must envisage the scheme which will produce the greatest development value. A zoning in a development plan will be helpful as section 38 of the Planning and Compulsory Purchase Act 2004 states that any development proposed must be made in accordance with the development plan unless there is a "material consideration" which suggests otherwise (and this is unlikely). The valuer in these circumstances is presented with a use which is overwhelmingly the most likely, and should value the site on that basis. It is more problematic if the type of development on the site is unclear, perhaps as a result of an imminent revision to the development plan. The valuer is seeking to establish the value of the site from the viewpoint of a hypothetical developer, and any bidder would discount the site value for the period of time that it will reasonably take to achieve planning consent. Using judgment and market knowledge this is taken into account in the valuation, and an appropriate discount employed to reduce the site value. It may be that an alternative development may result in an immediate planning consent and this produces a higher value. In this case the alternative development would be used in the valuation. All these assumptions must be stated in the valuation report and a perusal of the development plan documents together with a meeting with the planning officer dealing with the site will provide evidence of such matters as plot ratio, site coverage, car parking requirements and landscaping. Fortunately the frequent unavailability of planning officers has been offset by the ready availability of development plan documents via the internet.

Other matters to be considered include the possibility of improvements being necessary to off site roads by way of an agreement under section 278 of the 1980 Highways Act. Similarly, it may be probable that the developer will be obliged to build for community uses under Section 106 of the Town and Country Planning Act 1990 in order to achieve planning consent. For a residential site, the planning authority will require some social or affordable housing to be constructed and this must be allowed for in the valuation. If the local authority require a particularly high provision of the housing to be affordable, say more than 20%, this may affect site value. Again, local market knowledge will be used to adjust the probable bid for the site. In essence, the effect of adjacent cheap homes on the market value of the more expensive dwellings must be estimated, and this will depend on the location of the site, the type of neighbourhood, and the degree of integration on site.

# Time scale

A residual valuation is prepared using present day values and costs with no allowance for increases during the development period. Any cost associated with the proposed development is charged to short term interest until the scheme is completed and notionally sold. The valuation assumes that the site is purchased, the scheme built and let, and the resulting investment sold to repay capital and short term interest. Money remaining is the developer's profit. In preparing the valuation, every type of cost must be predicted and charged to interest. This will require the preparation of a programme, as it is assumed that no capital or interest is repaid until the scheme is eventually sold. Short term interest is said to "roll up" in the development period. In Study 1, below, a simplistic programme is adopted to illustrate the technique, but the study must be read in conjunction with the notes that follow it in order to appreciate the full complexity of the calculation. Development projects are planned using network analysis which will show the logical progression of activities and the way in which activities influence the start and finish dates of other parts of the project. The time scale adopted for study 1 is based on a simple bar chart as a cost loaded network can only be properly prepared with a computer programme.

## Study 1

A site of 0.512 ha (1.28 acres) on the outskirts of a major town in the south east of England is to be valued using the residual method. The site is zoned for B1 (offices) development in the local plan and discussion with the local planning authority reveals that a plot ratio of 2:1 would be appropriate. A possible pre-let tenant for the whole of the building has been identified. A site investigation report prepared for the existing freehold owner shows that that the sub strata are clay, thanet sand and chalk. The substructure of the building will require piling but the investigations have revealed no unusual problems in the ground. All mains services are readily available.

| Residual valuation | £ | £ |
|---|---:|---:|
| Rental income (1) | | |
| 8,280 m² at £300 per m² | 2,484,000 pa | |
| YP perp at 7% (2) | 14.29 | |
| Gross development value (GDV) | | 35,496,360 |
| *Less* Costs of disposal at 5.5% (3) | | 1,850,521 |
| Net Development Value (NDV) | | 33,645,839 |

| Expenditure | £ | |
|---|---:|---|
| Construction and fees on construction (4) | | |
| 9,936 m² at £1,610 per m² | 15,996,960 | |
| Architects and QS fees at 12.5% | 1,999,620 | |
| Sub total | 17,996,580 | |
| Contingency at 5% | 899,829 | |
| Site preparation and surveys (5) | 200,000 | |
| Statutory costs (6) | 300,000 | |
| Marketing (7) | 100,000 | |
| Letting agents fees at 15% of rent (8) | 372,600 | |
| Legal costs on lettings etc. (9) | 100,000 | |
| Developer's profit at 20% of GDV (10) | 7,099,272 | |
| Short term finance (11) | | |
| Construction + contingency | | |
| ¹/2 x £18,896,409 at 8% for 1 year | 755,856 | |
| Site preparation and surveys | | |
| £200,000 at 8% for 1.25 years | 20,000 | |
| Statutory costs £300,000 at 8% for 1 year | 24,000 | |
| Marketing £100,000 at 8% for 1 year | 8,000 | |
| Sub total Costs | 27,876,137 | |

Interest in disposal period (12)

| | | |
|---|---:|---:|
| £27,876,137 at 8% for 3 months | 557,523 | |
| Total Cost | 28,433,660 | −28,433,660 |
| Gross site value (13) | | 5,212,179 |
| (NDV less total cost) | | |
| PV £1 in 1.25 years at 8% | | 0.90878 |
| Net site value (14) | | 4,736,724 |
| Say (15) | | £4,600,000 |

## Notes

Rental income (1) is assessed by estimating the net internal area of the building to be constructed. If an architect's sketch scheme is available measurements can be taken using the RICS Code of Measuring Practice. Present day rents will be used, with no allowance for rental growth during the development period. When the rent has been calculated it is then capitalised into perpetuity using a yield and year's purchase derived from market evidence (2). Although the site value is rarely derived from the comparative method, rental income and yield are, and the valuer must allow for variations presented by any comparable evidence. Matters such as the proposed floorplate, specification, massing on site, lot size and, of course, location, will all be influential.

The gross development value (GDV) derived from the rent and yield is the present day value of the building which will be constructed. From this is deducted the costs involved with selling the building at the end of the project (3). The residual assumes that long term funding will be provided by disposing of the property on the investment market. Obviously if the property is one which is institutionally acceptable (in terms of location, tenant, lease terms and specification) the gross development value will be higher. Costs to be deducted are taken at 5.5% of the GDV and include agents' and solicitor's fees involved with the disposal, as well as the purchaser's structural engineer's fees. If the purchaser is assumed to be an institution, it is conventional for the developer to take responsibility for the fees of the purchaser's consultants.

It may not be the developer's intention to sell the building on completion. Other methods of long term funding may be envisaged such as a commercial mortgage or profit erosion. The residual is, however, prepared for a hypothetical developer and the value of the completed building is the starting point.

Expenditure involved with constructing the building is now calculated. In practice it is likely that the developer will know a quantity surveyor who will provide an estimated building cost and time scale in the expectation of being appointed to the project should it proceed. Alternatively a publication such as Spons may be used or figures provided by the Building Cost Information Service (BCIS). The figures provided will be taken on the gross internal area of the building and valuers often use a spot figure per $m^2$ if little, if any, design has taken place. Building cost figures provided by the various publications are, however, difficult to interpret and there will be allowances for geographical location as well as a multiplicity of options for building services and specification. An estimate from a quantity surveyor is always to be preferred.

The fees of the professional consultants are conventionally taken as a percentage of the building cost as this is how the consultants' standard agreements calculate them. The 12.5% in the calculation represents the fees of the architect, quantity surveyor, services engineer and structural engineer but it may be apparent, even at a preliminary stage, that additional consultants will be required. A planning consultant may be necessary to prepare an environmental impact study or there may be issues with noise or effluent which require specialist consultants. If there is a probability that additional consultants will be required this should be reflected in the calculation.

Statutory costs (6) may be necessary if fees need to be paid to statutory authorities for foul water connection, surface water run off or other matters.

Marketing costs (7) will cover the preparation of a brochure and advertising although at this early stage they will be difficult to estimate. The costs associated with other similar schemes are an obvious starting point. The letting agents' fees (8) represent two joint sole agents, each being paid 7.5% of one year's rent. Again, this is conventional, but if the property will require specialist letting expertise in addition to the two agents, this should be allowed for. For example, a shopping centre in an uncertain letting market may need additional marketing advice from a specialist.

Legal costs on letting (9) are solicitor's costs involved with the occupational leases. Where complex pre-letting agreements are required these may be substantial.

Developer's profit (10) is normally included as a percentage of the GDV, or sometimes as a percentage of cost. The level of profit should represent the return that a hypothetical developer will require for undertaking the project and should be commensurate with the risk involved. The broad brush nature of the figures in the residual usually means that 20% is used as an acceptable return for all projects in the absence of a fully researched risk analysis. Levels of return are only meaningful as comparative figures, and it is suggested that the level to be included in the residual should relate to the risky nature of development and to the length of the project. The study envisages a project where construction takes one year, with a total project time of 1.25 years. 20% is therefore roughly the return over one year and could be compared with other investment media such as equities and gilts. If the project was of three years duration, the return should reflect the time period and should be more than 20%. If, for example, the gross redemption yield on short term gilts was 8% pa, a developer might wish to see a return of 12% or 16% pa depending on the risk of the project. If, for example, a figure of 20% of GDV was included for profit and the project duration was three years, the return pa is 6.266% (take 3rd root of 1.20 − 1 × 100) which would probably be regarded as inadequate. This point is further considered later in this chapter.

Short term finance (11) is the amount of interest that the developer will incur in borrowing all the finance required to construct the building with all associated costs. Finance on the site cost is reflected later in the calculation when the site value is reduced to account for the costs of holding the it (14). It is assumed that the developer borrows all the finance at the short term rate of interest (taken here at 8% pa), and repays it when he refinances by selling the complete, and let, building. The developer is assumed to be 100% debt financed but in reality this is never the case as no business that is funded in such a way can hope to survive. Although the developer is going to be partly financed by equity in terms of the valuation, it is appropriate to include finance in this way as the opportunity cost of using equity will approximate to the short term rate of interest. It will be seen from the calculation that simple interest (for the sake of simplicity) has been used when in reality the developer would agree a facility with a bank (or sometimes with the long term funder) which will allow the drawdown of funds as they are required up to a set limit. It is conventional for banks to charge interest at quarterly rests, so, for example, if the developer is quoted 10% p.a. this will equate to 2.4% per quarter (4th root of 1.10 − 1 × 100). The bank will charge interest every quarter at 2.4% on the total borrowings of capital and accumulated interest. For the calculation to be as accurate as possible, short term interest should be calculated in this way using a spreadsheet program with best estimates of capital drawdowns included. The total charge to interest can then be calculated.

For short term finance to be accurately calculated the first requirement is a draft program of activities. The residual is prepared before the site is acquired and the program, although based on the best information available, is likely to be very broad brush. In reality, matters such as contract strategy, pre-let tenant requirements, and contractual delays will have a major influence on the program and the interest charged. The program on which the study is based is illustrated in Figure 13.1.

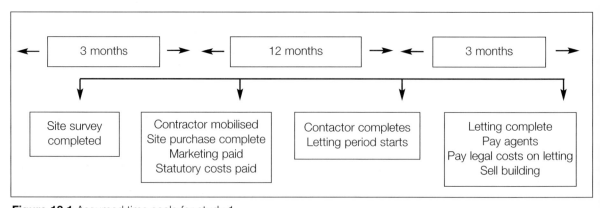

**Figure 13.1** Assumed time scale for study 1

It will be appreciated that the residual has been prepared using a time scale which is likely to change considerably in reality. No account has been taken of the following matters:

(a)  a deposit (10% of purchase cost) will be paid on exchange of contracts for the site with the balance on completion
(b)  a retention (usually 5%) is made from the monies paid to the contractor during the remediation period when defects to the building are corrected
(c)  there is likely to be a longer period following the letting of the building before disposal takes place. If the purchaser is an institution a structural survey will be required as well as inspections by other professionals employed by the fund
(d)  the way in which interest is charged on the building contract is approximated both for the contractor's payments and the design team fees and this is further discussed below
(e)  for the sake of simplicity, no allowance has been made for the cost of obtaining planning permission and building regulation approval.

Other matters are subject to change as the project proceeds but at the time the residual is prepared estimates of cost, rent and program are of necessity broad brush. The construction cost and the contingency are outstanding for one year but as the contractor is usually paid monthly, based on quantity surveyor's certificates, the interest is calculated for an approximation of half the total period. It is also assumed that the contingency (for matters which are unforeseen and unforeseeable) is spent in its entirety during the contract. The spend on a building contract will rise during the contract period as an S curve. At the commencement of the contract the contractor will erect huts, move plant onto site, construct builder's roads and clear the site. The middle part of the contract will be when the fastest spend occurs as the substructure and superstructure are completed. The final part of the contract is concerned with finishing trades, which in comparison require a slow rate of spend. Hence expenditure on the contact rises as an S curve but the interest calculation approximates this as a straight line at 45 degrees to the horizontal. It is also assumed that the construction professionals are paid as the contractor is paid, but this is not strictly the case. During the preliminary part of the design period the architect and the team will be working on a time basis and paid accordingly. It is common practice for this time based fee to be subsumed into the percentage fees which are paid for later design stages, so the team does not receive time fees in addition (unless work needs to be redone at the client's behest). During the preparation of the working drawings the team is paid on a percentage basis, and the residual assumes that they are paid at the same rate as the contractor. This is not exactly correct as consultants' percentage fees are front end loaded with the architect, for example, being paid 70% of his or her percentage fee before work begins on site.

When the total interest charge has been calculated this is added to the capital expenditure and the whole is charged to interest for three months to account for letting and disposal. (12) This total cost figure is then deducted from the net development value to give a gross site value (13). No account has been taken of the interest charge on the site purchase and therefore this figure overstates the site value. Interest on site purchase is calculated by discounting the site value for the period of time that the site is owned by the developer up to the disposal of the building. The short term rate of interest is used to give a net site value (14). This is conventionally approximated (15), "say", to a round figure which further emphasises the nature of the residual.

# Development appraisal

Once a site has been purchased, it is necessary to be able to predict the profit for the developer as the details of the project become more certain. A development appraisal is prepared to reflect the particular circumstances of the developer and is thus a calculation of worth rather than an open market valuation. In the development appraisal, all costs are deducted from the capitalized rental value of the building, but, as these costs now include the site cost, the residual figure is the profit from the scheme. Sometimes the development appraisal is called a residual appraisal for this reason. The appraisal is prepared with the benefit of better information than the residual valuation because the

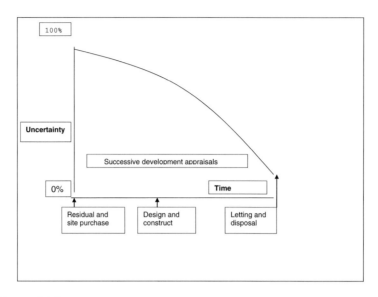

**Figure 13.2** Project uncertainty

scheme will have progressed. Design will be more certain, as will rents. Any unusual costs will have been recognised and a pre-let tenant may have signed an agreement to lease, with consequent changes to the scheme incorporated. A diagram illustrates the process (see Figure 13.2).

In Figure 13.2 project uncertainty is shown to reduce as the project progresses through the design stage. A residual valuation early in the process will provide the necessary reassurance to proceed to a site purchase. As uncertainty reduces successive development appraisals predict the eventual profit from the scheme. An example of a typical development appraisal is shown below in conventional format and is followed by an appraisal calculated by discounted cash flow using the same basic information.

## Study 2 Conventional Development Appraisal

The project in this case is an office building of 8,000 m² gross internal area and 6,400 m² net internal area. Other project details are as follows.

| | | |
|---|---|---|
| Rent from pre-let tenant: | £250 per m² | Note (1) |
| Yield: | 7.5% | Note (2) |
| Site purchase cost: | £6 million | Note (3) |
| Acquisition cost: | 5.75% | Note (4) |
| Construction cost: | £1,000 per m² | Note (5) |
| Contingency: | 5% | Note (5) |
| Section 278 works: | £30,000 | Note (6) |
| Preliminary surveys: | £10,000 | |
| Architect's fees: | 5% | |

| | | |
|---|---|---|
| Quantity surveyor's fees: | 2% | |
| Project Manager's fees: | 2% | |
| Planning fee: | £10,000 | |
| Engineer's fees: | 4% | |
| Agents' fees: | 12% | Note (7) |
| Estimated development period: | 10 months | Note (8) |

*Development appraisal*

| | | £ | £ |
|---|---|---|---|
| Rental income | | | |
| 6,400 m² at £250 per m² | | 1,600,000 | |
| YP perp at 7.5% | | 13.33 | |
| Gross development value | | | 21,333,333 |
| Less purchaser's costs at 5.437% | | | 1,159,968 |
| Net Development Value | | | 20,173,365 |

| | £ | | £ |
|---|---|---|---|
| Expenditure | | | |
| Site costs | | | |
| Purchase price | | 6,000,000 | |
| Acquisition cost at 5.75% | | 345,000 | |
| Total site cost | | 6,345,000 | |
| | | | |
| Development cost | | | |
| Construction cost | | | |
| 8,000 m² at £1,000 per m² | | 8,000,000 | |
| Contingency at 5% | | 400,000 | |
| Section 278 works | | 30,000 | |
| Preliminary surveys | | 10,000 | |
| Total construction cost | | 8,440,000 | |
| Fees on construction: | | | |
| Design team at 13% + | | | |
| £10,000 planning fee | | 1,102,000 | |
| Letting and legal fees: | £ | | |
| 12% of 1 year's rent | 192,000 | | |
| Sale agent's fee at 0.71% | 151,300 | | |
| Total letting cost | 343,300 | 343,300 | |
| Finance: | | | |
| Site etc at 6.21% for 12 months | 394,153 | | |
| Building cost and fees: | | | |
| 9,460,000 at 3.06% for 12 months | 291,910 | | |
| Total finance costs | 686,064 | 686,064 | |
| Total cost | | | −16,916,364 |
| Profit | | | 3,257,000 |

| | | |
|---|---|---|
| Profit on cost | 19.25% | Note (9) |
| Profit on NDV | 16.15% | Note (10) |
| Development yield | 9.46% | Note (11) |
| Profit on GDV | 15.26% | Note (12) |

Notes

(1)    The rent from the pre-let tenant has been agreed at this level. A pre-let tenant is often a pre-condition of obtaining short term finance, especially if a forward funding arrangement has been agreed as a package deal with the short term funder. Subject to negotiation, it is possible that the rent during the development period would increase, using an acceptable index, although the letting market would have to be in the landlord's favour before this would be agreed to.

(2)    The yield at 7.5% sets the capitalisation rate for the sale of the building upon completion. As in Study 1, the appraisal assumes that long term funding will be provided by selling the building to a long term investor. The yield is derived from market evidence.

(3)    The site cost was the unknown figure in the residual valuation, but in the appraisal the site has been purchased and the residual figure will be the profit.

(4)    Acquisition costs will be known and are included at their actual level.

(5)    The construction cost is low and reflects the extent of the fit out works to be undertaken by the tenant. In view of the uncertainty surrounding the tenant's works perhaps a 10% contingency rather than 5% should have been used. An agreement will have to be struck in the lease about the status of these works and whether they will be charged in the rent upon rent review.

(6)    These works result from an agreement with the Highways Authority for the developer to provide road improvements off site to cater for additional traffic caused by the scheme. The relevant legislation is section 278 of the Highways Act 1980.

(7)    Agents' fees now state the reality rather than the estimate that was used in the residual valuation of study 1.

(8)    The development period has been approximated to 12 months although in reality, as can be seen in the discounted cash flow below, the period is 10 months. Without the use of a spreadsheet it is not practicable to calculate finance on a monthly basis. Interest on construction expenditure has been taken at approximately one half of the short term rate of interest as the construction cost does not have to be borrowed as a lump sum. In the residual, the interest was charged for one half of the contract period but it will be appreciated that mathematically the calculation shows the same result.

(9)    Profit on cost shows the return on the capital employed in the scheme, and as it is a capital return achieved over one year it can be easily compared to competing investment media.

(10)  Profit on NDV shows the return in comparison with the capital sum raised, after costs of disposal.

(11)  The development yield (9.50%) shows income as a percentage of total cost and when the investment yield (7.50%) is deducted it represents the developer's annual profit (2.0%).

(12)  Profit on GDV shows the return in comparison to the actual value of the building created with no allowance for disposal costs.

For illustrative purposes some costs have been omitted from the development appraisal and some numbers rounded up. Nothing is included for marketing or a brochure for example and, again the timescale is very simplistic. The figures can now be used in a discounted cash flow which will provide a more accurate profit figure as timing can be taken more precisely into account.

## Study 3  Discounted Cash Flow Appraisal

| Month | Site acquisition | Construction | S278 + preliminaries | Fees | Agent + letting fees | NDV | Net cash flow | PV £1 | DCF |
|---|---|---|---|---|---|---|---|---|---|
| 0 | *6 million* *+345,000* | | | | | | 6.345 million | 1.0 | 6.345 million |
| 1 | | *840,000* | *10,000* | *110,200* | | | 960,200 | 0.994 | 954,431 |
| 2 | | *840,000* | | *110,200* | | | 950,200 | 0.988 | 938,816 |
| 3 | | *840,000* | | *100,200* | | | 950,200 | 0.982 | 933,175 |
| 4 | | *840,000* | | *110,200* | | | 950,200 | 0.976 | 927,568 |
| 5 | | *840,000* | | *110,200* | | | 950,000 | 0.970 | 921,994 |
| 6 | | *840,000* | | *110,200* | | | 950,200 | 0.964 | 916,454 |
| 7 | | *840,000* | | *110,200* | | | 950,200 | 0.959 | 910,948 |
| 8 | | *840,000* | | *110,200* | | | 950,200 | 0.953 | 905,474 |
| 9 | | *840,000* | | *110,200* | | | 950,200 | 0.947 | 900,034 |
| 10 | | *840,000* | *30,000* | *110,200* | *343,300* | 20.173365 million | 18.8498 million | 0.941 | 17.747391 million |
| | | | | | | **NPV** | **3,093,499** | | **3,093,499** |

Note: In this table, negative cash flows are shown in italics.

Notes
Some assumptions and changes have been made to the discounted cash flow (DCF) above in comparison with the conventional appraisal. Preliminary costs of £10,000 have been added in month 1, and as the cash flow has been constructed on a monthly basis it is assumed that the short term finance is borrowed with compound interest added every month rather than each quarter. The DCF discounts all flows of money back to the present day, and as such is a true indication of return. Whereas the conventional appraisal approximated the development period to 12 months, the DCF is able to show the return over a realistic 10 month period. The construction cost is assumed to rise as a straight line over the contract period but if the information was available this would be amended to show expenditure rising as an S curve.

The discount figure chosen is the short term rate of interest of 6.215 % pa, and as the cash flows are monthly this figure is reduced to a monthly rate of 0.6% over the 10 month period. To reduce the annual rate to a monthly rate the 10th root of 1.0621 is taken and from the resulting figure (1.0060) the 1 is deducted and the figure is raised to a percentage to give 0.60% per month. Although for illustrative purposes the cash flow is fairly simplistic, it will be seen that any amount of detail can be incorporated as the project becomes more certain. Cash flows can be exactly accounted for in the month that they occur rather than being taken on a cumulative basis in the conventional appraisal. If rental income is received during the development period this can be included and will influence the net deficit cash flow for the month received.

When cash flows are discounted by the short term rate of interest the cost of money is accurately reflected in the calculation, so that when the building is sold (which is assumed to be in month 10) the profit figure (the net present value or NPV) represents the present value of the profit that will be received in 10 month's time. It is also possible to calculate the true return from a project by constructing a cash flow with a discount rate which results in a zero net present value. In these circumstances the discount rate is an Internal Rate of Return (IRR) which would be a true monthly return on money expended. The discounted profit figure is usually more useful from the developer's viewpoint.

# Ground rent and partnership schemes

The residual method can be adapted to situations where a site owner wishes to let a ground lease rather than sell the freehold of a site. This is a frequent occurrence where a local authority owns the freehold of a town centre site and wishes to arrange for is development. Not surprisingly, it is usually

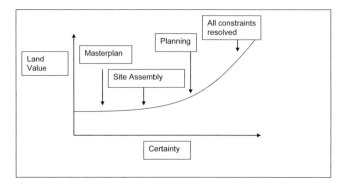

**Figure 13.3** Value and certainty

retail schemes that result from what are termed partnership arrangements, and there is a variety of ways in which the local authority can benefit financially from the scheme. Also, when a freehold owner disposes of a site to a developer it is frequently at the time of highest risk from the landowner's viewpoint. Planning consent may not yet be granted, and various other constraints may need to be resolved. When a local authority disposes of a town centre by way of a ground lease the constraints are resolved, and all information will be available in the development brief prepared by the authority. It would be expected, therefore, that the authority would benefit financially in comparison with a freehold disposal of a raw site, for, as certainty is achieved, site value rises, and this should be reflected in the rental and premium bids received. This is shown in Figure 13.3.

To illustrate the calculations for the grant of a ground lease the figures in Study 2 will be used. It is assumed that a local authority wish to dispose of a town centre site by way of a ground lease and will be seeking bids on the basis of a ground rent with or without a premium. It is assumed that the authority will wish to receive a geared ground rent. This means that the ground rent will always be geared to a percentage of the rack rental value of the building or, sometimes, the rent passing. This is further discussed below.

## Study 4  Ground Rent Calculation

| | | £ | £ pa |
|---|---|---:|---:|
| Income from scheme | | | 1,600,000 |
| | | | |
| Cost of scheme rounded up (from study 2) | | 16,916,364 | |
| *Less* total site cost (£6,345,000 + £394,153) | | 6,739,153 | |
| | | 10,177,211 | |
| | | | |
| Cost of long term finance | 7.5% pa | | |
| Developer's risk and profit | 2.0% pa | | |
| Total return required | 9.5% | 0.095 | |
| Rental return required by developer and fund | | 966,835  pa | 966,835 |
| | | | |
| Amount available for ground rent | | | 633,165 |
| Say £633,000 pa | | | |

The local authority, in disposing of the site to a developer, could expect to receive a ground rent which shows a return of 10.55% pa on a site value of £6 m if it is assumed that the site cost in Study 2 is a true reflection of value. The authority will carry out its own appraisal of site value, and it must be borne in mind that local authorities are subject to scrutiny by the district auditor when land is let or sold. From the figures above it would appear that the land value of £6 m is low, and a residual valuation would establish a higher value. The local authority would also wish to see the ground rent substantially exceeded by the occupational rent from the building.

It is unlikely that the site owner would receive an offer for ground rent at 40% of the rack rent of the building. If it is assumed that the building is institutionally acceptable as an investment, the opportunities for rental growth are reduced by the high percentage ground rent. For this reason funds do not find it acceptable to become involved with schemes where the ground rent exceeds 10% of the rack rental value of the investment. In Study 4 above this means that the ground rent would have to be no more than 10% of £1,600,000 (£160,000) with the remaining £473,165 (£633,165 less £160,000) being capitalised as a premium. In this case, the premium becomes a residual figure after the fund's and the developer's returns are allowed for. The new financial situation is shown in Study 5.

## Study 5  Ground Rent with Premium

|  |  |  | £ pa |
|---|---|---|---|
| Income from scheme |  |  | 1,600,000 |
| *Less* Ground Rent to local authority at 10% |  |  | 160,000 |
| Rental return to fund and developer |  |  | 1,440,000 |
| Long term finance at at 7.5% pa | 7.5% pa |  |  |
| Developer's return at 2% pa | 2.0% pa |  |  |
| Total return required | 9.5% |  |  |

|  |  | £ | £ |
|---|---|---|---|
| Maximum cost by fund and developer | $\dfrac{1,440,000}{9.5} \times 100$ |  | 15,157,895 |
| Cost of scheme (from study 2) |  | 16,916,365 |  |
| *Less* total site cost |  | 6,739,153 |  |
|  |  | 10,177,212 | 10,177,212 |
| Amount available for premium |  |  | 4,980,683 |

The local authority has received a premium which capitalizes its excess income (over £160,000) at 9.5% (£633,165 less £160,000 = £473,165 at 9.5% in perpetuity), and with the all risks yield at 7.5% this should be an acceptable premium.

Studies 4 and 5 above assume that the landowner is risk-averse and wishes to receive a ground rent well covered by the passing rent in addition to a premium. If the landowner is a local authority this is a reasonable assumption, but more creative arrangements may be negotiated if the landowner is prepared to be more flexible and the scheme is good enough. Study 6 below assumes that the landowner is happy to take 5% of the rack-rent as his "safe" income with any rental returns above the level of the developer's and fund's returns being shared between the parties to the scheme. The landowner's base return is guaranteed by calculating it as a percentage of the rent that should be paid by the occupational tenants under their leases (rent receivable) rather than on what is actually paid (rent received)

## Study 6  Partnership Scheme

|  | £ | £ pa | £ pa |
|---|---|---|---|
| Income from scheme, as above |  |  | 1,600,000 |
| 5% of rack rent to landowner |  |  | 80,000 |
| Net rent |  |  | 1,520,000 |
|  |  |  |  |
| Cost of scheme rounded up | 16,916,365 |  |  |
| *Less* total site cost | 6,739,153 |  |  |
|  | 10,177,212 |  |  |
|  |  |  |  |
| Return required by long term funder |  |  |  |
| 7.5% on | 10,177,212 |  |  |
|  | 0.075 |  |  |
|  | 763,291  pa | 763,291 |  |
| Return for developer |  |  |  |
| 2% on: | 10,177,212 |  |  |
|  | 0.020 |  |  |
|  | 203,544  pa | 203,544 |  |
| Total rent for fund and developer |  | 966,835 |  |
| Add base rent to landowner |  | 80,000 |  |
| Total |  | 1,046,835 |  |
| Occupational rent |  |  | 1,600,000 |
| *Less* |  |  | 1,046,835 |
| Amount available for profit share top slice |  |  | 553,165 |

Each of the three parties can share £553,165 as a top slice profit on a side by side basis.

Figure 13.4 shows the financial situation more clearly.

In Figure 13.4 the landowner grants the developer a licence to occupy the site during the development period, and when the scheme is completed the ground lease is granted to the developer and immediately assigned to the fund. The ground lease will provide for the freehold owner to receive a ground rent of 5% of the rent receivable by the fund from the occupational tenant, and in addition 33.33% of all rent received by the fund above £1,046,835 pa (£80,000 + £763,291 + £203,544). It would be usual for the fund to capitalise the developer's share of income when the scheme is completed and let, and the developer would then depart with a capital profit. The financial arrangements are given legal weight by means of a ground lease, a development agreement between developer/fund and landowner, and a funding agreement between developer and fund. Matters such as rent, timing of payments, and approvals are all defined in the various agreements. The relationship between the fund

| | | | |
|---|---|---|---|
| £184,388 pa<br><br>Fund | £184,388pa<br><br>Developer | £184,388 pa<br><br>Landowner | Rent paid by occupational tenant |
| £203,544 pa to developer | | | |
| £763,291 pa to fund | | | |
| £80,000 pa to landowner | | | |

and the developer is usually on the basis of profit erosion with priority yield, where the developer will receive a guaranteed capital fee payment and, in addition, further fee payments calculated as incentives. These incentive payments are based on multiples of rent above a certain level of return to the fund (the fund's priority yield). The fund's cost is the amount of money lent to the developer (capital and rolled up interest) in the development period and will include a capital sum payable to the developer if his or her rental share is capitalized as a profit. The developer's incentive is, therefore, to keep costs to a minimum level (as the fund also provides short term finance) and to let the building(s) at the highest possible initial rents. Retail schemes where the anchor tenants are secured through agreements to lease early in the development process fit well with this type of arrangement, as the developer can then let the standard units in the scheme at high rents (if the scheme is good enough).

## Rental slices

The local authority's base ground rent in Figure 13.4 (£80,000) would normally be guaranteed by the head lessee, the Fund. In the ground lease the local authority would be paid 5% of the rents receivable by the head lessee from the sub-tenants in the scheme. The base rent is therefore based on the amount of rent that the occupational sub-tenants should be paying the head lessee under the terms of their leases. If they do not pay, the head lessee still has to pay the base ground rent to the landowner. The next most secure slice is the rental shares to fund and developer although in practice the developer's slice would be capitalized into a lump sum profit. This would be paid by the fund to the developer, and the fund would then receive the developer's £203,544 pa as well as its own £763,291 pa. The least secure slice is the top slice to be shared between the local authority, the fund and the developer. In the ground lease and the finance agreement between developer and fund the three parties would be paid one third of all rents received by the head lessee above £1,046,835 pa (£80,000 + £763,291 + £203,544). Note the importance of the words "received" and "receivable" in this context. The scenario described above is the most usual in these types of agreement, but rental shares could be based on rack rent, rent passing or full rental value (rack rent) if this suits the wishes of the parties.

# Sensitivity analysis

Sensitivity analysis identifies those development variables which have the greatest influence on the developer's profit. During the development period items such as rent, construction cost and yield may fluctuate depending on market conditions, although it is possible to fix certain items. Rents may be set by pre-lets and yield may be fixed by a forward funding agreement with an institution. It may be expected, however, that certainty comes at a price, and a forward funder will seek a higher return from a scheme than the all-risks yield for the type of investment created. Similarly a pre-let tenant, especially a department store, will seek to negotiate a concessionary rent. A fixed price building contract is also likely to be more expensive than a contract where the employer accepts some risk of cost increase. An example of a sensitivity analysis is shown in Study 7.

## Study 7  Sensitivity Analysis

|  | Basic appraisal | Rent down 10% | Yield up 10% | Building cost up10% |
|---|---|---|---|---|
| Rent | 2,000,000 | 1,800,000 | 2,000,000 | 2,000,000 |
| YP 7% | 14.28 | 14.28 | 12.98 | 14.28 |
| GDV | 28,560,000 | 25,704,000 | 25,960,000 | 28,560,000 |
|  |  |  |  |  |
| Costs |  |  |  |  |
| Building cost | 10,000,000 | 10,000,000 | 10,000,000 | 11,000,000 |
| Site cost | 12,860,000 | 12,860,000 | 12,860,000 | 12,860,000 |
| Total cost | 22,860,000 | 22,860,000 | 22,860,000 | 23,860,000 |
| Profit | 5,700,000 | 2,844,000 | 3,100,000 | 4,700,000 |
|  |  |  |  |  |
| Effect of change on profit: |  | Down 50% | Down 46% | Down 18% |

Notes

The analysis in Study 7 starts with a basic appraisal which shows a profit of £5,700,000. The appraisal is deliberately skeletal to illustrate the affect of changing the main variables and it is assumed that the figures encompass all additional costs such as short term interest and fees. The column next to the basic appraisal shows what happens if the rent is decreased by 10% and the consequence is that profit falls by 50%. The next column increases the yield by 10% with the consequence that profit falls by 46%. The final column shows the result of a 10% increase in building cost, which is a fall in profit of 18%. On the basis of these figures, rent and yield are seen to be the most important variables as relatively small movements can cause a dramatic reduction in profit. Sensitivity analysis is not risk analysis as no probability figures are attached to the changing variables. The usefulness of the method is that it establishes which variables are most sensitive, and developers should therefore pay close attention to anything which may change rent or yield, even by a small amount as the results may be serious. If, for example, an increase in building cost, perhaps to accommodate an important tenant, will result in an increase in rental value, it is worthwhile running the analysis in the development appraisal as the cost increase may well be more than offset by increased profit.

## Study 8  Contractor's budget calculation

The residual method can also be used to establish a contractor's budget for a scheme. In this case the building cost and design team fees are the residual figures and all other costs are deducted from the NDV. This type of calculation is useful when a maximum cost is required when site has been acquired and the development strategy is being debated. An example of such a calculation is shown below. The figures, summarized into major headings, are the same as in Study 2 above.

All figures are from Study 2.

| | £ | £ |
|---|---|---|
| GDV | | 21,333,333 |
| NDV | | 20,173,365 |
| *Less* | | |
| Site cost and acquisition cost | 6,345,000 | |
| Site finance for 12 months | 394,153 | |
| Section 278 costs | 30,000 | |
| Survey | 10,000 | |
| Letting and sale fees | 343,300 | |
| Developer's profit at 20% GDV | 4,266,667 | |
| Sub-total cost | 11,389,120 | −11,389,120 |
| Amount available for construction, fees, finance and contingency | | 8,784,245 |

*Less* Interest on construction at 3.06% for one year:

$$8,784,245 \text{ less } \frac{8,784,245}{103.06} \times 100 \qquad -260,817$$

Amount available for construction, fees and contingency    8,523,428

*Less* Fee element at 12%:

$$8,523,428 \text{ } less \text{ } \frac{8,523,428}{112} \times 100 \qquad -913,225$$

Amount available for construction cost and contingency    7,610,203

In the above example the developer has been able to establish the maximum build cost and this can help determine the contract strategy for the scheme, the type of specification, and financial strategy in general.

It may be noted in passing that this approach can be used to enable any element to be deduced as the concluding, residual, value.

©Nigel Dubben, 2008

# Garages, Petrol Filling Stations and Motorway Service Areas 14

In the past the two types of property, garages and filling stations, were often on the same site and run by the same proprietor. This is becoming less common. The two types of business have been moving steadily away from each other in the last 20 years and in this sixth edition the two are dealt with separately with an additional section covering service areas. The chapter is divided into three parts:

- Part 1   Petrol filling stations
- Part 2   Motorway service areas
- Part 3   Garages

Both garages and petrol filling stations have one thing in common — they have seen a substantial reduction in their numbers in the last 30 years. The traditional garage has been steadily losing ground to a growing number of specialist fast-fit businesses which now dominate the market in tyres, exhausts, shock absorbers, batteries and windscreen replacement. In 1975 there were 10,500 franchised garages in the UK. A number of motor manufacturers have sharply reduced these numbers so that by the turn of the century there were fewer than 6,000 at which stage they were outnumbered by fast-fit centres.

Similarly the number of petrol filling stations in the UK has fallen sharply and the survivor sites tend to be larger, well laid out and capable of generating substantial non-fuel revenues and the oil companies are eagerly setting about transforming themselves into general roadside retailers. This might be a long-term, defensive strategy as the arrival of new forms of power generation for the car becomes more possible with pressure for zero-emission increasing.

A surveyor may be required to carry out valuations and conduct negotiations in connection with the purchase or sale of both categories of property and deal with new lettings and rent reviews. In the past it was rare that petrol filling stations were bought for investment purposes. Because their value is related to annual throughput they have been regarded as vulnerable to bypasses and traffic management schemes which can rob them of important traffic flow. However, more conventional types of investment properties suffered greatly in the recession of the early 1990s and the financial standing of tenants became much more important to investors so that there is more interest now in petrol filling stations as investments, particularly when the tenant is a major oil company. The same cannot really be said of garages. Even the best of the listed motor groups have market capitalisations which are tiny by comparison with the oil companies. Some motor groups have nevertheless arranged sale and lease-backs on some of their garages. This has been achieved usually by accepting artificial rent review provisions so that the rent at review is raised in a formula related to the Retail Price Index or to another type of property such as off-centre retail stores.

# Part 1  Petrol filling stations

## *The industry*

The valuer specialising in petrol station properties should have a working knowledge of the oil industry. Petrol is a standard product and it is marketed internationally, most of the larger wholesalers in the United Kingdom being part of multi-national oil companies. Shell, BP and Exxon, Esso's parent company, are among the largest companies in the world. Petrol is one by-product of crude oil and in the industry oil is described as equity crude to mean the oil produced by an oil company itself, and crude which is oil purchased from a third party in the open market. Equity crude as well as crude itself has to be valued at international market prices and not in relation to production costs, if for no other reason than that it could be sold at such prices, but also because it is hardly possible to identify the production costs of petrol alone as part of the oil-refining process. Bulk trade in northwest Europe is referred to as the Rotterdam market but the Rotterdam market is not an organised market with prices established in open-floor trading but rather a market where direct bargains are struck, although they are then reported as Rotterdam prices. Refiners themselves sometimes buy or sell from Rotterdam when they have supply difficulties. It should be noted at this stage that the price of crude oil is now quoted in terms of spot market prices rather than the official listed prices, and this is because of the supply situation.

  For the purposes of definition, the functions of the international oil industry can be divided between upstream activities and downstream activities. The former description covers exploration and production. However, in this chapter we are concerned only with downstream activities, or refining and marketing and chiefly with the latter, though the activities of an international, vertically-integrated oil company will have considerable influence on the market place for petrol filling stations. Petrol is one of a range of products produced by refining crude oil, the others including fuel oil, lubricants, kerosene, naphtha, etc, and in the course of the process a decision has to be made regarding the proportions of the various products demanded by prevailing market conditions, and, as already remarked, it is for this reason that it is difficult to specify the cost of refining petrol alone.

## *History of the petrol business*

Petrol was first imported from the USA and marketed in Britain at the turn of the century and at that time it was sold in sealed two-gallon (9.092 litres) cans. After the first world war, private motoring and commercial road transport developed quickly and there was a consequent increase in the number of petrol retailers. In 1920 hand-operated petrol pumps came into use in conjunction with underground storage tanks of about 500 gallons (2,273 litres) capacity and premises where petrol had been sold in cans then became pump sites. By 1938 there were over 35,000 outlets with an average throughput of 24,000 gallons (109,104 litres) a year and there was only a fraction of today's traffic on the roads. After the first world war Shell and Esso controlled the market, but between the wars various other suppliers established themselves. There was a rapid growth in demand and the establishment of refineries in the UK quickly followed.

  The Petroleum Board, which controlled petroleum distribution and prices during the second world war, was dissolved in 1948 and Anglo-American (Esso) introduced the tied garage or solus trading system from the USA. Considerable capital investment by oil companies was needed for the construction of refineries and it was therefore particularly necessary for the companies to secure assured outlets for their products. When petrol rationing ended, Esso, followed by Shell-Mex and BP and then by others, started to acquire outlets to secure ties and within a very short period all the leading suppliers had introduced similar arrangements, so that by 1953 eight out of 10 stations were subject to exclusive supply arrangements of one sort or another. There were by then five supplying groups, namely Shell-Mex, BP and National Benzole; Anglo-American (Esso) and Cleveland; Regent; Vacuum (Mobil) and Fina (Petrofina) and they controlled virtually the whole market. These companies invested heavily in the improvement of existing petrol stations and the development of new stations, either directly or through their dealers. The then existing system would not have been adequate to cope with the increased demand and the oil companies provided the basis and the finance for change. Indeed, at the beginning of this period many petrol stations had no forecourts as we know them today and were simply prepared to sell petrol, over the pavement as it were, as an adjunct of their garage and repair businesses.

At the end of the last war there were about 30,000 stations still in existence, mostly of poor quality, and the number of stations hardly altered over the next 20 years, although average sales per station doubled more or less in line with the increase in motor traffic. For some years pump prices were fairly stable, but Jet entered the market in 1958 and began to cut prices, to be followed by others.

By 1968 there were 39,958 stations in the UK with an average annual throughput of 85,789 gallons (389,996 litres). The number of stations was virtually halved by 1989 and the average annual throughput rose to 343,158 gallons (1,559,996 litres).

By 1997 the number of stations fell to 14,748 and the average annual throughput rose to 435,547 gallons (1,979,996 litres). In the period from 1968 to 1997 the demand for motoring fuel had virtually doubled. In 2007, after sustained disinvestment by the oil companies, the number of stations in the UK had fallen to approximately 10,000. And the average annual throughput is estimated at closer to 500,000 (2,273,000 litres).

## Present market

While the total number of stations has been reducing steadily in the last 30 years, a major factor in the recent closure rate has been the arrival in the market of the leading grocers like Tesco, Sainsbury and Asda. In a decade this sector has gone from nowhere to operating approximately 900 stations and they now account for approximately 26% of the UK market. Earlier superstore stations were not usually planned to pick up business from the passing motorist. They relied instead on their grocery customers and throughputs were often price-induced. However, a number of recent superstore developments have stations planned so that they can also be used easily by the passing motorist.

BP have merged their downstream interests in the UK with Mobil and they, with Shell and Esso, can be described as the majors. Between them they own or supply approximately 6,000 stations in the UK and enjoy approximately 40% of the market.

The Burma chain has been sold to Save, and Shell acquired the Gulf chain which earlier in 1997 was involved in an abandoned merger with Elf and Murco. Total, Elf and Fina have joined forces in the UK and trade as Total.

There is a healthy demand for high throughput stations, but the supply is limited. It is estimated that in a typical year there are no more than six opportunities to acquire sites with a proven or prudently projected annual throughput of a million gallons (4.546 million litres).

## Petrol tie and the law

The simple early method of obtaining a tie was to enter into a short-term solus agreement under which the dealer agreed to sell the oil company's petrol to the exclusion of other petrols, usually in return for special wholesale price rebates on petrol sold and sometimes in return for loans. These agreements were sometimes secured by a mortgage, but the type of arrangement that developed very considerably was a purchase and lease-back transaction, because in that way the dealer-owner could obtain the highest price for his freehold and then lease back for a tied rent which was low in relation to value, the lower because a rebate on petrol purchases was usually arranged as well, and on the other hand, because the oil company could obtain a secure tie. On top of these transactions, the petrol companies themselves acquired successful stations by outright purchase, and they purchased greenfield sites or run-down stations and then carried out development. From a legal point of view all these arrangements seemed at the time to be suitable ways of securing a tied outlet, but the position has been conditioned by case law and if the valuer has to deal with a tied station he is well advised to consider the precise nature of the tie and the question of whether it is binding or not.

The security of tenure provisions of the Landlord and Tenant Acts apply and a tenant is able to claim a new lease at a rent suitably adjusted to allow for the continued imposition of a tie, although of course petrol companies are able to oppose claims for new leases on the grounds specified in the Acts. As between the 1927 Act and the 1954 Act there is no great alteration in the provisions relating to the terms of a new tenancy and, while the court has a wide discretion, it is difficult to imagine a case where the petrol company would be required to grant a new lease without the tie, since the tie was the company's only object in acquiring the property in the first place. On the other hand, and having regard to one aspect of the reasoning in the restraint of trade cases discussed below, it is just conceivable that the courts might choose not to reimpose a tie on renewal, but in such circumstances it would seem to follow that the new rent would have to be at the free rental value level and would thus be beyond the reach of a company other than an oil company. Rent review clauses in tied tenancy agreements should make the tied basis of valuation clear but often do not.

Covenants on the part of a tenant to deal with his landlord alone and to take his goods have been accepted for many years, subject only to the implied term that the landlord will supply the goods at a fair and reasonable price. Such covenants have been held to be legal and binding in equity on an assignee.

The practice of arranging for the operation of oil company stations by licensees or by wholly-owned operating companies is now fairly common, though there was a tendency in the late 1990s for oil companies to take stations back into direct management. In the case of licensee operations an extra-statutory code is applied, which does give the dealer some protection and security. The normal practice is to grant a licence for a three-year period, renewable on terms and with provision for compensation in the alternative. For obvious reasons the valuer will not be greatly concerned with this type of arrangement

Contracts contrary to public policy are unlawful and agreements in restraint of trade have been held to be contrary to public policy and against the public interest. In the past the term has been taken to include arrangements restricting business activity and limiting competition. Over a period of seven years or so, starting in the mid-1960s, four important petrol tie cases were decided: *Petrofina (GB) Ltd v Martin* (1966); *Esso Petroleum Co v Harper's Garage (Stourport) Ltd* (1968); *Cleveland Petroleum Co v Dartstone* (1969) and *Total Oil (GB) Ltd v Thompson Garages (Biggin Hill) Ltd* (1972). The valuer should study these cases. *Petrofina v Martin* more or less destroyed the simple solus agreement method of obtaining a tie. *Esso v Harper's Garage* largely destroyed the mortgage coupled with a sales agreement method. The *Dartstone* case tended to validate lease-and-leaseback or purchase-and-purchase back types of tie arrangements and the position was strengthened by the *Total* v *Thompson* case.

In a normal rent review situation the valuer will find that he is required to arrive at the open market rent on the assumption that the premises are offered to let with vacant possession and he therefore values on a free of tie basis. From the point of view of the tenant this may produce unfortunate results because, as will be appreciated, the free rent could be much higher than the tied rent, although in such circumstances the tenant's subsequent remedy would probably be to sublease the premises to an oil company at the free rent that he himself would otherwise have to pay, arranging to lease the premises back at a tied rent.

## *Direct management, licensees, leases and dealers — the modern retailing operation*

Sites owned by wholesalers are either operated by them under a form of direct management, run by licensees or operated by tied tenants. On the other hand, independent sites are normally the business of solus dealers who operate under exclusive supply agreements with particular oil companies. Nowadays these agreements usually run for three or five years. Hypermarkets and supermarkets are also supplied under solus agreements and the larger the operation the greater the bargaining power of the retailer in terms of prices paid for supplies.

Oil companies prefer to lease out premises where there are workshops and car sales businesses, or even to sell off such parts, but simple petrol stations are more often found to be under direct management. It should be noted of course that a lessee operation gives the tenant the benefit of the lease renewal provisions of the Landlord and Tenant Act 1954, although section 30 of that Act and in particular subsections (f) and (g) enable an oil company landlord to obtain possession at the end of a lease for reconstruction or direct occupation, so the protection of the Act is to some extent illusory. The licensee type of occupation deserves examination case by case because it is well established in law that the mere use of the word licensee does not preclude the possibility that the occupier is in fact a tenant with all the advantages that flow from that.

Major petrol stations, while usually the property of oil companies, are sometimes found to be owned by the major petrol retailing chains. Smaller wholesalers might take on lesser sites in rural areas, but low volume and badly located sites are not attractive and are often sold for alternative development, sometimes with restrictive covenants against petrol sales where the vendor fears redevelopment and resulting competition. Of course low-rated sites can be improved, particularly in terms of layout and the provision of forecourt shops and indeed are improved if the potential for growth exists.

When petrol stations are acquired by oil companies it is, as has already been remarked, for the purpose of securing a tied outlet for fuel, but oil companies do not enter the market where the petrol sales potential is too low to make acquisition worthwhile. In these latter circumstances the market is restricted to dealers who, as tenants or as owners, in the normal course of business make solus agreements with oil companies for the supply of particular brands of petrol; they are the solus dealers.

Under a solus agreement a station proprietor agrees to buy his fuel from a wholesaler — a refiner or a supplier — at scheduled prices less a rebate and in addition he will sometimes receive what is known as selective price support on a temporary basis, this being a form of support enabling a dealer to reduce his prices to meet or to try to beat local competition. A recent development has been the introduction of supply agreements which give the dealer what has been called a guaranteed margin. With the exception of hypermarkets there are usually found to be only small overall variations in retail prices, although prices do tend to be higher in areas where there is little competition and lower where there is a real competition — from hypermarkets for example.

## Modern filling stations

Thirty years ago most filling stations had small sales kiosks where the motorist could pay for his petrol and at the same time buy from a very limited range of goods, mainly cigarettes and confectionery. This has been changing steadily since then so that today oil companies and petrol retailers are striving to improve their retailing skills and enlarge their facilities in order to offer a much wider range of goods to the motorist.

Shell UK Ltd published some seven years ago a study called *Night-time Convenience Shopping and the 24 hour Society*. The key facts were revealing. It was estimated that the UK forecourt convenience market was worth £2.8 billion. Shell's own retail business, Select, did 20% of its business between 10.00 pm and 6.00 am and 80% of the 850 shops were open 24 hours a day and seven days a week. There were 2,500 to 3,000 product lines and Select was the country's fourth largest own-label sandwich supplier. Forty four of the shops had off-licences. Sixteen per cent of customers arrived on foot and 65% of all customers bought something other than fuel.

An E K Williams survey at the time for *Forecourt News* suggested that typical profit margins vary from 35.1% on ice-creams down to 8.1% on cigarettes. The margin of 20.7% for groceries will be the envy of grocers generally. Since cigarettes will account typically for over a third of takings, average overall margins should be in the region of 15 to 20%. It is now becoming more common for a successful filling station to have non-fuel shop takings of £8,000 a week and many will have commission from lottery ticket sales which will be worth, typically, £5,000 a year.

It is likely that the oil companies will continue to look for new sources of profit and new ways of attracting customers. In late 1997 Texaco signed a five-year contract with the Bank of Scotland to launch a nationwide cash machine service at 200 of its stations by the end of 1998. It will be seen, therefore, that non-fuel activities are rapidly becoming a fundamental source of revenue at a station. Since the motorist is encouraged to buy non-fuel items before paying for his petrol, the lingering time at the pump will be longer. This will tend to make greater demands on ideal site areas so that a station standing on much less than half an acre will not have enough space to extend the facilities to include a convenience store. Planning applications by oil companies to extend their shops to 1,000 sq ft (93 m²) or even 2,000 sq ft (186 m²) have caused some concern among planning officers, who fear that established high streets and shopping centres close by could be adversely affected by the growth of convenience stores attached to filling stations. However, resistance to the increasing public demand for this type of retailing and the trend for larger shops will be difficult to justify, particularly at a time when many councils are looking hard at ways of reducing the amount of traffic in the centre of their towns and cities.

Adding to the demands on space is the importance to a successful station of a car wash where gross receipts might typically be £40,000 a year and for a simple hand-held lance wash £10,000 a year. The profit margins are remarkable. Returns of approximately 30% are not uncommon. Strong traffic volumes are essential to a successful station. The valuer can usually obtain traffic counts from the highways departments of local authorities or the Highways Agency. In regard to an established station with a proven throughput, the valuer must have regard to factors which might affect traffic volume and therefore trade. He will make careful enquiries to establish that there are no road improvements proposed which could affect the traffic volume. He should also discover whether there are any proposed developments of new stations in the area and not necessarily on the same route. This latter point is particularly relevant in the case of superstores with stations. He should look too at other stations which are already trading in the area, and compare their prospects for survival. With new petrol stations the valuer will be called on to estimate the potential mature throughput. In some cases the mature throughput can be reached within a year of the start of trading but in other cases it is only reached after about three years. He should look carefully at competing stations in the area which are trading poorly and are unlikely to survive. This calls for some fine judgment but is more apparent in the case of stations which have severely restricted site areas or difficult access and egress. Such a station might have a modest annual throughput of say 300,000 gallons (1,363,800 litres) but its closure could lead to most of its throughput going to the next station along the route and will rarely benefit more than two other stations in the area.

As a rule of thumb, on trunk roads the turn-in proportion is usually about 3 to 4% of passing motorists. On busy roads most of the throughput would come from traffic which is passing on the same side of the road as the station. The average purchase in 2004 was approximately 5.5 gallons (25 litres).

By way of illustration, in the case of a good site with a frontage of 150 ft to 200 ft, a depth of no less than 100 feet, and a two-way 24 hour traffic flow of 40,000 private vehicles and vans, the estimated throughput calculation would look like this:

## Assessment of throughput

| | | | |
|---|---|---|---|
| Total two-way 24 hour traffic flow for cars and vans | | 40,000 | |
| Less half for those vehicles on the side of the road opposite to the station | | 20,000 | |
| Effective daily number of potential motoring customers | | 20,000 | |
| Average turn-in rate | | 3% | |
| Estimated number of daily customers | (vehicles) | 600 | |
| Average fuel purchase | (gallons) | 5.5 | (25 litres) |
| Average daily throughput | (gallons) | 3,300 | (15,001 litres) |
| Effective number of days in the year | | 313 | |
| Estimated annual throughput | (gallons) | 1,032,900 | (4,695,563 litres) |

It is important to note that this example assumes that the traffic flow only applies to six days of the week. This will vary enormously from one road to another. It is never the case that a road as busy as the one in the example will be totally without Sunday traffic, and a large residential population in the immediately vicinity of a station would have a considerable bearing on the calculation. The 24-hour day seven-day week contemplated in the Shell study is of special relevance to the future. So too is the increasing number of major sporting events taking place on Sundays.

In 1995 the Department of Transport estimated traffic growth of 58–92% between then and 2025. Despite this, in late 1996 the Department of Transport abandoned over 100 longer-term road schemes though they did say that they would consider smaller-scale improvements to tackle local problems. It is possible that in the longer term the rate of petrol sales growth may decrease as saturation point is reached in terms of car ownership, and because motor manufacturers have made great progress in terms of economic fuel consumption. These matters will have a considerable bearing on the future growth of throughputs and non-fuel turnover.

## Basis of valuation

As is the case with the valuation of hotels, public houses, restaurants, cinemas and theatres, a close knowledge of the trade carried out on the premises is a necessary concomitant to this type of work.

Oil companies do not purchase for investment but to obtain an outlet for their products. Their petrol profits arise from the refining and wholesaling process — and to some extent from rents from tied outlets — whereas dealers' profits arise from retailing. The concept of a tied rent and a free rent must be understood. The free rent is the rent which an oil company is prepared to pay and is the total of the tied rent that it can expect to receive from its dealer plus an overbid, or special value element, which in fact is met out of refining and wholesaling margins. The dealer pays his rent out of his retailing profits whereas the oil company can and does add to the rent received from the dealer a proportion of its own earnings. In practice oil companies will be found in most cases to purchase rather than lease their outlets.

It must be borne in mind that this special aspect of property valuation involves projections in an imperfect market; this in the sense that where substantial petrol throughputs are involved, the oil companies and petrol distributors are virtually the only buyers and, furthermore, their interest varies from month to month because of marketing or budgeting factors rather than for property investment reasons. There is a limited number of possible purchasers and they are not all in the market at the same time; nor are they necessarily in the market in the same region of the country. In the circumstances, consistency will be found to be lacking and it is probably more true to say in this case than most that the value of the property is what it will fetch.

The valuer must be cautious in terms of analysis and synthesis when dealing with individual transactions because it is occasionally the case that excessively high prices are paid for large volume outlets. The oil companies usually operate within budget parameters and the availability of funds as well as marketing needs will condition offers. It is of vital importance that the valuer should consider all factors likely to affect petrol sales and the trade generally such as possible road diversions and a variation in traffic volume, the possibility of new stations being established, of old stations being enlarged and improved and of stations being closed. Valuable information can usually be obtained from the planning and highway authorities. The Freedom of Information Act has been particularly helpful to the valuer.

Directly comparable evidence is of great assistance, but it is not always available and valuations may well have to be constructed from first principles. To be useful, a comparable must be directly analogous and, by way of example, if in search of such evidence the valuer analyses the sales at, say, six stations on a busy 30-mile stretch of road with little variation in traffic volume, he will nevertheless find great variations in sales owing to such things as comparative accessibility, visibility, lay-out, dealer operation standards and, of course, pricing policy. It is after all from throughput and other turnover, and the consequent profit, that forecourt rent derives. Competitive pricing is vital but standards of operation are equally so and, surprising though it is, a good dealer might possibly sell twice as much petrol as a bad dealer at a given station. National advertising and special promotions by major oil companies can also affect throughputs.

It must be made clear that stations on other routes are no more directly comparable than would be rental evidence taken from shops in positions other than the one to which a valuation relates. It is true though that there is a general pattern of values and it is this pattern that the specialist valuer must build up in his mind in the course of practice. With particular reference to forecourts, it is clear that oil companies apply specific rates per gallon (or litre) on a sliding scale adjusted for throughput when buying or leasing.

# Forecourt values

Tied rents and licence fees, on the one hand, and open market rents, on the other, are to be distinguished from each other, tied rents being artificial or closed market rents paid to suppliers and conditioned by retail profit margins. Private landlords — non-oil company landlords — will let premises at open market rents but these rents will be affected by oil company demand only if the throughput is sufficient to attract it. The oil company wholesaler will of course be found to pay an open market rent and then, if he subleases, to do so at a lower tied rent.

Open market forecourt rents, that is to say free rents or the rents an oil company might be expected to pay, are related to the cost of producing petrol, Derv and lubricants and the profits of marketing them, whereas tied rents are related to the profits derived from retailing these products. Dealer rents, in the sense of rents paid to private landlords — non-oil company landlords — or low throughput stations of insufficient interest to attract oil companies are directly derived from likely trading profits.

Oil company interest is clear at about 900,000 gallons (4,091,400 litres) pa or above, but there is an area between that and about 600,000 gallons (2,727,600 litres) pa where the interest of oil companies is variable and depends on a variety of marketing and extraneous factors. It can be taken for granted that at the present time even a minor oil company would probably not be interested in the acquisition of a station selling 600,000 gallons (2,727,600 litres) pa or less. This does not mean that there is no demand for small throughput stations but it is a demand from dealers, retail chains or speculators. Whether for capital value or rental value the valuer will use great caution when dealing with a station with a throughput of less than 500,000 gallons (2,273,000 litres) and particularly so if the throughput is falling

By and large the property investor acquires a hedge against inflation and his yield, in the case of freeholds anyhow, is a net yield in the sense that he should not have to build up a sinking fund to compensate for the deleterious effects of inflation on the capital investment. However, petrol filling stations are bought by oil companies and oil companies are not investors in the sense in which the word has been used here; they buy property to secure an outlet for their fuel products. Of course, good, well-sited stations will still be a hedge against inflation but the central reason for their purchase is to obtain a property from which petrol can be sold or a property that can be let subject to a tie to secure the sale of petrol. It will thus be apparent that normal investment considerations are irrelevant.

Most petrol station transactions are purchases, leasing being a least preferred alternative, and it follows from this that most real evidence of value arises from capital transactions. The price paid for petrol filling stations can be closely correlated with throughput and rises steeply in terms of the amount paid per gallon as throughput increases.

The table overleaf (p388) is an attempt by way of synthesis to give an approximate indication of free of tie capital values in relation to throughput:

It is important to stress that these values apply to the whole station and assume non-fuel and car-wash receipts usually found at stations with the sort of fuel throughputs indicated.

Helpful to the specialist valuer is the *UK Pump Price Report* which is published twice a month by the *Petroleum Times*. This lists current pump prices for all forms of motor fuel at 39 towns and cities throughout the United Kingdom. There is a difference of approximately 25p a gallon (5.49p a litre) between the dearest and the cheapest unleaded fuel. In the Highlands and Islands this difference can sometimes double. There is evidence too of a well established trend for cheaper fuel along a line following the M62 between Liverpool and Hull. There is a similar band of cheaper fuel following the M8 between Edinburgh and Glasgow. Logically, the valuer should experience corresponding variations in value when dealing in the open market. In fact, he will find it difficult to discern such a pattern. He could find that he gets a better price for a site in Edinburgh than he does for a site in Inverness even

| Annual through-put in gallons | Capital value per gallon £ | Capital Value £ | Annual through-put in litres | Capital value per litre | Capital Value £ |
|---|---|---|---|---|---|
| 500,000 | 70p | 350,000 | 2,273,000 | 15.398 | 350,000 |
| 600,000 | 80p | 480,000 | 2,722,760 | 17.597 | 480,000 |
| 700,000 | 100p | 700,000 | 3,119,200 | 21.997 | 700,000 |
| 800,000 | 120p | 960,000 | 3,636,800 | 26.396 | 960,000 |
| 900,000 | 140p | 1,260,000 | 4,091,400 | 30.796 | 1,260,000 |
| 1,000,000 | 150p | 1,500,000 | 4,546,000 | 32.996 | 1,500,000 |
| 1,250,000 | 160p | 2,000,000 | 5,682,500 | 35.195 | 2,000,000 |
| 1,500,000 | 170p | 2,550,000 | 6,819,000 | 37.395 | 2,550,000 |
| 1,750,000 | 190p | 3,325,000 | 7,955,500 | 41.794 | 3,325,000 |
| 2,000,000 | 200p | 4,000,000 | 9,092,000 | 43.994 | 4,000,000 |

though Inverness pump prices are stronger. This is not entirely explained by the fewer number of oil companies active in the Highlands. In the last few years major oil companies have sold many sites in the North, though remaining willing to enter into supply agreements with independent retailers.

## From capital value to rental value

Capital values being the yardstick, rental value often has to be derived from the level of figures given in the last section, or rather from their market equivalents, and this is the reverse of the normal investment valuation process. The distinction between an investment purchase and a purchase made to secure an outlet for petrol sales will have become clear.

It follows that since the determination of the rental value of a petrol station often involves the ascertainment of capital value as a first step — the opposite of the normal investment valuation process — the years' purchase used for decapitalisation purposes must be constant. Decapitalisation at a higher years' purchase for a better station would produce an incongruous result in trading terms, since the more valuable station would then have a rental value proportionately lower than the less valuable station, and as a matter of strict logic it could be suggested that the better a petrol filling station is in trading terms the lower should be the decapitalisation years' purchase figure, although this would result in the initially surprising conclusion that, for example, a 500,000 gallons (2,273,000 litres) pa station's capital value should be decapitalised on an 8% basis and a 1,000,000 gallons (4,546,000 litres) pa station on a 10% basis. If it is argued that varying rates should be employed as in the conventional investment valuation process then it would follow that the rate of values per gallon must rise at the lower end of the scale and this would not represent the real nature of the market. The oil company's target is throughput and the higher the throughput the more valuable the station. It cannot be the case that an oil company would be prepared to pay a higher rent in proportion to capital value for a smaller station.

Of course the conventional investor might be prepared to pay a proportionately higher price in relation to rent for the better station but that would follow the initial free-rent equation and the increment could be regarded as the premium a conventional investor would be prepared to pay for security of income in terms of covenant and perhaps for growth prospects, although he would have to possess a specialised knowledge of the oil industry to judge the latter. An investor might also take into account the possibility of redevelopment for alternative uses, a factor usually ignored by oil companies, though that could change in the long term.

It is now necessary to turn to the question of rental yield in relation to capital value and as far as possible to base the conclusions on empirical evidence. In the normal investment market the valuer's analysis is derived from evidence of lettings followed by sales, or from sales followed by lettings, so the relationship of capital and rental value in terms of yield is immediately clear, whereas in the case of petrol stations the property will have been bought by an oil company and, if let, let at a tied rent related to the tied tenant's expected retail profits not to the wholesale profits that fix the capital value of a free station, and the same conditions apply if an oil company leases a station. In short, one factor in the equation, rent or capital value, is likely to be hypothetical, or at least not derived from the same transaction, simply because oil companies do not buy a station and then lease it at a free rent.

From the evidence available it would appear that in terms of today's market the inherent rent yield lies between 7% and 9% and can fairly be said to average 8%, and it will have become evident from the foregoing that the decapitalisation rate must be constant throughout the range of capital values.

In practice it is suggested that in determining rental value the practitioner should start from capital value unless, that is, he has equally good or better evidence of rental value, in which case he should use both approaches. A table of indicated free rental values derived from the table given above and based on a decapitalisation rate of 8% follows:

| Annual throughput in gallons | Rental value per gallon | Annual throughput in litres | Rental Value per litre |
|---|---|---|---|
| 500,000 | 5.6p | 2,273,000 | 1.251p |
| 600,000 | 6.4p | 2,722,760 | 1.407p |
| 700,000 | 8.0p | 3,119,200 | 1.759p |
| 800,000 | 9.6p | 3,636,800 | 2.111p |
| 900,000 | 11.2p | 4,091,400 | 2.463p |
| 1,000,000 | 12.0p | 4,546,000 | 2.639p |
| 1,250,000 | 12.8p | 5,682,500 | 2.815p |
| 1,500,000 | 13.6p | 6,819,000 | 2.991p |
| 1,750,000 | 15.2p | 7,955,500 | 3.343p |
| 2,000,000 | 16.0p | 9,092,000 | 3.519p |

As with capital values these rents are for the whole station and the same assumptions about non-fuel and car-wash receipts are made.

These figures are perhaps unduly precise but, if nothing more, they will establish the fact that a free rent could hardly be paid out of the retailing profits of a tied tenant.

The values of petrol stations have risen more or less in line with commercial and industrial property values, possibly doubling between 1984 and 1989, and it is believed that it was in the period from 1986 to 1987 that capital prices of £1 per gall were first realised for prime stations. Major oil companies claim that their return on the capital investment in service stations is as low as 6%, but this must reflect the fact that tied rents as an element in the return are artificially low, the difference between the tied rent and the open market rent, or the annual equivalent of the purchase price in the open market, being the premium or special value element that petrol companies are prepared to pay for the tie.

## The Harewood case

If there is a lack of evidence about petrol throughput in lease renewal arbitrations and in new tenancy cases under the Landlord and Tenant Act 1954 — for example if the tenant or supplier refuses to supply throughput figures — the valuer has in reserve the possibility of applying for an order for disclosure as part of the arbitration or court process. Such an order can usually be obtained if a valuation depends on evidence of trade done or of profits, as is invariably the case with petrol stations, throughput being a vital element in determining the rental value of a forecourt. Disclosure might, of course, include valuation reports and general accounts as well as petrol sales figures of course. The leading case is *Harewood Hotels Ltd* v *Harris* (1957), which was referred to in *WJ Barton Ltd* v *Long Acre Securities Ltd* (1982). The *Harewood* case made it clear that evidence of the financial results of a business was admissible if needed to show what rent premises were likely to fetch and the second case in limiting this judgment nevertheless made it clear that in the case of petrol stations trading evidence was relevant and could be called for.

It would seem to follow from the foregoing that if in negotiations a valuer acting for a wholesaler or for a retailer declines to produce petrol throughput figures he is encouraging the lessor to go to arbitration or to the court and that this might as well be recognised from the beginning. Petrol stations can only be valued on the basis of actual or potential throughput, although, of course, surveys and checks can be carried out to determine throughput and indeed to estimate potential throughput. There is at least one firm specialising in this type of work.

## Fast food caterers

There is a growing trend for oil companies to acquire sites where they can trade alongside fast food caterers, particularly those with strong brand images. The two activities are complementary and a popular caterer can typically improve throughputs by as much as 10%.

## Shops and car washes

Shops and car washes are usually valued with the forecourt and as far as shops are concerned it would appear on analysis that rents charged by oil companies to their tenants often relate directly to throughput in a ratio of something in the region of £2 per sq ft per 100,000 gallons pa (£21.528 per m²), in other words £20 per sq ft (£215.28 per m²) at 1,000,000 gallons pa, but this would not apply in an urban situation where there was non-motorist trade as well, this probably justifying a higher rent. The size of the shop is critical of course and would normally be in the region of 500 sq ft to 800 sq ft (46.45 m² to 74.32m²) though in appropriate locations shops will be much larger. In the last few years the Esso and Shell shops are more than double the top of this range.

In the last few years the sector has seen remarkable changes in the oil companies' approach to retailing at filling stations. Shops have become larger and larger and the amount of trade which walks onto the station can be remarkably high in places where the filling station is surrounded by houses or offices. Shops with areas of close to 2,000 sq ft (185.8 m²) are in appropriate cases the objective of the oil companies. They are called convenience stores. The range of goods carried in these shops has expanded dramatically, so that one can buy fast food, groceries, freshly baked bread and fresh flowers. The businesses have expanded to include lottery ticket sales, electricity tokens and few new filling stations presently being planned are without a cash dispensing machine. Often premium pricing policies are applied to certain goods such as cigarettes.

The oil companies have clearly aimed at establishing themselves as roadside retailers. Esso have joined forces with Tesco, and Shell with Sainsbury's, in an attempt to develop this type of business.

It might well be that the valuer would do well to deal with the rental value of shops or convenience stores on a profits basis and expect a rent of approximately 15% to 25% of the net profitability of the shop. He will need to be satisfied that he has all the necessary information about the tenant to carry out this task. It should be noted that as shop sizes become bigger then the profitability per sq ft or m² will fall as the range of goods is extended to include items with lower profit margins and goods with slower turnovers.

A fully equipped car wash in a permanent building would cost in the region of £100,000 to build and might produce a gross return of £60,000 pa with running costs of £20,000 pa, leaving a net return of £40,000 pa. An oil company landlord would probably look for something in the region of a third of the net profit as rent, so in the case quoted the rent would be around £13,333 pa.

## Study 1

By way of example the following is a rental valuation of a modern petrol filling station free of tie, with an annual motor fuel throughput of 1,000,000 gallons, annual shop sales of £400,000 excluding VAT and car wash annual sales of £60,000 and in an area of average pump prices.

| | | | |
|---|---|---|---|
| Annual throughput | | 1,000,000 | gallons 4,546,000 litres |
| Rent per gallon | | 9p | 1.97976 say £90,000 |
| | | £90,000 | |
| Non fuel turnover | £400,000 | | |
| Profit Margin | 20% | | |
| | £80,000 | | |
| Attributable to rent | 25% | £20,000 | |
| Car wash turnover | £60,000 | | |
| Cost of running | £20,000 | | |
| Return | £40,000 | | |
| Attributable to rent | 33.33% | £13,332 | |
| Annual Rental Value | | £123,332 | |

## Cost of construction and site values

At present the cost of constructing a typical fully-equipped petrol service station or petrol filling station stands between £700,000 and £1,200,000. A good modern station will be found to have a forecourt with a frontage of at least 150 ft (45 m) and a depth of about 150 ft (45 m) as well, and will be equipped with three or perhaps four multi-hose pump islands under a canopy. The storage tank capacity of a station selling upwards of 1,000,000 gallons pa (4,546,000 litres) usually consists of two or often three 12,000-gallon (54,552 litre) tanks or six 5,000 gallon (22,730 litre) tanks. Deliveries are nowadays made by 6,000-gallon tankers and if three deliveries a week are assumed it will be seen that it is possible to calculate the required tank capacity when peak sales are taken into account.

A typical station construction cost would relate to a station comprising a forecourt with storage tanks and interceptors, a canopy, a shop and an area of hard-standing, together with electrical and mechanical services. The provision of a car wash would increase the cost by between £50,000 and £100,000.

The residual method of valuation is commonly used by oil companies to determine site value and, speaking in very general terms, the value of a prime site with a potential throughput of 1,000,000 gallons pa would be in the region of £600,000 to £1,000,000, this for a site with an area of about half an acre (2,023.41 m$^2$).

It appears to be the case that certain oil companies which follow USA accounting principles prefer short-term building leases, for example leases for 20 years, and are prepared to pay rents equivalent to those normally paid under long-term leases. This follows from the usually fallacious reasoning that rent is a liability, this conclusion ignoring the fact that a lease is usually regarded as an asset. It will sometimes be found that building leases for normal long terms contain break clauses in favour of the tenant at, for example, 20-year intervals and the valuer will perceive the reason for this in the circumstances just outlined.

# Part 2  Motorway service areas

Prior to 1992 the Department of Transport was responsible for identifying, acquiring by compulsory purchase where necessary, and promoting the development of, motorway service areas.

These sites were leased, mainly to oil companies and large catering organisations. Originally turnover rents were paid but later the practice was for the Department of Transport to grant 50-year leases at a premium with nominal ground rents.

In 1992 the Department of Transport opened up this sector to private initiative so that oil companies, catering companies and developers have since then identified what they saw as potentially good sites, often filling in large gaps between existing sites.

Obtaining planning permission for these service areas is a daunting and expensive task. Often a multi-site public inquiry is held and can last for many months. It is usual for planning consent to be granted subject to a Signs Agreement from the Department of Transport. This permits the erection of signs on the approach section of the motorway which are helpful to the motorist and to the business of the service area, and they are important to road safety. They enable the motorist to manage his exit manoeuvre in safety in what are known as weaving distances. In exchange for a signs agreement the Department of Transport will create a rent charge on the motorway service area under the terms of which they can continue to enforce the regulations which apply to these sites in connection with car parking capacity, HGV parking capacity, provision of facilities, a ban on alcoholic drink and a limit to operations which could be reasonably described as destination activities. A valuer will approach this subject with the greatest care and a knowledge of the scarce comparable evidence is fundamental.

The Department of Transport has a general rule that motorway service areas will be available at roughly 30-mile intervals and that additional 'infill' services will be allowed, but only exceptionally and where a clear and compelling safety case can be established. For this reason, establishing need is fundamental to obtaining planning permission. However, motorway traffic use has increased at a much greater rate than trunk road traffic use, and pump prices and catering is stronger on motorways so that the profitability is greater. Assuming that a site is big enough to provide a large petrol filling station, an HGV service area, a large amenity and catering building and a motorists' lodge, then the developed value of such a site will be in the region of £6 to £12 per annual gallon of throughput, ignoring bunkered derv throughput. Site values are very difficult indeed to estimate. Infrastructure costs vary considerably from site to site. These can include the provision of services, drainage, road works and even the cost of building crowned roundabouts. The early motorway service areas such as Knutsford on the M6 occupied twin sites of not much more than 13 acres whereas the later service areas like Norton Canes on the Birmingham North Relief Road occupy over 60 acres on a single site serving both traffic flows.

The principles applied to the valuation of an ordinary petrol station apply to a motorway service area but the scale is substantially different. This is less marked at the extreme sections of some motorways such as the M6 north of Preston and the M4 west of Newport. In areas like these, and particularly in Scotland, values are more easily related to values on trunk roads. The difference is explained entirely by the great difference in traffic flows.

In 2007 the Department of Transport started a re-examination of motorway service area policy and at the time of writing the outcome of this study is not yet known. There has been some pressure for a relaxation of retail space which can be described as a destination activity. There may be greater pressure still on the number of motorway service areas, so that gaps between them will be influenced more by travelling time than by distance and proximity to important trunk road junctions.

The nature of motorway service areas has already changed greatly. Recent developments are changing motorway service areas from being businesses to being straightforward property investments, and first class sites will produce annual rents of well over £1,000,000 and in a few cases as much as £5,000,000. With the present strong demand for property investments this will lead to capital values of £20,000,000 to £120,000,000. Motorway service areas will become attractive to institutional investors as more sophisticated forms of leases are developed, providing the investor with a base rent plus an annually reviewable turnover rent. Though franchise arrangements are a common feature of motorway retailing and catering, the covenant strength available, particularly from the oil companies, will be attractive.

# Part 3  Garages

In 1975 there were approximately 31,000 petrol filling stations in Great Britain. By 1999 there were less than half that number. In 2007 there were approximately 10,000. The change has been dramatic.

The change in the motor trade has been just as great. The traditional business of a garage has been greatly eroded by the development of Fast Fit Centres which have enjoyed a meteoric rise. Thirty years ago the traditional motor trade captured 80% to 90% of all car tyre sales. In 1998 this share had fallen to 8% to 10%. In the same period the trade's share of exhaust sales has fallen from 60% to 30%. These trends have continued into the 21st century.

The Fast Fit businesses were able to succeed because they concentrated on jobs that could be done quickly and easily by semi-skilled staff. Motorists could go in without making an appointment and sit down and wait while the job was done. They did not need to book an appointment for three days' time. These businesses were also perceived by the public as being cheaper, although this was not always the case. Their advantage came from menu pricing so that the motorist knew in advance what he was going to pay.

The valuer will usually be faced by the need to arrive at the capital value of a franchised garage — that is to say a garage which holds a franchise from a motor manufacturer with a prescribed dealer territory in which to operate. A franchise agreement ties the proprietor of the garage to sell only the new products and parts of a particular manufacturer. In approaching the valuation of a franchised garage or dealership, the most important thing is to distinguish between the value of the property and the value of the franchise. Although the Block Exemption Rules have become a little kinder to garage proprietors since 2002, a sound property valuation should ignore the advantage or disadvantage of a particular franchise.

Retaining a franchise is not guaranteed. Neither is the popularity of any manufacturer's range of cars. Clearly there are advantages to certain franchises but these advantages should be found in the profit and loss account of the particular business and not in a valuation of the property itself.

It is important too to ignore the profitability of a business carried on from the property to be valued. It is not uncommon to see businesses in poor property with less than desirable franchises but nevertheless making good profits. In such a business the proprietor will have special strengths and skills such as used car sales. The profitability of a business should not, therefore, be confused with the value of the property.

# Valuation

## Study 2

By way of example the following valuation is of a provincial dealership in a good position with attractive buildings erected 20 years ago.

| | Sq ft | Rent £ pa per sq ft | m² | Rent £ pa per m² | Rent £pa |
|---|---|---|---|---|---|
| Showroom | 3,000 | 18.00 | 278.706 | 193.752 | 54,000 |
| Sales Office | 300 | 8.00 | 27.870 | 86.114 | 2,400 |
| General Office | 1,000 | 10.00 | 92.902 | 107.640 | 10,000 |
| Reception | 200 | 8.00 | 18.580 | 86.114 | 1,600 |
| Workshops | 6,000 | 5.00 | 557.413 | 53.820 | 30,000 |
| Parts Storage | 200 | 4.00 | 18.580 | 43.057 | 800 |
| Mezzanine | 500 | 2.00 | 46.451 | 21.528 | 1,000 |
| Rental Value of Buildings | | | | | 99,800 |
| *Plus* | | | | | |
| Open used car stances | 25 stances | 500 | | 12,500 | |
| Car parking spaces | 40 spaces | 150 | | 6,000 | 18,500 |
| Total Rental Value | | | | | 118,300 |
| YP in perpetuity at 7% | | | | | 14.28 |
| | | | | | £1,689,324 |

Open Market Value, say £1,600,000

The valuation of a franchised dealership is based on the capitalised estimated rental value, as shown in the example. Considerable skill and experience is required in the valuation of the whole and the constituent parts.

# Showroom

There is a relationship between the rental value in the example of £18 per sq ft and good quality, off-centre retail rents in the area. Had the showroom been of 4,000 sq ft and not 3,000 sq ft, it would still have been reasonable to adopt the same rent per sq ft, but had it been of 10,000 sq ft the rent to be adopted would have been considerably reduced. Showrooms are a manufacturer's requirement. Unless proportionately increased sales can be made, having showrooms which are too large merely reduces the amount of retained profit per vehicle sold. The showroom has to bear rent, or occupancy costs, rates, cleaning, lighting and decorating. The number of new cars a garage proprietor can sell is influenced to a large extent by the market share of the particular franchise he holds.

## Offices

It is sometimes the case that a garage is also the head office of a larger business with other branches. In the example, offices of 1,000 sq ft are appropriate to a garage of this size. Had the general offices been 3,000 sq ft it would have been sensible to adopt a lower rent on the assumption that a typical willing buyer would be unlikely to have a need for so much office space. There is little relationship between the value of ancillary offices in a garage and the value of purpose-built offices in the same town. Typically an ancillary office rent should not exceed two-thirds of the value of showroom space and is more usually half.

## Workshops

The rent of £5 per sq ft allocated to the workshops represents the value of stand-alone workshop space of a comparable quality in the same town. The principle of marginal utility applies to workshops just as it does to showrooms and offices. The servicing and repair of vehicles has changed enormously in the last 20 years so that much less workshop accommodation is needed.

## Used car stances

Valuing used car stances calls for some fine judgment. The annual rental value of one stance should be related approximately to the average retained profit on one vehicle. A good used car operation as part of a franchised garage should turn over in a year approximately 10 times the number of cars displayed. The difficulty for the valuer is that some businesses do far better and many more do far worse. Here again the principle of marginal utility is important. Had there been 100 used car stances in the example it would be wrong to assume that the willing purchaser might sell 1,000 cars a year.

Such a volume could only be achieved by a specialist used car operation because a typical franchisee in this type of property would be hard pressed to generate more than 300 used car sales a year from part exchange, and the only way to get to a much higher number would be to have a very skilled used car buyer. It would be wrong for the valuer to assume that the typical willing buyer would have those skills.

If the garage in the example had no used car spaces or a very limited number of stances then the valuer's approach to the other constituent parts of the valuation would be different. The property would lack a vital source of profit. The increased reliability of used cars has changed the business considerably and is likely to lead to an increase in the establishment of massive, specialist used car operations.

Most garages are owner-occupied and open market lettings are rare, so good comparable evidence is difficult for the valuer to find.

## Capitalising the rent

This is the most difficult stage for the valuer. In other types of property there is an investment market in which institutions, property companies and individuals are prepared to be landlords. There is only a tiny investment market in garage properties.

Some sale and lease backs have been achieved by garage groups, though usually they have guaranteed rental growth and/or they have, for rent review purposes, artificially related the rental value to other types of property, such as off-centre retail space.

The yield adopted for motor trade property should have a closer relationship to the actual cost of money at the time. This can vary considerably. For example, a substantial motor group listed on the London Stock Exchange might buy the garage in Study 2 for £1,600,000. As we start the 21st century the trend has been for lower inflation and lower interest rates. The motor group might borrow the money to buy this property and pay 6% interest, which in this case would be £96,000 pa — considerably less than the estimated rental value.

## Location

The location of a garage is very important to the valuer. As the number of franchised dealerships has been falling steadily for the last 25 years, the trend is likely to be the establishment of larger, flagship dealerships in big towns and cities enjoying much larger dealer territories. It will be increasingly difficult to value garages in smaller towns. Some will survive as unfranchised garages, but many will go out of the trade completely.

## Fast Fit premises

While franchised garages generally tend to be owner-occupied, this is less so in the case of Fast Fit properties and the valuer will find it comparatively easy to obtain evidence of comparable transactions in the area. Rental values for such premises are not directly related to the value of workshop space in the area. Rents can often be as much as twice the level of ordinary workshop rents, particularly when the property is in a prominent location.

## Recommended reading

*Automotive Management.* EMAP Automotive, Media House, Lynchwood, Peterborough PE2 6EA. (Published fortnightly.)

*Forecourt.* Retail Motor Industry Federation, 201 Great Portland Street, London W1W 5AB. The magazine of the Petrol Retailers Association. (Published monthly.)

*The Motor Industry of Great Britain World Automotive Statistics.* Society of Motor Manufacturers and Traders Ltd, Forbes House, Halkin Street, London, SW1X 7DS. (Published annually.)

I am grateful to my colleague of many years, Peter Squire, and to Andrew Long of Swayfields for their help and guidance.

# Lease Renewals and Rent Reviews of Commercial Property

*Note: This chapter is arranged in three parts, consisting of the text and two appendices, one an example of an arbitration award and the other of an independent expert determination.*

This chapter deals with the issues involved in the rental valuation of commercial properties both for rent review purposes and also lease renewals. Deliberate emphasis is placed on practical matters, knowing that Professor Del Williams has dealt with procedural and legal issues so comprehensively in Chapter 4.

At any one time, there are literally hundreds of thousands of commercial rent reviews and lease renewals being negotiated throughout the UK. These are mainly dealt with by professionally qualified chartered surveyor valuers but it needs to be recognised that unqualified and less experienced people are also involved in the process including some in-house property managers and lay principals or directors of small businesses. Ultimately the rent paid by a business occupying commercial property is but one of its costs and the valuation process needs to recognise this.

Fortunately most cases are successfully negotiated and agreed but where negotiations between landlords and tenants in rent review cases break down the remedy, usually provided by the lease, is to refer the matter to an independent surveyor acting either as an arbitrator or an independent expert. The main issues involved in this process are dealt with later on in this chapter.

Where lease renewals are not capable of being agreed, the final arbiter is the county court or, in some cases, the High Court. The costs of this process are relatively high and the timescales involved generally lengthy and as a result a joint initiative by the RICS and the Law Society introduced a scheme known as PACT which offers the parties an opportunity to contract out of the county court procedure. Again, this will be dealt with in a little more detail later on.

So the following pages look at the subject of lease renewals and rent reviews of commercial property essentially from a practical point of view in the hope that they will inform and guide advanced students in the subject and perhaps, just as importantly, remind practitioners of their roles and responsibilities which are all too often forgotten or overlooked when the pressure is on.

## Best practice

Over recent years expert witnesses in all spheres have been making headlines, all too often for the wrong reasons, particularly where courts or tribunals have taken decisions based upon poor expert witness evidence. In his report *Access to Justice*, Lord Woolf devoted a whole chapter to expert witnesses and in 1996 he said:

> Those responsibilities (of the Expert Witness) should be made clear in the guidance issued by relevant professional bodies.

It was this that spurred the RICS to publish its *Practice Statement and Guidance Note for Surveyors Acting as Expert Witnesses* in rent review and lease renewal cases which became mandatory in March 1997. The fundamental point that the practice statement and guidance note underlines is that a witness has a primary duty to the court or the tribunal to be truthful, honest as to opinion, objective and unbiased. It is vital that the expert witness advises his client before taking on the role of this duty and explains that he may have to give evidence that includes facts and other material that is contrary to his client's best interests in fulfilling this duty.

Under the practice statement, a chartered surveyor may only accept instructions to appear as an expert witness:

1. if he advises the client in writing that the practice statement applies and offers to supply a copy of it on request (the practice statement is published and obtainable from the RICS Book Shop)
2. if he keeps a written record of matters on which the expert evidence is required.

Any changes in instructions or supplemental instructions from the client must also be recorded in writing. Where an inspection of the property is required, it must always be carried out to the extent necessary to produce an honest opinion for the tribunal, which is professionally competent, having regard to its purpose and the circumstances of the case. In producing the report, the surveyor must:

- personally sign and date it
- make a declaration that:

    (a) he believes it to be accurate and it includes all facts relevant to his opinion (ie it includes all comparables, even those which may not support his case)
    (b) the report complies with the 1997 RICS Practice Statement and Guidance Note
    (c) the expert understands his duty to the judicial body and has complied with it.

At the time of writing, new practice statements and guidance notes governing expert witnesses and advocates are about to be published by the RICS, but the above principles are not expected to be altered.

The policing of the procedure however is somewhat difficult although in cases which are in the public arena, such as Lands Tribunal decisions and court cases, the contents of the expert witnesses report may come directly to the attention of the RICS and attract disciplinary proceedings.

In the case of arbitration and independent expert determinations which are essentially private procedures, the policing role mainly falls upon the arbitrator or independent expert and there is ongoing debate as to whether arbitrators or independent experts should 'report' their colleagues who do not comply with the mandatory requirements of the practice statements. It is, of course, open to arbitrators and independent experts to refer to any shortcomings of the evidence in their awards or determinations and to that extent the parties to the proceedings, including the other side's surveyor, may well become aware of the Tribunal's view which may open up the opportunity of some other person referring the matter back to the RICS.

More importantly, however, the whole process of chartered surveyor expert witnesses' behaviour and to that extent compliance with the practice statements and guidance note is more a matter of education. A good expert witness should be able to accept and deal with information and evidence which is contrary to his case. If he does this successfully his credibility before the court or tribunal will be much enhanced.

One further area of confusion which is worth referring to at the present time is the prospect of chartered surveyors taking on the dual role of both expert witness and advocate. This more often arises in valuation tribunals, rent assessment committees and planning appeals although in many cases, in the arbitration and independent expert world, the expert witness slips into an advocacy role without even noticing it himself.

The 2008 practice statements and guidance notes will address this area and highlight the difference between the two notes. In particular, the introduction of a practice statement for advocacy is expected to beef up the responsibilities of surveyors who take on this duty.

Valuers should consider very seriously the desirability of taking on both roles in a single case, and at the very least they must make it clear to the judicial body which one they are adopting at each and every stage. This, however, can be a dangerous and messy business and, if at all possible, it is far better to stick to one or the other.

In the case of *Multi Media Productions Ltd* v *Secretary of State for the Environment* [1988], it was said:

> The expert who has also played the role of advocate, should not be surprised if his evidence was later treated by the court with some caution.

In practice many chartered surveyors do appear in the dual role, which is both necessary and cost effective not only for the client but for the whole dispute resolution process. They need to be especially aware not only of their duties and responsibilities under the practice statements and guidance notes but also, and just as importantly, to the tribunal. An arbitrator who is confused as to whether he is receiving expert evidence or advocacy is just as likely to attach little weight to it or reject it altogether.

# Lease renewals

As previously indicated statutory rights and legal procedures will not be dealt with in detail here because these have been set out so effectively in Chapter 4. However, the surveyor dealing with lease renewals needs to have background knowledge in order to be able to advise his client and carry out an effective valuation.

Lease renewals in respect of commercial properties are essentially governed by Part II of the Landlord and Tenant Act 1954 and supplementing this there is a considerable body of case law which defines what a business tenancy is. The definition is far wider than might otherwise be expected and for example a tennis club and a strip of land used for training race horses have both been held to be business uses as defined under section 23 of the Act.

The underlying purpose of Part II of the Landlord and Tenant Act 1954 is to provide security of tenure for the business tenant and to put in place a mechanism whereby an application can be made to the county court for a new tenancy at the expiry of the present one. The main difficulties arise where the landlord wishes to oppose the grant of a new tenancy and there are seven grounds specified under section 30(1) of the Act, which may enable the landlord to do this successfully. Two of the most commonly used grounds upon which a landlord can successfully prevent a tenant obtaining a new lease are that the landlord intends to redevelop the property or needs it for his own occupation.

Clearly any surveyor advising a client in connection with a lease renewal has to have knowledge of this process and even if the grant of a new tenancy is opposed, the question of assessing an interim rent and therefore the consideration of rental values arises.

Where a landlord wants to ensure that he can obtain possession at the end of a new lease that is being negotiated he may protect himself by registering any new business tenancy in the court under the provisions of the Law of Property Act 1969 as being outside the terms of Part II of the Landlord and Tenant Act 1954. This procedure is known in practice as 'contracting out' and in effect means that the tenant has no 1954 Act rights of renewal. Clearly this can have a significant effect upon the valuer's involvement and the way in which either the tenancy is negotiated or some form of temporary or continued occupation is agreed.

The valuer normally first gets involved with the lease renewal when the necessary statutory notices have to be served either by the landlord or the tenant. The valuer should point out to his client that a lease is coming to an end and the appropriate statutory notices and counter notices should be served. Because these notices lead directly to county court involvement there are considerable advantages in allowing the client's solicitors to deal with them and ensure that the strict timescales laid down by the Act are complied with. However, the valuer needs to be aware that any failure to respond to a notice or deal with the counternotices within strictly prescribed time-limits will significantly affect the rights of either the landlord or the tenant.

Perhaps the most important point of principle that the valuer should be aware of at the present time is that the Civil Procedure Rules (CPR) have introduced the need for valuers' reports at a much earlier stage in the lease renewal process than existed before. For example, if a tenant intends to serve a notice requiring a new tenancy, CPR lays down that a valuer's report setting out the rent and the terms upon which the tenancy is proposed needs to be prepared at the outset in a form that can be presented to the court, if necessary. This means that the report may have to be in the form of an expert witness report reflecting all the requirements and obligations set out earlier.

In practice, however, this strict regime is not always applied, mainly because the whole process of lease renewals is largely consensual. In most cases the valuer's role will probably be limited to helping the client and the solicitor establish the terms of the new lease being sought so that a court application can be made under section 34. Very often the landlord or indeed the tenant may seek to vary the terms of the original lease, for example, by seeking to change the repairing obligations or altering the length of the new term. The valuer's input will be required in this area to ensure that the new tenancy is in line with market circumstances. Once this initial formal process has been completed the valuer's job of negotiating the terms of the new lease can begin.

When dealing with old leases the language used or the terms expressed in the lease may be out of date and may not reflect current market conditions or current legal thinking. Section 35 of the Act gives the court the right to impose such terms and conditions as it thinks fit, having regard to the current tenancy and all the relevant circumstances. The valuer giving evidence must therefore be prepared to include in his report (and be prepared to be cross examined upon it if necessary) any changes that might be made by the court and to express an expert view as to how these changes might influence the rent to be determined under the new tenancy.

In February 1982 the House of Lords in *O'May v City of London Real Property Co Ltd* [1982] gave a judgment that is a good indication as to how far the courts will go to protect the tenant from imposed variations to the terms of the new lease. The facts of the case were that the landlords had sought to impose a full repairing lease (where one did not exist previously) upon the tenants and the High Court determined that this was a reasonable variation. The High Court was overturned by the Court of Appeal and the House of Lords upheld the judgment of the Court of Appeal. It was an agreed fact between the valuers that the transfer of the burden could be adequately compensated by a reduction in rent from £10.50 psf to £10.00 psf. The High Court set out certain tests that should be applied in these cases and neither the Court of Appeal nor the House of Lords materially dissented from them. There are essentially three main tests to be applied, as follows:

1. Has the party seeking a variation in the terms of the lease shown any reason for doing so?
2. If such a change in term is granted, can it be adequately reflected by a reduced rent determined under section 34?
3. Will the proposed change materially impair the tenant's security in carrying out his business or profession?

With all three questions in mind the court should consider whether each one is fair and reasonable between the parties.

The acid point of the case was the transfer of risk. It was not just sufficient for the landlord to show that the reduction in rent was adequate compensation for the additional burden imposed on the tenants. The transfer of risk itself had to be fair and reasonable as well.

## PACT

It is probably fair to say that most practitioners recognise that the county court does not offer a perfect solution for dealing with disputed lease renewals. The principal problem is that a county court judge dealing with the case is not likely to be an expert on leases or lease renewals and consequently the court judgment can lead to a somewhat unexpected answer which may not necessarily satisfy either party.

As a result, a joint initiative in 1997 by the Royal Institution of Chartered Surveyors and the Law Society instigated a service known as Professional Arbitration on Court Terms (PACT).

The scheme is available as a means of determining disputes relating to unopposed lease renewals under the Landlord and Tenant Act 1954 and offers an opportunity for these disputes to be resolved without the necessity of going to the county court. Both the terms of the lease and the rent can be decided by a surveyor or solicitor acting as either an arbitrator or independent expert.

The key parts of the service, which can be found on the RICS web site at *www.rics.org/drs* are as follows:

- the parties to decide whether it would be advantageous to them to refer aspects of the lease renewal to a third party solicitor or surveyor rather than a judge
- identify which issues or aspects of the renewal (if any) are agreed
- decide which aspects are to be referred to the third party, ie the new rent, interim rent, other terms of the lease, the detailed drafting of the terms or any particular clause or any combination of these issues
- choose which aspects are to be resolved by a third party solicitor and which by a third party surveyor
- choose third party's capacity — either arbitrator or independent expert
- draft the court application making use of or adapting the PACT model orders
- apply to the court for consent for the identified issues to be dealt with under the PACT procedure
- apply for the appointment of the third party arbitrator or independent expert. For a surveyor appointee, the application is to the RICS Dispute Resolution Service; a solicitor appointee, the application is to the Law Society

- proceed with the adjudication
- receive the decision of the third party appointee(s) subject to any cooling off rights and, in arbitration, any right of appeal.

# Rent reviews

As is the case with lease renewals, the valuer will either be acting for the landlord or the tenant. Irrespective of whether he is negotiating for one of the parties or whether he is preparing a proof of evidence for arbitration or independent expert determination, the fundamental issue that he has to determine is the basis of the rental valuation to be carried out. The golden rule in the case of rent reviews is that the lease will invariably set out in detail how the valuation has to be made and the valuer needs to read the appropriate rent review clause and interpret the wording strictly.

If there is some uncertainty or ambiguity about the meaning of the rent review clause or part of it, it may be necessary to seek an order of the court to determine the true meaning before a valuation can be carried out. If the dispute is before an arbitrator or independent expert it may be possible for that third party also to determine the issue. In the case of an independent expert the agreement of the parties will be necessary if the terms of that expert's appointment within the lease do not include his ability to determine points of law. Either way the use of an arbitrator or independent expert to deal with points of law or interpretation of the rent review clause may well be more speedy and cost effective than referring the matter to the court.

In recent years a great deal of debate has taken place on the question of the interpretation of certain parts of rent review clauses. The underlying approach the court will take in clarifying ambiguous wording is to adopt a stance equating, as far as possible, with the actual position of the parties at the time of review. This has commonly become known among valuers and solicitors as 'the presumption of reality'.

A good example to illustrate this point is where the length of the hypothetical term to be valued is in doubt. The issue normally centres on the argument as to whether the term should be the whole term (ie the original term of years granted) starting on the review date or the date of the commencement of the term, ie the unexpired residue as at the review date in question. The reality which the parties face at the review date is that the lease only has the unexpired residue to run and unless it is absolutely clear from the wording that the valuers are required to adopt the whole term starting on the review date, the presumption of reality would lead to the adoption of the unexpired residue.

The valuers need to be alert, however, to rent review clauses that do specifically provide for an 'unreal' hypothetical situation to be adopted. The presumption of reality argument does not extend so far as to overturn clear and precise instructions which nevertheless are out of line with the actual circumstances. This might, for example, include a presumption that the building to be valued is a different size from the actual building (subject to some formula adjustment afterwards) or is located in a different position or, perhaps, a combination of both.

The rent review clause will normally require the value to be determined by reference to such expressions as 'the open market value', 'the open market rental value' or the 'rack rental value' of the property. There is a very wide range of wording used to describe what generally boils down to open market rental value and again the valuer needs to be alert to the exact meaning and definition of value that he has to interpret.

Despite the fact that the Royal Institution of Chartered Surveyors and the Law Society have suggested standard appropriate wording to define rental value there remains a wide range of wording in use.

Coupled with this definition of the revised rent virtually all leases contain a series of assumptions and disregards which the valuer has to take into account. The principal assumptions are as follows:

- willing landlord and/or willing tenant.
- hypothetical term
- fit for immediate occupation and use
- to be valued as a whole — (or in parts?)
- terms of the lease to be the same as the existing lease (other than the amount of the revised rent)
- tenant has complied with its obligations under the lease — ie repairs.

The fact that there has to be assumed either a willing landlord or willing tenant (or both) does not mean that the rent should be at an artificially high figure. This assumption is normally included to counter any argument that at a particular review date there is no willing tenant (or landlord) in the market place. From the willing tenant's point of view such an assumption does not automatically mean that the tenant is so willing that he would pay more than the open market rate. The willing tenant will pay the correct rent and no more, just as a willing landlord is entitled to receive an amount in line with market value and no less.

As indicated above, the length of the hypothetical term to be valued is usually defined. In recent years this one issue has been the source of many disputes, particularly where the hypothetical term is relatively long and the market evidence at the review date indicates that tenants prefer to take shorter term leases. In these cases there is a divergence between the majority of open market evidence and the basis for the longer term set out in the lease. The valuer needs to be aware of this and make a suitable allowance in his valuation to reflect it. Very often this issue is interlinked to the 'whole term versus unexpired residue' argument referred to above.

The words 'fit for immediate occupation and use' are usually included simply to indicate that the premises are in a state to be occupied immediately and that the hypothetical tenant does not expect any delay before he can use the premises for his business. However, this does not extend to any time needed by the tenant for specialised fitting out works such as are usually required in retail shop units. These periods are normally regarded as part of the tenant's occupation.

Most rent review clauses contain the assumption that the property has to be valued as a whole. This is to ensure that the situation faced by the tenant in reality is followed. Sometimes, when the property is partly sublet or it is a large rambling unit, the value of the constituent parts may be more than the whole, and this is where the tenant's valuer needs to ensure that his client is not over charged. In a small number of cases the rent review clause specifically includes a provision for the revised rent to be the higher of either the value of the whole or the sum of the constituent parts. On the face of it this may appear unfair on the tenant, but it usually goes back to the original negotiations when the lease was drawn up and he will be bound by it.

More often than not most rent review clauses also provide that the hypothetical lease contains the same terms as the actual lease other than the rent to be reviewed and it also sets out the obligation for both the landlord and the tenant to have complied with their obligations under the lease. The important point to bear in mind here is that if the building is in a poor state of repair this needs to be disregarded, assuming of course that there are clear repairing covenants in place.

Usually the standard disregards under section 34(1) of the Landlord and Tenant Act 1954 are also provided for, i.e. tenant's occupation, goodwill and improvements carried out with the consent of the landlord. This latter item usually has the most influence on value particularly if the improvements are substantial. If they have not been correctly identified and excluded from the rental assessment then a significant over valuation can result. This is something that the tenant's valuer must take particular note of at an early stage of the rent review process in order to ensure that he does not need the support of his professional indemnity policy later on.

# Methods of valuation

There are three basic methods of valuation that are generally adopted for commercial property rent reviews, namely open market rental value, the profit's test and capital values (depreciated replacement building costs).

## *Open market rental value*

With the vast majority of standard business premises such as shops, offices and industrial units the normal method of valuation adopted is by way of comparison with similar properties. Evidence of values of similar types of units is gathered and any adjustments, in order to compare like with like, are made to support the valuation of the subject property.

The amount that the hypothetical tenant might pay in any particular case (or the hypothetical landlord might accept) must in the long run reflect the fact that the tenant needs to make a profit out of his occupation of the property and the landlord should fall short of asking for a rent so high that the tenant is unwilling to pay because he could not continue trading. If the tenant is drawn into paying more than an economic rate that his business will support there is a good chance that the business will fail and the landlord will end up with an empty property that he will have to relet. In such circumstances the cost to the landlord of the void period before a new tenant is found usually exceeds the small cost of a slightly more moderate rent in the first place. This is something that a landlord's valuer needs to have in mind at all times. This whole circumstance is reflected in the willing landlord and/or willing tenant assumption referred to earlier.

In considering the best evidence to adopt in an open market rental valuation assessment, it is worth reminding ourselves as to the 'hierarchy of evidence'.

The following is the generally accepted order of preference:

(a)   open market lettings
(b)   lease renewals
(c)   rent reviews
(d)   independent expert's determinations
(e)   arbitrator's awards.

Most arbitrators and independent experts will place weight on comparable evidence reflecting this order of preference, coupled with the fact that the proximity of the comparable transaction to the subject property also needs to be taken into account, ie an open market letting within the same building or same development should attract more weight than an open market letting at a more distant location. There is some argument as to whether arbitrator's awards should be used at all, reflecting the fact that they are primarily based upon evidence put before the arbitrator and unlike an independent expert's determination his answer may not reflect his own true opinion of the value of the property. In addition, the arbitrator's award in isolation without seeing the evidence which led to that award may be quite meaningless.

## *Profit's test*

There are a number of properties where reference to open market rental evidence is not possible and a test against profitability of the tenant may be the most appropriate approach. Examples of these are public houses, cinemas, theatres, clubs, night clubs, football clubs, etc. The valuer will make reference to certified or audited accounts of the tenant and needs to be able to analyse those accounts and calculate the true profits of the business. Once this is done, a notional apportionment of those profits between the landlord and tenant has to take place in order that an assessment of the amount of rent can be made.

It should be borne in mind that the auditing of accounts as a true and correct record by a chartered accountant is only the point at which the valuer's work begins. Most company accounts are prepared for inland revenue tax purposes and the skilled valuer needs to interpret the figures in order to arrive at a fair profit assessment upon which the rental apportionment can be made.

## *Capital values*

Finally, there is a small group of commercial properties where the valuer can obtain no assistance by way of market rental evidence and there is no profitable business being carried on there. In these cases, the valuer has to resort to the use of capital value. The type of property where this arises is normally a building used for a particular purpose such as a church, school or public building where there are no open market transactions to rely upon or no commercial business in existence on which to base a profit's test calculation. A more traditional property might fall into this category if the tenant is tied to a very restrictive user clause that has to be adopted under the rent review mechanism.

The valuer would only use this approach as a very last resort and may feel the need to bring in other expert advice to deal with such matters as the condition of the property and its ongoing maintenance and running costs.

# Methods of resolving disputes

This section concentrates on arbitration and independent expert determinations in relation to commercial rent reviews because these two methods of dispute resolution apply in the vast majority of cases. Other methods such as mediation, adjudication and PACT are also adopted but in a relatively small number of situations. Mediation is a different skill set altogether and is a process which tends to encourage the parties to reactivate any stalled negotiations. Both mediation and adjudication are perhaps more commonly used in construction industry disputes rather than with commercial rent reviews. PACT is specifically an alternative method of dealing with lease renewal disputes, as indicated above.

So this section will concentrate on trying to identify the differences between the roles of arbitrator and independent expert and help the reader decide how each process might be appropriate for any particular dispute.

An excellent starting point is to look at the Arbitration Act 1996.

# The Arbitration Act 1996

The Arbitration Act 1996 now completely governs the process of arbitrations and with effect from 31 January 1997 introduced a whole raft of new concepts which, over the 10 years since its introduction, are now well established. Only the key parts of the Act can be touched upon here and it is recommended that a copy of the Act itself should be on the bookshelf of all those involved in the rent review process. In addition *The Handbook of Rent Review* by Kirk Reynolds and Guy Fetherstonhaugh is regarded by most arbitrators as the 'bible' on the subject. A handier sized paperback with an excellent commentary on each section of the Act is *The Arbitration Act — a Commentary* by Bruce Harris, Rowan Planterose and Jonathan Tecks.

Section 1 of the Act sets out very clearly the basic principles to be adopted:

1(a)     the object of arbitration is to obtain the fair resolution of disputes by an impartial tribunal without unnecessary delay or expense

1(b)     the parties should be free to agree how their disputes are resolved, subject only to such safeguards as are necessary in the public interest;

Section 1(a) is quite self explanatory but rent review surveyors should be particularly aware of the duty imposed on the arbitrator to proceed 'without unnecessary delay or expense' and they should endeavour to work with him to promote this objective.

A principal feature of the Act set out in section 1(b) is party autonomy. Throughout its various sections, the Act provides that if the parties agree on a particular procedure then the arbitrator will be bound by it. This has come to be seen as the parties' ability to take ownership of their dispute and how it is handled. In some instances such agreements may not necessarily produce the best or most cost effective way of dealing with the dispute and if this occurs the arbitrator may wish to step in and point this out to the parties, reflecting his duties under section 33 as follows:

33(1)    The Tribunal shall—

(a)     act fairly and impartially as between the parties, giving each party a reasonable opportunity of putting his case and dealing with that of his opponent, and

(b)     adopt procedures suitable to the circumstances of the particular case, avoiding unnecessary delay or expense, so as to provide a fair means for the resolution of the matters falling to be determined.

As can be seen from this section the arbitrator's duties are very clearly and specifically set out. If the arbitrator does detect a conflict between the expressed wishes of the parties and these duties he will need to bring it to the attention of the parties and if necessary impose his preferred procedure. A good example of this conflict is where the parties require the matters to be dealt with in a particularly short timescale but the arbitrator believes it will compromise his duty under section 33(1)(a) to give each party a reasonable opportunity of putting his case and dealing with that of his opponent.

Section 34 sets out that the arbitrator has the power to deal with all procedural and evidential matters unless the parties have agreed otherwise.

Section 34(2) sets out a list of eight items which the arbitrator has specific power to decide and all rent review surveyors should be aware of these as follows:

(a)     when and where any part of the proceedings is to be held;

(b)     the language or languages to be used in the proceedings and whether translations of any relevant documents are to be supplied;

(c)     whether any and if so what form of written statements of claim and defence are to be used, when these should be supplied and the extent to which such statements can be later amended;

(d)     whether any and if so which documents or classes of documents should be disclosed between and produced by the parties and at what stage;

(e)     whether and if so what questions should be put to and answered by the respective parties and when and in what form this should be done;

(f)     whether to apply strict rules of evidence (or any other rules) as to the admissibility, relevance or weight of any material (oral, written or other) sought to be tendered on any matters of fact or opinion, and the time, manner and form in which such material should be exchanged and presented;

(g)     whether and to what extent the tribunal should itself take the initiative in ascertaining the facts and the law;

(h)     whether and to what extent there should be oral or written evidence or submissions.

The fundamental basis of arbitration has always been that the arbitrator will base his award on the evidence presented to him and decide the issues by weighing that evidence and coming to a conclusion in favour of one party or the other, or somewhere in between. It is therefore theoretically possible, if this is applied strictly, for the arbitrator to come to a totally valid award based on the evidence before him but not producing the correct valuation.

Section 34(2)(g) introduces an important concept into arbitration which did not exist prior to the 1996 Act and gives the arbitrator the power to ' ascertain the facts and the law' relating to the case. In practice this means the arbitrator has the power to make investigations himself if he is not satisfied with the quality of evidence put before him. In the event that the arbitrator decides to do this, he needs to advise the parties of his intentions and make sure that any evidence he unearths is put before them and they are given an opportunity of commenting upon it before he uses it in making his award.

Where two competent valuers appear before an arbitrator on a traditional rent review case this procedure is probably unnecessary but it may be of considerable benefit where, for example, one of the parties is unrepresented and is unfamiliar with arbitration procedure and rental values. In these types of case it will enable the arbitrator to bring his own skill and knowledge of the market to the dispute, thus giving a level of protection to the unrepresented party and ensuring that the correct answer is obtained.

By bringing his own expert knowledge and skill to bear it enables the arbitrator to adjust his role somewhat towards that of an independent expert while retaining the formal authority of an arbitrator. However, this procedure should only be used with a great deal of caution and it is certainly not something that should be adopted automatically or indeed should be expected to be so used by the parties.

Having set out the powers and duties of the arbitrator it is worth reminding ourselves that the parties also have specific responsibilities set out under section 40. In short they are required to do all things necessary for the proper and expeditious conduct of the proceedings including complying with any directions of the arbitrator.

Unlike an independent expert, who may be sued for negligence if he comes to a wrong answer, an arbitrator has immunity under section 29 and cannot be held liable for anything done in the discharge of his duties unless it can be shown that he acted in bad faith. He has the power under section 30 to make a ruling on his own substantive jurisdiction subject only to a challenge and application to the court under section 67. Perhaps more worryingly from the arbitrator's point of view his actions can be challenged under section 68 if he has been guilty of serious irregularity and the section sets out nine grounds upon which a successful challenge can be made. The court has the power either to remit the award back to the arbitrator for reconsideration or can set it aside or declare it to be of no effect. Finally here it is worth mentioning appeals which can only be made on a point of law under section 69. Most importantly it should be noted that any agreement by the parties to dispense with reasons in the award will mean that their ability to appeal under this section will be lost.

# Arbitration procedure

The vast majority of all commercial rent review arbitrations are dealt with by way of written submissions rather than a hearing. This is probably driven by the preference of the surveyors representing the parties to make written representations rather than to expose themselves to the stressful experience of cross examination by solicitors, barristers or even their contemporaries on the other side.

However, it is worth saying that hearings have a number of distinct advantages and in some cases can be quicker and more cost effective than written submissions which can easily drag on and need to be supplemented by further correspondence.

To repeat an important point made earlier, the rent review surveyor's duty is to the tribunal and not to his client once arbitration proceedings have commenced. This is something that is very often overlooked and the written submissions turn out to be a slugging match between valuers, sometimes reducing to personal insults rather than supplying useful evidence in order to assist the arbitrator in his task. Surveyors and valuers who proceed in this way are easily spotted by the arbitrator and inevitably this kind of approach ends up being to the detriment of their client's case rather than enhancing it.

In written submission procedures, being straightforward, open and honest is just as important as when giving oral evidence at a hearing.

Once the arbitrator has issued directions it is important that the valuers adhere to them, not least because it contravenes the RICS Practice Statement if they do not, but perhaps more importantly because the patience of the arbitrator will be worn thin if he has to continually bring a wayward valuer back on course. As has been mentioned before, section 40 sets out the duties of the parties which includes doing 'all things necessary for the proper and expeditious conduct of the arbitral proceedings' and the giving of truthful and complete evidence is certainly part of that duty.

Most arbitrators expect responsible valuers to present a statement of agreed facts including such basic information as the floor area of the property to be valued, rateable values, planning information and the factual details of comparables to be introduced. As far as evidence is concerned, the majority of arbitrators will now accept evidence in a relatively informal format, certainly in small cases. This is usually in the form of proformas signed by the party responsible for the transaction in question. While technically this may not be strictly proven, most arbitrators adopt this procedure because the Civil Evidence Act 1995 itself allows the introduction of hearsay evidence subject to certain rules and notices. An experienced arbitrator with knowledge of the market place within which he is working is usually able to weigh this type of evidence satisfactorily rather than simply rule it out as inadmissible.

The parties to an arbitration should bear in mind that unless there is a written agreement between them to the contrary, the arbitrator's award will be given with reasons. These reasons should be expected to explain to the parties how the answer has been arrived at and they should ensure that, at the very least, the losing party understands why they have lost. An arbitrator's award can be appealed on a point of law under section 69 only if reasons are given and such appeal must be lodged within a period of 28 days from the date of publication of the award. On matters of fact, including valuation issues, an arbitrator's award is final and binding unless it can be shown that he has been guilty of serious irregularity (section 68) or he has exceeded his jurisdiction (section 67).

Finally, it is worth noting that an arbitrator must deal with the question of costs in order to make the award final. This can either be included within the award itself or if he has had a request to defer matters or has been made aware of the presence of a Calderbank offer his directions should provide for the making of a first award (as an interim award) with the question of costs being reserved to a later date. A subsequent procedure should be set in place for the parties to make submissions on the issue of costs once the first award has been delivered.

The whole question of costs in arbitrations is a matter which many surveyors and even some arbitrators find difficulty in dealing with. It is worth reminding ourselves that the fundamental principle in English law that governs the award on costs is that costs follow the event. In simple terms this means that the party who has won the arbitration can expect to have their costs paid by the loser. In rent review cases, however, the event is always the award of the revised rental figure and this is where some of the difficulties begin simply because most awards lie somewhere in between the parties' submissions rather than exactly and neatly in line with one or other of them. It is for this reason that the parties are advised to make a Calderbank offer — a process somewhat equivalent to a payment into court — to protect their positions on costs. In coming to his decision on costs, the arbitrator has to act judicially in exercising his discretion as to who has won and who has lost and certainly needs to take into account any Calderbank offers made.

One of the biggest dilemmas in this area is for the arbitrator to work out which, if any, of the Calderbank offers is close enough to his award for it to be considered a winner, and this introduces the much debated concept of 'near miss' Calderbanks. The *RICS Guidance Note for Surveyors Acting as Arbitrators & Independent Experts in Commercial Rent Reviews* is essential reading on this subject. In paragraph 3.10.6 the guidance is given to the effect that the Arbitrator needs to make a ruling of fact as to whether the Calderbank offer in question is within a negligible amount of the award. If it is, then in exercising his discretion, he may regard it as successful and make an award on costs reflecting that. If the difference between the offer and award is what is described as a 'non-negligible' amount it should be regarded as unsuccessful.

The difficulty that exists is establishing the difference between a negligible and a non-negligible amount and unfortunately different arbitrators approach the question differently. It is suggested that a straight forward percentage of the award figure is not the right approach as it may not reflect the circumstances of the particular case and where large sums are involved may produce a difference that is actually an unacceptably large figure. In practice most arbitrators tend towards a relatively small sum but beyond that it is not possible to be more definitive.

In some cases the parties try to get the arbitrator's prior agreement as to how he will deal with this problem or even put to him some sort of agreed formula. The difficulty here is that the arbitrator will be signing away his discretion which is such an essential ingredient in his judicial decision making process.

Finally on costs, section 63 of the Act deals with the recoverable costs of the arbitration and sets out the principle that they shall be based on a reasonable amount in respect of all costs reasonably incurred. Section 65 gives the arbitrator the power to limit in advance the amount of recoverable costs and this is a useful tool to be applied in cases where the arbitrator believes one side intends to spend an excessive amount in prosecuting its case in relation to the amount in dispute. As an example, this might typically apply where a wealthy landlord employs an expensive barrister in a small case against an unrepresented tenant where the amount of rent in dispute is quite small. This principle of cost capping therefore goes to the root of the arbitrator's duty under sections 1 and 33 of the Act to provide a fair means of resolving the dispute and avoiding unnecessary expense.

# Independent expert determinations

In practice many surveyors and valuers tend to blur the edges between the differing roles of independent expert procedure and arbitration and they make submissions which they expect the independent expert to deal with similarly to an arbitrator. First and foremost, the independent expert is obliged to use his own expertise and market knowledge in arriving at his assessment of the revised rent.

He may be required to invite submissions from the parties under the terms of the lease but ultimately he is quite entitled to disregard what he reads if he believes it to be unhelpful or incorrect. This is a fundamental difference between the two roles and it is vitally incumbent upon the independent expert to arrive at the correct answer reflecting his own expertise and skill. If he fails to do this he is liable to be sued for negligence by one or other of the parties whereas an arbitrator is immune from this risk.

Over recent years there has been a great deal of debate about the need for independent experts to give reasoned determinations. The early thinking was that the giving of reasons was an opportunity for the parties to challenge the determination and particularly for the losing party to cause trouble. A competent independent expert should not be afraid of giving at least an explanation for his determination and certainly a breakdown of his valuation so that the parties can see how he has arrived at his answer. It may be good practice for him to avoid the use of the word 'reasons' in his procedural instructions to the parties in order to help differentiate his role from that of an arbitrator.

The *RICS Guidance Note for Surveyors Acting as Arbitrators & Independent Experts in Commercial Property Rent Reviews* is vital reading to both rent review surveyors and independent experts on this subject. There is a particularly helpful section on this whole question of reasons and under what circumstances an independent expert should give reasons when requested to do so by one or other of the parties with or without an objection from the other side.

The principal differences between the roles of arbitrator and independent expert are set out in table form below. These may be a useful quick reference to help the reader decide which role to opt for if there is an opportunity to do so under the terms of the lease or, in the case of a new lease being drawn up, which role to incorporate.

## The Differing Roles

| Arbitrator | Independent Expert |
|---|---|
| Process supported by a statutory framework | Contractual appointment — 'an expert employed by both parties' |
| Party autonomy | Investigatory duty rather than adversarial |
| Arbitrator's positive powers | Binding decision based on own knowledge and skill |
| Duty to act fairly and impartially | No procedural restrictions |
| Ability to ' tailor make' procedures | Expert not bound by the 'evidence' |
| Reasons must be given (if required) | No right of appeal |
| Appeal on a point of law | No power of disclosure |
| Arbitrator relies on evidence | Reasons may not be given |
| Arbitrator not liable for negligence | No power to determine costs — unless under terms |
| Power (and duty) to determine costs | of appointment |
| Award enforceable as a judgement | Duty of care — negligence |

# Evidence

Both in formal arbitration procedures and court cases, the introduction of evidence is now governed by the Civil Evidence Act 1995 which came into force in 1997.

Essentially the Act allows the introduction of hearsay evidence to be given if the party wanting to produce it has given notice. If requested, the party must give particulars of or relative to the evidence.

Section 4 of the Act lays down what considerations an arbitrator (or the court) should bear in mind in weighing hearsay evidence and deciding its weight and importance as follows:

. In estimating the weight (if any) to be given to hearsay evidence in civil proceedings, the Court shall have regard to any circumstances from which any inference can reasonably be drawn as to the reliability or otherwise of the evidence.

. Regard may be had in particular to the following:

(a) Whether it would have been reasonable and practicable for the party by whom the evidence was adduced to have produced the maker of the original statement as a witness.

(b) Whether the original statement was made contemporaneously with the occurrence or existence of the matter stated.

(c) Whether the evidence involved is multiple hearsay.

(d) Whether any person involved had any motive to conceal or misrepresent matters.

(e) Whether the original statement was an edited account or was made in collaboration with another or for a particular purpose.

(f) Whether the circumstances in which the evidence is adduced is hearsay such as to suggest an attempt to prevent proper evaluation of its weight.

## Proof of evidence

Under arbitration procedures the parties are free to agree whether strict rules of evidence apply or not. In practice most commercial rent review arbitrations proceed on the basis of strict rules not applying. In other words, some form of hearsay or evidence that has not been strictly proved is usually admitted and dealt with as a matter of weight by the arbitrator. As a practical point in arbitration, the arbitrator is normally happy to accept details of comparable evidence as being proved and fully admissible if both parties agree the information as a matter of fact. In the High Court or formal arbitration proceedings where strict rules are being applied, the proof that is required is the production of the actual documents or certified copies.

One interesting area that sometimes proves difficult in arbitration cases is the question of post review date evidence. In an early Australian case, *Melwood Units Pty Ltd* v *Commissioner of Main Roads* [1979], PC the Privy Council held the Land Appeal Court wrong to reject evidence of a June 1966 sale in relation to a compulsory purchase order of September 1965. A 1982 case, *Duvan Estates Ltd* v *Rossette Sunshine Savouries Ltd* [1982] expressed a contrary view and in *Segama NV* v *Penny le Roy Ltd* [1984] it was held that an arbitrator was entitled to admit evidence as to rents agreed after the relevant date.

The weight that an arbitrator attaches to post review date evidence is therefore a matter of his judgment and experience just as it is with any other evidence before him. Post review date evidence is therefore probably correctly admissible not just in relation to a trend in values, as was highlighted in the *Melwood* case, but also of actual values. However, the greater the period between the valuation date and the date of the transaction, the less weight the evidence is likely to carry with the arbitrator. If there has been some economic or local event which has significantly affected market values in between the two dates, this too will reduce the weight to be attached to the evidence, perhaps even to zero.

## Arbitrator's award

The arbitration process is concluded by the arbitrator making an award containing his decision and the reasons leading to that decision.

In general terms an arbitrator is bound by the evidence put before him whether it be evidence of comparable properties or opinion evidence of the experts before him. If in what are genuinely exceptional cases, he has introduced evidence himself under the provisions of section 34(2)(g) of the Arbitration Act 1996 and given the parties the opportunity of commenting upon it, he needs to include a reference to that evidence and information in his award.

Although it is quite possible for the parties to agree to dispense with reasons under The Arbitration Act 1996, it is a very rare occurrence mainly because they are, quite naturally, not prepared to dispense with their right of appeal at the outset of a case. Apart from that it is perhaps the single most important part of the whole arbitration process for the parties to understand how the arbitrator has reached his decision and what evidence and comparables he has found most (and least) helpful. In difficult cases it may also be necessary and beneficial for the performance of the valuers giving evidence to be commented upon and any deficiencies highlighted.

In Appendix I to this chapter there is a sample award which the reader should not take as a definitive format as to how an award should be written. It is intended, however, to show how the arbitrator should marshal the facts and evidence and deal with the information given to him. Most importantly, it indicates how he has addressed the evidence and the stance of the parties and come to his decision based upon that evidence.

As a contrast Appendix II contains a sample independent expert determination which is intentionally drafted in a more informal fashion. It demonstrates the essential expert skill of the independent expert by the use of such phrases as 'I am of the opinion that ...' and 'I am of the view that ...' This can be contrasted with the arbitrators words such as 'I agree with Mr Jones on this point' or 'I prefer Mr Brown's evidence on this issue to that of Mr Green'.

# Appeals and challenges

Since the introduction of the Arbitration Act 1996 the courts have proved extremely reluctant to overturn arbitrators' awards unless there is something manifestly wrong. Section 70 of the Act specifically provides that a challenge cannot be made without the aggrieved party first having exhausted any available process under the Act itself either by way of appeal, review or even a correction of an error by the arbitrator under section 57. Subject to that, challenges can be made upon:

(a)  the substantive jurisdiction of the arbitrator (section 67)
(b)  grounds of serious irregularity (formerly known as misconduct) (section 68)
(c)  an appeal on the question of law, arising out of the award (section 69).

Under section 69 an appeal needs to be lodged within 28 days of the publication of the award.

## Appendix I  An arbitration award

There is no standard or approved format for an arbitration award under the Arbitration Act 1996. The award may be as lengthy or as brief a document as is necessary, depending upon the arguments advanced and the complexities of the issues of the case. A true test of the quality of the award is not only that the parties, and in particular the losing party to the arbitration, should understand how the decision has been reached but it should also contain sufficient information for a judge to follow the arguments advanced and the reasons for the arbitrator's conclusion in the event of a challenge or an appeal.

A 1981 High Court decision in *Gleniffer Finance Corporation* v *Guardian Royal Exchange Assurance Ltd* [1981] 14 April before the Hon. Mr Justice Goff concluded his judgement with the very useful guidance and wording as follows:

> The arbitrator's award was a full and accurate one in which he reviewed all the evidence and on the basis of that he exercised his judgment based upon his skill and experience. The exercise of an arbitrator's judgment on this basis is not an appropriate case for review under the Arbitration Act 1979.

> I therefore dismiss this application with costs.

The sample award that follows is fictitious as it is not possible to reproduce an actual award which is a private document between the parties concerned. Those of you that have read awards from the past will have noticed a number of 'Whereas' and 'Wherefores' which you will find are noticeably absent from this sample. In recent years arbitrators have been coached and encouraged to produce awards that contain plain and simple English leading logically through from the appointment, any background information, the facts of the case, the evidence and eventually to the conclusion.

Arbitration Award
of
AB Potter, FRICS, MCIArb

in respect of premises known as

Unit 3 Station Square Shopping Centre
Solihull
West Midlands

Dated this 20th day of August 2007

AB Potter, FRICS MCIArb
Arbitrator
Hillman Smith & Potter
Chartered Surveyors
99 Old Street
Birmingham B1 3PL

In the matter of the Arbitration Act 1996
and
In the matter of an Arbitration
between
Wealthy & Wise Investments Ltd (Claimants)
and
Bill Smith Butchers Ltd (Respondents)
this is the Partial Award
of
AB Potter FRICS MCIArb

## 1.0 Preliminaries

1.1 By a lease dated the 19 March 2001 between *Apex Assurance plc (Landlords)* and *West Midland Meats Ltd (Tenants)* the premises known as Unit 3 Station Square Shopping Centre, Solihull, West Midlands were demised by the Landlord to the Tenant for a term of 10 years from the 19 March 2001 upon the terms and conditions specified therein.

1.2 The annual rent reserved by the lease commenced at £22,500 pa and is subject to review under the terms of the lease as set out in Clause 9.

1.3 The parties were unable to agree the revised rent, as defined in the lease with effect from the 19 March 2006 and I was appointed by the President of The Royal Institution of Chartered Surveyors to act as Arbitrator on 1 March 2007.

1.4 I have been supplied with a verified copy of the lease which I have read and have given the parties the opportunity to make representations to me. Both sides have requested that the matter be dealt with by written submissions.

1.5 I issued draft Directions to the parties on the 17 May 2007 and following receipt of their comments I issued my Directions dated 8 June 2007 in an agreed format.

1.6 I have received and read the submissions and counter submissions from both sides and inspected the subject property and the comparables put forward.

1.7 As confirmed in my Directions, my Award is given with reasons and at the request of the Tenant's representative I am issuing a Partial Award initially and reserving the question of costs to a later date.

1.8 In their initial responses both parties indicated to me that they considered that there were no points of law at issue and from their submissions and counter submissions I find this to be the case. The dispute, therefore, revolves entirely around the question of fact and rental valuation of the subject property at the review date.

## 2.0 The Parties' Representatives

2.1 The Landlord is represented by Mr C Smith of Smith & Partners, Chartered Surveyors, and the Tenant is represented by Mr J Brown of Brown & Green, Estate Agents and Valuers.

## 3.0 Rent Review Provisions

3.1 Clause 9 of the lease sets out the mechanism for reviewing the rent which shall be revised at each review date to whichever is the greater of the existing rent prior to the review date or the sum determined under Clause 9 of the lease.

Clause 9.1 defines the revised rent as

… the yearly rent at which the Premises let as a whole might be expected to be let in the open market without any fine or premium for a term of years equivalent to the residue of the Term or 10 years (whichever is the greater) commencing on the relevant Review Date as between a willing landlord and a willing tenant with vacant possession upon a lease in the same terms as this Lease apart from the amount of rent thereby reserved (but including the provisions for rent review) on the assumptions that at the relevant Review Date:-

9.1.1 the Premises are fully fitted out and equipped to the requirements of the willing tenant and are ready and fit for immediate occupation and use

9.1.2 no work has been carried out to the Premises by the Tenant or any predecessor in title of the Tenant which has diminished the rental value of the Premises

9.1.3 if the Premises have been destroyed or damaged they have been fully re-instated

9.1.4 the covenants by the Tenant and the Landlord have been fully observed and performed (but without prejudice to the Landlord's rights against Tenant if there has been any breach)

There then follows the usual disregards for Tenant's occupation, goodwill and improvements. There is a further disregard for any rent-free or concessionary rental period that might be allowed to an incoming Tenant to enable it to fit out the premises.

## 4.0 Floor Areas

4.1 The Surveyors representing the Landlord and the Tenant have agreed the factual floor areas with regard to the subject property and I therefore find that the area of the shop ITZA is 566 sq ft. The first floor storage area is agreed at 487 sq ft.

## 5.0 The Parties' Valuations

5.1 I now turn to the valuation of the two Surveyors wherein the dispute lies.

5.2    Mr Smith, on behalf of the Landlord, values the property at £36,500 pa based upon a Zone A rate of £61.85 and £3.00 psf on the first floor storage. Mr Brown, on the other hand, values the shop at £41.30 Zone A but adopts the same figure of £3.00 psf on the first floor. He arrives at a valuation of £24,835 pa.

## 6.0    Comparable Evidence

6.1    In support of their respective valuations, the two Surveyors have submitted a number of comparables to me which I have considered carefully. As the majority of comparables are agreed, I shall list them briefly as a single group and then indicate which I prefer to rely upon.

6.2    <u>Unit 20 Station Square Shopping Centre</u>
Rent review with Sports Gear Ltd with effect from the 26 July 2003 at £30,500 pa devaluing at £47.50 Zone A.

6.3    <u>Unit 1 Station Square Shopping Centre</u>
Rent review with Cartelephones Ltd with effect from the 26 July 2003 at £29,500 pa at a Zone A rate of £40.30.

6.4    <u>Unit 7 Station Square Shopping Centre</u>
Rent review with Chatterleys with effect from the 26 July 2003 at £35,750 pa devaluing at a Zone A rate of £47.50.

6.5    <u>Unit 19 Station Square Shopping Centre</u>
Rent review with Jones Opticians with effect from the 26 July 2003 at £23,000 pa, devaluing at £48.91 Zone A.

6.6    <u>Unit 5 Station Square Shopping Centre</u>
An open market letting to Photo Supplies with effect from the 21 August 2003 at a rent agreed at £23,000 pa, devaluing at £42.30 Zone A.

6.7    <u>Unit 4 Market Street</u>
An open market letting to Phones For All with effect from the 29 September 2004 at a rent agreed at £24,000 pa, devaluing at £47.50 Zone A.

6.8    <u>Unit 8b Station Square Shopping Centre</u>
An open market letting to Games & Toys Ltd with effect from the 17 October 2005 at a rent agreed at £47,000 pa. The area ITZA is 685 sq ft and the Zone A rate devalues at £64.25.

6.9    <u>Unit 8a Station Square Shopping Centre</u>
Open market letting to Card Greetings with effect from the 4 November 2005 at a rent agreed at £23,500 pa. The area ITZA is 463 sq ft and the Zone A rate devalues at £70.00.

6.10   <u>Unit 4 Station Square Shopping Centre</u>
An open market letting to Barkers with effect from the 30 June 2006 at a rent agreed at £36,000 pa. The area ITZA is 556 sq ft and the Zone A rate £61.85.

6.11   <u>Unit 6 Station Square Shopping Centre</u>
This is an open market letting to Spot Off for a 5 year term outside the 1954 Landlord & Tenant Act with effect from July 2007. The rent is stated to be £23,500 pa devaluing at £68.85 Zone A. However, this letting is still in Solicitor's hands and has not yet concluded. In addition, it is a kiosk size unit significantly smaller than the remaining shops and the subject property.

## 7.0    Valuation Issues

7.1    The difference between the two Valuers in this case is relatively straightforward. Mr Smith on behalf of the Landlord follows the level of rent achieved by the 2003 and 2004 open market lettings in respect of Units 8a, 8b and 4. Although the Unit 4 letting to Barkers is a post review date transaction, he adopts the £61.85 psf for his valuation of the subject property, placing particular emphasis on the fact that Barkers is the immediate next door unit.

7.2    On the other hand, Mr Brown relies almost entirely on the raft of 2003 rent reviews to support his £41.30 figure. He criticises the open market lettings of Unit 8a and 8b as being out of line with the established tone of value and draws my attention to the fact that such traders as card shops and computer/computer games retailers may pay excessive rents to gain representation.

7.3    I have considered these two arguments very carefully and have come to the conclusion that I cannot follow Mr Brown's approach of relying almost entirely upon a group of July and August 2003 rent review transactions almost three years before the subject review date. Whilst I accept there may be some anomaly in the lettings to Card Greeting and Games & Toys Ltd, I cannot accept his complete disregard of these two lettings which are not only open market transactions but are relatively close to the subject review date. In addition, they are supported by the Barkers letting on the immediately adjoining unit to the subject property and to that extent I agree more with Mr Smith's approach than Mr Brown's.

7.4    Reference has also been made to the Phones For All letting at 4 Market Street in September 2004. I place less weight on this transaction because it is in a different location at the far end of the main shopping street and I find it less persuasive than the three open market lettings in the immediately vicinity.

## 8.0    Conclusions

8.1    Having effectively disregarded Mr Brown's group of historic rent reviews I look a little more closely at the three open market lettings, Units 8a, 8b and 4. I am happy to accept Mr Brown's contention that the 8a and 8b lettings are probably somewhat excessive and reflect special purchaser situations. In particular, the fact that the £70.00 Zone A figure on the Card Greetings letting is significantly higher than any of the others tends to support this argument. However, I do find that they indicate demand in this locality and that this is supported by the subsequent letting to Barkers on the adjoining unit to the subject property.

8.2    I accept Mr Brown's argument, albeit based on the 2003 rent review evidence, that there is a 'tailing off' of values towards the Station Street end of the mall where the Cartelephones Ltd unit is.

8.3    I also accept his argument that the lettings at Units 8a and 8b reflect a somewhat superior location than the subject property being within the main mall. I am also inclined to agree with his view that these two lettings reflect an overbid of between 5% and 10% for location and 5% and 10% for special purchaser, although because of the significant difference between the two Zone A figures, I do not think it is possible to try and analyse these more scientifically. However, I do find that they support the level achieved, albeit some three months after the review date, on the Barkers unit of £61.85 psf.

8.4    As I have already indicated I am happy to accept Mr Brown's argument that there is a tailing off between Units 7 (Chatterleys) and 1 (Cartelephone Ltd), which is at the far end of the mall. This leads me to the conclusion that the rent on the subject property, which is one unit closer to the Station Street end of the mall from Barkers, should be slightly below their rent of £61.85.

8.5    After careful consideration of all these arguments therefore and placing most weight on the open market letting of the Barkers unit immediately next door, I have come to the conclusion that the appropriate Zone A figure to apply to the subject property is £60.00 Zone A.

8.6    I then turn to the first floor accommodation where both Valuers agree on the size and the rate to adopt and I therefore follow their approach and use £3.00 psf for this area.

## 9.0    Valuation

9.1    My valuation of the subject premises therefore is as follows:

| | |
|---|---|
| Agreed area ITZA 566 sq ft at £60.00 | £33,960 |
| FF Storage — agreed area 487 sq ft at £ 3.00 | £1,461 |
| Total | £35,421 |
| Say   £35,400 pa | |

## 10.0    Award

10.1    I THEREFORE HEREBY AWARD that the revised rent of the subject premises, as defined for rent review purposes in Clause 9 of the Lease, with effect from the 19 March 2006, is the sum of **£35,400 (Thirty Five Thousand, Four Hundred Pounds) per annum**.

**11.0 Fees**

11.1 My fees for this Partial Reasoned Award amounts to £——- plus VAT in accordance with my Directions Letter of the 8 June 2007.

**12.0 Costs**

12.1 This Award is a Partial Award and at the request of the Tenant I defer the question of costs upon which I shall invite the parties to make submissions to me at a later date.

**13.0 Seat of Award**

13.1 The Seat of this Arbitration Award is England & Wales.

**Dated this 20th day of August 2007**

**Signed** _____

       **A B Potter FRICS MCIArb**

       **Arbitrator**

# Appendix II An Independent Expert Determination

This sample independent expert determination is in an informal letter format deliberately to highlight the difference between it and an arbitrator's award. It is quite acceptable for a determination to be set out more formally like an award particularly for larger properties where a substantial amount of rent may be in dispute or where detailed reasons are appropriate. However there is no specifically required format so long as the basics in line with the dozen or so sub headings in the following sample are covered. If there are further issues that affect the valuation being prepared then they may need to be included but an unnecessarily long document at increased cost is certainly not needed. If the determination is being given without reasons then a recital of the facts and background information is probably required but paragraphs 8–11 in the following sample may be either excluded or much reduced. This sample includes a complete recital of the rent review clause which is intended to indicate to the parties that the independent expert is aware of its details and has taken it into account. However this may not always be necessary so long as the independent expert makes it clear that his valuation has correctly reflected the precise requirements of the lease.

Our ref : ABP KE AB0007

20 August 2007

F Smith Esq  
Frederick Smith & Company  
Chartered Surveyors  
3 High Street  
Birmingham

T Jones Esq  
Jones Green & Barker  
Surveyors and Valuers  
100 Station Street  
Birmingham

Gentlemen

**Unit 99 Airport Industrial Estate Coventry**

This is my Determination as Independent Expert in respect of the rent review relating to the above property, with effect from the 23 August 2006.

**1.    Appointment**

    I was appointed on the 10 April 2007 by the President of The Royal Institution of Chartered Surveyors to act as Independent Expert under the provisions of the lease, dated 23 August 1996 between *Coventry Estates*

*Ltd* and *Plastic Parts Trading Ltd* in respect of premises described in that lease as Unit 99 Airport Industrial Estate, Coventry, West Midlands.

## 2.    Rent Review Provisions

The lease provides that the rent shall be reviewed on the 23 August 2006. The definition of the revised rent is contained in Paragraph 2.2 of Part 2 of the Fourth Schedule to the lease.

2.1    The revised rent shall be the higher of the rent passing prior to the review date or the

... annual rent at which the demised premises might reasonably be expected to be let at the relevant review date which might be granted in the open market on a new letting of the demised premises by a willing lessor to a willing lessee.

2.2.1    On the following assumptions at the review date:-

2.2.1.1    That the demised premises are available to let on the open market by a willing lessor to a willing lessee and without a fine or a premium or value in the nature of a fine or a premium and with vacant possession for a term ten years or the unexpired residue of the Term (whichever is the longer) beginning on the relevant review date.

2.2.1.2    That the demised premises are to be let as a whole subject to the terms of this Lease (other than the amount of the rent reserved by this Lease but including these provisions for a review of the rent at 5 yearly intervals throughout the whole of the proposed term).

2.2.1.3    That the demised premises are fit and available for immediate occupation and use by an incoming tenant for the purpose of beginning its fitting out works.

2.2.1.4    That the covenants contained in this Lease have been fully performed and observed.

2.2.1.5    That no work has been carried out to the demised premises which has diminished their rental value except to the extent that it has been carried out pursuant to any statutory requirements or the requirements of any local authority or other public body and that in case the demised premises have been destroyed or damaged they have been fully restored.

2.2.1.6    That no reduction is to be made to the rental value to take account of any rental concessions which on a new letting with vacant possession might be granted to the incoming tenant for a period in which its fitting out works might take place.

There then follows the usual disregards for tenant's occupation, goodwill and improvements.

## 3.    The Premises

3.1    The premises comprise a self contained two bay steel portal frame warehouse unit built in 1993 and incorporating two storey offices to both front and rear elevations. It has concrete floors throughout and the external elevations are of brick construction to 2m approximately, with lined PVC coated profile metal cladding above. The office sections have brickwork to full height. The height of the warehouse eaves is approximately 6.0m.

3.2    Internally, the warehouse provides an open floor area with four roller shutter loading doors to the side elevation and two to the rear where there is a self contained secure yard area.

3.3    The offices are fitted to a good standard with suspended ceilings and gas fired central heating system supplying radiators throughout. The warehouse has sodium light fittings and there is natural light provided by intermittent roof lights.

3.4    Externally the property has a total of 40 car parking spaces and a shared side entrance and parking area with the adjoining unit (No 101).

## 4.    Floor Areas

The Surveyors representing the Landlord and the Tenant have agreed the floor area at 26,777 sq ft gross internal area. At my request they have confirmed to me that they wish me to adopt this area for my valuation as a term of my instruction.

## 5.    Reasons

I have been requested by both parties to give a reasoned Determination in this matter and I therefore set out below an explanation as to how I arrive at my rental valuation together with a breakdown of it.

**6. Submissions Received**

6.1 I invited and received submissions and counter submissions from each side and have inspected the subject property and viewed the comparables put forward as evidence.

**7. Tenant's Improvements to be Disregarded**

7.1 My attention has been drawn to the fact that certain improvements have been carried out by the Tenant and these are to be disregarded for rent review purposes.

7.2 In brief, these comprise:-

1. Construction of the trade counter area with doorway at the side of the unit.
2. Extension and alteration of first floor offices.
3. Construction of a secure store and warehouse manager's office, staff rest rooms, etc. within the warehouse area.
4. Mezzanine floor within warehouse.

**8. The Parties' Valuations**

Mr Jones, on behalf of the Landlord, contends for a revised rent of £128,800 pa based on a rate of £4.81 psf on the agreed floor area of 26,777 sq ft. Mr Smith on the other hand applies a rate of £4.42 to which he makes an adjustment of 10% against one of his comparables and a further adjustment of 2.5% for the hypothetical term against the same comparable. He arrives at a net figure of £104,450 pa, as his opinion of the open market rental value of the unit.

**9. Valuation Issues**

9.1 The two Surveyors have supplied me with a significant number of comparables with Mr Jones mainly relying on evidence on Airport Industrial Estate itself.

9.2 I do not propose to set out details of the comparables but would simply say that the transactions I found most useful are the open market lettings on Unit 44 Airport Industrial Estate and Units B2 and B3 Ringway Estate, both in 2006. I have also had regard to the open market letting on Unit 101 Airport Industrial Estate in 2004.

9.3 A number of the comparables relate to historic transactions in 2000 and 2001 and by and large I have found those of little help, bearing in mind that later evidence is available. Furthermore, several of these simply recite the fact that the unit is currently vacant and on the market to let and whilst I have noted the quoted terms given, again I have not found them of significant help except to reinforce the state of the industrial and warehouse market in the locality with which I am familiar in any event.

9.4 Mr Smith has principally relied upon Units B2 and B3 Ringway Estate, although he has approached his analysis of that transaction in a way that I believe is inconsistent with other parts of his valuation. More importantly to me, the answer to which he comes of £3.90 psf (for the subject unit) seems to fly in the face of the actual evidence available.

9.5 From my own perspective the open market letting on Unit 101, which is immediately next door to the subject property and of similar specification albeit somewhat larger, shows a rate of £4.57 psf dating from November 2004 and I regard this to be an absolute base value from which to work.

9.6 I turn now to the two comparables that I think are of most help, namely Unit 44 on the estate and Units B2 and B3 at Ringway Estate.

9.7 As far as Unit 44 is concerned this is an open market letting at £4.74 psf with effect from 1st August 2006. I believe that this is a helpful comparable because it is a letting at a date very close to the subject review date on a building of similar specification although in a slightly inferior location and somewhat smaller than the subject property.

9.8 Mr Jones has made the point that the specification is inferior but I have taken a look inside the unit and do not believe that there is much difference, certainly as far as it affects rental value. The significant points of difference are that the lease on Unit 44 contains a tenant's 5 year break option which I think is advantageous to a tenant, whereas on the other hand the unit is semi-detached and does not have a dedicated and secure external yard.

9.9    I have considered these points very carefully and have come to the conclusion that in valuation terms these two factors effectively cancel each other out. On a like for like basis, therefore I accept that this comparable supports a rent of around £4.74 psf for the subject property.

9.11   I turn now to Units B2 & B3, Ringway Estate. I find this transaction to be of help because it is an open market letting with effect from 24th June 2006 at £4.90 psf. Whilst this date is reasonably close to the review date the transaction differs from the hypothetical terms of the rent review on the subject property in a number of ways and both valuers have made a series of adjustments in their analysis of the transaction in order to compare like with like. I have considered their figures and as I am not happy with some of them I have carried out my own exercise.

9.13   My own view about the adjustments that have been incorporated is that they should be approximately as follows:

| | | |
|---|---|---|
| Valuation Term | — | Minus 1.5% |
| Incentive | — | Minus 7.5% |
| Service Charge | — | Plus 1% |
| Age and Specification | — | Plus 3% |
| Total | — | Minus 5% |

9.14   By way of explanation, this letting was subject to a tenant's break option at 4 years coupled with a 3 month rental penalty whereas the subject property has to be valued on the basis of a 10 year hypothetical term. My view is that the adjustment for this factor should be in line with the rate of 1.5% used by Mr Jones in his analysis of the other transactions having 4 year terms. I consider this to be quite sufficient to reflect the small disadvantage of the 10 year assumed term on the subject property at the review date.

9.15   With regard to incentive, my own enquiries have revealed that there was in fact no specific agreement as to whether the 9 months rent-free given in this case reflected partly fit-out period and partly incentive or all of one or the other. My own judgement is that it would have been part fit-out, reflecting perhaps a slightly longer than the normal 3 month period due to the amount of works required, with the remainder being true incentive. As a result I believe that the figure of 7.5% used by Mr Jones is fair and reasonable for this factor and I have adopted it.

9.16   As far as the service charge is concerned, I have noted what Mr Jones has said but I am not sure that this would have anything but a marginal effect upon rent. As a result, therefore, I am only inclined to apply a token adjustment of say plus 1% to reflect the fact that there is no service charge on the subject property.

9.17   As far as age and specification is concerned, I accept Mr Jones's 3% figure reflecting in my own mind 2.5% adjustment for the yard and say 0.5% adjustment for the inferior car parking provision.

9.18   This then gives me a total adjustment of minus 5% which leads me to an adjusted rent of £4.65 based on the starting figure of £4.90.

## 10.   Conclusion

In conclusion I agree with both valuers that there is no direct comparable for this property and it is therefore necessary for me to do the best I can with the evidence I have. As a result I have come to the view that the appropriate rate to adopt for the valuation of the subject property lies in the range of £4.65–£4.74 psf reflecting my two preferred comparables to which I apply equal weight. I take the view that the adjustments which both valuers have adopted and which I have taken on board in respect of both units is an imprecise science and I therefore choose to adopt a figure which is in the middle of the range indicated. As a result I have decided to value the subject property at £4.70 psf. As I have indicated above I am instructed to adopt the agreed floor area of 26,777 sq ft used by both valuers.

## 11.   Valuation

My valuation of the subject property, therefore, is as follows:

| | | |
|---|---|---|
| 26,777 sq ft at £4.70 | = | £125,851 |
| **Say** | = | **£125,850 pa** |

## 12. Determination

I THEREFORE HEREBY DETERMINE that the revised rent, as defined in Paragraph 2.2 of Part 2 of the Fourth Schedule to the lease, with effect from the 23 August 2006, is the sum of **£125,850 (One Hundred & Twenty Five Thousand, Eight Hundred & Fifty Pounds) per annum**.

## 13. Fees

13.1 My fee for this Determination amounts to the sum of £——— plus VAT in accordance with my procedural instruction letter of the 7 June 2007.

13.2 There has been correspondence relating to my power to deal with the apportionment of my fee and I have determined that Paragraph 2.4 of Part II to the Fourth Schedule of the lease gives me power as Independent Expert to determine the apportionment or otherwise of my fee between the parties.

13.3 Mr Smith has asked me to issue my Determination on an interim basis, reserving the question of the apportionment of my fee and I confirm that this Determination is issued on this basis. If I am requested to deal with the apportionment of my fee at a later date, I will invite submissions and counter submissions from the parties in due course.

**Dated this 20th day of August 2007**

**Signed** _____

**A B Potter FRICS MCIArb**
**Independent Expert**

# Valuations for Financial Statements

16

## Introduction

Financial statements, sometimes also known as financial reports, are produced by an entity (ie any incorporated or unincorporated organisation) to provide information about its financial position, performance and changes in its financial position that is useful to management, owners and other stakeholders in making economic decisions. A financial statement will typically will include a balance sheet, profit and loss statement and a statement of cash flows. The term financial statement specifically relates to accounting information published for the benefit of the public or investors, whereas the word accounts is generic term that can include other types of document, for example, management accounts for internal use only.

Most financial statements are produced to meet the regular reporting requirements of entities. All companies in the UK have to produce annual financial statements by law, and listed companies are required to produce interim results. However, there are some specific situations where statements may be required at another time, for example in order to justify or defend a take-over bid, or where a company is making a public offering of shares. Some types of business or entity are also subject to specific rules. These special applications are discussed later in this chapter.

From the beginning of 2005, the consolidated financial statements of all companies listed on a stock exchange within the European Union have to be prepared in accordance with International Financial Reporting Standards (IFRS). In the UK, non listed companies may elect to adopt IFRS, or may continue to follow UK Generally Accepted Accounting Principles (UK GAAP). Public sector accounting in the UK still reflects UK GAAP, although the government is currently considering the adoption of International Public Sector Accounting Standards (IPSAS), which are based on IFRS.

Furthermore, the policy of UK Accounting Standards Board is to converge UK GAAP with IFRS, so that by 2010 there are likely to be few material differences between the two sets of standards. However, at the time of writing there are distinct differences between the valuation requirements of the two regimes, which are discussed later in this chapter. It will be appreciated that it will be of fundamental importance that the valuer establishes which accounting standards the client is using before undertaking a valuation for inclusion in a financial statement.

# International Financial Reporting Standards (IFRS)

Until 2004 these were known as International Accounting Standards. IFRS are issued by the International Accounting Standards Board (IASB). In 2007 IFRS were either required or permitted by law in over 90 countries, including all those in the European Union. At this time the most notable countries that did not permit the use of IFRS without reconciliation to domestic standards are the United States, Japan and Canada. However, the US standard setter, the Financial Accounting Standards Board (FASB), and IASB have agreed a project to converge their standards and align their future agendas. This will mean that within a few years there should be little material difference between the US standards and IFRS. Canada is also committed to adopt IFRS by 2010.

Consequently, although IFRS are rapidly becoming the globally recognised standard, valuers need to be aware that individual countries' generally accepted accounting principles (GAAP) may still apply.

IFRS is a generic term that encompasses a framework for the preparation and presentation of financial statements, thirty seven individual standards on different topics and interpretations published by an associated committee of the IASB. The thirty two standards first published before 2001 retain the prefix IAS; those published since are prefixed IFRS.

Many of the individual standards permit or require valuation. Those that are most likely to be encountered by a valuer of fixed assets are:

| | |
|---|---|
| IAS 2 | Inventories |
| IAS 16 | Property, plant and equipment |
| IAS 17 | Leases |
| IAS 36 | Impairment of assets |
| IAS 40 | Investment property |
| IAS 41 | Biological Assets |
| IFRS 3 | Business combinations |
| IFRS 5 | Non current assets held for sale and discontinued operations. |

While a valuer does not need an in depth knowledge of each of these standards, an appreciation of their basic valuation requirements is important. It is also important to recognise that although most of these standards require the assessment of fair value, which is discussed later, the assumptions on which the value is assessed may be different under different standards. For example, the fair value of an asset that is being used, and will continue to be used in the enterprise, will be normally be assessed on the assumption that it is sold in combination with other assets as part of the going concern. On the other hand, an asset that has been declared surplus to requirements, or which has been impaired, will be assessed on the assumption that it is sold separately from the other assets employed in the business.

It is therefore of fundamental importance that the valuer discusses the future utilisation of the asset with the entity before preparing the valuation in order to ensure that the appropriate valuation assumptions are made.

# UK generally accepted accounting principles

The Accounting Standards Board (ASB) produces standards for the UK. Originally these were published as Statements of Standard Accounting Practice (SSAPs) but as these have been reviewed and updated since the mid 1990s they have been replaced with Financial Reporting Standards (FRSs). In this paper the collective term of UK GAAP is used for both FRSs and SSAPs.

Although listed UK companies have had to use IFRS since 2005 onwards, the unlisted sector may continue to use UK GAAP, and UK GAAP still prevails in the public sector. However, as ASB completes its programme of convergence, it is likely that there will be few significant differences between UK GAAP and IFRS by 2010.

The UK standards most likely to be encountered by valuers are FRS 15 — *Tangible Fixed Assets*, SSAP 19 — *Investment Property* and SSAP 21 — *Leases and Hire Purchase Contracts*. The Chartered Institute of Public Finance and Accounting (CIPFA) produces a Statement of Recognised Practice (SORP) that applies the principles of UK GAAP to the specific situations encountered in the public sector. Awareness of the asset identification and classification criteria in the SORP is important for any valuer valuing public assets.

# Special applications

In this section a number situations are discussed where although the valuation may be included in a financial statement prepared in accordance with either IFRS or UK GAAP, there are further regulations governing either the valuation process or the valuer. Although the requirements of these regulations are summarised here, readers are recommended to refer to the UK Section of the RICS Red Book in order to find a more detailed explanation and to ensure that the requirements of the most current regulations are reflected.

## *Share listings and company circulars*

Valuations for incorporation or reference in listing particulars and circulars have to be accordance with the Financial Services Authority (FSA) Listing Rules. These were revised in 2005 and provide that a property company issuing a prospectus for new shares must include a valuation of the properties, although this report may be in a condensed form. No description is given in the listing rules of what is considered to be an acceptable 'condensed form' of report, but after consultation with the FSA the RICS has produced a framework for an acceptable report that is included in the UK section of the Red Book.

When any UK listed company proposes the acquisition or disposal of property where the value of the property exceeds 25% of the value of the company, shareholder approval must be obtained. The circular to shareholders must include a valuation by an expert valuer. Valuers undertaking this type of work must be aware of the Criminal Justice Act 1993, Part V, Insider Dealing, and also with the EU Market Abuse Directive 2003, which was implemented in the UK with effect from July 2005. Most transactions referred to in this and the following section on mergers are price-sensitive, and members must ensure compliance with the law.

The valuations must be by an independent expert, which normally will be an external valuer who has also disclosed such of the following matters that are applicable:

- ownership of securities issued by the company issuing the prospectus or circular
- former employment by the company
- membership of any of company's bodies
- any connection with the intermediaries involved in the offering or listing.

## Takeovers and mergers

Valuations in connection with a takeover or merger may not necessarily be incorporated in a financial statement, although if intended for publication or in support of a bid or defence they will share many of the characteristics, and from the valuer's perspective it is advisable to follow the follow the requirements of the appropriate accounting standard. If the merger or takeover goes ahead, under IFRS 3 *Business Combinations* there will be a requirement to value tangible assets of the acquired business in the first financial statements of the acquiring entity after the transaction.

The Takeover Panel regulates the City Code on Takeovers and Mergers, which is concerned with takeovers and mergers involving companies whose shares are listed in the UK and held by the public. The code seeks to ensure good business standards and fairness to shareholders and to maintain fair and orderly markets. The panel is a non-statutory body, but the code has been endorsed by the Financial Services Authority under the Financial Services and Markets Act 2002. Valuers must conform to the spirit of the code, as well as the letter.

The valuer must understand that he and his colleagues involved in the valuation task are ranked as associates as defined in the code. When undertaking a valuation of this nature, a register of shareholdings of the valuer, his colleagues, spouses and dependent children must be maintained and dealings are forbidden before or during any offer. This longstanding provision in the code is now reinforced by the regulations introduced in 2005 under the EU Market Abuse Directive, which has a wider application. The valuer is in possession of price-sensitive information and must be careful not to give share tips to others — this is an offence under the Company Securities (Insider Dealing) Act 1985.

Valuations under the code should be by a named independent valuer, which the code defines as a valuer meeting the definition of an external valuer in the RICS Red Book. It recognises that in some situations it is not possible for the valuer to complete a full valuation of every property and, in exceptional circumstances a restricted valuation of a representative sample may be permissible, subject to conditions. Finally, any special assumptions should not normally be made, but if these are necessary they should be agreed with the panel and fully explained in the report.

If the valuer gets drawn into discussions on profit forecasts he may become a financial adviser and become caught by the Financial Services and Markets Act 2000 — in which case he would need to be a member of a regulated professional organisation under the Act. He should be very careful about following such a path and, if in doubt, should seek legal advice before proceeding.

## Pension funds

Valuations for the financial statements of pension funds shall be in accordance with the SORP issued by the Pensions Research Accountants Group (PRAG). Occupational pension scheme are governed by the Pension Scheme Act 1993 and the Pensions Act 1995. The Occupational Pensions Regulations 1996[1] impose obligations with regard to audited accounts.

The SORP recommends that property valuations be carried out by independent valuers at least as often as actuarial valuations, but in any case not less than triennially. In other cases properties may be included on the basis of an annual valuation by an internal or external valuer. Less stringent

---

1    Occupational Pension Schemes (Requirement to obtain Audited Accounts and a Statement from the Auditor) Regulations 1996 (SI 1996 No 1975).

requirements apply where the proportion of property assets is low, and more stringent requirements where significant development properties are owned.

The SORP does not define the criteria for an independent valuer. The trust deed under which the fund is administered may have specific criteria, and the valuer will need to establish these, and confirm compliance, before proceeding. Otherwise an external valuer as defined in the RICS Red Book can meet the requirements of the SORP.

The current SORP published in 1996 states that property should be valued at 'Open Market Value' in accordance with the Red Book. Open market value is no longer a recognised definition in the Red Book and until the SORP is updated valuers should report market value, but may need to explain that this produces the same figure as the former definition of open market value.

The valuer needs to be aware of related parties as defined under the SORP. Related parties may be employer-related, trustee-related or manager-related. It is recommended in the SORP that related parties must be separately advised where:

- there is a transfer of a property interest between a related party and the fund
- in any negotiations between the related party and the fund affecting value — for example disposals, lettings and rent reviews
- where the related party and the fund have different legal interests in a property.

This may well give rise to circumstances where a valuer to the fund is prohibited from advising on a transaction with a related party.

The valuation report needs to identify total values by type or by class or category of property, related party transactions and projected rental income reflecting reversionary increases or decreases.

Finally, although the valuer will usually be employed by and report to the fund manager, their responsibility is equally to the trustees and beneficiaries. Copies of the report should therefore be provided to the trustees.

## Collective investment schemes

These schemes are subject to the provisions of the Financial Services and Markets Act 2002 and include authorised unit trusts and investment companies with variable capital. The detailed regulations are contained in the FSA *New Collective Investment Schemes Sourcebook* that came into effect in April 2004. The details relating to both the acquisition and regular valuation requirements for property, known in the regulations as an immovable, are set out in UK Appendix 2.3 of the Red Book.

Particular note should be taken of the difference between an appropriate valuer, who is a valuer engaged to value an immovable on behalf of the trustees at the time of acquisition, and a standing independent valuer who is engaged to undertake a full valuation at least once a year and to review the last full valuation at least once a month.

## Unregulated property unit trusts

Unregulated property unit trusts are a form of collective investment scheme permitted under section 75 of the Financial Services Act 1986 where assets are held in trust for participants who do not have day to day management involvement. They may not be marketed to the general public and are thus distinguished from authorised unit trusts.

There is no regulatory requirement for an independent valuer, but most trust deeds do provide for one. If so, the valuer must check the independence requirements and confirm they can be met.

Valuations, which should be to market value, are critical to the pricing of units and will normally be required to be reviewed frequently. The valuer must judge trends from the most recent transactions even if those trends may be short term.

The valuer is usually employed by and reports to the fund manager, but copies of the report should be provided for the trustees.

# Bases of value

This section examines the bases of value of which the valuer will need to be aware of when undertaking valuations for use in a financial statement. A basis of value is defined in the Red Book as a statement of the fundamental measurement principles of a valuation on a specified date. A basis should be distinguished from the method used to estimate the valuation figure, and from descriptions of the state or condition of the asset involved in the transaction. It describes only the nature of the assumed transaction and the relationship between the parties. If the value is likely to be affected by other factors, eg the state of the property or the degree to which it is combined with other assets in the transaction, additional assumptions may need to be made to further qualify the basis used.

There are three valuation bases that a valuer is likely to encounter when undertaking a valuation for inclusion in a financial statement. These are fair value, market value and existing use value. The latter is only relevant under UK GAAP, and is likely to be phased out as convergence with IFRS progresses.

## Fair value

Fair value is defined in the IVS as:

> the amount for which an asset could be exchanged between knowledgeable willing parties in an arms length transaction.

Fair value and market value (see below) are not always synonymous. Fair value simply requires there to be a hypothetical exchange at arm's length. No mention is made of the exchange taking place in the market. It is perfectly possible for a settlement to be reached at arm's length that is fair to the parties involved but which does not reflect what might happen if the asset or liability were offered in the market place. Equally, it is possible that the price obtainable in the market may not be regarded as fair by either or both parties to a transaction.

However, although fair value is defined in similar terms in IFRS, in so far as the valuation of property, plant and equipment is concerned, further conditions are required in addition to those in the definition. IAS 16 — *Property, Plant and Equipment*, states that:

> the fair value of land and buildings is usually determined from market based evidence by appraisal that is normally undertaken by professionally qualified valuers. The Fair Value of items of plant and equipment is usually their market value determined by appraisal.

This statement clearly indicates that for the purposes of this standard, fair value and market value are effectively the same thing, as regard must be had to the market, not just what two parties may deem to be fair. In a similar vein, IAS 40 — *Investment Property*, stipulates that the fair value of an investment property shall reflect market conditions as at the balance sheet date.

The question might be legitimately asked why IFRS stipulate that fair value may be used when it is clear from the commentary that it is really market value that is required. There is no clear answer to this, although one explanation commonly offered is that the term fair value was adopted by the original draftsmen of the international accounting standards as they were concerned that there was a proliferation of different market value definitions, some of which were deemed unsuitable. As there are also many different definitions of fair value in use around the world this reasoning seems somewhat tenuous. The fact of the matter is that fair value has become well established around the world as an accounting term. As accounting standards became accepted internationally well ahead of valuation standards, and have a much broader economic impact, it appears that valuers will just have to accept the anomaly in terminology.

## Market value

Market value is defined in the IVS as:

> The estimated amount for which a property should exchange on the date of valuation between a willing buyer and a willing seller in an arm's-length transaction after proper marketing wherein the parties had each acted knowledgeably, prudently and without compulsion.

It will be seen that the principal difference between market value and fair value is the condition that the exchange takes place after proper marketing. Regardless of the intentions or wishes of the owner, it requires an assumption that a sale is taking place and consideration of whether the price obtainable is fair to either or both parties is irrelevant. Market value is also supported by a detailed conceptual framework, which provides a detailed explanation of each element of the definition.

The amount of the valuation is the best price which can be obtained by widely offering the property for sale, using an appropriate marketing method and taking adequate time to conduct all negotiations to bring about the best price. However, providing a proper period has been allowed, the actual time taken to conclude marketing, negotiations and a legal formalities is irrelevant. The valuation is as at the date of exchange. In this context it should be noted that exchange does not have the same meaning as exchanging contracts under UK law, which may proceed the actual transfer of the property interest by a significant period. It describes a situation where a binding contract is struck, with an immediate transfer of the interest to follow.

The conceptual framework that supports market value in the IVS makes it clear that the estimated price in the definition excludes any element of special value, ie any additional bid that may be made by a purchaser who has a special interest in the property. Such bids can come from various sources, such as a developer seeking to assemble a site, or perhaps the adjoining owner of a shop wishing to expand, or the owner of a superior or inferior interest in the same property to release marriage value.

The property is to be sold in its existing state. It is legitimate to include any hope value for, say, a more valuable use or redevelopment, but only to the extent that it is generally recognised by the market and ordinary purchasers.

## Existing use value

This is a basis that is used only for owner occupied property valued under UK GAAP, (FRS 15 — *Tangible Fixed Assets*). The current definition of existing use value (EUV) in the UK section of the Red Book (UK PS 1.3) is the same as that for market value with the addition of the following words:

> ... assuming that the buyer is granted vacant possession of all parts of the property required by the business, and disregarding potential alternative uses and any other characteristics of the property that would cause its Market Value to differ from that needed to replace the remaining service potential at least cost.

The assumption of vacant possession refers to the passing of legal and physical control to the buyer. It does not, as some people think, require an assumption that the business in occupation has been closed and the property is therefore empty. Indeed, the vacant possession assumption is followed in the definition by qualifying words that it relates to 'all parts required by the business', which make clear the underlying assumption that the business is continuing.

Under FRS 15, an entity has the option of carrying property at either historic cost less depreciation or at current value. The concept of EUV is based on the theory that for as long as a property is gainfully occupied for the purpose of the business, any revalued figure should reflect what the entity would have had to pay at the date of valuation to replace what it actually needs. It is a reflection of the remaining service potential of the property to the entity on the basis that it will be utilised indefinitely for the purpose of the continuing business. This replicates the decision it would have made when originally acquiring the property, so in theory the historic cost option and the EUV option reflect the similar criteria, which supports the rationale of offering them as alternative approaches.

EUV requires alternative uses or any other characteristics that would cause the value of the remaining service potential to differ from the market value to be disregarded. Alternative uses are disregarded for a number of reasons. First, when the entity initially acquired the property it would not have paid extra for features, such as a more valuable alternative use, that would be irrelevant to its requirements. Second, carrying a property in the balance sheet at a figure reflecting an alternative use that was incompatible with the continued use of the property by the business would create an inconsistency in the accounts. Third, a depreciation charge is made in the profit and loss account based on the carrying amount in the balance sheet. If the balance sheet figure relates to a more valuable use, the depreciation charge would be reduced. This would increase the apparent profitability of the enterprise for reasons unconnected with the operation of the business, and distort the entity's performance in comparison to its peers.

Examples of other characteristics that may cause the market value to differ from the EUV include:

- the presence of contamination that is of no consequence to the current operation but which would severely depress the value for an alternative use
- where a property is oversized for its location, which would depress the market value, but where the cost of replacing the service potential to the business would be significantly greater
- where the existing buildings are old and would have a limited market value if vacant but where the replacement cost of the remaining service potential to the occupier would be considerably higher.

Clearly knowledge that an asset is worth significantly more or less than its value for the current use is relevant to management and investors alike, even if it conflicts with the underlying assumption that the accounts are prepared on the basis that the business will continue as a going concern. FRS15 provides that where the EUV is materially different from the market value, the latter should be disclosed in the notes, supported by the reasons for the difference. The Companies Act 1985 also requires a similar disclosure where the carrying amount differs significantly from the market value.

A detailed explanation of EUV and the underlying accounting concepts is provided in UK section of the Red Book.

Many properties are fitted out by the owners to meet their particular needs and to make them suitable for occupation. Shop fronts for retail units and partitioning for offices are obvious examples. They often add little or nothing to the market value. The valuer needs to discuss with the client whether he should disregard them in arriving at his valuation. If disregarded by the valuer, the cost of such adaptation works is treated separately in the accounts either by writing off in the year of expenditure or by assessing their current value on a depreciated replacement cost basis.

## Which bases to use?

The Red Book directs that RICS members undertaking valuations for inclusion in a financial statement produced under IFRS shall follow the International Valuation Standards. These provide that any estimate of fair value required under the standards listed at the beginning of this chapter is met by the valuer adopting market value. This reflects the clarifying conditions for the application of fair value in the standards themselves, in particular IAS 16 — *Property Plant and Equipment* and IAS 40 — *Investment Property*. The IVS also reflect the disclosure requirements in these two accounting standards, by requiring the valuation report to disclose the method adopted, and in particular, the extent to which the valuation was derived from market based evidence.

The IASB has recognised that although fair value is permitted or required in many of its standards, there is no clear framework for its application and a lack of clarity as to the assumptions that should be made in establishing the valuation figure. This does create some practical difficulties for valuers, their clients and auditors alike. An example is the lack of guidance on whether a particular asset or liability has to valued as an individual item without regard to the remainder of the business or as part of the overall combination of assets and liabilities of the business. At the time of going to press the IASB has produced a consultation document seeking views on some of these issues and the role of valuation in financial reports. In the meantime the International Valuation Standards simply provide that the valuer should discuss and agree the appropriate assumptions with the entity and clearly state these in the report.

Under UK GAAP, the UK Section of the Red Book follows closely the requirements of FRS 15 — *Tangible Fixed Assets*, and SSAP 19 — *Investment Property*. Surplus property and investment property are valued to market value. Owner occupied property is to be valued to existing use value. Specialised property is to be reported at its depreciated replacement cost. The categorisation of properties for valuation purposes is normally straight forward, but may need to be discussed with the entity if the valuer is in doubt as to how any particular property should be treated.

## Depreciated replacement cost

No discussion about valuations for financial statements would be complete with out examining depreciated replacement cost (DRC). This is a method of valuation that is used where it is impractical to use a sales comparison or income approach because of the specialised nature of the asset.

DRC is defined in the IVS as:

> The current cost of replacing an asset with its modern equivalent asset less deductions for physical deterioration and all relevant forms of obsolescence and optimisation.

The use of this approach valuations for financial statements is endorsed in IAS 16 — *Property, Plant and Equipment*, which provides that where there is no market based evidence of fair value because of the specialised nature of the item fair value may be estimated using a depreciated replacement cost approach.

DRC is an approach that is only applicable to specialised property, which is defined in the IVS as:

> Property that is rarely sold in the market, except by way of a sale of the business or entity of which it is part due to uniqueness arising from its specialised nature and design, its configuration, size, location or otherwise.

Examples of specialised property include refineries, power stations, docks and specialised manufacturing facilities. There are also many examples of specialised property in the public sector such as schools, hospitals, town halls, museums and so on. The definition is not exclusive to real estate; in the IVS property is used in a broad sense to include not only real estate but also plant, equipment and personal property.

A detailed discussion on the application of the DRC approach is beyond the scope of this chapter, although an example is included later. However, the approach has been criticised in the past, mainly because of a misapplication of its principles by some practitioners. The most common errors have been basing the replacement cost on the actual asset rather than its modern equivalent, and failing to reflect economic and functional obsolescence adequately. The valuer should always remember that DRC is based on the economic principle of substitution; if a willing buyer is offered an asset in the market, the price he would be willing to pay would reflect the price of the available alternatives. In the case of a specialised asset the only alternative is likely to be the purchase or procurement of a new asset. The price he will pay will therefore reflect the relative disadvantages of the actual asset compared to the modern equivalent. In 2007 RICS published a Valuation Information Paper *The Depreciated Replacement Cost Method of Valuation for Financial Reporting* which discusses the principles in detail.

There is a requirement in IAS 16 — *Property, Plant and Equipment* to disclose in the accounts the extent to which the fair value of an item was determined by reference to observable prices in an active market or by using other valuation techniques. The use of a DRC or an income approach must therefore be disclosed in the financial statement, alerting users to the fact that the figure is not derived from direct market evidence. However, the underlying valuation objective remains the same, regardless of the method used. The International Valuation Standards therefore provide that the valuer reports the market value, but discloses that this has been derived using the DRC approach.

Under UK GAAP, FRS 15 — *Intangible Fixed Assets*, adopts a slightly different approach. Unlike IAS 16, there is no general requirement for the valuation approach to be disclosed in the entity's accounts. However, it does stipulate that specialised properties should be valued separately on the basis of depreciated replacement cost. This means that the valuer reports the DRC figure as such. However, this is merely a different presentation; there should be no difference in approach between a DRC valuation carried out for use under IFRS and one for use under UK GAAP.

There is a further reporting requirement that is common to both the International Valuation Standards and the UK section of the Red Book where DRC has been the prime or sole method used to arrive at the valuation. In the case of a private sector entity the valuation must be reported as being subject to adequate profitability. In the case of a public sector or other not for profit entity, the valuation must be stated to be subject to adequate service potential, which is defined as 'The capacity of an asset to continue to provide goods and services in accordance with the entity's objectives'.

Some are confused by these statements, believing that they invite the client to make some further adjustment to the valuation. This is not the case. These caveats simply alert the reader of the report to the fact that the DRC valuation is based on a method that implies that there is a willing buyer who would purchase the specialised property in order to continue the same use. If this were not the case then the whole premise of the valuation approach becomes invalid as the buyer would not establish the price by reference to a modern equivalent asset, as he would have no need of such an asset. Instead, the price would be established by reference to the potential for an alternative use or the asset's scrap or salvage value.

In a similar vein, when reporting a DRC valuation, the Red Book additionally requires the valuer to state the market value for any readily identifiable alternative use that would result in a higher value, or to indicate if it is believed that the market value would be materially lower if the current business was extinguished.

# Depreciation accounting

Under IAS 16 an entity is required to depreciate any part of an item that has a cost that is significant in relation to the total cost of the item. Depreciation has to be made, regardless of whether the asset is carried at either its historic cost or at a subsequent valuation in the balance sheet. In relation to land and buildings, different buildings on a site may well be depreciated separately and it is also possible that elements of a single building, for example the services, may be depreciated separately. There is a specific requirement in the standard to account for land and buildings separately, as land is not depreciated, except in special cases such as quarries or landfill sites.

Before considering these depreciation requirements further, it should be made clear that the calculation of the depreciation charge has absolutely nothing to do with the calculation of depreciation in the depreciated replacement cost method of valuation. The first is an exercise to establish the charge that should be made against an entity's income to reflect the consumption of a fixed asset during a specified period. The second is an alternative valuation approach that can be used to help establish the capital value of an asset in the absence of direct market evidence.

In order to calculate the depreciation charge, it is necessary for the entity to first assess the future useful life of the asset, or part of the asset. This is not necessarily the same as the future economic life as would be recognised by a valuer. It is based on the length of time for which the entity believes the asset will be of use for its purposes, which may well be shorter than the economic life in the open market.

Once the useful life is determined, the residual value is assessed. This is where the valuer is most likely to get involved as the residual value is the estimated amount that could be currently obtained for the asset after deducting the costs of disposal, if the asset were already of the age and in the condition expected at the end of its useful life. In other words, armed with the entity's decision on the useful life, the valuer has to consider the likely age and condition of the asset at the end of this period, but assume that it was already in this state at the date of valuation. Sometimes it will be difficult to make this judgment without assumptions as to future maintenance and repair, and these would always need to be discussed with the entity before proceeding.

In assessing the residual value of land and buildings it is first necessary to deduct the value of land element from the market value, or in the case of owner occupied property accounted for under UK GAAP, the existing use value. This will provide a notional current value of the building element. The valuer then needs to estimate what the equivalent value would be if the building were already of the age and condition that could reasonably be anticipated at the end of its useful life. This Residual Value is then deducted from the current value of the building in order to calculate the depreciable amount.

IAS 16 states that the useful life and residual value shall be reviewed at least annually and, if expectations differ from those made previously, appropriate adjustments should be made. It also states that an increase in the value of the land on which a building stands does not affect the calculation of the depreciable amount of the building. At first sight these two provisions seem to conflict. However, the author believes that the intention is that once a depreciable amount and useful life has been determined for a building, increases in land value due simply to market trends should not be grounds for altering those figures. However, where the increase in the value of the land arises because it has potential for a more valuable alternative use, an entity undertaking a review of the useful life of the building for its existing purpose would have good grounds to reduce it. From the perspective of the valuer it is advisable in such cases to provide values for the land based on the alternative assumptions that

(a)  it can be used for the more valuable alternative use or
(b)  that the existing use will continue.

It is for the entity in conjunction with its auditors to adopt a residual figure that is consistent with the assumptions made in preparing the financial statements generally.

There is no requirement to depreciate investment property in IAS 40, and therefore no requirement to consider residual value for this class of property. Property that is surplus will be at the end of its useful life and therefore will no longer need to be depreciated by the entity. It will be accounted for under IFRS 5 — *Non Current Assets held for Sale and Discontinued Operations* and the valuer may need to advise on the Market Value, but there will be no need to consider residual value

UK GAAP (FRS 15) contains similar provisions for depreciation accounting to IAS 16. There are subtle differences, eg 'useful life' in IAS 16 becomes 'useful economic life' in FRS 15, although the meaning is the same. Also there is no specific requirement to depreciate different parts of an asset that have a cost that is significant in relation to the total cost of the item, although nothing to suggest that such an approach would be incorrect, and in practice this is often done.

As is the case under IFRS, there is no need under UK GAAP to depreciate property that is either held as an investment or surplus. Apportionments between the land and buildings and residual valuations will only be required for owner occupied property. Since this class of property is valued on the basis of EUV, it follows that the land element must also be considered on this basis.

# Working studies

Before reading these studies it is important to realise that the preparation of a valuation for inclusion in a financial statement does not require the use of techniques or methods that are any different from those used in valuation for other purposes. The sales comparison, income and depreciated replacement cost approaches can all be used, and the criteria for their use is identical to those that determine which approach should be used for any other purpose. The principal differences between valuations that are intended for use in a financial report and other types of valuation are the specific reporting requirements and, in particular, the need to identify and disclose appropriate assumptions. The case studies therefore mainly focus on the presentation of the valuation, rather than its calculation.

## Study 1 Reporting value of owner occupied property

A very large owner occupied office in on the outskirts of a regional city is fully utilised by the occupier. However, if the occupier were to vacate the valuer considers that there would be little or no demand from other owner occupiers. The most likely interest would be from developers looking to either break up the building into smaller suites or to demolish and redevelop. How is this valued for inclusion in the owner's accounts?

If valuing under UK GAAP, the appropriate basis to use is existing use value (EUV). This requires the assumption that the business in occupation is continuing, at that there is no intention to liquidate or curtail its operations. This means that a valuation of the property on the assumption that it was empty and disused would be entirely inappropriate. Although the valuer must not base the valuation on what the actual occupier would pay to acquire the property, as this would not provide the required degree of objectivity, it can be assumed that there is at least one buyer with an identical requirement in the market. The valuer has to ascertain what they would pay. This could include looking at the availability and cost of alternatives, which in the case of a very large building may be elsewhere in the region or even the country. While some people advance the theory that an owner occupier with a large and unusual requirement would only need to pay £1 more than a speculator or someone buying to knock down and redevelop in order to secure the property, in the opinion of the author this is based on a unbalanced interpretation of the EUV definition. Not only is the buyer deemed to be acting knowledgeably and prudently, so is the seller. The hypothetical seller of a large and unusual building is deemed to be aware of the alternatives, or lack of them, available to the buyer and would negotiate a price that reflected this.

The amount by which the EUV of such a property exceeds its market value on the assumption it was empty and surplus to requirements cannot be expressed as a specific percentage or other mathematical calculation. It is a matter for the valuer's judgment having regard to all the material factors, such as the degree of specialisation and the nature of possible alternatives. However, if the EUV exceeds the valuer's estimate of the price that could be obtained in the absence of a party requiring the whole building as part of an ongoing business by a significant margin, then the Red Book does require this difference to be disclosed and an explanation provided of the reasons.

In circumstances such as this, an appropriate phrasing for the presentation of the valuation could be:

We are of the opinion that the Exiting Use Value of the freehold interest is £50,000,000 (Fifty million pounds)

Because this property is unusually large for its location, if it were to become surplus to your requirements and offered for sale vacant there would be very little or no demand from other owner occupiers. The only demand would be from prospective buyers who would need to either demolish the buildings or heavily adapt them. We are of the opinion that Market Value in such a situation would be only £30,000,000 (Thirty million pounds)

If the same property were being valued under IFRS, EUV is not available as a recognised reporting basis. However, IFRS also requires financial statements to be produced on the assumption of a going concern and therefore the same distinction between the Market Value of a property fully utilised as part of the continuing business and its Market Value if surplus still arises.

Suitable words for the phrasing of the first sample paragraph given above valuation would be:

We are of the opinion that the Market Value of the freehold interest on the assumption that it were sold as part of a sale of the continuing business is £50,000,000 (Fifty million pounds)

The second paragraph could remain unaltered.

## Study 2 Calculation of residual amount

The following is an illustration of a simple calculation of the depreciable amount of a property developed with a single building, where the useful life is less than the economic life. This example shows an apportionment of the market value, which would be the appropriate basis for an owner occupied property under IFRS. However, exactly the same procedure is used for calculating the depreciable amount under UK GAAP where the apportionment would be of the EUV.

| | | |
|---|---|---|
| **Market Value:** | | £1,000,000 |
| Apportioned Value of land | | £400,000 |
| Apportioned Value of building | | £600,000 |
| Useful life — as determined by entity | 15 years | |
| Future economic life — estimated by valuer | 35 years | |
| **Residual Value:** | | |
| Valuer's estimate of current value of building assuming it was 15 years older with a remaining life of 20 years | | £250,000 |
| **Depreciable Amount** of building: | | £350,000 |

The entity then has to allocate £350,000 over 15 years, either on a straight line basis or using some other systematic approach, and the resulting annual amount is charged to the profit and loss statement.

## Study 3 Depreciated replacement cost

A general hospital in a city centre is housed in a listed building dating from the mid 19th century. A valuation is required for inclusion in the health authority's annual accounts. The valuer considers that because of the specialised nature of the building there are no relevant sales or lettings of hospital buildings. According a depreciated replacement cost approach (DRC) is used.

The DRC approach involves comparing the actual asset with a modern equivalent. The valuer's first task is then to establish the characteristics of a modern equivalent hospital. It certainly would not be located in a listed building; modern design and construction would also probably mean that the equivalent accommodation could be provided in a building that was smaller than the actual building. The size of actual building to be valued is therefore largely irrelevant; the valuer's job is to establish the size that would be required to provided the same functionality and facilities in a new building at the date of valuation. The fact that the actual building is listed should not be allowed to distract the valuer. The historic features are incidental to the primary purpose of the building as a hospital, and would not be feature in any replacement. Only where the historic features are intrinsic to the service provided by the building should allowance be made for a replacement incorporating exceptional architectural features.

Exactly the same concept applies to the land element of the modern equivalent building. The amount of land is not automatically that included with the actual hospital, but the amount required to accommodate the modern equivalent. However, a further problem the valuer then has to address is where this equivalent site is deemed to be located? The theoretical answer is that the site would be where the hospital managers would choose to locate a new hospital. In practice there are obvious difficulties. A city centre site may be too remote from the main residential areas, and difficult to access because of traffic congestion and lack of parking. It may therefore be in a suburban location. On the other hand, a modern equivalent might still need to be in the city centre as this only location capable of servicing a citywide catchment. These issues have to be considered and weighed by the valuer.

Another problem to be addressed in valuing the equivalent site, and connected to the decision as to where it would be located, is the assumed alternative use or uses. The site would be being acquired for hospital use but most suitable sites in the market would have to be bid away from other competing uses in the locality. What are those uses deemed to be? If a new hospital were actually to be built then it is probable that land would be specifically identified in the local development plan thus limiting potentially competing uses. However, the acquisition price would probably still need to reflect the value of any alternative development that would permitted if it were not for the hospital project. The correct valuation approach is to determine the most probable location for a modern equivalent hospital and then to establish the lowest cost of acquiring a suitable site in that location.

Having established the size, specification and location of a modern equivalent hospital the valuer then has to calculate the cost of providing that facility. Due allowance has to be made for fees and financing the costs during the construction period. A typical calculation might be:

|  | £ |
|---|---:|
| Cost of new building of 10,000m² @ £2,000/m² | 20,000,000 |
| Design and other professional fees | 2,600,000 |
| Finance Costs @ base plus 2% on 50% costs over 3 years | 2,734,600 |
| **Total Construction Cost:** | **25,334,600** |
| Land cost 2.5 hectares @ £500,000 | 1,250,000 |
| Finance cost of land 3 years at base plus 2% | 30,200 |
| **Cost of Modern Equivalent Asset:** | **£26,614,850** |

This calculation is simplified for illustration and obviously can be finessed as the situation requires.

Having established the gross cost of providing a modern equivalent asset, the valuer then has to begin the process of comparing this notional building with the actual building. The thought process is to replicate what a hypothetical buyer would do in formulating a bid for the actual building when its only real alternative would be to buy a site and construct the modern equivalent. This is the depreciation part of DRC. Allowances have to be made for all types of obsolescence, including physical, functional and economic. The old buildings will have a shorter life expectancy than the new equivalent, they will cost more to maintain, they will cost more to run, the layout may make the wards and other facilities difficult to utilise efficiently. All of these are reasons why a buyer would bid a much lower price for the current building than the modern equivalent.

Although it is customary to break down the depreciation elements, there is nothing wrong with the valuer taking an all encompassing allowance, other than taking such an approach can make it difficult to understand what factors have been reflected if subsequent analysis or justification of the figure is required. In this example the following might be a suitable approach to depreciation:

|  |  | £ |
|---|---:|---:|
| **Modern Equivalent Asset:** |  | **26,614,850** |
| Economic life of modern equivalent: | 60 years | |
| Remaining economic life of actual building: | 20 years | |
| Depreciation on straight line basis* : | 66% | (£17,725,490) |
| Additional depreciation: |  | |
| for cost of maintaining listed building: | 5% | (£1,330,742) |
| inefficient layout | 5% | (£1,330,742) |
| **Depreciated Replacement Cost:** |  | **£6,227,876** |
| Say £6,250,000 |  | |

*This example has used straight line depreciation over the life of the building — other depreciation calculations include reducing balance or 'S — curve'. The valuer should select the basis that would most closely replicate how the market depreciates broadly similar assets.

The objective of the DRC method is to estimate the value in exchange, ie EUV under UK GAAP or market value under IFRS. Consequently the valuer should stand back and look at the conclusion reached by applying the DRC method and ask whether it is broadly credible having regard to the sort of price differential between new and 150 year old buildings seen in the market for other types of property? While direct price comparisons cannot be drawn (otherwise DRC would not have been used in the first place) the relativities for other types of property can provide a useful credibility check.

Confusion is often caused when the actual site is identified as having a higher alternative use value. In the current example the listed status of the building would preclude demolition and redevelopment, but a common example is where a specialised industrial property in a town centre location has a far higher site value for an alternative use.

However, this is irrelevant for the purpose of the DRC calculation, which must always reflect the cost of a site for a modern equivalent of the actual buildings or structures. If having calculated the DRC the valuer considers that the alternative use value is higher, then this is the market value, and should be reported as such. However, the valuer is valuing the site using a sales comparison method, not DRC. Since this Market Value could only be realised by closure of the existing operations the Red Book does require the valuer to report both figures, on the appropriate alternative assumptions, see Study 1 above

# Public Houses

## Property

The character of public houses has evolved over the centuries from two distinct roots — rudimentary drinking houses supplying ale or cider to a regular, local trade; and taverns, somewhat grander premises, offering food, drink and accommodation to travellers. Merged into this mix were the gin palaces of the 19th century, built with imposing façades and opulent interiors by gin merchants competing with brewers.

Until the 1960s, other than inns, public house trade consisted almost entirely of liquor and tobacco. In most, food was limited to crisps, pickled eggs and possibly sandwiches. Although the other function of a pub is to provide for social encounters, and pubs have always been closely associated with games, in the main these were restricted to customer activities such as darts, dominoes, cribbage and shove ha'penny. Some public houses may have had a piano or bar billiards; few had snooker or billiard tables, skittle alleys or crown bowls.

Beer consumption in pubs has been declining nationally since the middle of the 20th century. To respond to population movements, more competition, the growth of home consumption and drink driving laws, and the increase in customers' leisure time, disposable income and life expectancy, many public houses have had to adapt to survive. This has involved growth of the business by developing new income streams and expanding their customer base.

Most public houses now sell food, either snacks or meals in the bar or in a dedicated restaurant area. Some truly specialise as gastronomic pubs where the trade emphasis has been re-orientated to offer exceptional quality food.

Although only a small minority, the number of public houses with letting accommodation has grown. Rooms over bars have been utilised or purpose built lodges constructed on adjoining land, offering modern and convenient facilities for travellers.

The nature of pub entertainment has changed. Besides traditional activities, the vast majority have now had gaming machines (amusement with prizes/AWPs) and juke boxes for many years. A considerable number of pubs show sporting events and other programmes on big screen TV, particularly BSkyB Premiership football. Some have pool tables and video games. Others offer live music or performances, and music and dancing are not uncommon, particularly in town/city centre venues attractive to a younger audience.

Some rural pubs have needed to diversify in other ways to survive and additional uses such as the village shop and post office or community use are not unusual.

With the change in fashions and growth of large public houses in town centres, those public houses unable to adapt, or adapt successfully, have been closing. This may be due to physical characteristics, location or competition rendering them no longer economically viable, or where the underlying value for an alternative use for which planning permission is available is greater.

The evolution of the public house has led to fusing with other uses, such as restaurants, hotels and nightclubs. This chapter deals with the valuation of the broad range of public houses but does not explore additional factors that it may be necessary to take into account when valuing licensed properties at the interface. Practitioners in Scotland and Northern Ireland should be aware that there may be differences in both law and markets in comparison with England and Wales.

# Planning

In England and Wales, prior to 21 April 2005, planning law grouped public house and restaurant use together (the sale of food and drink for consumption on the premises) under class A3 Town and Country Planning (Use Classes) Order 1987 (SI 1987 No 764).

From 21 April 2005, in England, the 1987 Use Classes Order was divided into three new separate classes, the new A3, A4 and A5 classes by the Town and Country Planning (Use Classes) (Amendment) (England) Order (SI 2005 No 84). Similar provisions have not been made in Wales where the 1987 Order continues to apply.

The new class A3 is 'restaurants and cafes', designed for places where the primary purpose is the sale of food and drink for consumption on the premises. A restaurant where the trade is primarily in-house dining but which has ancillary bar use will fall within this class. The new class A4 is 'drinking establishments' including use as a public house or wine bar. This caters for places where the primary purpose is the sale and consumption of alcoholic drink on the premises. (The new class A5 is 'hot food takeaways' where different environmental issues arise.)

When reading leases for properties in England, it is important to establish whether the permitted use relates to the original 1987 Order or 2005 amendment.

The Town and Country Planning (Use Classes) (Scotland) Order 1997 (SI 1997 No 3061 (section 195)) applies north of the border. Broadly, class 3 is similar to the new A3 class in England — pubs and hot food fall outwith this class, and under article 3(5) are treated *sui generis*.

# Licensing

Before alcohol can be offered for the sale by retail on any premises, it is also necessary for both retailer and property to be licensed.

The provisions of the Licensing Act 2003 took effect fully from 24 November 2005, the second appointed day, and apply only in England and Wales. The rigid hours under the former regime have been swept away and replaced by a more flexible, customised approach balancing the desires of a business operator with the wishes and needs of the neighbouring community and the regulatory authorities, within the overall objectives set out in the Act.

In order to sell intoxicating liquor for consumption on or off the premises, there must now be a premises licence allowing the holder of a personal licence to use specified premises for licensable activities. Besides the sale of alcohol, the premises licence will also cover provision of regulated entertainment. Previously, the requirement under the provisions of the Licensing Act 1964 was for a justices' licence for the sale by retail of intoxicating liquor, granted to a suitable person in respect of a specific property, and where applicable, a separate public entertainment licence obtained from the local authority under the Local Government Act 1963.

An application for the transfer of an existing justices' licence (prior to the second appointed day), a new licence or a variation must be made to the licensing authority in accordance with the Licensing Act 2003 (Premises Licences and Club Premises Certificates) Regulations 2005 (SI 2005 No 42). A copy of the application should be sent to the responsible authorities and also displayed for 28 days on the premises, in a prominent place visible from outside the property such as a window, and advertised in a local paper for at least one day.

The application must set out, on the operating schedule, the various details as to how the holder proposes to operate the premises, including the licensable activities to be carried out and the proposed hours during which they are to take place. The licensing authority do not have the power as a body itself to amend the details but must grant the application on the terms sought, (subject only to imposition of mandatory or other conditions that are consistent with the operating schedule), unless there are 'relevant representations' made by an interested party or responsible authority.

An interested party is defined in the 2003 Act, section 13(3), as meaning a person living, or involved in a business, in the vicinity of the premises or a body representing such a person. A responsible authority is defined in section 13(4) as including the chief of police; the fire authority; the local planning authority; the local authority itself and any recognised competent bodies in the area responsible for the welfare of children.

If the licensing authority accepts such representations and a compromise between the parties cannot be reached, the application falls to be heard and determined by the licensing committee.

Once a premises licence is granted it will continue indefinitely, unless revoked or suspended under section 52 or was only granted for a specified period that has since expired, subject only to the death, incapacity, insolvency etc of the licence holder or to its voluntary surrender.

Nevertheless a premises licence may be reviewed at any time if either the holder applies to the licensing authority for a variation of the original terms, subject to the same rights of representation by interested parties and responsible authorities, or at the behest of any interested party or responsible authority calling for a review.

Applications for a new personal licence, or 10 yearly renewal, have to be made to the licensing authority in accordance with the Licensing Act 2003 (Personal Licences) Regulations 2005 (SI 2005 No 41). Personal licences must be granted where the licensing authority is satisfied applicants are aged over 18, they have attended an accredited course and gained a personal licensing qualification (other than licensees who acquired grandfather rights arising from the old regime) and have not had a personal licence forfeited in the last five years. Applicants are however required to disclose any convictions they have for relevant offences in order that the licensing authority can forward the application to the police authority, which has fourteen days in which to lodge an objection.

The licensing of individuals separately from the premises gives the holder of a personal licence mobility. A personal licence is not needed to work in a premises selling alcohol, but a designated premises supervisor (DPS), who must hold a personal licence, is required under a premises licence. Although the DPS does not have to be present at all times, all sales have to be authorised ultimately by a personal licence holder.

Different licensing laws apply in Scotland and Northern Ireland.

# Ownership

The last decade or so of the 20th century saw a rapid transformation of the industry. Prior to this time, local brewers with tied estates bought or merged with competitors until they too, in turn, were taken over. This led to the emergence of a number of regional brewers and the big six national brewers — Allied, Bass, Courage, Grand Metropolitan, Scottish & Newcastle and Whitbread.

The Monopolies & Mergers Commission Report (MMC), *The Supply of Beer* (Cm 651), published in March 1989 acted as a catalyst for radical change, halting the ongoing practice of takeovers and mergers among large national brewers. By the Supply of Beer (Tied Estate) Order 1989 (SI 1989 No 2390) breweries with over 2,000 public houses had to:

(1)   dispose of their brewery business or
(2)   dispose of the excess of houses over 2,000 or
(3)   release half of the excess over 2,000 from tie clauses.

This resulted in the abandonment of the traditional vertically integrated structure within the industry (farmers, brewers, transporters, property owners and retailers) by the big six national brewers and by other companies to concentrate either on brewing or running estates of tied tenancies or managed houses.

With so many public houses coming to the market, many brewers and independent retail companies (pubcos) took the opportunity of improving the quality of their tied estates by acquiring properties that met the core criteria of their businesses, and in the process selling off the poorer houses or those that no longer fitted their portfolios.

There are some 56,500 public houses in England, Wales and Scotland (as at the end of 2006), over half owned by pubcos, 16% owned by brewers and 30% independent. In the pubco estate one quarter of pubs are directly managed, three quarters leased or tenanted. The ratio for brewer owned houses is similar at approximately 30% : 70%. Additionally there are some 90,000 other establishments, besides shops and off-licences, licensed to sell alcohol including hotels, wine bars, restaurants and clubs.

# Landlord and Tenant Reform

Prior to 1991, lessees in England and Wales had no right to a new lease under the Landlord and Tenant Act 1954 since public houses were specifically excluded from the provisions of Part II. One recommendation in the 1989 MMC Report, *The Supply of Beer*, was to repeal this exclusion, contained in section 43(1)(d) of the 1954 Act.

This was implemented in the Landlord and Tenant (Licensed Premises) Act 1990 (which took effect on January 1 1991), and lessees of public houses, like other commercial and business premises, now have a right to a new lease and, where it is refused, to statutory compensation.

There are no similar provisions in Scotland.

# Market

Purchasers in the market for public houses fall within a number of categories — pubcos, brewery companies, institutional and other investors, property companies and individual traders. As it is not a single homogeneous market, the valuation approach for any particular type of public house is influenced by who is most likely to be the potential purchaser or lessee.

Generally, the market is dominated by pubcos that either run pubs under management or lease them out on full repairing and insurance terms. Pubcos are often able to outbid the individual retailer by reason of wholesale profits derived from their power to obtain high discounts.

When looking at a public house from the point of view of a brewer, the wholesale profits are a factor — houses with sound trades and large beer volumes are more attractive. The two super regionals, Greene King and Marston's, other regionals and local brewers are still adding to their tied estates by purchasing good trade public houses.

Besides individual sales and lettings, many public houses change hands in package deals, that is acquisition and disposal of portfolios of properties involving pubcos or brewers, or as a result of company takeovers. These transactions involve additional considerations beyond those dealt with in this chapter.

Institutional and other investors are interested in pubs let on leases to sound lessees and look at the quality of the rental stream. Investment markets seek portfolios of public houses with rents in excess of £100,000 pa. Small investment companies will look at public houses rented at less, but usually not below £30,000 pa. Before capitalising the rent it is necessary to check this against sustainable net profit. The years' purchase adopted will depend upon the covenant, risk and desirability of the public house.

The emergence of real estate investment trusts (REITs) into the UK market has provided a liquid, transparent and properly managed investment vehicle for all. With the emphasis on income projection rather than net asset value, changes in market perception are likely to filter through to public house valuations as more property use classes attain investment grade status, and specialised REITs develop in the leisure sector. Legislation sets out the framework for REITs, giving rise to tax efficient investments, but only income arising from 'the exploitation, as a source of rents or other receipts, of any estate, interest or rights in or over land' will qualify. To be able to fit into the REIT regime, pubcos will have to restructure, most likely by splitting into a property group and a pub operator.

Individual retailers and small retail groups rely normally on straightforward valuations based on accounts and receipts, and little or no regard is paid to discounts as an additive although they too are now able to obtain significant discounts.

Some public houses, particularly in country areas, may be attractive to people who view the living quarters as highly desirable and who are seeking a change in lifestyle, for example upon retirement or following redundancy. As licensees do not have to over-stretch themselves, with profits earned merely topping up a pension or other income, the pub may be less than business focused. Sale prices of such properties are often driven by and reflect the residential market, as opposed to being solely rooted in accounts, even where planning permission for wholly residential use is unlikely to be granted.

In recent years there has been a significant growth in the leasehold market, which historically had been restricted to traditional public houses, generally let to brewers by landed estates that seldom changed hands. This development has arisen from brewers, followed by pubcos, granting tied leases on their fully fitted public houses on full repairing and insuring terms and as properties they themselves leased became surplus to requirements.

Owing to the unsuitability of these leases for managed house companies or tied lease/tenancy operators, or due to the high tied rents, only individuals or small groups of managed operators will be in the market.

# Trade

If a public house is to be offered for sale or to let, one of the first questions that will be asked by the majority of prospective takers is: what is the trade? The trade is an influential factor in valuing licensed premises. It does not form the basis of a separate goodwill valuation as in shops and other businesses but is integral to the valuation of the property. The premises have the benefit of a premises licence and the trade is regarded, in most cases, as being an inherent attribute of the licence. Hence, goodwill, if any, is normally considered to be included in the price.

In what form can information about trade be obtained by the valuer? Audited and certified accounts are the most reliable source, embracing the whole business operation, and should be sought wherever these have been prepared.

If it is a tied house (that is, tied to one brewer or pubco) a certificate of liquor (wet) trade may be available. The brewers and pubcos keep records of the draught and bottled beers delivered to each house and issue a statement annually showing the barrelage, expressed in barrels (36 gallons). Details of any other deliveries will also be shown — wines and spirits (in gallons or in litres) and minerals expressed in terms of ready to drink litres (RTDLs). Some operators measure premium packaged spirits (PPS) (alcopops) on the same basis as beer due to the low level of margin. It must be remembered that in many cases this statement is unlikely to comprise the total trade as wine, spirits and minerals are now free of tie for a substantial proportion of houses and licensees can, and often do, purchase from other suppliers.

In a free house, selling different brews, an alternative source of information is the accounts or the receipts and purchases. If a publican employs a stocktaker periodically, say monthly or quarterly, the stocktaker's statements can be referred to for the receipts and purchases.

An experienced valuer may also find a cellar count a useful tool as part of the overall picture where little information is available, or to reach a judgment where a question as to the veracity of the accounts arises.

Besides liquor, income from food, letting accommodation, machines and other sources will also need to be ascertained or determined. In particular for let properties, where machines are not rented directly, the income needs careful scrutiny due to the host of differing arrangements between pubcos/brewers and their lessees/tenants. These usually involve the landlord taking a share of the machine profit, but very rarely is any clue as to the magnitude of these amounts found in the accounts.

In some circumstances, little or no trade or financial information will be available. The public house may be new, substantially altered or recently reopened after a period of closure. A previous tenant or lessee may have been buying outside the tie, have disappeared or been declared bankrupt. The valuation may be required in connection with a proposed takeover where the books have not been opened. The valuer will have to rely on personal knowledge, experience and database records held by specialist valuers, together with inspections of the public house and comparables as well as verbal enquiries to licensees, staff and customers. The valuer should also consider and assess any business plan researched and drawn up by a potential purchaser or lessee.

All data considered and research undertaken should be recorded.

# Methods of valuation

Public houses are located where they are as a result of historic accident. While populations shift away from rural areas to towns and cities, most land uses evolve and properties are redeveloped, reflecting changing demand and market fashions (eg out of town shopping centres and leisure parks), by and large public houses have evolved and survive despite the numbers falling to redevelopment accelerating in recent years. This may be of positive or negative value, but due to a pub's unique characteristics it is unlikely to correlate with rental values of other property types or be capable of analysis on a floor area basis owing to the value being dependent upon customer demand for intoxicating liquor at the particular outlet in its existing location.

Public houses are not homogeneous. Rarely will two houses be identical, even when they have the same overall floor area and are situated in the same locality. Not only are a pub's individual characteristics, such as nature and character, attractiveness, age, facilities, style and customer base variable but their proximity to transport links, car parks, other property uses and occupiers will also vary, if only slightly.

Within similar sized footprints layouts will differ. Some public houses may have a restaurant area, while others have a dance floor, function rooms, games rooms, children's room, indoor sporting facilities etc. Generally, these are not interchangeable without a capital injection, which may or may not be economically viable. Other pubs may have external facilities such as pavement drinking area, beer garden, children's play area, sporting facilities etc. All these features affect the relative values of public houses even if they share the same overall floor area.

Trade profiles will also vary with most operators aiming at a specific market sector, such as office workers, tourists, local residents, younger clientele etc. These attach to the property and take time to alter, often involving considerable sums of expenditure, although operators will not all wish to aim for the same particular niche in the market, since were they to do so the available trade would be shared and those whose taste is not satisfied will go elsewhere.

Each operator will have regard to the competition in the market, and the characteristics of its pub. They will position their business accordingly so as to achieve maximum profitability from their optimum customer base. Different styles of operation will often be complementary to one another rather than in direct head to head competition.

Due to these factors similar levels of trade are unlikely, even for pubs with similar overall floor areas within the same locality. When valuing public houses, market practice regards the level of trade as the major determinant of value, rather than a physical measure such as floor area. Furthermore the level and type of trade will determine the type of public house operator likely to be in the market. Failure to identify this correctly could result in the property being valued erroneously.

The RICS *Valuation Standards* (The Red Book) Guidance Note 1 (GN1) 'Trade related property valuations' indicates that special consideration must be given to the application of market value to properties, such as public houses, that are normally bought and sold on the basis of their trading potential.

Valuation Information Paper 2, *The Capital and Rental Valuation of Restaurants, Bars, Public Houses and Nightclubs in England and Wales*, considers the general approach to be adopted for capital and rental valuations.

Valuations should include:

- land and buildings
- trade fixtures, fittings, furniture, furnishings and equipment and
- the market's perception of the trading potential excluding personal goodwill, together with an assumed ability to obtain/renew existing licences, consents and certificates (often referred to as inherent or transferable goodwill).

A valuation on the basis of market value should only reflect inherent or transferable goodwill, related to the trading potential of the business. Any personal goodwill, attributable to the present licensee or operator, that would not be passed on is to be excluded.

The Red Book (GN 1.2.2.5) defines transferable goodwill as:

> That intangible asset that arises as a result of property-specific name and reputation, customer patronage, location, and similar factors, which generate economic benefits. It is inherent to the specialised trading property, and will transfer to a new owner on sale.
>
> (IVS GN 12 para 3.3.3)

and personal goodwill (GN 1.2.2.5.1) as:

> The value of profit generated over and above market expectations, which would be extinguished upon sale of the specialised trading property, together with those financial factors related specifically to the current operator of the business, such as taxation, depreciation policy, borrowing costs and the capital invested in the business.
>
> (IVS GN 12 para 3.3.2)

Past trade will reflect the business acumen, or lack of it, as well as the practices of the actual operator, including the hours the pub traded within those permitted under its premises licence. The valuer is required to reach a judgment as to the hours and activities a reasonably efficient operator would choose within the terms of the premises licence, consider whether or not there is any potential to extend these or to use the property in a more profitable way so as to assess the future sustainable trading potential: see The Red Book GN 1.4.4.6.

A reasonably efficient operator is defined in The Red Book (GN 1.2.2.4) as:

> A market-based concept whereby a potential purchaser, and thus the valuer, estimates the maintainable level of trade and future profitability that can be achieved by a competent operator of a business conducted on the premises, acting in an efficient manner. The concept involves the trading potential rather than the actual level of trade under the existing ownership so it excludes personal goodwill.
>
> (IVS GN 12 para 3.4)

In order to report accurately, and to make an informed professional judgment, it is essential for the valuer to inspect the public house and its environs. Indeed, PS 5.1 requires that 'Inspections and investigations must always be carried out to the extent necessary to produce a valuation which is professionally adequate for its purpose'. It is also recommended, where possible, that the licences, approvals, consents, permits and certificates relating to the property are inspected (GN 1.4.4.6).

Valuations based on trading potential, excluding personal goodwill, reflect the value of land, buildings, goodwill and, where the property is owner–occupied, trade inventory. Such valuations are valid for loan security purposes as this reflects the workings of the market: see The Red Book UK Appendix 3.1.4.6. The constituent parts cannot be valued separately, but if requested to do so for accounting purposes the valuer may make an informal apportionment: see The Red Book GN 1.5.2.

Where the value for an alternative use exceeds the market value of the property determined on the basis of trading potential, this should be stated: see The Red Book Appendix 4.1 (IVA 1.6.6.3).

# Capital valuations

Public houses are generally valued as fully equipped operational entities, having regard to trading potential and this should be clearly set out in the standard terms of engagement. The following methods are based on the assumption that the property is, or will be, used for the foreseeable future only as a public house. The possibility of potentially more valuable alternative uses is not explored but the valuer should always comment on this where it is believed such value may arise.

## Profits method

The profits, or accounts, method (sometimes referred to as receipts and expenditure method) is the primary valuation approach, particularly for free houses.

The starting point is the actual accounts net of VAT — the quality of these is important and the valuer should be satisfied as to their accuracy and dependability. Wherever possible, audited or certified accounts should be sought. Accounts prepared by an unqualified person, or unsigned accounts, should be treated with reserve. It should be borne in mind that accounts are not usually drawn up for property valuation and will be presented in a variety of forms. The purpose of the accounts, together with the accounting practice adopted, needs to be established and understood by the valuer at the outset.

The accounts will show how the pub has traded under its present ownership — there can be no guarantee this performance is maintainable, or alternatively that it cannot be bettered under a different management style or if the property is adapted to a more profitable style of operation.

Wherever possible at least three years' accounts leading up to the valuation date should be examined to identify trends and eliminate any distortions. These may include receipts encompassing a one-off event in the locality, which will not be repeated, or be for a period during which the property was closed, resulting in actual receipts shown in the accounts being higher or lower than those normally to be expected. Gross profit may be understated due to poor stock control, pilfering, excessive wastage, personal consumption or, (depending upon the accounting method adopted), to stock levels being higher at the end of the year. Conversely, if stock levels are run down during the year the profit will be overstated.

Working expenses typically include the following items, although these may be shown under more or less heads and described in different ways — wages, salaries, rent, rates, water rates, refuse collection, light heat and power, repairs and renewals, general insurance, machinery and equipment costs, glasses and bar requisites, cleaning, motor expenses, telephone, postage, stationery and printing, advertising, newspapers/journals, subscriptions, licences, accountancy costs, bank charges, sundry expenses and miscellaneous sales costs etc. The actual expenses may be higher or lower than other operators, eg due to excess wages, personal expenses being included in the accounts, or the depreciation policy adopted. Or the sum shown may be exceptional, either due to a cost that will not recur annually, such as re-thatching a roof, or because payment does not represent the true outgoing, such as rates payable being reduced by a rebate covering a number of years.

### Study 1

A typical set of free public house accounts may be adjusted by adding back to, or deducting from, the net profit.

| *Example* | £ | £ |
|---|---|---|
| Receipts | | 625,000 |
| *Less* cost of sales | | 250,000 |
| Gross profit (60%) as shown in accounts | | 375,000 |
| *Less* operating expenses | | 250,000 |
| Net profit (20%) as shown in accounts | | 125,000 |
| *Add back* interest paid and depreciation | 26,500 | |
| *Adjust* re stock in hand of consumable goods | −1,500 | |
| | | 25,000 |
| Adjusted net profit (24%) | | £150,000 |

The object of adding back interest is to exclude the personal circumstances of the publican. A prospective buyer may be financially independent and require no loans or mortgage. An adjustment for levels of stock in hand of consumable goods, reflecting the difference between the beginning and end of the financial year, may be necessary so that the adjusted net profit represents the profit emanating solely from the licensed premises and is not over- or understated due to running down or building up stock levels.

A valuer should be cognisant of trading conditions and be able to examine previous and current years' trading performance, as well as analyse and adjust accounts proficiently so as to provide a medium for valuation. Projections for future years should include any potential to expand permitted hours under the Licensing Act 2003 and reflect the likely impact of the smoking bans recently extended to Wales (2 April 2007) and England (1 July 2007). This is necessary in order to be able to assess the fair maintainable trade and future sustainable profits that can be achieved by a reasonably efficient operator, and likely to be envisaged by a potential purchaser at the valuation date.

Having determined the fair maintainable trade, the valuer will deduct the cost of sales and working expenses envisaged in running the operation. (Unlike for most other classes where the profits method is used, it is not common practice in public house valuations to make an allowance for the cost of replacing trade furnishings and equipment.) This figure is the fair maintainable net profit generated by the business or earnings before interest, tax, depreciation, amortisation and rent (EBITDAR).

## Study 2

Valuation of the freehold interest with vacant possession in a fully operational public house having regard to trading potential (excluding personal goodwill) inclusive of trade fixtures, fittings, furnishings and equipment, looked at in Study 1:

| *Example* | £ |
|---|---|
| Fair maintainable turnover | 625,000 |
| *Less* cost of sales | 231,250 |
| Gross profit (63%) | 393,750 |
| *Less* operating expenses (33%) | 206,250 |
| Fair maintainable net profit (EBITDAR) (30%) | 187,500 |
| YP | 8 |
| Capital Value of freehold interest | £1,500,000 |
| (Equivalent to a multiple of 2.4 × receipts £625,000) | |

There is a wide practice of offering the freehold interest for sale to include the trade inventory. In all cases an inventory should be prepared by the vendor and adjustments agreed for deficiencies, if any, on day of practical completion.

Wet and dry stock, consumables such as fuels and cleaning materials, and glassware should be excluded from all valuations. Payment for these is always made or adjusted at the change on the day of ingoing. Where an upgrade is required to achieve the FMT, there are outstanding repairs or the inventory is poor, the value will be less as it is necessary to make a deduction for the expenditure required.

The years' purchase will depend upon evidence derived from comparable transactions and the risk and desirability of the public house. For freeholds, this generally ranges from 6.5 up to 10 where a purchaser is actively building a brand. Leasehold interests are valued in the same way and for leases of up to 25 years the range one to four years' purchase will be the norm. Additional factors to consider include whether there is a special purchaser or any prospect of synergistic value (marriage value); the underlying value; the fierceness or lack of competition; the extent and character of the accommodation including domestic and other facilities; and the security of future trade, including the long term impact of the smoking bans.

## Comparative method

In Study 2, it will be seen that the capital value has been devalued as a multiple (2.4) by reference to the receipts as a means of comparison. Using receipts as a valuation tool is a comparative method of valuation, not a shortened profits method as it is sometimes referred to. Valuers may be forced to use the receipts in a primary way because the only available evidence of trade is the receipts, the accounts for the previous year not being prepared and certified at the date the property is advertised for sale. Valuations such as this (which take one line) are attractive as time-saving exercises, but it is essential to fully appraise the age, type of property, location and type of trade for both the public house being valued and those from where evidence is derived. This approach can be used reliably only by valuers with extensive licensed property market experience, practical knowledge and a dependable database.

Preferably, the comparative method should be regarded as a check method, but as explained above this is not always its role, or as a tool for analysing rental or capital transactions when only receipts information is known.

In looking at receipts, valuers should establish whether they include VAT. Receipts extracted from certified accounts will be exclusive of VAT but if they are taken direct from till records or cashbooks, they may include VAT which should be deducted for valuation purposes. The valuer should then consider the annual figure in the light of comparables, and only if satisfied that this represents the fair maintainable turnover adopt this for the valuation and apply a years' purchase.

## Study 3

|  | £ |
|---|---|
| Estimated annual receipts based on till records | 734,375 |
| *Deduct* VAT (17.5%): £734,375 ÷ 6.714 | 109,375 |
| Receipts ex VAT | 625,000 |
| Fair maintainable turnover | 625,000 |
| Multiple (per Study 2) | 2.4 |
| Capital Value | £1,500,000 |

The multiple to turnover for most public houses generally falls within the range of 1.5 to 2.5, but this may fall to 1 or rise to 3 depending on the profitability of the property and its desirability in the market. A valuer will need to exercise judgment based upon experience, a comprehensive database and knowledge of comparables.

## Investment method

Where public houses are let, the majority of goodwill transfers to the lessee on grant of the lease with the inherent goodwill shared between the lessor and lessee by way of rent. Besides the investment value to the freeholder or head lessee, the lessee also has a beneficial interest. As the appropriate approach to the valuation will differ dependent upon whether or not the lease is free of tie, it is necessary to ascertain this at the outset, and to consider who is in the market to buy the property and what return they require.

In the past, few public houses were sold subject to leases free of tie. Broadly, this market comprises individual public houses owned by corporations and landed estates or forming part of a commercial development. As part of the recent changes in the industry major managed operators, seeking to raise finance, have sold packages of good quality public houses to institutional investors as part of a sale and leaseback transaction. Other packages of poorer-quality leased public houses, free of tie, have come to the market.

The valuer should make an appraisal of the rent payable under a lease to determine whether it is at, above or below a sustainable level, assess the level of rent achievable if the tenant defaults and weigh up the prospect for future rental growth.

If an open market rent, free of tie, is being paid, under a full repairing and insuring lease, and has been recently fixed, a years' purchase may be applied to the rent, in common with other types of commercial property. The yield will be dependent on the strength and quality of the covenant, rental appraisal and prospect for alternative uses should a pub become unviable. This will generally fall between 10–17 years' purchase.

### Study 4

Valuation of the freehold interest in a fully operational public house (Study 2), but subject to a 20 year full repairing and insuring lease, free of tie. The rent of £87,500 pa, recently fixed at the first five yearly review, is considered to be at a sustainable level for the business.

| | |
|---|---:|
| Rent paid under lease | £ 87,500 pa |
| YP, say | 16 |
| Capital Value | £1,400,000 |

A similar approach may be adopted where the estimated current rental value, based upon a full appraisal of the sustainable net profit, is greater than the rent passing under the lease.

## Study 5

Estimate the capital value of a public house (Study 4) on the assumption the lease expires in four years' time. The rent passing is £50,000 pa and the current rental value is £87,500 pa.

| | £ pa | £ |
|---|---|---|
| *First 4 years* | | |
| Rent under lease (well secured) | 50,000 | |
| YP 4 years at 5% | 3.546 | 177,300 |
| *After 4 years* | | |
| Revised rent | 87,500 | |
| YP in perp deferred 4 years at 6.25% | 12.555 | 1,098,562 |
| | | £1,275,862 |
| Capital Value, say | | £1,275,000 |

If a lease is terminating within a few years, prior to the Landlord and Tenant (Licensed Property) Act 1990, it was often the practice to consider a reversion to capital value instead of a revised lease rent. This will now only arise where landlords wish to use a public house for their own occupation and have owned the property for longer than five years.

## Study 6

Compare the capitalisation of rent (Study 5) with the present capital value of the fully operational public house with vacant possession, estimated to be £1,500,000 (Study 2). The value of the tenant's inventory is £29,500.

| | £ | £ |
|---|---|---|
| *First 4 years as before* | | 177,300 |
| *After 4 years* | | |
| Reversion to capital value at the end of the lease | 1,500,500 | |
| Defer 4 years at 6.25% | 0.785 | 1,177,500 |
| | | 1,354,800 |
| *Less:* | | |
| Section 37 payment (see Study 10) 1 × RV | −51,000 | |
| Purchase of tenant's inventory (lease term) | −29,500 | 80,500 |
| | | £ 1,274,300 |
| Capital Value, say | | £1,275,000 |

The payment of compensation under section 37 of the Landlord and Tenant Act 1954 for the refusal to grant new lease is set at once times rateable value where the period of occupation is less than 14 years, twice rateable value where 14 years or longer: see Chapter 4 on the Landlord and Tenant Acts.

Additionally, there may be a requirement under the terms of lease for the landlord to purchase the tenant's inventory. Depending on the facts in any particular case, a further deduction may therefore be necessary before the capital value is determined.

Different valuation considerations apply to public houses let on tied leases or tenancies. Unlike other commercial valuations linked to rental income, brewers and pubcos with tied estates also receive wholesale profits on the supply of beer and other products supplied to their tenants plus a share of receipts from gaming and other machines. Valuations are therefore closely linked to sustainable profits.

As the income generated by leased and tenanted public houses is considered more secure than the retail profits earned by owner-occupiers or managed house operators, the years' purchase multiple i likely to be higher than that adopted when valuing as a fully operational public house with vacan possession (Study 2).

An overall years' purchase figure, generally between 6 to 10, may be applied to the total income, o differential rates applied to reflect the security of the individual income streams. Whichever approach is adopted it must mirror the methodology used to analyse evidence of comparable transactions.

## Overbids

Until quite recently overbid methods were adopted for two kinds of valuation — in valuing a public house from the point of view of a brewer supplying liquor, or of a pubco able to command high discounts off purchases.

A brewery company charges a rent for its freehold pub and also receives wholesale profits on the beers supplied and is able to use the potential wholesale profits and receipts from AWPs to substantiate an overbid. If the barrelage of a pub is not known but the valuer is aware of the receipts, a similar brewer's overbid might be estimated by reference to gross turnover.

If the brewery company intends to run a public house under management it would make a net trading profit as well as a wholesale/manufacturing profit, and if it wanted to acquire the property might be prepared to make an overbid from an amalgam of both sources of profit. Likewise a pubco able to obtain a significant discount on its purchases of beer and certain other consumable goods is able to use part of its discounts to make an overbid.

Although significant discounts are currently available these are not guaranteed, even though there is overcapacity in the brewing industry. This would pose additional risk, however slight in today's market conditions, to tiny retail groups or individual freetraders who, in any case, would be unable to make an overbid without sacrificing a slice of their net trading profit.

As fair maintainable net profit (EBITDAR) is now similar for most operators competing for any particular type of public house, and is of more importance to a pubco or brewer in determining the value, any overbid will be reflected in a higher years' purchase, depending upon the desirability of the public house to a prospective purchaser.

As a consequence of the growth of securitisation of assets over the past decade the yields achievable for public houses forming part of the estate are generally more favourable than for an individual property. This enables a managed house company or tied lease operator to make an overbid where a property comes on to the market that would fit into its portfolio.

An overbid may also be paid where a special purchaser is in the market for a fully operational public house and is willing to pay a higher multiple of EBITDAR than any of the competitors, or that can be supported by analysis of comparable transactions.

## Price per barrel

In the past, when up to 90% of pubs were tied to breweries and wet trade was the only significant income stream, the method of valuing at a price per barrel — draught and bottled beer expressed in barrels (36 gallons) — was normal. It has ceased to be regarded as a reliable method for general use as public houses no longer sell mainly liquor from the one brewery company — many purchase guest beers, wines and spirits elsewhere making it difficult or impossible to obtain accurate sums of barrelage. Furthermore, many practise extensive catering while others derive significant receipts from amusement machines and other sources.

Converted or composite barrelage — draught and bottled beer expressed in barrels plus converted units of wines and spirits (gallons divided by the relevant conversion factor used in practice, currently 14) is rarely used beyond free trade discounts.

Some valuers may still use barrelage as a check or for analysing comparables.

## Kennedy method

This method, based on brewer's profit, is called the Kennedy method after the judgment of Kennedy J in an earlier case (*Ashby's Cobham Brewery Co Ltd, re The Crown, Cobham and Ashby's Staines Brewery Co Ltd, re The Hand and Spear, Woking* (1906)) on redundancy compensation. The reasoning behind the Kennedy method is that it goes directly to what the brewery is making from the public house: first, the brewery makes wholesale profit from supplying goods and, second, it charges its tenants a tied rent. Or, if the pub is managed, the brewer makes a net retail profit in lieu of a tied rent.

In a development land tax case, *Inland Revenue Commissioners* v *Allied Breweries (UK) Ltd* (1982), concerning a back-street pub in Birmingham, two traditional methods of valuation were in collision. The valuation prepared by the Revenue's licensed property valuer was based on the Kennedy method, and updated, whereas the brewer's valuer depended on a valuation per converted or composite barrel — the price per barrel being derived from a scrutiny and analysis of comparable sales over a wide area.

The Lands Tribunal amended the valuation in the 1982 case so that, at the end of the day, it looked like this.

### Study 7

*Malt Shovel PH, Birmingham*
Current use value at 14.12.1976

| | £ | £ |
|---|---:|---:|
| *Wholesale profits* | | |
| 356 barrels draught beer at £3.90 | | 1,388 |
| 36 barrels bottled beer at £6.50 | | 234 |
| 135 gallons wines and spirits at 90p | | 122 |
| | | 1,744 |
| *Add*: 25% to allow for increase over tone of the 1973 valuation list | | 436 |
| Wholesale profits in 1976 estimated at | | 2,180 |
| *Tied rent* (1974) | 1,560 | |
| *Add*: 15% to update | 234 | 1,794 |
| Gross income | | 3,974 |
| *Deduct:* for repairs and insurance | | 374 |
| Net income | | 3,600 |
| YP | | 10 |
| Current use value | | £36,000 |

The Lands Tribunal uplifted the brewers' wholesale profits, established as the tone for 1973 rating assessments, by 25% to reflect the position at the relevant valuation date in 1976.

The Lands Tribunal looked selectively at the comparables and reinforced its decision by using a check valuation at £ per converted barrel. It is interesting to note that it expressed no preference for either method. As it has done in other cases where methods of valuation have come into conflict, it rationalised the evidence and showed how the answers on both methods could confirm each other.

Since the MMC Report in 1989, the licensed property market has radically altered as both the number of brewers with tied estates, and the proportion of tied houses operating under direct management, has significantly reduced, rendering both these methods of little significance in today's market.

## Building costs

The cost of construction or rebuilding plus land value will seldom equate to the market value of a public house. An assessment based upon reinstatement cost in the event of destruction will however be necessary for insurance purposes. It is recommended that this should be undertaken by a qualified building or quantity surveyor, or building cost consultant.

# Rental valuations

The first and most important step for a valuer involved in rental valuations is to read the lease. The length of lease, term unexpired, tenant's obligations including tie, rent review cycle, rental assumptions, section 34 disregards (Landlord and Tenant Act 1954) and user clauses will all affect value.

The physical characteristics and most likely hypothetical lessee are also important. For town centre public houses rented as fully fitted, the regular cycle of refurbishment required to maintain market share will need to be reflected whereas this will be of less importance for older established public houses. Where the lessee is likely to be a pubco, the cost of installing a manager to run the public house must be taken into account, whereas a private individual's remuneration will be encompassed within the tenant's share of the divisible balance.

## Profits method

A profits method of valuation, by reference to accounts, is the primary approach used to determine rents under tied leases and tenancies (where, due to the requirement to buy most products from the landlord or nominated suppliers future sustainable profits are effectively controlled) as well as for fully fitted free-of-tie traditional pubs. This method can also be used to calculate ground rents.

The use and adjustment of accounts in a profits valuation is considered above for capital purposes. While it is important for the valuer to understand the business, it should be remembered that it is not the business of the actual occupier, but that of the hypothetical lessee, that is being valued.

Any rent payable under the current or former lease, as shown in the accounts, must be added back since the object of the exercise is to ascertain the profit emanating from the licensed premises in order to determine what rent is affordable. If depreciation is shown, this should also be added back but reflected when considering the sum allowed for interest on tenant's capital.

As a profits valuation is seeking to establish the net sustainable annual profit, turnover and expenses must be projected, either upwards or downwards, rather than merely adjusting the preceding year's accounts. At the time of writing the impact of the smoking ban is open to conjecture. Some public houses will be more affected than others — where appropriate this should be reflected in the fair maintainable trade adopted.

Having determined the fair maintainable turnover, the valuer will deduct the cost of sales, working expenses envisaged in running the operation and interest on tenant's capital. This encompasses allowances for the cost of replacing trade furnishings and equipment, carrying stock of consumable goods and the cash required to run the business. A further deduction, of interest on tenant's capital, is made so that the fully adjusted net profit represents the profit emanating solely from the licensed premises.

This adjusted net profit is often referred to as the divisible balance. It falls to be shared between the tenant as reward for operating the business and the landlord as a return on his capital by way of rent. The valuer should test the end result against comparable transactions, differentiating according to any trading restrictions, eg liquor tie impacting upon operating profit, and treatment of gaming machine receipts.

## Study 8

A summary rental valuation, based on the figures from Study 1, but assuming the public house is let on a tied lease:

|  | £ | £ |
|---|---|---|
| Receipts |  | 625,000 |
| *Less* cost of sales |  | 250,000 |
| Gross profit (60%) as shown in accounts |  | 375,000 |
| *Less* operating expenses (36.2%) |  | 226,000 |
| Net profit (23.8%) as shown in accounts |  | 149,000 |
| *Add back* interest paid on loans | 2,500 |  |
| Adjust operating expenses to reflect savings | 18,500 |  |
|  |  | 21,000 |
| Adjusted net profit (27.2%) |  | 170,000 |
| *Deduct* interest on tenant's capital: |  |  |
| Trade inventory of furniture, fixtures and fittings valued at ingoing 3 years ago at £32,000. Allow for depreciation and additions since, say | 29,500 |  |
| Stock in hand of consumable goods (actual) | 10,900 |  |
| Cash required to run business, allow 2 weeks' purchases of consumable goods, say | 9,600 |  |
|  | 50,000 |  |
| Interest at, say, 10% on £50,000 |  | −5,000 |
| Divisible balance |  | 165,000 |
| Tenant's bid at 50% (Rent) |  | £82,500 |

Rent equates to 13.1% of fair maintainable trade.

If any improvement works to the property are required or the inventory needs upgrading, in order to achieve the FMT, a deduction to reflect the amortisation of these costs should be made.

The quantum of the tenant's bid will depend on the risk and desirability of the property. This may rise to 60% for a good public house in the City of London but only be 30% for a public house/nightclub where regular, periodic refurbishments are required and the trade is more fickle.

Although written for rating, a guidance note produced by the Joint Professional Institutions' Rating Valuation Forum and the Valuation Office — *The Receipts and Expenditure Method of Valuation for Non-Domestic Rating* — contains useful advice including a page on the 'shortened' method (see below).

Full and detailed accounts valuations will be found in Chapter 12 (Rating) and in Chapter 18 (Hotels)

## Comparative method

The comparative or shorter receipts method is in essence a one line valuation calling for a percentage to be applied to the turnover to produce a rental value, although this approach is more useful as a check or as a tool for analysing rents when only receipts information is known.

While valuations based on full accounts are always preferable, if turnover is to be used as the primary valuation the rental percentage adopted must follow a full appraisal. It is essential that the valuer carrying this out is actively involved in the public house market, fully appreciates the factors impacting upon the sustainable net profit and has a comprehensive database of comparables. This approach should not be used when advising tenants or potential tenants as to the rent they should pay under a new lease or at review.

Rental values can vary greatly, for a fully fitted, free of tie pub the range is 12 to 17%. Shell rents are likely to fall between 9 to 12% and for ground rents 4 to 6%. The norm for tied houses is generally from 9 to 13% according to extent of the tie and whether the pub is let as a tenancy, or as is now more normal, on a full repairing and insuring lease. Other factors affecting the rental level include the quality of the house and trade, and treatment of machine income. Where a pub is trading above expectation the landlord may accept a lower rental percentage.

## Floor area

Unlike most other types of commercial property, valuations of public houses do not generally depend on floor area appraisals or zoning, as size will not be the major factor influencing value. However, since about the mid 1990s, pubcos and brewers have competed with restaurateurs and other users for units suitable for conversion to public house, restaurant and bar use, particularly former retail premises in town and city centres attractive to a wide range of A3/A4, and possibly A1/A2 and D2, users. Without a previous trading history, where such properties are let as shells and user provisions in the lease allow, initial rents are likely to be determined on a floor area basis, and are therefore more correctly valued on this basis to reflect competition and the workings of the market. A check should however be made using a profits method based on projected trade.

Similarly, direct comparison by using floor area is also appropriate when considering shell rents for pubs, restaurants and other A3/A4 units, both on leisure or retail parks and in secondary retail parades where the unit is readily adaptable for commercial purposes. Such an approach is not appropriate for purpose built public houses where the tenant is required to fit out.

For rent review purposes, when valuing such properties it is essential to establish the exact provisions under the lease. Often this involves valuing for alternative uses — if A1 or A2 use is stipulated a valuer with the prerequisite commercial expertise should undertake the valuation. If the rent review is restricted to the public house value, the valuer should revert to the traditional profits approach.

The appropriate measure of floor area will vary according to the characteristics of the property, competition and local market practice. Where the most likely purchaser is A3/A4, either bar space (public areas) or gross internal area (GIA) are likely to be adopted. If a pub is competing with retail use,

GIA or zoning, depending on size of unit, may be more appropriate. There is however no universal practice and there are wide variations in the way GIA or net internal areas are measured, although the RICS Code of Measuring Practice recommends standards for measurement of such properties.

Likewise, there are a number of ways in which measurement may be used, including overall; support areas at 50% of public area rate; basement/first floor at 50% of ground floor overall rate and basement/first floor at 50% of ground floor rates for public area/support areas. Adjustments to reflect the type of multi-level property, the nature and user friendliness of additional space and quantum of support areas, compared with total area, are becoming more sophisticated. It is however important that whatever method is adopted to analyse comparables is also used when valuing similar properties. For properties with differing characteristics it may be necessary to make valuation adjustments to reflect factors such as superior or inferior layouts.

See Chapter 7 (Retail) for details regarding methodology.

# Rating valuations

The following relates to the position in England and Wales only. Reference should also be made to Chapter 12, Rating, for the general principles of this tax, which is based upon the annual value of property.

For 50 years public houses were assessed on the Direct Approach Method or Direct Method as it was shortly known. This was introduced following the House of Lords decision in *Robinson Bros (Brewers) Ltd v Durham County Assessment Committee* [1938], where it was held that brewers (who then owned most of the public houses) should be considered also as hypothetical tenants of public houses and, furthermore, the effect on values of competition between brewers to acquire public houses should be taken into account. In skeleton it had this form:

## Study 9

| | |
|---|---|
| Tied rent | £ |
| *Add*: Brewer's wholesale profits on tied supplies of beers and wines & spirits | £ |
| Total brewer's net income | £ |
| Brewer's rental bid between 30% and 50% | £ = Rental value for rating |

Prior to this case, valuations had been estimated on the basis of what a retailer would pay, and for the 1995 and subsequent rating lists valuations have reverted to a receipts based rental method. For the 1990 revaluation, the rateable values were assessed on an overbid method based on tied rents plus a percentage of brewer's wholesale profits.

The basis of rating was revised by the Local Government Finance Act 1988, removing domestic accommodation from rateability. The Local Government Finance Act 1992 introduced Council Tax, based on the capital value of domestic property, with effect from 1 April 1993.

Responsibility for valuing domestic and non-domestic hereditaments, and maintaining Council Tax Valuation Lists and Non-Domestic Rating Lists, lies with Valuation Officers of the Valuation Office Agency, an executive agency of HM Revenue & Customs.

The antecedent valuation date (AVD) for the 1990 and each subsequent five yearly non-domestic rating revaluation has been set two years prior to the lists coming into force. For the 2005 revaluation, the AVD is 1 April 2003 and the date the lists took effect is 1 April 2005.

Notices requesting information of rents and turnover, frequently referred to as forms of return (FORs), are served by valuation officers (the statutory officers) mainly in the AVD year.

Rents agreed in the market for public houses are adjusted to accord with the definition of rateable value (Local Government Finance Act 1988, Schedule 6 para 2(1)) and analysed by valuers with reference to turnover. The evidence of public house rents devalued on this basis, expressed as rental percentages, can then be applied directly to value other similar public houses where the turnover is known, or used to create a scheme of valuation as is done for rating purposes.

Public house assessments are determined by applying rental percentages to fair maintainable receipts (FMR). The use of receipts and expenditure analyses, which may well have been used to inform the actual tenant and landlord, is therefore unnecessary as the task of rating valuers is to interpret rather than set the market, and to arrive at a valuation that will accord with levels of value produced in the market.

As the majority of public houses have living accommodation, subject to a separate assessment for council tax, rental evidence has to be adjusted to reflect the value of the living accommodation. The appropriate methodology for this and other rental adjustments, the evidential weight to be given to rents under differing leases and tenancies, and the actual rental analysis were examined in detail, by both the Valuation Office Agency and valuers acting on behalf of the industry, and this formed the basis of detailed negotiations.

Following consultation with interested parties, and agreement with the British Beer & Pub Association (formerly the Brewers' Society, then Brewers & Licensed Retailers Association), the Valuation Office Agency has issued Approved Guides for assessing the rateable values of public houses in the 1995, 2000 and 2005 Rating Lists. These Approved Guides contain tables and graphs showing the results of rent analysis in terms of rental percentages, according to the level of turnover, for the liquor (plus gaming machines), catering and letting income streams. In most cases, the sum of rateable value determined by applying these percentages to the FMR for each separate income stream, plus the value attributable to any other part of the property that is not reflected in the FMR, will provide the total rateable value.

For liquor sales, there are three broadly based geographical areas — central London, outer London and provincial England and Wales, although before determining either of the London options, which reflect higher levels of rents, it is also necessary to consider the trading position of the pub including its exact location, nature and type of trade.

There are three valuation bands for liquor sales (plus gaming machine receipts) for the appropriate geographic area. The underlying concept is to identify types of property and trading locations where there is a potential for a pub to charge higher prices and trade at its optimum level, and those pubs that must compete in, and be responsive to, a market with fickle or restricted demand. Pubs producing higher or lower gross profitability than average are likely to prove more or less attractive to potential tenants, and this will be reflected in their rental bids.

The correct percentage to be adopted from within the band range should reflect the nature and quantum of outgoings required to operate the pub having regard to its individual characteristics and the impact of these outgoings on the bottom line. At any given level of turnover, a modern, single bar, brick and tile built pub with high gaming machine receipts will generate higher net profits than a similar sized, older, multi-bar pub with low gaming machine receipts. Other factors affecting the net profit, and impacting upon rent, will include the size of the pub and whether or not it is able to trade at its optimum potential, the level of prices charged, staffing costs having regard to level and scope of service, provision of entertainments, security, marketing, incentives, maintenance, insurance etc in relation to the FMR.

For the 2005 Rating Lists, in the provinces, the minimum rental percentage on liquor and gaming machine receipts at an annual turnover of £100,000 is 2.5% for pubs with the least attractive physical and locational characteristics, rising up to a maximum of 12.5% on receipts in excess of £500,000 pa for good quality, well located pubs.

There are two valuation bands for food. Besides trading locality and physical characteristics, it is necessary to consider the ability of a pub to conduct food trade and the style and profitability of the food operation.

For food sales the percentages vary from 0%, where turnover is low with little or no profit — often a loss-leader to attract custom — to a maximum of 10.5% for high turnover operators with moderate to low outgoings. For pubs with an annual food turnover of between £50,000–£100,000 the percentages range from 4% to 7%.

Rental percentages for lodges, determined according to receipts per room, fall between 11% (receipts less than £5000/room) to 17.5% (receipts £18,000 plus/room). For other letting bedrooms the rental percentages will be up to 3% higher than the adopted food percentages, subject to a minimum range of 6% to 9%.

The valuation basis for liquor sales and gaming machine takings will generally reflect all other incidental receipts, such as tobacco sales, juke boxes, other machines (except AWPs), telephones, ATMs (automatic teller machines), pool tables, room hire etc. Where, due to advantageous characteristics of the actual property, receipts generated from other sources fall outside the norm, such as admission charges, room hire or those arising from adjoining land or buildings (eg used as caravan/camping site), and return a secure level of profit on a regular basis, this income stream will generally be valued separately at an appropriate rental percentage.

The approved guide states:

> To ascertain the correct rateable value the first and most important consideration is to determine the 'fair maintainable receipts' (FMR) of the property excluding VAT. These will be split between gross receipts for liquor, food, accommodation and other sales; and net receipts from gaming machines.
>
> The figure of receipts adopted should represent the annual trade considered to be maintainable at the antecedent valuation date (AVD), 1st April 2003, having regard to the physical nature of the property and its location as at 1st April 2005 when the new rating lists come into force (or subsequently following a material change of circumstances) on the assumption that the business will be proficiently carried out by a competent publican responding to the normal trading practices and competition of the locality.

Vacant and to let, unless it is a new property, potential tenants would be aware of the actual trade carried out at the property. The valuer's estimate of FMR should, wherever possible, have regard to this when adhering with the principles endorsed by Mr Justice Thompson in *Watney Mann Ltd* v *Langley (VO)* [1966]:

> He (the hypothetical tenant) will, as the prospective tenant would, endeavour to estimate what trade could reasonably be expected to be done by the ordinary tenant if he were the licensee in the premises as they now are, in the area in which they are located. While I do not doubt that such a prospective tenant would consider in his mind whether he could make as great or a greater success of the house than his predecessor, the base from which he would ponder on his prospects would, I have no doubt whatever, be the actual trade his predecessor had in fact done.

## Study 10

Assess the rateable value of Study 1 *et seq*. This is a roadside public house and restaurant, with large adjoining ca▮ park, built inter-war with residential and neighbourhood shopping close by and no direct competition. The turnove▮ for the financial year ending 31 May 2002 has been projected to estimate FMR at AVD at £400,000 liquor, £90,00▮ food and £15,000 net income from gaming machines.

| | | |
|---|---|---:|
| Liquor including AWPs: | £415,000 at 11.0% | 45,650 |
| Food: | £90,000 at 6.0% | 5,400 |
| | | £51,050 |
| Rateable Value say | | £51,000 |

## Study 11

Assess the same pub (Study 10) following the construction of a 40 bedroom lodge. The first 12 months' trading▮ receipts for the financial year ending 31 May 2003 are £455,000 liquor, £105,000 food, £335,000 accommodation▮ and £18,000 gaming machines. These are regarded as representative of the FMR.

| | | |
|---|---|---:|
| Liquor including AWPs: | £473,000 at 11.25% | 53,212 |
| Food: | £105,000 at 6.35% | 6,667 |
| Accommodation: | £335,000 at 12% | 40,200 |
| | (£8,375/room) | |
| | | £100,079 |
| Rateable Value say | | £100,000 |

## Study 12

Assess the rateable value of a character public house with good kitchen facilities. It runs a quality catering operation▮ employing specialist staff with varied à la carte menus. Receipts for liquor are £105,000, food £180,000 and £2,000▮ net income from gaming machines. These represent the FMR.

| | | |
|---|---|---:|
| Liquor including AWPs: | £107,000 at 6.45% | 6,901 |
| Food: | £180,000 at 7.25% | 13,050 |
| | | £19,951 |
| Rateable Value say | | £20,200 |

## Study 13

Assess the rateable value of a public house, close to the prime retail location in the town centre, comprising a modern, well-designed single open plan bar, in a former shop. It achieves good trade throughout the day. The FMR is £1,525,000 — liquor £1,125,000; food £320,000; gaming machines £80,000.

| | | |
|---|---|---:|
| Liquor including AWPs: | £1,205,000 at 12% | 144,600 |
| Food: | £320,000 at 9.5% | 30,400 |
| | | £175,000 |
| Rateable Value say | | £175,000 |

## Study 14

Assess the rateable value of a public house (as Study 13) situated within a popular public house circuit forming part of a distinct market for which there is relevant market evidence. The floor area of the bar space (public areas including servery but excluding kitchens, beer stores, cellars, office, WCs etc) is 440 m². The total floor area measured to GIA is 770 m².

An example of the floor area comparative basis:

| | |
|---|---|
| Bar space: 440 m² at £437.50 m² | £192,500 RV |
| or | |
| GIA: 770 m² at £250 m² | £192,500 RV |

The appropriate unit of measurement will depend on local market practice and the methodology adopted to analyse rents. Valuations must be carried out by reference to the same criteria. Unless all units are similar, it is unlikely that market transactions will have been informed by reference to zone A since, unlike retail, the rental value for licensed use is not unduly sensitive to frontage/depth ratios, nor to other relevant factors likely to impact upon retail rents.

In this study, if the bar space is 400 m², the rateable value based on the tone value of £437.50 m² would be £175,000, the same as a receipts based valuation (Study 13). If the floor area is only 360 m² the rateable value would be £157,500.

Where a floor area tone is established from rental evidence of bar style public houses within a defined locality, this evidence should be used to determine the rateable values of similar properties, irrespective of whether the resultant figure is higher or lower than a valuation based on receipts.

## Study 15

Assess the rateable value of a small, unmodernised, terrace public house with poor kitchen facilities and no special features or attractions. The pub is in a small village with several competing public houses, little passing traffic and it is only used by locals. A wet turnover of £105,000, dry trade of £1,750 and net income from a gaming machine of £1,500 represent the FMR.

| | | |
|---|---|---|
| Liquor including AWPs: | £106,500 at 4.75% | 5,059 |
| Food: | £1,750 at 0% | Nil |
| | | £5,059 |
| Rateable Value say | | £5,050 |

A valuation using the approved guide is a comparative rental valuation, not a valuation on accounts. Furthermore, it should not be viewed as an arithmetical calculation. Valuations must be considered by a competent valuer within the terms of the approved guide having regard to the individual nature of each property, its trading location and style of trade.

If a valuer is acting for a ratepayer and considers an assessment based on the approved guide does not truly reflect the rental value of the property, especially when valuing a hybrid or other public houses at the interface with other uses, it may be advantageous to look at a receipts and expenditure accounts valuation. Summaries in Studies 1 and 8 above provide a guide. The principle is also illustrated for an hotel in Chapters 12 (Rating) and 18 (Hotels).

# Further reading

*Law & Valuation of Leisure Property* (Marshall, H., Williamson, H., eds) Estates Gazette

*Rating List 2005 Approved Guide to Valuation of Public Houses* Valuation Office Agency (available on website *www.voa.gov.uk*)

*Receipts and Expenditure Method of Valuation for Non-Domestic Rating* RICS Books (1998)

*RICS Appraisal and Valuation Standards* 6th ed (the 'Red Book') RICS Books

*The Capital and Rental Valuation of Restaurants, Bars, Public Houses and Nightclubs in England, Wales and Scotland* Valuation Information Paper No 2 (2nd ed) RICS Books (2006)

Valuation — Public Houses chapter, published by RICS (*www.isurv.com*)

# Hotels

## Introduction

The term hotel encompasses a wide spectrum of property types from larger units having up to or maybe in excess of 1,000 letting bedrooms, to smaller units maybe having as few as 10 or even less. They range from modern, purpose-built properties, incorporating the latest design techniques, to converted old manor houses or coaching inns often with parts listed. Services provided vary considerably between hotel types. Budget hotels may offer letting bedrooms only, three, four and five star hotels may in addition provide a wide range of facilities such as restaurants, bars, conference and banqueting rooms, leisure clubs, golf courses, business centres, outdoor pursuits, etc. and top five star hotels may even provide the services of a butler.

Whichever category a particular property falls in, one factor that most hotels have in common is that they have been built or adapted for their particular use and have limited potential for conversion to alternative uses.

Hotels are a class of property which normally changes hands in the open market as fully operational businesses at prices based primarily on their trading potential. A property will therefore, be normally sold inclusive of all trade fixtures, fittings, furniture, furnishings and equipment, and with the benefit of all licences, permits, certificates and trading potential, but excluding stock in trade, leased and badged items, some of which may be subject to a separate agreement. The whole principle of valuation is therefore based upon potential turnover, net profit and the return on capital required by prospective purchasers in the market. However, while larger hotels are generally operated by hotel companies to make a profit, this may not be the primary consideration for some, mainly smaller hotels where the owners are motivated by other reasons such as the provision of a family home, quality of life etc.

## RICS *Valuation Standards*

Reference to the 2008 edition of the RICS's *Valuation Standards* (The Red Book) is essential before undertaking the valuation of any hotel. The valuation basis generally adopted will be market value (MV) which is defined in Practice Statement PS3.2 as:

> The estimated amount for which a property should exchange on the date of valuation between a willing buyer and a willing seller in an arm's-length transaction after proper marketing wherein the parties had each acted knowledgeably, prudently and without compulsion.

The valuation basis may require special assumptions when applied to hotels, valued by reference to their trading potential.

### GN 1

Guidance Note 1 (GN1) of the Red Book relates to trade related property valuations and goodwill, and sets out the general approach and special considerations to be adopted for the capital valuation of properties which are normally bought and sold on the basis of their trading potential.

The contents of GN 1 are not set out in full here, but particular attention is drawn to the following.

### GN 1.2.3

The operational entity will usually comprise:

- the legal interest in the land and buildings
- the plant and machinery, trade fixtures, fittings, furniture, furnishings and equipment
- the market's perception of the trading potential, excluding personal goodwill, together with an assumed ability to obtain/ renew existing licenses, consents, certificates and permits
- the benefits of any transferable licences, consents, certificates and permits.

### GN 1.4: The valuation approach
### GN 1.4.3

The valuer should distinguish between the market value of an operational entity and the value to the particular operator (its worth to that operator). ... While the present operator will be one potential bidder in the market, to come to an opinion of value the valuer will need to understand the requirements and the achievable profits of all other potential bidders, and the dynamics of the open market.

### GN 1.3.2

Where the property is trading the correct basis would be:

Market Value as a fully-equipped operational entity, having regard to trading potential.

### GN 1.3.5

Where a property is vacant or if it is a new property with no existing trade the correct basis would be:

Market Value of the empty property having regard to trading potential.

### GN 1.4.5

Specialised trading related properties are considered as individual trading concerns and typically are valued on the basis of their potential earnings before interest, taxes, depreciation and amortisation (EBITDA) on the assumption that there will be a continuation of trading.

### GN 1.4.6

The task of the valuer is to assess the fair maintainable level of trade and future profitability that could be achieved by a reasonably efficient operator, of the business upon which a potential purchaser would be likely to base an offer.

*GN 1.2.4*

A reasonably efficient operator is defined as:

> a market-based concept whereby a potential purchaser, and thus the valuer, estimates the maintainable level of trade and future profitability that can be achieved by a competent operator of a business conducted on the premises, acting in an efficient manner. The concept involves the trading potential rather than the actual level of trade under the existing ownership so it excludes personal goodwill.

# Valuation Information Paper no 6

Valuation Information Paper no 6 — *The Capital and Rental Valuation of Hotels in the UK*, agreed with the British Association of Hospitality Accountants, was produced by the Trading Related Valuation Group of the RICS in 2004 and considers:

- the general approach to be adopted for the valuation of hotels
- the valuation of trading potential and goodwill
- licences, consents, certificates and permits and
- the valuation of hotels for loan security purposes.

The paper recommends that when valuing hotels as fully equipped and operational entities the valuer should reflect the same methodology, or combination of methodologies, as adopted by prospective purchasers in the market and broadly divides these into:

(i)   the direct comparables approach
(ii)  by reference to the capitalisation of a stabilised year's assessment of maintainable profit
(iii) the discounted cash flow (DCF) approach.

Again, it is essential that the valuer makes full reference to this paper before undertaking valuations of hotels. The case of *Corisand Investments Ltd* v *Druce & Co* [1978] is also worthy of note.

# Inspections

The details that need to be taken into account on inspection differ from many other types of property because floor area is not the essential ingredient that determines value. The information required relates to the trading potential of the property.

Location, as with any property, will provide an hotel's most important advantage or disadvantage. What is a good location will depend upon the type and class of hotel under consideration, but the valuer must always have regard to the economic prosperity of the surrounding locality and the hotel's proximity and accessibility to sources of business and the target market sector. City centres are always in need of an appropriate supply of hotel bedrooms, and major airports are another prime location where high occupancy can be expected, provided there is not an over-supply. Many cities have seen a vast expansion in room numbers in recent years. Easy road access, closeness to motorway junctions or public transport facilities will generally be of benefit to a hotel's trade, but in other instances a remote, attractive countryside location may hold particular appeal. Some hotels may benefit from their proximity to large industrial or commercial establishments or tourist attractions.

Even if a full structural survey is not being carried out, it will be important, as with all property inspections, to note the age, type and quality of construction, and the general condition of the buildings and accommodation both internally and externally. Consideration needs to be given to the fair annual cost of repairs and maintenance, and account taken of any outstanding works or necessary items of expenditure, either current or likely to occur in the near future.

Supply to the property of gas, electric, water and drainage should be confirmed together with the nature and extent of any heating systems, air conditioning etc.

Hotels commonly provide a number of different services within the same building which are often available for the use of both residents of the hotel and non-residents. They include the provision of letting bedrooms, food and beverage services in restaurants, bars and banqueting rooms, conference/ meeting rooms and leisure facilities. Full details are required of all the accommodation, facilities and services offered by the hotel, always having regard to the overall layout and ease of operation, staffing and management. The valuer needs to formulate an idea not only of the potential turnover of the hotel but also of the running costs necessary to achieve that turnover.

The letting bedrooms will normally provide the most profitable source of income. It is not necessary to measure all the rooms, but the valuer needs to know the total number of available rooms; the size and specification of the rooms in general terms; the mix between different room types, ie doubles, twins, singles, suites etc.; the total sleeper capacity; whether the rooms have en-suite bathrooms or showers; the heating or air conditioning systems; the quality, condition and standard of furnishings and fittings; and the facilities provided including such items as televisions, radios, fax machines, modem points/ internet access, in-house movies, direct dial telephones, hairdryers, mini-bars, safes, tea and coffee making facilities, trouser presses, etc. The general comfort of the rooms in terms of noise, views, etc. should also be noted. A certain degree of noise may be expected in city centres for instance, but this can be greatly alleviated by the provision of secondary or double glazing and air-conditioning. Some hotels may charge higher rates for rooms with attractive views, perhaps sea facing. The advertised room rates should be noted, but in the knowledge that these will likely differ significantly from the actual achieved room rates. Breakfast may or may not be included in the room rates. Rooms may have different tariffs to reflect different size or location of room or simply to reflect the facilities offered in the room such as bathrobes, better quality toiletries or possibly the use of an executive lounge. The provision of lifts and disabled facilities are important factors.

The net internal floor areas of bars, restaurants, banqueting and conference rooms may be taken, but of greater importance in estimating potential revenue are factors such as the capacities of the facilities, their general layout and location within the hotel, their flexibility and ease of use, modernity and character/ambience. For instance, a bar or restaurant with direct access to the street may attract a good level of non-resident trade, whereas it can be extremely difficult to attract such trade to facilities above ground-floor level. Public rooms situated on more than one floor may necessitate the staffing and management of two kitchens whereas in other circumstances one would easily be able to cope with the same level of trade. Far greater use can be made of large conference rooms that are capable of being subdivided than those that are not. Again, the general comfort of the rooms should be noted in terms of natural light, heating, air conditioning, etc and whether the facilities offered meet modern standards and requirements. Record should be made not only of the facilities offered but also of the type of product and services available, functions catered for and pricing structures.

Leisure facilities, as separate profit centres, historically may have added relatively little to the overall profits of an hotel, but their provision is becoming ever more essential to attract trade and improve room rates, particularly in the higher 3 star and above hotels. Many hotels now establish viable clubs in their own right in addition to use by hotel guests. Equipment is constantly being updated, and as well as recording the facilities offered and their quality, it is important to determine the level of outside demand, whether there is a separate membership scheme and if so, the joining and annual membership fees, how many members there are, and whether there is a waiting list.

The availability and ease of car parking, preferably secure and directly controlled by the hotel is another important factor. Well tended, landscaped grounds may add to the appeal of an hotel but will also increase running costs. Golf course hotels have become a separate category in their own right, and it is essential to record the quality of the golfing facilities, including the course on offer, separate membership schemes etc.

The valuer must confirm whether all facilities are operated directly by the hotel or whether any are operated under a separate lease or franchise arrangement.

Note should be made of ancillary accommodation, its suitability, adequacy, and location in the hotel. This will include management and administration offices, kitchens, stores, plant rooms, cellars, staff accommodation, and delivery points. Record should also be made of expansion opportunities.

Hotels should ideally have in place a planned refurbishment programme. As well as any structural alterations to a property, the valuer should ascertain when any recent refurbishments were carried out. Outstanding or catch up works may require considerable capital investment which will impact upon the valuation.

While not carrying out an inventory valuation, the valuer during his inspection, should ascertain which, if any, items are leased or hired. He should confirm the existence and validity of all necessary licences and certificates which may include among others a Justices' liquor licence, gaming licence, licences for civil weddings and a public entertainment licence. The fire certificate should be checked, and confirmation sought that there are no outstanding statutory notices. Compliance with all appropriate Acts of Parliament must be ascertained and these may include among others, Health and Safety at Work, Food, Food Safety, Building, Disability Discrimination, Transfer of Undertakings, Environment and Consumer Protection. If it is not possible to inspect all necessary licences a report on title should be obtained from solicitors.

It is always worth obtaining a copy of the hotel brochure and any available conference packs etc. These will provide a summary of location, style of hotel, facilities offered and pricing structures.

It is essential to meet the manager or proprietor to discuss the hotel, its trading performance and potential. He should be questioned about such factors as occupancy levels; average achieved room rates; the main sources of business and principal business generators; average length of stay; trading profile; competitors; reasons for past fluctuations in trade; trade outlook; and any known factors likely to affect the hotel's future trading prospects including new competition. Tenure must be confirmed together with details of any sub-lettings or franchise arrangements.

A SWOT (strengths, weaknesses, opportunities and threats) analysis is a good means of summarising all relevant factors.

From the inspection and discussions with the manager or proprietor, the valuer will be able to give consideration to the style and class of the hotel, its suitability to the market sector at which it is aimed, and begin to make an assessment of its trading potential and security and the necessary costs involved in running the hotel to achieve that potential.

# Definitions

It is worth noting here several terms commonly used in the hotel industry.

- *Room occupancy*: the total number of letting bedrooms occupied during the year as a percentage of the total number of rooms available.
- *Bed occupancy*: the total number of bed spaces occupied during the year as a percentage of the total number of bed spaces available.
- *Double occupancy*: the percentage of total rooms sold occupied by two guests.
- *Average achieved room rate (AARR)*: the average rate achieved for every bedroom that is let. AARR is calculated by dividing the rooms revenue by the total number of bedrooms occupied during the year. This will differ from the advertised rack rates because of various discounts offered to guests.
- *Revenue per available room (REVPAR) (also referred to as average daily rooms yield)*: the average rate achieved for all letting bedrooms available throughout the year. Rooms yield is calculated by dividing the rooms revenue by the total number of rooms available during the year. This is also the product of room occupancy multiplied by the average achieved room rate.

# Investigation of market sector, demand and competition

Having completed the inspection of the particular hotel and on-site enquiries, the valuer will need to investigate the market sector at which the hotel's trade is aimed, demand for the services offered and competition.

It is paramount that an hotel competes in a class of business and market sector to which it is suited. The valuer needs to understand and appreciate the economic profile of the surrounding locality and the sources, nature and level of demand for the hotel's services and facilities, particularly the letting bedrooms as these will normally contribute the greater proportion of profits. Sources of business vary considerably between hotels and can include travelling businessmen, residential conferences, day conferences or meetings, functions such as weddings, dinners etc; tourists, holiday makers, coach parties and leisure breaks. The valuer must be particularly careful if an hotel is heavily reliant upon one client for its trade because any breakdown in that relationship would have a significant effect on the hotel's trade and consequently value. The type of trade coming to the hotel will determine the rates that can be charged for the facilities provided.

The size, nature and location of competitor hotels needs to be established together with the facilities they offer and their relative advantages and disadvantages. Depending upon the type of hotel in question, competitors may be confined to other hotels within a relatively small area, or may be spread throughout the country or indeed internationally. Efforts should be made to try and ascertain an idea of the relative trading performances of competitor hotels. The nature of trade may determine that other establishments such as restaurants and leisure clubs are also considered to be major competitors. All competition, whether in existence or planned, has to be investigated and any possible effect on the trade of the subject hotel properly taken into account.

The valuer needs to formulate his opinion of the fair level of trade at the hotel in question, its security and future prospects for improvement or possible downturn.

# Trading potential/fair maintainable trade

The first stage is to then assess the fair maintainable trade or annual turnover. From his inspection of the property and other enquiries and analyses, the valuer will be able to formulate a reasonable estimate of the trading potential of the hotel initially by reference to average room occupancy and average achieved room rates. However, wherever possible, the valuer should obtain the actual trading accounts for at least the last three years, the current year and projections for future years. It must be remembered that accounts can be in different formats and may be prepared for a variety of purposes. Hence it is difficult to generalise, but the valuer must carefully analyse whatever accounts are available and be satisfied that the information provided is full, accurate and reliable. Whatever accounts are available, the valuer must establish their accuracy and reliability. He needs to ensure that all relevant items of income and expenditure are included. Where a company operates more than one property particular care is required to ensure that all the items shown in the individual accounts relate only to the property being valued. On the other hand, he must also ensure that no revenue or costs are taken centrally and hence are not shown in individual accounts.

The accounts of a particular property will only show how that property is trading under the particular management at the time, and may or may not represent the market's perception of the fair maintainable level of trade. The task of the valuer is to use his experience, knowledge of the market and analysis of the trading accounts of comparable properties, to form an opinion of the fair maintainable trade achievable not by the best or worst operator, but by the average or reasonably efficient operator. Assistance may be gained in this respect from the local Tourist Board and statistics published by specialist hotel consultancies or the British Hospitality Association.

# Goodwill

This essentially takes two forms:

(i)   inherent or transferable goodwill — attaches to and passes with the property
(ii)  personal goodwill — extinguished on sale of a property.

It must be emphasised that goodwill can only have a market value to the extent that it is transferable. Value which attaches to the building and passes with the property by virtue of circumstances such as its location, design, licences, planning permission and occupation for it's particular use is reflected in its trading potential which can enhance the value of buildings but does not form a separate element in the saleable value of the business. A hotel which is open and trading will probably have a higher value than if it has been closed and empty for some time. A newly built hotel will likely attract a lower value than one with a proven trading history.

This needs to be differentiated from personal goodwill which is attributable to the personal skill, expertise and reputation of the present owner or management. It may also arise as a result of access to a central reservations system or a particular brand name. Personal goodwill attaches to the operator and will disappear when he leaves the property. However, it may be replaced or even enhanced by the personal goodwill of the new operator. The valuer must exclude any element of trade attributable to personal goodwill, but include all or any additional potential that would be realised by a reasonably efficient operator. It is therefore, essential to be able to identify the likely potential purchaser (excluding a special purchaser). For example, if a hotel is sold by a major chain operator to another similar operator, it may be considered that there will be little or no effect on trading levels. On the other hand, a private purchaser may suffer a drop in trade if he does not have the benefit of national marketing, central reservations etc. An internationally well known operator may be able to attract a higher level of trade to a particular hotel than a purely national operator. Consequently, it is essential for the valuer to have detailed knowledge of the market. He must consider who the potential purchasers are and address the following:

(i)   How, and under what brand, would another operator trade from the hotel?
(ii)  What is the anticipated level of turnover and profitability which a purchaser would anticipate?
(iii) What costs would be incurred by the purchaser in re-branding the hotel?

# Accounts

It must always be remembered that a valuation is being carried out of a property and not of a particular operator or business, but the actual accounts should be obtained and analysed if at all possible. Actual accounts may be in different formats, but the valuer must ensure that not only is all income shown, but also that all necessary and relevant items of expenditure are included. Items of a non-recurring nature should be discounted as should any items unrelated to the trading activities of the hotel. If for instance, a leisure club is operated separately under a franchise arrangement, it may be appropriate to replace a franchise payment or licence fee shown in the accounts with an estimate of the club's total revenue. Accounts for a new hotel may include joining fees for a leisure club and care needs to be taken to discount these to the extent that they would not be receivable on a year-to-year basis.

Expenses such as repairs and maintenance may vary significantly from one year to the next, and these will require adjustments to reflect a fair year on year cost. Depreciation policies can vary between companies and actual depreciation figures should be added back into the accounts and later adjustments/allowances made. Each item should be considered carefully on its reasonableness and adjustments made if necessary. Head office costs can be substantial and these have to be allowed for where appropriate. Companies may take some income and costs at head office level rather than attributing them to individual hotels and the valuer may have to make adjustments for these. Directors' remuneration is not normally allowed as an outgoing when assessing net profit, the directors' return coming from the profit of the business. However, if directors are carrying out duties that would normally be carried out by an employee, then a deduction for appropriate salary costs should be made.

Where actual accounts are not available, the valuer will have to construct his own shadow or hypothetical accounts. Trade profiles will vary significantly between hotels and different classes of hotel, and this together with a variety of other factors including, *inter alia*, location, age, size, design, condition, layout of accommodation, facilities and services offered and required staffing levels will affect profitability.

When considering the trading potential of any hotel, the income derived from the letting bedrooms will normally prove to be the most profitable source of income. Total rooms revenue is the product of:

$$\text{total rooms available} \times \text{room occupancy} \times \text{average achieved room rate}$$

or

$$\text{total rooms available} \times \text{revenue per available room (REVPAR)}$$

For example:

| | | | |
|---|---|---|---|
| 100 bedroom hotel | = | 36,500 available rooms pa |
| × 70% occupancy | = | 25,550 let rooms pa |
| × £80.00 AARR | = | £2,044,000 total rooms revenue |

or:

| | | | |
|---|---|---|---|
| available rooms | = | 36,500 available rooms pa |
| 70% occupancy × £80.00 AARR | = | £56.00 |
| × 36,500 | = | £2,044,000 total rooms revenue. |

Reference is often made to occupancy levels and average achieved room rates, but it is misleading to consider either in isolation. The product of the two gives the REVPAR or average daily rooms yield, and this is the principal indicator of a hotel's performance. For example:

| Hotel | Room occupancy | Average room rate | Rooms yield |
|---|---|---|---|
| A | 80% | £85.00 | £68.00 |
| B | 60% | £112.00 | £67.20 |

When considering these two hotels, the use of either the occupancy levels or average achieved room rates in isolation would give a false comparison. Their true performance, shown by the rooms yield, is very similar.

Having considered the potential rooms revenue, the valuer should turn to the other sources of income. These will vary considerably depending upon the type of hotel in question. For a standard 3/4 star hotel in central London, where most guests tend to go out to eat and drink, rooms revenue will typically account for between 65% and 80% of total revenue. In provincial hotels, where guests are generally more captive, this will usually fall to between 40% and 60% for modern hotels and between 30% and 45% for smaller, older hotels where greater reliance may be placed upon local restaurant and bar trade.

Accounts should always be shown net of VAT, and the following show examples of how hypothetical turnover figures may be calculated:

## Study 1

A modern purpose-built 4 star provincial hotel, having 125 double letting bedrooms with well planned, flexible accommodation including restaurant, bar, conference rooms and leisure club with 400 members. Easily accessible just off a motorway junction and with good car parking facilities. Reliant on corporate business and conference trade during the week with some leisure based trade at weekends.

| Advertised tariff: | Double | £150.00 per night incl VAT. |
|---|---|---|
| | Single | £120.00 per night incl VAT. |

Assumptions:
(a) Room occupancy       75%
(b) Double occupancy       35%
(c) Average achieved room rate at 70% of advertised tariff.
(d) Room revenue 55% of total revenue.

1.  Total rooms available      =   125 × 365    =   45,625
    75% room occupancy                         =   34,219 let rooms
    Double occupancy at 35%                =   11,977 rooms let
    Single occupancy at 65%                =   22,242 rooms let

2.  Average achieved room rates:
    Double room £150.00 at 70%   =   £105.00    =   £91.87 net of VAT
    Single room £120.00 at 70%    =   £84.00     =   £69.30 net of VAT

3.  Total room revenue can then be calculated as:
    11,977 double rooms at £91.87            =   £1,100,327
    22,242 single rooms at £69.30            =   £1,541,370
    Total room revenue                              £2,641,697
    Say                                           £2,640,000

    Note:
    AARR = £77.15
    Room yield = £57.86

4.  Room revenue accounts for 55% of total revenue, which therefore amounts to £4,800,000. Adopting reasonable assumptions for this class of hotel for the other sources of income, this may be made up as follows:
    Say:

    | | | |
    |---|---:|---:|
    | Rooms | £2,640,000 | 55.00% |
    | Food | £1,200,000 | 25.00% |
    | Beverage | £480,000 | 10.00% |
    | Telephone | £48,000 | 1.00% |
    | Other | £432,000 | 9.00% |
    | Total | £4,800,000 | 100% |

## Study 2

An old manor house extended and converted to hotel use and providing 65 letting bedrooms mainly in 1970s wings, character restaurant and bar and several small meeting rooms. Set in attractive landscaped gardens in a rural location, and heavily reliant upon non-resident restaurant trade and its local reputation.

Advertised tariff:       Double          £130.00 per night incl. VAT.
Single                            £115.00 per night incl. VAT.

Assumptions:
(a)   Room occupancy            65%
(b)   Double occupancy         40%
(c)   Average achieved room rate at 65% of advertised tariff.
(d)   Room revenue 35% of total revenue.

1.  Total rooms available      =   65 × 365    =   23,725
    65% room occupancy                         =   15,421 let rooms
    Double occupancy at 40%                =   6,168 rooms let
    Single occupancy at 60%                =   9,253 rooms let

2.  Average achieved room rates:
    Double room £130.00 at 65%    =   £84.50    =   £69.71 net of VAT
    Single room £115.00 at 65%    =   £74.75    =   £61.67 net of VAT

3.  Total rooms revenue can then be calculated as:

    | | | |
    |---|---|---|
    | 6,168 double rooms at £69.71 | = | £429,971 |
    | 9,253 single rooms at £61.67 | = | £70,632 |
    | Total room revenue | | £1,000,603 |
    | Say | | £1,000,000 |

    Note:
    AARR = £64.85
    Room yield = £42.15

4.  Room revenue accounts for 40% of total revenue which therefore amounts to £2,500,000. Adopting reasonable assumptions for this class of hotel for the other sources of income, this may be made up as follows:
    Say:

    | | | |
    |---|---|---|
    | Rooms | £1,000,000 | 40.0% |
    | Food | £ 937,500 | 37.5% |
    | Beverage | £ 475,000 | 19.0% |
    | Telephone | £ 25,000 | 1.0% |
    | Other | £ 62,500 | 2.5% |
    | Total | £2,500,000 | 100.0% |

Having arrived at the fair maintainable level of turnover, the valuer then needs to calculate his valuation net profit. He needs therefore, to turn his attention to the outgoings and expenses necessary to maintain the turnover. Actual accounts can be presented in different formats. Reference may be made in this respect to the Generally Accepted Accounting Principles (GAAP), the International Accounting Standards (IAS) and the Uniform System of Accounts for the Lodging Industry. Whether using actual accounts or hypothetical accounts the valuer needs to make appropriate and sufficient deductions for expenses on a year by year basis to arrive at his valuation or adjusted net profit. Profit lines in accounts may be drawn at different levels, those often referred to being:

*   gross operating profit (GOP) — revenue less purchases
*   earnings before interest tax depreciation and amortisation (EBITDA)
*   earnings before interest and tax (EBIT)
*   income before fixed charges (IBFC).

It is essential that the valuer is fully aware of the profit line he is considering in his valuation and when drawing comparison between different properties. EBITDA is very often referred to in analyses but this does not represent the net cash flow because it omits cash required to fund the replacement of old equipment.

In simple terms valuation or net profit can be described as revenue less all purchases and operating expenses necessary to maintain that revenue.

Revenues and departmental costs and expenses are often shown on a department by department basis according to the principal activities of an hotel and, as shown in the above examples, include rooms, food, beverage, telephones, other operated departments, rentals and other income. Departmental costs and expenses only include items which can be specifically identified and related to each department, and include, *inter alia*, wages, costs of food and beverage, licences, travel agents' commissions and central reservations.

Total operated department income is calculated by deducting the total departmental costs and expenses from the total revenues.

Undistributed operating expenses are those outgoings which cannot be specifically related to a particular department, and include administrative and general costs, marketing, property operation and maintenance, energy costs, etc.

If accounts show a management fee deduction this should be added back before arriving at the profit and rental. outgoings need to be considered carefully under the terms of any lease.

An accounts summary may be as follows:

| *Revenues* (as calculated in Study 1 above) | £ | % of total |
|---|---|---|
| Rooms | 2,640,000 | 55.0 |
| Food | 1,080,000 | 25.0 |
| Beverage | 600,000 | 10.0 |
| Telephones | 48,000 | 1.0 |
| Other Departments | 432,000 | 9.0 |
| Total Revenues | 4,800,000 | 100.0 |
| | | |
| *Departmental costs and expenses* | | |
| Rooms | 765,600 | 29.0 |
| Food & Beverage | 1,021,440 | 60.8 |
| Telephones | 19,968 | 41.6 |
| Other Departments | 192,672 | 44.6 |
| Total Costs & Expenses | 1,999,680 | 41.7 |
| Total Operated Department Income | 2,800,320 | 58.3 |
| | | |
| *Undistributed operating expenses* | | |
| Administration and general | 451,200 | 9.4 |
| Marketing | 120,000 | 2.5 |
| Property operation and maintenance | 172,800 | 3.6 |
| Energy | 158,400 | 3.3 |
| Total undistributed expenses | 902,400 | 18.8 |
| EBITDA | 1,897,920 | 39.5 |

Notes

Figures and percentages shown are for illustration purposes only. Accounts may not be presented in this format.

Further deductions may still need to be made to determine the adjusted valuation net profit, or net free cash flow, which will then be capitalised. Depending upon the nature of the interest being valued, these may include; tenant's return on capital: insurance of the property and it's contents; uniform business rates; finance charges for leased items; a renewals fund to allow a yearly sum for the replacement and maintenance of the furniture, furnishings, fixtures and equipment at a level necessary to maintain the adopted level of trade; rents; and service charges.

# Valuations

The valuer must always have regard to the interest being valued, the nature and purpose of the valuation and the appropriate method of valuation to be adopted. The approach used by the valuer should reflect the approach used by the market for the particular type of property.

Hotels are a type of property which generally change hands in the open market as fully operational businesses at prices based on their trading potential for their existing use. Properties will normally be sold inclusive of all trade fixtures, fittings, furniture, furnishings and equipment, with the benefit of ll licences, permits, certificates and trading potential, but excluding stock in trade, any leased items nd equipment and badged items. The whole principle of valuation is therefore based upon potential urnover, net profit and the return on capital required by prospective purchasers bidding against each ther in the market.

The primary method of valuation is by capitalisation of income/profit, the two approaches to which re:

    capitalisation of maintainable profit (earnings multiple)
    discounted cash flow (DCF)

Historically, the RICS favoured the use of the capitalisation of maintainable profit approach using omparable market transactions whereas the British Association of Hospitality Accountants (BAHA) avoured DCF. In November 1993 BAHA published its *Recommended Practice for the Valuation of Hotels* o which the RICS responded in August 1994. The debate between the two approaches was highlighted in 1993 as a result of a wide disparity in group valuations of Queens Moat Hotels by two irms of hotel valuers. The two professional bodies sought to prepare an agreed statement of best practice but this was not forthcoming. However, Valuation Information Paper 6 produced by the Trading Related Valuation Group of the RICS was agreed by BAHA. The paper recognises that the primary method of valuation for most 3 star and below hotels and many 4 star hotels will be the capitalisation of maintainable profit.

# Capitalisation of maintainable profit (earnings multiple approach)

This approach, also known as the accounts or profits method of valuation, has been traditionally adopted by hotel valuers. The valuer formulates his opinion of the adjusted maintainable level of valuation net profit (net free cash flow) after allowing for all items of income and expenditure (excluding finance costs) and taking into account all factors which will impact on future levels of profitability. To this extent the approach mirrors that undertaken during a DCF analysis. The valuation net profit is then capitalised by a multiplier to arrive at the property's capital value to include all trade fixtures, fittings, furnishings and equipment, statutory consents, licences, certificates and permits, but excluding stock in trade and any leased or badged items, some of which may be subject to a separate valuation or agreement.

The multiplier to be applied is derived from the rate of return or yield which an hotel operator would require in order to purchase the property taking into account the security and growth prospects of the income. The valuer must assess how the market would view the risk or security associated with the particular hotel and its profit potential, and establish the level of the multiplier from his experience, regular contact with operators and his analysis of any market transactions on comparable properties.

This approach represents a mature situation, and is based upon the maintainable level of net profit in the normal stabilised year of trading. In the case of new hotels for example, this would ignore the build-up of earnings over the first few years of trading, and in such cases a deduction would need to be made from the capital value to reflect any shortfall in earnings prior to the stabilised year. This may commonly take one to three years to reach, but is very dependent on individual circumstances. For example, the rapid growth in budget hotels in recent years has seen many reach their maximum trading potential within a matter of months. Particular care must be taken in the choice of yield or multiplier where a hotel has no proven track record. A deduction would also need to be made to reflect any capital expenditure necessary to maintain the level of income adopted.

### Study 3

|  | | £ pa | £ pa |
|---|---|---|---|
| EBITDA (as calculated above) | | 1,897,920 | |
| Less: | | | |
| | (a) Uniform business rates say | 190,000 | |
| | (b) Insurance say | 75,000 | |
| | (c) Renewals fund at 4% of FMT | 192,000 | |
| | | 457,000 | 457,000 |
| Adjusted valuation: | | | |
| | Net profit/net free cash flow | 1,440,920 | (30.0%) |
| | YP in perp at 8.5% | 11.76 | |
| | Capital value | £16,945,219 | |
| | But say £16,950,000 | | |

Notes

(1) Rates deduction is calculated by multiplying the Rateable Value by the Uniform Business Rate. This may be affected by transitional arrangements.
(2) Renewals Fund: annual sum to provide for the replacement of fixtures, fittings and furnishings necessary to maintain the adopted level of trade.

# Discounted Cash Flow approach

DCF is a well respected financial model used extensively in investment appraisals. It's underlying principle is the time value of money — income receivable in the short term is more valuable than that receivable in the longer term. It is a particularly appropriate method in the consideration of new projects or where a property is to be extensively refurbished, extended or improved. It is also a suitable technique to take into account major changes in the market or in the economy as a whole.

DCF involves creating a cash-flow model of a property's projected financial performance over a time scale of typically five–10 years which is then capitalised by a discount rate which reflects the rate of return an investor would require given the inherent risk attaching to the projected cash-flow and the alternative investment possibilities available. To this is added the residual value where appropriate. The approach simulates a series of cash-flows representing the anticipated receipts, expenses and operating performance of a property together with any required capital expenditure and the effects of inflation.

The discount rate adopted contains three elements:

(i) compensation for the effects of inflation
(ii) a risk free rate of return
(iii) a premium compensating for the inherent risk associated with the industry and the individual property.

| Year | Total Revenue £ | Net Free Cash Flow £ | % of Revenue | PV of £1 @ 15% | Net Present Value |
|------|-----------------|----------------------|--------------|----------------|-------------------|
| 1    | 4,800,000       | 1,440,920            | 30.0         | 0.87           | 1,253,600         |
| 2    | 5,040,000       | 1,512,966            | 30.0         | 0.76           | 1,149.854         |
| 3    | 5,292,000       | 1,588,614            | 30.0         | 0.66           | 1,048,485         |
| 4    | 5,556,600       | 1,668,045            | 30.0         | 0.57           | 950,785           |
| 5    | 5,834,430       | 1,751,447            | 30.0         | 0.50           | 875,723           |
| 6    | 6,126,151       | 1,839,019            | 30.0         | 0.43           | 790,778           |
| 7    | 6,432,459       | 1,930,970            | 30.0         | 0.38           | 733,768           |
| 8    | 6,754,081       | 2,027,518            | 30.0         | 0.33           | 669,080           |
| 9    | 7,091785        | 2,128,895            | 30.0         | 0.28           | 596,090           |
| 10   | 7,446,375       | 2,235,340            | 30.0         | 0.25           | 668,835           |
|      |                 |                      |              |                | £8,625,002        |

| | | | | |
|---|---|---|---|---|
| Add residual value | | £2,235,340 pa | | |
| YP in perp. @ 10% | 10.00 | | | |
| PV £1 in 10 years  @ 10% | 0.39 | 3.9 | | £8,717,826 |
| | | | | £17,344,828 |

In normal circumstances, the minimum level of return an investor will expect is the yield obtainable from long-term government bonds. The discount rate is built-up, therefore, from the current risk-free rate of return, ie the yield on long-term gilt-edged securities.

Adopting the net free cash-flow calculated above of £1,440,920 as the starting point, the following shows a DCF model for the purposes of which it is simply assumed that income and expenditure will increase by 5% pa (see table above).

## Comparison of the two approaches

Valuations are intended to show the price which an interest in property could be expected to fetch in the market place. The fundamental factor common to the capitalisation of maintainable profit and DCF approach is that both value hotels by reference to recent and current trading performance and future trading potential.

Adopting the capitalisation of maintainable profit approach and using comparable market transactions, the valuer must consider the past and anticipated trading performance of the hotel. This gives greater weighting to achieved levels, and thus the method is less reliant on the potential inaccuracy of future projections. However, it must always be remembered that past performance is not necessarily a true indicator of future potential.

The valuer will have regard to all available market evidence, and the multiplier applied will be derived from the valuer's opinion of how the market will view the subject hotel and its profit potential compared to the multipliers generally applied in the hotel sector. This market led approach results in hotel valuations reflecting the volatility of market conditions. Consequently it can be argued that, at certain times in the economic cycle, these estimations of price may not be fully supported by the hotel's underlying earnings potential but they will reflect market value.

The discount rate applied in a DCF is intended to approximate the weighted average cost of capital to the likely investor but it is not derived from the hotel market. Many companies use DCF in taking investment and dis-investment decisions but companies have their own fixed rates of return which tend to vary over time and are not necessarily derived from the hotel market. The method is used to calculate worth to the owner and this may vary between owners according to the criteria each company has used in setting its own fixed rate of return.

In many cases the residual value in a DCF may account for up to 50% of the total value. There may be considerable uncertainty as to the accuracy of estimations of profit potential say 10 years ahead, which casts doubt over the accuracy of the valuation as a whole.

DCF may be considered to be more responsive to anticipated changes in future earnings, expenditure, market conditions etc: and demands a more rigorous, detailed and disciplined approach. The method relies more heavily on general investment yields, inflation, perceived risk and liquidity, and the characteristics and potential of the hotel being valued, than it does on the sentiment influencing the hotel investment market.

BAHA considered the capitalisation of maintainable profit to be an indicative method with market transactions as reasonability checks. It is considered a major flaw that while its heavy reliance on market generated yields reflects the market's volatility, it can result in valuations which may be in line with market sentiment but may not be supported by a detailed analysis of the hotel's earnings potential and risk. DCF, it is argued, is a more objective approach which relies on benchmarks set by the financial markets. It must be noted, however, that these are not infallible.

In all cases, it has to be stressed that hotels are a specialist class of property, and it is essential that the valuer has the appropriate experience and expertise to carry out the valuation required. Whichever method of valuation is adopted, it may also be wise to carry out a valuation by the other method as a secondary check. The capitalisation of maintainable profit is probably more appropriate where a valuation is being carried out of a property with an established trading history, and DCF where this is not the case — when valuing new hotels/projects, and properties which are going to be subject to major investment in alterations and/or additions.

The end valuation should be checked by reference to analysis of comparable market transactions wherever possible. In some cases this may even provide sufficiently reliable evidence to be adopted as a method of valuation in its own right.

It should also always be pointed out that the value of an hotel could be lower if:

(i)   the business is closed
(ii)  the inventory has been removed
(iii) the justices' or other licenses, consents, certificates and/or permits are lost or are in jeopardy or
(iv)  the property has been vandalised.

The end valuation should be checked by reference to analysis of comparable market transactions wherever possible though it may be difficult to find two hotels exactly comparable in all respects. In some cases however, this may provide sufficiently reliable evidence to be adopted as a method of valuation in its own right.

# Comparable approach and evidence of open market transactions

For most property types, valuations are generally based on evidence of open market transactions in similar properties. Valuers must not ignore the evidence of the market. A valuation however, is an exercise in judgment and should represent the valuer's opinion of the price which would have been obtained if the property had been sold at the valuation date on the terms of the appropriate basis of valuation. The valuer is not bound to follow evidence of market transactions unquestioningly, but should take account of trends in value and the market evidence available to him, adjusting such evidence as appropriate, and attaching more weight to some pieces of evidence than others. In a rapidly rising or falling market, undue weight should not be given to historic evidence which may have become outdated, but care is needed where acquisitions may be based purely on speculation and hope value. It is essential that the valuer has the relevant skill and experience to analyse transactions correctly.

For hotels in particular, market knowledge is generally imperfect. The parties to a transaction normally treat the relevant details upon which the sale is based as confidential, and very often sales take place through corporate transfers. Consequently, very rarely will the valuer have available to him all the relevant accounting and other background information to enable a full analysis of other transactions.

Comparison may be made on physical factors, most commonly a price per bedroom or equivalent double bedroom. This does allow an analysis to be carried out on information that is readily available, but if adopting such an approach, it is essential that the property being analysed is truly comparable in all respects. This method may be appropriate when considering the valuation of certain categories of hotel such as terraces of privately owned small hotels/guest houses often found in seaside resorts and not necessarily always operated to maximise profits, lifestyle and trophy hotels. However, unless all relevant factors are known, care has to be taken because comparison on physical aspects alone can be misleading and unreliable.

In a depressed market, a significant proportion of sales may be by vendors who are obliged to sell, such as liquidators and receivers. The valuer should establish whether or not these sales took place after proper marketing for a reasonable period. Liquidators and receivers are normally under a duty to obtain the best price and their sales may then be regarded as genuine open market transactions. However, there can be a stigma attached to such transactions which can depress the values achieved and they are therefore, not always reliable evidence.

In other instances a purchaser may pay a seemingly high price for a trophy property or for a property which will improve the profile of a portfolio as a whole. An operator may be especially keen to gain a presence in a particular locality, and be prepared to pay in excess of what might be considered the norm to secure that presence. Particular care is needed to identify such purchases when analysing transactions.

Often a portfolio of properties may be sold, and when attempting an analysis of such a transaction, it must be remembered that the acquisition may comprise, in broad terms, four elements of value:

- the aggregate of the individual existing use values of the properties
- a lotting premium reflecting the ability of the purchaser to make a single purchase of a large portfolio of properties with much reduced acquisition costs and a significantly shorter period within which the global acquisition would be earnings enhancing
- reduced head office costs and possible improvements in buying power having absorbed the acquisition into the estate

- the acquisition of a strong brand which enables the purchaser to reposition a large part of its existing estate.

In the case of *Alliance & Leicester plc* v *Harrison Robertshaw* (2000) a non specialist hotel valuer was held not to be incompetent when valuing a hotel in Bournemouth by reference to a price per bedroom backed up by comparable evidence.

There may be growing merit in certain circumstances to attempt to analyse transactions by reference to floor areas and to use this analysis as further comparable evidence.

## Rental valuations

Historically, relatively few hotels were held leasehold. Generally, considerable investment is required in fixtures, furnishings and equipment, and operators undertaking such expenditure preferred the freedom and flexibility of a freehold interest. From a landlord's point of view, the success of his hotel and consequently his income and the capital value of his asset, is not in his own hands, but is dependent upon the abilities of his tenant, over whom he may have little control. As a result, there was a paucity of open market rental evidence. In recent years there has been a spate of sale and leaseback transactions although these have largely been replaced by sale and management deals. Great care is required when considering sale and leaseback rents. They may provide evidence of open market rental levels or they may be considered to be financing arrangements used as a vehicle to raise capital or entered into for taxation or other reasons. Where rents are available for analysis, care has to be taken to ensure that they are genuine open market rents. Some rents may have simply been maintained under upward only review provisions where the fair rental at the time may actually be lower than the rent passing.

Full consideration must be given to the terms of the lease and particular regard paid to any unusual or onerous terms and conditions. For many property types there is an established pattern for rent reviews, but this historically has not been so evident for hotels, and each has to be considered on its own merits.

Rental values, like capital values, are therefore determined primarily by reference to trading potential, and the amount of profit that will be paid as rent. Whether or not actual accounts are available, the valuer has to go through the same process to arrive at the EBITDA based upon the estimated trading performance of a reasonably efficient operator. Deductions are then made to arrive at the level of fair maintainable profit or net free cash flow (often referred to as the divisible balance), upon which the tenant will base his rental bid.

### Study 4

| | £ | £ pa | £ pa |
|---|---|---|---|
| EBITDA (as previously) | | | 1,897,920 |
| Less: | | | |
| (a)  Uniform business rates | | 190,000 | |
| (b)  Insurance | | 75,000 | |
| (c)  Renewals fund at 4% of FMT (1) | | 192,000 | |
| (d)  Interest on tenant's capital: | | | |
| (i)  Stock | 100,000 | | |
| (ii)  Working capital | | | |

|  | Say 2 weeks turnover | 185,000 |  |  |
| (iii) | Fixtures and fittings | | | |
|  | at £15,000 per room | 1,875,000 | | |
| Total | | 2,160,000 | | |
| at 7% (2) | | 0.07 | 151,200 | |
| | | | −608,200 | −608,200 |
| | | | | £1,289,720 |
| Divisible balance | | | | |
| Tenant's rental bid (3) at say 55% = | | | £709,306 | |
| Rental (4), say £710,000 (14.8%) | | | | |

Notes

(1) Renewals fund: annual sum to provide for the replacement of fixtures, fittings and furnishings necessary to maintain the adopted level of trade.

(2) Interest on tenant's capital: a debt has to be serviced on working capital in the form of a cash float, food and drink stock and trade creditors. This has traditionally been accepted but in some cases these items may no longer be appropriate. An argument can be made that allowing for a renewals fund and interest on tenant's capital in the form of his outlay on fixtures and fittings is allowing the tenant too much and artificially reduces the calculated rental value. These allowances must, therefore, be considered carefully and in conjunction with the adopted tenant's rental bid.

(3) Tenant's rental bid: this will vary depending upon a range of factors that a tenant would take into account including location, the level and security of the income, the lease terms, alternative opportunities, the hotel's design constraints and future developments that might benefit or adversely affect the property. It is generally accepted that the tenant's rental bid will normally fall in the range of 40% to 50% of the divisible balance. Circumstances however, may warrant a bid outside this range, perhaps as low as 35% or as high as 60%. The percentage to be adopted has to be considered in the light of the deductions made to arrive at the divisible balance, and it must be ensured that the tenant receives an appropriate return for running the hotel and the risk involved.

(4) When carrying out a rental valuation care must be taken to ensure that the actual rent passing is not included as a deduction before arriving at the Divisible Balance.

# Tenant's improvements

Tenant's improvements may fall to be disregarded for rent purposes at review. This poses the problem of how and where the effect of the improvements is reflected in the rental calculations.

Ideally, the valuer should ignore the improvements when formulating his trading projections and arrive at the net free cash flow for the hotel in its unimproved condition. However, depending upon the nature of the works to be disregarded, this may prove to be an extremely difficult or even impossible task. This being the case, the return that the tenant would expect on his expenditure should be treated as a deduction in the rental calculations This deduction may be made from the calculated rental, as above, or as an additional item to be taken into account before arriving at the Divisible Balance on which the rental bid will be based. The two approaches may give significantly different end answers as demonstrated below:

| (a) | Calculated rental as before | £710,000 | |
|---|---|---|---|
| | *Less:* | | |
| | Tenant's improvements cost | | |
| | Say £1,000,000 at 7% return | £ 70,000 | |
| | Rental value | £640,000 | |

| (b) | EBITDA as before | £1,897,920 | |
|---|---|---|---|
| | *Less:* | | |
| | (i)  Deductions as before | £ 608,200 | |
| | (ii) Tenant's improvements cost | | |
| | Say £1,000,000 at 7% | £ 70,000 | |
| | | −£678,200 | −£678,200 |
| | Divisible valance | | £1,219,720 |
| | Rental value at 55% | | £670,846 |
| | Say  £670,000 pa | | |

In such cases the question has to be asked if and why the tenant's improvements should be treated any differently than for instance, the deduction for renewal of fixtures and fittings. Generally, the second approach will be the most appropriate to adopt.

Valuations of leasehold interests will follow the same valuation principles as freehold valuations except that the rent will be a deduction to make to arrive at the net free cash flow or valuation net profit to be capitalised for the appropriate number of years.

## Investment valuations

Hotel investment valuations will follow normal valuation principles, ie capitalisation of the rental income. It can be seen above that the net free cash flow comprises the two elements of rent and residual profit.

An operator for the hotel will make his bid on both of these and therefore, to arrive at market value the two elements are capitalised. However, when considering an investment valuation, it has to be remembered that the residual profit is only available to the tenant or operator, and therefore investment valuations will only capitalise the rental income. Consequently, unlike many other properties, the value of the freehold interest held as an investment may be significantly lower than the existing use value.

## Property company/operating company valuations

In recent years, the hotel industry has been at the forefront of transactions divorcing the ownership of property assets from the operation of the business either by means of sale and leasebacks or sale and management agreements. On the sale of the property assets, the operator will typically take back a lease or management agreement for a term of between 25 and 35 years. Rents are generally set by reference to turnover and profitability perhaps with a guaranteed minimum payment. By these means, operators are able to raise immediate capital and concentrate on what they are perceived to do best, running the business. They do however, sell off the company's assets.

Such a structure necessitates separate valuations of the property company and operating company interests. The property company receives a rental income and the operating company receives the residual profit. Where a property asset or assets is suitable for such a structure the valuer may need to consider a Property Company/OpCo valuation. Depending upon the appropriate yields such a structure may maximise value.

## Valuations for loan security purposes

These are referred to in Practice Statement 6.2 (Secured Lending) of the RICS Appraisal and Valuation Standards. The valuer must confirm whether the value based on trading potential excludes any personal goodwill and must stress the sensitivity of the value to future changes in trading potential.

## Conclusion

Hotels are a specialist class of property which normally change hands in the open market as fully operational businesses at prices based on their trading potential. It is essential that the valuer has appropriate experience and a detailed knowledge of the hotel market.

©Brian D Scott 2008

# Leisure Properties 19

## Introduction

With the turn of the new Millennium, the leisure market appeared to have come of age as a separate property sector. The demand for an increasingly wide range of leisure property formats was high; institutional investors had begun to move in and there was a corporate hunger for expansion. The previous decade had seen the rise of leisure as an integral part of the retail offer and the genesis of leisure parks. All this impacted on the approach to leisure property valuation.

Earlier decades had witnessed the growth of the service industries and consequent changes in both the specification of, and investor attitudes towards, offices and retail premises; the 1990s saw the manifestation, in property terms, of social and economic changes which resulted in enormous growth of commercial leisure activity. By the end of the century the proportion of average consumer spending devoted to leisure had risen significantly so that, for the average household, it now forms one of the major items of expenditure. The growth in spending on leisure had been disproportionately higher for those people in higher income brackets, and as people devoted more money to leisure their expectations of better quality facilities were realised with the emergence of better specified units. In employment terms too, leisure gained increasing importance, but its association with low pay has been a persistent theme.

Since the year 2000 the leisure dream of increased values and continuing strong demand has not been realised. A number of world events such as 9/11 and the Iraq War played adversely upon confidence and tourism levels, but more locally, the trading projections upon which many developments were predicated simply remained unfulfilled. The consequence has been that some sub-sectors of the leisure market, notably the health and fitness area, have been characterised by company failures and corporate re-structures. Additionally, the property investment community turned greater attention to the traditional bulk classes of shops and offices. Finally, a plethora of legislative and planning changes have had an impact on the operation of the markets. In turn, these changes have reinforced the necessity for those charged with the valuation of leisure properties to have a good understanding of the market context and of the fundamental drivers of value to the leisure operator.

It must also be borne in mind when valuing leisure properties that not all leisure activities are based upon commercial enterprise. Leisure embraces a range of non-commercial activities and this presents a challenge to valuers in terms of methodology, as will be outlined briefly below. To illustrate the problem: a multiplex cinema is clearly a commercial enterprise built with the intention of providing profit in the hands of both the operator and the property owner. A community theatre on the other hand, is operated for reasons of civic pride and commitment to the enterprise, and not for any financial gain to the owner, although the operator will almost certainly require a profit. Similarly, and perhaps presenting even more of a difficulty for the valuer, a private sector health and fitness club which is a purely profit driven operation may, at first glance, look very similar to a local authority owned scheme which offers similar facilities, but closer observation of the pricing policies may reveal very different ownership agendas.

It can be seen from this introduction that some special issues arise in relation to the valuation of leisure properties. These relate not only to the wide variety of properties that come under the umbrella heading of leisure but include the choice of valuation method appropriate, which may depend on whether the property is, or is not, owned for commercial purposes.

# Types of leisure property

The question of what are leisure properties is almost unanswerable. To say that they are simply those used for leisure purposes is far too all-embracing to be useful to the valuer. Instead, it is helpful to consider them under certain categories:

- *sport and recreation*: including health and fitness, sports stadia and golf; increasingly the sport and recreation sector has become associated with town centre developments and full profit-driven facilities
- *entertainment*: including cinemas, bingo, bowling, family entertainment centres and nightclubs; of these family entertainment centres proved to be a short-lived phenomenon while clubbing has experienced some growth
- *food, drink and hospitality*: including restaurants, bars, public houses, hotels (including new formats such as aparthotels, which comprise suites with their own catering facilities appropriate for longer-stay guests) and holiday villages.
- *culture and heritage*: including art galleries, museums and historic houses.

This list is certainly not exhaustive, and it can be seen that within each of these main categories a wide variation of property types exist, both commercial and non-commercial. Therefore, as well as considering what basic type of property is to be valued, which is necessary if economic analysis is to be undertaken, it is also appropriate for the valuer to review whether the property is owned primarily for:

- *commercial reasons*: such as a bowling alley or amusement arcade
- *social reasons*: such as a local authority swimming pool or municipal golf club
- *furtherance of a self-help group*: such as a local cricket club
- *cultural and educational purposes*: such as a museum or
- *free public enjoyment*: such as a public park.

By considering both the type of property and the ownership motivation, the valuer has begun to determine the most appropriate method of valuation. However, before considering the method of valuation, it is vital to consider the purpose for which the valuation is prepared. In practice, the valuer will normally be called upon to value for:

- market transaction or letting

- loan security
- asset valuation
- special purposes, such as rating, taxation or compulsory purchase.

Specific comment is made in relation to these later in the chapter, but first the normal approach to valuing for market transactions is considered.

If the premises are owned commercially, the valuation method chosen will usually be based upon comparable rental and yield evidence where this is available, or the valuer may adopt the profits method. If the premises are held for any other purpose, evidence to support the use of these methods might be limited, and a fundamental approach using all available methods, including a cost approach, might have to be considered.

In this chapter the emphasis is placed on commercial entertainment properties as this sub-category is the one with which the practising valuer is most likely to be concerned. No specific reference is made to public houses or hotels, for which see Chapters 17 and 18. It must be pointed out however that the boundaries between types of leisure property — notably between nightclubs, bars and restaurants is not always entirely clear and this may have a bearing on the valuation.

# Planning

Under planning legislation, critical factors in relation to the valuation of leisure properties are found not only in the main statutory provisions but also in the statutory instruments, and in the planning policy statements (PPSs). The planning laws were altered in 2004, by the introduction of a new regime under the Planning and Compensation Act 2004. The regime brought in by this Act places a much greater emphasis on the environmental impact of schemes, and may act to reduce the likelihood of some marginal schemes gaining consent or being viable to develop. However, since the boom of commercial leisure developments during the last years of the old millennium, the pace of leisure developments has slowed down markedly. The main driver of value lies now in the demand side with supply remaining relatively constant. Therefore of greater influence on value has been the changes brought about by the Town and Country Planning Use Classes Order 1987 (SI 1987 No 764), which has to be read in conjunction with the Town and Country Planning (General Permitted Development) Order 1995 (SI 1995 No 418), both as amended.

The use classes order (UCO) was changed significantly in 2005, and the changes introduced mainly affected leisure property. Following a long series of proposals and consultations, during which many in the leisure industry had hoped for a clear definition of leisure to emerge, the order makes it clear that there continues to be no one categorisation of leisure property. The aim of the UCO was to deal with the myriad changes in the way property sectors, and in particular leisure properties, had changed in nearly two decades but the resultant position is still less than clear.

Some leisure properties continue to be included within the retail use class A, while others are in class C (hotels and boarding houses). The majority of entertainment properties, such as cinemas, concert halls, bingo hall, dance halls and swimming pools come under the assembly class, class D as D2 (assembly and leisure) while still others (for example theatres, amusement centres, casinos and nightclubs) are *sui generis*, that is, to be considered as individual cases. It should be noted that prior to 2005 casinos were part of D2.

The most significant changes related to restaurants and to premises selling food and drink. These have long been regarded as leisure properties, possibly because they often generate their income from leisure spending and because many are located within leisure parks or are regarded as leisure units within shopping schemes. However, for planning purposes strictly they are still treated as being within the retail category. While prior to 2005 they were categorised together as A3, this grouping has been sub-divided as follows:

- A3: restaurants and cafés: includes restaurants, snack bars, cafés
- A4: drinking establishments: includes pubs ad bars
- A5: hot food takeaways: includes hot food takeaways only.

Under the UCO any property with A4 or A5 consent can change to A1 (general retail), A2 (financial services) or A3 as long as no material physical change is involved, but there is no permitted change to either A4 or A5 from any other class or sub-class. The sub division of food and drink into three sub-groups may be problematic for the valuer particularly as the boundary between A3 and A4 is blurred. The distinction revolves around the amount of food offer — but there is an unresolved question at to whether this to be assessed in terms of value or of floor area. Additionally, the balance can change in value terms very quickly and it is therefore difficult in practical terms to see a clear dividing line. It might be possible that even within one branded format one outlet may be more food orientated than another, and for them to fall into different sub-classes would appear to defeat any sense of reason. Furthermore, given that pubs often now trade late (see licensing, below), the distinction between a drinking establishment and a nightclub is less obvious than it was some years ago. This is particularly the case since the introduction of so-called chameleon operations which operate as restaurants by day and as nightclubs by night. Yet, despite this blurring of planning boundaries and greater categorisation, nightclubs are still *sui generis* — ie each has to be treated as an individually on a case by case basis.

It is important that the valuer of any leisure property has a good appreciation of the planning situation and, in the case of an A3/A4/A5 unit, an awareness of whether or not a premium value attaches to the sub-use. This will depend on market conditions, and in many locations and in certain circumstances an A1 or A2 use may generate a higher level of rental value. If this is the case, the alternative use should be considered subject to consideration of, for example, tenure constraints. The valuer should also bear in mind that many current leases were granted prior to the current UCO and this may need careful interpretation by the valuer as to the assumptions made regarding the permitted use of the premises.

## Licensing

Many leisure enterprises rely on a variety of licences in order to trade. Often these will be licences to sell alcohol and for gaming but there are other licences upon which the trade depends and the terms of the licence can be critical. In the past such licences for the sale of alcohol were issued by magistrates, but consequent on the Licensing Act 2003, which came into effect in 2005, they are now part of the remit of local authorities. The passing of the 2003 Act was intended to simplify the regime under which licensed premises operated and in many ways this has been the case. No longer is there the grant of a full on licence; instead there are two types of licence: the personal and the premises licence. The restricted hours rules have been relaxed, and operators can now apply to operate during hours which suit their format. Although it was originally feared that this might have a negative social impact, with pubs open almost round the clock, such fears have not materialised. What has resulted has been the emergence of late night bars and pubs competing more directly with nightclubs, and the valuer must always consider the potential for competition to the unit being valued, competition which a combination of changes to the planning and licensing laws has made more likely. The quality of the management of any club, whether owner-operated or not, is crucial to its success; consequently, changes in the management can have swift and substantial effects on the pattern and strength of competition within the locality, and this is something of which the valuer must be aware.

Associated with the changes to the licensing laws have been changes to gaming provisions, and the passing of the Gambling Act 2005 introduced new controls over gambling. While licences for gaming fall within the remit of the local authority, the Secretary of State has reserved rights to make regulations over the provision of casinos. At the time of writing [June 2007] the matter of super casinos is very much a live debate, as is the potential impact on the value of gaming premises of the rise of internet gambling. When valuing any property subject to licences, and this includes most leisure properties, the valuer must be familiar with the relevant regulatory and legislative framework, and that the premises and business comply with the requirements as any failure as regards compliance will adversely affect value.

## Other legislative matters

Leisure properties are often affected by legislation that is aimed at social or welfare improvements and there are several recent important examples of this. The application of the Disability Discrimination Act 1995 to access to premises occurred in 2005. This Act required reasonable adjustments to be made to premises to ensure accessibility and use by people suffering from a range of disabilities. By now all premises should have been adjusted as far as practicable; however, it is likely that some premises are not yet fully compliant and the valuer should be aware of this when making inspections.

Many leisure properties have been constructed since the mid-1990s which was a period of intense development activity; however, many leisure enterprises can trade successfully in older buildings, where the character of the building can add to the appeal of the enterprise. These older buildings will almost certainly not comply with current building regulations, and, indeed, many premises constructed only five years ago may now not comply with the latest energy requirements under Part L of the regulations. Therefore, if plans for future refurbishment are being included in the valuation, the valuer should be very aware that the costs of bringing the building into compliance can be very high. With mounting concerns over environmental issues and carbon saving, the need to upgrade environmental standards with each refurbishment is likely to remain.

The Smoke-Free (Premises and Enforcement) Regulations 2006 (SI 2006 No 3368) under the Health Act 2006 are due for implementation on 1 July 2007. This will introduce a ban on smoking in public places. Such a ban has been hard fought by the leisure and licensing sector, but implementation of similar bans elsewhere has not proved as negative as first thought. Clearly, the impact will be variable with some premises suffering an initial drop in trade while others might benefit.

## Tenure restrictions

At the time of taking instructions the valuer should check what interest is to be valued and how far any investigations as to title should be undertaken (for further information see RICS requirements and guidance). Although the question of tenure will normally be a matter of fact that the valuer can take as given, it is important that appropriate checks as to title are undertaken, either before the property is inspected or afterwards. Additionally, when inspecting, the valuer should be alert to any potential problems such as rights of way or access difficulties.

Where the property is subject to a lease this should always be scrutinised carefully and particular note made of any onerous conditions such as restrictions on assignment or change of use. The position in relation to licences, and the assumptions at rent review should also be examined carefully. Given the recent changes to the UCO, the user clause should be examined very carefully as there may well be a difference between the planning position under legislation and under the lease.

The rent review clause should also be carefully scrutinised. With the growth of the leisure parks and the number of multiple operators, so greater standardisation of lease terms has occurred. It is not unusual for leisure properties to be let on terms of 25 years, or even longer, and on standard institutional terms (that is on full repairing and insuring terms with five-yearly upwards-only rent reviews). It should be noted that, due to concerns over the availability or otherwise of comparable evidence and the lack of investment return data, many institutions have insisted on additional protection to the landlord at rent review. Therefore, if the property is let on a lease granted in the 1990s it may well have a review clause which ties the reviewed rent to the higher of open market rental value or the initial rent plus a 3% pa compound increase. Where such a property is being valued, care should be taken to ensure that the onerous nature of the clause is taken into account as such clauses have resulted in many properties being over-rented (ie the passing rent exceeds market rent). The use of such onerous clauses is unlikely to be found in more recent leases, as the level of comparables available has grown and operators have resisted their inclusion in a less heated market.

Leisure properties frequently require fitting out to meet operational needs. Depending on the nature of the business this can involve very high capital cost. Accordingly, particularly in a weak letting market, tenants may require any or all of the following:

- a substantial rent-free period to amortise the costs
- the right to make subsequent alterations to the premises
- a lease of length sufficient to allow for full benefit to be gained from the work and
- disregard of the value of the fit out at rent review.

These factors may have significant effects on both rental levels and yields. The issue of how the costs should be amortised can be a matter of debate during rent review negotiations.

## RICS regulatory framework

All valuers are required to be aware of the provisions of the RICS *Valuation Standards* (the Red Book) and the Practice Statements contained in it. These provisions apply to the valuation of leisure properties as they do to other valuations and therefore need no further comment here. The Red Book, as it is colloquially known, is of global application and is concerned primarily with process, not methodology. Instead, the RICS publish valuation information papers designed to supply valuers of specific types of properties. Of these three are germane to the valuation of leisure properties:

- *Valuation Information Paper 1*: The Valuation of Owner–Occupied Property for Financial Statements
- *Valuation Information Paper 2:* the Capital and Rental Valuation of Restaurants, bars, public houses and nightclubs in England, Wales & Scotland (2nd ed 2006)
- *Valuation Information Paper 6*: The Capital and Rental Valuation of Hotels (1st ed 2004).

The mandatory practice statements (PSs) are of global application; additionally there are UK Practice Statements that relate to specific UK applications, such as valuations for secured lending (UK PS 3) and for financial statements (UK PS 1). It is essential that all valuers are familiar with the Red Book and how it applies to the valuation under question. Among the most important provisions are those relating to the taking of instructions (contained in PS 2) and those relating to inspections (PS 4 and PS5).

Further commentary and explanation relevant to leisure properties will be found in Chapters 17 (Pubs) and 18 (Hotels), and reference should be made to Chapter 16 regarding valuations for financial statements. The following points, in particular, should be borne in mind.

- *Taking instructions*: care must be taken when taking instructions to clarify the assumptions in respect of fixtures, fittings and licences as these are often important in terms of value. It is also important to point out to the client that personal goodwill will normally be excluded.
- *Value*: Where valuing under the UK practice statements, owner-occupied property valued for the balance sheet is normally valued to existing use value. This is defined in UK PS 1 as:

  The estimated amount for which a property should exchange on the date of valuation between a willing buyer and a willing seller in an arm's-length transaction, after proper marketing wherein the parties had acted knowledgeably, prudently and without compulsion, assuming that the buyer is granted vacant possession of all parts of the property required by the business and disregarding potential alternative uses and any other characteristics of the property that would cause its Market Value to differ from that needed to replace the remaining service potential at least cost.

- *VAT*: When reporting the valuation, if based on trading, the sources of the figures (actual or projected) should be quoted exclusive of VAT.
- *Goodwill*: Many trading assets change hands at prices which include goodwill. The guidance to valuers is that, if conducting a valuation of the asset, only transferable goodwill should be included. This is defined as 'that intangible asset that arises as a result of property-specific name and reputation'. Personal goodwill is excluded, and this should be spelled out to the client.
- *Operator's competence*: When valuing using a trading approach the valuer should assume the concept of a reasonably efficient operator; this may require an adjustment from the average trading figures realised.

Although the Red Book addresses process, not method, there are some specific references to valuation practice methodology. In relation to leisure property, the valuer should take careful note of the provisions of guidance note GN 1 *Trade related property valuations* which applies, internationally, to all property that is normally bought and sold on the basis of its trading potential. Reference to its contents, and to those of the relevant information papers are made through the remaining sections of this chapter.

It will be noted that the information papers and the guidance notes relevant to leisure property primarily address the concept of property valued according to its trading potential. This should not be read to mean that all leisure properties are valued using a profits method. This is not the case at all. The valuer should always consider the market for the asset being valued and the purpose for which the valuation is being undertaken, and form a view as to which approach is most appropriate given the circumstances. While it remains the case that many types of leisure property do continue, as before, to be traded on a freehold basis between operators as occupational units, many more units are now let and valued on an investment method using comparable evidence, possibly backed up by trading accounts. Similarly, for hotels, there is increasing use of the discounted cash flow (DCF) approach to valuation, and in many cases, where the valuer is advising on worth as opposed to market value, a DCF is the most appropriate approach. A full discussion of this in the context of hotels, with a worked example, will be found in Chapter 18. A fuller explanation of DCF itself, together with a discussion as to its applicability, and further examples, will be found in Chapter 11.

Where properties are to be valued with reference to their trade, the information papers (if they apply to the type of property concerned) should be consulted. While the adoption of their advice is not mandatory, in the event of any subsequent dispute regarding the valuation, adherence to both the Red Book and its guidance and the associated information papers is important as a failure or omission to follow guidance may lay the valuer open to a charge of negligence.

# Collating the evidence and referencing

Fundamental to any valuation is appropriate research. The Red Book gives clear guidance on the matters to be investigated when carrying out any valuation. As many of the newer leisure developments have been constructed on brownfield sites, particular cognisance should be taken of the need to ensure that appropriate investigations concerning possible site contamination have been undertaken. For the valuation of a standing property, this will normally be a desk-based exercise.

As well as the standard enquiries, two particular sets of factors should be taken into account for commercially-trading units:

* those which affect profitability of the enterprise in the hands of an operator and
* those which determine whether the property is likely to be of interest primarily to investors or to owner–occupiers or, indeed, whether the nature, layout or location of the property render it of potential interest only to non-profit motivated owners.

## Physical factors

The physical specification of the property is important and the valuer must be aware of the likely operator requirements based on the type of property and location, as well as any requirement to comply with legislation (such as disabled access).

## Unit size

The individual requirements of leisure operators are extremely variable and in the past it has been almost impossible to standardise the required specification. However, changes during the 1990s meant that greater standardisation was achieved as leisure parks were developed. One discernible trend has been towards much bigger units, with recent years seeing the advent of both the super-pub comprising properties of up to 1,390 m² and super clubs with floor areas of up to 2,320 m². Restaurants, too, particularly aiming at the family budget, are seeking more floor space. Large units provide the operator with the ability to increase the ratio of income producing area to service area.

It should be noted that operator requirements change rapidly due to the fast evolving nature of the leisure markets. Valuers should ensure that they are fully conversant with the latest trends and requirements.

# Location

The location must be appropriate for the use and the operator. Most forms of commercial leisure property require ease of access and prominent visibility, but of equal importance may be proximity of complementary uses. For a commercial leisure unit to operate successfully it is often of paramount importance that it can synergise with other adjacent uses; hence the growth in popularity of the purpose built leisure parks, many of which are located adjacent to, or integrated with, retail schemes. While planning restrictions and a change of trading conditions have meant that few of these are now being developed, there has been increasing importance attached to the proximity of similar uses. One trend within existing town and city centres has been the development of the leisure quarter and the drinking circuit, with operators seeking to locate close to competitors. Such a trend has been facilitated by later opening hours, which have resulted in drinkers often going out less frequently but visiting a number of venues in the course of an evening. This trend has placed great pressure on the viability of solus units in terms of market appeal, and hence profitability. On the other hand, some types of leisure operation can trade successfully in non-purpose built units and, indeed, unusual elements to the fabric can enable a niche trade to become established. These units may be dependent on the strength of an individual operator, and the valuer should be very careful to exclude personal goodwill in such circumstances.

The presence, or otherwise, of car parking should be noted together with the proximity and adequacy of public transport facilities. Car parking will be particularly important where either public transport links are weak or where the unit relies on late night trade, although the introduction of more late night buses may act as a counter-balance.

The valuer should also be alert to any proposed changes to traffic planning policies at national or local level which could adversely affect trade.

# Trading potential

Whether or not the property is to be valued by reference to a full profits approach valuation, any competent leisure valuer should assess the trading potential of the property in the hands of a 'reasonably efficient operator' in order to estimate the maintainable trading potential. Valuation Information Paper (VIP) 2 quotes the International Valuation Standards (Guidance Note 12) in defining the reasonably efficient operator as:

> ... a market-based concept whereby a potential purchaser, and thus the valuer, estimates the maintainable level of trade and future profitability that can be achieved by a competent operator of a business conducted on the premises, acting in an efficient manner. The concept involves the trading potential rather than the actual level of trade under the existing ownership so it excludes personal goodwill.

The VIP goes on to state that 'the valuer needs to be actively involved in the market for the class of property because a practical knowledge of the types and classes is essential'. It therefore makes the case that familiarity with matters affecting trading is vital to a successful valuation, and all valuers are reminded of the requirements of the Red Book regarding competence to act (PS 1). In particular VIP 2 points out the need to be familiar with the licensing situation, with the configuration appropriate to the use, and with trading aspects.

Establishing trading potential begins with the inspection. The valuer should note carefully all details of the existing trade, such as admissions price (if applicable) and general tariff together with those factors which affect the ability to build revenue. These include layout, facilities and potential for expanding business within the existing envelope as well as by development. Note should also be taken of any disadvantages, such as poor configuration, which could adversely affect trade, and of matters which will adversely impact on future fit-out, for example failure to comply with current building regulations.

Where the property is trading, it is normal for access to trading accounts to be supplied, but the valuer should always treat these with caution, as, even in the case of correctly audited accounts, levels of trade in the past may not be a good indication of future revenue flows. Additionally, when inspecting accounts, the valuer should take careful note of items that might be allowable for accounting purposes but which are personal to the operator and so would not necessarily be replicated by other operators. These include items such as depreciation, bank charges and proprietor's drawings. Conversely, the accounts may not show an accurate picture of all items that are necessary to cost the business in full. For example, in the case of family run businesses, it is very common that inadequate amounts are entered in the accounts to reflect the true wage costs. At all times, in analysing the accounts the valuer should use experience and judgment to determine whether the current operation is what could be achieved by an efficient operator. All elements of either under- or over-trading should be disregarded. It is the maintainable trading position and its profitability which is sought. Verification of information supplied should be sought, and suitable adjustments should also made for appropriate levels of likely repair and maintenance particularly when these have not been allowed for in the accounts.

Increasingly, leisure units are operated by either franchisees or by multiple operators as part of branded operations. In these cases, the question of trading potential becomes especially problematic because the trading profile may be closely linked to the quality of the operator and the brand image rather than to the intrinsic property quality — yet it is the latter that the valuer must try to establish by stripping out any element of trade that is deemed to relate to the brand. However, if the valuer thinks that the property is one that would be likely to attract a range of different, competing branded operations (as may be the case with cinemas, nightclubs, bingo halls, etc.) then the element of turnover which relates generically to the ability of the property to attract a multiple operator can be included. This is, perhaps, one of the most difficult dilemmas facing the valuer in assessing the trading potential, and it should not be underestimated.

It follows, therefore, that when inspection takes place it is important that the valuer obtains as much information as practicable in relation to the trade in order that a true assessment of potential is made.

In addition to careful inspection of the property, the valuer ought to take into account any technical changes which could affect the design and trade of the unit in the future. For example, in the case of nightclubs, the valuer must be aware of the latest technology used in sound systems so that a judgment can be made as to how the property compares with those with which it is in competition.

It must be remembered that not all leisure properties trade at a profit. The market is fashion driven and some types of units face a high risk to their profitability, so it is important to determine the sustainability of both turnover and profit. An over-estimation as to the future trade based on continuance of current fads can lead to optimistic valuations which will result in future loss. It it therefore important in undertaking the valuation that the valuer recognises the assumptions as to the future that the market is making and, if supplying a full appraisal, points out to the client that the trade assumptions may not be sustainable.

There are many units that are owned and/or operated for social reasons. For example, local authorities own many leisure properties, particularly those that fall into the C1 and D2 use classes — such as theatres and leisure centres; also many sporting and recreational clubs operate leisure properties. The common theme for these is that the bottom line is not the reason for the ownership, and the trading and tariff policies reflect this. Where a valuer is instructed in relation to any non-profit orientated properties, great care must be exercised, and consultations with the client should take place in respect of valuation assumptions in relation to trade. In such cases, a valuation for an alternative use may be required, depending on the purpose for which the valuation is being prepared.

## Comparable evidence

One of the basic tenets of every valuation is that all comparable evidence must be obtained and analysed. This applies to leisure properties as to any other class of property, even if a decision has been made to value by reference to trade potential. However, the collection of comparable evidence may be more difficult than with, for example, a shop or office, as the amount of stock is less and its location more diverse. For example, there are unlikely to be more than one or two multiplex cinemas in any one town — barring the major conurbations. Similarly, it would be unusual to find bingo halls located close to each other, so comparable evidence from the near vicinity may be impossible to obtain. However, as drinking venues and night clubs are increasingly situated on leisure circuits or in leisure quarters, good local comparables may exist. At the time of writing, comparable evidence has been scarce for many leisure properties, given both the decline in the number of new builds to provide new letting evidence and the strong over-renting situation, which has meant that many rent reviews of units such as cinemas have not been activated.

Therefore comparable evidence may need to be gathered from a wider geographical base, if it can be gleaned at all. This can make analysis difficult as so much of the trading potential of any unit relates to the demographic and economic profile of the area. In some cases it may be possible to obtain rental evidence from within a single scheme or from other schemes offering a similar range of outlets. A typical leisure park scheme would be anchored on a multiplex cinema with bingo, bowling, health and fitness and eating and drinking units dominating the rest of the park.

For other types of property, there is still a dearth of comparable evidence, and for some types of property that appeal to the international market, comparability may take on a cross-national perspective. Where comparable evidence analysed on a floor area basis is to form the main evidence base for the valuation, the valuer will often be well advised to do a double check against a trading approach.

## Measurement

For most commercial properties, it is usual to measure the property in accordance with the current edition of RICS *Code of Measurement Practice*. This best practice guide sets out various methods of measurement: gross external (used for insurance purposes), gross internal (used for industrial premises), and net internal (used for retail and offices). The fifth edition points out that for many leisure properties the floor area may lack relevance as it is the configuration into bed spaces, restaurant covers etc. which may be most important. However, it does state that there are occasions on which the value does relate more directly to overall size. Where this is the case it is recommended in the code that measurement is to gross internal area (GIA) but that, depending on local market practice, the following should be stated separately (Code of Measuring Practice, S30):

- internal load-bearing walls and columns
- fire escape stairs and corridors
- in the case of purpose-built multiplex cinemas the floor levels providing raised projection boxes and the stepped flooring providing the auditoriums seating

- for restaurant premises the public seating areas, kitchens and stores
- in measuring the effective drinking area for licensed premises, the trading area must exclude the servery.

It follows that before undertaking any valuation the valuer should ascertain any local practice and ensure that all analysis is carried out on the same basis as is proposed for the valuation.

## Appropriate method of valuation

Until fairly recently the valuation method adopted for commercial leisure properties was almost universally the profits or accounts method, or, as detailed in the Red Book, a valuation in relation to trading potential. However, the growth of an investment market in some forms of commercial property, principally for urban entertainment uses such as bingo halls, cinemas, and food and drink units, and the greater availability of rental evidence has led to a change towards the use of the investment method where that appears more appropriate.

This change in context means that before undertaking either a capital or rental valuation, the valuer should determine:

- whether the property is operated for profit
- the level and validity of comparable capital value or rental evidence
- whether the property is likely to be of interest to an investing owner and
- whether it is, or is likely to be, let to a major covenant.

If the answer to these enquiries demonstrates that a valuation based on direct market evidence is available, then this may be the more appropriate method, but if there is some doubt then either a profits approach should be used or both methods should be adopted. Although the investment approach is gaining in popularity, and DCF is increasingly used to establish worth, it should be noted that at the time of writing some recent third party dispute cases arising in relation to rent reviews, have led to a preference for the profits or accounts approach. It may be relevant that these cases were in relation to restaurants and the same outcome may not result in the case of different types of property.

If the property does not trade at a profit, whether due to poor market conditions or to the pricing and management policy of the owner as, for example, in the case of many publicly owned premises, the valuer may, in the absence of any market evidence, have to choose between adopting some form of revenue approach while accepting that a low value will result, or contemplate the use of a cost approach. This however is very much a specialist area, and valuers should only attempt such a valuation if they have the requisite professional experience. It is stressed that the use of a cost approach should be regarded as a last resort, although its use has been confirmed by the Lands Tribunal in relation to the rating of public sector leisure centres, see *Eastbourne Borough Council and Wealden District Council* v *Allen (VO)* (2001). However, although the decision was in 2001, the case dated back to the 1990 and 1995 Rating Lists and should therefore not be seen as necessarily creating a precedent for current best practice.

# Accounts approach to the valuation of leisure properties

The accounts or profits approach is used to establish both rental and capital values for a wide range of trading leisure properties. VIP 2 and VIP 6 both give appropriate guidance, but only relating to some types of leisure property.

## Full accounts approach to rental value

For rental valuations, the first task is to establish the fair maintainable trade in the hands of the reasonably efficient operator. Often the starting point will be the actual detailed trading accounts. Ideally, trading forecast figures will also be available. In the case of properties which are new build, recently refurbished, or vacant due to default or other reasons, trading accounts will not be available. The valuer will then have to rely on making a personal assessment. Even where figures are available, it should be remembered that it is not the actual business that is being valued but the ability of the property to support a business in the hands of an efficient operator. Therefore great care should be taken at inspection to ensure that all evidence of under- or over-trading is noted — for example whether the tariff is appropriate and the management of high quality. In forming an opinion as to the maintainable trade, every opportunity should be taken to consult staff and managers. In addition, competing operations should be inspected, and it must be remembered that the competing units may either add trade to the unit being valued or, alternatively, that competing under-trading property may lead to a future downturn in the profits of the subject premises.

Once an estimate of the maintainable level of trade has been established which the valuer considers to be a fair reflection of all the available evidence, verified as far as possible (see Red Book, PS 5), the next step is to examine the expenditure which relates to the estimated turnover. Again, the actual accounts can be the best guide to the maintainable level of outgoings, but a critical assessment should be made so that any extraordinary items, or items personal to that individual operator, are excluded from the calculation: they should be added back (see below).

The accounts should also be adjusted, if necessary, to allow a realistic amount for wages, as in the case of, for example, a family run operation. It is helpful in estimating this to consider the expenditure under two main headings: variable costs and fixed costs.

- *Variable costs* are those items which will vary directly with the level of trade achieved, such as the cost of food and drinks purchased in the case of a restaurant. Therefore, where the valuer has made adjustments to the level of turnover recorded in the accounts, care should be taken to adjust the variable costs accordingly. It is also important that unit costs are appropriate to the style of operation being undertaken.
- *Fixed costs* are those which the business must carry whatever the level of trade, such as rates, heating, repairs etc. In examining the fixed costs, the valuer should ensure that the costs that are likely to be incurred in an average year are used, and that any extraordinary items, such as periodic major repairs, are excluded from the calculation but included, nominally, in the annual average.

Two other major adjustments should be undertaken by the valuer. These are adding back and the adjustment for a sinking fund.

## Adding back

To the notional accounts, a series of add back adjustments are normally made. These include any personal items related to the particular operator which are not connected with the notional business — such as interest, loan and bank charges, depreciation costs, director's remuneration and, in the case of multiple operations, head office costs — provided that the equivalent solus operation costs are included. Wages and salaries should also be adjusted — either up or down — to provide figures which would be appropriate to the operation of the business on a purely commercial basis. This can often be difficult, as many small operations would not be viable if full labour costs were included. In this case, the valuer must exercise judgement: if a likely hypothetical tenant would bid for the premises working on the assumption that family members would help to run the business unpaid, then it is reasonable to build this assumption into the figures. However, if it is necessary to do so to reach a figure of viability, then the long term future of the enterprise must be seriously questioned.

## Adjustment for sinking fund

The notional accounts should be further adjusted to include a reasonable amount to cover:

- a sinking fund to allow for the periodic replacement of trade fixtures, fittings and equipment and likely future capital expenditure to ensure compliance with legislation and
- a return on the notional capital invested in the business (that is to allow for the acquisition of the fixtures and fittings and to supply working capital).

The fully adjusted accounts will reveal a level of adjusted net profit. This figure is that which a normally competent operator could normally expect year on year to be sustained as at today's trading performance and as at today's prices. It represents the amount that is available for:

- rent and
- profit in the hands of the operator.

If the valuer is seeking to establish a rental value, the next step is to decide how much of the adjusted net profit would be bid away as rent. Typically this may be in the order of between 35% and 60% of the profit level, but in the case of sale and leaseback deals, of which there have been many as the investment market in leisure has matured, the rents may reach in the order of 70% or even more. However, such high levels of rent are not likely to be sustainable and are normally more the reflection of a complex financial deal than a true indicator of rental value. In determining the percentage to be adopted as the rental value, the valuer should consider:

- the likely level of demand for the unit from other potential operators
- the future prospects, based on trends in performance
- the risk of profit being eroded. (Note that where the fixed costs form a high percentage of the total costs, the profit is vulnerable in a downturn but potential is great in an upswing) and
- the total level of profit and its relationship to turnover.

In relation to the last point, it should be borne in mind that different types of operation have different norms of level of net profit as a percentage of turnover. Any operation where the net profit falls below, ay, 15% of the turnover could be regarded as highly risky, and this risk should be reflected in the rental bid.

## Alternative approaches to establishing rental value

In many ways the estimation of a rental value from a fully adjusted set of accounts is the truest approach to establishing a rental value, as the relationship between the occupier's benefit from use and the value ascribed is transparent. However, it is very subjective and, in the hands of an inexperienced valuer, can be misleading. Therefore, where it is possible to establish rental value by analysing comparable transactions, this will normally be preferred, although as stated above, the profits approach has been preferred at arbitration. The best approach is to look first for rental evidence and to back this up with accounts information and interpretation.

Even though evidence may be found, great care should be taken to establish that true comparability does exist in terms of specification, location, access and in lease details. If comparables are used, the valuer should make every effort to read the lease as there are considerable variations in the types of leases used for leisure properties.

The following clauses are critical:

- *Term length*
  It is not unusual for leisure properties to be let on terms of 25 or more if the property is to be let out in shell condition and the tenant has to pay a high level of fit out costs such as would be incurred for a cinema or bowling alley. Where fit out is cheaper, as in the case for example of a restaurant, then terms may be shorter. Given that practice for other commercial properties is for short leases, this represents a significant advantage to the landlord and potential burden to the tenant, especially if assignment requirements are onerous. If they are, the rental level could well be depressed.

- *Repair*
  Modern practice will normally dictate that the lessee is responsible for all repairs and redecoration work but the valuer should carefully examine the clause and ensure that no unduly onerous provision is included.

- *Improvements and fit-out*
  For many leisure properties the cost of fitting out can be substantial; there is also a need to refurbish more frequently than with many other types of property. Not only will the fabric require on-going commitment to alter and update as fashion trends demand, but also the operator will need to spend on fixtures and fittings. For this reason the well-drafted leisure lease should make adequate provision for the lessee to carry out works of improvement and alteration to the premises as trading conditions require, but care should be taken to ensure that the value of such improvements are excluded at rent review. Therefore, the valuer should check the provisions regarding alterations and improvements carefully, and note any inter-relationship with the rent review clause. Due to the costs of fitting out, it is common practice for lessees to be granted either rent-free periods or capital contributions at the commencement of the term. In such cases the value of the work may well be included at rent review, and the valuer should check this position.

- *User clause*
  As stated above, leisure comes under a number of different use classes. Care should always be taken when assessing a rental value to check the permitted planning use and note whether or no this is related to any particular planning use class. Given that there have been recent significan changes to the use classes order, the valuer should take great care in analysing the implications of the user clause.

- *Licensing and other consents*
  As with user clauses, the valuer should check whether the lease places any special requirement on the lessee in relation to the gaining and keeping of licences or other consents, as these can materially affect value. These clauses should be checked against the recent changes in licensing and gaming law.

- *Rent review provisions*
  The rent review provisions in any lease will be critical in establishing rental value, both initially and at review. The development of standard lease drafting has led to many review clauses being five yearly upward only to open market rental value, or more recently market rent as defined in the Red Book. With many commercial leisure properties, concerns about the likely availability of comparable evidence at review led to the inclusion of special provisions during the leisure boom years of the late 1990s. This was especially so with Class D2 and *sui generis* properties. In these cases the review clause may include reference to the greater of either Open Market Rent or a prescribed increase, typically 3% pa compound. Other formulae sometimes used are to: open (A1 retail use; a percentage of turnover; some other prescribed use; or by reference to other retail or office rents in the locality. All these formulae, of which the escalator clause appears to be the most common, reflect concerns about lack of evidence. If any such clause is present, the valuer should consider whether it is sufficiently onerous to reduce the rental value.

Where a rental value has been established by reference to comparable letting evidence it is advisable to check that the rent is sustainable by reference to profits — if it is feasible so to do. Another approach sometimes adopted to find rental value is by reference to gross turnover, with the rental value being expressed simply as a percentage of the take. This approach has several advantages:

- it is simple to ascertain
- is not open to argument as it can be related to audited accounts and it is difficult to manipulate in a non-transparent way and
- it can form a basis for rent review which encourages the lessee to maximise profit — as the containment of costs will not lead to higher rent.

However, set against these advantages are several disadvantages:

- if costs escalate due to inflation but without profitability improving, this will act against the tenant
- the relationship between profit and turnover is not static over time or between types of operation and thus it is at best a crude measure and
- it does not take into account issues of improvements, or lease terms generally.

## Establishing capital value

There are several approaches to establishing the capital value of commercial leisure units. These are:

- the profits approach
- the investment method or
- capital comparison.

## The profits approach

The profits approach can be used either where the property is owner occupied or likely to transfer between occupying operators or in the event of let premises. It is the normal approach when the valuation is for the purpose of sale from one operator to another and it is also used extensively where the proposed purchase is equity investor funded.

There are three variants in use:

- the single earnings multiplier
- the dual capitalisation approach and
- the discounted cash flow approach.

## Single earnings multiplier

The single earnings multiplier is the approach most commonly adopted for commercial leisure properties that are sold to an operator rather than an investor. It assumes that the capital value is related directly to the profitability of the premises in the hands of an efficient operator. It builds on the full profits approach to finding rental value (see p497). The starting point is the adjusted net profit.

The method appears deceptively simple as in essence it is the net profit multiplied by a single figure to find a capital value, as set out below:

| | |
|---|---|
| Maintainable Net Operating Profit (NOP) Say (arrived at after full adjustment of accounts) | £300,000 pa |
| YP, say | 8 |
| Capital Value | £2,400,000 |

It is obvious from this valuation that the choice of YP is extremely critical as a slight error at this stage can lead to enormous error in the capital valuation. This is probably the most important single reason that leisure valuations are regarded by many as being the province of specialist valuers.

It follows that it is important to understand what drives the choice of YP. The rationale for the YP is that it reflects the prospects for the profitability of the business moving forward. It is based on observations and the analysis of market transactions, and it is important that the multiplier chosen reflects the volatility of the sub-sector as well as the individual characteristics of the property being valued. However, as the information in the public domain on a sale will not include the adjusted net profit, it is critical that the valuers involved do understand in some detail the likely relationship between audited turnover and adjusted profitability. Otherwise they will only be in a position to analyse deals of which they have first hand information.

The valuer, in choosing the multiplier, should have regard to a number of factors:

- *Comparable evidence*
  Due note should be taken of any comparable market evidence, although unless the valuer has first hand information of the deal he is unlikely to be aware of the level of adjusted net profit to which a capitalisation factor was applied. This limits the use of comparables when applying a profits approach to capital value.

- *Likelihood of sustainability of the profit*
  To establish this, a careful note should be taken of the profit trends, analysis of competition and other external factors. It is also important that the valuer assesses any issues of branding and considers whether the level of maintainable profit is independent of any brand imaging considerations.

- *Level of demand*
  The valuer should consider the state of the market and the availability of other premises on the market. The demand for the unit will also be affected by the supply of alternative units, so an analysis of supply factors should be undertaken in order to assess demand.

- *Development potential*
  The valuer should assess whether the possibility exists to exploit the trading potential of the unit further, either with, or without, major capital expenditure, and should adjust the multiplier accordingly. Following the recent changes to planning and other regulations, it is important to consider the extent to which the property's physical characteristics and trading identity may be altered without the need for fresh consent.

- *Profitability of the underlying operation*
  This is very important as the value of any trading leisure property relates to the profitability of the underlying activity. For example, at the time of writing, cinema admissions have undergone very significant growth over a fairly long period. This has been reflected in both the rents attainable and in capital values. However, for profitability to continue, not only must the property's specification be right, but also the type of films screened must be capable of attracting audiences. A downturn in the film industry and shortage of blockbusters could quickly and adversely affect property confidence and, hence, capital values.

In making a judgment of which capitalisation rate to use, the valuer should, at all times, ensure that no element of double counting has resulted from any adjustments made to the accounts. For example, if the level of turnover has been adjusted upwards to reflect the presence of an inefficient operator and the potential for increased trade, it would be important that the capitalisation factor is not increased — nor the yield decreased — to reflect this fact. Similarly, where the expectations of revenue have been muted due to the likelihood of a new competing development depressing trade, care should be taken before building this factor into the capitalisation. At all times it is important to reflect on the trading figures, and ensure that they are used to guide the valuer to an appropriate figure, and not simply used as the only evidence of value. For leisure properties where the single earnings multiplier is used, it is important that valuation checks (see below) are carried out to add confidence in the resultant figure.

## The dual capitalisation approach

The dual capitalisation approach builds directly on the full profits approach to finding rental value. It is favoured by many valuers as providing the truest method of establishing the market value because it separates out the element of value attributable to the property itself from the element that is attributable to the trade. It is normally used where the market is dominated by owner-operators because the very construction of the model implies that there is an addition to the capital value over and above that which an investing owner would pay for the premises.

The starting point is to take the two elements of the divisible balance (rent and residual profit) and to capitalise each separately. The resultant figures are then added to give the capital value. So, taking the example used above, the calculation becomes:

**Stage 1: The divisible balance**

|  | £ pa | £ |
|---|---|---|
| Maintainable Net Operating Profit (NOP) Say | 300,000 | |
| (arrived at after full adjustment of accounts) | | |
| *Less* Annual rental value at say 60% of NOP | 180,000 | |
| Residual profit | 120,000 | |

**Stage 2: Capitalisation**

|  | £ pa | £ |
|---|---|---|
| Annual rental value | 180,000 | |
| YP in perpetuity at say 8.5% | 11.76 | |
| Value of rental slice | | 2,116,800 |
| Residual profit | 120,000 | |
| YP at say | 2.5 | |
| Value of residual profit slice | | 300,000 |
| Capital Value | | 2,416,800 |

But say  £2,400,000

The rationale for splitting the two elements is that the rental income slice is more secure and is capable of being capitalised at a property yield, whereas the residual profit is less secure. In assessing the yield to be applied to the bottom tranche of income, the valuer should take cognisance of all the factors that have been detailed above in relation to the quality of the property as an investment, but excluding consideration of the volatility of the business in the hands of the operator. This method of valuation is used primarily where the anticipation is that the most likely purchaser in the open market would be an operator. If an investment sale of the freehold interest is in contemplation then, unless the investing owner will be able to access a share of the business, the use of the dual capitalisation approach will be inappropriate because the property owner will only receive the rental income. If the property is to be let and sold on an investment basis, the investor will seek to capitalise only the rent.

In assessing the top slice multiplier (not yield) which is typically in the region of 2 to 5 YP, the factors listed above become paramount in relation to the profit stream's potential for growth and the likelihood of sustainability of the profit in the hands of an efficient operator. It must be stressed that the residual value in the hands of the operator does not represent personal goodwill; it is the residue that an efficient operator would expect to make over and above the amount bid away as rent. As such, the valuer should acknowledge that it is part of the property valuation and not something that adheres to the occupier — it runs with the land.

There is one situation in which the valuer may need to make an adjustment to the valuation. In the event that the premises to be valued are new, or for some other reason have no recent trading record the value of the projected residual value in the hands of the operator may be depressed, because the incoming operator will have to establish the trading reputation of the premises. This additional risk factor can be taken into account by making an adjustment to the multiplier used to capitalise the residual profit. In these circumstances, the use of the dual capitalisation approach is more flexible than the single earnings multiplier because it enables the valuer to reflect accurately within the valuation both the quantum and the nature of the trading risk.

Theoretically, the capital valuation arrived at by the use of the dual capitalisation approach should be reconciled with that achieved by use of the single earnings multiplier. In practice the single earnings multiplier can be considered to be a blended yield where the valuer makes a judgment as to the total risk involved. This obviously requires much knowledge and fine judgment, and as such the dual capitalisation approach provides a more transparent way of assessing value.

While the use of the dual capitalisation and single capitalisation approaches remain the first choice for many leisure and licensed premises, where an institutional market based on standard leases has developed, such as in the case of multiplex cinemas and health and fitness premises, the conventional investment method will normally be adopted.

## Discounted cash flow

Valuers are now making increasing use of full DCF approaches to the capital valuation of leisure properties. This has been facilitated by their having access to either industry standard software (such as Excel™) or bespoke property programmes (such as Circle Investor). Properties valued by DCF tend to be those of higher value.

Whenever a DCF approach is used two important principles should be recognised:

- The quality of a DCF calculation is only as good as the inputs on which it is based. If these are objective and market based, DCF is simply a more flexible method by which capital value can be derived.
- If the inputs are taken with reference to a specific client's requirements, and without full reference to the market or to the factors set out above in relation to the profits approach, the result will be in the nature of an investment worth calculation, as defined by the Red Book, and not a market valuation. That is to say, it becomes a calculation that represents an evaluation of the net monetary worth of the asset to the individual client, based on that client's requirements, and not something which is necessarily reflective of market opinion.

In order to carry out a DCF appraisal, not only must the NOP (Net Operating Profit) be calculated but a number of other issues must be considered. VIP 6 contains a list of considerations, and the issue set out below has been adapted from this:

- the period over which specific cash flows are projected; typically this will be in the range of 5–10 years only due to the difficulty of projecting accurately for longer periods
- the capitalisation rate or exit yield to be applied at the end of the period
- whether to discount on a monthly, quarterly or annual basis
- the discount rate to be adopted

the need to build in for possible refurbishment and refitting costs

whether to inflate the incomes and outflows for inflation or not — which may have an impact on the discount rate chosen

the projected trading profile and

the initial and running yields as well as the terminal values.

DCF can easily be adjusted to allow for monthly or quarterly cash flow periods. The cash flow should build in specific projections for changes in profit levels. Where a market value is required (as opposed to a worth calculation), any growth built in to reflect, for example, inflation, must be related to the discount rate which is chosen, derived from the analysis of external objective indicators, and should represents the valuer's best opinion of the overall return relating to that class of property which is required by investing owners active in the market.

Having set the discount rate and made the explicit assumptions for the cash flow, the valuer should then determine the period over which the calculation will be run. In general, due to market uncertainty, it is unusual to take a time frame longer than say 10 years. This is because the specific projections on which a DCF calculation is based are difficult to establish with any degree of accuracy for periods longer than that. At the end of the time period involved it is necessary to establish an end value, either by taking a capital sum direct or by calculating the last estimated profit at a realistic market yield or exit yield. The matter of the exit yield to be applied is one of valuer judgment. It will often be taken at slightly above that which was applied as a market capitalisation rate for the dual income approach. This is because at the end of the set period the unit may be suffering from obsolescence. However, such an adjustment is not always justified, particularly if allowance for refurbishment during the period has been made.

Once all the inputs have been established the discounted cash flow for each year (or period) can be computed and the sum of all of these will represent the net present value or capital value of the asset. Worked examples of DCF calculations are to be found in Chapters 11 and 18, and a discussion in the latter.

The DCF approach, as with any variation of the profits method, requires great care in execution, particularly as the very nature of the calculation means that it is difficult to obtain good objective comparable evidence. It should not be undertaken by the inexperienced valuer. In many cases its use is best suited to the establishment of worth. It is, however, useful to run a DCF calculation alongside another approach to gain a better overall picture of value.

## Investment method

The use of the investment method to establish the capital value of a leisure property has become increasingly widespread as institutional investors have bought into the market, generally into units which have a low hospitality content (i.e. where the sale of food, drink and alcohol is incidental to the main offer) and which are purpose built modern units. This means that generally their interest has been restricted to:

- high street A3 and A4 users where there is good comparable letting evidence
- well specified and let leisure parks — preferably integrated with or adjacent to retail schemes
- well let multiplexes and
- other large commercial D2 units with strong covenants and tried and tested brand product.

Where such a unit is to be valued for capital purposes, the usual approach will simply be to capitalise the rental value, in the case of a rack rented unit, at an appropriate all-risks yield or, where reversionary value is to be established, to capitalise using a hardcore approach and equivalent yield.

The yields that might be appropriate for prime leisure properties have declined from around 9% in the mid 1990s, when standard leisure boxes were a very new concept, to yields relating more closely to retail warehouses and shopping centres. A yield of 6% or even less would now not be unrealistic in a strong market. However, unlike offices, retail and industrial property, the yield movements in the leisure market are not routinely monitored by organisations such as the Investment Property Databank (IPD). This is because the numbers sitting within the portfolios of institutional investors are too small. It must, however, be stressed that it is only the very best well-let schemes that will be attractive to investors, and the leisure sector is still, in many ways, a misnomer as it comprises a wide ranging series of sub-sectors.

The principal factors that determine the attractiveness of a leisure property to investors are set out below.

- *Depth of the investment market*
  The investor will wish to be assured that there are sufficient units of this type within similar localities to ensure that comparability to support valuations exists, and to ensure liquidity in the event that they wish to sell.

- *Quality of the covenant and potential occupational demand*
  Not only will the investor need to establish that the present or, in the case of a pre-let scheme, the prospective, tenant provides a satisfactory covenant, but they will also require assurance that there is the strength of tenant demand to ensure long-term performance in the event of a proposed assignment. Given that recent years have seen a succession of both business failures and corporate mergers, this is of real concern to investors.

- *Planning and environmental considerations*
  The fast moving nature of the leisure market means that there may be a need for the occupier to refresh the image and brand during the life of the lease. It is important that, as far as possible, the planning considerations are not likely to hinder this process. At a macro-level, consideration of potential policy decisions (for example in relation to environmental and other sustainability issues) should be factored in. At the time of writing, significant changes to the planning system have been put forward and the introduction of energy performance certification is imminent. These measures are likely to have differential impacts on both tenant demand and investor preference.

- *Design and obsolescence factors*
  Leisure properties which are purpose built are particularly susceptible to obsolescence, in much the same way as shopping centres. This in turn will reflect in poor rental growth prospects. The valuer should be aware that the very nature of leisure means that functional and design obsolescence are real problems that can quickly lead to falling values. However, for some other types of leisure property, such as heritage schemes, obsolescence is not an issue and may add value! Currently, however, heritage schemes are not usually of any attraction to institutional investors, who seek prime modern stock.

*Location, accessibility and footfall*

The location, accessibility (including visibility) and ability to generate footfall are all issues which will underpin the long-term viability of a commercial leisure property. Therefore these should be considered, and the analysis of economic and demographic trends undertaken before a yield is chosen.

*Brand image and tenant mix*

The brand image can be important to an investor as it will be influential in creating an image for the whole investment. The presence of recognised brand names in a leisure scheme is just as important as in a shopping centre where the brand image of the anchor tenant and the appropriate trading mix are crucial factors to securing a good long-term investment. For a leisure scheme the mix of users that will create synergy is constantly changing. When making investment choices, consideration should be given to the riskiness of the format. For example, the health and fitness sector, which only five years ago looked to be low risk due to increased media attention to health matters and rising real wealth, has been subject to significant covenant failures.

*Lease terms*

Imperative to institutional investors is a lease which provides the terms that have become established over three decades for other commercial premises. The nature of these has been discussed above and can be summarised as 15 to 25 years in length, depending on fit out requirements, with five yearly rent reviews, upward only. However, whereas a ratchet was commonplace some years ago this is no longer the case. There will normally be full repairing and insuring tenant liability. It is sufficient here to emphasise that a property which is not so let is likely to be valued at a considerably higher yield and lie in the portfolio of a small or specialist investor.

It must be pointed out that a valuation carried out on the investment method will be less likely to be adopted when the valuer is presenting an existing use value (EUV), given that the use of EUV is now restricted by the Red Book to the valuation of owner-occupied property for accounts purposes under UK accounting standards (UKPS 1).

## Capital comparison

Most commercial leisure properties will be valued using either the profits approach or, in the case of let schemes, an investment method. To an extent both of these rely on some measure of comparable evidence. It is unusual for a leisure property to be be valued only by direct capital comparison. This is because it will either be let — in which case the investment method is the preferred method — or it will be owner occupied — in which case it will be the trading potential rather than the physical specification and size which will drive the valuation. However, there are instances where capital comparison can be used. If evidence of transactions involving similar properties having comparable trading potential does exist, then a simple analysis and comparison should be used, although a check against a profits approach is advisable.

# Valuation for various purposes

## General

The reader is reminded of the requirements of the RICS regulatory regime described earlier in thi chapter which it is not proposed to repeat here. Valuers should ensure that they are familiar with the definitions, and adherence to the statements is mandatory unless there are specific, justified an clearly articulated reasons for not so doing.

The most common purpose for which a valuation of leisure property is required is for marke transaction, but frequently valuations are required for the purposes of either loan security or the balance sheet. In all these cases the appropriate basis is normally market value but if the the valuation is for balance sheet purposes, owner-occupied property is to be valued to existing use value.

## Valuations for loan security

There are no specific rules contained in the practice statements in relation to loan security valuation: for properties that are to be valued as trading operational entities. However, UKPS 3 does addres: some particular concerns in relation to valuations for secured lending, though without specific reference to trading assets. Additionally, section 9 of Valuation Information Paper 2 makes provisior regarding valuations for the purposes of secured lending.

The purpose of including this assessment is to give information to the lender which will assist ir determining the amount of loan that will be granted against the security of the property. It follows tha the valuer is instructed to give the lender information concerning the sources of trading evidence, and in particular whether the assessment of value is founded on actual, estimated or projected accounts.

The value of a trading leisure property is very dependent upon the availability (or otherwise) of trading records, and it follows that there is greater risk attaching to the security offered by a property so valued as compared to, say, an office investment property where the property's value is less dependent on the nature and trade of the occupier. Therefore, due to the inherent risks that go with trading properties, the VIP guidance note makes it clear that the valuer should not include elements of personal goodwill in the valuation. It also states that the valuer should ensure that the valuation details the fair maintainable trade assessment, so that the lender can assess the serviceability of the loan. Furthermore, the valuer is advised to emphasis the factors that may impact on value and the 'risk associated with the future trading potential at the property'.

It may sometimes be the case that the nature of the operation is one for which the potential demand is particularly fickle. Where this is the case, and the valuer considers that the market demand for the property in its existing use may change significantly over the period of the projected loan, the matter should be reported to the client. If appropriate, a valuation on the basis of alternative use may be advised.

Where a valuation is required for loan security purposes of a property that is let as an investment, the usual considerations for any investment property apply.

## Valuations for balance sheets

The specific requirements for valuations for balance sheets and other financial statements are covered in UK PS 1 and UK PS 2. This statement makes clear that market value and existing use value (EUV) are the recognised bases.

In all cases the choice of the appropriate basis should be determined by the valuer in agreement with the client. Market value is the basis used where the property is either let as an investment or is operational land surplus to requirements. EUV is only used where the property is operational land occupied for the purposes of the business and for which there would be a market for sale to a single occupier.

Properties that are tenanted will be valued on the basis of market value, normally using the investment method. However, for those leisure properties that are owner-occupied, the basis will normally be EUV, in which case the trading approach will normally be used, but the VIP notes that where premises are capable of occupation but not trading, appropriate adjustments should be made to the valuation. These adjustments may include the cost of refitting. Where there is no intention that the property will be re-occupied in the future for the purposes of the business, the property will normally be declared surplus and valued to market value.

# Non-commercial leisure properties

The valuation methods and considerations detailed above relate to trading leisure properties that are owned and operated primarily for commercial purposes. There are, however, very many leisure properties that are not so held. For example, local authorities and charitable institutions often hold a wide range of leisure properties within their portfolios. Typical local authority portfolios contain leisure and sports centres, municipal golf courses, theatres, community centres, museums, playing fields, parks and gardens. Charitable institutions may hold heritage assets, such as castles and gardens. Many of these are revenue producing and, indeed, some have very high turnovers; others are free access. The common thread is that they are provided as a service to the local populace in furtherance of political, cultural, educational and/or social aims. The terms under which some are held allow lease or sale; others are inalienable.

In general such properties have seldom been sold on the open market, and until the introduction of changed accounting procedures in the mid 1990s there was no general requirement to place a monetary value on them. Where valuations were prepared (for example for rating purposes) the general approach has been to use a cost-based method, and this approach has been supported in the Lands Tribunal. For asset valuation purposes, the choice is between a trading approach and depreciated replacement cost (DRC). Guidance on the latter approach is found within the Valuation Information Paper 10. A DRC approach will be taken when, after appropriate research, it is considered that the owner (ie the public body) would be the only possible purchaser or tenant for the property. For properties deemed to be heritage assets (ie those held for cultural and/or educational purposes) there is only a limited requirement to value, although at the time of writing the matter is under review.

It follows that until the mid 1990s, the question of the basis of valuation was largely academic because properties were seldom valued other than for rating, and even here the valuations did not carry real financial significance for the owners. However, the valuation of such properties is no longer academic. In rating terms, the decisions matter; additionally local authorities are now required to produce capital valuations of most of their assets for capital accounting purposes.

While many publicly owned leisure properties are still valued on the basis of DRC for capital accounts and rating purposes, market changes and changes in the philosophy of public provision are leading to a slow change in the methodology adopted for valuation, with a growing reluctance to use DRC as there is now more evidence on which to base valuations.

Additionally, many public sector assets are now privately managed on a commercial basis, which again points to the adoption of a trading valuation approach.

For valuing any non-commercial leisure property, whether held in the public or charitable sector, the valuer should consider:

- the purpose of the valuation
- the attraction of the property to other potential occupiers or owners
- the appropriateness of the trading policy and whether this is driven by social or political reasons and
- the effect of grant funding on both the original capital provision of the property and its revenue expenditure.

It must be stressed that the methods to be adopted in relation to the type of properties detailed above are very much subject to detailed discussion with the client, as the choice of method may significantly affect the end result.

## Study 1  Freehold Valuation of a Leisure Park Investment

You are instructed to value the freehold interest of a leisure park investment for possible sale. The property is situated in the south-east of England and the vendors have been the owners since its inception some seven years ago. The land abuts an edge of town retail scheme which is fully let, mainly to multiple tenants. It has adequate on-site parking, there is good visibility, and vehicular access from a main road. A motorway junction is within two miles, but public transport links are limited. The scheme is anchored on an eight screen multiplex cinema and has two other D2 users (bingo and bowling) and four food and drink outlets, all of which have leases indicating A3 use only, but which are now, in planning terms, three A3 units, and one A4 unit.

The leases are as follows:

- *Cinema*
  Let on a 30 year term with a five-yearly upward-only rent review clause to open market rental value (OMRV) or 3% pa compound. At the time of letting the tenant received a rent-free period of six months to allow for fit out. The user clause restricts the use to a cinema.

- *The other D2 units*
  Let on 25 year terms with five-yearly upward-only rent review clauses to OMRV or 3% pa compound. The user clauses permit any D2 use.

- *Food and drink units*
  Let on 20 year terms with five-yearly upward-only rent review clauses to OMRV. The user clauses are restricted to any A3 use.

There is little evidence from recent lettings of other leisure units in the region, as few new schemes have been built out within the last three years. The rent reviews that have fallen due for competitor schemes demonstrate that rents have changed little recently. The rents on the subject scheme were reviewed in 2005. At that time the rents of the cinema and the D2 units were increased in line with the escalator clause. Reviews to market value were exercised in relation to the food and drink units. Since that time, cinema and bowling rents in the area have recovered sufficiently so that the units are no longer regarded as over-rented. There has been little growth in restaurant rental values since the rents were last reviewed.

There is also evidence of yields from other investment sales. The adequacy of the location and access, combined with the quality of the covenants, which are all multiple traders, lead you to rank this as semi-prime. The age of the scheme, although only seven years, means that the design is not state of the art, and the poor public transport links lead you to question its ability to achieve rental growth in line with the best of new schemes.

## Valuation

| Note: All rents are MRV | | £ pa | £ |
|---|---|---:|---:|
| *8 screen multiplex cinema* | | | |
| Rent passing 3,000 m$^2$ at £150 per m$^2$ | | 450,000 | |
| *Two D2 Units* | | | |
| *Unit 1* | | | |
| Rent passing | | 100,000 | |
| Ground floor: 1,000 m$^2$ at £100 per m$^2$ | | | |
| *Unit 2* | | | |
| Rent passing | | 72,000 | |
| Ground floor: 600 m$^2$ at £100 per m$^2$ | | | |
| First floor: 300 m$^2$ at £40 per m$^2$ | | | |
| *A4 Unit* | | | |
| *Unit 1* | | | |
| Rent passing 500 m$^2$ at £185 per m$^2$ | | 92,500 | |
| *A3 Units* | | | |
| *Unit 2* | | | |
| Rent passing 640 m$^2$ at £170 per m$^2$ | | 108,800 | |
| *Unit 3* | | | |
| Rent passing 200 m$^2$ at £175 per m$^2$ | | 35,000 | |
| *Unit 4* | | | |
| Rent passing 180 m$^2$ at £175 per m$^2$ | | 31,500 | |
| Total Rent Roll | | 988,800 | |
| YP in perp at 6.75% | | 14.81 | |
| Gross Capital Value | | | 14,644,128 |
| *Less* Purchaser's Costs | | | |
| Stamp Duty at 4.0% | 585,766 | | |
| Agents and legal fees at 2.0% | 292,883 | | |
| | | | 878,649 |
| Net Capital Value | | | £13,765,479 |
| Say £13,770,000 | | | |

## Study 2

You are instructed to act on behalf of a tenant in a rent review negotiation of a 20-lane bowling alley. The property is located in a town centre and it is let to an independent trader. It is not in a prime pitch. The building, of which it comprises part of the ground floor, is now some 30 years old. The remainder of the building comprises a restaurant unit with four floors of offices above. The unit still trades well but due to the pitch and the age of the building it is unlikely to attract a multiple operator. The premises have a net internal area of 2,137 m$^2$.

The lease is for a term of 25 years with five-yearly reviews, and this is the second rent review. The lease is on the equivalent of full repairing terms as there is a full service charge and the rent review clause specifies a review to market rental value to the effect that any rent free periods are to be ignored. Furthermore, it is to be assumed that the premises are in good condition and fully fitted.

After inspection, you consider that the tenant is trading to the unit's full potential. In addition to the income derived from lane hire the tenant has maximised the opportunities for food and drink sales and from machine revenue.

The audited accounts show that the trade for the last three years has increased slightly above inflation, but you are also aware that a new leisure park located about one mile away is due to open in a few months. This enterprise includes a 36-lane ten pin bowling centre. You therefore consider it appropriate to reflect the potential competition in preparing adjusted accounts on which to advise as to rental value.

The accounts show a current turnover of just over £1,200,000 and a profit of £420,000. However you consider that the turnover is unsustainable in view of the impending competition and examination of the accounts reveals that a large sum is included for annual depreciation. The tenant is also showing in the books the cost of running a small catering business which is run from the premises. These items you exclude by the add-back process to produce an estimated maintainable trade.

There is little rental evidence of bowling alleys in the immediate area — this unit has a local monopoly position. However, you are aware that in neighbouring towns bowling alleys have commanded rents in the region of £105 per m².

### Estimated Maintainable Trade

| *Income* | £ pa |
|---|---|
| Bowling (including shoe hire) | 459,250 |
| Machine revenue | 167,000 |
| Beverage sales | 133,600 |
| Food sales | 33,400 |
| Merchandising/vending/other income | 41,750 |
| Total maintainable sales | 835,000 |
| | |
| *Cost of sales* | |
| Food and drink purchases | 58,450 |
| Other purchases pins, balls etc. | 66,800 |
| Total cost of sales | 125,250 |
| | |
| Gross profit | 709,750 |
| | |
| *Outgoings* | |
| Wages | 270,000 |
| Heating and lighting | 40,000 |
| Promotion | 10,000 |
| Service charge | 19,000 |
| Repairs | 30,000 |
| Other operating expenses | 50,000 |
| Total outgoings | 419,000 |
| | |
| Gross profit | 709,750 |
| *Less* Total outgoings | 419,000 |
| Adjusted Net Profit | £290,750  (35%) |
| | |
| Rental bid at say 47.5% of profit | 138,106 |
| Say | 140,000 |

This bid represents some 16.7% of turnover.

To check the valuation an analysis is carried out against floor area. The suggested rental bid represents a figure of £95.20 per m². This is lower than the comparable evidence suggests, but appears reasonable given the poor quality of the building and the future competition.

# The use of valuation checks

The use of valuation checks is widespread in practice, but there is a danger in their use by those not practised in such properties. In a sense, every rental value found by reference to a price per m² is a check method in that, underlying the rate adopted, will be the ability of the premises to be profitable at that figure if in the hands of an efficient operator.

Common units of comparison are per seat (theatres and cinemas), per court (sports such as squash) per cover (restaurants) and per bedroom (hotels). However, it is advocated that, however experienced the valuer, the use of checks is confined to just that purpose: providing a check against a more detailed calculation.

# Summary

The leisure property sector is extremely disparate. It comprises both commercial and non-commercial property; it includes properties located in urban positions and those in rural locations. Leisure properties are found in the occupational portfolios of both public sector bodies and private sector organisations, and they are increasingly acceptable within institutional investment portfolios.

All these facts are reflected in the wide range of valuation approaches that may be appropriate. The maturing of the commercial leisure market has increased the applicability of the investment method. The rental values of such investment properties are found by analysing comparable transactions on a m² basis, usually based on gross internal areas, although where the unit is in a high street alongside retail units, a net basis may be used. However, a great number of leisure properties are still primarily traded for owner-occupation or for occupation under franchise or financial management contracts. For these, the profits or trading approach continues to be the most frequently adopted method, and the use of the full profits approach still provides the best method of establishing the maximum rental value that an operator could afford to pay.

Within the field of social provision, the use of cost approaches to valuation is still widespread whenever it can be argued that the property is not one for which there would be a ready market. With the growth of private sector provision and the consequent availability of market evidence, this method is becoming less relevant.

# Easements and Wayleaves for Sewers, Pipelines and Cables

## General

### Introduction

While the underlying principles of easement and wayleave valuations remain largely the same, a number of new Acts have been introduced since the last edition which reflect the changing nature of many of the parties involved with the subject-matter of this chapter and which seek to regulate them.

Practitioners will be aware of the multiplicity of the parties now entitled to acquire rights, ranging from the established companies, through what were previously public utilities to the new entrants in communications in its many facets. Throughout this chapter they will collectively be referred to as the promoter and their works will often be referred to as pipeline or scheme irrespective, except in specific cases, of the actual nature of the work. In the same way those from whom the rights are being acquired or obtained, whether owner or occupier, lessee or tenant, are referred to as the claimant.

The majority of easements and wayleaves involve agricultural land and, except where otherwise indicated in relation to sewers in particular, the practical implications are considered from an agricultural point of view.

### Legal background

In the majority of cases it is unnecessary for the valuer to study the detail of the statute, if any, under which a particular scheme is being promoted. A brief background to the legislation can be found in four advisory handbooks produced by the Country Landowners and Business Association (CLA): *Telecommunications Wayleaves* (CLA 39 October 2005); *Electricity Wayleaves* (CLA 40 December 2005); *Water Pipes and Sewers* (CLA 43 November 2006); *Gas and Oil Pipelines* (CLA 44 2007).

In general, most sewers and some water mains are laid under statutory powers without any further legal documentation beyond the original notice and without any easement being acquired. Others, such as gas, oil and other pipelines, are installed by voluntary deed of grant, agreements being negotiated within the general compensation code because the compulsory powers are there, although they are rarely used. The electricity companies install apparatus under a combination of voluntary deeds and statutory powers and the burgeoning telecommunications industry tends to do likewise, although, in the case of many telegraph lines, no deed of grant is entered into. Under the Electronic

Communication Code established by the Communications Act 2003 there are certain circumstances in which equipment can be installed with only the agreement of the occupier and without obtaining the consent of the owner.

The valuer will need to give consideration to the terms of any proposed deed, to ensure that it suits the claimant's circumstances with or without amendment, and because the terms may have a bearing upon the valuation. The claimant's solicitor should be consulted on the wording of the deed at as early a stage as possible. It should be noted, however, that the large companies are reluctant to agree any substantive amendments

Exceptionally, in the case of sewers, private individuals may have indirect access to statutory powers. Generally, however, such individuals can install equipment in the land of another only if the owner and any occupier agree when, in the interests of all parties, a formal deed of grant should be prepared by the respective solicitors to be attached to the title deeds of each. This should be a comprehensive document incorporating not only the terms commonly used by commercial promoters, but also the additional undertakings that such promoters usually volunteer, including an arbitration or alternative procedure in the event of disputes, and confirmation that the promoter will reimburse professional fees incurred by the claimant as well as the usual list of undertakings with regard to working arrangements.

## Role of the valuer

The initial role of the valuer depends on the nature of the claimant's interest. For a landlord there is only the capital compensation for permanent loss, and possibly for the long term effect of the route of the pipeline on the property overall. For an owner occupier there is the same capital compensation to be considered as well as compensation for surface damage arising from the works. For an occupier, whether tenant or lessee, there is only the surface damage compensation to be considered, except in the case of long leases where the lessee may be joined in the deed and may in fact be entitled to part or even all of the capital payment.

Strictly, for an owner occupier, the two heads form part of the same claim, but in practice the capital payment valuation will be made before work commences and the surface damage element is formulated during and after the scheme is implemented.

On larger schemes, the basis for the capital payment will have been agreed by the promoter with the CLA and National Farmer's Union (NFU), or their Scottish and Welsh equivalents. Only in exceptional circumstances will the promoter be prepared to vary these, although it has been known for a group of valuers acting for clients on a scheme collectively to persuade the promoter to amend the proposals, understandably upwards.

There are other factors a valuer might have to consider, such as whether interest is payable and from when, what the promoter's proposals are with regard to the claimant's fees, and the disputes procedure if any particular aspect of compensation cannot be agreed.

On larger schemes in particular, there may well be problems continuing for several years which result in annual run-off claims.

Depending on the nature of compensation offered, it might be necessary for the valuer to seek the advice of the claimant's accountant.

In the case of surface wayleaves, particularly for electrical apparatus, the annual payments should be checked from time to time by the claimant or his valuer to take account of changes to field boundaries and land use.

# Compensation for permanent damage

## General

The first of the two categories of compensation is that for the permanent or long-term presence of the equipment in or on the land. This is referred to variously as easement consideration, the claim for freehold damage, a recognition payment, the freehold claim, or the landlord's claim.

The amount of compensation for permanent damage depends in part upon the restrictions imposed on the claimant by the presence of the equipment and in part on the rights granted to the promoter for inspection and subsequent re-entry on to the land.

All deeds should include clauses covering losses in the future if the presence of the pipeline interferes with, or prevents, the development of any adjoining land. They should ensure that compensation must be assessed at the date of the loss.

The claim should include compensation for surface structures such as manholes, even if buried, for cathodic protection equipment and for injurious affection to any other property belonging to the claimant.

Where the deed includes the right to install further equipment at a later date, care should be taken that the relevant clauses are worded so as to entitle the claimant to full compensation at values then current.

Specific provisions should apply to land containing minerals, where the precise legal position must be clarified.

In common with the general law of compulsory purchase, promoters using statutory powers can deduct betterment from any compensation that would otherwise be due, although this rarely applies.

In *Mercury Communications Ltd* v *London & India Dock Investments Ltd* [1994], the court declined to award a ransom element within the compensation for freehold damage, and in *Northern Electric plc* v *Addison* [1997] a similar attitude was taken by the Court of Appeal on a rent review for an electricity sub-station.

See also *Kettering Borough Council* v *Anglian Water Services plc* (2005) where the Lands Tribunal held that where a developer successfully requisitioned a sewer connection across a third party's land, the third party was not entitled to claim a ransom value from the undertaker installing the sewer. *Andrew Logan* v *Scottish Water* (2005) reiterated this ruling.

## Advance payments

Note: This part of the chapter is dealing with permanent damage: see p532 regarding surface damage.

Generally the claimant will be entitled to an advance payment of 90% of the compensation as agreed or estimated, either on the date notice is served, or the date of entry provided title has been proved. The claimant's valuer should apply for such advance payment as soon as a relevant notice is served or well in advance of the expected date of entry and instruct solicitors to prove title without delay. Promoters are understandably reluctant to make any advance payment before title has been proved. Outstanding compensation normally attracts interest although it does not always do so.

On some schemes the final amount of compensation will not be assessed until detailed as laid plans have been produced. These should be checked by the valuers, and should form part of the deed of grant.

Valuers should not omit to provide for the payment of statutory interest. See p540 below.

## Sewers and water pipes

Originally the power to construct such pipes was given under various Public Health Acts which were consolidated in the Water Industry Act 1991. This Act gave the various Water Service plc's powers to enter land to lay pipes and sewers, subject to the service of statutory notices as set out in section 159. Only rarely will water companies offer better terms than those available to them under the Act. Such occasions can arise if the project is particularly urgent, or if there has been a problem with the service of the appropriate notice. However, such occasions are rare and generally the compensation code applies

The Lands Tribunal has refused to lay down rigid formulae for the calculation of compensation for permanent damage, although such formulae are widely used in practice. The Tribunal's refusal is because of the need to judge each individual case on its merits.

In *St John's College, Oxford* v *Thames Water Authority* [1990], the Tribunal expressed an opinion that such valuations should ideally be made on a before and after basis, but conceded that no such evidence was available. Their award was actually based on 50% of the freehold value discounted by 60% to reflect that the land was tenanted.

In *Felthouse* v *Cannock Rural District Council* and *Markland* v *Cannock Rural District Council* (1973), which remain the definitive cases, the Tribunal referred to four main factors:

1. the trend of payments for sewer pipelines
2. the relationship of sewer pipeline payments to capital land values
3. the trend of capital land values
4. the trend of payment for other types of pipeline.

The common formula for agricultural land involves three elements and is calculated per metre run of pipe:

1. the value of the land
2. a percentage of that land value
3. a notional width.

The land value used should be the value to the owner of the farm as equipped as a working farm, but should exclude any uplift for amenity or non-agricultural value unless such uplift is itself affected by the scheme. Promoters tend to adopt a standard value for land of a similar quality within each scheme and to vary that figure only where circumstances are markedly different. Traditionally the vacant possession value of the land has been taken, even where it is let on an agricultural tenancy, although the Lands Tribunal adopted the tenanted value in the *St John's College* case.

Almost universally for sewers and water pipelines, 50% of freehold value is used and has been adhered to generally by the Lands Tribunal, despite other promoters using higher percentages. However, where the route of a pipeline was altered at the request of a claimant so that it followed a longer route around the edge of a field, the Lands Tribunal reduced this to 40%: see *Abercrombie* v *Derwent Rural District Council* (1971).

Usually the greatest scope for discussion concerns the width affected. To a certain extent this is determined by the size and depth of the sewer but typically widths of six to eight metres are common, with six tending to be the norm. The width used in such calculations tends to be less than the working width employed by the promoters, and sometimes a multiplier of two-thirds of that working width is used, but in the *St John's College* case the Lands Tribunal award was calculated on the full 20 m working width necessary to lay a 750 mm water pipe at a depth of 2 to 2.5 m. An argument in favour of a larger width is that, on the rare occasions when major problems subsequently occur, necessitating the re-excavation of a sewer or water pipe, a far greater width is required than that when the pipe was first laid. On the other hand, a direct link with the working width might encourage promoters to reduce that width, which is in neither party's interests for reasons explained below. In addition, it should be noted that other pipeline promoters base their capital payments on an easement width which is invariably less than the working width required for construction.

Besides agricultural land, water pipes and sewers are occasionally laid through house gardens. Compensation is rarely paid for the permanent presence of such pipes if the house concerned benefits from the scheme through the provision of a connection to the public sewer, if this does away with reliance on a cesspool or septic tank: see *Collins and Collins v Thames Water Utilities Ltd* [1994]. Otherwise, there is no formula and much will depend upon the skill and experience of the valuers. Obviously, the rights of inspection and re-entry assume great significance in such circumstances where they involve the loss of privacy within a garden, although these rights are rarely exercised.

If the sewer or pipeline interferes with a building plot, by preventing or reducing the size of the potential development, compensation will be based on the usual before and after valuations.

In addition to the payment for the sewer or pipeline, separate payments are made for surface chambers. The amount will depend upon the precise location, with more being paid for chambers within the cultivated area than for those within tracks or fence lines.

There is no general established payment across the industry, and there is quite a wide variation in the methods adopted by different companies.

Compensation is also payable, but at a lesser rate, for buried manholes: see *Felthouse v Cannock Rural District Council* and *Markland v Cannock Rural District Council* (1973).

No form of payment is made to agricultural tenants to cover the permanent presence of sewers or water pipes, despite such payments being common for pipelines carrying other materials. In any case most tenancy agreements except and reserve all wayleave and easement payments to the landlord. Occupiers are of course entitled to compensation arising out of the construction of the sewer or pipeline.

Special provisions apply to land containing minerals. These are set out in Schedule 14 to the Water Industry Act 1991.

## Gas and other pipelines

The valuer's role in the calculation of compensation for permanent damage for gas and other pipelines tends to be more limited than for sewers and water pipes. The same type of formula is used but often the land value is subject to a generous minimum level. The percentage value is usually around 80% and the width is specified in the deed of grant, so there is less scope for disagreement, especially as the promoters of such schemes invariably agree terms beforehand with the CLA and NFU. These are on an industry-wide basis in the case of gas, now predominantly National Grid, the latest general revision for which was in 2005; and are agreed for each individual scheme in the case of oil and chemical pipelines.

Nevertheless, the valuer should consider whether the terms are adequate for his client and should pay particular regard to the possible loss of development potential and safety restrictions. Most promoters include the right for the claimant to be awarded additional compensation in the future if development, including mineral extraction, is prevented by the pipeline, but such clauses need careful drafting. For safety reasons, some development is being prevented and additional restrictions imposed by the Health and Safety Executive over a greater width even than that envisaged by the promoter. In cases where such additional restrictions are possible, the parties should seek the views of the Health and Safety Executive at an early stage and should base the compensation and the restrictions in the deed of grant thereon: see below: Preparing for the scheme.

Areas required for pressure-reducing stations or valve chambers tend to be purchased on a freehold basis.

The promoters of gas and other pipelines normally make a payment in addition to the easement payment on a per metre run basis to both owners and occupiers. The owner's payment tends to be nominal, the occupier's is based on the size of the pipe. Owner–occupiers of course get both payments, both of which are subject to a minimum. Usually this is stated as being in return for prompt co-operation with the project rather than as compensation, and does not attract fees for the valuer.

## Electricity pylons, poles and fibre optic cables

While the promoters of electricity projects have extensive compulsory powers they prefer to negotiate private agreements with claimants on either an annual or permanent basis.

Annual wayleaves are more common, and are the norm for smaller apparatus and are subject to annual payments of rent and compensation which are revised each year by agreement between the electricity industry and the CLA and NFU. The rental payments are made to the landowner and depend upon the size and type of apparatus concerned. The annual compensation is paid to the occupier and is dependent, not only on the size and type of apparatus, but also on the type of land, with different rates for grassland and arable, with a further uplift for various specialist categories such as orchards, hop gardens and double-cropped market garden land, paddocks and strip grazing. There are also reduced payments for smaller apparatus in hedgerows, irrespective of the crop growing on the adjoining land. These payments are based on five criteria as set out in the ADAS 1994 Study, as amended, and are subject to change based on the overall profitability of farming.

The five criteria are:

1. loss of output on the uncropped area
2. time wasted when using machinery around the structure
3. wasted seeds, fertiliser and sprays
4. yield loss in the vicinity of the structure
5. weed control around the structure.

In practice the current payments are generally slightly above the ADAS rates. Owner-occupiers receive both rental and compensation payments. Certain landlords receive the compensation in addition to the rent under the terms of their tenancy agreements. Annual wayleaves are usually subject to not more than 12 months' notice.

Permanent or capital wayleaves apply where the parties agree that the equipment shall remain on the land in perpetuity in exchange for a one-off capital payment.

In *Clouds Estate Trustees* v *Southern Electricity Board* [1983], it was accepted that the annual payments agreed with the CLA and NFU were based on agricultural values with no allowance for any loss of amenity. Accordingly, where the claimant's property suffers loss of amenity, such as that caused by a large pylon close to a dwelling, the valuer should negotiate higher rental payments to reflect such loss. If agreement cannot be reached, the claimant should consider serving notice terminating the annual wayleave, whereupon the promoter will have to apply for compulsory powers and then the usual compensation will be payable for injurious affection.

The annual nature of such wayleaves is relevant when the apparatus is found to interfere with subsequent proposals for development. Again, the claimant should consider serving a termination notice, whereupon the promoter either will have to move the apparatus so that it does not interfere with the proposed development or will have to pay compensation for the loss in development value. As an alternative, a permanent wayleave can incorporate a clause providing for future compensation for loss of development value, but such clauses need careful drafting: see *Turris Investments Ltd* v *Central Electricity Generating Board* [1981] and *Mayclose Ltd* v *Central Electricity Generating Board* [1987].

While the capital sum obtained for a permanent wayleave is attractive to claimants, it will be seen from the foregoing that annual wayleaves offer flexibility which can be valuable, particularly where there is the possibility of future development. However, a capital claim may be more appropriate where amenity has been damaged.

Because the annual compensation payments are based on the use to which the land is put, from time to time they need to be checked by the claimant, who should notify the promoter of any changes. For instance, the annual compensation due to the claimant should be increased where hedges have been removed with the result that apparatus which was previously located in the hedge is now in the centre of an enlarged field; or where grassland has been ploughed and converted to arable; or where arable land has been planted with orchards or double cropped with market-garden produce.

## Telecommunications

The Telecommunications Act 1984 was amended and updated by the Communication Act 2003, section 106 of which introduced the Communication Code. There are now well in excess of 100 Communication Code System Operators, all of whom are licensed under the code. This gives them compulsory powers to acquire rights to install their equipment. Disputes are resolved by application to the county court.

It should be noted that where the Act deals with wayleaves it is concerned almost exclusively with the occupier rather than the owner. The occupier's valuer should check the tenancy agreement, which will in most cases prohibit the tenant from entering into such arrangements. If it does, he should consult with the owner or his agent.

There are provisions for the promoter and the claimant to opt out of most, but importantly not all, of the provisions of the code in arriving at an agreement. There are other provisions covering the removal of the equipment at the request of the claimant and for the removal of the equipment on its abandonment. There are special provisions for BT equipment, including annually agreed compensation payments for it which are based on the use of the land on which it is situated. There are similar payments for other promoter's equipment. Annual payments can normally be commuted to capital payments but care should be exercised if there is any possibility of the equipment interfering with any long term development of the land. The right to have the equipment removed is normally lost if a capital payment is agreed.

There were separate provisions for fibre optic cables on electricity apparatus, cable television and other communication companies, and cables laid by 186k[1] and its successors. However, there does not

---

1     186k was the then British Gas subsidiary laying loops of fibre optic cables linking major cities following existing high pressure pipelines.

now seem to be much additional work being undertaken by these organisations. Where there are ren
review provisions in the existing agreements, claimants should be encouraged to seek professiona
advice to ensure they achieve comparable rates to those currently on offer from BT and other promoters

## Private easements and wayleaves

There is a wide variety of circumstances in which a private individual or company might require ar
easement or wayleave over the land of another, ranging from a private individual wishing to lay a smal
water pipe through a neighbour's property to a large company requiring an extensive range of services

Where a private individual is able to requisition a public sewer, compensation for permanen
damage will be assessed on the basis described previously. Otherwise compensation for permanent
damage is a matter between the parties. In certain cases, for smaller projects, the claimant might settle
on a neighbourly basis for 50% of the land value, as for public sewers and water pipes, or even less
In others the valuer should consider a ransom strip valuation similar to that in *Stokes* v *Cambridge
Corporation* (1961). In all cases the valuer must consider the precise terms of the deed of grant to ensure
that the rights and liabilities granted and imposed are reflected in the figures.

So far as is practical the terms and conditions for both the easement and the construction work
should be based on those for commercial undertakings because, at the end of the day, the rights being
acquired and the potential for damage during installation are the same.

# Preparing for the scheme

## General

The valuers for both the promoter and the claimant have roles to play during the inception and
planning stages.

In the case of the water, sewerage and telecommunications industries, codes of conduct or practice
are required by statute, and the gas industry, in particular, periodically negotiates such an agreement
with the CLA and NFU or their Scottish and Welsh equivalents. This also includes specific codes in the
form of letters of undertakings which apply to all their pipelines. These undertakings will cover most
of the points described below, but the valuer must still have regard to the particular circumstances of
each claimant and to any amendments that are required to the relevant undertakings as a result.

While such preparations are necessary to mitigate the physical damage resulting, their primary
purpose is to anticipate and avoid problems to the mutual benefit of both promoter and claimant. The
promoter's valuer will need to work with the agricultural liaison officer (ALO) who is usually
appointed to advise the engineers on practical agricultural matters, or the engineers direct, if one is
not. However, it must be accepted that from the engineers' and promoter's point of view the point of
the exercise is to get the pipe in the ground or the cable in the air and often such advice as may be
given is ignored for economical or practical reasons. This can give rise to continuing compensation
matters long after the engineers have completed the project and are no longer involved.

The claimant's valuer will have a similar role when advising an occupier, who may well have little
experience of civil engineering schemes. However, close liaison between the valuers is essential from
the outset, continuing through the construction period and beyond

The most important points that valuers need to consider are listed below. They apply in whole or in
part to all schemes whether involving the installation of above or below ground equipment. Any

eference in these pages to pipeline services or equipment should be taken as the same, *pari passu*. On major schemes, great attention will have to be paid to all or the majority of these, together with others which might be specific to that particular scheme or the land affected. On others, certain points may not be relevant, but the valuer should be wary as small schemes can give rise to major problems, and the degree of experience and commitment of promoters and their engineers may not be as great as with major projects.

This aspect of the valuer's role should not be underestimated, for it is generally far better to anticipate and prevent a problem than to pay compensation for something which could have been prevented. For instance, most farmers would prefer a live cow rather than compensation for its death following a fall into a pipeline trench because of damaged fencing.

In addition to the specific preparations listed, the valuer must bear in mind those items which will be need to be included in the compensation claim at a later date, for certain of those will require monitoring as the project proceeds. For instance, the difficulties and costs involved in farming small areas of land which are wholly or partly cut off from the remainder of the farm must be considered.

# Route

The promoter will normally have planned the route for the scheme before the claimants are approached. This will have been on the basis of the shortest way between any two given points, subject to various environmental or planning constraints. These will be with particular regard to SSSIs and similar designations, known badger sets, the presence of newts and similar environmental considerations. However, variations may be necessary for a variety of reasons.

Even on the smallest schemes it is necessary to avoid dwellings, buildings and other structures and also locations where such development is expected in the future. On the larger schemes this assumes greater significance, as the deed of grant may impose a limitation, if not a total prohibition, on building within the easement width. The Health and Safety Executive may oppose future development on an even greater width than the easement in the deed of grant. Provision to eliminate such impositions can be made by increasing the thickness of the pipes or taking alternative security precautions, such as placing concrete slabs over the pipe, all of which should be discussed with the promoter before consent is given. However, bear in mind such alternatives add to the initial cost of the project, and the valuer will only persuade the promoter if there is a realistic prospect of such developments taking place in the foreseeable future.

If possible, woodland, fruit orchards and trees, particularly if subject to a preservation order, should be avoided and care should be taken to lay pipelines at sufficient distance to avoid damage to root systems. The deed of grant may limit the size of trees that can be replanted, or may prohibit such replanting altogether within specified distances from the pipeline.

A claimant may persuade a promoter to change the route for other reasons, such as the need to avoid existing drainage systems, but this is less likely on the larger schemes, where the costs of increasing the length, or even installing bends to avoid such obstacles, can be considerable. Also it must be remembered that gravity sewers have to be laid to a gradient and that promoters prefer to avoid laying services in public roads because of the costs involved, and the risks involved from other promoters damaging the service when installing or working on their own installations.

A large number of underground pipelines are now constructed of plastic which is difficult to detect, thereby giving rise to problems if the precise position needs to be determined at a later date. This is no less important on the smaller schemes where a leaking joint between two lengths of alkathene water pipe can be hard to find.

## Surface structures

Consideration needs to be given to the location of surface structures such as manholes, marker post aerial marker posts, stiles and gates.

Sewer manholes and surface chambers on water pipelines have to be provided at regular interval or at particular locations determined by contours and need to be borne in mind when the route under discussion. Ideally they should be located in or adjoining farm roads, tracks, hedges or fenc lines. However, if a claimant is faced with a significant length of pipeline he may have to becom resigned to a number within farmable land. While such structures appear intrusive when positione in the middle of a field, they can cause less hindrance there than in headlands, and particularly in fiel corners where agricultural machinery passes and turns, and especially so if insufficient width is le for machinery to pass between the structure concerned and the field edge.

Some promoters are happy to bury manholes, which is of obvious benefit to claimants, althoug triangulation points should be considered to enable easy relocation. Otherwise there is merit in surfac structures being raised well above ground level and identified by tall markers which can be spotte by the operators of farm machinery when the field is cropped, thereby avoiding damage by or to th machinery concerned.

Marker posts, whether for ground or aerial detection, are more easily accommodated, as are th stiles and gates which some promoters use to indicate the presence of underground pipelines and t facilitate inspections. These can be located within field boundaries, although consideration must b given to the possibility of such boundaries changing in the future, when the claimant should contac the promoters to arrange for their removal or re-location.

## Promoter's representative

Before the project starts, claimants should be provided with the name, address and telephone numbe of the promoter's representative. On larger schemes ALOs may be provided, but otherwise th representative is usually the resident engineer. Provision should be made for resolving problems tha arise overnight and at weekends.

## Trial investigations

Before the scheme begins, it may be necessary for the promoter to carry out trial borings or trial pits tc determine the type of underground strata or to locate existing pipelines or underground services. Such borings should be carried out only after consultation between the parties, with a record of conditior being prepared and with compensation being paid for all damage caused. Such compensation is usually confined to surface damage, but some promoters will make a payment for such investigations, particularly if lengthy accesses are involved.

## Timing of the work

The valuer should establish the likely date of entry and the probable duration of the scheme. This should have been set out in the initial offer letter. Provision should be made for the promoter to serve a notice of intention to commence work, which should be of not less than 14 days duration with the ideal being nearer 28 days or as long as is reasonably practical. On the larger schemes the occupier should be made aware of the promoter's detailed programme, which should refer to all movements of equipment, machinery and vehicles required for construction purposes, and should be kept in touch with progress and changes this should be part of the role of the ALO, if there is one.

Ideally, schemes should be carried out when the weather is likely to be dry, between April and October, thereby minimising the risk of long term damage to the soil.

## Single farm payments

The promoters should lodge a copy of the proposed route of all schemes with the appropriate regional service centres. However, this does not always happen and it is incumbent on the claimant and or his valuer to ensure that the appropriate regional service centre is aware of the proposals and how they will impinge on the individual holding. DEFRA have produced guidance notes, which claimants should obtain at an early opportunity if these are not provided by the promoters. Depending on when entry takes place, claimants must conform to any regulations as to the timing of formal notification to the service centre of the work commencing. Derogation from some of the regulations concerning cross compliance must be sought as necessary. As with many government department, regulations are subject to alteration and it cannot therefore be stressed too much that contact with the service centres should be made at the earliest opportunity and then maintained throughout the scheme. Should the scheme affect the amount of payment, full documentation of all correspondence with the appropriate authority must be presented to the promoter if a claim is to be successful.

## Land drainage

This is potentially one of, if not the most, important facets of pipeline construction so far as the claimant is concerned. How it is addressed can hold the key to the long term effect the pipeline has, not only on the working areas, but also on the adjoining land.

In nearly all large scale projects pre-entry drainage should be agreed with the promoter, and to assist with this copies of any plans of existing drainage should be supplied at the earliest opportunity.

It is now accepted practice to allow for cut off, header, drains to be laid on the high side as close as possible to where the working width fence will be, or even outside it. The purpose of these drains in to connect to all existing drains with properly manufactured connections, thereby maintaining the drainage of the land outside the working area, and at the same time to prevent the working area becoming waterlogged which is something that would only add to likely long term problems with soil structure. The outlet for such drains should be carefully considered so as to locate it within the claimant's ownership wherever possible. If this cannot be achieved, it will be necessary for the promoter to secure the grant of any necessary easements by the adjoining owners to permit such drainage across their land.

Depending on the width of the working area, additional post-construction drains will need to be laid parallel to the pipeline to discharge into the same outlet. This work should be undertaken before the top soil is reinstated and the ground is subsoiled. Care must be taken to ensure the subsoiling does not damage the new drains. All such drains, including the initial header drains, should have clean permeable aggregate over them to the top of the subsoil to assist with the working of the new drains and to enhance the effect of the subsoiling.

When it is necessary for any drains to cross the excavated trench, great care must be exercised to ensure they are properly supported, preferably on a concrete lintel cut well back into unaffected ground on either side of the excavation.

Although it is now normal practice for the promoter to accept responsibility for the function of drains or drainage systems which can be shown to have been affected by the construction of a pipeline, it is essential for the valuer to ensure that there are no time constraints placed on this undertaking. Drainage problems can manifest themselves several years after construction, and it is most important that there should be no time-limits on the promoter's liabilities.

On smaller schemes and those not requiring a formal deed the valuer should ensure that the promoter at best follows the above procedure, or at least undertakes to reinstate all affected drains or drainage systems. Where header drains are not installed, it is essential that the position of all drains cut by the excavation should be marked by pegs or similar on both working width fences. If they are to be re-connected over the excavation they must be adequately supported on some form of bridging, adequately cut back into unaffected ground.

When land is let on anything less than a full secure tenancy, the owner as well as the occupier should be given the opportunity to be involved in all matters relating to drainage.

At all times the claimant must be given the opportunity to inspect all drainage work as it progresses.

The valuer should require the promoter to provide, on completion of the scheme, a fully annotated plan indicating the position of all drains connected to the header drain and the position of all new drains.

It is also recommended that the valuer ensures there is an effective disputes resolution procedure if agreement cannot be reached on the method of reinstating drains following construction of the scheme.

## Soil protection

This, or rather the lack of it, is probably the second most important consideration — after land drainage — with regard to the long term effects of pipeline construction. Although techniques and working procedures have greatly improved over the years, the greatest care must be taken by the claimant to ensure no short cuts are taken.

It is, or should be, standard practice on all but the smallest schemes, to strip whatever depth of topsoil there is and store it on the edge of the working area. This should be the first action by the contractor once the fences have been erected and any pre-construction drainage installed. Subsoil from the excavation must be kept separately, preferably on the other side of the excavation to prevent any risk of the two being mixed or contractor's machinery and vehicles coming into contact with the topsoil, or worse still, running on it.

Wet weather cannot necessarily be avoided, but every effort should be made to avoid the need to do anything with the topsoil when it is wet. This may in extreme conditions mean postponing the final spreading of the top soil until the following summer. Working with wet or cold topsoil or subsoil gives rise to more long term soil structure problems than any other factor; whereas if it is left until conditions are dry, such problems can largely be avoided.

Whenever the final reinstatement takes place, the working area should be subsoiled before the top soil is replaced. This should be done at a depth of at least 450mm. To be effective it should only be done when the ground is dry and when all construction traffic has ceased to use the working area.

The protection of the top soil is also important when trenchless or narrow trench methods are used for small diameter pipes, such as private water supplies. In such schemes, as well as the installation of small overhead or underground electricity supplies, weather conditions should be the overriding factor in determining when the work should be undertaken to minimise damage, as the topsoil is not normally nor necessarily stripped.

Other than when the excavation is through rock which must be removed from site, it should not be necessary for, nor should the claimant agree to, the removal of any subsoil from the site. Any surplus not compacted into the excavation can be lost over the remainder of the working area before final reinstatement takes place.

Provision should also be made by the promoter to ensure that the contractor sprays any weed growth on the top soil mound to prevent seeding taking place, otherwise the problem will be perpetuated after reinstatement.

## Depth

It is obviously important that pipelines, buried manholes, cables and other underground equipment should be laid at sufficient depth to avoid damage to or by normal agricultural operations. This is generally regarded as being a minimum of 1.1 m cover, measured from ground level.

Both parties should have regard to the possibility of ground levels changing with erosion, and the claimant should take care to avoid subsequent operations which materially affect the cover, which would be in breach of the normal covenants in the deed.

In rare situations, where a claimant has installed or is planning an unusually deep tile drainage or irrigation system, a greater depth might be necessary, but generally an excessive depth would be resisted by the promoter.

## Ditches

Problems can arise where underground equipment passes beneath a ditch if the ditch is subsequently cleaned, enlarged or deepened. The sound bottom of the ditch, which is often below the apparent bottom, should be located. The equipment should be protected by a concrete raft at least 150 mm thick with its top being a minimum of 300 mm below the true clean bottom of the ditch giving minimum cover of 450 mm from the bottom of the ditch.

## Stacking and protecting turf and plants

Subject to the likely timetable and weather conditions, where a scheme involves non-agricultural land, particularly domestic gardens and sports fields, it is usually best for the turf and as many of the plants as possible to be dug up beforehand to be stored on one side, preferably on plastic sheeting, tended and watered, for replacement after the promoter has finished. Such action is possible only if the scheme can be carried out swiftly, but it does enable mown grass to be left with a more uniform appearance than if turfs are brought in, and enables gardens to be left with their original plants rather than with less mature replacements. Nevertheless, some losses may occur, particularly in dry weather. By and large however, promoters do endeavour, if possible, to avoid crossing gardens and playing fields.

## Working widths

It is important that adequate width be granted to enable the scheme to be executed properly and it is rarely in either the promoter's or claimant's interest for this to be reduced. The width must accommodate, separately, the excavated top soil, the subsoil, any base rock, provide access for the promoter's machinery, space to assemble the pipes, cables or other equipment, and space for the machinery to lower the equipment into the ground or raise it on to poles. If insufficient width is allowed, there is a risk of damage to crops and soil outside the width caused by materials falling through the fences, and damage to the excavated and stored top soil either through compaction by vehicles running on it or by mixing with subsoil.

In certain places, particularly adjoining roads, railways and rivers, additional working areas might be required. These should be negotiated beforehand, and where they are they should be treated as part of the agreed working area, attracting surface damage compensation accordingly. Where such areas are requested after construction has begun — normally at the request of the contractors — a rent should be paid in addition to surface damage compensation.

## Fences

The limits of the working area and any access points should be fenced to a specification agreed between the parties beforehand. This is essential where the land on either side is grazed by livestock.

Fences are equally important in arable land and should be provided to ensure that the promoter's contractors stay within the working area. While a lesser specification may suffice, such fences should comprise at least post and wire. Normally in a mixed farming area the fence should be to the same standard as for grassland to prevent, in the event of livestock escaping on to the working area, damage being caused to crops growing outside the fences.

Where existing field boundary fences are affected, provision must be made for straining posts on either side of the breach to maintain tension. Again, this is important in fields grazed by livestock, but it applies none the less in arable fields, as fences deteriorate once tension is lost.

Where horses graze the land, special fencing consisting of at least a top wooden rail for both visibility and strength should be used. Some promoters avoid barbed wire on such fences, but a strand just above the rail stops it being damaged by rubbing or being chewed.

As part of the environmental survey during the planning stage it will have been necessary for the promoters to establish if the route may affect any known colonies of newts. If it does it will be necessary to construct and control special newt fencing to protect the newts from being affected by construction works.

## Access points for promoter

On most projects the promoter will have sufficient access to the working area from public highways. Any additional access points should be the subject of prior agreement, incorporating the same terms as apply to the working width, and promoters should be obliged to pay accordingly for such accesses. Access beyond what is agreed should be prohibited because of the risk of damage occurring.

## Access for claimant

Where a field is severed by a scheme, the parties must decide how the cut-off portion is to be dealt with if there is no alternative means of access. If the cut-off area is small, it will probably not be in the interest of either party to have access across the working area, and compensation should be paid for any resulting loss of crop arising on the isolated portion. Where this is the case, agreement should be eached before the fences are erected. Alternatively, access points might be left to enable the claimant to cross the working area so that he can continue to farm the cut-off portion. The provision of a supply of water will be an important consideration in livestock areas, and if difficult could be a reason for not providing access across the working area. These points need careful planning, particularly if the land is to be used by livestock, in which case a system of gates will be required, or if unusually large or heavy machinery, such as combine harvesters, are to cross, an appreciable width and substantial ridging of the trench might be necessary. Additionally, there will be times when even these accesses need to be closed, and it is important for the parties to liaise over such closure in advance to prevent substantial losses occurring.

## Continuity of supplies

If the scheme crosses the line of any services, such as water, electricity, gas, drainage or telephone, arrangements must be made for these to be protected or reinstated without delay. Their precise position needs to be located beforehand whenever possible. The existence of water supply pipes in particular is often unknown, and their severance can pass unnoticed and/or unrecorded. To guard against this, claimants should be encouraged to check their water meters at frequent intervals as excavations continue and afterwards.

## Record of condition

Shortly before the scheme commences, the promoter's valuer should make a record of condition of the working width, access points and any features nearby which risk being damaged. This record should be as full as possible and should include photographs. The record and photographs should be sent to the claimant's valuer, ideally before work commences, for approval with or without amendments and further photographs.

It would be in the promoter's interest for detailed schedules of condition to be prepared for residential or other buildings which lie close to the scheme. The valuer should arrange for a building survey where such buildings are old, of great value or have a known history of structural problems.

## Trees

Trees should be removed only after prior consultation with the owner and both parties should ensure that there is no breach of any tree preservation order nor of the legislation regarding trees in conservation areas. A felling licence may be required. Prior agreement should be reached as to whether the promoter is to remove the tree completely, paying compensation for its timber value, cut it up for firewood, or is to leave it for the claimant to dispose of.

In order to comply with current legislation, and in particular the Wildlife and Countryside Act 1981 and Hedgerow Regulations 1997 (SI 1997 No 1160), trees and hedges cannot be removed between certain dates depending on location, but generally between March and August. This means that they may well have to be removed in advance of the construction work and the erection of the working width fences. Subject to agreement as to means of access, sympathetic consideration must be given to the needs of the promoter as to the timing of this work.

## Working hours

It is usually in the interests of both promoter and claimant that the scheme be completed and the lan reinstated as quickly as possible, even when this involves work at weekends and after dusk. Howeve where all or part of a scheme is within sight or sound of domestic dwellings, the promoter should b encouraged to cease or curtail activities during unsociable hours.

## Reinstatement of roads and paths

Generally, the excavated width across roads and paths will be reinstated. However, a precise match c materials is rarely possible and in particularly sensitive areas of high amenity value, usually involvin residential rather than agricultural property, it might be necessary to resurface the entire roadway o path concerned.

## Control of dust

In dry periods, the promoter's machinery will probably raise dust which can have a deleterious effec on domestic property, market garden crops — including fruit — and on grass for silage and grazin by livestock. On larger schemes, the promoter should provide water bowsers and sprinklers to dam down such dust.

## Noise, vibration and explosives

The noise and vibration caused by the promoter's machinery can be a problem, especially durin unsociable hours, when close to residential property. In such cases the promoter should reduce nois and vibration by restricting working hours and by the use of alternative machinery. Explosives shoul be used only when the parties have agreed beforehand, and in particular when any third parties, suc as neighbouring residents, have been notified.

## Support structure

Particular care should be taken in urban areas with regard to pipes to ensure that suitable temporary support is given, and that other protective measures are taken, for buildings or structures adjacent tc the pipe trench, both during and, most importantly, subsequent to, construction to guard agains subsidence.

## Infectious diseases

In the (however unlikely) event of a scheme or part of a scheme falling within an area declared to be infected by foot and mouth disease, fowl pest, bird flu, swine fever, rhizomania or any other notifiable disease, the scheme should only continue with the approval of DEFRA.

## Wet weather working

Most promoters endeavour to plan their construction programme for the drier summer period, normally April to October.

It is not always possible, for a variety of reasons, to complete a project within this time-scale and if it is necessary to over run beyond October most promoters will agree, once the engineering works are complete, to postpone the final reinstatement until the following summer. This is, of course, very much in the long term interest of the claimant and usually of the promoter as well.

Wet weather can be encountered during the normal construction period. Whereas only in the most exceptional circumstances would the promoter agree to suspend work, provided the top soil is protected and final reinstatement only takes place when ground conditions are dry and warm, then the long term effect will be minimised.

## Pollution of water supplies

Promoters should be wary of polluting water supplies, particularly in water catchment areas and in marshy and drained land. Noxious liquids deposited on the land can often seep through the soil into water and be carried long distances by drains. Such pollution can spread over a wide area, posing a threat particularly to livestock and to a wide variety of arable crops through spray irrigation, and can lead to highly damaging pollution of watercourses. Such events are actionable at law and under certain circumstances may lead to criminal prosecution.

## Soil-borne pests and diseases

The promoter should be required to take care to avoid the spread of soil-borne pests and diseases such as potato cyst, eel worm and verticillium wilt, and should be alerted by claimants where such a risk exists.

## Fossils and artefacts

Prior agreement should be reached that any fossils, coins or other objects of value discovered by the promoter will remain the property of the claimant, unless found to be treasure as defined by the Treasure Act 1996.

## Security and the prohibition of certain activities

The promoter should be prohibited from using the working area for purposes or activities other than the scheme itself. Caravans or site offices should not be permitted. Dogs should not be allowed, except in the case of properly controlled guard dogs. Poaching should be prohibited and firearms or shotguns should not be allowed on the working area for any reason. The use of metal detectors should be confined to the locating of underground services. On larger schemes, where numerous people will be employed by the promoter, adequate means of identification should be carried at all times.

## Diary

The promoter's representative and the claimant should keep diaries and photographs of problems that arise as the project proceeds. The diary should include a detailed record of incidents as they occur including dates, times, names of people and the description and registration numbers of vehicles involved, to form the evidence for complaints or to support compensation claims. The claimant should report more serious problems immediately to the promoter's representative and to his own valuer. The importance of notification of serious problems cannot be over stressed. This helps prevent disputes arising, and assists with the eventual agreement on compensation.

## Normal farming practice

The valuer should recommend to the claimant that he should continue to farm his holding in accordance with his normal practice. Schemes can be delayed, deferred or abandoned. Promoters will be very reluctant to consider any claim, unless the principle has been agreed beforehand, for any losses arising from any alteration to the normal farming practice which has resulted from the anticipation of a scheme.

# Surface damage compensation

## General

This covers all aspects of compensation arising from the construction works associated with the scheme.

A very wide variety of items can be included within this heading, some being described in more detail below, and the valuer should take care to ensure that no item of loss or damage is omitted.

The claimant is under an obligation to mitigate his loss, and as a general rule of law the costs of attempted mitigation are recoverable. However, care needs to be taken, particularly during the initial planning of a project; and steps towards mitigation should not be taken until the promoter has given written confirmation of a precise date for entry, and even then only after agreement with the promoter's valuer if the scope for subsequent disputes is to be minimised. If a claimant adjusts his cropping in anticipation of a scheme which is then cancelled or postponed, he could be left with losses for which he is not entitled to compensation.

In exceptional circumstances, compensation is payable where losses result from construction works not on the claimant's land but in an adjoining highway: see *George Whitehouse Ltd (trading as Clarke Bros (Services))* v *Anglian Water Authority* [1978] and *Leonidis* v *Thames Water Authority* (1979).

## Advance payments

Note: This section of the chapter is dealing with surface damage arising from the scheme. Compensation for permanent damage, advance payments (see above, p517), covers the same for the capital payment, but they are separate items.

Claims cannot realistically or practically be formulated until the scheme is complete, or, on larger schemes, at least one year's crop has been lost.

No loss of crops can be claimed before entry has been taken as none has been suffered by the claimant. Similarly, a loss is not automatically suffered if an arable crop is not actually sold direct off the field but stored for future sale or use.

Promoters are reluctant to make advance payments without evidence of an actual claim, but as soon as a claim has been submitted a payment on account should be requested. It is often agreed by the promoter that such a payment should be made if agreement cannot be achieved within a reasonable period of time from submission, normally two to three months.

Interest on compensation claims normally only runs from the date of agreement of the claim.

## Monitoring

The valuer will initially need to check the record of condition prepared by the promoter's valuer. He should subsequently inspect the progress of the scheme as necessary, and should keep in touch with his client so that he is aware of problems that arise. Notes should be taken during these inspections, together with photographs. Problems should be communicated immediately to the promoter's valuer, preferably by fax or post so there will then be a record on file. Joint inspections by the two valuers might be necessary to agree the extent of losses before the evidence is lost, and such agreement should be confirmed in writing. For instance, damage to standing crops by straying stock because of inadequate or damaged fencing should be agreed before those crops are harvested. Photographic records should be taken at the time. If agreement cannot be reached, the claimant's valuer must assemble sufficient evidence to enable him subsequently to prove the loss.

## Loss of arable crops

Loss of arable crops on entry is an obvious head of claim. The area lost may well include strips of at least a metre on either side of the working width fences because of difficulties manoeuvring harvesting equipment, although this is not always the case and will depend on the angle of the fences relative to the rest of the field. It will in any case only apply while the fences remain. In addition, there may be losses outside the fences caused by having to alter the direction of tram lines to accommodate the fenced area.

The yield of the crops in the field will need to be recorded, as will the price obtained for them. The valuer should, in the case of cereal crops, make allowance for the loss of straw as well as of grain.

Depending on the time of entry, it is possible that there will be certain saving in costs such as sprays, fertilisers and harvesting costs. If entry is taken before the crop is sown, or if the scheme extends into a second season, there could in fact be as much as a complete saving of all variable and machinery costs associated with a crop. These savings should be reflected in the amounts claimed. All such costs are normally based on those produced by the Central Association of Agricultural Valuers (CAAV) or farm management books produced by among others Professor John Nix of the Imperial College London, Wye Campus.

Special circumstances arise if schemes such as sewers affect gardens. The valuer would have to ensure that great care is taken by the promoter to minimise the effect of construction works and allowance should be made for the replacement, at the promoter's cost, of all affected plants and turf.

## Loss on grassland

On dairy farms it matters little, so far as actual production is concerned, whether schemes affect land used for summer grazing or the production of silage or hay for winter consumption as the two go together to produce the gross margin achieved from the forage area. Obviously, schemes affecting summer grazing potentially carry a far higher risk of causing disruptive inconvenience and a resulting reduction in production. Only if there is a relatively large scheme affecting a small-holding will it be necessary for the claimant actually to reduce his stocking rate, although it may be necessary to temporarily adjust the number of followers. Provided production figures and costings are accurately recorded, it should be possible to agree the claim based on the gross margins achieved from the forage area. Where they are not, a more simplistic approach will have to be taken on average production figures for the locality. There is a school of thought which does not accept claims based on gross margins for dairy farms, but this is mainly on account of the sums involved. However, if the valuer is to claim on the gross margin basis, full supporting evidence must be available. In extreme cases extensive supplementary feeding might be required and the cost included in the claim, but the promoter's valuer should be informed in advance.

Where less intensive forms of production are affected, such as beef, sheep or suckler cows, the claim should again be based on production figures achieved in the locality, and again it would be reasonable to base these on average gross margins for each enterprise. The value of winter grazing, particular for hill sheep, should be taken into consideration where appropriate.

## Reinstating arable land

Irrespective of how well the promoter may have reinstated the land, additional costs will inevitably be incurred in restoring the soil structure to a similar condition as the unaffected land. Additional subsoiling may be necessary as well as extra cultivations, which will all have to be in addition to the normal cultivations in the field. Depending on the nature of the ground, stone picking, to a greater or lesser extent, is nearly always necessary following the construction of a scheme. There may also be a weed problem if the top soil mound was not correctly sprayed, and this will result in additional spraying of the working area in isolation to the rest of the field. Additional farmyard manure will also probably be required, and all these operations will have to be carried out on a relatively small area in what may be a substantially larger one, giving rise to a lot more turning of machinery. All such additional costs should be recorded and included in the claim.

It should also be borne in mind that it might not be possible to carry out all the necessary operations at the same time as the rest of the field is being prepared and a crop sown. The timing of when the scheme is completed is therefore critical as to whether the work can be done and a crop established. It is better to lose another year's crop on the pipeline than seriously compromise the crop on the whole field. If this is likely, agreement with the promoter's valuer should be sought as a matter of urgency.

## Reseeding grassland

It should be established at an early stage in the scheme whether the claimant will be expected, if he is capable of undertaking the work, to reinstate and re-seed the working area. If he is not, then either the promoter should employ an agricultural contractor to do the work or agree to the claimant employing one directly, at the promoter's expense.

Again, the timing of the work is critical and should not be undertaken too late in the year, nor, for that matter, too early. If the promoter is to organise the work, agreement must be reached as to the seed mixture and fertilisers, and the rates of application. It is normally recommended that at least 50% more seed than usual should be applied to ensure successful germination. It must also be borne in mind that all operations will have to be undertaken in a confined space, as it is essential that the working width fences remain to protect the new grass from stock on the adjoining land.

Very much the same operations will be necessary as for arable land. While the same costings can be used as a base, the confined working area will inevitably give rise to an appreciable uplift. Care will have to be taken in achieving a blending in of the edges of the fenced areas with the undisturbed grassland. Provision should be made for the fences to remain until the grass has been established and will not be damaged by grazing. It is normal practice for the fencing materials to be left for the claimant to remove and have the use of any residual value.

## Additional costs of working severed areas

Fields which are divided by a project invariably take more time to cultivate, tend and harvest than the original undivided field. This is due partly to the extra turning time required, especially for larger machinery, and is increased where the only access to severed portions is across the working width. The extra costs based on the time involved should be included in the claim. Where particular difficulties are envisaged, the claimant's valuer must consider whether he should mitigate his claim by abandoning the crop in the severed areas, where its value may well be less than the extra costs involved. If this appears likely to be the case, agreement with the promoter's valuer should be sought at the earliest opportunity, preferably before construction commences, so that the promoter will not have to provide an access for the claimant over the working area.

## Future losses

However well a project is executed and the land reinstated, some shortfall is inevitable in succeeding crops and, where soil structure has been badly affected, these have been recorded for many years. Also, weed problems such as wild oats, sterile broom, black grass or chick weed can occur. The valuers should preferably inspect the land together each year before harvest, so that the extent of the losses can be agreed.

The valuer should in most circumstances wait at least to the joint inspection in the year following reinstatement before assessing whether a full and final settlement would be appropriate. Promoters should be obliged to pay compensation for losses in succeeding years where these are quantifiable. Equally they are normally amenable to agreeing a lump sum payment to cover future losses were appropriate. Before agreeing to a lump sum payment, the claimant's valuer should check with his client's accountant as to how such a sum would be treated for tax purposes.

Where land has suffered no undue damage and has been properly reinstated, and when it is agreed such a method would be appropriate, a rule of thumb for the calculation of such a lump sum is a 50% loss in the first year after reinstatement, 25% in the next year, and 12.5% in the final year.

In *Smith, Stone & Knight Ltd* v *Birmingham City District Council* [1988] in Official Referee's Business in the High Court, it was held that damage caused by subsidence occurring more than six years after the initial laying of a sewer was not statute-barred.

Valuers should consider, in areas where potatoes (in particular) are grown, that, depending primarily on climatic conditions, significant losses above on pipelines can still occur several years after construction. In these cases it would not be unreasonable for the promoter to agree to a full and final settlement for all crops other than those specifically referred to to cover this eventuality.

In most pipeline deeds there is also provision for further claims arising from drainage problems that can be shown to have arisen from the construction of the pipeline.

## Field boundaries

It should be incumbent on promoters, and is in fact normal practice, that all field boundaries should be replaced like for like on completion of the scheme.

Before major schemes commence, one of the environmental surveys which will have been carried out should have noted the various types of plants making up each hedge, and the same proportion should be re-planted. However, wherever possible the claimant should request that no elder should be replaced. Although elder has become established in many hedges, it really is a weed species and can soon dominate a newly planted hedge as it grows far faster than normal hedge plants. Care must be taken to ensure only healthy plants are used and that planting takes place in the winter months only, even if this means the work has to take place well after the scheme has otherwise been completed.

The replacement hedges should be protected with stock proof fences and, where rabbits are in evidence, rabbit netting should be used. It should be taken across both ends of the fence to prevent rabbits from gaining access down the hedge line. This may seem an obvious precaution but is often overlooked. Hedges should be maintained by the promoter for at least three years after planting, this is easily done with modern granular weed killer. If the promoter is not prepared to carry this out, an appropriate sum should be included in the final claim to compensate the claimant for doing the work himself. This should be in addition to the normal claim for maintaining the hedges after the initial period.

Special provision should be made when hedges have been allowed to grow tall in order to provide wind breaks for orchards, specialist crops or privacy for gardens. The method of replacement should be discussed and agreed with the promoter before the scheme commences.

Care must also be taken by valuers where stone walls have been affected. The stone should have been placed alongside the working width fences when being taken down. The promoter should employ an experienced dry stone wall contractor to rebuild walls. Quite often there is insufficient stone available at the end of the scheme, and the claimant should insist on only suitable local stone being imported to make up any deficiency. It is important that the contractor use sufficient throughs to ensure the wall is stable. If a re-built wall moves when kicked it needs more stability, but be warned: use the sole of your boot for this, not the toe.

## Disturbance to animals

Certain types of livestock, most notably poultry and to a lesser extent dairy cows, and horses, are susceptible to disturbance caused by noise and unfamiliar activities. Losses in egg and milk production can result and, in extreme cases of panic, livestock can be injured or killed. Such disturbance may be difficult to prove, so care must be taken to collect evidence from experts such as vets when advising the promoter's valuer. The quantum of such loss is often hard to calculate without a detailed study of production over the months and years prior to, and following, the project.

# Deficiency of top soil

Top soil can be lost as a result of removal by the contractor, which strictly speaking should never happen but unfortunately sometimes does, mixing it with subsoil or subsidence in the trench are the most common causes. Usually, supplies of top soil for making good are limited and of suspect quality, but such supply and spreading is an item of claim none the less. It is, however, better to oblige the promoter to carry out the necessary importation. In all cases the claimant must be given an opportunity to inspect the source of the soil to ensure it is suitable. It should also be certified as free of any soil borne diseases. In extreme cases, where large amounts are required, the promoter's valuer might argue that the cost of such work exceeds the amount of the claimant's loss. Such an argument should be resisted as, invariably, such problems arise from subsidence or promoters failing to keep to their undertakings.

In *Lucey (Personal Representative of)* v *Harrogate Corporation* (1963), where a deficiency of top soil was alleged, the Lands Tribunal held that the occupier might please himself whether he restored the top soil, claiming the cost thereof, or whether he tolerated the physical damage, claiming compensation for the resultant losses. However, it would be unwise for a tenant to adopt this view as he could be liable for dilapidations at the end of his tenancy.

# Claimant's time

Even where a claimant employs a valuer, he will spend time on matters related to the project, and on major schemes this can be considerable. On properly managed schemes, the promoter through the ALO and claimant will establish a close working relationship which will operate to their mutual benefit because it should reduce the number and size of problems that arise, and the amount of the compensation claim for surface damage. Accordingly, it is in neither party's interest for such time to be stinted.

During the preparation for the scheme, the claimant will attend meetings with both the promoter and his own valuer to consider the points referred to above. Once work has begun, he will need to inspect the working fences at frequent intervals and keep an eye on the promoter's activities. He will need to be careful to differentiate between time necessarily and properly spent and time when watching engineering works which may be new to him and of interest in their own right. Furthermore, time will be taken on matters such as checking the stripping of soil, gaining access across the working width, checking land drains and their replacement and checking water meters. At the end of the scheme discussions will take place concerning items of final reinstatement, including the reinstatement of field boundaries. Throughout, the claimant will have to liaise with his valuer and provide much of the evidence on which the surface compensation claim is based. It will greatly assist in proving a claim if a detailed diary is kept.

An item should be included for such extra effort on the claimant's part, in accordance with the decision of the Court of Appeal in *Minister of Transport* v *Pettit* (1968). Wherever possible, the heading for the claimant's time, should be supported by evidence such as photographs, diaries, time sheets and similar records: see *Rush & Tompkins Ltd* v *West Kent Main Sewerage Board* (1963).

# Miscellaneous

The claimant's valuer must ensure that a claim is submitted for every item of loss resulting from the scheme. Certain of these might be unusual or unique to the particular claimant concerned.

Compensation has been paid for

- loss of shooting days or a lack of game owing to the construction of a pipeline
- loss of fruit and vegetables rendered unmarketable by a coating of dust
- weed infestation in severed land owing to access being unavailable during the crucial spraying period
- additional travelling time when the main access to a farm was blocked, necessitating a longer alternative
- reduction in crop resulting from an inability to irrigate because of the pollution of a water course
- losses because of interference with livery businesses and farm rides
- loss of profits in a public house during a period when much of the car park was excavated
- loss of privacy within a house when its garden was crossed by a scheme.

## Full and final settlement

Unless a scheme has gone without any hitches, which in itself is rare, and the land has been reinstated in ideal conditions and or the scheme is relatively small, claimants should be wary of initially agreeing compensation in full and final settlement of all claims. Particularly on larger schemes it is recommended that for the year of construction a separate claim should be submitted. Then on the evidence of the quality of reinstatement, and how the crops grow in subsequent years, annual claims should be submitted accordingly. Once the quantifiable losses have clearly reduced and no other problems have manifested themselves, a full and final settlement could be considered. Even then, if possible, reservations should be made with regard to losses arising from subsidence or subsequent drainage problems, unless this is covered in the deed. A further point to bear in mind is that some crops, when grown over pipelines, can be seriously affected several years after construction, this resulting from climatic conditions such as drought or very heavy rainfall affecting soil conditions at critical times in their growth. Crops such as potatoes and soft fruit are particularly susceptible. A proviso should be sought where these crops are likely to be grown to reserve the right to make further claims should they be affected.

# Miscellaneous information

## Betterment

Where schemes are executed under compulsory powers, betterment can be set off against any compensation that would otherwise be payable, whether for permanent or surface damage. Strictly calculated, this can result in a nil figure, and frequently does in the case of sewer schemes affecting residential property. Nevertheless, most promoters are willing to pay some compensation towards surface damage. The Lands Tribunal took a similarly benevolent view towards limited compensation for permanent damage in the case of *Raper* v *Atcham Rural District Council* (1971), and the promoter did not pursue such an argument in the case of *Rush & Tompkins Ltd* v *West Kent Main Sewerage Board* (1963).

# Future planning consents

The claimant's valuer should ensure that provision is made in the deed to cover future planning permissions that may be affected by the scheme. This is a standard clause in most pipelines promoters' deeds, but is not available for sewers nor for most water mains. The clause should allow for compensation to be assessed on the date planning permission is granted — in some older deeds it was as at the date of deed. The promoter will reserve the right to divert the pipeline if this is more economical and can be accommodated within the claimant's ownership, this will involve a free easement for the diverted pipeline, which in itself might cause restrictions to the development.

# Minerals

The working of minerals in land through which a pipeline is constructed is usually covered in the deed by reference to the Mining Code laid down by Part II of the Mines (Working, Facilities and Support) Act 1923. The code sets out a formal procedure which has to be followed or the claimant's rights may be seriously compromised.

# Cathodic protection

Pipelines constructed of metal need cathodic protection to avoid decay caused by ionisation. The equipment required should be the subject of separate negotiations and should be referred to in the deed of grant or be covered by a separate deed. The promoter should take steps to ensure that existing buildings, structures and other installations are protected in accordance with the appropriate British Standard Code of Practice.

# Abandonment

Both parties need to agree the action to be taken in the event of the equipment being abandoned in the future. This should be covered in the deed.

# Contractor's liability

Promoters invariably employ civil engineering contractors to install their equipment and it is not uncommon for such contractors to cause damage to the claimant's property by negligence, nuisance or trespass, and in breach of their contract with the promoter.

While they may have a claim against the contractors concerned, claimants are advised to pursue all such claims against the promoter alone, for in *Smith, Stone & Knight Ltd* v *Birmingham City District Council* [1988] in Official Referee's Business in the High Court, it was held that the promoter could not deny the validity of a statutory claim by seeking to establish that the damage was outside the statute because it was attributable to the negligence of its contractors.

Claimants and their valuers should bear in mind at all times that their privity of contract is with the promoter not the contractor, and should avoid any dealings with the contractors in matters directly related to the scheme. Otherwise there is a risk of them making an inadvertent contractual arrangement under which their claim will lie against the contractor rather than the promoter. Specifically, complaints arising during the scheme should be directed to, and any meetings arranged with, the promoter's representative and not the contractors. By the same token, requests from the contractors for additional working areas or additional access points should be refused unless redirected through the promoter.

It might appear to be to the claimant's financial advantage to negotiate a site office, pipe dump or similar facilities for the sole use of the contractor, but such arrangements should be regarded as quite separate from the scheme. The claimant should ensure that his position is protected, as he will not have any claim against the promoter for losses that result. It is recommended that any such arrangement should only be entered into by the claimant if a legally binding agreement is drawn up by his solicitors and signed before entry is taken, with all rents or similar moneys together with fees paid in advance, and with a bond against subsequent claims to be held by a stakeholder. An agricultural tenant will in all probability be prevented from entering into such an agreement by the terms of his tenancy. If approached, he should inform his landlord and seek his agreement. The landlord would be advised to be a party to any such agreement and can expect to get part of the consideration.

## Interest

Whether this applies and the date it is payable from varies depending on the type of claim as detailed in the Planning and Compensation Act 1991, section 80, and Part 1 of Schedule 18.

It was established in *Taylor* v *North West Water Ltd* (1995), that the Lands Tribunal had the power to award interest on compensation from the date of entry as the acquisition of a pipeline creates a legal interest in the land.

For most schemes, provision is made for interest at the statutory rate on any outstanding compensation for permanent damage calculated from the date of entry. The claimant's valuer should confirm that this is the case, particularly where private easements and wayleaves are being negotiated, and should consider the insertion of a commercial rate of interest in place of the statutory rate wherever possible.

Generally, interest is not paid on compensation for surface damage from the date of entry. Sometimes it is from the date of the submission of the claim. In most instances, however, interest is paid from the date of settlement. If interest is not applicable by statute or deed, valuers should consider the inclusion of such an item to run from the date of the submission of the claim or from a reasonable date thereafter, in order to encourage prompt negotiation and settlement by promoters. This is likely to be resisted. Where possible, the final settlement of compensation should include an agreement that interest will run if payment is not made by a specified date.

## Professional fees

It is normal practice for promoters to agree to pay, by way of a contribution, the reasonable fees for valuers and solicitors in connection with schemes, whether constructed under statute or deed. However, the valuer should at the outset — and before taking any other action — get this confirmed in writing. If it is not, he will have to advise the claimant accordingly and agree with him on the basis to be used. Fees strictly form part of the claimant's claim, either for permanent or surface damage losses and any shortfall in the promoter's contribution would have to be made up by the claimant.

Ryde's Scale was not revised by the valuation officer after the 1996 revaluation because it was said to infringe free competition. However, for many years it formed the basis of fees for valuers advising on wayleaves and easements. Many promoters continue to contribute to fees based on the 1996 Scale subject to a variety of uplifts depending on the nature of the claim.

There always was provision within the scale for fees to exceed it where circumstances were appropriate for a *quantum meriut* basis. This still applies but it is essential that claimants and their valuers are clear at the outset what the basis of fees will be, and these should be agreed in advance with the promoter.

Where the claimant is registered for value added tax (VAT), the promoter's contribution should be to the fee net of tax which the claimant will be able to reclaim in the normal course of business. Where the claimant is not so registered, the promoter's contribution should include VAT.

## Disputes

The Lands Tribunal ultimately remains the forum for the resolution of all claims for permanent loss, and also for surface damage claims on schemes executed under statutory powers. There are now, however, methods other than formal arbitration open for the resolution of claims. The two most commonly used are either independent expert or mediation. In the former, an expert agricultural valuer looks at the evidence and makes a determination. The expert could be appointed by prior agreement between the parties, although the CAAV do offer an appointment service. For mediation it is normal to apply to the RICS Disputes Resolution Service for the appointment of a mediator. The mediator will hear the parties' cases independently of each other and will then endeavour to resolve the differences between them by discussion with them individually. Most promoters provide for disputes on compensation to be resolved by any one of these methods, and the claimant's valuer should ensure such provision is included in the deed of grant.

Neither the independent expert nor the mediator have the power to award costs, and both parties should expect to meet their own, although it has been known for promoters to contribute towards the claimant's costs. The Lands Tribunal and full arbitrations award costs in favour of the successful party, unless the behaviour of the successful party is such that it would be inequitable to do so.

It must also be borne in mind that the Lands Tribunal is one of the most expensive forms of dispute resolution, and that route should only be considered if a most important point of principle is involved and/or if the claimant has the backing of a national organisation. The Lands Tribunal can award only one lump sum to cover past and future losses, which may be a disadvantage if surface damage is likely to continue.

It must be borne in mind, if an application is made to the Lands Tribunal or arbitration, that both demand a far higher standard of proof, with considerable more and exhaustive detail being required, than would be called for if simply negotiating with the other party's valuer, or through mediation or independent expert.

Claimants and their valuers are, however, to be urged always to endeavour to settle disputes without recourse to third party resolution because unless the sums involved are substantial the costs can well outweigh any possible gain. The claimant's valuer should always be aware that the burden of proof in a dispute lies with the claimant.

Claimants should be advised that the larger commercial promoters offer compensation for permanent damage on a higher basis than that awarded by the Lands Tribunal, see *Wells-Kendrew* v *British Gas Corporation* (1974) and *Sanders* v *Esso Petroleum Co Ltd* (1962).

# Formulating the claim

## Permanent damage

It is only necessary to prepare a formal claim where the rights are being taken under statute or other than by deed. In all other cases the promoter will normally, as part of his initial approach, make a formal offer for the easement rights required, or in the case of the electricity industry, the annual payment pertaining to the equipment. These offers can of course be negotiated, but a formal claim as such will not be necessary.

Where a claim has to be submitted it should clearly set out the length of the pipeline on the claimant's property quoting as necessary the Ordnance Survey numbers of the fields involved, supported if possible, by a plan.

It should show the easement width claimed and the vacant possession value placed on the land, as well as the percentage of that value on which the claim is based. It would be advantageous if these could be agreed with the promoter's valuer beforehand.

Particularly in the case of sewers, the position of manholes should be recorded and the type of field they are in, arable, grassland, track, yard or field boundary, and if substantial, their dimensions together with a schedule of what payments are claimed setting out the amount for each one.

The total amount of the claim should be given at the end. It is surprising how often valuers fail to give a total figure, which is aggravating for the promoter's valuer.

The basis upon which fees on the settled figure, are to be claimed should be set out, as well as whether or not the claimant is registered for VAT. The fee should be submitted when the compensation is agreed for a deed and not delayed until it is completed, which could be many months or years down the line.

## Surface damage compensation

The claim for the period of construction should be itemised field by field. It should clearly show the area claimed and how it is arrived at, and give separately the areas severed if appropriate.

The cropping should be given and how the claim is formulated, crop production per hectare including both grain and straw for cereals, together with the price claimed or stocking rate, if in grass, and how gross margins, if claimed, are arrived at.

Whether a silage crop had to be taken early to facilitate the start of the scheme, should also be included if appropriate.

There should always be sufficient information provided to enable the promoter's valuer to follow precisely how the amounts claimed have been arrived at. He may not agree with the figures but it should not be left to him to have to work out how the figures claimed had been formulated.

Where items are included which refer to dealings with a third party, documentary evidence should be included.

If claims are made for additional purchases of feed stuffs or particular sprays, copies of the appropriate invoices should be included.

If advice has to be sought from other professionals full details should be included with copies of any relevant correspondence and invoices.

Where appropriate any savings in costs should be clearly set out.

Items should be included for the claimant's time in connection with the scheme, preferably supported by a dairy and any specific items of inconvenience.

An item should be included for the future cost of maintaining any replanted hedges after the promoter's initial period of maintenance.

The cost of reinstating the working area including additional cultivations, additional harvesting or spraying costs should also be included.

If the scheme has progressed in such a way that a full and final claim can be formulated, it must clearly show what the future cropping of each field will be and what the percentage loss in each year will be.

If it is decided to claim — at least initially — on an annual basis, this should be submitted following, preferably, a joint inspection with the promoter's valuer, or, if this is not practical, after the claimant's valuer has made his own inspection to assess the likely losses. Again, how the claim is formulated should be clearly shown.

All claims should clearly show the basis on which it is expected fees will be paid by the promoter.

Claims should also contain reservations for any items which subsequently come to light but which are omitted because they are not evident; examples are such things as losses from drainage problems, soil structure problems or subsidence.

Again the total amount of the claim should be given.

Finally the claim should be signed and dated.

# Assessments for Insurance Purposes

The assessment or valuation of buildings and plant and machinery for insurance purposes is a skill that is not regarded with sufficient importance. To be able to deal with the subject competently a valuation surveyor needs to have, *inter alia*, a good knowledge of the asset to be valued and the basis upon which an insurance policy will compensate an insured party if the asset is destroyed.

Although a building society or bank might require an insurance valuation of a residential property, as part of a mortgage valuation, so that insurance is on the basis of an accurate sum insured, such instructions are normally relatively straightforward. The only exception to this would be where the subject property to be assessed for insurance is a listed building where recognition would need to be given to particularly expensive features, perhaps of some antiquity. The valuation of listed buildings for insurance is beyond the scope of this chapter because such properties require an assessment by specialists including a quantity surveyor. A list of references for those seeking material on the insurance of listed buildings is given at the end of the chapter. In particular, *Fire Insurance Law & Claims* by RM Walmsley is an excellent manual used by insurers and loss adjusters alike.

The most problematical valuations for insurance tend to be those where the subject-matter is in the commercial or industrial sector. The valuation of commercial and industrial buildings can be complex and potentially risky to the valuation surveyor because of the large sums involved. If a mistake is made it could be costly, particularly if average applies. (Average is explained later in the chapter.)

When valuing any asset for insurance purposes it is good advice to look carefully at the result of the calculations to see if they feel right. Bear in mind that once the valuation report has been sent to the client and his advisers it is likely that they will act on it and the valuation amount will become the sum insured. If the valuation was incorrect, the sum insured insufficient and the building destroyed by a fire then the insurance company will only wish to pay an amount up to the sum insured. In such circumstances there is every likelihood that your client will look to you to make up any shortfall.

## Reinstatement cost and indemnity value

In commercial insurance there are two main methods of insuring assets against loss by fire or similar perils. These are commonly referred to as reinstatement cost and indemnity value.

Most insurance policies today are arranged on a reinstatement basis, effectively compensating the insured on the basis of a new asset for an old one. The concept of indemnity value is still in existence but it is usually as a basis to calculate the amount of a loss where the insured does not intend to reinstate. This will be examined later.

# Reinstatement memorandum

Insurance arranged on a reinstatement basis provides the optimum level of cover but, to be effective the insurance policy must contain a reinstatement memorandum. The memorandum provides that the amount payable in respect of any destroyed item will be the cost of reinstatement of the item destroyed or damaged subject to certain conditions.

Reinstatement is defined in the memorandum as:

The carrying out of the after-mentioned work, namely:

(a)  Where property is destroyed the rebuilding of the property, if a building, or, in the case of other property its replacement by similar property, in either case in a condition equal to but not better or more extensive than its condition when new.

(b)  Where property is damaged, the repair of the damage and the restoration of the damaged portion of the property to a condition substantially the same as but not better or more extensive than its condition when new.

A reinstatement policy goes further than an indemnity policy, which places the insured in the same position as he was before the fire, since it allows him to rebuild, replace or repair a property to a condition equal to, or the same as, its condition when new.

# Reinstatement cost

When instructions to value an old factory are received and the valuation surveyor is asked to prepare a valuation on a reinstatement basis it is easy to make the wrong assumptions as to the correct approach. A common fault in a valuation of this type of property is to assume that, because the building to be valued has stone walls, wooden floors and a slated roof, a reinstatement or replacement would take the form of a modern building. In the practical sense if the stone building were to be destroyed, it is certainly likely that it will be replaced by a modern structure. However, unless you are directed by your client, supported by his insurance broker and preferably the insurers of the building as well, a prudent valuation surveyor would prepare a calculation of the reinstatement cost of the building based on its existing materials and to its existing design. The reason for this is simple. Most buildings affected by a fire are damaged not destroyed. It is likely that the damaged stone walls and roof will be repaired using existing materials. In fact it could be impossible to introduce modern, cheaper materials, such as steel or cladding, into the repair. This being the case, the valuation should be assessed on a reinstatement basis as originally constructed. If a surveyor recommends a reinstatement cost based on modern materials at lower cost, it is likely that insurers, faced with paying for a repair of the building using existing more expensive materials, will reduce the payment to account for this under-insurance.

## Study 1

Bentham Factory was built in about 1880. It is three storeys high and has walls of squared blocks of sandstone. The roof is slated and is supported by pitch-pine king post trusses. Floors are flagged and supported by cast iron pillars and brick vaulted fireproof ceilings. The factory is still in full use for manufacturing purposes.

*Reinstatement cost based on original design and original materials:*

|  |  | £ |
|---|---|---|
| Floor area on a gross internal area basis | 3,840 m² |  |
| Estimated rebuilding cost | ×£1.515 | 5,817,600 |
| Demolition and debris removal costs, 5% |  | 290,880 |
|  |  | 6,108,480 |
| Professional fees, 15% |  | 916,272 |
|  |  | 7,024,752 |

|  | £ |  |
|---|---|---|
| Other costs: |  |  |
| Planning fees | 13,410 |  |
| Building Regulations Charge | 17,290 | 30,700 |
|  |  | 7,055,452 |
| Estimated reinstatement cost, say |  | £7,055,000 |

*Reinstatement cost based on original design but using modern materials, including a steel frame and profile steel cladding:*

|  |  | £ |
|---|---|---|
| Floor area on a gross internal area basis | 3,840 m² |  |
| Estimated rebuilding cost using modern materials | ×£600 | 2,304,000 |
| Demolition and debris removal costs, as above |  | 290,880 |
|  |  | 2,594,880 |
| Professional fees, 10% (1) |  | 259,488 |
|  |  | 2,854,368 |

|  | £ |  |
|---|---|---|
| Other costs: |  |  |
| Planning fees | 13,410 |  |
| Building Regulations Charge | 7,665 | 21,075 |
|  |  | 2,875,443 |
| Estimated reinstatement cost using modern materials, say |  | £2,880,000 |

Note

1) Professional fees will vary depending on building type. It will be apparent that professional fees have been set at 15% of the rebuilding cost for the replacement of Bentham Factory to its original design and using original materials. A lower amount for professional fees has been included for the modern replacement building where the allowance is only 10%. The reason for the higher percentage is that the replacement of Bentham Factory in its original form will require an architect, quantity surveyor and structural engineer because of the complexity of the project. The modern replacement building, being much simpler, would probably not require the services of an architect but only a building surveyor and structural engineer.

The contrast between the reinstatement cost of £7,055,000 to replace the original building and £2,880,000 for a modern alternative could not be starker. If the stone factory was insured for £2,880,000 and subsequently damaged in a fire requiring a repair using original materials at a cost of £1,000,000, the insured would not be paid in full by insurers despite the fact that the sum insured was greater than the amount of the repairs. Because of the application of the average clause the calculation to arrive at a payment would be:

$$\frac{£2,880,000}{£7,055,000} \times £1,000,000 = £408,221$$

The insured would be paid only £408,221 by the insurer and would be forced to find £591,779 out of his own funds to make up the total cost of £1,000,000.

If a property owner sustains a serious loss, let us say a total loss of his building, and it has been properly and adequately insured on a reinstatement basis, he would be entitled to ask insurers to agree to pay for the reinstatement of the building. This is provided that he has complied with all policy conditions. However, the policy terms ensure that he would not be entitled to a bigger or better building, other than by paying the additional cost himself. He would not be able to ask for the reinstatement cost to be paid in cash without actually rebuilding and incurring the cost.

It is often overlooked at the time of a valuation that the reinstatement memorandum does not allow an insured to obtain settlement of a claim for replacement or repair on a reinstatement basis if the act of replacement or repair is not carried out. This and other controls on the insured are exercised by the special provisions. These are effectively conditions which form part of the reinstatement memorandum and which, because of their importance, deserve some mention.

The work of reinstatement may be carried out on another site provided that the liability of the insurer is not increased. This means that the insurers would not object to the insured's decision to relocate to Manchester a factory destroyed in London provided that the cost to insurers would not be greater.

The special provisions provide that the work of reinstatement must be carried out with reasonable despatch, which is taken to mean that there must be no unreasonable delays in the rebuilding or repair of a property which would result in insurers being asked to pay a higher amount.

Special provision 3 is of vital importance and shows that there would seem to be no point in valuing and then insuring a building on a reinstatement basis where there is no intention to rebuild or repair. It states:

> 3. No payment beyond the amount which would have been payable under the policy if this memorandum had not been incorporated therein shall be made until the cost of reinstatement shall have actually been incurred.

If the cost of reinstatement is not incurred, the insured will only be able to recover settlement of the claim on an indemnity basis.

Where there would be no intention to reinstate or repair a building following damage, and the building owner knows this at the time of organising the insurance, he should seek an alternative type of insurance perhaps where only the costs of demolition and debris removal are insured.

## Indemnity value

It is unusual now to find a commercial insurance policy covering buildings and plant and machinery arranged on an indemnity value basis because reinstatement policies offer a superior method of insuring against the loss of a building or the machinery within it.

Many years before the introduction of reinstatement insurance, cover to protect buildings and machinery was provided by a Standard Fire Policy and it gave compensation against loss on an indemnity basis.

A Standard Fire Policy allows the insurer to:

> Pay to the insured the value of the property at the time of the happening of its destruction or the amount of such damage or at its option reinstate or replace such property or any part thereof.

Insurers rarely ever take the option to reinstate or repair an insured's property but it is a device available to them if they are dealing with a difficult or intransigent insured. It is an option, however, which is of doubtful advantage to the insurer since there would be the possibility of a continuing dispute with a dissatisfied claimant.

In dealing with an indemnity value on behalf of an insured or insurer it is necessary sometimes to deal with a hybrid. There was a case recently where a company, part of a large group, decided that it did not wish to rebuild its modern factory, which included clean room facilities, because this part of the business was to be sold as a going concern. They simply could not delay the sale of this part of the group by waiting for the completion of rebuilding. The clean room was a valuable part of the property and it contributed greatly to the earning capacity of the business. The factory was on a relatively new trading estate with a number of identical premises, some of which were for sale. It was decided, therefore, that the correct measure of indemnity in respect of the factory was diminution in market value since the market value could be easily and accurately identified from sales of identical neighbouring premises. It was not possible however, to establish a market value for the extensive clean room which accounted for more that 50% of the floor space. As clean rooms are seldom, if ever, sold in the open market there was no transactional evidence to support a market value. In the circumstances the clean room was valued on the basis of depreciated replacement cost taking into account age, wear and tear.

Where an insured would be entitled to reinstate a destroyed building but decides not to because, for example, it is difficult to find a tenant or the location is no longer suitable, the policy might then dictate that the insurer should "... pay to the insured the value of the property at the time of the happening of its destruction". Unfortunately what is meant by "value" is not defined and the Courts have decided that it is the Tort definition: "place the insured by payment or otherwise in the same position as it was in before the destruction, neither better nor worse".

In the 1980s quite a number of cases were heard where there was a dispute between the insurer and insured as to the correct method of assessment of an indemnity value where rebuilding was not to take place. It was a time when insurers felt strongly that the correct method of assessment should be diminution in market value. However, the courts did not fully concur with insurers and seem to have laid down some broad terms of reference for the treatment of claims dealt with on an indemnity basis:

- where a property is owner-occupied and in use — depreciated replacement cost
- where a property was for sale before destruction — diminution in market value
- where a property is owned by a property company — diminution in market value
- where the property is of a type not normally sold in the open market — depreciated replacement cost.

Obviously it is not possible to give a precise interpretation and each case must be treated on its merits but the following are worthy of consideration.

- In the case where a property is owner-occupied it is said that it would usually be impossible to find an identical replacement in terms of construction, layout and location in the open market. This being the case, depreciated replacement cost should be adopted. In valuation terms location is of fundamental importance to the value of a property, as is layout. Recent valuation cases relating to fire loss seem to dictate that if a property has been destroyed and cannot be replaced with a more or less identical one, in terms of construction type, location and layout, the courts will not favour an attempt to calculate a market value for it. However, in practice there may be exceptions to this, for example, where location and layout are not of such importance to the insured.

- If a property was for sale before destruction, the insured was obviously prepared to accept a market value from the sale. In these circumstances he should not object to an indemnity value payment post-loss based on the same criteria, namely market value.
- Most properties owned by property companies will be tenanted and subject to leases with covenants to insure and rebuild so that the prospect of an indemnity, other than by reinstatement, is remote. Where indemnity other than by reinstatement is an issue, it is generally felt that a property company should be entitled to a payment derived from a calculation of diminution in market value.
- If an asset such as a cleanroom, is not normally sold in the market place, there will be little or no evidence of market prices. In the circumstances such assets should be valued on a depreciated replacement cost basis.

## Study 2

The contrast between an indemnity value calculated on a depreciated reinstatement cost basis and one on a diminution in market value basis in respect of a modern freehold single storey factory on an established industrial estate in South Humberside can be significant.

*Depreciated reinstatement cost:*

| | | £ |
|---|---|---|
| *Estimate of cost:* | | |
| Floor area (gross internal area) of a modern factory used for pharmaceutical products | 2,328 m² | |
| Estimated reinstatement cost | ×£833 m² | 1,939,224 |
| Demolition costs | | 100,000 |
| | | 2,039,224 |
| Professional fees, 12.5% | | 254,903 |
| | | 2,294,127 |
| | £ | |
| Other costs: | | |
| Planning fees | 13,410 | |
| Building Regulations charge | 6,457 | 19,867 |
| | | 2,313,994 |
| Estimated reinstatement cost, say | | £2,310,000 |

*Depreciation:*
*Straight line basis*

| | |
|---|---|
| Estimated expected life of the building structure: | 40 years |
| Actual age of the structure | 11 years |
| Depreciation $\dfrac{11}{40} \times 100 =$ | 27.5% |
| Reinstatement cost, as above | £2,310,000 |
| *Less* | |
| Depreciation at 27.5% | 635,250 |
| | 1,674,750 |
| Depreciated reinstatement cost, say | £1,675,000 |

*Diminution in market value:*

|  |  | £ pa | £ |
|---|---|---|---|
| Floor area (gross internal area): | 2,328 m² |  |  |
| Current net rental value: | £43 m² |  |  |
| Current net rent |  | 100,104 |  |
| YP in perp at 8% |  | 12.5 | 1,251,300 |
| *Less* |  |  |  |
| Market value of the cleared freehold site: |  |  |  |
| 0.6 hectares @ £370,000 per hectare |  |  | 222,000 |
|  |  |  | 1,029,300 |
| Diminution in market value, say |  |  | £1,030,000 |

It will be apparent that there is a considerable difference between the indemnity value of £1,675,000 calculated by reference to depreciated rebuilding costs, and indemnity value of £1,030,000 calculated on the basis of diminution in market value. The criterion to be adopted must be this: if, under an indemnity policy, there is an intention to rebuild, the basis of the indemnity calculation must be depreciated replacement cost in the sum of £1,675,000. It does not follow that if the insured expresses an intention not to rebuild in the event of a serious loss, he should insure the building on a diminution in market value basis. The diminution in market value figure for an old building is likely to be much lower than one calculated on a depreciated replacement cost basis. If a building is insured on a market value basis and is seriously damaged but not destroyed, the insured will probably wish to carry out repairs. In these circumstances the cost of repairs could be greater than the sum insured calculated on a market value basis and the amount of compensation would be reduced by insurers to recognise this under insurance.

# Calculation of rebuilding costs
## Method of measurement

To enable a client to fix a sum insured, the method to be adopted in assessing building costs for insurance purposes requires that industrial and commercial properties be measured to obtain the gross internal floor area. This is attained by taking measurements from inside external walls, including all areas within the envelope of the building, and multiplying the result by the number of floors in the building.

Houses are measured to attain the gross external area by taking measurements from outside external walls and multiplying the result by the number of floors.

The RICS Code of Measuring Practice provides guidelines for the measurement of properties for various valuation purposes.

## Property type

The unit cost of a building will depend on many factors but the surveyor will need to establish a basic description of the property, whether it is single- or multi-storey, a factory, warehouse or office building and, in the case of a house, detached, semi-detached, terraced or bungalow.

# Construction type

As complex foundations can add considerably to the cost of reconstruction, it will be necessary to ascertain details of the site and any particular problems which it might present. Principal construction details will be required such as whether the building to be assessed is of traditional brick and slate construction, a light steel-framed metal-clad or a reinforced-concrete-framed brick building. A surveyor will need to describe internal features of the building such as partition walls, suspended ceilings and services such as lighting, heating, fire and intruder protections. The survey will encompass details of thickness of external walls, quality of floor and roof finishes, types of windows and doors, quality of internal and external finishes and the extent of internal and external services. Adjustments are also made for non-standard features such as high ceilings, additional bathrooms, dormers, large chimneys, outbuildings and suchlike. Walls, fences, gates, patios, pathways and landscaping should also be allowed for in the valuation.

# Method of assessment

The calculation of rebuilding costs to fix a sum insured is often by applying unit cost rates to superficial measurements of a building, adjusting the rates to particular types of building. In certain cases it will be preferable to produce priced full or elemental bills of quantities but these are labour intensive exercises and therefore costly. However, following destruction of a building which is to be rebuilt there is probably no alternative to the provision of a specification and full bill of quantities using the services of a quantity surveyor and others.

# Sources of cost information

Surveying practices involved with building projects will be able to analyse details of costs and record them for future assessment use. Other sources of building costs are the Building Cost Information Service of the RICS, which provides, *inter alia*, a comprehensive online service, quarterly reviews of building prices taken from actual tenders, and technical books such as Spons and Laxtons.

The *Guide to House Rebuilding Costs* published by the Building Cost Information Service and by the Association of British Insurers is a useful guide to the cost of domestic construction costs.

# Additions to the basic assessment

So far we have reached the stage of establishing the cost of rebuilding a particular property but there are several additions required to complete an assessment suitable for insurance purposes, as shown in the earlier examples.

In assessing the replacement cost of a particular structure a valuer needs to consider what effect current regulation would have on design, materials and methods if the building were to be destroyed by a fire. For example, it is possible that an existing old factory with a steel frame and corrugated iron walls and roof, without proper insulation, could continue in use indefinitely. If, however, it were destroyed by fire it would be subject to current planning policy and building regulations. If the owner of the property intends to rebuild it in a modern form using modern materials (and perhaps altering the size and height) a new planning permission would be required. If the owner's new plans do not conform with current planning policy, permission could be refused. This could force the owner to move to a different, more suitable site which, if ground conditions are poor, could increase the cost of building the sub-structure compared with the original site. While insurers would pay for increased costs for complying with building regulations they will not wish to pay for increased sub-structure costs except in extenuating circumstances.

While the replacement of an old factory on an existing site following a fire may be acceptable to the planning authority, it is almost certain that building regulations will force the owner to improve the replacement building, requiring possibly more substantial foundations, better insulation, lighting and means of escape. This too will impact on cost and the pre-fire cost assessment must take account of such requirements.

A valuer preparing a cost assessment for insurance purposes needs a crystal ball to foresee the future and to reflect in his assessment the cost of satisfying the requirements of local and statutory authorities and those of the European Union. However, in commercial property insurance, to ensure that the insurance policy will compensate the factory owner for increased construction costs, it might need to be amended to incorporate a number of additional valuable clauses. (Such clauses are usually automatically included in a policy insuring a domestic property.)

Further additions to the basic assessment of building costs will be required. First, the damaged building will require demolition and the valuer needs to add for this and site clearance. Second, when the insured contemplates a new factory to replace the one destroyed, he will do so probably with the assistance of a construction team composed of one or all of the following: architect, quantity surveyor and engineer. He will also be required to appoint a planning supervisor under the Construction (Design and Management) Regulations 2007 (SI 2007 No 320) which impose legal obligations on the construction team in the area of health and safety. As will be seen later a surveyor will also make allowances for debris removal cost and professional fees.

## Public authorities and European Union clause

This clause, which must be specifically included in the policy to gain its benefits, covers the additional cost incurred in reinstatement of damage to comply with building or other regulations made in pursuance of any Act of Parliament, bye laws or other relevant laws. The costs to be incurred under this item are not separately insured but are to be included in the overall sum insured covering the buildings.

## Debris removal clause

As with the Public Authorities Clause, it may be necessary to extend the policy to include for debris removal. The term removal of debris is meant to refer to the costs of demolition, the removal of debris, and of any shoring or propping up necessarily incurred by the insured with the consent of the insurer. As far as buildings and machinery are concerned the cost of removal of debris should be included within the cost assessment which forms the relevant sum insured. However, the removal of debris left from destroyed stock and materials-in-trade has to be insured as a separate item with a separate sum insured.

It should be borne in mind that environmental and health and safety issues figure prominently in the demolition of a property damaged by fire, particularly an old industrial one. Costs can be high, especially if asbestos is present.

## Architects', surveyors' and consulting engineers' fees clause

Professional fees are not always automatically insured and special provision by means of a relevant clause in the policy may need to be made for them. They will normally form part of the overall sum insured on buildings and relate to fees necessarily incurred in reinstating or repairing the damage, not for preparing a claim. Professional fees will normally include the cost of appointing a construction design and management co-ordinator in accordance with the Construction (Design & Management) Regulations 2007 (SI 2007 No 320).

## Fees for building regulation and planning applications

An allowance for the cost of fees for building regulations approval, planning applications, listed building, conservation area and other consents should be made in an assessment at appropriate levels. However, such costs are likely to represent only a small proportion of the total rebuilding costs.

## Value Added Tax

There are two reasons why VAT may not need to be taken into account when assessing value for insurance purposes. First, the building works may be zero-rated (VAT free) and second if VAT is chargeable on the building works, it may be reclaimable from HM Revenue & Customs by the insured business using the property.

### Zero-rating is available for certain works to the following types of buildings:

(a)  new dwellings
(b)  charitable buildings (for non-business use or as a village hall or similar)
(c)  residential buildings (communal residences eg nursing homes, student accommodation, hospice)

For these buildings, new building works in the event of complete reconstruction following a fire or other damage are zero-rated. Repair work to such buildings will be standard-rated. It has been accepted by insurance companies that the sum insured on the building of a dwelling does not need to include VAT, despite the fact that claims for partial damage will attract VAT. In such cases insurers will reimburse the cost of repairs plus VAT.

### Recovery of VAT on building works

An insured business that is registered for VAT and owns a property will be able to reclaim VAT incurred on works to the building if it is occupied by the business for the purposes of making taxable supplies (zero-rated, lower-rated or standard-rated). For example, a VAT registered manufacturing business will be able to reclaim VAT on works to the factory that it owns and occupies to manufacture its products for sale. Consequently, insurers exclude VAT payments from claims by such businesses. Assessments of buildings and plant and machinery belonging to such VAT registered businesses need not, therefore, include VAT.

However, certain businesses eg financial institutions and private schools, which are exempt for VAT purposes, will not be able to reclaim VAT on works to the buildings that they own and occupy. Exempt businesses do not have to charge VAT to their customers, and are not able to reclaim VAT on their expenses. This means that the complete reconstruction of buildings owned and occupied by such organisations will attract VAT that they will not be able to recover from HM Revenue & Customs. In these circumstances, the assessment of the cost of buildings and contents of such organisations for insurance purposes must include VAT.

In some instances businesses may have some income which is liable to VAT and some which is exempt. In these cases, the business is partly exempt and a proportion of the VAT on the works will be recoverable if the business is registered for VAT.

Where property is held as an investment to provide rental income, the VAT situation will depend on the type of property and whether or not the investor has elected to waive exemption. The letting of dwellings, charitable property and communal residential property referred to above is exempt from VAT and therefore no VAT is recoverable on any works to such property.

The letting of commercial buildings is exempt unless the investor has elected to waive the exemption. If an election has not been made, no VAT is recoverable on works to the building and this must be taken into account when assessing the cost of building for insurance purposes. However, if the investor has elected to waive the exemption, this usually means that VAT is chargeable on the rent and therefore VAT is recoverable on works to the building. Where an effective election is in place, no VAT need be taken into account when assessing the cost of the buildings for insurance purposes. There are some instances however, when even though an election has been made, the election is disapplied (eg for supplies between connected parties when certain anti-avoidance provisions may apply). In these circumstance VAT on works to the building will not be recoverable. If a building is completely destroyed the investor's election on the original building will end and it will be necessary for the investor to make an election in respect of the new building if VAT is to be recoverable on the construction of the new building.

# Average conditions

## Ordinary pro rata average

Most valuers will be aware of the punitive effect of the average clause which is now incorporated into most classes of fire and special perils insurance. The pro rata condition of average operates where there is under-insurance and provides that the insurer will be required to pay only that proportion of the loss which the sum insured bears to the true indemnity value of the property at the time of the fire. The pro rata condition of average affects assessments calculated on an indemnity basis. It is an attempt to ensure that the indemnity value is correct at the time of the fire, not at the time of reinstatement.

### Study 3

The indemnity value of a factory in Birmingham is calculated to be £1,100,000. Let us assume that the insured decided not to insure the property to its full value but fixed the sum insured at £500,000. Shortly after the valuation the property was damaged by fire and the cost of reinstatement was estimated to be £12,000. The repairs would provide some improvement, so that an equitable assessment of the loss on an indemnity basis, ignoring under-insurance, would be £10,000. However, it is clear that there is under-insurance and that average will apply on the following basis:

Pro rata condition of average formula:

$$\frac{\text{Sum insured} \times \text{Loss}}{\text{True indemnity value at the time of the fire}}$$

$$\frac{£500,000}{£1,100,000} \times £10,000$$

Insurer's liability £4,545

In this example the insured would receive only £4,545, being his own insurer for the balance, namely £5,455.

## Reinstatement average

Policies containing the reinstatement memorandum are also subject to average but its application is slightly different from that of pro rata average. As we have seen, pro rata average requires that the sum insured should be equal to the indemnity value of the property at the time of the fire. Reinstatement average, which is much more common, determines that the sum insured should be equal to the value of the property at the time of reinstatement which, in a large building contract, could be two or three years after the destruction.

# Combating inflation

During periods of high inflation it is difficult for an insured to ensure that the level of insurance to cover his assets remains at adequate levels. If it does not, a claim could be reduced by the application of the average clause, as described earlier. In the last few years building and other costs have risen less steeply so that some of the devices to combat inflation are not as common. The surveyor dealing with insurance assessment should be aware of the following.

## Day-one reinstatement scheme

With the day-one scheme, which is now the most popular anti-inflation device, the insured declares the total rebuilding cost at prices ruling at renewal each year. Cover for inflation is provided by reason of the fact that the sum insured is stated to be a certain percentage more than the declared value at the beginning of the insurance year. However, the declared value must be accurate because it will be used as the basis for the average calculation at the time of the loss. In some circumstances an additional premium is levied by insurers to provide payment for the inflation element.

## 85% Reinstatement memorandum

As a temporary amendment to the reinstatement memorandum insurers decided to restrict the operation of the average clause to those cases where, at the time of reinstatement, the sum insured is less than 85% of the full reinstatement cost. For example, if the sum insured represents 90% of the reinstatement cost the average clause will not be applied to reduce the amount of the claim (unless of course the property is totally destroyed, in which case the insured will have to stand 10% of the loss himself). Where, however, the sum insured at the time of reinstatement represents less than 85% of the full reinstatement cost, any claim will be subject to average as if the 85% concession did not exist.

Where a day-one reinstatement scheme is in place, it cannot contain an 85% reinstatement memorandum.

# Inflation indices for building cost assessments

Every valuer needs to have access to inflation indices and there are several good sources of information.

The Building Cost Information Service calculates, *inter alia*, the Tender Price Index, which indicates the movement of tender prices for new building work in the United Kingdom. Because the data is derived from actual tenders, the indices reflect the conditions prevailing at the time of pricing. The indices measure not only changes in building costs such as labour rates and materials prices but also the influence of economic conditions operating at the time in a competitive market.

The index gives details of actual changes in tender prices and also forecasts likely changes in the immediate future, the latter being of assistance to valuers in giving recommendations on inflation projections. Inflation data and projections are used by insurers and insurance brokers when deciding upon the method and basis of protecting assets after the date of inception of the policy up to the next renewal, usually one year later. During this period, with some policies, the sum insured would otherwise remain static which could lead to under-insurance.

# Insurable interest

All property whether material, or represented as a right, is insurable provided that the party seeking insurance has an insurable interest in the property or right for which insurance is sought. The insured need not, however, have an insurable interest at the time insurance is effected but he must have such an interest at the time of the loss. For example, a freeholder has an insurable interest in the property which he owns because if it is destroyed he stands to sustain a loss. However, he will not have an insurable interest in an adjoining property which he does not own or lease even though its destruction could cause him loss in terms of, say, loss of support for his own buildings. In such circumstances the freeholder may have a right of recovery against the owner of the property in which the fire started but only if it can be proved that the fire was caused as a result of a negligent act. If negligence cannot be proved, compensation would only be available from the owner's own insurance.

# Utmost good faith

Insurance is based on the doctrine of *uberrima fides* or utmost good faith. This means that the insured, the policyholder, is required to disclose to the insurer any information which is likely to cause the insurer to review the basis upon which he is insuring a property or refuse to insure it at all. Failure to disclose material facts would make the policy voidable. For example, failure by a policyholder to declare that a building, for which insurance is sought, shows cracks in the brickwork, even though they may be old, could make the policy voidable.

# References in connection with listed building insurance

The following publications will be found helpful:

*An Insurance Manual for Historic Houses.* 1996, Historic Houses Association.

*Approach to Property Insurance and Insurance Valuations for Historic Buildings.* A Guidance Note, 1994 RICS.

*Approach to Property Insurance and Insurance Valuations for Historic Buildings.* Sutch and Davey. 1994 RICS Information Sheet No 1.

*BCIS Guide to House Rebuilding Costs.* Current Edition, RICS.

*Fire Insurance Law & Claims* R M Walmsley, Chartered Institute of Loss Adjusters 1997.

*Guide to Carrying Out Reinstatement Cost Assessments.* 1999, RICS.

*Insuring your Historic Building: Houses and Commercial Buildings.* 1994, English Heritage & RICS.

*Property Insurance: Some Points to Consider in Relation to the Proper Cover of Risks.* 4th Edition, 1995, RICS.

*Reinstatement Cost Assessment and Insurance Claims*, 2nd Edition. 1985, RICS.

*RICS Appraisal and Valuation Manual*, The Red Book. RICS. This does not specifically deal with listed buildings insurance but has relevance as indicating best practice etc.

*The Code of Measuring Practice.* 6th Edition, RICS.

©Arthur Broadhurst FRICS 2008

# The Valuation of Care Homes

## Background

Much has changed in the care sector since the start of the 21st century and the last edition of this book. There has been a plethora of new legislation and regulatory bodies, a boom in values and rapid consolidation of the sector. The fundamental approach to valuation has, however, remained unchanged.

*Trade Related Property Valuations* provides guidance on the criteria that need to be considered when valuing properties such as care homes that are normally bought and sold on the basis of their trading potential. Supplementing this is *Valuation Information Paper 11* by the Trade Related Valuation Group. The various banks also set out via service agreements and guidance notes, the content they require in a care home valuation report over and above that required by the Red Book. The comments made in this chapter should therefore be read in conjunction with these documents. For consistency, a care home in this chapter is taken to be the same as that defined in VIP 11 and concerns properties which are registered with an appropriate regulatory body to provide nursing or personal care. Personal care homes are homes in which people live on a permanent or respite basis. They provide accommodation, meals and personal care (such as help with washing and eating). Nursing homes are the same as personal care homes but they also have registered nurses who can provide care for more complex health needs.

Care homes provide care and accommodation to a wide variety of service users. This chapter concentrates on the elderly, although the approach used here can be applied equally to other client categories.

## Approach to valuation

Paragraphs 4.1 to 4.16 of GN1 set out the valuation approach to be adopted. The following are of particular note:

4.5   Specialised Trading Related properties are considered as individual trading concerns and typically are valued on the basis of their potential earnings before interest, taxes, depreciation and amortisation (EBITDA) on the Assumption that there will be a continuation of trading.

4.6   The task of the valuer is to assess the fair maintainable level of trade and future profitability that could be achieved by a Reasonably Efficient Operator, of the business upon which a potential purchaser would be likely to base an offer.

2.4    A Reasonably Efficient Operator is defined as 'a market-based concept whereby a potential purchaser, and thus the valuer, estimates the maintainable level of trade and future profitability that can be achieved by a competent operator of a business conducted on the premises, acting in an efficient manner. The concept involves the trading potential rather than the actual level of trade under the existing ownership so it excludes personal goodwill' (IVS GN 12).

4.7    When assessing future trading potential, the valuer should exclude any turnover and profit that is attributable solely to the personal skill, expertise, reputation and/or brand name of the existing owner or management. However, in contrast, the valuer should include any additional trading potential which might be realised under the management of a Reasonably Efficient Operator taking over the business at the date of Valuation.

4.8    When valuing properties by reference to trading potential, the valuer will need to compare trading profitability with similar types and styles of operation. Therefore a proper understanding of the profit potential of those property types, and how they compare to one another, is essential.'

VIP 11 elaborates on how this may be achieved.

5.7    An analysis and review of the current and past years trading performance of the business is helpful in forming an opinion, as at the date of valuation, of future sustainable trading potential achievable by a competent operator. This will include consideration of the fee income and likely sustainable occupancy of the business, the adequacy, suitability and cost of staffing and all other expenditure in the context of the type of building and its location.

5.8    With the exception of closed or start-up enterprises, actual accounts will generally be used as a valuation tool in formulating the valuer's opinion of the future sustainable trading potential, however they will show only how a particular enterprise is trading under the particular management at the time. The valuer will need to interpret the accounts to understand how bidders in the current market will view the future potential of the business and the returns expected from it. Current and past trading is no guarantee as to the future fair maintainable trade and operating profit.

5.9    Access to data on accounting information from which to derive benchmark fees, occupancy, staff and all other running costs relating to this class of business forms an essential element in providing a competent assessment of sustainable profitability and value.

In the following sections the key drivers of the care sector are examined and through the use of a hypothetical example, a set of accounts analysed noting divergences from industry norms and a calculation of the fair maintainable trade undertaken.

# Key drivers

The key income drivers in a care home are occupancy and fee rates.

Underpinning these are standards of accommodation and care and the supply of, and demand for, care beds.

The cost drivers can be divided into staff costs and non-staff operating costs.

# Standards of accommodation

The care sector is tightly regulated. While the aims and objectives of the legislation are the same throughout the United Kingdom, the Acts and regulations vary from one country to another. It is important to understand the differences in the legislation and how they are interpreted in each country as the differences can have an impact on the calculation of trading potential. It would not be practical here to detail the legislation as it applies in each part of the United Kingdom. Instead, in the worked study described later, it is assumed that the care home is in England and consequently the regulatory context as it applies to England is summarised below.

Care homes in England are governed by the Care Standards Act 2000 as amended and its associated regulations. Under the Act, a care home cannot operate unless it is registered by the Commission for Social Care Inspection (CSCI). In order for a care home to continue to trade on a sale, it is necessary for the purchaser to be approved as the registered provider. Most sales are therefore conditional on registration being transferred to the prospective purchaser. This can be time consuming and while it is only a part of the process, it is one of the main reasons why it is not unusual for care home sales to take up to 12 months to complete.

The Care Homes Regulations 2001 (SI 2001 No 3965) as amended are generalised in so far as the required facilities and services in a care home are concerned. For example, paragraph 16(2)(c) of the regulations requires the registered person, having regard to the size of the care home and the number and needs of service users to provide in rooms occupied by service users, adequate furniture, bedding and other furnishings, including curtains and floor coverings, and equipment suitable to the needs of service users, and screens where necessary. There is one specific, however. Under paragraph 4(1) the registered person has to compile a 'statement of purpose' and this should include a statement as to the number and size of rooms in the care home. In reality, providers interpret their obligations under 4(1) differently but valuers should request a copy of the statement of purpose and check the information contained therein.

Under section 23(1) of the Care Standards Act 2000 the Secretary of State for Health published national minimum standards in 2001 and subsequently amended them in 2003. The standards cover all facets of running a care home. Section 5 (standards 19–26) deals with the environment. A number of the standards apply equally to all homes. However, there are some environmental requirements that are higher for new homes than so called pre-existing homes ie those which were in existence prior to April 2002. Essentially, pre-existing homes are required to maintain the physical standards that they achieved as at 31 March 2002 and not fall below them while new care homes for the elderly have to achieve certain minimum criteria. These include having:

- 12 m$^2$ single en-suite rooms (with the en-suite in addition to the 12 m$^2$)
- 4.1 m$^2$ of day space for each service user
- one assisted bath/shower for every eight service users
- doorways with a clear opening of 800mm for wheelchair users
- the height of the window enables the service user to see out of it when seated or in bed
- service users wishing to share are offered two single rooms for use, for example, as bedroom and sitting room.

All homes, whether new or pre-existing, have to be suitable for their stated purpose, be accessible, safe and well maintained. They should meet service users' needs in a comfortable and homely way. Buildings should comply with the requirements of the local fire service and environmental health department. Outdoor space should be accessible to those in wheelchairs and be kept tidy, safe and attractive. Lighting in communal areas should be domestic in character and sufficiently bright and positioned to facilitate reading and other activities. Furnishings of communal rooms should be domestic in character and of good quality. Toilets should be available close to lounge and dining areas and within close proximity of a service user's bedroom. Service users should have access to all communal and private parts of the home via ramps and passenger lifts if necessary or stair lifts where appropriate. Aids, hoists, assisted toilets and baths should be available which are capable of meeting the assessed needs of service users.

Single rooms accommodating wheelchair users should have at least 12 m² of usable floor space excluding any en-suite. Rooms should have sufficient dimensions and layouts to ensure that there is sufficient space on either side of the bed to enable access for carers and any equipment needed. Where rooms are shared, they should be occupied by no more than two service users who have made a positive choice to share with each other. When a shared place becomes vacant, the remaining service user should have the opportunity to choose not to share, by moving into a different room if necessary. Rooms that are currently shared should have at least 16 m² of usable floor space excluding any en-suite. Adjustable beds should be provided for service users receiving nursing care. Doors to service users' rooms should be fitted with suitable locks. Screening should be provided in shared rooms to ensure privacy. Pipe work and radiators should be guarded or have guaranteed low surface temperature surfaces. Water should be stored at a minimum of 60°C, distributed at a minimum of 50°C and delivered via thermostatic valves at close to 43°C. Rooms should be centrally heated and the heating controllable in the room. Emergency lighting should be provided throughout the home.

The above, while covering the main physical requirements, is by no means a complete list of all the environmental standards. There are various others and the practitioner should be familiar with them.

**Table 22.1** Single rooms as a percentage of all bed spaces and the percentage of bed spaces which have en-suite wcs in for profit homes for older people March 2006

| Area | Care homes providing nursing care | | Care homes providing personal care only | |
|---|---|---|---|---|
| | Single places (%) | Places with en-suite wc (%) | Single places (%) | Places with en-suite wc (%) |
| North | 93 | 61 | 94 | 65 |
| Yorkshire & Humberside | 85 | 71 | 83 | 49 |
| North West | 88 | 55 | 84 | 48 |
| West Midlands | 84 | 67 | 82 | 42 |
| East Midlands | 77 | 58 | 85 | 55 |
| East Anglia | 85 | 72 | 84 | 58 |
| Northern Home Counties | 83 | 67 | 86 | 58 |
| Greater London | 95 | 86 | 75 | 47 |
| Southern Home Counties | 83 | 64 | 84 | 55 |
| South West | 84 | 65 | 90 | 65 |
| Wales | 85 | 68 | 81 | 41 |
| Scotland | 85 | 82 | 91 | 69 |
| N Ireland | 84 | 49 | 73 | n/a |
| United Kingdom | 85 | 66 | 85 | 54 |

Source: Laing & Buisson

# Standards of care

In order to assess compliance with the regulations and the extent to which the national minimum standards are met, the CSCI regularly inspect care homes. Their reports are published and made available on the CSCI web site. Inspection reports assess some or all of the 38 standards against a scale of 1 to 4 (major shortfalls to commendable). Inspection reports may set out statutory requirements and recommendations. The requirements should have a time scale for action and the registered provider must comply with the given time scale. Recommendations are seen as good practice for the registered provider to consider carrying out. Failure by a registered provider to comply with statutory requirements can lead to the home being placed on special measures and the ultimate sanction is for registration to be withdrawn and any residents moved from a home. The loss of registration or even the threat of losing registration can have a detrimental impact on capital value. Poor inspection results can impact adversely on a home's reputation and in turn on levels of admissions and occupancy. The difficulty, the skills required and the time and cost involved in turning around a failing home should not be underestimated.

It is possible for a home devoid of residents to maintain its registration and the sale of such a home still be subject to the transfer of registration. However, should the registration be withdrawn or rescinded and then an application to register the premises be made subsequently, it is probable that the property would be assessed against the standards for new homes. The cost of upgrading a pre-existing home with, for example, 10 m² rooms, into a home with all 12 m² single en-suites, could be prohibitive. It some cases it would not be feasible. In others it would result in a much lower registration being granted than had previously been in place. Lenders in particular need to be made aware of the possible impact on capital value associated with the loss of registration.

The CSCI are proposing to introduce a star rating scheme in 2008. Homes which are awarded the top grade could benefit significantly from this scheme by having their reputation enhanced, while those found wanting could experience more intensive inspections and fewer enquiries and admissions. In addition to the CSCI initiative there are other quality schemes in existence including those operated by individual local authorities and RDB star rating. Under these schemes homes that are assessed as having high standards can be rewarded by receiving quality premiums from local authorities.

The government are proposing to review the national minimum standards although at the time of writing, no indication of likely changes has been received.

# Standards driven upwards by market forces

The purpose of regulation is to protect vulnerable people and ensure that minimum standards are met. Regulation does not mean that all homes offer the same standard of accommodation or care. As in any market, those homes which are perceived to offer high standards and value for money will tend to generate more business than those which do not. Notwithstanding the fact that government does not require pre-existing homes to improve their physical standards beyond those which existed on March 31 2002, market forces continue to drive up standards. The most widely used benchmarks for standards of amenity are the single room ratio and the proportion of rooms with an en-suite wc. The position in early 2006 is summarised in Table 22.1.

Environmental standards are likely to continue to rise in response to consumer demands, with the preference being for large single rooms with en-suite showers and wcs. When appraising a particular home it is pertinent to compare the standards that it offers with those offered by its competitors, both locally and nationally. While there is an element of subjectivity involved in the process, some firms such as GLP Taylors have introduced grading tools in order to enable more objective comparisons.

# Demand for care beds

Most service users in care homes for the elderly are women. Most also suffer from one or more clinical conditions and are not simply frail. Most have no partner who could have looked after them in their own home. In homes providing personal care only 8% are aged under 75 years and in nursing homes some 18% are aged under 75 years, these being mainly physically disabled. The average age of residents in care homes for the elderly is in the mid 80s. Generally the demand for care increases with age. Research published by the Alzheimer's Society in 2007 found that the prevalence of dementia, one of the main causes of disability in later life increases with age, doubling with every five-year increase across the age range 65–95. The proportion of those with dementia living in care homes also rises steadily with age rising to 60.8% of those aged 90 and over.

The population of the UK aged 75 and over is expected to rise by 15.7% to 5.385m over the 10 year period to 2016. The numbers aged 85 and over in the UK are expected to rise even more in percentage terms (28.4% by 2016). As the numbers of elderly grow, so will their demand for health and social care. However, the fact that the UK's 85+ population is forecast to rise by 28.4% will not necessarily lead to a pro-rata increase in the demand for care beds. The demand for care beds will depend not only on demographic factors but also on a variety of other factors including advances in medicine, the availability of alternative types of care provision, government policy, government funding, individual preferences and telecare.

Laing & Buisson project that the demand for care beds in the UK will in fact only increase from 421,000 occupied places in 2006 to 445,000 by 2016 an increase of 0.57% pa or 2,400 places a year.

# Supply of care beds

In the 10 years to 2006, over 100,000 care beds were lost in the U.K. The reduction in the stock of beds has now halted, at least in the independent sector, and while the number of local authority care beds may continue to fall as they move away from their supplier role, the independent sector is unlikely to incur further significant contraction in the short term..

## *Competitive pressures*

In Table 22.2 below, the national supply of care beds has been analysed in relation to the population aged 75 and over. There are 9.9 persons aged 75 and over in the United Kingdom per care bed, 26.6 persons aged 75 and over per nursing bed and 17.1 persons aged 75 and over per personal care bed.

**Table 22.2** UK population aged 75 and over per care bed 2006

| | |
|---|---|
| UK population aged 75 and over per nursing home place | 26.6 |
| UK population aged 75 and over per personal care place | 17.1 |
| UK population aged 75 and over per total care places | 9.9 |

Note: Total places include nursing, personal care and long stay NHS places

Analysing local rates of supply and comparing them to the national average is particularly useful when considering new developments and extensions to existing units. In an urban area, the primary catchment of a care home for the elderly might only extend to a two to three mile radius. In a rural area, the primary catchment may extend for many miles. Homes accommodating specialist categories may have regional or even national catchments. Each home needs to be assessed individually and a calculation of its primary catchment area made based on an investigation of its existing/proposed market. Once the primary catchment area has been identified, local rates of supply and hence competitive pressures can be calculated.

Let us assume, for instance, that the care home which is the subject of the valuation is one catering for the general elderly and it lies in the centre of a market town. Its elderly residents largely come from the town or surrounding villages or have moved to the home to be close to relatives who live in the town. The valuer's investigations lead him to conclude that the vast majority of residents derive from within a three mile radius of the home. Within this area is a population (estimated from census data and population projections) of 18,095 people of whom 1,719 (9.5%) are aged 75 and over. The 9.5% is higher than the national average (7.7%) and reflects the town's popularity with the retired. The 75 and over age group has been chosen since we know that the average age of care home residents is in the mid 80s and relatively few are aged under 75.

The supply of beds within the same three mile radius can be calculated by reference to the CSCI database or to a number of proprietary databases. In this example there are a total of 172 care beds for the elderly and physically disabled across all categories, nursing, personal care, local authority and independently owned homes. The position is summarised in Table 22.3.

**Table 22.3** Rate of Supply of Beds relative to the National Average

|  | UK | Within a three mile radius of property |
|---|---|---|
| Population aged 75 and over | 4,656,000 | 1,719 |
| Care beds | 468,000 | 172 |
| Population aged 75 and over per care bed | 9.9 | 10.0 |
| Difference relative to the national average |  | 2 care beds |

Table 22.3 shows us that the rate of supply within a three mile radius of the target property is only two care beds different from the national average. This situation could change rapidly with the opening of new homes and extensions or closures of existing ones. Closures can be monitored on the CSCI web site and new developments monitored by subscribing to one of the proprietary services such as ABI.

Having analysed competitive pressures it is important to then understand how the particular care home stands in its competitive environment.

Is the property in a suitable location for its existing/proposed use as a care home? Accessibility by foot, public transport and car should be considered in relation to service users, their friends and relatives, and staff. Visibility is not important in one sense since care homes do not attract passing trade, but kerbside appeal is. The outside presentation of a home including the standard of its signs and the neatness of its gardens can say much about a home and its standards internally. Neighbouring land uses should also be considered in relation to whether they conflict with or complement the care home use.

How do the facilities compare at the subject home to those in competitor units? Are they above average, average or below average? While much can be achieved from an analysis of published data, there is no substitute for actually visiting competitor units in order to obtain a definitive view.

## Occupancy rates

Occupancy levels result from the relative forces of supply and demand. According to healthcare analysts Laing & Buisson, they averaged just over 90% in March 2006. Occupancy levels have recovered from a low of 86% in the mid 1990s as the supply of beds has contracted. Occupancy rates can vary significantly from area to area and from home to home. It is important to understand the reasons why a particular home may be under or over performing in its local market and in the light of this to assess whether the under/over achievement is likely to be a short or longer term phenomenon.

**Table 22.4** Occupancy rates in private and voluntary care homes for older people, by area, March 2006

| Area | Care homes with nursing (%) | Care homes providing only personal/residential care (%) |
|---|---|---|
| North | 85.3 | 89.2 |
| Yorkshire & Humberside | 91.6 | 92.6 |
| North West | 92.2 | 87.8 |
| West Midlands | 91.7 | 88.7 |
| East Midlands | 89.9 | 90.1 |
| East Anglia | 89.6 | 91.1 |
| Northern Home Counties | 88.6 | 91.9 |
| Greater London | 94.0 | 92.5 |
| Southern Home Counties | 90.9 | 89.0 |
| South West | 92.8 | 91.8 |
| Wales | 91.2 | 93.5 |
| Scotland | 92.1 | 92.5 |
| N Ireland | 98.1 | 93.2 |
| United Kingdom | 91.3 | 90.2 |

Source: Laing and Buisson

Colliers CRE found in their Spring 2007 Healthcare Review, average occupancy rates to be slightly higher at 93.1% (nursing) and 91.9% (personal care) although their survey did exclude Northern Ireland.

## Fee rates

Nationally, 60% of care home fees are payable by local authorities. The balance is made up by private payers (32%) and the NHS (8%). However, it is increasingly common for the relatives of a local authority funded person to 'top up' the local authority contribution.

People wishing to enter a care home are entitled to apply to their local authority for financial help towards the cost of their care. Eligibility for funding depends upon an assessment by the local authority of a person's care and financial needs.

Local authorities as the dominant purchaser of care beds are in a powerful negotiating position with independent sector providers. During the mid 1990s when occupancy rates were lower, local authorities used their bargaining power to limit fee increases to levels sometimes below the prevailing rate of inflation. It meant that some homes became insufficiently profitable to continue trading and they closed. As the stock of beds declined and occupancy rates started to recover, local authorities became increasingly concerned at their ability to secure the number of places they needed. Consequently the independent sector became in a better negotiating position to secure higher fees. Local authorities in turn were able to award higher fees because they were receiving better funding from central government. Personal social services received 5–6% annual increases in cash allocations over the period 1998–99 to 2002–03 and 9% annual increases in the three year period to 2005–06. Looking forward, with tighter restrictions resuming on government spending, it is likely that future increases in personal social services spending and hence fee rates will be set at more modest levels once again.

Each local authority decides its own commissioning strategy, fee rates and contracts. Most local authorities determine fees for the elderly at fixed rates in contrast to other groups eg adults with learning disabilities for whom fees are typically set according to an individual needs assessment. Elderly care fees are usually reviewed every year with any increases coming into effect in April. It is important to be aware of the fee policies of individual authorities as any changes can have a significant impact on a home's fee income and hence trading potential. GLP Taylors monitor all local authority fees and their preliminary results for 2007 have shown a wide variation, but on the whole lower increases than in 2006–07. There can be a major realignment of fees when a local authority has commissioned an independent review into the fair rate for care.

While spot contracting is the norm, some local authorities also enter into block contracts with independent providers. Where these exist, they do offer the provider greater security of income, but this is often at the expense of a lower fee than the provider might otherwise have been able to negotiate.

The 8% of care home fees paid by the NHS is in relation to nursing care and the majority of this is payable in respect of the registered nursing care contribution (RNCC). From the care home owners' point of view the RNCC contribution is, strictly speaking, revenue neutral.

Private payers and families who top up local authority fees are an increasingly important source of fee revenue. Not surprisingly, the number of private payers varies widely in relation to socio-economic factors. Overall, an increasing number of elderly are asset rich (due to the rising capital value of their housing) even though they may be income poor. Private pay residents generally pay the highest fees and can be the most discerning in relation to standards of service.

The funding position in Scotland is slightly different from that south of the border as residents of Scotland are entitled to free personal care as well as nursing care. This is increasingly contentious.

The average fees payable are an amalgam of local authority, NHS and private pay fees. The results for 2006 are summarised in Table 22.5.

In their Spring 2007 Review, Colliers CRE found from their survey of homes in Great Britain that in the first half of 2006, nursing home fees averaged £575 per week while those in personal care homes averaged £520 per week. The results for nursing homes are in line with those obtained by Laing & Buisson but the results for personal care homes are well in excess of those shown in Table 22.5.

The average weekly fee for a single room in a nursing home has grown by 45.6% in the five years to March 2006 and this spectacular growth has been the single most important driver behind improved EBITDA levels in the elderly care sector.

**Table 22.5** Average weekly fees in for profit homes for the elderly, by area, March 2006

| Area | Care Homes with nursing single rooms £ | Care Homes with nursing twin rooms £ | Care Homes providing single rooms £ | Care Homes providing twin rooms £ |
|---|---|---|---|---|
| North | 489 | 474 | 359 | 340 |
| Yorkshire & Humberside | 491 | 453 | 376 | 347 |
| North West | 519 | 489 | 371 | 344 |
| West Midlands | 561 | 489 | 371 | 346 |
| East Midlands | 534 | 486 | 386 | 372 |
| East Anglia | 590 | 570 | 400 | 409 |
| Northern Home Counties | 750 | 636 | 472 | 422 |
| Greater London | 766 | 617 | 564 | 482 |
| Southern Home Counties | 695 | 578 | 454 | 389 |
| South West | 621 | 542 | 420 | 390 |
| Wales | 490 | 477 | 360 | 343 |
| Scotland | 508 | 454 | 413 | 367 |
| N Ireland | 503 | 472 | 321 | 310 |
| United Kingdom | 581 | 509 | 403 | 369 |

Source: Laing and Buisson

Armed with information on a home's fee income, the source of that income, the socio-economic nature of the area, prevailing local authority fees and regional benchmarks, it should be possible to assess whether a reasonably efficient operator would be capable of sustaining and increasing fees or alternatively, struggle to maintain them.

## Staff costs

Staffing is the single largest operating cost in a care home. Staffing costs are a function of wage rates and hours worked. In turn wage rates are determined by market forces whilst the hours worked are largely determined by the needs of service users and they are monitored by the CSCI to ensure that they are satisfactory.

Standard 27.3 of the National Minimum Standards states that:

> The ratios of care staff to service users must be determined according to the assessed needs of residents, and a system operated for calculating staff numbers required, in accordance with guidance recommended by the Department of Health.

The government has been slow to put specific guidance in place. The Residential Forum published a staffing matrix in 2002 which proposed a range of 16 (for low dependency) to 20 (for high dependency) care assistant hours per resident per week for older people with additions for ancillary activities, training, holidays and sickness. However, in most instances staffing levels still tend to be determined in relation to the staffing notices issued by former Health Authorities and Local Authority Staffing Guidelines. As these vary from one authority to another, no one formula prevails.

In practice however, there are benchmarks. In all homes there should be at least two people on duty at all times of the day and night. In care homes for the elderly providing personal care only, a typical staffing rota provides 11 care hours per resident per week (for low dependency service users) rising to 13 (for high dependency frail elderly) and 15 (for dementia sufferers). Night time staffing levels can vary significantly, but typically are within the range of one member of care staff to every 10-16 service users. Ancillary hours spent on cooking, cleaning, laundry, maintenance, administration and management are in addition.

In nursing homes there should be a minimum of one registered nurse on duty at all times of the day and night. Staffing ratios are typically one member of staff to every five residents during the day falling to one member of staff per 10 residents at night. The percentage of skilled nurses to care assistants can vary widely from 20 to 50%. Overall care hours provided by RGNs and care assistants usually equate to three and a half per resident per 24 hour period for homes accommodating the general elderly, rising to four in homes accommodating those with dementia. As with personal care homes, ancillary hours are in addition.

Nursing homes may provide care to a mix of service users, some of who may require nursing and others who may require only personal care. Care staffing levels in these cases will be a mix of the above, dependent upon the relative numbers of nursing and personal care clients accommodated.

Staffing levels not only vary in accordance with the numbers accommodated and their dependency levels, but also according to the layout of the home. An older home with a sprawling layout and several levels is likely to require higher staffing levels than a modern home with an efficient layout.

There are no definitive staffing levels for ancillary staff. Typical levels for cooking, cleaning and laundry are six per resident per week. The hours worked on gardening, maintenance, financial administration, marketing, training and resident activities are in addition. These will vary according to the size of the home. Some activities eg payroll and gardening may be subcontracted out to self employed contractors and their cost will appear not under wages but under different headings in the profit and loss account. In homes with over 20 residents, the registered manager is normally supernumerary to the care staffing rota.

In small homes it is common for the owner to also be the registered manager and for other family members to work in the home. In larger homes it is common for the owner to appoint a manager to take day to day control. Corporate operators may undertake a number of functions centrally, such as marketing and payroll. It is important when comparing staffing costs between homes to make appropriate adjustments to ensure that the basis of comparison is consistent.

The wage rates payable in any particular home are determined by its local labour market, subject to rates not falling below minimum levels. Minimum rates are enshrined in the minimum wage legislation and the rise in the national minimum wage has been a major factor influencing the rising costs faced by care home owners. Currently it stands at £5.35 per hour, an increase of 5.9% on the previous year. It is set to rise to £5.52 in October 2007, a rise of 3.2%. The government have also proposed to add the eight bank holidays to the statutory four weeks annual leave allowance, which will add to wage costs.

Care home owners have often struggled to find sufficient staff of the quality that they require and some have had to turn to agencies. This can add enormously to staff costs. Some operators looked overseas for staff. The influx of overseas staff has certainly been a major help in minimising agency costs and in keeping wage rate rises lower than they would otherwise have been, but staffing a home with large numbers of overseas staff can create difficulties.

Average hourly rates of pay are given in Table 22.6.

**Table 22.6** Hourly wages of staff in UK private care homes, February 2006

| | Weighted average across all shifts £ per hour |
|---|---|
| Level 1 Nurses | 10.81 |
| Level 2 Nurses | 10.25 |
| Care assistants without NVQ | 5.51 |
| Care assistants with NVQ2 and above | 5.77 |
| Senior carers | 6.33 |
| Cleaning, laundry and catering staff | 5.44 |
| Chefs/Cooks | 7.11 |
| Manager | £30,811 pa |

Source: Laing & Buisson

GVA Grimley in their Spring 2006 Bulletin also surveyed average wage rates. The results for level 1 nurses were similar. However, the rates for care assistants were generally slightly lower than those in Table 22.6, albeit they covered the year to 30 September 2005.

Larger homes generally benefit from economies of scale when it comes to staffing. However because there are no fixed rules on staffing levels, there is no one registration size which can be said to be the optimum. Laing & Buisson have calculated the cost of operating an efficient larger scale care home for older people, which is fully compliant with all the English National Minimum Standards for homes constructed since April 2002 and their wage analysis is detailed in Table 22.7.

**Table 22.7** Reasonable Wage Costs in Care homes in 2006–07 by type and location

| | Nursing care for older people or people with dementia £ | | Residential (personal) care for older people £ | | Residential (personal) care for people with dementia £ | |
|---|---|---|---|---|---|---|
| | Provincial Location | Outer London and Environs | Provincial Location | Outer London and Environs | Provincial Location | Outer London and Environs |
| Qualified nurses | 100 | 104 | n/a | n/a | n/a | n/a |
| Care staff | 138 | 144 | 113 | 118 | 142 | 148 |
| Domestic staff | 43 | 46 | 43 | 46 | 43 | 46 |
| Management and Administration | 30 | 33 | 30 | 33 | 30 | 33 |
| Agency staff allowance | 4 | 5 | 2 | 2 | 2 | 2 |
| Training backfill | 3 | 3 | 2 | 2 | 2 | 2 |
| Total including on-costs per registered bed per week | 318 | 335 | 190 | 201 | 219 | 231 |

Source: Laing & Buisson

accordingly the annual wage cost in a 49 bed nursing home in the provinces would, using the figures in Table 22.7, equate to £810,264.

Wage costs can also be analysed as a percentage of fee income. In nursing homes they typically range from 50 to 60% of fee income while in residential (personal) care homes they typically range from 40 to 50% of fee income.

# Non-staff operating costs

Non-staff operating costs typically cover:

- food and other consumables eg cleaning materials and medical supplies
- utility costs — gas, electricity, water
- insurance — property and business
- repairs and maintenance
- clinical waste and other refuse disposal costs
- telephone
- postage, stationery and printing
- advertising
- training
- motor expenses
- registration fees
- audit and other professional fees
- council tax
- sundries.

Laing & Buisson have estimated reasonable non-staff operating costs to be around £67 per registered bed per week with maintenance capital expenditure of £14 per bed per week in addition. Non-staff operating costs usually range between 11% and 18% of income with 15% being a typical average.

The main problem when analysing non-staff operating costs is to extract one-off or non-recurring costs from those necessary for the continuing operation and extracting costs which are incurred as a result of the particular provider's style of operation and which would not be incurred by the reasonably efficient operator.

Some owners have items of plant and equipment or vehicles on hire purchase, leasing or rental agreements. An assumption is normally made that any such items that are necessary for the continuing successful operation of the business are transferred to the purchaser, unencumbered.

# EBITDAM

The earnings from a specific freehold property are generally expressed in terms of EBITDAM (earnings before interest, taxation, depreciation and management charges). For management charges read head office costs (in the case of a group), directors' drawings (in the case of a company), or proprietor's drawings (in the case of a sole trader or partnership).

It is not surprising, given variations in occupancy, average fees, staff and non-staff operating costs that levels of EBITDAM can vary widely. That said, overall levels of profitability, driven largely by fee increases, have on the whole improved dramatically in the last five or so years. Typical levels of

EBITDAM in 2002 were between £4,000 and £6,000 per bed. Today [2007] typical levels are betwee
£6,500 and £9,000 per bed, with EBITDAM levels in London and the South East commonly exceedir
£10,000 per bed.

As a proportion of revenue, EBITDAM in nursing homes averaged 29.2% in the first half of 200
while those in residential (personal) care homes averaged 33.9% according to research undertaken k
Colliers CRE. A typical range would be 25–38% for nursing homes and 28-41% for personal care home

## Study 1 Worked example

Let us take the example of The Old Vicarage nursing home in the Southern Home Counties which is registered f
49 and accommodates the general elderly. It comprises an original manor house which was converted into a ca
home in the mid 1980s and extended on subsequent occasions, the last being in 2002 when a purpose built wir
was added. Like many homes, therefore, the standard of accommodation at The Old Vicarage is variable, with th
most recent extension being built to meet the national minimum standards as they apply to new homes. The hom
is on two floors connected by an eight person passenger lift. Unfortunately, four rooms (two twins and two single
in the original part of the home at first floor are at a different level to the rooms in the new extension. This means th
service users in the four rooms have to climb several stairs to access the lift (which is in the new extension). Thes
four rooms are consequently not accessible to service users with mobility problems.

The home has 39 single rooms and five twins. Twenty nine of the rooms are en-suite. All are over 10 m$^2$ and mo
are in excess of 12 m$^2$ excluding the en-suite. There is ample day space. The building is well maintained ar
presented. The standard of plant, fixtures, fittings and equipment is good and the large garden is a particular
attractive feature. As indicated above in the section on competitive pressures, an analysis of the rate of supply of bec
within a three mile radius found it to be in line with the national average. An inspection of competing homes confirme
that they are of a similar standard but have no better facilities than the subject. Enquiries did not indicate any closure
or proposals for new homes or major extensions to existing units in the vicinity.

The Old Vicarage has been the subject of two CSCI inspections in the last 12 months. The outcome of both wa
good. The home scored levels 3 and 4 on all the standards inspected and no requirements were issued following th
last inspection. The manager was picked out for particular praise.

An analysis of the audited accounts for the three years to 31 March 2006 produced the following:

| Item | Year ended 31.03.04 £ | Year ended 31.03.05 £ | Year ended 31.03.06 £ |
|---|---|---|---|
| Income | 1,327,658 | 1,406,890 | 1,479,202 |
| Staff Costs | 736,746 | 776,500 | 781,191 |
| Non-staff costs | 229,500 | 286,331 | 302,200 |
| EBITDAM | 361,412 | 344,059 | 395,811 |
| EBITDAM per bed | 7,376 | 7,022 | 8,078 |
| EBITDAM as a % of income | 27.2 | 24.5 | 26.8 |
| Staff costs as a % of income | 55.5 | 55.2 | 52.8 |
| Non-staff costs as a % of income | 17.3 | 20.3 | 20.4 |

A list of fees payable by the 46 residents at the time of inspection (May 2007) was obtained. This showed a tota
weekly fee income being received of £30,267, a weekly average of £657.98. There was no sundry income in addition
The provider was able to produce evidence to support his claim that occupancy levels had been consistently a
around this level for the past 12 months or so. 56.5% of the residents had their fees partially paid by the local authority
with the balance in the form of third party top-ups. 43.5% were wholly private pay residents. The fees, while well ir

excess of the local authority baseline rate were found to be in line with local and regional levels. A socio-economic analysis of the local area confirmed the good private pay potential evident by the service user mix at the home. Fees of local authority residents were increased in April 2007 in line with the increase in the local authority baseline rate (2.5%). Private payers have their fees reviewed on the anniversary of their admission.

An analysis of staffing levels showed them to be in line with industry norms and the latest inspection report confirmed them to be satisfactory to CSCI. Wage rates were also in line with the rates payable by other homes in the locality. The last wage review had taken place in October 2006 and the next was due in October 2007.

Non-staff operating costs were discussed in detail and one-off and unusual costs identified.

From monthly management accounts for the year to 31 March 2006 and current trading information, the following analysis was possible.

| Item | Year ended 31.03.07 £ | May 2007 Run Rate (annualised) £ |
|---|---|---|
| Income | 1,514,845 | 1,573,884 |
| Staff Costs | 820,552 | 836,678 |
| Non-staff costs | 225,439 | 234,580 |
| EBITDAM | 468,854 | 502,626 |
| EBITDAM per bed | 9,568 | 10,258 |
| EBITDAM as a % of income | 31.0 | 31.9 |
| Staff costs as a % of income | 54.2 | 53.2 |
| Non-staff costs as a % of income | 14.9 | 14.9 |

In financial terms the home has gone from strength to strength. The May 2007 run rate was supported by trading records seen and the key components seen to be in line with industry benchmarks. Non-staff costs were unusually high in the historic accounts and fell in the management accounts. Investigation of these costs found that major refurbishment costs had been allocated to the profit and loss account in the accounts for the years ended 31.03.05 and 31.03.06.

The owner was confident of achieving a fee income in the coming year of £1,632,000 and an EBITDAM of £550,000 with tighter cost controls.

In calculating the fair maintainable trade there were a number of issues to consider including:

- the future marketability of the twin rooms
- the four beds not easily accessible from the lift and
- whether a reasonably efficient operator would be able to sustain levels of profitability at their current levels should the existing manager leave.

If it is concluded that the four rooms on the different floor level have a limited future earning potential, are there ways of compensating for this by building a second lift or extending the property further? What would the capital cost be? What are the chances of obtaining planning permission? In this example, it is assumed that planning permission is in place to build six extra en-suite rooms at a cost of approximately £250,000 including VAT and fitting out costs, there being no requirement for additional day space or service areas. The VAT is included as it is levied on care home extensions (unless it is a listed building) and cannot usually be reclaimed.

The six new beds could replace the six beds in the four rooms which are not easily accessible. The Old Vicarage would then have a more sustainable 43 singles and 3 twins with 35 of the rooms being en-suite (excluding the 4 rooms). With the extension built, the calculation of the fair maintainable trade might look as follows:

| | |
|---|---:|
| Registration | 49 |
| Rooms | 46 |
| Assumed Average Occupied rooms | 45 |
| Average bed occupancy (%) | 91.8 |
| Average weekly fee (£) | 695 |
| Annual Fee Income (£) | 1,626,300 |
| Staff costs (£) | 865,000 |
| Non-staff costs (£) | 240,000 |
| EBITDAM (£) | 521,300 |
| EBITDAM per bed (£) | 10,639 |
| EBITDAM as a % of income | 32.0 |
| Staff costs as a % of income | 53.2 |
| Non-staff costs as a % of income | 14.8 |

A marginally lower occupancy than that currently being achieved is assumed in the light of regional occupancy levels and the better than average manager, notwithstanding the better room structure and layout which the home would offer. However, a higher average fee is assumed, again in line with benchmark data, assuming that the six new rooms would appeal to private payers at premium prices.

While the new rooms are being built there is no reason why the home should not in the short term continue to trade in line with the May run rate (adjusted for inflation).

The future is, of course, uncertain and particularly so in healthcare where so many factors combine to determine levels of profitability. It is therefore important when undertaking a valuation for loan security purposes not only to investigate and identify the level of fair maintainable trade, but also to comment on those risk factors, as the ability to repay interest and capital is likely to be dependent on the profits generated.

The bank should be able to understand how volatile levels of profitability are likely to be and the reasons for that volatility. Given that commercial mortgages are now over terms of up to 25 years, the task of the valuer is to highlight both short and longer term risk issues. The identification of these issues should then help to frame the valuer's comments and recommendations on the suitability of the property for the level and term of the proposed loan.

# Capital value

Capital value is calculated using the income capitalisation approach assuming that the use as a care home is to continue and that care home use is the most valuable use of the property. The income capitalisation approach estimates the value of an interest by calculating the present value of anticipated future benefits. The two most common methods are discounted cash flow and capitalisation of income. In healthcare capitalisation of income is the more common method. This requires the estimation of a capitalisation rate.

Paragraph 7.1 of VIP11 notes that:

> The determination of the capitalisation factor applied in arriving at the market value of healthcare properties relies upon the knowledge and experience of the valuer. It reflects the valuer's opinion of the market's perception of the risk or security associated with the subject business, the relevant market sector's approach to value, the availability and likely cost of development funding for new enterprises and funding for transfers of this category of enterprise and its current and future trading potential. It will also take account of and be supported by all available market evidence, intelligence and general economic factors.

Capitalisation rates for care home properties have risen in recent years in response to falling interest rates, falling property yields and the increasing popularity of the sector with investors. The highest multiples are paid for properties which comply fully with the national minimum standards since the view is taken that there is a lesser risk attached to the EBITDAM stream from these properties than that from poorer quality properties. The argument is broadly as follows:

1.  those homes that already offer the best spatial standards will be at least risk from future regulatory changes
2.  modern well presented homes should, all other things being equal, be more attractive to clients than less well presented units and therefore able to win market share at the expense of their competitors
3.  modern purpose built homes are more operationally efficient, should require less future maintenance and have a longer economically useful life than rambling conversions with old plant and equipment.

As the future is uncertain, the flexibility of a building to adapt to changing market circumstances might also be deemed an asset.

In the case of the Old Vicarage, based on comparable transactional evidence a multiplier of 8 has been adopted. The calculation of capital value is as follows:

|  | £ pa | £ |
|---|---|---|
| Fair Maintainable EBITDAM | 521,300 | |
| Capitalisation factor | 8 | |
|  | 4,170,400 | 4,170,400 |
| *Less* anticipated cost of 6 rooms | | 250,000 |
| Capital Value | | 3,920,400 |

Say £3,920,000

A useful cross check can be undertaken on a per bed and a per room basis. In this case it equates to £80,000 per bed and £89,091 per room based on the present format. The value of the planning permission has not been separately identified in the above calculation as it is incorporated into the capitalisation factor. An alternative would be to utilise a lower capitalisation factor and apply this to the May 2007 run rate EBITDAM and then add an additional sum for the value of the planning permission. The result should be similar. The space occupied by the four soon-to-be-redundant rooms on the first floor has not, in the above calculation, been attributed any value which in practice it would have if it could have an alternative economic use, for example, as staff accommodation.

## Trends in capital values

The care home market is currently buoyant. Whether care home values have peaked remains to be seen. Growth in EBITDAM levels is likely to be more sluggish than in recent years but still positive. Capitalisation rates are at an all time high and show no signs of falling, boosted in part by acquisition activity in the corporate market.

At the corporate level, private equity is a key driver of consolidation through leveraged buyouts supported by banks and property investment funds that have become willing to accept yields on care homes similar to those being achieved on mainstream commercial properties. The opco/propco model is currently seen as the best approach to maximising value. In this model, the EBITDA of good quality purpose built properties can be split, part going to pay rent and part being retained as residual operating profit after rent. Using the opco/propco split, the weighted average of these two income streams can generate an Enterprise Value for the whole entity of 12–14+ times earnings, well in excess of the multiples achieved only a few years ago. This aspect at present is really restricted to the larger corporates, although its influence is being felt throughout the sector.

Take for example a modern purpose built home generating say £8,000 per bed EBITDARM. A simplified calculation of its enterprise value using the opco/propco split is as follows:

| | |
|---|---|
| EBITDARM | £8,000 |
| Rent cover ratio | 1.67 |
| Rent payable | £4,790  pa |
| Residual operating profit | £3,210 |
| Yield on property investment | 6% |
| Prop-co value | £79,833 |
| Opco capitalisation rate | 5.0 |
| Opco value | £16,050 |
| Enterprise value | £95,883 |
| Blended capitalisation rate | 12.0 |

Note: The prop-co yield will depend partly on the perceived quality of the property and partly on the perceived strength of the tenant covenant. Leases are typically of 25–35 year duration with upward only rent reviews geared to inflation with caps and collars. The opco capitalisation rates can vary widely and there is little market evidence of the sale of single unit leased assets. The rent coverage ratio is set at a level which gives a sufficient margin to both the opco and propco from the available EBITDARM.

© Philip CL Hall 2008

# References

Care Standards Act, 2000

Care Homes Regulations 2001 (SI 2001, No 3965) as amended by Care Standards Act 2000 (Establishments and Agencies)(Miscellaneous Amendments) Regulations 2002 (SI 2002, No 865)

*GN1 Specialised Trading Property Valuations and Goodwill.* RICS Appraisal and Valuation Standards (5th ed)

*Valuation Information Paper No 11: The Valuation and Appraisal of Private Care Home Properties in England, Wales and Scotland.* Trade Related Valuation Group (Exposure Draft), 2007

*Annual Survey of Local Authority Fees,* Taylors Business Surveyors and Valuers Ltd, 2007

*Care Homes for Older People: National Minimum Standards.* 3rd ed, Department of Health, 2003

*Care Homes Review,* Colliers CRE, March 2007.

*Care of Elderly People,* UK Market Survey, Laing & Buisson, 2007.

*Dementia UK* — A Report into the prevalence and cost of dementia prepared by the Personal Social Services Research Unit (PSSRU) at the London School of Economics and the Institute of Psychiatry at King's College London, for the Alzheimer's Society, 2007.

*Population Trends No 123,* Office for National Statistics, Spring 2006

*Solutions,* GVA Grimley, Spring 2006

# Negligence and Valuations

The spectre of a claim for professional negligence is something which haunts every practising valuer. Quite apart from the financial implications of liability (which will normally be cushioned, at least partially, by professional indemnity insurance cover), there is the risk of the adverse publicity, for both the individual valuer and the firm, which inevitably follows a high-profile lawsuit.

Such lawsuits occur most frequently in the aftermath of a sharp fall in the property market. It is then that a lender, faced with a defaulting borrower and a property which is no longer adequate security for the outstanding loan, is most likely to turn on those professionals (solicitors as well as valuers) who advised on the making of that loan. The full extent of such claims is difficult to assess, since professional indemnity insurers are notoriously reluctant to divulge information about claims; however, anecdotal evidence from both the profession and the insurance industry suggests that, where surveyors' liability is concerned, valuations account for the lion's share.

In order to succeed in an action for professional negligence against a valuer, a claimant must satisfy the court on three issues: that the valuer owed them a legal duty to exercise care and skill; that this duty was breached; and that the breach of duty resulted in loss or damage of a kind recognised by the law. Assuming that these three elements are present, the valuer may still be able to avoid or limit liability by relying on one or more of three specific defences. These defences (the onus of establishing which lies on the valuer) are exemption clauses or disclaimers, time-limits and the partial defence of contributory negligence.

## Duty of care

The duty of care issue answers the question: "To whom can a negligent valuer be held liable?" Unless a legal duty of care exists, it does not matter whether or not the valuer's work has fallen below the appropriate standard; there can be no liability in respect of it.

### Liability in contract and tort

The most straightforward example of a duty of care is that which arises by virtue of a contract between the parties. It has long been established that, in carrying out professional work, a valuer owes a duty of care to the client. This duty arises primarily as an implied term of the contract under which the work is done, a principle confirmed by section 13 of the Supply of Goods and Services Act 1982; however should it be relevant (which would normally be in the context of statutory time-limits — see p.000 the client is also entitled to bring a claim in the tort of negligence.

Of course, it is only a contracting party who can be held liable for breach of contract. This mean that, in the vast majority of cases, a client's action for negligence will be brought against the firm which contracts to supply valuation services rather than against the individual member of that firm who actually carries out the work. However, the case of *Merrett* v *Babb* [2001] demonstrates that i certain circumstances the individual valuer may also be at risk. In that case a mortgage valuation wa carried out by the defendant, who was employed by a firm of valuers as manager of one of it branches. By the time the house purchaser sought to bring a claim for negligence, the owner of the firm had become bankrupt and was not worth suing. Worse, from the point of view of employees the owner's trustee in bankruptcy had cancelled all professional indemnity insurance cover. In highly controversial decision, the Court of Appeal held by a 2–1 majority that the individual valuer who had signed the mortgage valuation in his own name and added his professional qualifications had assumed responsibility for it and was therefore personally liable to the purchaser in the tort o negligence.

For many years, the law insisted that liability for negligent advice leading to financial loss coul only arise in contract, which of course meant that a valuer could be liable only to the client. However in *Hedley Byrne & Co Ltd* v *Heller & Partners Ltd* [1964] it was ruled by the House of Lords that provided there was a special relationship between the parties, liability for negligence could also arise in tort. This is not the place for an extended discussion of what amounts to a special relationship fo this purpose; however, it appears to require, not just foreseeability by the defendant that the claimant may rely on the defendant's advice, but also some sense of proximity or closeness between them.

An example of what is meant by proximity, and thus of where the boundaries of liability might lie may be found in the case of *Shankie-Williams* v *Heavey* [1986]. The defendant there, who traded as a dry rot surveying specialist, inspected and report on the ground floor flat in a property which had recently been converted into three flats. This task was commissioned and paid for by the developer who had carried out the conversion work, but the defendant knew that the latter wanted the report to show to potential purchasers. It was held that the defendant, who negligently failed to find clear evidence of dry rot, owed a duty of care to the purchasers of the ground floor flat. However, no duty was owed to the purchasers of the flat above, who had been shown the report and had concluded from it that their flat too was free from dry rot.

## Duty of borrower's valuer to lender

Before agreeing to lend money on the security of commercial property, the lender will almost invariably require a valuation of that property. It would seem sensible for the lender to commission such a valuation but, rather oddly perhaps, a practice has grown up whereby the lender is content to rely on a valuation commissioned by either the borrower itself or by an intermediary, such as a broker, acting on the borrower's behalf.

Provided that the valuer is aware, both of the purpose of the valuation and of the identity of the lender for whom it is intended, it is clear that the valuer will owe that lender a duty of care. This was the decision in *Banque Bruxelles Lambert SA* v *Eagle Star Insurance Co Ltd* [1994], where it was also held that the valuer's duty of care extended to an insurance company which had relied on the valuation in agreeing to issue a mortgage guarantee indemnity policy on the property in question.

Where the valuer knows that the valuation will be shown to a lender, but is unaware exactly who that lender will be, the legal position is rather less clear. On the one hand, it was held in *Assured Advances Ltd* v *Ashbee & Co* [1994] that ignorance of the lender's identity was irrelevant: the valuer would owe a duty of care to anyone who ultimately lent in reliance on the valuation, even if the valuation "passed from hand to hand until a lender could be found".

A very different approach emerges from the case of *Omega Trust Co Ltd* v *Wright Son & Pepper* [1997], where valuers who had produced a valuation of commercial property for prospective borrowers agreed that it could be readdressed (for a fee) to a named lender. The resulting loan was made partly by this lender and partly by another company (a Belgian bank) with which the named lender was closely associated. The Court of Appeal held unanimously that, because of a disclaimer in the valuation report (see p.000), the valuers were not liable to the Belgian bank. However, what is important for present purposes is that two members of the court clearly thought that, even without such a disclaimer, there would have been no duty of care to the second, unidentified, lender.

## Duty of lender's valuer to borrower

Where residential, rather than commercial, property is concerned, lending practices are very different. What almost invariably happens is that the lender, whether a building society, bank, local authority or other institution, will commission an inspection and mortgage valuation of the property before agreeing to lend. This valuation, which may be undertaken either by an independent valuer or by one employed in-house by the lender, is intended solely to assist the lender in deciding whether or how much to lend and on what terms. It is emphatically not designed for the benefit of the purchaser of the house or flat in question, although many mortgage lenders routinely provide the purchaser with a copy of the valuer's report.

Against this background, the question which was bound to arise sooner or later was whether a purchaser who relied on a mortgage valuation report in deciding to purchase could recover damages for negligence (usually for overlooking defects which a reasonable and competent inspection would have revealed) from the valuer concerned. In *Yianni* v *Edwin Evans & Sons* [1981], it was held that this was indeed possible; in carrying out an inspection of this kind, the valuer owed a duty of care, not only to the lender, but to the purchaser as well. The decision was a highly controversial one, but it was resoundingly approved by the House of Lords in *Smith* v *Eric S Bush*; *Harris* v *Wyre Forest DC* [1989]. In explaining why it was fair and reasonable to impose a duty of care upon the mortgage valuer, the judges regarded two factors as crucial: the evidence showed that valuers generally knew that purchasers were very likely to rely, directly or indirectly, on the mortgage valuation report, and the underlying economics of the mortgage transaction were such that the purchaser actually paid the valuer's fee for carrying out this work.

*Smith* v *Bush*, as its full name suggests, was actually two separate cases which were heard by the House of Lords at the same time. They were selected from a number of other contemporary claims on the basis that, between them, they covered a wide range of factual possibilities. In one the lender was a building society, in the other a local authority; in one the valuation was carried out in-house, in the other by a local firm of chartered surveyors; in one the valuer's report was disclosed to the purchaser, in the other it was not; and in one the mortgage loan was a mere 25% of the value of the house, in the other it was 90% (the maximum then permitted by statute for a local authority loan). Given that, in both cases, the House of Lords ruled in favour of the purchasers, it can be taken that none of these factors makes a difference to the valuer's liability.

The House of Lords in *Smith* v *Bush* clearly regarded its decision as a matter of consumer protection, emphasising more than once that both cases concerned very modest residential properties, of a kind whose purchaser might reasonably be expected not to commission their own survey, but to rely instead on the mortgage valuation (for which, after all, they had directly or indirectly paid). The judges suggested strongly that the same principle could not be extended to commercial property, and even expressed some doubt as to whether it would apply in the case of very expensive residential properties; here, it was thought, purchasers might reasonably be expected to employ and rely on their own professional advisers.

In accordance with these suggestions, it was held in the Scottish case of *Wilson* v *DM Hall & Son* [2005] that a valuer, in appraising a residential development project for a lending bank, owed no duty of care to the developer, who relied on the valuer's report in fixing a selling price for the complete flats. However, in the unreported case of *Qureshi* v *Liassides* (22 April 1994), which concerned valuation of a small shop with a greengrocer's business, it was held that the lending bank's valuer owed a duty of care to the purchaser, to whom his report was shown; the judge clearly regarded purchasers at this level of commercial property as just as much in need of consumer protection as those in the residential field.

## Other cases

There are two other situations in which the existence of a duty of care has been considered by the courts. The first is where the parties to an actual or potential dispute agree to be bound by the decision of a valuer acting as an independent expert. A common example of this would be in the context of a rent review, since leases frequently provide that, if landlord and tenant are unable to agree on the new rent, the matter shall be referred to such an independent third party. It is now well established that, in carrying out this task, the valuer will owe a legal duty of care to both parties (*Zubaida* v *Hargreaves* [1995]). The position of the valuer in such a case would appear to be a very exposed one, since it is highly likely that at least one of the parties will disagree with the decision; however, it is noteworthy that, to date, no claimant has succeeded in establishing negligence on the part of an independent expert.

The second situation arises where a mortgage lender repossesses and resells the mortgaged property after the borrower has defaulted on the loan. It has long been the law that, in exercising its power of sale, the lender owes the borrower a legal duty to take all reasonable steps to obtain the best available price for the property. In cases where a sale at an undervalue is based on negligent advice from the lender's valuer, what normally happens is that the lender is held liable to the borrower for any loss suffered and that the lender then recoups this loss in an action for negligence against the valuer concerned. Judges in a number of cases appeared to accept that the borrower might short-circuit this procedure by claiming directly for negligence against the valuer, but this was decisively rejected by the Court of Appeal in *Raja* v *Austin Gray* [2003]. This was because there was insufficient "proximity" between the valuer and the borrower; to hold otherwise, and thus put the valuer under a duty to both lender and borrower, would create an unacceptable conflict of interest.

# Breach of duty

Assuming that a duty of care and skill is held to exist, the next question is whether or not it has been breached. This might be regarded as the most fundamental issue of all, for it rests on whether or not the defendant's conduct can be described as negligent.

# Defining the valuer's task

t is important to remember that negligence is not an abstract concept, but a failure to apply an adequate degree of care and skill to a specific task. This means that, in any professional negligence action, a court's first task is to define precisely what it was that the defendant undertook to do. Only when this has been done will it be possible to decide whether or not the task has been carried out to an appropriate standard.

A simple example of this principle in operation may be seen in the context of mortgage valuations, where the courts readily accept that the valuer's inspection of the property will be more limited than that expected on a full building survey or a Homebuyer Survey and Valuation. In *Lloyd* v *Butler* [1990], for example, the judge described a mortgage valuation as "an inspection which on average should not take longer than 20 to 30 minutes. It is effectively a walking inspection by someone with a knowledgeable eye, experienced in practice, who knows where to look ... to detect either trouble or the potential form of trouble". In consequence, a surveyor might escape liability for overlooking a defect on a mortgage valuation, where it would have been regarded as negligent to overlook the same defect in the course of a building survey.

Even in the course of a limited inspection, however, a valuer who detects signs of hidden trouble must "follow the trail" of suspicion, even if this will involve going beyond what would otherwise be expected, for example by moving furniture or lifting carpets (*Roberts* v *J Hampson & Co* [1988]). Thus in *Sneesby* v *Goldings* [1995], a mortgage valuer, having noticed that a chimney breast in the kitchen of a converted property had been removed, checked the room above for visible signs of distress but found none. Had the valuer looked under the cooker hood or in the kitchen cupboards, he would have realised that the chimney had been left without adequate support. Such inspections would not normally be expected on a mortgage valuation, but the Court of Appeal held that, given the trail of suspicion in this case, the valuer was negligent in not following it through.

The need to define the professional task to be carried out applies equally in the context of pure valuations. In *Arab Bank plc* v *John D Wood (Commercial) Ltd* [2000], for example, the claimants were approached to lend on the security of a large and complex industrial estate. Having obtained a valuation of the property from one firm of valuers, they then asked another leading firm to carry out a franking or review valuation, but without making clear what this meant. It was held by the court that the second firm's task fell between a full standalone valuation and a mere desktop exercise; they were entitled to accept the factual content of the earlier valuation report (such matters as floor areas, physical condition, lease terms and geological survey) but were expected to make their own assumptions and carry out their own calculations based on this material.

In *Predeth* v *Castle Phillips Finance Co Ltd* [1986] a mortgage lender repossessed a house on the borrower's default and, having asked a local valuer for a crash sale valuation, sold the property for slightly more than the figure provided. The borrower successfully sued the lender for failing to obtain the best available price when exercising its power of sale, whereupon the lender in turn sued the valuer for failing to advise it as to the open market value of the property. This claim was rejected by the Court of Appeal, on the ground that the valuer had done precisely what the client had requested.

It should be noted that the client in the *Predeth* case was a commercial mortgage lender, which should be well aware of its legal obligations and could not expect the valuer to warn it of the dangers of selling at an undervalue. By contrast, the case of *McIntyre* v *Herring Son & Daw* [1988] was one in which specialist rating valuers were instructed by a private client to seek a reduction in the rateable value of the client's leasehold house. The valuers, having successfully negotiated what was agreed by expert witnesses to be a satisfactory reduction, advised their client that it would not be worth taking legal proceedings to secure a further reduction, since the chances of success were very slim. It was held that the valuers were negligent in not alerting their client to the fact that a comparatively slight further reduction would have brought the property within the Leasehold Reform Act 1967, thus giving the client a valuable right of enfranchisement. It should be said that the decision is a highly controversial one, since the client's instructions had nothing whatsoever to do with leasehold enfranchisement however, it can perhaps be explained on the basis that the client was a private individual and the defendants were specialists in the field.

## The objective standard

Once the task has been identified, the valuer's duty is to carry it out with reasonable care and skill This is an objective standard, based on what could have been expected from the hypothetical ordinarily competent member of the profession. In *Watts v Savills* [1998], for example, it was alleged that a valuer had negligently failed to recognise the potential of certain land for housing development The Court of Appeal, reversing the decision of the trial judge, held that the valuer would only be liable if it could be said that no competent valuer, acting carefully, could have shared his opinion as the lack of development potential.

Useful guidance on the objective standard is provided by the case of *Kenney v Hall, Pain & Foster* [1976], where it was held that this standard will be applied even to work for which no fee is charged, and even where the person undertaking the work lacks formal qualifications or relevant experience. The latter aspect also underpins the decision in *Baxter v FW Gapp & Co Ltd* [1938], where a valuer who undertook to value a house in an area with which he was unfamiliar was held negligent for overlooking comparables which would have been well known to someone with knowledge of the local residential property market.

## The professional knowledge required

A valuer, like any other professional adviser, is expected to possess a reasonable body of relevant professional knowledge. Some at least of this professional knowledge is to be found in RICS Practice Statements and Guidance Notes; evidence as to whether or not a valuer complied with these is likely to prove highly persuasive in an action for professional negligence although, as was pointed out in *PK Finans International (UK) Ltd v Andrew Downs & Co Ltd* [1992], it is certainly not conclusive. Examples of reliance being placed on RICS guidance include *Craneheath Securities Ltd v York Montague Ltd* [1994], where a valuer was held entitled to base mortgage valuation of restaurant on open market value, and *Allied Trust Bank Ltd v Edward Symmons & Partners* [1994], where a valuer was held entitled to include hope value in a mortgage valuation of a country house with clear development potential.

Valuers are not required to know the law to the same degree as would be expected of barristers or solicitors. However, they are expected to have at least a broad understanding of the legal principles which underpin different types of valuation. Thus in the old case of *Jenkins v Betham* (1855), the defendant's valuation of dilapidations in a rectory was based on rendering the premises habitable (the normal leasehold standard) rather than putting them into good and substantial repair (the standard applicable to tenancies of ecclesiastical properties). The defendant was held to have been negligent, since any valuer holding himself out as competent to value such property should be aware of the standard to be applied. Similarly, in *Corisand Investments Ltd v Druce & Co* [1978], a hotel valuer was held negligent in failing to take account of the work which would be required under the Fire Precautions Act 1971 to obtain a fire certificate.

In so far as valuers are expected to have legal knowledge, they are also expected to keep that knowledge up to date. In *Weedon* v *Hindwood, Clarke & Esplin* [1975], for example, the defendants were representing a client in compulsory purchase negotiations with the local authority. Shortly before negotiations began, a ruling by the Court of Appeal altered (in favour of the client) the law as to the date on which property should be valued for compulsory purchase and, before negotiations ended, this change in the law was confirmed by the House of Lords. The defendants were held negligent for agreeing a price with the local authority which was clearly based on the old law.

This need to keep up to date applies, not only to legal knowledge, but to all aspects of a valuer's professional expertise. In *Izzard* v *Field Palmer* [2000] a mortgage valuer, in reporting on a maisonette in a four-storey block, failed to point out that it was constructed under a system which combined large concrete panels and timber cladding. The valuer was held to have been negligent, since the risks of structural problems and high maintenance charges inherent in such buildings had been well documented in the professional literature at the time of the defendants' inspection.

## Specific valuation issues

In the absence of specific instructions, a valuer's task is to provide a view of the open market value of a property at the relevant date. It was made clear by the Court of Appeal in *Banque Bruxelles Lambert SA* v *Eagle Star Insurance Co Ltd* [1994] that it is not the valuer's responsibility to predict future market movements, either nationally or locally; such movements are relevant only in so far as they are already reflected in current prices and therefore in the open market value.

Of course, valuers are frequently instructed to value on the basis of certain assumptions which depart from open market value and, where this is the case, they must follow their instructions. However, this in itself will not necessarily lead to a different result; whether or not it will do so depends upon the prevailing market conditions. In *UCB Corporate Services Ltd* v *Halifax (SW) Ltd* [2000], for example, the defendants were instructed by lenders to value a new unit in a small commercial development on two bases: open market value and a sale within 90 days by a mortgagee in possession. The valuers took the view that, as the market then stood, the two values were identical, and the Court of Appeal held that they were not negligent in coming to this conclusion.

As far as valuation methodology is concerned, the courts have traditionally been reluctant to interfere, repeatedly emphasising (as in *Love* v *Mack* (1905)) that there is no single correct approach, departure from which will automatically amount to negligence. It was acknowledged in *Corisand Investments Ltd* v *Druce & Co* [1978], for example, that a valuer might quite legitimately base a valuation simply upon comparables, without any further calculation, or even upon the valuer's own experienced awareness of and instinct for the property market. Nevertheless, a more prescriptive approach has occasionally been in evidence, as for example in *Singer & Friedlander Ltd* v *John D Wood & Co* [1977], where the judge described valuation as a four-stage process: collecting of relevant information about the property; analysing that information; checking one's findings by an appropriate alternative valuation method; and reporting to the client. Such willingness to examine the details of valuation process may also be seen in *Merivale Moore plc* v *Strutt & Parker* [1999], which concerned the valuation of a development property held on a lease with 46 years to run. The Court of Appeal held the valuers negligent, not in their choice of yield, but for failing to warn their developer clients that, in view of the dearth of suitable comparables, the choice of yield was a more than usually risky exercise and might well turn out to be incorrect.

A number of cases have considered the impact on valuations of a recent transaction involving the subject property. The courts have emphasised that such a transaction is potentially the most cogent evidence of current market value and have accordingly held that a valuer's failure to take it into account will almost always amount to negligence, unless there is clear evidence to suggest either that the transaction did not reflect market value or that the valuer has been specifically instructed to disregard it (*Banque Bruxelles Lambert SA* v *Eagle Star Insurance Co Ltd* [1994]). Indeed, it has been held that a valuer who is aware that such a transaction has taken place, but is unaware of its terms, will be guilty of negligence in failing to take reasonable steps to find out more (*Interallianz Finanz AG* v *Independent Insurance Co Ltd* [1997]).

## Margin of error

In all the cases discussed above, the issue of whether a valuer has been negligent has been decided on the basis of evidence as to how the valuer carried out the particular valuation; the question is then whether the hypothetical reasonably competent valuer would have acted in the same way. In recent years, however, the courts have accepted that there is an alternative approach to this issue, one which focuses on the amount by which the allegedly negligent valuation diverges from the what the court, on the basis of expert evidence, decides was the true value of the property on the date on which the defendant valued it. According to this approach, a valuation may be judged negligent by reference, not to the process adopted, but simply to the figure arrived at.

This margin of error or bracket approach is ultimately derived from the courts' recognition that every valuation contains an element of subjective opinion and that merely being "wrong" does not necessarily amount to negligence. Thus: "valuation is a matter of opinion" (*Baxter* v *FW Gapp & Co Ltd* [1938]); "valuation is an art, not a science" (*Singer & Friedlander Ltd* v *John D Wood & Co* [1977]); and "valuation is not an exact science; it involves questions of judgement on which experts may differ without forfeiting their claim to professional competence (*Zubaida* v *Hargreaves* [1995]).

Unfortunately for valuers, judicial recognition that valuation is a matter of opinion carries another implication, first expressed in *Singer & Friedlander Ltd* v *John D Wood & Co* [1977]. The judge in that case was faced with evidence put forward by expert witnesses to the effect that, as a general principle, any competent valuation should fall within 10% (or, in exceptional circumstances 15%) of the property's true value. Having accepted this evidence, the judge concluded: "There is ... the permissible margin of error, the 'bracket' as I have called it. What can properly be expected from a competent valuer using reasonable skill and care is that his valuation falls within this bracket." As developed in subsequent cases (notably the decisions of the Court of Appeal in *Merivale Moore plc* v *Strutt & Parker* [1999] and *Arab Bank plc* v *John D Wood (Commercial) Ltd* [2000]), this has come to mean that a valuation which falls outside the appropriate bracket raises a presumption of negligence, which it is then for the valuer to disprove.

The margin of error approach may operate in favour of a valuer. It was held in *Mount Banking Corporation Ltd* v *Brian Cooper & Co* [1992] that, where a valuation falls within the appropriate bracket, the valuer cannot be held liable for negligence, however erroneous the method by which that valuation was arrived at. This principle was approved by the Court of Appeal in *Merivale Moore plc* v *Strutt & Parker* [1999], although it was said in *Arab Bank plc* v *John D Wood (Commercial) Ltd* [2000] that there may be an exception to this where the valuer is guilty of a discrete error (such as a failure to diagnose and make a specific deduction for the presence of asbestos) which feeds through into the final figure; here there may be liability for negligence in respect of that error, even though the final figure is within the overall bracket.

As mentioned above, the judge in the *Singer & Friedlander* case accepted expert evidence that the appropriate bracket was 10%, extending in exceptional circumstances to 15%. In subsequent cases there has been much discussion and debate over the appropriate bracket to adopt, which is said to depend upon such factors as the type of property, the availability of suitable comparables, the sensitivity of the type of valuation involved and the state of the property market. Having taken such factors into account, judges (strongly influenced by expert witnesses) have adopted brackets ranging from a mere 5% (on a standard house on an estate) to 20% (where a valuer was asked to review an earlier valuation by another firm). However, in most cases the figure has been between 10% and 15%. Although the margin of error approach appears well accepted by the courts, it has been subjected to heavy academic criticism (see Crosby, Lavers & Murdoch, "Property Valuation Variation and the Margin of Error' in the UK" (1998) 15 *Journal of Property Research* 305). This is on the ground that, while expert witnesses have routinely advised judges as to what would be an appropriate bracket, they have never offered any empirical support for the figure chosen; moreover, their unsupported assertions are frequently at variance with published research on valuations, which suggests that variation between valuers is actually much greater than the brackets adopted. In short, the margin of error approach means that the issue of valuation negligence is reduced to a single question: whether or not the defendant's opinion as to value falls within an arbitrary and unsupported distance of another matter of purely subjective opinion, namely the hypothetical true value of the property, as determined by the judge on the basis of (often wildly conflicting) opinions put forward by those same expert witnesses!

## The role of expert evidence

In seeking to decide whether or not a valuer has acted with reasonable care and skill, a court will frequently pay close attention to evidence provided by expert witnesses. Such witnesses are normally called by the parties, although the court has jurisdiction under rule 35 of the Civil Procedure Rules to direct that evidence shall be given by a single expert, to be appointed either by agreement between the parties or by the court. In any of these cases, an expert witness must possess relevant expertise; thus in *Whalley* v *Roberts & Roberts* [1990], where it was alleged that a mortgage valuation had been negligently carried out, the court refused to take any account of evidence put forward on this issue by an engineer and an architect.

It is important to appreciate that the duty of a valuer who appears as an expert witness is to assist the court on matters within his or her expertise, and that this duty overrides any obligation to the client. This principle is laid down in the Civil Procedure Rules; it is also spelled out in no uncertain terms by the RICS Practice Statement and Guidance Notes, 2006: "Surveyors acting as Expert Witnesses". According to that Practice Statement (which applies to appearances before arbitrators, tribunals and independent experts as much as in court):

> The overriding duty of the surveyor is to the judicial body to whom the evidence is given. It overrides the contractual duty to the client ... This duty to the judicial body, which is ongoing, is to set out the true facts and give honest, impartial opinions, which should cover all relevant matters ... The surveyor's evidence must be independent, impartial, objective and unbiased. Special care must be taken to ensure that it is not biased towards those who are responsible for instructing or paying the surveyor. The evidence should be the same whoever is instructing the expert or is paying him.

A practical effect of this overriding obligation appears to be that an expert witness cannot be sued by the client in respect of anything said in court or contained in a report prepared for the purpose of court proceedings (even if those proceedings do not in fact take place). This was laid down by the Court of Appeal in the non-valuation case of *Stanton* v *Callaghan* [2000], but a word of caution may be sounded:

the reasons given by the court for its decision in that case are very similar to those formerly used to justify the immunity of barristers against claims in negligence, an immunity which has since been removed by the House of Lords.

## Extent of liability

The third essential element in an action for negligence is proof that the negligence in question has caused the claimant to suffer loss of a kind recognised by law. This means that if the loss would have occurred anyway, irrespective of the defendant' negligence, then the defendant cannot be held legally responsible for it. Thus in the non-valuation case of *Thomas Miller & Co* v *Richard Saunders & Partners* [1989], surveyors acting for the tenant at a rent review arbitration failed to submit certain relevant evidence to the arbitrator. The surveyors were held to have been negligent, but they were nevertheless not liable to their clients as they had not caused them any loss; this was because the court was satisfied that, even if the evidence in question had been submitted, the rent fixed by the arbitrator would have been exactly the same.

In cases involving negligent valuations, the question of causation is most likely to focus on the question of whether or not the claimant has relied on the valuation in deciding to enter into a particular transaction. This is a question of fact, but the courts appear fairly ready to assume that, where a person receives professional advice before entering into a transaction, they have relied on that advice. In *Charterhouse Bank Ltd* v *Rose* [1995], for example, where a lending bank asked one valuer to review an earlier valuation by another valuer, it was held that the bank had relied on both valuations in reaching its decision to lend on the property. Likewise, where a lender's practice is to obtain two independent valuations of a property before agreeing to lend, it appears that there is sufficient reliance on the higher valuation, whether the loan is in fact based on that valuation (as it was in *Cavendish Funding Ltd* v *Henry Spencer & Sons Ltd* [[1998]) or, as is normal, on the lower one (as in *Housing Loan Corporation plc* v *William H Brown Ltd* [1997]).

The decision of the Court of Appeal in *Western Trust & Savings Ltd* v *Strutt & Parker* [1998] must be regarded as a very borderline one. The defendants there carried out a valuation for lending purposes of certain property which was in the process of being redeveloped to provide holiday accommodation. When a loan was made, it was secured on most but not all of the property valued. Despite this, it was held that the lenders had sufficiently relied on the defendants' valuation to allow them to recover damages in negligence.

Despite the judicial tendency to lean in favour of claimants on this issue, it is not always possible to establish reliance. In *Shankie-Williams* v *Heavey* [1986], for example, where a person who traded as a dry rot surveying specialist prepared a report on the timbers in a flat, he was held not liable to the purchasers of that flat, since it was clear on the evidence that they had not been shown his negligently prepared report. Again, in *Banque Bruxelles Lambert SA* v *Eagle Star Insurance Co Ltd* [1994], it was held in respect of one property that the bank was in fact highly sceptical about the defendants' valuation and had based its decision to lend on purely commercial factors which were entirely unconnected with it.

## Liability to purchaser

The assessment of damages in an action for professional negligence brought by a purchaser against a valuer or surveyor has given rise to considerable argument. However, the legal principles now appear to be settled, although it should be pointed out that they have not yet been confirmed by the House of Lords.

## The basic measure of damages

A purchaser who relies on negligent advice in deciding to purchase a property will normally be entitled to damages reflecting the difference between what has been paid and what the property is truly worth. Where the advice in question is a pure valuation, there can be little argument that this is the appropriate way to measure the claimant's loss (*Patel* v *Hooper & Jackson* [1999]). However, in cases involving building surveys, there has for many years been a strong body of opinion in favour of awarding damages based on the cost of repairing those defects which the surveyor has negligently overlooked. This view is based on the belief that, where a prospective purchaser commissions a building survey, they are concerned not so much with the value of the property as with a desire for reassurance that they will not be saddled with large and unforeseen repair costs.

In *Watts* v *Morrow* [1991], the leading case in this area, such arguments convinced the trial judge that the purchasers of an old farmhouse should be awarded the cost of repairing negligently overlooked defects (£34,000) rather than the £15,000 difference between what they paid for the property (£172,500) and what it was actually worth in its defective condition (£162,500). However, this award was overturned by the Court of Appeal, who held that £15,000 was the correct award; the higher figure, it was said, could only have been awarded if the surveyor had given a contractual guarantee that the property was free from defects not mentioned in the survey report.

The justification for awarding damages reflecting the difference between the price paid and the true value of the property is that the purchaser has relied on the negligent valuation or survey in agreeing to pay that price. It might therefore be thought that, where the purchaser pays more than the value placed on the property by the valuer or surveyor (for example because the vendor refuses to reduce the price to that level), the damages should be based on the valuation rather than the purchase price; it is difficult to see how the purchaser can claim to have relied on the report in paying more than the value given in it. However, this logic was not followed in *Oswald* v *Countrywide Surveyors Ltd* [1996], where the claimant purchased a house for £225,000 after the defendants had valued it "in the region of £215,000". Its true value, due to defects which the defendants had failed to detect, was held to be £165,000. The trial judge, whose decision was upheld by the Court of Appeal, ruled that the correct measure of damages was £60,000 rather than £50,000, on the basis that the defendants could be said to have caused the claimants to pay £225,000 for the property.

The difference in value measure of damages carries further implications. First, if the evidence establishes that, defects notwithstanding, the property is worth what the claimant has paid for it, then the claimant is entitled only to nominal damages for breach of contract (*Smith* v *Peter North & Partners* [2002]). Second, there may be circumstances in which this measure of damages works in the purchaser's favour. In *Gardner* v *Marsh & Parsons* [1997], for example, the claimants purchased a recently converted maisonette for £114,000, relying on a survey report which failed to disclose serious structural defects. As a result of these defects the property was worth only £85,000 at the date of purchase. By the time the claimant's action for negligence against the surveyors came to court, the landlord who had converted the property rectified the defects at his own expense (under threat of legal action by the claimants), as a result of which the claimants now had a property worth what they had paid for it. It was nevertheless held by the Court of Appeal that the claimants were entitled to damages of £29,000 from the surveyors; what had happened between the claimants and their landlord was a completely separate issue and had no bearing on the case.

# Incidental losses

In addition to the basic measure of damages described above, a purchaser may be entitled to compensation for certain incidental losses. These will include the legal costs involved in purchasing and reselling the defective property and, if it is uninhabitable, the cost of alternative accommodation while this is done. If the purchaser chooses to remain in the property and to repair it, compensation may also include the cost of temporary accommodation while repairs are carried out, although not of course the cost of repair itself.

Where the negligence of a valuer or surveyor leads to a purchaser suffering inconvenience, the purchaser may be entitled to a modest amount of compensation for this. Inconvenience for this purpose means physical discomfort rather than mere distress, vexation or frustration and, as a result, such an award will only be made in respect of residential property. It was held in *Watts v Morrow* [1991] that purely mental suffering can only be the subject of damages where there is a breach of a contract specifically designed to provide pleasure or avoid displeasure; a normal contract with a surveyor or valuer does not fall into either of these categories.

# Liability to vendor

Where a valuer's negligence results in a vendor selling property at an undervalue, the principles on which damages are assessed are in effect a mirror image of those governing claims by purchasers. The basic measure of damages consists of the difference between what the property is sold for on the basis of the negligent valuation, and its true market value at the date of sale. This principle was applied in the case of *Weedon v Hindwood, Clarke & Esplin* [1975], where valuers who were conducting compulsory purchase negotiations for a landowner negligently advised their client to accept too low a price from the local authority.

In the unusual circumstances which arose in *Kenney v Hall, Pain & Foster* [1976], the defendants were held liable to a vendor client for a negligent overvaluation of his property. Relying on the defendants' assurances as to the price for which his house could be sold, and leaving what appeared to be a sensible safety margin, the claimant took a bridging loan in order to acquire and renovate two other properties. When the house, which had been negligently overvalued, proved very difficult to sell, the substantial interest charges incurred by the claimant brought him to the point of bankruptcy. It was held that the defendants were responsible for all the losses suffered by the claimant, apart from those which he would have suffered in any event had he been properly advised and had entered into a much more modest scheme.

# Liability to lender

Where a mortgage lender has lost money through relying on a negligent valuation, a court seeking to assess damages must first of all answer, on the basis of the available evidence, the following preliminary question: if the valuer had not been negligent, and had therefore provided an accurate valuation of the property, would a mortgage transaction, albeit a smaller one, still have taken place? If the answer to this question is yes, then the case is known as a smaller transaction case, and the basic measure of damages for the lender will consist of whatever has been lent and lost, less whatever would still have been lent and lost on the (hypothetical) smaller transaction, for example due to a fall in the property market (*Corisand Investments Ltd v Druce & Co* [1978]). This is because the valuer's negligence cannot be treated as having caused losses that would have been suffered, even if there had been no negligence. In such a case the lender is not normally entitled to recover the cost of repossessing and reselling the mortgage property, since it is to be presumed that these costs would equally have been incurred under the smaller transaction.

If the answer to the preliminary question is no, that without the negligence there would have been no lending transaction at all, then the case is known as a no transaction one. In such a case, lenders would argue that the negligent valuer should be liable for all their losses since, in the absence of negligence, they would have lost nothing. Such an argument was accepted by the Court of Appeal in *Banque Bruxelles Lambert SA* v *Eagle Star Insurance Co Ltd* [1994]. However, this ruling was overturned by the House of Lords in *South Australia Asset Management Corporation* v *York Montague Ltd* [1996]. The court there held that a valuer's duty is to exercise reasonable care and skill in attempting to provide accurate information; as a result, if the valuer is negligent, liability for that negligence is limited to "the consequences of the information being wrong". What this means in practical terms is that the lender's damages cannot exceed the difference between the value negligently attributed to the property by the defendant and its true value at that date. In short, there is a cap on the valuer's liability, equivalent to the amount by which the property has been negligently over-valued.

The effect of this formulation can be seen in the facts of two of the three cases which were reported together as *South Australia Asset Management Corporation* v *York Montague Ltd* [1996]. In one (the *South Australia* case itself), where the valuer's error was £10 million, the lender was held entitled to recover its entire loss of £9.75 million. In another (*Nykredit Mortgage Bank plc* v *Edward Erdman Group Ltd*), where the valuer's error was £1.5 million, it was held that the lender could recover no more than this, despite having lost a total of £3 million.

The ruling in *South Australia* means that, subject to the cap, the lender is entitled to recover its entire loss. This will normally consist of the full amount of the mortgage loan together with interest on that capital. Such interest will be assessed at normal commercial rates; it was held by the House of Lords that a lender who specialised in loans to high-risk borrowers at exorbitant rates of interest could not hold a negligent valuer liable for damages based on those interest rates (*Swingcastle Ltd* v *Alastair Gibson* [1991]). The lender's loss will also include costs incurred in repossessing and reselling the mortgage property following the borrower's default. Against these losses must be set off whatever is realised on a resale of the subject property, any loan repayments made by the borrower and (according to a controversial ruling by the House of Lords) whatever value can be placed on the borrower's covenant to repay the loan (*Nykredit Mortgage Bank plc* v *Edward Erdman Group Ltd (No 2)* [1988]).

Quite apart from the part played by interest in working out the loss suffered, a lender, like any other litigant, may be entitled to simple interest on damages for the period between the date on which the cause of action arose and the date of judgment (Supreme Court Act 1981, section 35A). In the context of a no transaction case, it has been held by the House of Lords that interest of this kind runs from the date on which the lender's total loss (including interest on the mortgage loan at normal commercial rates) reaches the level of the "cap" (*Nykredit Mortgage Bank plc* v *Edward Erdman Group Ltd (No 2)* [1988]).

# Defences

Assuming that the three issues (duty of care, breach of duty and resulting damage) discussed above are established to the satisfaction of the court, the claimant is in principle entitled to win the case. However, there remains the possibility that the defendant can raise one of a number of defences which, if proved, will negate or at least reduce liability. It should be noted that, in relation to all these defences, the onus of proof is on the defendant; this is why simply proving that there has been no negligence is not treated as a defence, because the onus of proving negligence is always on the claimant.

## Exemption clauses and disclaimers

As a matter of legal principle, there is nothing to prevent a valuer from excluding or restricting liability for negligence, to either a client or a third party, by means of a contract term or an adequate notice of disclaimer. As far as clients are concerned, valuers do not in practice seek to exclude liability altogether; they would presumably regard it as highly unprofessional to take a fee for providing a service, while at the same time insisting that they are not liable for providing it badly. However, it is not uncommon for valuers, especially those who carry out valuations of very expensive commercial properties, to seek to restrict or limit liability to the client in some way, usually by providing a financial limit which may be related to the amount of professional indemnity insurance cover carried by the valuer.

Whether such a restriction is valid will depend on the legal principles governing contractual exemption clauses generally and, in particular, the rules contained in the Unfair Contract Terms Act 1977. According to these statutory rules, a contract term which seeks to exclude or restrict liability for negligently causing financial loss (which is the type of loss for which a negligent valuer is normally liable) can only be relied upon if it is regarded by the court as a fair and reasonable term. It is quite possible that a limitation of liability related to professional indemnity insurance cover would be regarded as fair and reasonable, although no court has yet been faced with this question.

Although quite prepared to accept liability to clients, valuers are usually very reluctant to be held responsible to third parties, and routinely attempt by means of disclaimers to ensure that they will not be. A standard form of disclaimer found in valuation reports, for example, is a clause to the effect that the report is for the use only of the person to whom it is addressed (and sometimes also of that person's professional advisers), and that no responsibility is undertaken towards any other person. Similarly (though, as we shall see, less effectively), building society mortgage documentation frequently contains warnings to purchasers that they should not rely on the mortgage valuation report, since it is intended solely for the use of the lender, and that the valuer does not assume any legal responsibility towards them.

As a general principle (and subject to the Unfair Contract Terms Act), it seems that such disclaimers can be perfectly effective to exclude a duty of care in tort. In the unreported Scottish case of *Commercial Financial Services Ltd v McBeth & Co* (15 January 1988), for example, valuers prepared a report on a proposed development for the developer, knowing that the latter intended to show it to potential lenders, but nevertheless inserting in the report a standard clause stating that the report was for the use only of the client and that "no responsibility is accepted to any third party". This clause was held effective to exclude any liability for negligence to a lender who was shown and relied on the valuers' report.

It is important to note that, in order to be effective, a disclaimer must be brought to the attention of the third party at the appropriate time. This requirement proved fatal to a mortgage valuer in the Scottish case of *Martin v Bell-Ingram* [1986], where notices purporting to exclude the valuer's liability to a house buyer were not shown to the buyer until after he had relied on the mortgage valuation report in deciding to purchase the house in question.

Disclaimer notices, like contractual exemption clauses, are subject to the Unfair Contract Terms Act 1977. However, the provisions of that Act operate slightly differently in relation to disclaimers; here it is the defendant's reliance on the notice, rather than the term itself, that must be shown by the defendant to satisfy the statutory test of reasonableness.

In dealing with this question of reasonableness, the courts have adopted two widely differing approaches, depending on whether the case concerns commercial or residential property. As far as commercial property is concerned, the prevailing judicial attitude has been a very robust one, exemplified by the decision of the Court of Appeal in *Omega Trust Co Ltd* v *Wright Son & Pepper* [1997]. As noted earlier, the valuers in that case agreed, in return for a fee, that their valuation of their client's property (three small supermarkets held on short leases) could be readdressed to a named lender. The valuation report contained a term stating that it was confidential to the addressee and was not to be relied upon by third parties for any purpose whatsoever without the defendants' express written permission. The Court of Appeal held unanimously that this term was effective in principle to exclude any duty of care to a second company, associated with the addressee, which advanced part of the mortgage loan. The court further ruled that it was perfectly fair and reasonable for the defendants to rely upon this term: the parties were of equal bargaining strength, it would have been a simple matter for the associated company to obtain its own valuation, and a professional valuer in a commercial context ought to be entitled to refuse to assume legal responsibility to unknown lenders, or to insist on a fee for assuming such responsibility.

On turning to residential property, a very different picture emerges. In *Smith* v *Eric S Bush*; *Harris* v *Wyre Forest DC* [1989], which concerned mortgage valuations of modest houses, the House of Lords was faced with statements in the documentation provided by lenders to purchasers which clearly sought to exclude any liability on the part of either the lender or the mortgage valuer in respect of the valuation report. The judges accepted that, in terms of general legal principle, these statements would be effective to prevent any duty of care from arising, but held unanimously that they failed to satisfy the statutory test of reasonableness and should therefore be struck down.

In considering this issue, the House of Lords noted that the parties (house purchaser and mortgage lender) were not really of equal bargaining power; that requiring purchasers to obtain their own reports would place a considerable financial strain on people at the lower end of the housing market; that mortgage valuations were "work at the lower end of the surveyor's field of professional expertise"; and that the availability of professional indemnity insurance meant that the risk of liability for negligence could easily be spread and its impact thus lessened. Taking all these factors into account, the judges agreed that it would be wholly unreasonable to allow the valuer's duty of care to be excluded.

## Statutory time-limits on liability

For almost four centuries, English law has insisted that every civil (as opposed to criminal) legal action must be commenced within a certain period of time. In a succession of statutes starting in 1623 (the current legislation is contained in the Limitation Act 1980), a set of time limits have been laid down for different types of claim. If a claimant fails to start legal proceedings against the defendant within the relevant limitation period, then the claim is statue-barred and will fail. The intention is to strike a balance between preserving the rights of claimants (whose claims are just as meritorious, whether or not they are pursued within the time-limit) and protecting defendants against the possibility of being hounded by stale claims, at a time when the memories of witnesses may have dimmed and the evidence necessary to mount a successful defence may have disappeared.

In some areas, this balance results in time-limits being based upon a claimant's awareness that they have a potential claim. However, the requirement of awareness is by no means a universal principle (except in special circumstances, for example where a defendant has deliberately concealed evidence); in many types of claim, a time-limit may expire long before the claimant is, or could possibly have been, aware of their rights. In such circumstances, the defence of limitation appears a harsh one; nevertheless, it is one which has frequently been utilised by negligent valuers.

In considering the provisions of the Limitation Act 1980 which are relevant to professional negligence claims against valuers, it is important to remember that those claims may be framed either as a breach of contract or in the tort of negligence. A third party (who by definition has no contract with the valuer) can obviously sue only in tort. Clients, by contrast, have a choice; it was held in *South Australia Asset Management Corporation* v *York Montague Ltd* [1996] that, notwithstanding their contractual relationship with the valuer, they are perfectly entitled to sue in tort in circumstances where this will provide them with a longer limitation period.

## Claims in contract

As far as claims for breach of contract are concerned, section 5 of the Act provides that legal proceedings must be commenced within six years of the date on which the contract is breached; in the unlikely event that the contract between valuer and client is made in the form of a deed, the relevant period, for purely historical reasons, is 12 years (section 8). In the valuation context, the date of breach would normally be when the valuer commits a negligent act, for example by inadequately inspecting the subject property or by making an unacceptable error in calculation. However, a claimant can probably argue that the submission of a negligently prepared valuation report to the client is also a breach of contract, so that a legal action commenced within six years of that date will be in time. In either event, it should be noted that these time-limits apply whether or not the client is or should be aware of the breach in question.

## Claims in tort

The principles which govern claims in the tort of negligence for financial loss are complex, although one fundamental principle is clear and straightforward. If more than 15 years have elapsed since the breach of duty on which the claim is based, then that claim is statute-barred (Limitation Act 1980, section 14B). If this final longstop has not been reached, then the claimant is offered a choice of two separate limitation periods and, so long as one or both of these is still running, it is not too late to start legal proceedings.

The first of these tort periods, laid down by section 2 of the 1980 Act, consists of six years from the date on which the cause of action accrues. In the present context, this means the date on which the claimant first suffers loss as a result of the defendant's negligence. Here, as with claims for breach of contract, the defendant's actual or constructive knowledge is irrelevant; it is the fact of loss which is important.

As to when a claimant first suffers loss, this will depend on both the nature of the negligent advice provided and the action taken by the claimant in reliance on it. In *Gulf Oil (Great Britain) Ltd* v *Phillis* [1998], for example, a tenant brought a negligence action against the valuer who, acting as an independent expert, had determined the new rent under a rent review provision in the lease. It was held that the limitation period on this claim began when the tenant first paid rent at the new level.

In the common case of a purchaser who pays too much for a property because it has been negligently over-valued, the limitation position is clear; it was held by the Court of Appeal in *Byrne* v *Hall Pain & Foster* [1999] that the six-year tort period starts to run on the date that the purchaser exchanges contracts for the acquisition of the property, since it is at this moment that the purchaser is locked into a loss-making transaction.

The most complex and controversial application of the first tort period is in relation to claims brought by mortgage lenders. It might be thought that the law, in seeking to identify the date on which the lender suffers loss, would choose one of two possibilities: either the date on which the mortgage is executed and the lender is committed to lending on an over-valued security, or the date on which, following the borrower's default, the property is repossessed and resold at a figure insufficient to cover the outstanding loan. Unfortunately, however, it was laid down by the House of Lords in *Nykredit Mortgage Bank plc v Edward Erdman Group Ltd (No 2)* [1988] that neither of these dates (either of which would be very easy to identify) is appropriate: instead what matters is the first date on which the total value of the lender's rights, which for this purpose include not only the value of the mortgaged property but also the value of the borrower's covenant to repay the loan, is insufficient to cover the outstanding mortgage debt (including accrued interest). Identifying this date will therefore require a retrospective valuation of the property at the relevant date (six years before legal proceedings were launched) plus a highly speculative estimate as to the value of the borrower's covenant at that date.

### Latent damage

The second, alternative period within which a claimant is permitted to start proceedings in the tort of negligence is related to the claimant's awareness of his position. Section 14A of the Limitation Act 1980 provides in effect that, where a claimant's cause of action accrues before they are aware of it, an action may be commenced within three years of the date on which they have knowledge of the material facts about the loss which has been suffered. Material facts for this purpose are defined as such facts as would lead a reasonable person to consider it sufficiently serious to justify starting legal proceedings against a solvent defendant who did not dispute liability.

Knowledge in this context includes, not only actual knowledge, but also knowledge which the claimant might reasonably be expected to acquire from those facts which they actually know, with the help of appropriate expert advice where it would be reasonable to seek this. Moreover, the courts have made it clear that knowledge does not mean "know for certain and beyond possibility of contradiction". It means knowing enough to consider litigation and to take preliminary steps. In short, vague and unsupported suspicion is not enough, but reasonable belief will be.

## Contributory negligence

The third defence which may be available to valuers is somewhat different from those already considered in that it operates, not to extinguish the defendant's liability altogether, but rather to reduce it. The Law Reform (Contributory Negligence) Act 1945 applies where a claimant's loss results partly from the fault of the defendant and partly from the claimant's own fault. In such circumstances, the Act provides that the damages payable shall be reduced "to such extent as the court thinks just and equitable", having regard to the claimant's share in the responsibility for the loss. The court has complete discretion in deciding how great the reduction in the damages shall be, but will attempt to produce a result that seems broadly fair, given the amount of blame that can be attributed to each party. In the context of a negligent valuation, this will usually mean that, despite fault on the part of the claimant, the valuer will bear the majority of the loss; this is perhaps on the basis that causing loss to others is more blameworthy than merely failing to take proper care of one's own interests.

In assessing the amount by which damages are to be reduced, account should be taken of a highly controversial decision of the House of Lords. In *Platform Home Loans Ltd v Oyston Shipways Ltd* [1999] a mortgage lender, relying on a negligent over-valuation of the subject property, had lost a total of £611,000. However, the lender's damages in respect of this loss were capped at £500,000 under the ruling in *South Australia Asset Management Corporation v York Montague Ltd* [1996] (see p.000). The Court of Appeal, holding the lender 20% responsible for the loss, awarded damages of £400,000 (ie 80% of £500,000) but the House of Lords ruled that the correct figure was £489,000 (ie 80% of £611,000). The result, which seems a very odd one, is that the lender lost a mere 2.2% of their damages on the basis of what had been held to be 20% responsibility for their loss!

It may be noted that the 1945 Act is not applicable to claims for breach of contract as such. However, the courts have ruled that, where an action might equally well be brought in contract or tort (as will invariably be the case in a client's claim for professional negligence), the defence will be available. This means that the client cannot evade the consequences of their own carelessness by saying: "I am choosing to sue for breach of contract, not tort". As a result of this, the defence of contributory negligence has in recent years been routinely raised (with considerable success) by valuers, most commonly in actions brought by mortgage lenders. In such cases the question to be decided by the court is whether the lender has acted in accordance with the standards to be expected of a reasonably prudent and competent banker, and whether any failure by the claimant to achieve this standard has contributed to the loss suffered from the defendant's negligent valuation.

The 1945 Act gives no guidance as to what kind of conduct will amount to contributory negligence, but the cases involving negligent valuations have effectively fallen into three main groups. The defendant may allege that the claimant has in some way contributed to the inaccuracy of the valuation itself; that it was unreasonable for the claimant to believe in or place reliance on the valuation; or that the claimant's decision to enter into a transaction on the basis of the valuation was imprudent for some other, unconnected, reason. While recognising that these are not separate categories of case in any legal sense, it is convenient for the purposes of discussion to deal separately with them.

## Contribution to erroneous advice

Very few of the valuation cases so far reported have concerned this type of contributory negligence, although it is not difficult to envisage circumstances in which it might arise. After all, a valuation is only as good as the information on which it is based, and a valuer is often dependent on the client for at least some of that information. One case where it did arise was *Craneheath Securities Ltd v York Montague Ltd* [1994], which concerned a valuation by the defendants of a restaurant business. It was alleged that the valuers had been negligent in basing their valuation on an unrealistic view of turnover. However, it emerged at the trial that, while the defendants had not seen any recent accounts of the business, the clients themselves had obtained a recent set of accounts but had inexplicably not made these available to the valuers. On the facts, it was held that the defendants had not been negligent and were therefore not liable at all; however, the judge made it clear that, had they been liable, the clients would almost certainly have been held guilty of contributory negligence.

Another way in which a client can be partly to blame for the poor advice received is where the instructions given to the valuer are inaccurate or misleading. In *South Australia Asset Management Corporation v York Montague Ltd* [1995], which concerned a major development site in London Docklands, it was clear that the parties had been in some considerable confusion as to the precise service to be carried out by the valuers, in particular as to whether they were to provide an open market valuation. This confusion arose at least partly because the claimants, in breach of their own internal lending rules, failed to provide direct and explicit guidance to the defendant valuers as to what they required. This failure was held to constitute contributory negligence.

A similar decision was reached by the Court of Appeal in *Western Trust & Savings Ltd* v *Strutt & Parker* [1998], where the defendants, in valuing a scheme to develop holiday accommodation, failed to discover and take account of a potential planning problem. It was held that the claimant lenders were partly responsible for this failure, because of the way in which they had allowed the borrower to instruct the valuers.

## Unreasonable reliance on the valuation

The most common allegation of contributory negligence by a valuer against a claimant, and the one which has most frequently been successful, is that the claimant's uncritical reliance on the valuation was unreasonably imprudent. At first sight, this might appear a difficult argument to support, since it involves the assertion that a client who has paid for professional advice is not simply allowed to act on it, but is expected to second-guess that advice and to make further enquiries if there is any reason to doubt its accuracy. Indeed, permitting a valuer to raise a defence of this kind carries considerable irony: the greater the valuer's negligence (and thus the more inaccurate the valuation), the stronger the defence will be and the more likely that the valuer will avoid at least part of the responsibility for the consequences of that negligence!

The first case in which a valuer successfully raised a defence of contributory negligence against a mortgage lender was *Banque Bruxelles Lambert SA* v *Eagle Star Insurance Co Ltd* [1994], where certain recently-acquired commercial properties were valued at up to 70% more than their purchase prices, without any explanation being offered for this startling discrepancy. The trial judge had no hesitation in holding that the valuers had been negligent; indeed, the valuers themselves did not deny this. However, in finding that the lenders were just as aware as the valuers of the differences between the valuations and the purchase prices, the judge held that the lenders' failure to demand an explanation from the valuers amounted to contributory negligence, justifying a reduction of 30% in the damages payable.

A defence of this kind was first accepted by the Court of Appeal in *Cavendish Funding Ltd* v *Henry Spencer & Sons Ltd* [1998]. There mortgage lenders, a secondary bank which specialised in short-term loans without checking the borrower's finances, obtained two independent valuations of the subject property (in this case a Grade I listed country house). Having adopted this prudent practice, however, the lenders then inexplicably based their lending decision on the higher of the two valuations, despite the fact that this exceeded the other by more than 50%! The court held that the lenders' failure to investigate the discrepancy between the two valuations amounted to contributory negligence and that they were responsible for 25% of their loss.

The principle laid down in *Banque Bruxelles Lambert* has been extended somewhat by the courts in subsequent cases. First, in *Nyckeln Finance Co Ltd* v *Stumpbrook Continuation Ltd* [1994], a lender was held guilty of contributory negligence despite having made an attempt (albeit a rather token one) to challenge the valuation provided by the defendants. Second, and more controversially, it was held in *Interallianz Finanz AG* v *Independent Insurance Co Ltd* [1997] that a lender who knew that the subject property had recently changed hands was contributorily negligent in failing to enquire about the purchase price, in order to see whether there was a serious discrepancy between that price and the valuation. This approach might be thought to place an unreasonably high burden upon lenders, since it apparently requires them to check up on a valuation even where they have no reason to doubt it.

As mentioned earlier, this type of contributory negligence has been commonly and successfully alleged by valuers against mortgage lenders. It seems highly unlikely that a court would regard a lay person (such as a house purchaser) as negligent for placing unquestioning reliance on a valuation. The defence was raised in *Yianni* v *Edwin Evans & Sons* [1981], where a mortgage valuer was sued by a house purchaser, on the ground that the purchaser failed to have an independent survey of the property; did not read the building society documentation; and generally took no steps to discover the true condition of the house. In holding that none of these failures amounted to contributory negligence, the judge held that the purchaser was entitled simply to rely on the defendants to make competent inspection and valuation.

## Lender's independent imprudence

The third group of cases consists of those where the valuer alleges that, while the claimant had no reason to doubt the valuation in question, their decision to enter into a specific transaction based on was imprudent and negligent for other reasons. Two particular allegations have arisen in the cases (all of which have so far related to mortgage lending): that the claimant's lending policy was based on an unreasonably high loan to value ratio (which reduces the lender's cushion of protection), and that the lender had taken insufficient steps, or no steps at all, to check the borrower's credit status before agreeing to lend.

Where both these factors are present, the courts have been fairly ready to treat the lender as guilty of contributory negligence. In *Platform Home Loans Ltd v Oyston Shipways Ltd* [1996], for example, the claimant lenders had a policy, when approached for a non-status or self-certification loan (that is, one where no check is made on the financial status claimed by the borrower) of obtaining two independent valuations of the subject property and then lending up to 70% of the lower of these. This practice was held to be one which a reasonably prudent lender would not adopt; the claimants were therefore contributorily negligent and were responsible for 20% of their losses.

Although the two factors together may well amount to contributory negligence, it seems that neither taken alone is likely to do so. In *Banque Bruxelles Lambert SA v Eagle Star Insurance Co Ltd* [1994], a loan to value ratio of 90% in a case involving commercial property was held to be a reasonable one for lender to adopt. Moreover, it was made clear in *HIT Finance Co Ltd v Lewis & Tucker Ltd* [1993] that where there is no particular reason to doubt the integrity or solvency of a proposed borrower, a lender is not automatically required to make enquiries on this front, but is perfectly entitled to rely exclusively on the security provided by the property. Of course, the position will be different where the borrower shows obvious signs of lack of integrity or substance; here a lender who agreed to lend without making further enquiries would almost certainly be regarded as imprudent and thus contributorily negligent.

# Index

## H

# N

**W**